Handbook of Exact Solutions for Ordinary Differential Equations

Andrei D. Polyanin
Valentin F. Zaitsev

CRC Press
Boca Raton New York London Tokyo

Library of Congress Cataloging-in-Publication Data

Polyanin, A. D. (Andrei Dmitrievich)
Handbook of exact solutions for ordinary differential equations /
Andrei D. Polyanin, Valentin F. Zaitsev.
p. cm.
Includes bibliographical references and index.
ISBN 0-8493-9438-4 (alk. paper)
1. Differential equations — Numerical solutions. I. Zaitsev,
V.F. (Valentin F.) II. Title.
QA372.P725 1995
515′.352—dc20 95-10217
CIP

Direct all inquiries to CRC Press, Inc., 2000 Corporate Blvd., N.W., Boca Raton, Florida 33431.

International Standard Book Number 0-8493-9438-4
Library of Congress Card Number 95-10217
Printed in the United States of America 1 2 3 4 5 6 7 8 9 0
Printed on acid-free paper

FOREWORD

Exact solutions have always played and still play an important role in properly understanding the qualitative features of many phenomena and processes in various fields of natural science.

Equations of applied and theoretical physics often contain parameters or functions which are found experimentally and therefore are not stringently fixed. At the same time, equations that model real phenomena and processes must be sufficiently simple to make possible their analysis and solution. It is natural to adopt, as one of feasible criterions of simplicity, the requirement that the model equation admits a solution in a closed form.

It should be noted that even exact solutions of nonlinear equations (including those without a clear physical sense and which do not correspond to real phenomena and processes) play an important role of "test" problems for verifying the correctness and assessment of accuracy of various numerical, asymptotic, and approximate methods. Moreover, the model equations and problems admitting exact solutions serve as the basis for the development of new numerical, asymptotic, and approximate methods, which, in turn, enable us to study more complicated problems having no analytical solution.

This book contains nearly 5000 ordinary differential equations and their solutions. The table below compares data presented in this work with those of currently available handbooks concerning the general number of concrete second- and higher-order ordinary differential equations analyzed.

The order of equations	E. Kamke (1976)	M. Murphy (1960)	This book
Second order	249	315	1227
Third order	13	22	587
Fourth order	3	3	75
Higher order	3	9	160
Total number of equations	268	349	2049

When selecting the material, the authors gave preference to the following two types of equations:

1. Equations that traditionally attracted the attention of many researchers: those of the simplest appearance but involving the most difficulties for integration (Abel equations, Emden—Fowler equations, Painlevé equations, etc.).
2. Equations that encountered in various applications (in the theory of heat and mass transfer, nonlinear mechanics, hydrodynamics, the theory of nonlinear oscillations, the theory of combustion, chemical engineering science, etc.).

Special attention is paid to equations containing arbitrary functions. The other equations contain one or more arbitrary parameters (i.e., actually, this book deals with whole *families* of ordinary differential equations) which can be fixed by a reader at will. Many solutions have been obtained just recently with the aid of new (discrete group) methods described in other books by the authors (1993, 1994).

When compiling this book, the handbooks by E. Kamke (1976), M. Murphy (1960), and D. Zwillinger (1989) were partly used in which one can find basic notions and definitions of the theory of ordinary differential equations, apart from concrete equations. In these handbooks, classical and some new methods of solving differential equations are described as well—see also the books by E.L. Ince (1964), P.J. Olver (1986), and N.H. Ibragimov

(1993). The latter books give a great number of references to the original papers and books by other authors, which are devoted to exact solutions and methods of the theory of ordinary differential equations.

In addition, when describing solutions of linear ordinary differential equations, which are connected to higher transcendental functions (Bessel, Legendre, Mathieu, hypergeometric, etc.), the handbooks by G. Beitmen and A. Erdeii (1953–1955), M. Abramowitz and I.A. Stegun (1964) were used.

In some sections of this book, asymptotic solutions of some classical equations of nonlinear mechanics and theoretical physics are also given, which are discussed in the books by J.D. Cole (1968), M.V. Fedoryuk (1983), and A.H. Nayfeh (1973, 1971) in detail.

The detailed table of contents enables a reader to quickly navigate through this book in searching for desired equations.

The authors hope that this book will be helpful for a wide range of scientists, lectures, engineers, and students engaged in the fields of mathematics, physics, mechanics, and chemical engineering science.

Andrei D. Polyanin
Valentin F. Zaitsev

Some Remarks and Notation

1. In this book, in the original equations the independent variable is denoted by x, and the dependent one is denoted by y. In the given solutions, the symbols C, C_0, C_1, C_2, … stand for arbitrary integration constants.

2. The following notation is used for derivatives: $y'_x = \dfrac{dy}{dx}$, $y''_{xx} = \dfrac{d^2y}{dx^2}$, $y'''_{xxx} = \dfrac{d^3y}{dx^3}$, $y''''_{xxxx} = \dfrac{d^4y}{dx^4}$, and $y^{(n)}_x = \dfrac{d^ny}{dx^n}$ with $n \geq 5$.

3. In some cases, we use the operator notation $\left(f\dfrac{d}{dx}\right)^n g$ which is defined by the recurrence relation

$$\left(f(x)\frac{d}{dx}\right)^n g(x) = f(x)\frac{d}{dx}\left[\left(f(x)\frac{d}{dx}\right)^{n-1} g(x)\right].$$

4. In some sections of the book (see, for example, 1.3, 2.3–2.6, 3.2–3.4), for the sake of brevity, solutions are represented as several formulae containing the terms with the signs "$\pm$" and "$\mp$." By this is meant two formulae—one correspond to the upper signs, and another corresponds to the lower signs. For example, the solution of equation 1.3.1.6 can be written in the parametric form

$$x = af^{-1}\exp(\mp\tau^2),\quad y = af^{-1}\bigl[\exp(\mp\tau^2) \pm 2\tau f\bigr],\quad \text{where } f = \int \exp(\mp\tau^2)\,d\tau - C,\quad A = \mp 2a^2.$$

This is equivalent to that the solutions of equation 1.3.1.6 are given by the formulae

$$x = af^{-1}\exp(-\tau^2),\quad y = af^{-1}\bigl[\exp(-\tau^2) + 2\tau f\bigr],\quad \text{where } f = \int \exp(-\tau^2)\,d\tau - C,\quad A = -2a^2$$

and

$$x = af^{-1}\exp(\tau^2),\quad y = af^{-1}\bigl[\exp(\tau^2) - 2\tau f\bigr],\quad \text{where } f = \int \exp(\tau^2)\,d\tau - C,\quad A = 2a^2.$$

5. When referencing to a particular equation, the notation like "4.1.2.5" stands for "equation 5 in Subsection 4.1.2."

6. The book includes two supplements that provide a reader with useful information on some elementary and special functions which appear in solutions of the differential equation outlined.

7. References that may be helpful for a reader are given at the end of the book.

THE AUTHORS

Andrei D. Polyanin, Ph.D., D.Sc., is a noted scientist in the fields of ordinary differential equations, engineering and applied mathematics, heat and mass transfer, nonlinear mechanics, and chemical engineering science.

Professor Polyanin graduated from the Faculty of Mechanics and Mathematics of the Moscow State University in 1974 and received his Candidate of Sciences (Ph.D.) degree in 1981 (at the Institute for Problems in Mechanics of the U.S.S.R. Academy of Sciences). His Ph.D. thesis was devoted to the asymptotic analysis of the problems of heat and mass transfer. In 1986, Professor Polyanin received his Doctor of Sciences degree; his D.Sc. thesis was dedicated to the mass and heat exchange between reacting particles and flow.

Since 1975, Professor Polyanin is a member of staff of the Institute for Problems in Mechanics of the Russian Academy of Sciences.

Professor Polyanin has made important contributions to new approximate analytical methods in the theory of heat and mass transfer, hydrodynamics, and chemical engineering science, as well as to new methods of the theory of ordinary differential equations. In 1991 he was awarded a Chaplygin Prize of the U.S.S.R. Academy of Sciences for his research in mechanics.

Professor Polyanin has published more than 100 research papers and seven books. He is also an author of three patents.

Valentin F. Zaitsev, Ph.D., D.Sc., is a noted scientist in the fields of ordinary differential equations, mathematical physics, and nonlinear mechanics.

Professor Zaitsev graduated from the Radio Electronics Faculty of the Leningrad Politechnical Institute (now Saint-Petersburg Technical University) in 1969 and received his Candidate of Sciences (Ph.D.) degree in 1983 (at the Leningrad State University). His Ph.D. thesis was devoted to the group approach to the study of some classes of ordinary differential equations. In 1992, Professor Zaitsev received his Doctor of Sciences degree; his D.Sc. thesis was dedicated to the discrete-group analysis of the ordinary differential equations.

Since 1971, Professor Zaitsev is in the Research Institute for Computational Mathematics and Control Processes of the St.-Petersburg State University. He is also a Professor at the Russian State Pedagogical University (St.-Petersburg), the Orel State Pedagogical Institute, and Orel State Politechnical Institute.

Professor Zaitsev has made important contributions to new methods in the theory of ordinary differential equations. He is an author of more than 70 scientific publications, including five monographs and the patent.

Zaitsev V.F. also read the theoretical course at the Leningrad Conservatory, later participating in development of mathematical methods in musical sciences.

CONTENTS

Chapter 1
First Order Differential Equations

1.1. Simplest Equations with Arbitrary Functions Integrable in a Closed Form*

1.1.1. Equations of the Form $y'_x = f(x)$

Solution:** $\displaystyle y = \int f(x)\,dx + C.$

1.1.2. Equations of the Form $y'_x = f(y)$

Solution: $\displaystyle x = \int \frac{dy}{f(y)} + C.$

Particular solutions: $y = A_k$, where A_k are roots of the algebraic equation $f(A_k) = 0$.

1.1.3. Separable Equations $y'_x = f(x)g(y)$

Solution: $\displaystyle \int \frac{dy}{g(y)} = \int f(x)\,dx + C.$

Particular solutions: $y = A_k$, where A_k are roots of the algebraic equation $g(A_k) = 0$.

Remark. The equation of the form $f_1(x)g_1(y)y'_x = f_2(x)g_2(y)$ is reduced to the form 1.1.3 by dividing both sides by $f_1 g_1$.

1.1.4. Linear Equation $g(x)y'_x = f_1(x)y + f_0(x)$

Solution:

$$y = Ce^F + e^F \int e^{-F} \frac{f_0(x)}{g(x)}\,dx, \qquad \text{where} \quad F(x) = \int \frac{f_1(x)}{g(x)}\,dx.$$

---

* Special cases of equations 1.1.1–1.1.5 for concrete functions f, f_0, f_1, f_n, and g are not discussed in this book; such cases can be readily recognized by the appearance of equations investigated, and the solution can be obtained using the general formulae given in Section 1.1.

** Hereinafter we shall often use the term "solution" to mean "general solution."

1.1.5. Bernoulli Equation $g(x)y'_x = f_1(x)y + f_n(x)y^n$

Here, n is an arbitrary number.

The substitution $w(x) = y^{1-n}$ leads to a linear equation:

$$w'_x = (1-n)f_1(x)w + (1-n)f_n(x).$$

Solution:

$$y^{1-n} = Ce^F + (1-n)e^F \int e^{-F}\frac{f_n(x)}{g(x)}\,dx, \qquad \text{where} \quad F(x) = (1-n)\int \frac{f_1(x)}{g(x)}\,dx.$$

1.1.6. Homogeneous Equation $y'_x = f(y/x)$

The substitution $u(x) = y/x$ leads to an equation with separation of variables: $xu'_x = f(u) - u$.

Solution: $\displaystyle\int \frac{du}{f(u)-u} = \ln|x| + C.$

Particular solutions: $y = A_k x$, where A_k are roots of the algebraic equation $A_k - f(A_k) = 0$.

1.2. Riccati Equations: $g(x)y'_x = f_2(x)y^2 + f_1(x)y + f_0(x)$

1.2.1. Preliminary Comments

For $f_2 \equiv 0$, we obtain a linear equation (see 1.1.3), and for $f_0 \equiv 0$, we have the Bernoulli equation (see 1.1.4 with $n = 2$), whose solutions were given previously. Below we discuss equations with $f_0 f_2 \not\equiv 0$.

1. Given a particular solution $y_0 = y_0(x)$ of the Riccati equation, the general solution can be written as

$$y = y_0(x) + \Phi(x)\left[C - \int \Phi(x)\frac{f_2(x)}{g(x)}\,dx\right]^{-1},$$

where

$$\Phi(x) = \exp\left\{\int \left[2f_2(x)y_0(x) + f_1(x)\right]\frac{dx}{g(x)}\right\}.$$

To the particular solution $y_0(x)$ corresponds $C = \infty$.

2. The substitution

$$u(x) = \exp\left(-\int \frac{f_2}{g}y\,dx\right)$$

reduces the general Riccati equation to a second order linear equation:

$$g^2 f_2 u''_{xx} + g\left[f_2 g'_x - g(f_2)'_x - f_1 f_2\right]u'_x + f_0 f_2^2 u = 0,$$

which often may be easier to solve than the original Riccati equation.

3. Let $g = f_2 = 1$, $f_1(x)$ and $f_0(x)$ be polynomials. If the degree of the polynomial $\Delta = f_1^2 - 2(f_1)'_x - 4f_0$ is odd, the Riccati equation can not possess a polynomial solution.

If the degree of Δ is even, the equation involved may possess only the following polynomial solutions:

$$y = -\tfrac{1}{2}\left(f_1 \pm \left[\sqrt{\Delta}\,\right]\right),$$

where $\left[\sqrt{\Delta}\,\right]$ denotes an integer rational part of the expansion of $\sqrt{\Delta}$ in decreasing powers of x (for example, $\left[\sqrt{x^2 - 2x + 3}\,\right] = x - 1$).

4. The general Riccati equation, with the aid of the transformation

$$x = \varphi(\xi), \quad y = \frac{1}{F_2}w - \frac{1}{2}\frac{F_1}{F_2} + \frac{1}{2}\left(\frac{1}{F_2}\right)'_\xi, \qquad \text{where} \quad F_i(\xi) = \frac{f_i(\varphi)}{g(\varphi)}\varphi'_\xi, \tag{1}$$

is reduced to the canonical form

$$w'_\xi = w^2 + \Psi(\xi), \tag{2}$$

where function Ψ is defined by the formula

$$\Psi(\xi) = F_0F_2 - \frac{1}{4}F_1^2 + \frac{1}{2}F_1' - \frac{1}{2}F_1\frac{F_2'}{F_2} - \frac{3}{4}\left(\frac{F_2'}{F_2}\right)^2 + \frac{1}{2}\frac{F_2''}{F_2}.$$

(prime denotes differentiation with respect to ξ).

Transformation (1) is determined by function $\varphi = \varphi(\xi)$ which may be arbitrary. For a specific original Riccati equation, different functions φ in (1) will generate different functions Ψ in equation (2).

In the special case where the original equation has the canonical form

$$y'_x = y^2 + f(x),$$

transformation (1) is written as

$$x = \varphi(\xi), \quad y = \frac{1}{\varphi'_\xi}w - \frac{1}{2}\frac{\varphi''_{\xi\xi}}{(\varphi'_\xi)^2},$$

and the transformed equation (2) is determined by function Ψ:

$$\Psi(\xi) = (\varphi'_\xi)^2 f(\varphi) - \frac{3}{4}\left(\frac{\varphi''_{\xi\xi}}{\varphi'_\xi}\right)^2 + \frac{1}{2}\frac{\varphi'''_{\xi\xi\xi}}{\varphi'_\xi}.$$

If the original Riccati equation is integrable by quadrature, we may obtain, specifying different functions φ, a variety of different integrable equations of the form (2). In Subsection 1.2.9, some useful transformations are given for specific functions φ.

5. The transformation (φ, ψ_1, ψ_2, ψ_3, and ψ_4 are arbitrary functions)

$$x = \varphi(\xi), \quad y = \frac{\psi_4(\xi)u + \psi_3(\xi)}{\psi_2(\xi)u + \psi_1(\xi)}$$

reduces the general Riccati equation to the Riccati equation.

1.2.2. Equations Containing Power Functions

1. $y'_x = ay^2 + bx + c.$

For $b = 0$, we have an equation of the form 1.1.2. For $b \neq 0$, the substitution $bt = bx + c$ leads to an equation of the form 1.2.2.4: $y'_t = ay^2 + bt$.

2. $y'_x = y^2 - a^2x^2 + 3a.$

Particular solution: $y_0 = ax - x^{-1}$.

3. $y'_x = y^2 + ax^2 + bx + c.$

This is a special case of equation 1.2.2.9 with $\alpha = 0$, $\beta = 0$.

4. $y'_x = ay^2 + bx^n.$

Special Riccati equation, n is an arbitrary number.

Solution: $y = -\frac{1}{a}\frac{u'_x}{u}$, where

$$u(x) = \sqrt{x}\left[C_1 J_{\frac{1}{2q}}\left(\frac{1}{q}\sqrt{ab}\,x^q\right) + C_2 Y_{\frac{1}{2q}}\left(\frac{1}{q}\sqrt{ab}\,x^q\right)\right], \quad q = \frac{n+2}{2};$$

J_m and Y_m are Bessel functions, $n \neq -2$. With $n = -2$, see equation 1.2.2.36.

5. $y'_x = y^2 + anx^{n-1} - a^2x^{2n}.$

Particular solution: $y_0 = ax^n$.

6. $y'_x = ay^2 + bx^{2n} + cx^{n-1}.$

For $n = -1$, we have 1.2.2.36. For $n \neq -1$, the transformation

$$\xi = \frac{1}{n+1}x^{n+1}, \quad \eta = yx^{-n}$$

leads to an equation of the form 1.2.2.25:

$$\xi\eta'_\xi + a\xi\eta^2 + \frac{n}{n+1}\eta = b\xi + \frac{c}{n+1}.$$

7. $y'_x = y^2 + ax^n y + ax^{n-1}.$

Particular solution: $y_0 = -1/x$.

8. $y'_x = y^2 + ax^n y + bx^{n-1}.$

The substitution $y = -u'_x/u$ leads to a second order linear equation of the form 2.1.2.42:

$$u''_{xx} - ax^n u'_x + bx^{n-1}u = 0.$$

9. $y'_x = y^2 + (\alpha x + \beta)y + ax^2 + bx + c.$

The substitution $y = -u'_x/u$ leads to a second order linear equation of the form 2.1.2.28:

$$u''_{xx} - (\alpha x + \beta)u'_x + (ax^2 + bx + c)u = 0.$$

10. $y'_x = y^2 + ax^n y - abx^n - b^2$.

Particular solution: $y_0 = b$.

11. $y'_x = ax^n y^2 + bx^{-n-2}$.

Solution:

$$\sqrt{ab}\ln x = \int \frac{du}{u^2 + \beta u + 1} + C, \qquad \text{where} \quad u = \frac{\sqrt{a}}{\sqrt{b}} x^{n+1} y, \quad \beta = \frac{n+1}{\sqrt{ab}}.$$

12. $y'_x = ax^n y^2 + bx^m$.

1°. For $n \neq -1$, the substitution $\xi = x^{n+1}$ leads to an equation of the form 1.2.2.4:

$$y'_\xi = \frac{a}{n+1} y^2 + \frac{b}{n+1} \xi^{\frac{m-n}{n+1}}.$$

2°. For $n = -1$ and $m \neq -1$, the transformation $\zeta = x^{m+1}$, $w = -1/y$ yields an equation of the form 1.2.2.4:

$$w'_\zeta = \frac{b}{m+1} w^2 + \frac{a}{m+1} \zeta^{-1}.$$

3°. For $n = m = -1$, the original equation is an equation with separation of variables. In this case we have

$$\ln|x| = \int \frac{dy}{ay^2 + b} + C.$$

13. $y'_x = y^2 + k(ax+b)^n (cx+d)^{-n-4}$.

The transformation

$$\xi = \frac{ax+b}{cx+d}, \qquad u = \frac{1}{\Delta}\left[(cx+d)^2 y + c(cx+d)\right], \quad \text{where } \Delta = ad - bc,$$

leads to an equation of the form 1.2.2.4: $u'_\xi = u^2 + k\Delta^{-2}\xi^n$.

14. $y'_x = ax^n y^2 + bmx^{m-1} - ab^2 x^{n+2m}$.

Particular solution: $y_0 = bx^m$.

15. $y'_x = -(n+1)x^n y^2 + ax^{n+m+1} y - ax^m$.

Particular solution: $y_0 = x^{-n-1}$.

16. $y'_x = ax^n y^2 + bx^m y + bcx^m - ac^2 x^n$.

Particular solution: $y_0 = -c$.

17. $y'_x = ax^n y^2 - ax^n(bx^m + c)y + bmx^{m-1}$.

Particular solution: $y_0 = bx^m + c$.

18. $y'_x = -anx^{n-1} y^2 + cx^m(ax^n + b)y - cx^m$.

Particular solution: $y_0 = (ax^n + b)^{-1}$.

19. $y'_x = ax^n y^2 + bx^m y + ckx^{k-1} - bcx^{m+k} - ac^2 x^{n+2k}.$

Particular solution: $y_0 = cx^k$.

20. $y'_x = (ax^{2n} + bx^{n-1})y^2 + c.$

The substitution $y = -1/w$ leads to an equation of the form 1.2.2.8:

$$w'_x = cw^2 + ax^{2n} + bx^{n-1}.$$

21. $xy'_x = ay^2 + by + cx^{2b}.$

The transformation $t = x^b$, $w = x^{-b}y$ leads to an equation with separation of variables: $bw'_t = aw^2 + c$.

22. $xy'_x = ay^2 + by + cx^n.$

The transformation $\xi = x^b$, $\eta = yx^{-b}$ reduces this equation to the special Riccati equation 1.2.2.4:

$$\eta'_\xi = \frac{a}{b}\eta^2 + \frac{c}{b}\xi^m, \qquad \text{where} \quad m = \frac{n}{b} - 2.$$

23. $xy'_x = ay^2 + (n + bx^n)y + cx^{2n}.$

The substitution $y = wx^n$ leads to an equation with separation of variables:

$$w'_x = x^{n-1}(aw^2 + bw + c).$$

24. $xy'_x = xy^2 + ay + bx^n.$

The substitution $y = -u'_x/u$ leads to a second order linear equation of the form 2.1.2.62: $xu''_{xx} - au'_x + bx^n u = 0$.

25. $xy'_x + a_3xy^2 + a_2y + a_1x + a_0 = 0.$

The substitution $a_3y = u'_x/u$ leads to a second order linear equation of the form 2.1.2.59: $xu''_{xx} + a_2u'_x + a_3(a_1x + a_0)u = 0$.

26. $xy'_x = ax^n y^2 + by + cx^{-n}.$

The substitution $w = yx^n$ leads to an equation with separation of variables: $xw'_x = aw^2 + (b+n)w + c$.

27. $xy'_x = ax^n y^2 + my - ab^2 x^{n+2m}.$

Particular solution: $y_0 = bx^m$.

28. $xy'_x = x^{2n} y^2 + (m - n)y + x^{2m}.$

Solution: $y = x^{m-n} \tan\left(\dfrac{x^{n+m}}{n+m} + C\right)$.

29. $xy'_x = ax^n y^2 + by + cx^m.$

The transformation $\xi = x^{n-b}$, $\eta = yx^b$ leads to the special Riccati equation1.2.2.4:

$$(n+b)\eta'_\xi = a\eta^2 + c\xi^k, \qquad \text{where} \quad k = \frac{m-n-2b}{n+b}.$$

30. $xy'_x = ax^{2n}y^2 + (bx^n - n)y + c.$

For $n = 0$, this is an equation with separation of variables. For $n \neq 0$, the solution is

$$n \int \frac{dw}{aw^2 + bw + c} = x^n + C, \qquad \text{where} \quad w = yx^n.$$

31. $xy'_x = ax^{2n+m}y^2 + (bx^{n+m} - n)y + cx^m.$

The substitution $w = yx^n$ leads to an equation with separation of variables:

$$w'_x = x^{n+m-1}(aw^2 + bw + c).$$

32. $(a_2x + b_2)(y'_x + \lambda y^2) + (a_1x + b_1)y + a_0x + b_0 = 0.$

The substitution $\lambda y = u'_x/u$ leads to a second order linear equation of the form 2.1.2.103:

$$(a_2x + b_2)u''_{xx} + (a_1x + b_1)u'_x + \lambda(a_0x + b_0)u = 0.$$

33. $(ax + c)y'_x = \alpha(ay + bx)^2 + \beta(ay + bx) - bx + \gamma.$

The substitution $t = ay + bx$ leads to a linear equation with respect to $x = x(t)$:

$$(\alpha at^2 + \beta at + \gamma a + bc)x'_t = ax + c.$$

34. $2x^2y'_x = 2y^2 + xy - 2a^2x.$

Particular solution: $y_0 = a\sqrt{x}.$

35. $2x^2y'_x = 2y^2 + 3xy - 2a^2x.$

Particular solution: $y_0 = a\sqrt{x} - \dfrac{x}{2}.$

36. $x^2y'_x = ax^2y^2 + b.$

Solution:

$$y = \frac{\lambda}{x} - x^{2a\lambda}\Big(\frac{ax}{2a\lambda + 1}x^{2a\lambda} + C\Big)^{-1},$$

where λ is a root of the quadratic equation $a\lambda^2 + \lambda + b = 0$.

37. $x^2y'_x = ax^2y^2 + bxy + c.$

The substitution $w = xy$ leads to an equation with separation of variables:

$$xw'_x = aw^2 + (b+1)w + c.$$

38. $x^2y'_x = x^2y^2 - a^2x^4 + a(1 - 2b)x^2 - b(b+1).$

Particular solution: $y_0 = ax + bx^{-1}.$

39. $x^2y'_x = cx^2y^2 + (ax^2 + bx)y + \alpha x^2 + \beta x + \gamma.$

The substitution $cy = -u'_x/u$ leads to a second order linear equation of the form 2.1.2.134:

$$x^2u''_{xx} - x(ax + b)u'_x + c(\alpha x^2 + \beta x + \gamma)u = 0.$$

40. $x^2y'_x = ax^2y^2 + bx^n + c.$

Having set $w = xy + A$, where A is a root of the quadratic equation $aA^2 - A + c = 0$, we arrive at an equation of the form 1.2.2.22: $xw'_x = aw^2 + (1 - 2aA)w + bx^n$.

41. $x^2y'_x = x^2y^2 + \dfrac{1 - n^2}{4} + ax^{2m}(bx^m + c)^n.$

The transformation

$$\xi = bx^m + c, \quad w = \frac{1}{bm}x^{1-m}y + \frac{1-m}{2bm}x^{-m}$$

leads to an equation of the form 1.2.2.4: $w'_\xi = w^2 + a(bm)^{-2}\xi^n$.

42. $x^2y'_x = ax^2y^2 + bxy + cx^m + s.$

The substitution $ay = -u'_x/u$ leads to a second order linear equation of the form 2.1.2.127: $x^2u''_{xx} - bxu'_x + a(cx^m + s)u = 0$.

43. $x^2y'_x = ax^2y^2 + bxy + cx^{2m} + sx^m.$

The substitution $ay = -u'_x/u$ leads to a second order linear equation of the form 2.1.2.128:

$$x^2u''_{xx} - bxu'_x + ax^m(cx^m + s)u = 0.$$

44. $x^2y'_x = cx^2y^2 + (ax^n + b)xy + \alpha x^{2n} + \beta x^n + \gamma.$

The substitution $cy = -u'_x/u$ leads to a second order linear equation of the form 2.1.2.141:

$$x^2u''_{xx} - (ax^n + b)xu'_x + c(\alpha x^{2n} + \beta x^n + \gamma)u = 0.$$

45. $x^2y'_x = (\alpha x^{2n} + \beta x^n + \gamma)y^2 + (ax^n + b)xy + cx^2.$

The substitution $y = -1/w$ leads to an equation of the form 1.2.2.44:

$$x^2w'_x = cx^2w^2 - (ax^n + b)xw + \alpha x^{2n} + \beta x^n + \gamma.$$

46. $(x^2 - 1)y'_x + \lambda(y^2 - 2xy + 1) = 0.$

The substitution $y = \dfrac{2\lambda - 1}{\lambda}x + \dfrac{1-\lambda}{\lambda}\dfrac{1}{u(x)}$ leads to an equation of the same form:

$$(x^2 - 1)u'_x + (\lambda - 1)(u^2 - 2xu + 1) = 0.$$

If $\lambda = n$ is a positive integer, then by using the above substitution, the original equation can be reduced to an equation of the same form, wherein $\lambda = 1$, i.e., to an equation of the form 1.2.2.49 with $a = 1$, $b = -1$.

47. $(ax^2 + b)y'_x + \alpha y^2 + \beta xy + \dfrac{b}{\alpha}(a + \beta) = 0.$

Particular solution: $y_0 = -\dfrac{a+\beta}{\alpha}x.$

48. $(ax^2 + b)y'_x + \alpha y^2 + \beta xy + \gamma = 0.$

The substitution $y = -\dfrac{a+\beta}{\alpha}x - \dfrac{1}{u(x)}$ leads to an equation of the same form:

$$(ax^2 + b)u'_x + \Big(\gamma - \frac{a+\beta}{\alpha}b\Big)u^2 + (2a + \beta)xu + \alpha = 0.$$

49. $(ax^2 + b)y'_x + y^2 - 2xy + (1 - a)x^2 - b = 0.$

Solution: $y = x + \Big(\displaystyle\int \frac{dx}{ax^2 + b} + C\Big)^{-1}.$

50. $(ax^2 + bx + c)y'_x = y^2 + (2\lambda x + b)y + \lambda(\lambda - a)x^2 + \mu.$

Particular solutions:

$$y_0 = -\lambda x + A, \qquad \text{where} \quad A = \tfrac{1}{2}\big(-b \pm \sqrt{b^2 - 4\mu - 4\lambda c}\big).$$

51. $(ax^2 + bx + c)y'_x = y^2 + (ax + \mu)y - \lambda^2 x^2 + \lambda(b - \mu)x + \lambda c.$

Particular solution: $y_0 = \lambda x.$

52. $(a_2x^2 + b_2x + c_2)y'_x = y^2 + (a_1x + b_1)y - \lambda(\lambda + a_1 - a_2)x^2 + \lambda(b_2 - b_1)x + \lambda c_2.$

Particular solution: $y_0 = \lambda x.$

53. $(a_2x^2 + b_2x + c_2)y'_x = y^2 + (a_1x + b_1)y + a_0x^2 + b_0x + c_0.$

Let λ and β be roots of the system of the quadratic equations

$$\lambda^2 + \lambda(a_1 - a_2) + a_0 = 0, \quad \beta^2 + \beta b_1 + c_0 - \lambda c_2 = 0,$$

that are solved consecutively (in the general case there are four roots). If some of roots satisfy the condition $2\lambda\beta + \lambda b_1 + \beta a_1 + b_0 - \lambda b_2 = 0$, the original equation possesses a particular solution: $y_0 = \lambda x + \beta$.

54. $(x - a)(x - b)y'_x + y^2 + k(y + x - a)(y + x - b) = 0.$

To the case of $k = 0$ corresponds an equation with separation of variables. To the case of $k = -1$ corresponds a linear equation. For $k \neq -1$ and 0, with the aid of the substitution $ku(x) = y + k(y + x)$, we obtain the general solution:

$$\frac{y + k(y + x - a)}{y + k(y + x - b)}\Big(\frac{x - a}{x - b}\Big)^k = C \qquad \text{if } a \neq b,$$

$$\frac{1}{y + k(y + x - a)} + \frac{1}{x - a} = C \qquad \text{if } a = b.$$

55. $(c_2x^2 + b_2x + a_2)(y'_x + \lambda y^2) + (b_1x + a_1)y + a_0 = 0.$

The substitution $\lambda y = u'_x/u$ leads to a second order linear equation of the form 2.1.2.166:

$$(c_2x^2 + b_2x + a_2)u''_{xx} + (b_1x + a_1)u'_x + \lambda a_0 u = 0.$$

56. $x^3y'_x = ax^3y^2 + (bx^2 + c)y + sx.$

The substitution $ay = -u'_x/u$ leads to a second order linear equation of the form 2.1.2.170:

$$x^3u''_{xx} - (bx^2 + c)u'_x + asxu = 0.$$

57. $x^3y'_x = ax^3y^2 + x(bx + c)y + \alpha x + \beta.$

The substitution $ay = -u'_x/u$ leads to a second order linear equation of the form 2.1.2.173:

$$x^3u''_{xx} - x(bx + c)u'_x + a(\alpha x + \beta)u = 0.$$

58. $x(x^2 + a)(y'_x + \lambda y^2) + (bx^2 + c)y + sx = 0.$

The substitution $\lambda y = u'_x/u$ leads to a second order linear equation of the form 2.1.2.177:

$$x(x^2 + a)u''_{xx} + (bx^2 + c)u'_x + \lambda sxu = 0.$$

59. $x^2(x + a)(y'_x + \lambda y^2) + x(bx + c)y + \alpha x + \beta = 0.$

The substitution $\lambda y = u'_x/u$ leads to a second order linear equation of the form 2.1.2.181:

$$x^2(x + a)u''_{xx} + x(bx + c)u'_x + \lambda(\alpha x + \beta)u = 0.$$

60. $(ax^2 + bx + c)(xy'_x - y) - y^2 + x^2 = 0.$

Solution: $\ln\left|\dfrac{y - x}{y + x}\right| = C + 2\displaystyle\int \frac{dx}{ax^2 + bx + c}.$

61. $x^4y'_x = -x^4y^2 - a^2.$

Solution: $y = \dfrac{1}{x} + \dfrac{a}{x^2}\tan\left(\dfrac{a}{x} + C\right).$

62. $x^2(x^2 + a)(y'_x + \lambda y^2) + x(bx^2 + c)y + s = 0.$

The substitution $\lambda y = u'_x/u$ leads to a second order linear equation of the form 2.1.2.206:

$$x^2(x^2 + a)u''_{xx} + x(bx^2 + c)u'_x + \lambda su = 0.$$

63. $ax^2(x - 1)^2(y'_x + \lambda y^2) + bx^2 + cx + s = 0.$

The substitution $\lambda y = u'_x/u$ leads to a second order linear equation of the form 2.1.2.205:

$$ax^2(x - 1)^2u''_{xx} + \lambda(bx^2 + cx + s)u = 0.$$

64. $a(x^2-1)^2(y'_x+\lambda y^2)+bx(x^2-1)y+cx^2+dx+s=0.$

The substitution $\lambda y = u'_x/u$ leads to a second order linear equation of the form 2.1.2.213:

$$a(x^2-1)^2u''_{xx}+bx(x^2-1)u'_x+\lambda(cx^2+dx+s)u=0.$$

65. $(ax^2+bx+c)^2(y'_x+y^2)+A=0.$

The substitution $y=u'_x/u$ leads to a second order linear equation of the form 2.1.2.220:

$$(ax^2+bx+c)^2u''_{xx}+Au=0.$$

66. $x^{n+1}y'_x = ax^{2n}y^2+bx^ny+cx^m+d.$

Having set $w=x^ny+A$, where A is a root of the quadratic equation $aA^2-(b+n)A+d=0$, we arrive at an equation of the form 1.2.2.22:

$$xw'_x = aw^2+(n+b-2aA)w+cx^m.$$

67. $x(ax^k+b)y'_x = \alpha x^ny^2+(\beta-anx^k)y+\gamma x^{-n}.$

The transformation $t=x^ny$, $z=x^{-k}$ leads to an equation with separation of variables:

$$[\alpha t^2+(\beta+bn)t+\gamma]z'_t=-k(bz+a).$$

68. $x^2(ax^b-1)(y'_x+\lambda y^2)+(px^b+q)xy+rx^b+s=0.$

The substitution $\lambda y = u'_x/u$ leads to a second order linear equation of the form 2.1.2.238:

$$x^2(ax^b-1)u''_{xx}+(px^b+q)xu'_x+\lambda(rx^b+s)u=0.$$

69. $(ax^n+bx^m+c)y'_x = cy^2-bx^{m-1}y+ax^{n-2}.$

Particular solution: $y_0=-1/x.$

70. $(ax^n+bx^m+c)y'_x = ax^{n-2}y^2+bx^{m-1}y+c.$

Particular solution: $y_0=x.$

71. $(ax^n+bx^m+c)y'_x = \alpha x^ky^2+\beta x^sy-\alpha\lambda^2x^k+\beta\lambda x^s.$

Particular solution: $y_0=-\lambda.$

72. $(ax^n+bx^m+c)(xy'_x-y)+sx^k(y^2-\lambda x^2)=0.$

Particular solutions: $y_0=\pm x\sqrt{\lambda}.$

73. $(ax^n+bx^m+c)(y'_x-y^2)+an(n-1)x^{n-2}+bm(m-1)x^{m-2}=0.$

Particular solution: $y_0=-\dfrac{anx^{n-1}+bmx^{m-1}}{ax^n+bx^m+c}.$

1.2.3. Equations Containing Exponential Functions

1. $y'_x = ay^2 + be^{\lambda x}$.

The substitution $t = e^{\lambda x}$ leads to an equation of the form 1.2.2.22: $\lambda t y'_t = ay^2 + bt$.

2. $y'_x = y^2 + a\lambda e^{\lambda x} - a^2 e^{2\lambda x}$.

Particular solution: $y_0 = ae^{\lambda x}$.

3. $y'_x = \sigma y^2 + a + be^{\lambda x} + ce^{2\lambda x}$.

The substitution $\sigma y = -u'_x/u$ leads to a second order linear equation of the form 2.1.3.5: $u''_{xx} + \sigma(a + be^{\lambda x} + ce^{2\lambda x})u = 0$.

4. $y'_x = \sigma y^2 + ay + be^x + c$.

The substitution $\sigma y = -u'_x/u$ leads to a second order linear equation of the form 2.1.3.10: $u''_{xx} - au'_x + \sigma(be^x + c)u = 0$.

5. $y'_x = y^2 + by + a(\lambda - b)e^{\lambda x} - a^2 e^{2\lambda x}$.

Particular solution: $y_0 = ae^{\lambda x}$.

6. $y'_x = y^2 + ae^{\lambda x}y - abe^{\lambda x} - b^2$.

Particular solution: $y_0 = b$.

7. $y'_x = y^2 + ae^{2\lambda x}(e^{\lambda x} + b)^n - \frac{1}{4}\lambda^2$.

The transformation

$$\xi = e^{\lambda x} + b, \quad w = \frac{1}{\lambda}\left(e^{-\lambda x}y - \frac{\lambda}{2}e^{-\lambda x}\right)$$

leads to an equation of the form 1.2.2.4: $w'_\xi = w^2 + a\lambda^{-2}\xi^n$.

8. $y'_x = y^2 + ae^{8\lambda x} + be^{6\lambda x} + ce^{4\lambda x} - \lambda^2$.

The transformation

$$\xi = e^{2\lambda x}, \quad w = e^{-2\lambda x}\left(\frac{y}{2\lambda} - \frac{1}{2}\right)$$

leads to an equation of the form 1.2.2.3: $w'_\xi = w^2 + (2\lambda)^{-2}(a\xi^2 + b\xi + c)$.

9. $y'_x = y^2 + axe^{\lambda x}y + ae^{\lambda x}$.

Particular solution: $y_0 = -1/x$.

10. $y'_x = ae^{\lambda x}y^2 + be^{-\lambda x}$.

Solution: $\displaystyle\int \frac{dz}{az^2 + \lambda z + b} = x + C$, where $z = e^{\lambda x}y$.

11. $y'_x = ae^{kx}y^2 + be^{sx}$, $\quad k \neq 0$.

The substitution $t = e^{kx}$ leads to an equation of the form 1.2.2.4: $ky'_t = ay^2 + bt^{s-k}$.

12. $y'_x = be^{\mu x}y^2 + a\lambda e^{\lambda x} - a^2be^{(\mu+2\lambda)x}.$

Particular solution: $y_0 = ae^{\lambda x}$.

13. $y'_x = ae^{\lambda x}y^2 + by + ce^{-\lambda x}.$

The substitution $z = e^{\lambda x}y$ leads to an equation with separation of variables: $z'_x = az^2 + (b+\lambda)z + c$.

14. $y'_x = ae^{\mu x}y^2 + \lambda y - ab^2e^{(\mu+2\lambda)x}.$

Particular solution: $y_0 = be^{\lambda x}$.

15. $y'_x = e^{\lambda x}y^2 + ae^{\mu x}y + a\lambda e^{(\mu-\lambda)x}.$

Particular solution: $y_0 = -\lambda e^{-\lambda x}$.

16. $y'_x = -\lambda e^{\lambda x}y^2 + ae^{\mu x}y - ae^{(\mu-\lambda)x}.$

Particular solution: $y_0 = e^{-\lambda x}$.

17. $y'_x = ae^{\mu x}y^2 + abe^{(\lambda+\mu)x}y - b\lambda e^{\lambda x}.$

Particular solution: $y_0 = -be^{\lambda x}$.

18. $y'_x = ae^{kx}y^2 + by + ce^{sx} + de^{-kx}.$

The substitution $t = e^{kx}$ leads to an equation of the form 1.2.2.42:

$$kt^2y'_t = at^2y^2 + bty + ct^{(k+s)/k} + d.$$

19. $y'_x = ae^{(2\lambda+\mu)x}y^2 + [be^{(\lambda+\mu)x} - \lambda]y + ce^{\mu x}.$

The substitution $w = e^{\lambda x}y$ leads to an equation with separation of variables:

$$w'_x = e^{(\lambda+\mu)x}(aw^2 + bw + c).$$

20. $y'_x = ae^{kx}y^2 + by + ce^{knx} + de^{k(2n+1)x}.$

The substitution $t = e^{kx}$ leads to an equation of the form 1.2.2.43:

$$kt^2y'_t = at^2y^2 + bty + ct^{n+1} + dt^{2(n+1)}.$$

21. $y'_x = e^{\mu x}(y - be^{\lambda x})^2 + b\lambda e^{\lambda x}.$

Particular solution: $y_0 = be^{\lambda x}$.

22. $y'_x = ae^{\lambda x}y^2 + bnx^{n-1} - ab^2e^{\lambda x}x^{2n}.$

Particular solution: $y_0 = bx^n$.

23. $y'_x = e^{\lambda x}y^2 + ax^ny + a\lambda x^ne^{-\lambda x}.$

Particular solution: $y_0 = -\lambda e^{-\lambda x}$.

24. $y'_x = -\lambda e^{\lambda x} y^2 + a x^n e^{\lambda x} y - a x^n.$

Particular solution: $y_0 = e^{-\lambda x}.$

25. $y'_x = a e^{\lambda x} y^2 - a b x^n e^{\lambda x} y + b n x^{n-1}.$

Particular solution: $y_0 = b x^n.$

26. $y'_x = a x^n y^2 + b\lambda e^{\lambda x} - a b^2 x^n e^{2\lambda x}.$

Particular solution: $y_0 = b e^{\lambda x}.$

27. $y'_x = a x^n y^2 + \lambda y - a b^2 x^n e^{2\lambda x}.$

Particular solution: $y_0 = b e^{\lambda x}.$

28. $y'_x = a x^n y^2 - a b x^n e^{\lambda x} y + b\lambda e^{\lambda x}.$

Particular solution: $y_0 = b e^{\lambda x}.$

29. $y'_x = -(k+1) x^k y^2 + a x^{k+1} e^{\lambda x} y - a e^{\lambda x}.$

Particular solution: $y_0 = x^{-k-1}.$

30. $y'_x = a x^n y^2 - a x^n (b e^{\lambda x} + c) y + b\lambda e^{\lambda x}.$

Particular solution: $y_0 = b e^{\lambda x} + c.$

31. $y'_x = a x^n e^{2\lambda x} y^2 + (b x^n e^{\lambda x} - \lambda) y + c x^n.$

The substitution $w = e^{\lambda x} y$ leads to an equation with separation of variables:

$$w'_x = x^n e^{\lambda x}(a w^2 + b w + c).$$

32. $y'_x = a e^{\lambda x}(y - b x^n - c)^2 + b n x^{n-1}.$

Particular solution: $y_0 = b x^n + c.$

33. $x y'_x = a e^{\lambda x} y^2 + k y + a b^2 x^{2k} e^{\lambda x}.$

Solution: $y = b x^k \tan\Bigl(ab \int x^{k-1} e^{\lambda x}\, dx + C\Bigr).$

34. $x y'_x = a x^{2n} e^{\lambda x} y^2 + (b x^n e^{\lambda x} - n) y + c e^{\lambda x}.$

Solution: $\displaystyle\int \frac{dw}{a w^2 + b w + c} = \int x^{n-1} e^{\lambda x}\, dx + C,$ where $w = x^n y.$

35. $(a e^{\lambda x} + b e^{\mu x} + c) y'_x = y^2 + k e^{\nu x} y - m^2 + k m e^{\nu x}.$

Particular solution: $y_0 = -m.$

36. $(a e^{\lambda x} + b e^{\mu x} + c)(y'_x - y^2) + a\lambda^2 e^{\lambda x} + b\mu^2 e^{\mu x} = 0.$

Particular solution: $y_0 = -\dfrac{a\lambda e^{\lambda x} + b\mu e^{\mu x}}{a e^{\lambda x} + b e^{\mu x} + c}.$

37. $y'_x = y^2 + 2a\lambda x e^{\lambda x^2} - a^2 e^{2\lambda x^2}$.

Particular solution: $y_0 = ae^{\lambda x^2}$.

38. $y'_x = ae^{-\lambda x^2} y^2 + \lambda xy + ab^2$.

Solution: $y = be^{\lambda x^2/2} \tan\Bigl(ab \int e^{-\lambda x^2/2}\,dx + C\Bigr)$.

39. $y'_x = ax^n y^2 + \lambda xy + ab^2 x^n e^{\lambda x^2}$.

Solution: $y = be^{\lambda x^2/2} \tan\Bigl(ab \int x^n e^{\lambda x^2/2}\,dx + C\Bigr)$.

40. $x^4(y'_x - y^2) = a + b\exp\Bigl(\dfrac{k}{x}\Bigr) + c\exp\Bigl(\dfrac{2k}{x}\Bigr)$.

The transformation $\xi = 1/x$, $w = -x^2 y - x$ leads to an equation of the form 1.2.3.3: $w'_\xi = w^2 + a + be^{k\xi} + ce^{2k\xi}$.

1.2.4. Equations Containing Hyperbolic Functions

1. $y'_x = \alpha y^2 + \beta + \gamma\cosh x$.

The transformation $x = 2t$, $\alpha y = -u'_x/u$ leads to the modified Mathieu equation 2.1.4.1:

$$u''_{tt} - (a - 2q\cosh 2t)u = 0, \qquad \text{where} \quad a = -4\alpha\beta, \quad q = 2\alpha\gamma.$$

2. $y'_x = y^2 - a^2 + a\lambda\sinh(\lambda x) - a^2\sinh^2(\lambda x)$.

Particular solution: $y_0 = a\cosh(\lambda x)$.

3. $y'_x = y^2 - \lambda^2 + a\cosh^n(\lambda x)\sinh^{-n-4}(\lambda x)$.

The transformation

$$\xi = \coth(\lambda x), \quad w = -\frac{1}{\lambda}\sinh^2(\lambda x)y - \sinh(\lambda x)\cosh(\lambda x)$$

leads to an equation of the form 1.2.2.4: $w'_\xi = w^2 + \lambda^{-2}\xi^n$.

4. $y'_x = y^2 + a\lambda - a(a+\lambda)\tanh^2(\lambda x)$.

Particular solution: $y_0 = a\tanh(\lambda x)$.

5. $y'_x = y^2 + 3a\lambda - \lambda^2 - a(a+\lambda)\tanh^2(\lambda x)$.

Particular solution: $y_0 = a\tanh(\lambda x) - \lambda\coth(\lambda x)$.

6. $y'_x = y^2 + a\lambda - a(a+\lambda)\coth^2(\lambda x)$.

Particular solution: $y_0 = a\coth(\lambda x)$.

7. $y'_x = y^2 - \lambda^2 + 3a\lambda - a(a+\lambda)\coth^2(\lambda x)$.

Particular solution: $y_0 = a\coth(\lambda x) - \lambda\tanh(\lambda x)$.

8. $y'_x = y^2 - 2\lambda^2\tanh^2(\lambda x) - 2\lambda^2\coth^2(\lambda x)$.

Particular solution: $y_0 = \lambda\tanh(\lambda x) + \lambda\coth(\lambda x)$.

9. $y'_x = y^2 + a\lambda + b\lambda - 2ab - a(a+\lambda)\tanh^2(\lambda x) - b(b+\lambda)\coth^2(\lambda x)$.

Particular solution: $y_0 = a\tanh(\lambda x) + b\coth(\lambda x)$.

10. $y'_x = \lambda\sinh(\lambda x)y^2 - \lambda\sinh^3(\lambda x)$.

Particular solution: $y_0 = \cosh(\lambda x)$.

11. $y'_x = a\sinh(\lambda x)y^2 + b\sinh(\lambda x)\cosh^n(\lambda x)$.

The transformation $\xi = \cosh(\lambda x)$, $w = \dfrac{a}{\lambda}y$ leads to an equation of the form 1.2.2.4: $w'_\xi = w^2 + ab\lambda^{-2}\xi^n$.

12. $y'_x = a\cosh(\lambda x)y^2 + b\cosh(\lambda x)\sinh^n(\lambda x)$.

The transformation $\xi = \sinh(\lambda x)$, $w = \dfrac{a}{\lambda}y$ leads to an equation of the form 1.2.2.4: $w'_\xi = w^2 + ab\lambda^{-2}\xi^n$.

13. $y'_x = [a\sinh^2(\lambda x) - \lambda]y^2 - a\sinh^2(\lambda x) + \lambda - a$.

Particular solution: $y_0 = \coth(\lambda x)$.

14. $y'_x = [a\cosh^2(\lambda x) - \lambda]y^2 + a + \lambda - a\cosh^2(\lambda x)$.

Particular solution: $y_0 = \tanh(\lambda x)$.

15. $2y'_x = [a - \lambda + a\cosh(\lambda x)]y^2 + a + \lambda - a\cosh(\lambda x)$.

Particular solution: $y_0 = \tanh\left(\dfrac{\lambda x}{2}\right)$.

1.2.5. Equations Containing Logarithmic Functions

1. $y'_x = y^2 + a\ln(\beta x)y - ab\ln(\beta x) - b^2$.

Particular solution: $y_0 = b$.

2. $y'_x = y^2 + ax\ln^m(bx)y + a\ln^m(bx)$.

Particular solution: $y_0 = -1/x$.

3. $y'_x = ax^n y^2 - abx^{n+1}\ln x\, y + b\ln x + b$.

Particular solution: $y_0 = bx\ln x$.

4. $y'_x = -(n+1)x^n y^2 + a x^{n+1}(\ln x)^m y - a(\ln x)^m.$

Particular solution: $y_0 = x^{-n-1}$.

5. $y'_x = a(\ln x)^n y^2 + bm x^{m-1} - ab^2 x^{2m}(\ln x)^n.$

Particular solution: $y_0 = bx^m$.

6. $y'_x = a(\ln x)^n y^2 - abx(\ln x)^{n+1} y + b\ln x + b.$

Particular solution: $y_0 = bx\ln x$.

7. $y'_x = a(\ln x)^k (y - bx^n - c)^2 + bnx^{n-1}.$

Particular solution: $y_0 = bx^n + c$.

8. $y'_x = a(\ln x)^n y^2 + b(\ln x)^m y + bc(\ln x)^m - ac^2(\ln x)^n.$

Particular solution: $y_0 = -c$.

9. $xy'_x = ay^2 + b\ln x + c.$

The substitution $x = e^t$ leads to an equation of the form 1.2.2.1: $y'_t = ay^2 + bt + c$.

10. $xy'_x = ay^2 + b\ln^k x + c\ln^{2k+2} x.$

The substitution $t = \ln x$ leads to an equation of the form 1.2.2.6 with $k = n - 1$: $y'_t = ay^2 + bt^k + ct^{2k+2}$.

11. $xy'_x = (ay + b\ln x)^2.$

Solution: $\ln x = \displaystyle\int \frac{dz}{az^2 + b} + C$, where $z = ay + b\ln x$.

12. $xy'_x = xy^2 - A^2 x\ln^2(\beta x) + A.$

Particular solution: $y_0 = A\ln(\beta x)$.

13. $xy'_x = xy^2 - A^2 x\ln^{2k}(\beta x) + Ak\ln^{k-1}(\beta x).$

Particular solution: $y_0 = A\ln^k(\beta x)$.

14. $xy'_x = ax^n y^2 + b - ab^2 x^n \ln^2 x.$

Particular solution: $y_0 = b\ln x$.

15. $xy'_x = a\ln^m(\lambda x) y^2 + ky + ab^2 x^{2k}\ln^m(\lambda x).$

Solution: $y = bx^k \tan\Big[ab\displaystyle\int x^{k-1}\ln^m(\lambda x)\,dx + C\Big]$.

16. $xy'_x = ax^n(y + b\ln x)^2 - b.$

Solution: $\dfrac{1}{y + b\ln x} + \dfrac{a}{n}x^n = C$.

17. $xy'_x = ax^{2n}(\ln x)y^2 + (bx^n \ln x - n)y + c \ln x.$

Solution: $\displaystyle\int \frac{dw}{aw^2 + bw + c} = \int x^{n-1} \ln x \, dx + C,$ where $w = x^n y.$

18. $x^2 y'_x = x^2 y^2 + a \ln^2 x + b \ln x + c.$

The transformation $\xi = \ln x$, $w = xy + \frac{1}{2}$ leads to an equation of the form 1.2.2.3: $w'_\xi = w^2 + a\xi^2 + b\xi + c - \frac{1}{4}$.

19. $x^2 y'_x = x^2 y^2 + a(b \ln x + c)^n + \frac{1}{4}.$

The transformation

$$\xi = b \ln x + c, \quad w = \frac{x}{b} y + \frac{1}{2b}$$

leads to an equation of the form 1.2.2.4: $w'_\xi = w^2 + ab^{-2}\xi^n$.

20. $x^2 y'_x = a^2 x^2 y^2 - xy + b^2 \ln^m x.$

The substitution $a^2 y = -u'_x/u$ leads to a second order linear equation of the form 2.1.5.24: $x^2 u''_{xx} + x u'_x + (ab)^2 \ln^m x \, u = 0$.

21. $x^2 \ln(ax)(y'_x - y^2) = 1.$

Particular solution: $y_0 = [x \ln(ax)]^{-1}$.

22. $(a \ln x + b)y'_x = y^2 + c(\ln x)^n y - \lambda^2 + \lambda c(\ln x)^n.$

Particular solution: $y_0 = -\lambda$.

23. $(a \ln x + b)y'_x = (\ln x)^n y^2 + cy - \lambda^2 (\ln x)^n + c\lambda.$

Particular solution: $y_0 = -\lambda$.

1.2.6. Equations Containing Trigonometric Functions

1. $y'_x = \alpha y^2 + \beta + \gamma \sin(\lambda x).$

The substitution $2t = 2\lambda x + \pi$ leads to an equation of the form 1.2.6.2: $\lambda y'_t = \alpha y^2 + \beta + \gamma \cos t$.

2. $y'_x = \alpha y^2 + \beta + \gamma \cos x.$

The transformation $x = 2t$, $\alpha y = -u'_x/u$ leads to the Mathieu equation 2.1.6.4:

$$u''_{tt} + (a - 2q \cos 2t)u = 0, \qquad \text{where} \quad a = 4\alpha\beta, \quad q = -2\alpha\gamma.$$

3. $y'_x = y^2 - a^2 + a\lambda \sin(\lambda x) + a^2 \sin^2(\lambda x).$

Particular solution: $y_0 = -a \cos(\lambda x)$.

4. $y'_x = y^2 - a^2 + a\lambda \cos(\lambda x) + a^2 \cos^2(\lambda x).$

Particular solution: $y_0 = a \sin(\lambda x)$.

5. $y'_x = y^2 + \lambda^2 + c\sin^n(\lambda x)\cos^{-n-4}(\lambda x)$.

This is a special case of equation 1.2.6.6 with $a = 0$, $b = \pi/2$.

6. $y'_x = y^2 + \lambda^2 + c\sin^n(\lambda x + a)\sin^{-n-4}(\lambda x + b)$.

The transformation

$$\xi = \frac{\sin(\lambda x + a)}{\sin(\lambda x + b)}, \qquad w = \frac{\sin^2(\lambda x + b)}{\sin(b - a)}\left[\frac{y}{\lambda} + \cot(\lambda x + b)\right]$$

leads to an equation of the form 1.2.2.4:

$$w'_\xi = w^2 + A\xi^n, \qquad \text{where} \quad A = c[\lambda\sin(b - a)]^{-2}.$$

7. $y'_x = y^2 + a\sin(\beta x)y + ab\sin(\beta x) - b^2$.

Particular solution: $y_0 = -b$.

8. $y'_x = y^2 + ax\sin^m(bx)y + a\sin^m(bx)$.

Particular solution: $y_0 = -1/x$.

9. $y'_x = y^2 + a\lambda + a(\lambda - a)\tan^2(\lambda x)$.

Particular solution: $y_0 = a\tan(\lambda x)$.

10. $y'_x = y^2 + \lambda^2 + 3a\lambda + a(\lambda - a)\tan^2(\lambda x)$.

Particular solution: $y_0 = a\tan(\lambda x) - \lambda\cot(\lambda x)$.

11. $y'_x = y^2 + a\lambda + a(\lambda - a)\cot^2(\lambda x)$.

Particular solution: $y_0 = -a\cot(\lambda x)$.

12. $y'_x = y^2 + \lambda^2 + 3a\lambda + a(\lambda - a)\cot^2(\lambda x)$.

Particular solution: $y_0 = \lambda\tan(\lambda x) - a\cot(\lambda x)$.

13. $y'_x = ay^2 + b\tan x\, y + c$.

Having set $ay = -u'_x/u$, we obtain a second order linear equation of the form 2.1.6.29: $u''_{xx} - b\tan x\, u'_x + acu = 0$.

14. $y'_x = ay^2 + 2ab\tan x\, y + b(ab - 1)\tan^2 x$.

The substitution $u = y + b\tan x$ leads to an equation of the form 1.1.2: $u'_x = au^2 + b$.

15. $y'_x = y^2 - y\tan x + a(1 - a)\cot^2 x$.

Particular solution: $y_0 = -a\cot x$.

16. $y'_x = y^2 - my\tan x + b^2\cos^{2m} x$.

Solution: $y = -b\cos^m x\cot\left(b\int\cos^m x\,dx + C\right)$.

17. $y'_x = y^2 - 2a\cot(ax)y + b^2 - a^2.$

Particular solution: $y_0 = a\cot(ax) - b\cot(bx).$

18. $y'_x = y^2 + my\cot x + b^2(\sin x)^{2m}.$

Solution: $y = -b\sin^m x \cot\Bigl(b\int \sin^m x\,dx + C\Bigr).$

19. $y'_x = y^2 - 2\lambda^2\tan^2(\lambda x) - 2\lambda^2\cot^2(\lambda x).$

Particular solution: $y_0 = \lambda\cot(\lambda x) - \lambda\tan(\lambda x).$

20. $y'_x = y^2 + \lambda a + \lambda b + 2ab + a(\lambda - a)\tan^2(\lambda x) + b(\lambda - b)\cot^2(\lambda x).$

Particular solution: $y_0 = a\tan(\lambda x) - b\cot(\lambda x).$

21. $y'_x = y^2 + ax\tan^m(bx)y + a\tan^m(bx).$

Particular solution: $y_0 = -1/x.$

22. $y'_x = y^2 - \frac{1}{2}\lambda^2 - \frac{3}{4}\lambda^2\tan^2(\lambda x) + a\cos^2(\lambda x)\sin^n(\lambda x).$

The transformation

$$\xi = \sin(\lambda x), \qquad w = \frac{y}{\lambda\cos(\lambda x)} + \frac{\sin(\lambda x)}{2\cos^2(\lambda x)}$$

leads to an equation of the form 1.2.2.4: $w'_\xi = w^2 + a\lambda^{-2}\xi^n.$

23. $y'_x = \lambda\sin(\lambda x)y^2 + \lambda\sin^3(\lambda x).$

Particular solution: $y_0 = -\cos(\lambda x).$

24. $y'_x = \lambda\cos(\lambda x)y^2 + \lambda\cos^3(\lambda x).$

Particular solution: $y_0 = \sin(\lambda x).$

25. $2y'_x = [\lambda + a - a\sin(\lambda x)]y^2 + \lambda - a - a\sin(\lambda x).$

Particular solution: $y_0 = \tan\Bigl(\frac{\lambda x}{2} + \frac{\pi}{4}\Bigr).$

26. $2y'_x = [\lambda + a + a\cos(\lambda x)]y^2 + \lambda - a + a\cos(\lambda x).$

Particular solution: $y_0 = \tan\Bigl(\frac{\lambda x}{2}\Bigr).$

27. $y'_x = [\lambda + a\sin^2(\lambda x)]y^2 + \lambda - a + a\sin^2(\lambda x).$

Particular solution: $y_0 = -\cot(\lambda x).$

28. $y'_x = [\lambda + a\cos^2(\lambda x)]y^2 + \lambda - a + a\cos^2(\lambda x).$

Particular solution: $y_0 = \tan(\lambda x).$

29. $y'_x = a\sin(\lambda x)y^2 + b\sin(\lambda x)\cos^n(\lambda x)$.

The transformation $\xi = \cos(\lambda x)$, $w = -\dfrac{a}{\lambda}y$ leads to an equation of the form 1.2.2.4: $w'_\xi = w^2 + ab\lambda^{-2}\xi^n$.

30. $y'_x = \lambda\sin(\lambda x)y^2 + a\cos^n(\lambda x)y - a\cos^{n-1}(\lambda x)$.

Particular solution: $y_0 = 1/\cos(\lambda x)$.

31. $y'_x = a\cos(\lambda x)y^2 + b\cos(\lambda x)\sin^n(\lambda x)$.

The transformation $\xi = \sin(\lambda x)$, $w = \dfrac{a}{\lambda}y$ leads to an equation of the form 1.2.2.4: $w'_\xi = w^2 + ab\lambda^{-2}\xi^n$.

32. $y'_x = \lambda\sin(\lambda x)y^2 + a\sin(\lambda x)y - a\tan(\lambda x)$.

Particular solution: $y_0 = 1/\cos(\lambda x)$.

33. $y'_x = \lambda\sin(\lambda x)y^2 + ax^n\cos(\lambda x)y - ax^n$.

Particular solution: $y_0 = 1/\cos(\lambda x)$.

34. $y'_x = -(k+1)x^k y^2 + ax^{k+1}(\sin x)^m y - a(\sin x)^m$.

Particular solution: $y_0 = x^{-k-1}$.

35. $y'_x = -(k+1)x^k y^2 + ax^{k+1}(\tan x)^m y - a(\tan x)^m$.

Particular solution: $y_0 = x^{-k-1}$.

36. $y'_x = a\sin^k(\lambda x + \mu)(y - bx^n - c)^2 + bnx^{n-1}$.

Particular solution: $y_0 = bx^n + c$.

37. $y'_x = a\tan^n(\lambda x)y^2 - ab^2\tan^{n+2}(\lambda x) + b\lambda\tan^2(\lambda x) + b\lambda$.

Particular solution: $y_0 = b\tan(\lambda x)$.

38. $y'_x = a\tan^k(\lambda x + \mu)(y - bx^n - c)^2 + bnx^{n-1}$.

Particular solution: $y_0 = bx^n + c$.

39. $xy'_x = a\sin^m(\lambda x)y^2 + ky + ab^2x^{2k}\sin^m(\lambda x)$.

Solution: $y = bx^k\tan\Big[ab\displaystyle\int x^{k-1}\sin^m(\lambda x)\,dx + C\Big]$.

40. $xy'_x = a\tan^m(\lambda x)y^2 + ky + ab^2x^{2k}\tan^m(\lambda x)$.

Solution: $y = bx^k\tan\Big[ab\displaystyle\int x^{k-1}\tan^m(\lambda x)\,dx + C\Big]$.

41. $\sin^{n+1}(2x)y'_x = ay^2 \sin^{2n} x + b\cos^{2n} x.$

The substitution $z = y\tan^n x$ leads to an equation with separation of variables: $2^n \sin(2x)z'_x = az^2 + n2^{n+1}z + b$.

42. $[a\sin(\lambda x) + b]y'_x = y^2 + c\sin(\mu x)y - d^2 + cd\sin(\mu x).$

Particular solution: $y_0 = -d$.

43. $[a\tan(\lambda x) + b]y'_x = y^2 + k\tan(\mu x)y - d^2 + kd\tan(\mu x).$

Particular solution: $y_0 = -d$.

1.2.7. Equations Containing Inverse Trigonometric Functions

► *In the equations 1–9, function* $\arccos x$ *may be substituted for* $\arcsin x$.

1. $y'_x = y^2 + \lambda(\arcsin x)^n y - a^2 + a\lambda(\arcsin x)^n.$

Particular solution: $y_0 = -a$.

2. $y'_x = y^2 + \lambda x(\arcsin x)^n y + \lambda(\arcsin x)^n.$

Particular solution: $y_0 = -1/x$.

3. $y'_x = -(k+1)x^k y^2 + \lambda(\arcsin x)^n(x^{k+1}y - 1).$

Particular solution: $y_0 = x^{-k-1}$.

4. $y'_x = \lambda(\arcsin x)^n y^2 + ay + ab - b^2\lambda(\arcsin x)^n.$

Particular solution: $y_0 = -b$.

5. $y'_x = \lambda(\arcsin x)^n y^2 - b\lambda x^m(\arcsin x)^n y + bmx^{m-1}.$

Particular solution: $y_0 = bx^m$.

6. $y'_x = \lambda(\arcsin x)^n y^2 + \beta mx^{m-1} - \lambda\beta^2 x^{2m}(\arcsin x)^n.$

Particular solution: $y_0 = \beta x^m$.

7. $y'_x = \lambda(\arcsin x)^n(y - ax^m - b)^2 + amx^{m-1}.$

Particular solution: $y_0 = ax^m + b$.

8. $xy'_x = \lambda(\arcsin x)^n y^2 + ky + \lambda b^2 x^{2k}(\arcsin x)^n.$

Solution: $y = bx^k \tan\Big[\lambda b \int x^{k-1}(\arcsin x)^n\,dx + C\Big].$

9. $xy'_x = (ax^{2n}y^2 + bx^n y + c)(\arcsin x)^m - ny.$

The substitution $z = x^n y$ leads to an equation with separation of variables: $z'_x = x^{n-1}(\arcsin x)^m(az^2 + bz + c)$.

▶ *In the equations 10–18, function* $\operatorname{arccot} x$ *may be substituted for* $\arctan x$.

10. $y'_x = y^2 + \lambda(\arctan x)^n y - a^2 + a\lambda(\arctan x)^n.$

Particular solution: $y_0 = -a.$

11. $y'_x = y^2 + \lambda x(\arctan x)^n y + \lambda(\arctan x)^n.$

Particular solution: $y_0 = -1/x.$

12. $y'_x = -(k+1)x^k y^2 + \lambda(\arctan x)^n(x^{k+1}y - 1).$

Particular solution: $y_0 = x^{-k-1}.$

13. $y'_x = \lambda(\arctan x)^n y^2 + ay + ab - b^2\lambda(\arctan x)^n.$

Particular solution: $y_0 = -b.$

14. $y'_x = \lambda(\arctan x)^n y^2 - b\lambda x^m(\arctan x)^n y + bmx^{m-1}.$

Particular solution: $y_0 = bx^m.$

15. $y'_x = \lambda(\arctan x)^n y^2 + bmx^{m-1} - \lambda b^2 x^{2m}(\arctan x)^n.$

Particular solution: $y_0 = bx^m.$

16. $y'_x = \lambda(\arctan x)^n(y - ax^m - b)^2 + amx^{m-1}.$

Particular solution: $y_0 = ax^m + b.$

17. $xy'_x = \lambda(\arctan x)^n y^2 + ky + \lambda b^2 x^{2k}(\arctan x)^n.$

Solution: $y = bx^k \tan\Big[\lambda b \int x^{k-1}(\arctan x)^n\,dx + C\Big].$

18. $xy'_x = (ax^{2n}y^2 + bx^n y + c)(\arctan x)^m - ny.$

The substitution $z = x^n y$ leads to an equation with separation of variables: $z'_x = x^{n-1}(\arctan x)^m(az^2 + bz + c).$

1.2.8. Equations Containing Arbitrary Functions

Notation: $f = f(x)$ *and* $g = g(x)$ *are arbitrary functions;* a, b, n, *and* λ *are arbitrary parameters.*

1. $y'_x = y^2 + fy - a^2 - af.$

Particular solution: $y_0 = a.$

2. $y'_x = fy^2 + ay - ab - b^2 f.$

Particular solution: $y_0 = b.$

3. $y'_x = y^2 + xfy + f.$

Particular solution: $y_0 = -1/x.$

4. $y'_x = fy^2 - ax^n fy + anx^{n-1}.$

Particular solution: $y_0 = ax^n.$

5. $y'_x = fy^2 + anx^{n-1} - a^2x^{2n} f.$

Particular solution: $y_0 = ax^n.$

6. $y'_x = -(n+1)x^n y^2 + x^{n+1} fy - f.$

Particular solution: $y_0 = x^{-n-1}.$

7. $xy'_x = fy^2 + ny + ax^{2n} f.$

Solution with $a > 0$: $y = \sqrt{a}\, x^n \tan\Bigl(\sqrt{a} \int x^{n-1} f\, dx + C\Bigr).$

Solution with $a < 0$: $y = \sqrt{|a|}\, x^n \tanh\Bigl(-\sqrt{|a|} \int x^{n-1} f\, dx + C\Bigr).$

8. $xy'_x = x^{2n} fy^2 + (ax^n f - n)y + bf.$

The substitution $z = x^n y$ leads to an equation with separation of variables: $z'_x = x^{n-1} f(x)(z^2 + az + b)$.

9. $y'_x = fy^2 + gy - a^2 f - ag.$

Particular solution: $y_0 = a.$

10. $y'_x = fy^2 + gy + anx^{n-1} - ax^n g - a^2 fx^{2n}.$

Particular solution: $y_0 = ax^n.$

11. $y'_x = fy^2 - ax^n gy + anx^{n-1} + a^2x^{2n}(g - f).$

Particular solution: $y_0 = ax^n.$

12. $y'_x = -f'_x y^2 + fgy - g.$

Particular solution: $y_0 = 1/f.$

13. $y'_x = fy^2 - fgy + g'_x.$

Particular solution: $y_0 = g.$

14. $y'_x = g(y - f)^2 + f'_x.$

Particular solution: $y_0 = f.$

15. $y'_x = \dfrac{f'_x}{g} y^2 - \dfrac{g'_x}{f}.$

Particular solution: $y_0 = -g/f.$

16. $f^2 y'_x - f'_x y^2 + g(y - f) = 0.$

Particular solution: $y_0 = f$.

17. $y'_x = y^2 - \dfrac{f''_{xx}}{f}.$

Particular solution: $y_0 = -f'_x/f$.

18. $y'_x = ae^{\lambda x}y^2 + ae^{\lambda x}fy + \lambda f.$

Particular solution: $y_0 = -\dfrac{\lambda}{a}e^{-\lambda x}$.

19. $y'_x = fy^2 - ae^{\lambda x}fy + a\lambda e^{\lambda x}.$

Particular solution: $y_0 = ae^{\lambda x}$.

20. $y'_x = fy^2 + a\lambda e^{\lambda x} - a^2 e^{2\lambda x}f.$

Particular solution: $y_0 = ae^{\lambda x}$.

21. $y'_x = fy^2 + \lambda y + ae^{2\lambda x}f.$

Solution with $a > 0$: $y = \sqrt{a}\,e^{\lambda x}\tan\Bigl(\sqrt{a}\int e^{\lambda x} f\,dx + C\Bigr)$.

Solution with $a < 0$: $y = \sqrt{|a|}\,e^{\lambda x}\tanh\Bigl(-\sqrt{|a|}\int e^{\lambda x} f\,dx + C\Bigr)$.

22. $y'_x = fy^2 - f(ae^{\lambda x} + b)y + a\lambda e^{\lambda x}.$

Particular solution: $y_0 = ae^{\lambda x} + b$.

23. $y'_x = e^{\lambda x}fy^2 + (af - \lambda)y + be^{-\lambda x}f.$

The substitution $z = e^{\lambda x}y$ leads to an equation with separation of variables: $z'_x = f(x)(z^2 + az + b)$.

24. $y'_x = fy^2 + gy + a\lambda e^{\lambda x} - ae^{\lambda x}g - a^2 e^{2\lambda x}f.$

Particular solution: $y_0 = ae^{\lambda x}$.

25. $y'_x = fy^2 - ae^{\lambda x}gy + a\lambda e^{\lambda x} + a^2 e^{2\lambda x}(g - f).$

Particular solution: $y_0 = ae^{\lambda x}$.

26. $y'_x = fy^2 + 2a\lambda xe^{\lambda x^2} - a^2 fe^{2\lambda x^2}.$

Particular solution: $y_0 = ae^{\lambda x^2}$.

27. $y'_x = fy^2 + \lambda xy + afe^{\lambda x^2}.$

Solution with $a > 0$: $y = \sqrt{a}\,e^{\lambda x^2/2}\tan\Bigl(\sqrt{a}\int e^{\lambda x^2/2} f\,dx + C\Bigr)$.

Solution with $a < 0$: $y = \sqrt{|a|}\,e^{\lambda x^2/2}\tanh\Bigl(-\sqrt{|a|}\int e^{\lambda x^2/2} f\,dx + C\Bigr)$.

28. $y'_x = f'_x y^2 + ae^{\lambda x} f y + ae^{\lambda x}$.

Particular solution: $y_0 = -1/f$.

29. $y'_x = f y^2 + g'_x y + a f e^{2g}$.

Solution with $a > 0$: $y = \sqrt{a}\, e^g \tan\Bigl(\sqrt{a} \int f e^g\, dx + C\Bigr)$.

Solution with $a < 0$: $y = \sqrt{|a|}\, e^g \tanh\Bigl(-\sqrt{|a|} \int f e^g\, dx + C\Bigr)$.

30. $y'_x = f y^2 - a \tanh^2(\lambda x)(af + \lambda) + a\lambda$.

Particular solution: $y_0 = a \tanh(\lambda x)$.

31. $y'_x = f y^2 - a \coth^2(\lambda x)(af + \lambda) + a\lambda$.

Particular solution: $y_0 = a \coth(\lambda x)$.

32. $y'_x = f y^2 - a^2 f + a\lambda \sinh(\lambda x) - a^2 f \sinh^2(\lambda x)$.

Particular solution: $y_0 = a \cosh(\lambda x)$.

33. $x y'_x = f y^2 + a + a^2 f (\ln x)^2$.

Particular solution: $y_0 = a \ln x$.

34. $x y'_x = f (y + a \ln x)^2 - a$.

Solution: $\dfrac{1}{y + a \ln x} + \displaystyle\int \frac{f(x)}{x}\, dx = C$.

35. $y'_x = f y^2 - a x \ln x\, f y + a \ln x + a$.

Particular solution: $y_0 = a x \ln x$.

36. $y'_x = -a \ln x\, y^2 + a f (x \ln x - x) y - f$.

Particular solution: $y_0 = \dfrac{1}{a(x \ln x - x)}$.

37. $y'_x = \lambda \sin(\lambda x) y^2 + f \cos(\lambda x) y - f$.

Particular solution: $y_0 = \dfrac{1}{\cos(\lambda x)}$.

38. $y'_x = f y^2 - a^2 f + a\lambda \sin(\lambda x) + a^2 f \sin^2(\lambda x)$.

Particular solution: $y_0 = -a \cos(\lambda x)$.

39. $y'_x = f y^2 - a^2 f + a\lambda \cos(\lambda x) + a^2 f \cos^2(\lambda x)$.

Particular solution: $y_0 = a \sin(\lambda x)$.

40. $y'_x = f y^2 - a \tan^2(\lambda x)(af - \lambda) + a\lambda$.

Particular solution: $y_0 = a \tan(\lambda x)$.

41. $y'_x = f y^2 - a \cot^2(\lambda x)(af - \lambda) + a\lambda$.

Particular solution: $y_0 = -a \cot(\lambda x)$.

1.2.9. Some transformations

Notation: f, g, and h are arbitrary composite functions of their argument which is written in parentheses following the function name (the argument is a function of x).

1. $y'_x = y^2 + a^2 f(ax+b)$.

The transformation $\xi = ax+b$, $u = y/a$ leads to the equation $u'_\xi = u^2 + f(\xi)$.

2. $y'_x = y^2 + x^{-4} f\left(\dfrac{1}{x}\right)$.

The transformation $\xi = 1/x$, $w = -x^2y - x$ leads to the equation $w'_\xi = w^2 + f(\xi)$.

3. $y'_x = y^2 + \dfrac{1}{(cx+d)^4} f\left(\dfrac{ax+b}{cx+d}\right)$.

The transformation

$$\xi = \frac{ax+b}{cx+d}, \quad w = \frac{1}{\Delta}[(cx+d)^2y + c(cx+d)], \quad \text{where } \Delta = ad - bc,$$

leads to a simpler equation: $w'_\xi = w^2 + \Delta^{-2} f(\xi)$.

4. $x^2 y'_x = x^4 f(x) y^2 + 1$.

The substitution $u = -\dfrac{1}{x^2y} - \dfrac{1}{x}$ leads to the equation $u'_x = u^2 + f(x)$.

5. $x^2 y'_x = x^2 y^2 + \dfrac{1-n^2}{4} + x^{2n} f(ax^n + b)$.

The transformation $\xi = ax^n + b$, $w = \dfrac{x^{1-n}}{an} y + \dfrac{1-n}{2an} x^{-n}$ leads to a simpler equation: $w'_\xi = w^2 + (an)^{-2} f(\xi)$.

6. $y'_x = f(x)y^2 + g(x)y + h(x)$.

The substitution $y = -1/w$ leads to an equation of the same form: $w'_x = h(x)w^2 - g(x)w + f(x)$.

7. $y'_x = y^2 - \dfrac{\lambda^2}{4} + e^{2\lambda x} f(e^{\lambda x})$.

The transformation $\xi = e^{\lambda x}$, $w = \frac{1}{\lambda} e^{-\lambda x} y - \frac{1}{2} e^{-\lambda x}$ leads to to a simpler equation: $w'_\xi = w^2 + \lambda^{-2} f(\xi)$.

8. $y'_x = y^2 - \dfrac{\lambda^2}{4} + \dfrac{e^{2\lambda x}}{(ce^{\lambda x} + d)^4} f\left(\dfrac{ae^{\lambda x} + b}{ce^{\lambda x} + d}\right)$.

The transformation

$$\xi = \frac{ae^{\lambda x} + b}{ce^{\lambda x} + d}, \quad w = \frac{(ce^{\lambda x} + d)^2}{\Delta \lambda e^{\lambda x}} y + \frac{c^2 e^{2\lambda x} - d^2}{2\Delta e^{\lambda x}}, \qquad \text{where } \Delta = ad - bc,$$

leads to a simpler equation: $w'_\xi = w^2 + (\Delta\lambda)^{-2} f(\xi)$.

9. $y'_x = y^2 - \lambda^2 + \sinh^{-4}(\lambda x) f\big(\coth(\lambda x)\big).$

The transformation

$$\xi = \coth(\lambda x), \quad w = -\frac{1}{\lambda}\sinh^2(\lambda x) y - \frac{1}{2}\sinh(2\lambda x)$$

leads to a simpler equation: $w'_\xi = w^2 + \lambda^{-2} f(\xi)$.

10. $y'_x = y^2 - \lambda^2 + \cosh^{-4}(\lambda x) f\big(\tanh(\lambda x)\big).$

The transformation

$$\xi = \tanh(\lambda x), \quad w = \frac{1}{\lambda}\cosh^2(\lambda x) y + \frac{1}{2}\sinh(2\lambda x)$$

leads to a simpler equation: $w'_\xi = w^2 + \lambda^{-2} f(\xi)$.

11. $x^2 y'_x = x^2 y^2 + f(a \ln x + b) + \frac{1}{4}.$

The transformation $\xi = a \ln x + b$, $w = \dfrac{x}{a} y + \dfrac{1}{2a}$ leads to a simpler equation: $w'_\xi = w^2 + a^{-2} f(\xi)$.

12. $y'_x = y^2 + \lambda^2 + \sin^{-4}(\lambda x) f\big(\cot(\lambda x)\big).$

The transformation

$$\xi = \cot(\lambda x), \quad w = -\sin^2(\lambda x)\Big[\frac{y}{\lambda} + \cot(\lambda x)\Big]$$

leads to a simpler equation: $w'_\xi = w^2 + \lambda^{-2} f(\xi)$.

13. $y'_x = y^2 + \lambda^2 + \cos^{-4}(\lambda x) f\big(\tan(\lambda x)\big).$

The transformation

$$\xi = \tan(\lambda x), \quad w = \cos^2(\lambda x)\Big[\frac{y}{\lambda} - \tan(\lambda x)\Big]$$

leads to a simpler equation: $w'_\xi = w^2 + \lambda^{-2} f(\xi)$.

14. $y'_x = y^2 + \lambda^2 + \sin^{-4}(\lambda x + b) f\left(\dfrac{\sin(\lambda x + a)}{\sin(\lambda x + b)}\right).$

The transformation

$$\xi = \frac{\sin(\lambda x + a)}{\sin(\lambda x + b)}, \quad w = \frac{\sin^2(\lambda x + b)}{\sin(b - a)}\Big[\frac{y}{\lambda} + \cot(\lambda x + b)\Big]$$

leads to a simpler equation: $w'_\xi = w^2 + [\lambda \sin(b - a)]^{-2} f(\xi)$.

TABLE 1.1
Solvable Abel equations of the form $yy'_x - y = sx + Ax^m$,
A is an arbitrary parameter

m	s	Equation	m	s	Equation
arbitrary	$-\dfrac{2(m+1)}{(m+3)^2}$	1.3.1.10	-1	0	1.3.1.16
-7	$15/4$	1.3.1.56	$-1/2$	$-2/9$	1.3.1.26
-4	6	1.3.1.54	$-1/2$	$-4/25$	1.3.1.22
$-5/2$	12	1.3.1.47	$-1/2$	0	1.3.1.32
-2	0	1.3.1.33	$-1/2$	20	1.3.1.55
-2	2	1.3.1.19	0	arbitrary	1.3.1.2
$-5/3$	$-3/16$	1.3.1.30	0	0	1.3.1.1
$-5/3$	$-9/100$	1.3.1.23	$1/2$	$-12/49$	1.3.1.53
$-5/3$	$63/4$	1.3.1.48	2	$-6/25$	1.3.1.45
$-7/5$	$-5/36$	1.3.1.27	2	$6/25$	1.3.1.46

1.3. Abel Equations of the Second Kind

1.3.1. Equations of the Form $yy'_x - y = f(x)$

Preliminary comments. For the sake of convenience, in Tables 1.1–1.4 are listed all the Abel equations discussed in Section 1.3. Tables 1.1–1.3 classify the Abel equations wherein functions f are of the same form; Table 1.4 gives the other Abel equations. In Table 1.1, equations are arranged in accordance with the growth of parameter m. In Table 1.2, equations are arranged in accordance with the growth of parameter s. In Table 1.3, equations are arranged in accordance with the growth of parameter p. In the rightmost columns of the tables are indicated equation numbers where the corresponding solutions are written out.

Below in this section are given all the Abel equations united into groups wherein all the solutions are expressed in terms of the same functions. A notation is given before each group.

In most cases the solutions are presented in the parametric form

$$x = F_1(\tau, C), \qquad y = F_2(\tau, C),$$

where τ is a parameter, C is an arbitrary constant.

1. $\boldsymbol{yy'_x - y = A}$.

Solution: $x = y - A\ln|y + A| + C$.

2. $\boldsymbol{yy'_x - y = Ax + B}, \qquad \boldsymbol{A \neq 0}$.

Solution in the parametric form:

$$x = C\exp\biggl(-\int\frac{\tau\,d\tau}{\tau^2 - \tau - A}\biggr) - \frac{B}{A}, \qquad y = C\tau\exp\biggl(-\int\frac{\tau\,d\tau}{\tau^2 - \tau - A}\biggr).$$

TABLE 1.2
Solvable Abel equations of the form $yy'_x - y = sx + \sigma A(\alpha x^{1/2} + \beta A + \gamma A^2 x^{-1/2})$, A is an arbitrary parameter

s	σ	α	β	γ	Equation
arbitrary $\neq 0$	arbitrary	0	arbitrary	0	1.3.1.2
$\dfrac{2(m-1)}{(m-3)^2}$	$\dfrac{2}{(m-3)^2}$	$m(m+3)$	$4m^2+3m+9$	$3m(m+3)$	1.3.1.12
$-1/4$	$1/4$	1	5	3	1.3.1.17
$-30/121$	$3/242$	21	35	6	1.3.1.29
$-12/49$	arbitrary	arbitrary	0	0	1.3.1.53
$-12/49$	$1/98$	25	41	10	1.3.1.25
$-12/49$	$6/49$	1	8	5	1.3.1.38
$-12/49$	$2/49$	5	34	15	1.3.1.24
$-12/49$	$4/49$	-10	27	10	1.3.1.31
$-12/49$	$1/49$	5	262	65	1.3.1.52
$-12/49$	$6/49$	-3	23	12	1.3.1.28
$-12/49$	$2/49$	1	166	55	1.3.1.58
$-12/49$	1	$3/49+3B$	$12/49-15B/2$	$15/196+75B/16$	1.3.1.64
$-6/25$	$2/25$	2	19	6	1.3.1.20
$-6/25$	$6/25$	2	7	4	1.3.1.39
$-28/121$	$2/121$	5	106	15	1.3.1.51
$-2/9$	arbitrary	0	arbitrary	arbitrary	1.3.1.3
$-2/9$	arbitrary	0	0	arbitrary	1.3.1.26
$-2/9$	6	0	1	2	1.3.1.11
$-10/49$	$2/49$	4	61	12	1.3.1.57
$-4/25$	arbitrary	0	0	arbitrary	1.3.1.22
$-4/25$	$1/50$	7	49	6	1.3.1.59
0	arbitrary	0	0	arbitrary	1.3.1.32
0	1	1	2	arbitrary	1.3.1.36
0	$n+2$	1	$2(n+2)$	$(n+1)(n+3)$	1.3.1.34
0	$n+2$	1	$2(n+2)$	$2n+3$	1.3.1.35
0	1	-1	2	0	1.3.1.37
0	2	1	4	3	1.3.1.4
0	arbitrary	0	arbitrary	0	1.3.1.1
2	2	-10	19	30	1.3.1.50
2	2	10	31	30	1.3.1.49
20	arbitrary	0	0	arbitrary	1.3.1.55

3. $yy'_x - y = -\frac{2}{9} + A + Bx^{-1/2}.$

1°. Solution in the parametric form with $A > 0$:

$$x = a\left[\frac{(2k-1)C\tau^k - (k-2)\tau - k - 1}{C\tau^k + \tau + 1}\right]^2, \quad y = -6a\frac{(k-1)^2 C\tau^{k+1} + k^2 C\tau^k + \tau}{C\tau^k + \tau + 1},$$

where $A = \frac{2}{3}a(k^2 - k + 1)$, $B = \frac{2}{3}a^{3/2}(2k-1)(k-2)(k+1)$.

TABLE 1.3
Solvable Abel equations of the form $yy'_x - y = sx + \alpha Ax^p + \beta A^2x^q$, A is an arbitrary parameter

p	q	s	α	β	Equation
-1	-3	arbitrary	1	-1	1.3.1.5
-1	-3	$\dfrac{2m+1}{4m^2}$	1	-1	1.3.1.13
-1	-3	0	1	-1	1.3.1.7
$-3/5$	$-7/5$	$-5/36$	arbitrary	arbitrary	1.3.1.62
$-5/11$	$-13/11$	$-33/196$	$286A/3$	$-770A/9$	1.3.1.69
$-1/3$	$-5/3$	$-3/16$	arbitrary	arbitrary	1.3.1.61
$-1/3$	$-5/3$	$-3/16$	3	-12	1.3.1.40
$-1/3$	$-5/3$	$-3/16$	5	-12	1.3.1.15
$-1/3$	$-5/3$	$15/4$	6	-3	1.3.1.60
$-1/5$	$-4/5$	$-10/49$	$13A/5$	$-7A/20$	1.3.1.68
0	$-1/2$	$-2/9$	arbitrary	arbitrary	1.3.1.3
2	3	$4/9$	2	2	1.3.1.14

2°. Solution in the parametric form with $A < 0$:

$$x = \xi[2\lambda e^{-\lambda\tau} - (C_1\lambda - 3C_2\omega)\sin\omega\tau - (3C_1\omega + C_2\lambda)\cos\omega\tau]^2,$$
$$y = 6\xi\{(C_1^2 + C_2^2)\omega^2 - [C_1(\lambda^2-\omega^2) - 2C_2\omega\lambda]e^{-\lambda\tau}\sin\omega\tau - [2C_1\omega\lambda + C_2(\lambda^2-\omega^2)]e^{-\lambda\tau}\cos\omega\tau\},$$

where $A = \frac{1}{9}a(3\omega^2 - \lambda^2)$, $B = \frac{2}{27}a\lambda(9\omega^2 + 5\lambda^2)$, $\xi = a(e^{-\lambda\tau} + C_1\sin\omega\tau + C_2\cos\omega\tau)^{-2}$.

3°. See 1.3.1.26 for the solution with $A = 0$.

4. $\boldsymbol{yy'_x - y = 2A(x^{1/2} + 4A + 3A^2x^{-1/2})}$.

Solution in the parametric form:

$$x = \tfrac{1}{4}a(3 \pm 2\tau L_\pm)^2, \qquad y = \pm aL_\pm(R_\pm^2 L_\pm + \tau), \qquad A = -\tfrac{1}{2}a^{1/2},$$

where

$$L_+ = \begin{cases} \displaystyle\int \frac{d\tau}{1+\tau^2} = \arctan\tau - C, & R_+ = \sqrt{1+\tau^2}, \\ \displaystyle\int \frac{d\tau}{\tau^2-1} = \frac{1}{2}\ln\left|\frac{\tau-1}{\tau+1}\right| - C, & R_+ = \sqrt{\tau^2-1}, \end{cases}$$

$$L_- = \int \frac{d\tau}{1-\tau^2} = \frac{1}{2}\ln\left|\frac{1+\tau}{1-\tau}\right| - C, \qquad R_- = \sqrt{1-\tau^2}.$$

5. $\boldsymbol{yy'_x - y = Ax + Bx^{-1} - B^2x^{-3}}$.

Solution in the parametric form:

$$x = \left(\frac{V}{W}\right)^{-1/2}, \qquad y = (\tau+1)\left(\frac{V}{W}\right)^{-1/2} - B\left(\frac{V}{W}\right)^{1/2},$$

TABLE 1.4
Other solvable Abel equations of the form $yy'_x - y = f(x)$

Function $f(x)$	Equation
$Ax^{k-1} - kBx^k + kB^2x^{2k-1}$	1.3.1.6 (particular solution)
$Ax^2 - \dfrac{9}{625\,A}$	1.3.1.44
$\dfrac{3}{4}x - \dfrac{3}{2}Ax^{1/3} + \dfrac{3}{4}A^2x^{-1/3} - \dfrac{27}{625}A^4x^{-5/3}$	1.3.1.66
$-\dfrac{6}{25}x + \dfrac{7}{5}Ax^{1/3} + \dfrac{31}{3}A^2x^{-1/3} - \dfrac{100}{3}A^4x^{-5/3}$	1.3.1.67
$-\dfrac{6}{25}x + ax^{1/3} + b + cx^{-1/3} + dx^{-2/3}$ (coefficients a, b, c, and d are related by an equality)	1.3.1.65
$-\dfrac{21}{100}x + \dfrac{7}{9}A^2\left(123x^{-1/7} + 280\,Ax^{-5/7} - 400\,A^2x^{-9/7}\right)$	1.3.1.70
$\dfrac{k}{\sqrt{Ax^2 + Bx + C}}$	1.3.1.63
$\dfrac{A}{\sqrt{x^2 + 4A}}$	1.3.1.18
$-\dfrac{3}{32}x + \dfrac{9a^2 - 6x^2}{64\sqrt{x^2 + a^2}}$	1.3.1.43
$\dfrac{3}{8}x + \dfrac{6x^2 + 5a^2}{16\sqrt{x^2 + a^2}}$	1.3.1.21
$\dfrac{3}{8}x + \dfrac{6x^2 + 9A}{16\sqrt{x^2 + A}}$	1.3.1.41
$\dfrac{9}{32}x + \dfrac{30x^2 + 33A}{64\sqrt{x^2 + A}}$	1.3.1.42
$A + B\exp\left(-\dfrac{2x}{A}\right)$	1.3.1.8
$A\left[\exp\left(\dfrac{2x}{A}\right) - 1\right]$	1.3.1.9
$a^2\lambda e^{2\lambda x} - a(b\lambda + 1)e^{\lambda x} + b$	1.3.1.73 (particular solution)
$a^2\lambda e^{2\lambda x} + a\lambda x e^{\lambda x} + be^{\lambda x}$	1.3.1.74 (particular solution)
$2a^2\lambda\sin(2\lambda x) + 2a\sin(\lambda x)$	1.3.1.75 (particular solution)

where

$$V = \begin{cases} (\tau^2 + \tau - A)\exp\Big[\dfrac{2}{\sqrt{-\Delta}}\arctan\dfrac{2\tau+1}{\sqrt{-\Delta}}\Big] & \text{if } \Delta < 0, \\ (\tau^2 + \tau - A)\exp\Big(-\dfrac{2}{2\tau+1}\Big) & \text{if } \Delta = 0, \\ (\tau^2 + \tau - A)\Big[\dfrac{2\tau+1-\sqrt{\Delta}}{2\tau+1+\sqrt{\Delta}}\Big]^{\frac{1}{\sqrt{\Delta}}} & \text{if } \Delta > 0, \end{cases}$$

$$W = \begin{cases} C + 2B\displaystyle\int \exp\Big[\dfrac{2}{\sqrt{-\Delta}}\arctan\dfrac{2\tau+1}{\sqrt{-\Delta}}\Big]\,d\tau & \text{if } \Delta < 0, \\ C + 2B\displaystyle\int \exp\Big(-\dfrac{2}{2\tau+1}\Big)\,d\tau & \text{if } \Delta = 0, \\ C + 2B\displaystyle\int \Big[\dfrac{2\tau+1-\sqrt{\Delta}}{2\tau+1+\sqrt{\Delta}}\Big]^{\frac{1}{\sqrt{\Delta}}}\,d\tau & \text{if } \Delta > 0, \end{cases}$$

where $\Delta = 4A + 1$.

6. $\boldsymbol{yy'_x - y = Ax^{k-1} - kBx^k + kB^2x^{2k-1}}$.

Particular solution: $y_0 = x - Bx^k - \dfrac{A}{kB}$.

7. $\boldsymbol{yy'_x - y = Ax^{-1} - A^2x^{-3}}$.

Solution in the parametric form:

$$x = a\tau^{-1}(\tau - \ln|1+\tau| - C)^{1/2},$$
$$y = a\Big[\frac{1+\tau}{\tau}(\tau - \ln|1+\tau| - C)^{1/2} - \frac{1}{2}\tau(\tau - \ln|1+\tau| - C)^{-1/2}\Big],$$

where $A = a^2/2$.

8. $\boldsymbol{yy'_x - y = A + Be^{-2x/A}}$.

Solution in the parametric form:

$$x = A\ln\left|\frac{\sqrt{\tau^2+AB}}{A\ln\big|\tau+\sqrt{\tau^2+AB}\big|+C}\right|, \qquad y = \tau\frac{A\ln\big|\tau+\sqrt{\tau^2+AB}\big|+C}{\sqrt{\tau^2+AB}} - A.$$

9. $\boldsymbol{yy'_x - y = A(e^{2x/A} - 1)}$.

Solution in the parametric form:

$$x = A\ln\left|\frac{\tau^2+1}{\tau}(\arctan\tau - C)\right|, \qquad y = \frac{A}{\tau}\big[\tau + (\tau^2-1)(\arctan\tau - C)\big].$$

▶ *In the solutions of equations 10–15, the following notation is used:*

$$E_{ml} = \int (1 \pm \tau^{m+1})^{\frac{1}{l-2}}\,d\tau - C, \qquad E_m = E_{m0} = \int (1 \pm \tau^{m+1})^{-1/2}\,d\tau - C,$$
$$R_m = \sqrt{1 \pm \tau^{m+1}}, \qquad F_m = R_m E_m - \tau.$$

10. $yy'_x - y = -\dfrac{2(m+1)}{(m+3)^2}x + Ax^m.$

Solution in the parametric form:

$$x = \frac{m+3}{m-1}aE_m^{\frac{2}{m-1}}\tau, \qquad y = aE_m^{\frac{2}{m-1}}\Big(R_m E_m + \frac{2}{m-1}\tau\Big),$$

where $A = \pm\dfrac{m+1}{2}\Big(\dfrac{m-1}{m+3}\Big)^{m+1}a^{1-m}$.

11. $yy'_x - y = -\frac{2}{9}x + 6A^2(1 + 2Ax^{-1/2}), \qquad A > 0.$

Solution in the parametric form:

$$x = A^2R^{-4}E^{-2}(R^2E \pm 6\tau^{1/2})^2, \qquad y = -12A^2R^{-4}E^{-2}(R^2E - 2\tau),$$

where $E = E_{-1/2,\,3/2}$, $R = R_{-1/2}$.

12. $yy'_x - y = \dfrac{2(m-1)}{(m-3)^2}x$

$$+\frac{2A}{(m-3)^2}[m(m+3)x^{1/2} + (4m^2+3m+9)A + 3m(m+3)A^2x^{-1/2}].$$

Solution in the parametric form:

$$x = \frac{a}{(m-3)^2}\tau^{-2}[(m-3)R_mE_m + 3\tau]^2,$$

$$y = \frac{a}{m-3}\tau^{-2}E_m[\pm(m-1)\tau^{m+1}E_m - 2E_m + 2\tau R_m],$$

where $A = -\dfrac{a^{1/2}}{m-3}$.

13. $yy'_x - y = \dfrac{2m+1}{4m^2}x + Ax^{-1} - A^2x^{-3}.$

Solution in the parametric form:

$$x = \frac{1}{a}\tau^{-1/2}R_m^{-2}E^{1/2}, \qquad y = \frac{1}{2ma}\tau^{-1/2}R_m^{-2}E^{-1/2}\{\tau - [1 \mp (2m+1)\tau^{m+1}R_m^2E\},$$

where $a^2 = -2mA$, $E = E_{m,\,3/2}$.

14. $yy'_x - y = \frac{4}{9}x + 2Ax^2 + 2A^2x^3.$

Solution in the parametric form:

$$x = \frac{1}{3A}\tau^{-1}F_3, \qquad y = \frac{1}{9A}\tau^{-2}E_3(\tau R_3 - E_3 \pm \tau^4E_3).$$

15. $yy'_x - y = -\frac{3}{16}x + 5Ax^{-1/3} - 12A^2x^{-5/3}.$

Solution in the parametric form:

$$x = a\tau^{1/2}E^{-3/2}F^{3/2}, \qquad y = \frac{a}{4}\tau^{1/2}E^{-3/2}F^{-1/2}(F^2 - 2\tau F - \tau^{-2/3}E^2),$$

where $A = a^{4/3}/24$, $E = E_{-5/3}$, $F = F_{-5/3}$.

▶ *In the solutions of equations 16–18, the following notation is used:*

$$f = \int \exp(\mp\tau^2)\,d\tau - C, \qquad g = 2\tau\left[\int \exp(\mp\tau^2)\,d\tau - C\right] \pm \exp(\mp\tau^2).$$

16. $yy'_x - y = Ax^{-1}.$

Solution in the parametric form:

$$x = af^{-1}\exp(\mp\tau^2), \quad y = af^{-1}\left[\exp(\mp\tau^2) \pm 2\tau f\right],$$

where $A = \mp 2a^2$.

17. $yy'_x - y = -\frac{1}{4}x + \frac{1}{4}A(x^{1/2} + 5A + 3A^2x^{-1/2}).$

Solution in the parametric form:

$$x = \frac{a}{16}\left[3 \pm 8\tau f \exp(\pm\tau^2)\right]^2, \quad y = af\exp(\pm\tau^2)\left[(2\tau^2 \pm 1)f\exp(\pm\tau^2) \pm \tau\right],$$

where $A = \frac{1}{4}\sqrt{a}$.

18. $yy'_x - y = \pm\dfrac{2a^2}{\sqrt{x^2 \pm 8a^2}}.$

Solution in the parametric form:

$$x = \pm a(fg)^{-1}(g^2 \mp 2f^2), \qquad y = a(fg)^{-1}\left[\exp(\mp\tau^2)g - 2f^2\right].$$

▶ *In the solutions of equations 19–21, the following notation is used:*

$$E = \sqrt{\tau(\tau+1)} - \ln\left|C\left(\sqrt{\tau} + \sqrt{\tau+1}\right)\right|,$$
$$R = \sqrt{\frac{\tau+1}{\tau}}, \qquad F = 1 - \sqrt{\frac{\tau+1}{\tau}}\ln\left|C\left(\sqrt{\tau} + \sqrt{\tau+1}\right)\right|.$$

19. $yy'_x - y = 2x + Ax^{-2}.$

Solution in the parametric form:

$$x = \frac{a}{3}E^{-2/3}\tau, \quad y = aE^{-2/3}\left(\frac{2}{3}\tau - RE\right), \qquad \text{where} \quad A = -\frac{243}{2}a^3.$$

20. $yy'_x - y = -\dfrac{6}{25}x + \dfrac{2A}{25}(2x^{1/2} + 19A + 6A^2x^{-1/2}).$

Solution in the parametric form:

$$x = a\tau^{-2}(5RE - 3\tau)^2, \quad y = 5a\tau^{-3}E[(2\tau+3)E - 2\tau^2 R], \qquad \text{where} \quad A = -\sqrt{a}.$$

21. $yy'_x - y = \dfrac{3}{8}x + \dfrac{3}{8}\sqrt{x^2+a^2} - \dfrac{a^2}{16\sqrt{x^2+a^2}}.$

Solution in the parametric form:

$$x = \frac{a}{2\sqrt{2}}\,\frac{E^2 - 2\tau^2 F}{\tau E\sqrt{F}}, \qquad y = \frac{a}{4\sqrt{2}}\,\frac{4\tau F^2 - E^2}{\tau E\sqrt{F}}.$$

▶ *In the solutions of equations 22–25, the following notation is used:*

$$P_2 = \pm(\tau^2 - 1), \qquad P_3 = \tau^3 - 3\tau + C, \qquad P_4 = \pm(\tau^4 - 6\tau^2 + 4C\tau - 3).$$

22. $yy'_x - y = -\frac{4}{25}x + Ax^{-1/2}$.

Solution in the parametric form:

$$x = 5aP_2^2 P_3^{-4/3}, \quad y = 4aP_3^{-4/3}(P_2^2 - \tau P_3), \qquad \text{where} \quad A = \pm\frac{4a}{5}\sqrt{5a}.$$

23. $yy'_x - y = -\dfrac{9}{100}x + Ax^{-5/3}$.

Solution in the parametric form:

$$x = 10aP_3^{3/2} P_4^{-9/8}, \quad y = 9aP_3^{-1/2} P_4^{-9/8}(P_3^2 - P_2 P_4), \qquad \text{where} \quad A = \pm 9a^2(10a)^{2/3}.$$

24. $yy'_x - y = -\dfrac{12}{49}x + \dfrac{2A}{49}(5x^{1/2} + 34A + 15A^2x^{-1/2})$.

Solution in the parametric form:

$$x = aP_2^{-4}(14\tau P_3 - 9P_2^2)^2, \quad y = 28aP_2^{-4}P_3(4\tau^2 P_3 - 3\tau P_2^2 \mp P_2 P_3),$$

where $A = -3\sqrt{a}$.

25. $yy'_x - y = -\dfrac{12}{49}x + \dfrac{A}{98}(25x^{1/2} + 41A + 10A^2x^{-1/2})$.

Solution in the parametric form:

$$x = aP_3^{-4}(21P_2P_4 - 16P_3^2)^2, \quad y = 21aP_3^{-4}P_4(9P_2^2 P_4 \mp P_4^2 - 8P_2P_3^2),$$

where $A = -8\sqrt{a}$.

▶ *In the solutions of equations 26–29, the following notation is used:*

$$S_1 = \exp(\sqrt{3}\,\tau) + C\sin\tau, \qquad S_2 = 2\exp(\sqrt{3}\,\tau) - C\sin\tau + \sqrt{3}\,C\cos\tau,$$
$$S_3 = 2\exp(\sqrt{3}\,\tau) - C\sin\tau - \sqrt{3}\,C\cos\tau, \qquad S_4 = 4S_1S_3 - S_2^2.$$

26. $yy'_x - y = -\frac{2}{9}x + Ax^{-1/2}$.

Solution in the parametric form:

$$x = 3aS_1^{-2}S_2^2, \quad y = 2aS_1^{-2}(S_2^2 - 2S_1S_3), \qquad \text{where} \quad A = 16(3a)^{3/2}.$$

27. $yy'_x - y = -\frac{5}{36}x + Ax^{-7/5}$.

Solution in the parametric form:

$$x = 48aS_1^{5/2}S_4^{-5/4}, \quad y = 5aS_1^{-1/2}S_4^{-5/4}(8S_1^3 - S_2S_4), \qquad \text{where} \quad A = (48a)^{2/5}a^2.$$

28. $\boldsymbol{yy'_x - y = -\frac{12}{49}x + \frac{6A}{49}(-3x^{1/2} + 23A + 12A^2x^{-1/2})}$.

Solution in the parametric form:

$$x = aS_2^{-4}(7S_1S_3 - 2S_2^2)^2, \quad y = -7aS_1S_2^{-4}(4S_1^2S_2 - 4S_1S_3^2 + S_2^2S_3),$$

where $A = \sqrt{a}/2$.

29. $\boldsymbol{yy'_x - y = -\frac{30}{121}x + \frac{3A}{242}(21x^{1/2} + 35A + 6A^2x^{-1/2})}$.

Solution in the parametric form:

$$x = aS_1^{-6}(11S_2S_4 - 64S_1^3)^2, \quad y = -11aS_1^{-6}S_4(S_4^2 - 5S_2^2S_4 + 32S_1^3S_2),$$

where $A = -32\sqrt{a}$.

▶ *In the solutions of equations 30–31, the following notation is used:*

$$T_1 = \tanh(\tau + C) + \tan\tau, \qquad T_2 = \tanh(\tau + C) - \tan\tau,$$
$$\theta_1 = \cosh\tau - \sin(\tau + C), \qquad \theta_2 = \sinh\tau + \cos(\tau + C), \qquad \theta_3 = \sinh\tau - \cos(\tau + C).$$

30. $\boldsymbol{yy'_x - y = -\frac{3}{16}x + Ax^{-5/3}}$.

Solution in the parametric form with $A < 0$:

$$x = 8aT_1^{-3/2}, \quad y = 3aT_1^{-3/2}(2 - T_1T_2), \qquad \text{where} \quad A = -12a^{8/3}.$$

Solution in the parametric form with $A > 0$:

$$x = 4a\theta_1^{3/2}\theta_2^{-3/2}, \quad y = 3a\theta_1^{-1/2}\theta_2^{-3/2}(\theta_1^2 - \theta_2\theta_3), \qquad \text{where} \quad A = 3a^2(4a)^{2/3}.$$

31. $\boldsymbol{yy'_x - y = -\frac{12}{49}x + \frac{4A}{49}(-10x^{1/2} + 27A + 10A^2x^{-1/2})}$.

Solution in the parametric form with $A < 0$:

$$x = a(10 - 7T_1T_2)^2, \quad y = 7aT_1(T_1^3 + 3T_1T_2^2 - 4T_2), \qquad \text{where} \quad A = -2\sqrt{a}.$$

Solution in the parametric form with $A > 0$:

$$x = a\theta_1^{-4}(7\theta_2\theta_3 - 5\theta_1^2), \quad y = -7a\theta_1^{-4}\theta_2(\theta_2^3 - 3\theta_2\theta_3^2 + 2\theta_1^2\theta_3), \qquad \text{where} \quad A = \sqrt{a}.$$

▶ *In the solutions of equations 32–43, the following notation is used:*

$$Z_\nu = \begin{cases} C_1J_\nu(\tau) + C_2Y_\nu(\tau) & \text{for the upper sign (Bessel functions)},\\ C_1I_\nu(\tau) + C_2K_\nu(\tau) & \text{for the lower sign (modified Bessel functions)}, \end{cases}$$

$$f_\nu = \tau(Z_\nu)'_\tau + \nu Z_\nu, \qquad Z = Z_{1/3},$$
$$U_1 = \tau Z'_\tau + \tfrac{1}{3}Z, \qquad U_2 = U_1^2 \pm \tau^2Z^2, \qquad U_3 = \pm\tfrac{2}{3}\tau^2Z^3 - 2U_1U_2.$$

Remark. *The solutions of equations 32–43 contain only the ratio Z'_τ/Z. Therefore, for symmetry, function Z is defined in terms of two "arbitrary" constants C_1 and C_2 (instead, we may set, for instance, $C_1 = 1$, $C_2 = C$).*

32. $yy'_x - y = Ax^{-1/2}$.

Solution in the parametric form:

$$x = a\tau^{-4/3}Z^{-2}U_1^2, \quad y = a\tau^{-4/3}Z^{-2}U_2, \qquad \text{where} \quad A = \mp\tfrac{1}{3}a^{3/2}.$$

33. $yy'_x - y = Ax^{-2}$.

Solution in the parametric form:

$$x = 2a\tau^{4/3}Z^2U_2^{-1}, \quad y = \pm 3a\tau^{-2/3}Z^{-1}U_2^{-1}U_3, \qquad \text{where} \quad A = -36a^3.$$

34. $yy'_x - y = A(n+2)[x^{1/2} + 2(n+2)A + (n+1)(n+3)A^2x^{-1/2}]$.

Solution in the parametric form:

$$x = aZ_\nu^{-2}[f_\nu - (\nu+1)Z_\nu]^2, \quad y = aZ_\nu^{-2}(f_\nu^2 - 2\nu Z_\nu f_\nu \pm \tau^2 Z_\nu^2),$$

where $A = \nu\sqrt{a}$, $\nu = \dfrac{1}{n+2}$.

35. $yy'_x - y = A(n+2)[x^{1/2} + 2(n+2)A + (2n+3)A^2x^{-1/2}]$.

Solution in the parametric form:

$$x = af_\nu^{-2}[\tau^2 Z_\nu \pm (2-\nu)f_\nu]^2, \quad y = \pm a\tau^2 f_\nu^{-2}[f_\nu^2 + 2(1-\nu)Z_\nu f_\nu \pm \tau^2 Z_\nu^2],$$

where $A = \mp\nu\sqrt{a}$, $\nu = \dfrac{1}{n+2}$.

36. $yy'_x - y = Ax^{1/2} + 2A^2 + Bx^{-1/2}$.

Solution in the parametric form:

$$x = A^2 Z_\nu^{-2}(\tau Z'_\nu - Z_\nu)^2, \quad y = A^2 Z_\nu^{-2}[\tau^2(Z'_\nu)^2 - (\nu^2 \mp \tau^2)Z_\nu^2],$$

where $B = (1-\nu^2)A^3$, prime denotes differentiation with respect to τ.

37. $yy'_x - y = 2A^2 - Ax^{1/2}$.

Solution in the parametric form:

$$x = a(Z'_0)^{-2}(\tau Z_0 \pm 2Z'_0)^2, \quad y = \pm a\tau(Z'_0)^{-2}[\tau(Z'_0)^2 + 2Z_0Z'_0 \pm \tau Z_0^2],$$

where $A = \sqrt{a}$, prime denotes differentiation with respect to τ.

38. $yy'_x - y = -\dfrac{12}{49}x + \dfrac{6A}{49}(x^{1/2} + 8A + 5A^2x^{-1/2})$.

Solution in the parametric form:

$$x = 3aU_1^{-4}(5U_1^2 - 7\tau^2Z^2)^2, \quad y = 28a\tau^2Z^2U_1^{-4}(3\tau^2Z^2 - ZU_1 - 3U_1^2),$$

where $A = 2\sqrt{3a}$; Z and U_1 are expressed in term of modified Bessel functions.

39. $yy'_x - y = -\frac{6}{25}x + \frac{6A}{25}(2x^{1/2} + 7A + 4A^2x^{-1/2}).$

Solution in the parametric form:

$$x = a\tau^{-4}Z^{-6}(U_1U_2 - 2U_3)^2, \quad y = 5a\tau^{-4}Z^{-6}U_2(U_2^2 - U_1U_3), \qquad \text{where} \quad A = -\sqrt{a}/2.$$

40. $yy'_x - y = -\frac{3}{16}x + 3Ax^{-1/3} - 12A^2x^{-5/3}.$

Solution in the parametric form:

$$x = a\tau^{-3/2}Z_{3/2}^{-3/2}f_{3/2}^{3/2},$$
$$y = \frac{3a}{4}\tau^{-3/2}Z_{3/2}^{-3/2}f_{3/2}^{-1/2}(f_{3/2}^2 - 2Z_{3/2}f_{3/2} - \tau^2Z_{3/2}^2),$$

where $Z_{3/2}$ and $f_{3/2}$ can be expressed in terms of modified Bessel functions, while $A = \frac{1}{8}a^{4/3}$.

41. $yy'_x - y = \frac{3}{8}x + \frac{3}{8}\sqrt{x^2 \pm b^2} \pm \frac{3b^2}{16\sqrt{x^2 \pm b^2}}.$

Solution in the parametric form:

$$x = -\frac{a}{4}\tau^{-1}Z^{-3/2}U_1^{-1/2}U_2^{-1}(2\tau^2Z^3U_1 \mp 3U_2^2),$$
$$y = \mp\frac{a}{8}\tau^{-1}Z^{-3/2}U_1^{-1/2}U_2^{-1}(3U_2^2 - 12U_1^2U_2 \pm 4\tau^2Z^3U_1),$$

where $b^2 = \frac{3}{2}a^2$.

42. $yy'_x - y = \frac{9}{32}x + \frac{15}{32}\sqrt{x^2 \mp b^2} \mp \frac{3b^2}{64\sqrt{x^2 \mp b^2}}.$

Solution in the parametric form:

$$x = -\frac{a}{2}\tau^{-1}Z^{-3/2}U_2^{-3/2}U_3^{-1/2}(2\tau^2Z^3U_3 \pm 3U_2^3),$$
$$y = \pm\frac{a}{4}\tau^{-1}Z^{-3/2}U_2^{-3/2}U_3^{-1/2}(3U_2^3 \mp \tau^2Z^3U_3 - 3U_3^2),$$

where $b^2 = 6a^2$.

43. $yy'_x - y = -\frac{3}{32}x - \frac{3}{32}\sqrt{x^2 + a^2} + \frac{15a^2}{64\sqrt{x^2 + a^2}}.$

Solution in the parametric form:

$$x = \frac{a}{2}U_2^{-3/2}U_3^{-1}(U_3^2 - U_2^3), \quad y = \frac{a}{24}U_2^{-3/2}U_3^{-1}(3U_3^2 - 12U_2^3 \pm 4\tau^2Z^3U_3).$$

▶ *In the solutions of equations 44–52, the following notation is used:*

$$E_1 = \tau^3\sqrt{\pm(4\wp^3 - 1)} + 3\tau^2\wp \mp 1, \qquad E_2 = \tau^2\wp \mp 1,$$
$$E_3 = \sqrt{\pm(4\wp^3 - 1)} \pm 2\tau\wp^2, \qquad E_4 = \tau\sqrt{\pm(4\wp^3 - 1)} + 2\wp.$$

Function $\wp = \wp(\tau)$ is given implicitly as follows:

$$\tau = \int \frac{d\wp}{\sqrt{\pm(4\wp^3 - 1)}} - C_2.$$

The upper sign in this formula corresponds to the classical elliptic Weierstrass function $\wp = \wp(\tau + C_2,\ 0,\ 1)$.

44. $\boldsymbol{yy'_x - y = Ax^2 - \frac{9}{625}A^{-1}}$.

Solution in the parametric form:

$$x = 5a\Big(\tau^2\wp \mp \frac{1}{2}\Big), \quad y = a\tau^2 E_4, \qquad \text{where} \quad A = \pm\frac{6}{125}a^{-1}.$$

45. $\boldsymbol{yy'_x - y = -\frac{6}{25}x + Ax^2}$.

Solution in the parametric form:

$$x = 5a\tau^2\wp, \quad y = a\tau^2 E_4, \qquad \text{where} \quad A = \pm\frac{6}{125}a^{-1}.$$

46. $\boldsymbol{yy'_x - y = \frac{6}{25}x + Ax^2}$.

Solution in the parametric form:

$$x = 5aE_2, \quad y = a\tau^2 E_4, \qquad \text{where} \quad A = \pm\frac{6}{125}a^{-1}.$$

47. $\boldsymbol{yy'_x - y = 12x + Ax^{-5/2}}$.

Solution in the parametric form:

$$x = a\wp^{-6/7}E_3^{-4/7}, \quad y = a\wp^{-6/7}E_3^{-4/7}(14\wp^2 E_4 - 3), \qquad \text{where} \quad A = \mp 147a^{7/2}.$$

48. $\boldsymbol{yy'_x - y = \frac{63}{4}x + Ax^{-5/3}}$.

Solution in the parametric form:

$$x = 2aE_3^{3/2}E_4^{-9/8}, \quad y = aE_3^{-1/2}E_4^{-9/8}(9E_3^2 \mp 16\wp E_4^2),$$

where $A = -\frac{128}{3}a^2(2a)^{2/3}$.

49. $\boldsymbol{yy'_x - y = 2x + 2A(10x^{1/2} + 31A + 30A^2x^{-1/2})}$.

Solution in the parametric form:

$$x = a\wp^{-2}\big[\tau\sqrt{\pm(4\wp^3 - 1)} - 3\wp\big]^2, \quad y = -2a\tau\wp^{-2}\big[\wp\sqrt{\pm(4\wp^3 - 1)} \pm 2\tau\wp^3 \pm \tau\big],$$

where $A = \sqrt{a}$.

50. $yy'_x - y = 2x + 2A(-10x^{1/2} + 19A + 30A^2x^{-1/2})$.

Solution in the parametric form:

$$x = aE_2^{-2}(E_1 - 6E_2)^2, \quad y = -2aE_2^{-2}(\pm 6E_2^3 - E_1^2 + 7E_1E_2), \qquad \text{where} \quad A = -\sqrt{a}.$$

51. $yy'_x - y = -\dfrac{28}{121}x + \dfrac{2A}{121}(5x^{1/2} + 106A + 15A^2x^{-1/2})$.

Solution in the parametric form:

$$x = a(22\wp^2E_4 - 5)^2, \quad y = \pm 44a\wp^2E_3(7\wp E_3 \mp 2\tau), \qquad \text{where} \quad A = \pm 2\sqrt{a}.$$

52. $yy'_x - y = -\dfrac{12}{49}x + \dfrac{A}{49}(5x^{1/2} + 262A + 65A^2x^{-1/2})$.

Solution in the parametric form:

$$x = aE_3^{-4}(28\wp E_4^2 \mp 15E_3^2)^2, \quad y = 56aE_3^{-4}E_4^2(6\wp E_2 + E_4), \qquad \text{where} \quad A = \mp 3\sqrt{a}.$$

▶ *In the solutions of equations 53–60, the following notation is used:*

$$I = \int \frac{\tau\, d\tau}{\sqrt{\pm(4\tau^3 - 1)}} \quad \text{— incomplete elliptic integral of the second kind in the form of Weierstrass}$$

$$R = \sqrt{\pm(4\tau^3 - 1)}, \quad I_1 = \tau(2I \mp \tau^{-1}R + C), \quad I_2 = \tau^{-1}(RI_1 - 1), \quad I_3 = 4\tau I_1^2 \mp I_2^2.$$

53. $yy'_x - y = -\dfrac{12}{49}x + Ax^{1/2}$.

Solution in the parametric form:

$$x = 7a\tau^2I^{-4}, \quad y = -2aI^{-4}(RI - 2\tau^2), \qquad \text{where} \quad A = \pm\frac{12}{49}\sqrt{7a}.$$

54. $yy'_x - y = 6x + Ax^{-4}$.

Solution in the parametric form:

$$x = a\tau^{-3/5}I_1^{-2/5}, \quad y = a\tau^{-3/5}I_1^{-2/5}(5RI_1 - 2), \qquad \text{where} \quad A = \mp 150a^5.$$

55. $yy'_x - y = 20x + Ax^{-1/2}$.

Solution in the parametric form:

$$x = aI_1^{-4/3}I_2^2, \quad y = -4aI_1^{-4/3}(I_2^2 \mp 9\tau I_1^2), \qquad \text{where} \quad A = \pm 108a^{3/2}.$$

56. $yy'_x - y = \frac{15}{4}x + Ax^{-7}$.

Solution in the parametric form:

$$x = aI_1^{1/2}I_3^{-3/8}, \quad y = \frac{a}{2}I_1^{-3/2}I_3^{-3/8}(I_2I_3 - 3I_1^2), \qquad \text{where} \quad A = \pm\frac{3a^8}{4}.$$

57. $yy'_x - y = -\frac{10}{49}x + \frac{2A}{49}(4x^{1/2} + 61A + 12A^2x^{-1/2})$.

Solution in the parametric form:

$$x = a(7RI_1 - 3)^2, \quad y = 14aI_1[\pm(10\tau^3 - 1)I_1 - R], \qquad \text{where} \quad A = \sqrt{a}.$$

58. $yy'_x - y = -\frac{12}{49}x + \frac{2A}{49}(x^{1/2} + 166A + 55A^2x^{-1/2})$.

Solution in the parametric form:

$$x = aI_2^{-4}(42\tau I_1^2 \mp 5I_2^2)^2, \quad y = \mp 84aI_1^2I_2^{-4}(3\tau I_2^2 + I_2 \mp 12\tau^2 I_1^2), \qquad \text{where} \quad A = \pm\sqrt{a}.$$

59. $yy'_x - y = -\frac{4}{25}x + \frac{A}{50}(7x^{1/2} + 49A + 6A^2x^{-1/2})$.

Solution in the parametric form:

$$x = aI_1^{-4}(5I_2I_3 - 16I_1^2)^2, \quad y = -5aI_1^{-4}I_3(\pm 3I_3^2 - I_2^2I_3 + 8I_1^2I_3), \qquad \text{where} \quad A = 8\sqrt{a}.$$

60. $yy'_x - y = \frac{15}{4}x + 6Ax^{-1/3} - 3A^2x^{-5/3}$.

Solution in the parametric form:

$$x = 2a\tau^{3/2}I_1^{3/2}I_2^{-3/4}, \quad y = a\tau^{-1/2}I_1^{-1/2}I_2^{-3/4}(2\tau I_2^2 + I_2 - 3\tau^2 I_1^2),$$

where $A = -\frac{1}{3}a(2a)^{1/3}$.

61. $yy'_x - y = -\frac{3}{16}x + Ax^{-1/3} + Bx^{-5/3}$.

The substitution $x = \tau^{-3/2}$ leads to the equation

$$yy'_\tau = -\frac{3}{2}\tau^{-5/2}y + \frac{9}{32}\tau^{-4} - \frac{3}{2}A\tau^{-2} - \frac{3}{2}B$$

coincident with equation 1.3.3.13 when $n = -1/2$, $c = 0$, $b = 3/4$, $d = 3A/2$, $a^2 = -3B$.

62. $yy'_x - y = -\frac{5}{36}x + Ax^{-3/5} - Bx^{-7/5}, \qquad B > 0$.

The transformation

$$x = \left(w + \frac{A}{B} - \frac{1}{3}\sqrt{\tau}\right)^{-5/4}, \qquad y = \frac{5}{6}x + \left(\frac{5}{3}B\right)^{1/2}\sqrt{\tau}\,x^{1/5}$$

leads to an equation of the form 1.3.1.3:

$$ww'_\tau - w = -\frac{2}{9}\tau + \frac{2A}{3B} + \left(\frac{5}{27B}\right)^{1/2}\frac{1}{\sqrt{\tau}}.$$

63. $yy'_x - y = k(Ax^2 + Bx + C)^{-1/2}$.

The transformation

$$x = \frac{4(b_2w^2 + b_1w + b_0)}{4A - w^2}, \quad y = \xi + \frac{4(b_2w^2 + b_1w + b_0)}{4A - w^2},$$

where parameters b_2, b_1, and b_0 are found from the relations $B = 4Ab_2 - b_0$ and $C = b_1^2 - 4b_0b_2$, leads to the Riccati equation:

$$\pm kw'_\xi = (-\tfrac{1}{4}\xi + b_2)w^2 + b_1w + A\xi + b_0.$$

For $C > 0$, we may set $b_2 = 0$, $b_1 = \sqrt{C}$, $b_0 = -B$.

In books by Zaitsev & Polyanin (1993, 1994) it is shown that the original equation is reducible to the degenerate hypergeometric equation.

64. $yy'_x - y = -\frac{12}{49}x + 3A(\frac{1}{49}+B)x^{1/2} + 3A^2(\frac{4}{49}-\frac{5}{2}B) + \frac{15}{4}A^3(\frac{1}{49}+\frac{5}{4}B)x^{-1/2}.$

The substitution $x = (\xi^2 + \frac{5}{4}A)^2$ leads to an equation of the form 1.3.3.13 with $n=3$, $a = 4/7$, $c = 0$, $b = A$, $d = 12A(\frac{2}{7} - B)$:

$$yy'_\xi = (4\xi^2 + 5A)\xi y - \Big[\frac{48}{49}\xi^4 + 12A\Big(\frac{2}{7} - B\Big)\xi^2 + 3A^2\Big]\xi^3.$$

65. $yy'_x - y = -\frac{6}{25}x$
$+ \frac{4}{75}B^2\big[(2-A)x^{1/3} - \frac{3}{2}B(2A+1) + B^2(1-3A)x^{-1/3} - AB^3x^{-2/3}\big].$

The transformation

$$x = w^{-3}, \qquad y = \bigg[\xi + \frac{B(3-2Bw)}{5(w^2 + \frac{1}{B}w)}\bigg]\Big(w + \frac{1}{B}\Big)^2 w^{-2}$$

leads to the Riccati equation:

$$\Big(\frac{2}{B}\xi^2 - \frac{2}{5}B^2\xi + \frac{4}{25}AB^5\Big)w'_\xi = \xi w^2 + \Big(\frac{1}{B}\xi - \frac{2}{5}B^2\Big)w + \frac{3}{5}B.$$

66. $yy'_x - y = \frac{3}{4}x - \frac{3}{2}Ax^{1/3} + \frac{3}{4}A^2x^{-1/3} - \frac{27}{625}A^4x^{-5/3}.$

The transformation

$$x = A^{3/2}f^{-3/2}, \quad y = 3A^{3/2}\Big(\xi - \frac{3}{25} - \frac{1}{2f} + \frac{1}{2f^2}\Big)\sqrt{f}, \qquad \text{where} \quad f = \frac{w}{\xi^2 - \frac{6}{25}\xi},$$

leads to an equation of the form 1.3.1.46: $ww'_\xi - w = \frac{6}{25}\xi - \xi^2$.

67. $yy'_x - y = -\frac{6}{25}x + \frac{7}{5}Ax^{1/3} + \frac{31}{3}A^2x^{-1/3} - \frac{100}{3}A^4x^{-5/3}.$

Denote $A = \frac{7}{100}a$ and perform the transformation

$$x = \xi^{3/2}, \quad y = \frac{7}{20}\Big(w + \frac{8}{7}\xi - \frac{3}{5}a - \frac{7}{50}\frac{a^2}{\xi}\Big)\sqrt{\xi}, \qquad \text{where} \quad \xi = z - \frac{3}{10}a.$$

As a result we obtain an equation of the form 1.3.4.30 with $n = \frac{1}{7}$, $c = -\frac{3}{10}a$:

$$\Big[\Big(z - \frac{3}{10}a\Big)w + \frac{8}{7}z^2 - \frac{9}{7}az + \frac{1}{7}a^2\Big]w'_z = -\frac{1}{2}w^2 + 2zw.$$

68. $yy'_x - y = -\frac{10}{49}x + \frac{13}{5}A^2x^{-1/5} - \frac{7}{20}A^3x^{-4/5}.$

Denote $A = 8a^{-2}$. The transformation

$$\sqrt{\tau} = a^3\Big(\frac{5}{112}x^{3/5} - \frac{1}{16}yx^{-2/5}\Big), \quad w = \frac{4}{7}\tau + x^{-3/5} - \frac{39}{42}a^2$$

leads to an equation of the form 1.3.1.64 with $B = -1/49$:

$$ww'_\tau - w = -\frac{12}{49}\tau + \frac{39}{98}a^2 - \frac{15}{784}a^3\tau^{-1/2}.$$

69. $yy'_x - y = -\frac{33}{196}x + \frac{286}{3}A^2x^{-5/11} - \frac{770}{9}A^3x^{-13/11}$.

Denote $A = \frac{8}{5}a^{-2}$. The transformation

$$\sqrt{\tau} = \frac{15a^3}{448}\left(x^{8/11} - \frac{14}{11}yx^{-3/11}\right), \quad w = \frac{3}{7}\tau + x^{-8/11} - \frac{39}{56}a^2$$

leads to an equation of the form 1.3.1.64 with $B = -1/49$:

$$ww'_\tau - w = -\frac{12}{49}\tau + \frac{39}{98}a^2 - \frac{15}{784}a^3\tau^{-1/2}.$$

70. $yy'_x - y = -\frac{21}{100}x + \frac{7}{9}A^2(123x^{-1/7} + 280Ax^{-5/7} - 400A^2x^{-9/7})$.

Denote $A = 1/a$. The transformation

$$x = \xi^{-7/4}, \quad y = \frac{35}{3a^2}\left(w + 4\xi + \frac{7}{5}a + \frac{3}{50}\frac{a^2}{\xi}\right)\xi^{-3/4}, \qquad \text{where} \quad \xi = z - \frac{21}{20}a,$$

leads to an equation of the form 1.3.4.30 with $n = 3$, $c = -\frac{21}{20}a$:

$$\left[\left(z - \frac{21}{20}a\right)w + 4z^2 - 7az + 3a^2\right]w'_z = \frac{3}{4}w^2 + 2zw.$$

71. $yy'_x - y = ax + bx^m$.

1°. For $m \neq 3$, the transformation

$$\tau = B^2\left[(m-3)\frac{y}{x} + 1\right]^2, \quad w = 2(m-3)B^2\left(bx^{m-1} - \frac{y^2}{x^2} + \frac{y}{x} + a\right)$$

leads to the equation

$$ww'_\tau - w = \frac{2(m-1)}{(m-3)^2}\{\tau - mB\tau^{1/2} + [2m-3-a(m-3)^2]B^2 + [2-m+a(m-3)^2]B^3\tau^{-1/2}\}.$$

2°. Let $m \neq 1$ and $a > -1/4$. Assume

$$a = -\frac{(n+2)(n+m+1)}{(2n+m+3)^2}, \qquad \text{whence} \quad n_{1,2} = \frac{1}{2}\left(\pm\frac{m-1}{\sqrt{1+4a}} - m - 3\right).$$

Then, the transformations

$$x = \xi^{\frac{n+2}{m-1}}w, \quad y = \frac{m-1}{2n+m+3}\xi^{\frac{n+2}{m-1}}\left(\xi w'_\xi + \frac{n+2}{m-1}w\right), \qquad n = n_{1,2}$$

reduce the original equation to the classical Emden—Fowler equation $w''_{\xi\xi} = A\xi^n w^m$, where $A = \left(\frac{2n+m+3}{m-1}\right)^2 b$, which is discussed below in Section 2.3.

72. $yy'_x - y = -\frac{m+1}{(m+2)^2}x + Ax^{2m+1} + Bx^{3m+1}$.

Assume $A = -\frac{am}{2(m+2)^2b^2}$, $B = -\frac{m^2}{2(m+2)^3b^2}$. The transformation

$$\sqrt{\tau} = -\frac{(m+2)^2}{m}byx^{-m-1} + \frac{m+2}{m}bx^{-m}, \quad w = \frac{2(m+1)}{m+2}\tau + x^m + \frac{m+2}{m}a$$

leads to the equation

$$ww'_\tau - w = \frac{2m(m+1)}{(m+2)^2}\tau + a + b\tau^{-1/2}$$

(see Table 1.2 with $\alpha = 0$ in Subsection 1.3.1).

73. $yy'_x - y = a^2\lambda e^{2\lambda x} - a(b\lambda + 1)e^{\lambda x} + b.$

Particular solution: $y_0 = ae^{\lambda x} - b.$

74. $yy'_x - y = a^2\lambda e^{2\lambda x} + a\lambda x e^{\lambda x} + be^{\lambda x}.$

Particular solution: $y_0 = ae^{\lambda x} + x + \dfrac{b}{a\lambda}.$

75. $yy'_x - y = 2a^2\lambda \sin(2\lambda x) + 2a\sin(\lambda x).$

Particular solution: $y_0 = -2a\sin(\lambda x).$

76. $yy'_x - y = a^2 f'_x f''_{xx} - \dfrac{(f+b)^2}{(f'_x)^3} f''_{xx}, \qquad f = f(x).$

Particular solutions: $y_1 = af'_x + \dfrac{f+b}{f'_x}, \quad y_2 = -af'_x + \dfrac{f+b}{f'_x}.$

1.3.2. Equations of the Form $yy'_x = f(x)y + 1$

1. $yy'_x = (ax + b)y + 1.$

The substitution $\xi = y - \frac{1}{2}ax^2 - bx$ leads to the Riccati equation with respect to $x = x(\xi)$: $x'_\xi = \frac{1}{2}ax^2 + bx + \xi.$

2. $yy'_x = (ax + b)^{-2}y + 1.$

The substitution $a\xi = -(ax + b)^{-1}$ leads to an equation of the form 1.3.1.33: $yy'_\xi = y + (a\xi)^{-2}.$

3. $yy'_x = \left(a - \dfrac{1}{ax}\right)y + 1.$

The substitution $\xi = y - ax$ leads to the Bernoulli equation with respect to $x = x(\xi)$: $\xi x'_\xi + a\xi x + a^2x^2 = 0.$

4. $yy'_x = (ax + b)^{-1/2}y + 1.$

The substitution $z = \dfrac{2}{a}(ax + b)^{1/2}$ leads to an equation of the form 1.3.1.2: $yy'_z = y + \frac{1}{2}az.$

5. $yy'_x = 3(ax^{3/2} + 8x)^{-1/2}y + 1.$

The substitution $z = \dfrac{12}{a}(ax^{1/2} + 8)^{1/2}$ leads to the equation of the form 1.3.1.10 with $m = 3$: $yy'_z = y - \dfrac{2}{9}z + \dfrac{a^2}{5184}z^3.$

6. $yy'_x = (ax^{-2/3} - \frac{2}{3}a^{-1}x^{-1/3})y + 1.$

The transformation $x = a^{3/2}w^3$, $y = \xi - w^2$ leads to the Riccati equation: $3a^{3/2}\xi w'_\xi = \xi - w^2.$

7. $\boldsymbol{yy'_x = ae^{\lambda x}y + 1.}$

The substitution $\xi = \dfrac{a}{\lambda}e^{\lambda x}$ leads to an equation of the form 1.3.1.16: $yy'_\xi = y + (\lambda\xi)^{-1}$.

8. $\boldsymbol{yy'_x = (ae^{\lambda x} + be^{-\lambda x})y + 1.}$

The transformation

$$\xi = y + \frac{b}{\lambda}e^{-\lambda x} - \frac{a}{\lambda}e^{\lambda x}, \qquad w = e^{\lambda x}$$

leads to the Riccati equation: $w'_\xi = aw^2 + \lambda\xi w - b$.

9. $\boldsymbol{yy'_x = ay\cosh x + 1.}$

This is a special case of equation 1.3.3.20 with $b = 0$, $c = 1$.

10. $\boldsymbol{yy'_x = ay\sinh x + 1.}$

This is a special case of equation 1.3.3.21 with $b = 0$, $c = 1$.

11. $\boldsymbol{yy'_x = a\cos(\omega x)\,y + 1.}$

The transformation

$$x = -\frac{2}{\omega}\arctan\frac{4u}{\omega}, \qquad y = \tau - \frac{8au}{16u^2 + \omega^2}$$

leads to the Riccati equation: $u'_\tau = -2\tau u^2 + au - \frac{1}{8}\omega^2\tau$.

12. $\boldsymbol{yy'_x = a\sin(\omega x)\,y + 1.}$

The substitution $x = \xi + \dfrac{\pi}{2\omega}$ leads to an equation of the form 1.3.2.11: $yy'_\xi = a\cos(\omega\xi)\,y + 1$.

1.3.3. Equations of the Form $yy'_x = f_1(x)y + f_0(x)$

Preliminary comments. With the aid of the substitution $\xi = \int f_1(x)\,dx$, these equations are reducible to the form

$$yy'_\xi = y + f(\xi), \qquad \text{where} \quad f(\xi) = f_0(x)/f_1(x), \tag{1}$$

and by means of the substitution $z = \int f_0(x)\,dx$ they can be reduced to the form

$$yy'_z = g(z)y + 1, \qquad \text{where} \quad g(z) = f_1(x)/f_0(x). \tag{2}$$

Concrete equations of the form (1) and (2) are outlined in Subsection 1.3.1 and 1.3.2, respectively.

1. $\boldsymbol{yy'_x = (ax+3b)y + cx^3 - abx^2 - 2b^2x.}$

The substitution $y = x^2t + bx$ leads to the linear equation with respect to $x = x(t)$: $(-2t^2 + at + c)x'_t = tx + b$.

2. $\boldsymbol{yy'_x = (3ax+b)y - a^2x^3 - abx^2 + cx.}$

The substitution $y = xw + ax^2$ leads to the Bernoulli equation with respect to $x = x(w)$: $(-w^2 + bw + c)x'_w = wx + ax^2$.

3. $\boldsymbol{2yy'_x = (7ax+5b)y - 3a^2x^3 - 2cx^2 - 3b^2x.}$

This is a special case of equation 1.3.3.11 with $m = 3/2$, $k = 1/2$.

4. $\boldsymbol{yy'_x = [(3-m)x - 1]y + (m-1)(x^3 - x^2 - ax).}$

The transformation

$$x = w/z, \quad y = -z^{m-1} + x^2 - x - a$$

leads to the equation $ww'_z = w + az + z^m$ whose solvable cases are outlined in Subsection 1.3.1 (see Table 1.1).

5. $\boldsymbol{yy'_x + x(ax^2+b)y + x = 0.}$

The substitution $z = -\frac{1}{2}x^2$ leads to an equation of the form 1.3.2.1: $yy'_z = (-2az + b)y + 1$.

6. $\boldsymbol{yy'_x = 3(ax+b)^{-1/3}x^{-5/3}y + 3(ax+b)^{-2/3}x^{-7/3}.}$

The substitution $w = \dfrac{1}{xy} + \dfrac{1}{3}\left(\dfrac{ax+b}{x}\right)^{1/3}$ leads to an equation with separation of variables: $w'_x = x^{-1/3}(ax+b)^{-2/3}(\frac{1}{9}a - 3w^3)$.

7. $\boldsymbol{3yy'_x = (-7Ax + 6s - 2\lambda)x^{-1/3}y}$

$$\boldsymbol{+\,2(Ax+5\lambda)(-Ax+3s+4\lambda)x^{1/3} + 6(\lambda sx - 1)x^{2/3},}$$

where $A = \lambda s(3s + 4\lambda)$.

The transformation

$$x = (\xi + \lambda s)^{-1}, \quad y = (w + 4\lambda + 3s - Ax)x^{2/3}$$

leads to an equation of the form 1.3.4.10 with $a = 1/3$:

$$[(\xi + \lambda s)w + (4\lambda + 3s)\xi]w'_\xi = \tfrac{2}{3}w^2 + 2(3\lambda + s)w + 2\xi.$$

8. $\boldsymbol{yy'_x = x^{n-1}[(1+2n)x + an]y - nx^{2n}(x+a).}$

The transformation

$$x = \frac{w}{z}, \quad y = -\frac{1}{z^n} + x^{n+1} + ax^n$$

leads to an equation with separation of variables: $w'_z = w^{-n} - a$.

9. $\boldsymbol{yy'_x = a(x-nb)x^{n-1}y + c[x^2 - (2n+1)bx + n(n+1)b^2]x^{2n-1}.}$

The substitution $\xi = ax^n\left(\dfrac{x}{n+1} - b\right)$ leads to an equation of the form 1.3.1.2: $yy'_\xi = y + (n+1)ca^{-2}\xi$.

10. $yy'_x = [a(2n+k)x^k + b]x^{n-1}y + (-a^2nx^{2k} - abx^k + c)x^{2n-1}.$

The substitution $y = x^n(w + ax^k)$ leads to the Bernoulli equation with respect to $x = x(w)$: $(-nw^2 + bw + c)x'_w = wx + ax^{k+1}$.

11. $yy'_x = [a(2m+k)x^{2k} + b(2m-k)]x^{m-k-1}y - (a^2mx^{4k} + cx^{2k} + b^2m)x^{2m-2k-1}.$

The transformation $z = x^k$, $y = x^m(t + ax^k + bx^{-k})$ leads to the Riccati equation with respect to $z = z(t)$:

$$(-mt^2 + 2abm - c)z'_t = akz^2 + ktz + bk. \qquad (1)$$

The substitution

$$z = \frac{mt^2 + c_0}{ak}\frac{w'_t}{w}, \qquad \text{where} \quad c_0 = c - 2abm,$$

reduces equation (1) to a second order linear equation:

$$(mt^2 + c_0)^2 w''_{tt} + (2m+k)t(mt^2 + c_0)w'_t + abk^2 w = 0. \qquad (2)$$

The substitution

$$\xi = \frac{t}{\sqrt{t^2 + (c_0/m)}}, \qquad u = (1-\xi^2)^{\mu/2} w, \qquad \text{where} \quad \mu = -\frac{m+k}{2m},$$

brings equation (2) to the Legendre equation:

$$(1-\xi^2)u''_{\xi\xi} - 2\xi u'_\xi + [\nu(\nu+1) - \mu^2(1-\xi^2)^{-1}]u = 0,$$

where ν is a root of the quadratic equation $\nu^2 + \nu + \dfrac{m^2 - k^2}{4m^2} - \dfrac{abk^2}{mc_0} = 0$.

12. $yy'_x = [(m + 2l - 3)x + n - 2l + 3]x^{-l}y + [(m + l - 1)x^2 + (n - m - 2l + 3)x - n + l - 2]x^{1-2l}.$

The transformation

$$x = \frac{\xi}{w}w'_\xi, \qquad y = A\xi^{n-l+2}w^{m+l-1} - x^{2-l} + x^{1-l}$$

leads to the generalized Emden—Fowler equation: $w'_\xi = A\xi^n w^m (w'_\xi)^l$, which is discussed in Section 2.5.

13. $yy'_x = [a(2n+1)x^2 + cx + b(2n-1)]x^{n-2}y - (na^2x^4 + acx^3 + dx^2 + bcx + nb^2)x^{2n-3},$

where a, b, c, d, and n are arbitrary numbers.

The substitution $y = x^n t + ax^{n+1} + bx^{n-1}$ leads to the Riccati equation with respect to $x = x(t)$:

$$(-nt^2 + ct - d + 2nab)x'_t = ax^2 + tx + b.$$

14. $yy'_x = [a(n-1)x + b(2\lambda+n)]x^{\lambda-1}(ax+b)^{-\lambda-2}y - [anx + b(\lambda+n)]x^{2\lambda-1}(ax+b)^{-2\lambda-3}$.

The substitution

$$y = \left[\frac{1}{w} + \frac{1}{x^n(ax+b)}\right]x^{\lambda+n}(ax+b)^{-\lambda}$$

leads to an equation of the form 1.3.4.5:

$$(w + ax^{n+1} + bx^n)w'_x = [anx^n + b(\lambda+n)x^{n-1}]w.$$

15. $yy'_x = (ae^x + b)y + ce^{2x} - abe^x - b^2$.

The transformation $x = \ln w$, $y = tw + b$ leads to a linear equation: $(-t^2 + at + c)w'_t = tw + b$.

16. $yy'_x = [a(2\mu+\lambda)e^{\lambda x} + b]e^{\mu x}y + (-a^2\mu e^{2\lambda x} - abe^{\lambda x} + c)e^{2\mu x}$.

The substitution $\xi = e^x$ leads to an equation of the form 1.3.3.10:

$$yy'_\xi = [a(2\mu+\lambda)\xi^\lambda + b]\xi^{\mu-1}y + (-a^2\mu\xi^{2\lambda} - ab\xi^\lambda + c)\xi^{2\mu-1}.$$

17. $yy'_x = (ae^{\lambda x} + b)y + c[a^2e^{2\lambda x} + ab(\lambda x + 1)e^{\lambda x} + b^2\lambda x]$.

The substitution $\xi = \dfrac{a}{\lambda}e^{\lambda x} + bx$ leads to an equation of the form 1.3.1.2: $yy'_\xi = y + c\lambda\xi$.

18. $yy'_x = e^{\lambda x}(2a\lambda x + a + b)y - e^{2\lambda x}(a^2\lambda x^2 + abx + c)$.

The substitution $y = e^{\lambda x}(\xi + ax)$ leads to a linear equation with respect to $x = x(\xi)$: $(-\lambda\xi^2 + b\xi - c)x'_\xi = ax + \xi$.

19. $yy'_x = e^{ax}(2ax^2 + 2x + b)y + e^{2ax}(-ax^4 - bx^2 + c)$.

The substitution $y = e^{ax}(\xi + x^2)$ leads to the Riccati equation with respect to $x = x(\xi)$: $(-a\xi^2 + b\xi + c)x'_\xi = x^2 + \xi$.

20. $yy'_x = (a\cosh x + b)y - ab\sinh x + c$.

The transformation $t = y - a\sinh x$, $\xi = e^x$ leads to the Riccati equation: $2(bt + c)\xi'_t = a\xi^2 + 2t\xi - a$.

21. $yy'_x = (a\sinh x + b)y - ab\cosh x + c$.

The transformation $t = y - a\cosh x$, $\xi = e^x$ leads to the Riccati equation: $2(bt + c)\xi'_t = a\xi^2 + 2t\xi + a$.

22. $yy'_x = (2\ln x + a + 1)y + x(-\ln^2 x - a\ln x + b)$.

The transformation $x = e^w$, $y = (\xi + w)e^w$ leads to a linear equation: $(-\xi^2 + a\xi + b)w'_\xi = w + \xi$.

23. $yy'_x = (2\ln^2 x + 2\ln x + a)y + x(-\ln^4 x - a\ln^2 x + b)$.

Performing the transformation $x = e^w$, $y = (\xi + w^2)e^w$, we obtain the Riccati equation: $(-\xi^2 + a\xi + b)w'_\xi = w^2 + \xi$.

24. $yy'_x = ax\cos(\omega x^2)\,y + x.$

The substitution $\xi = \frac{1}{2}x^2$ leads to the Abel equation of the form 1.3.2.11: $yy'_\xi = a\cos(2\omega\xi)\,y + 1.$

25. $yy'_x = ax\sin(\omega x^2)\,y + x.$

The substitution $\xi = \frac{1}{2}x^2$ leads to the Abel equation of the form 1.3.2.12: $yy'_\xi = a\sin(2\omega\xi)\,y + 1.$

1.3.4. Equations of the Form $[g_1(x)y + g_0(x)]y'_x = f_2(x)y^2 + f_1(x)y + f_0(x)$

Preliminary comments. With the aid of the substitution

$$w = \Bigl(y + \frac{g_0}{g_1}\Bigr)E, \qquad \text{where} \quad E = \exp\Bigl(-\int \frac{f_2}{g_1}\,dx\Bigr), \tag{1}$$

these equations are reducible to a simpler form:

$$ww'_x = F_1(x)w + F_0(x), \tag{2}$$

where

$$F_1 = \left[\frac{d}{dx}\Bigl(\frac{g_0}{g_1}\Bigr) + \frac{f_1}{g_1} - 2\frac{g_0 f_2}{g_1^2}\right]E, \qquad F_0 = \left(\frac{f_0}{g_1} - \frac{g_0 f_1}{g_1^2} + \frac{g_0^2 f_2}{g_1^3}\right)E^2.$$

Concrete Abel equations of the form (2) are outlined in 1.3.1–1.3.3. In the degenerate cases with $F_0 \equiv 0$ or $F_1 \equiv 0$, the variables in equation (2) are separable.

1. $(Ay + Bx + a)y'_x + By + kx + b = 0.$

Solution: $Ay^2 + kx^2 + 2(Bxy + ay + bx) = C.$

2. $(y + ax + b)y'_x = \alpha y + \beta x + \gamma.$

The substitution $y = u - ax - b$ leads to the equation

$$uu'_x = (a + \alpha)u + (\beta - a\alpha)x + \gamma - b\alpha$$

which is separable with $a = -\alpha$.

For $a \neq -\alpha$, the substitution $u = (a + \alpha)w$ yields an equation of the form 1.3.1.1 or 1.3.1.2:

$$ww'_x = w + \Delta^{-2}(\beta - a\alpha)x + \Delta^{-2}(\gamma - b\alpha), \qquad \text{where} \quad \Delta = a + \alpha.$$

3. $(y + akx^2 + bx + c)y'_x = -ay^2 + 2akxy + my + k(k + b - m)x + s.$

The substitution $y = z + kx$ leads to the Riccati equation with respect to $x = x(z)$:

$$[-az^2 + (m - k)z + s - ck]x'_z = akx^2 + (b + k)x + z + c.$$

4. $(y + Ax^n + a)y'_x + nAx^{n-1}y + kx^m + b = 0.$

Solution: $y^2 + \dfrac{2k}{m+1}x^{m+1} + 2(Ax^n y + ay + bx) = C.$

5. $(y + ax^{n+1} + bx^n)y'_x = (anx^n + cx^{n-1})y.$

The substitution $y = x^n(w-b)$ leads to the Bernoulli equation with respect to $x = x(w)$:

$$[-nw^2 + (bn + c)w - bc]x'_w = wx + ax^2.$$

6. $xyy'_x = ay^2 + by + cx^n + s.$

The transformation $\xi = x^{-a}$, $w = -\dfrac{a}{b}x^{-a}y$ leads to the equation $ww'_\xi = w + A\xi + B\xi^m$, where $A = -ab^{-2}s$, $B = -ab^{-2}c$, $m = (a-n)/a$ (see Subsection 1.3.1).

7. $xyy'_x = -ny^2 + a(2n+1)xy + by - a^2nx^2 - abx + c.$

The substitution $y = w + ax$ leads to the Bernoulli equation with respect to $x = x(w)$:

$$(-nw^2 + bw + c)x'_w = wx + ax^2.$$

8. $2xyy'_x = (1-n)y^2 + [a(2n+1)x + 2n - 1]y - a^2nx^2 - bx - n.$

The transformation $x = \xi^2$, $y = \xi t + a\xi^2 + 1$ leads to the Riccati equation:

$$(-nt^2 + 2an - b)\xi'_t = a\xi^2 + t\xi + 1.$$

9. $(Axy - Aky + Bx - Bk)y'_x = Cy^2 + Dxy + (B - Dk)y.$

The transformation $x = w + k$, $y = \xi w$ leads to a linear equation with respect to $w = w(x)$:

$$[(C - A)\xi^2 + D\xi]w'_\xi = A\xi w + B.$$

10. $[(3ax + \lambda s)y + (4\lambda + 3s)x]y'_x = 2ay^2 + 2(3\lambda + s)y + 2x.$

The substitution $w = ay^2 + (3\lambda + s)y + x$ leads to an equation of the form 1.3.3.3:

$$2ww'_y = (7ay + 5b)w - 3a^2y^3 - 2cy^2 - 3b^2y, \quad \text{where } b = s + 2\lambda, \quad c = \tfrac{1}{2}a(13\lambda + 6s).$$

11. $[(4ax + \lambda s)y + (4\lambda + 3s)x]y'_x = \frac{3}{2}ay^2 + 2(3\lambda + s)y + 2x.$

The substitution $w = \frac{3}{4}ay^2 + (3\lambda + s)y + x$ leads to an equation of the form 1.3.3.3:

$$2ww'_y = (7ay + 5b)w - 3a^2y^3 - 2cy^2 - 3b^2y, \quad \text{where } b = s + 2\lambda, \quad c = \tfrac{1}{8}a(60\lambda + 25s).$$

12. $(2Axy + ay + bx + c)y'_x = Ay^2 + Ak^2x^2 + my + k(ak + b - m)x + s.$

The substitution $y = z + kx$ leads to the Riccati equation with respect to $x = x(z)$:

$$[Az^2 + (m - ak)z + s - ck]x'_z = 2Akx^2 + (2Az + ak + b)x + az + c.$$

13. $\left[2xy + (1-m)Ay - \dfrac{2(m+1)}{m+3}x\right]y'_x = \dfrac{1-m}{2}y^2 + \dfrac{m-1}{m+3}y + x.$

The substitution $w = \dfrac{1-m}{2}y^2 + \dfrac{m-1}{m+3}y + x$ leads to an equation of the form 1.3.3.4:

$$ww'_y = [(3-m)y-1]w + (m-1)(y^3 - y^2 - ay), \qquad \text{where} \quad a = A - \frac{2(m+1)}{(m+3)^2}.$$

14. $x(2ay + bx)y'_x = a(2-m)y^2 + b(1-m)xy + cx^2 + Ax^{m+2}.$

The transformation $z = y/x$, $w = -Ax^m + amz^2 + bmz - c$ leads to the equation with separated variables: $ww'_z = m(2az+b)(amz^2 + bmz - c)$.

15. $(xy + x^2 + a)y'_x = y^2 + xy + b.$

Solution: $(x+y)^2 + a + b = C(bx - ay)^2$.

16. $(2Axy + Bx^2 + b)y'_x = Ay^2 + k(Ak + B)x^2 + c.$

The substitution $y = z + kx$ leads to the Riccati equation with respect to $x = x(z)$:

$$(Az^2 + c - bk)x'_z = (2Ak + B)x^2 + 2Azx + b.$$

17. $(Axy + Bx^2 + kx)y'_x = Dy^2 + Exy + Fx^2 + ky.$

The substitution $y = xz$ leads to a linear equation with respect to $x = x(z)$:

$$[(D-A)z^2 + (E-B)z + F]x'_z = (Az + B)x + k$$

18. $(Axy + Bx^2 + kx)y'_x = Ay^2 + Bxy + (Ab + k)y + Bbx + bk.$

This is a special case of equation 1.3.4.22.
Solutions: $y = Cx - b$ and $Ay + Bx + k = 0$.

19. $(2Axy + Bx^2 + kx)y'_x = Ay^2 + Cxy + Dx^2 + ky - C\beta x - A\beta^2 - k\beta.$

The substitution $y = \xi x + \beta$ leads to a linear equation with respect to $x = x(\xi)$:

$$[-A\xi^2 + (C-B)\xi + D]x'_\xi = (2A\xi + B)x + 2A\beta + k.$$

20. $(Axy + Bx^2 + kx)y'_x = Ay^2 + Cxy + Dx^2 + (k - A\beta)y - C\beta x - k\beta.$

The substitution $y = \xi x + \beta$ leads to a linear equation with respect to $x = x(\xi)$:

$$[(C-B)\xi + D]x'_\xi = (A\xi + B)x + A\beta + k.$$

21. $(Axy + Aky + Bx^2 + Bkx)y'_x = Cy^2 + Dxy + k(D-B)y.$

The transformation $x = w - k$, $y = \xi w$ leads to a linear equation:

$$[(C-A)\xi^2 + (D-B)\xi]w'_\xi = (A\xi + B)w - kB.$$

22. $(Axy + Bx^2 + a_1x + b_1y + c_1)y'_x = Ay^2 + Bxy + a_2x + b_2y + c_2.$

Jacobi equation.

1°. With the help of the transformation $x = \bar{x} + \alpha$, $y = \bar{y} + \beta$, where α and β are the parameters which are determined by solving the algebraic system

$$A\alpha\beta + B\alpha^2 + a_1\alpha + b_1\beta + c_1 = 0, \qquad A\beta^2 + B\alpha\beta + a_2\alpha + b_2\beta + c_2 = 0,$$

we obtain the equation

$$(A\bar{x}\bar{y} + B\bar{x}^2 + \bar{a}_1\bar{x} + \bar{b}_1\bar{y})\bar{y}'_{\bar{x}} = A\bar{y}^2 + B\bar{x}\bar{y} + \bar{a}_2\bar{x} + \bar{b}_2\bar{y},$$

where

$$\bar{a}_1 = 2B\alpha + A\beta + a_1, \qquad \bar{b}_1 = A\alpha + b_1,$$
$$\bar{a}_2 = B\beta + a_2, \qquad \bar{b}_2 = 2A\beta + B\alpha + b_2.$$

The transformation $z = \bar{y}/\bar{x}$, $\zeta = 1/\bar{x}$ leads to a linear equation:

$$[\bar{b}_1z^2 + (\bar{a}_1 - \bar{b}_2)z - \bar{a}_2]\zeta'_z = (\bar{b}_1z + \bar{a}_1)\zeta + Az + B.$$

2°. The original equation can be also rewritten in the form

$$(xy'_x - y)(n_3x + m_3y + k_3) - y'_x(n_1x + m_1y + k_1) + n_2x + m_2y + k_2 = 0.$$

The solution of this equation in the parametric form can be obtained from the solution of the following system of the constant-coefficient linear differential equations:

$$(x_1)'_t = n_1x_1 + m_1x_2 + k_1x_3,$$
$$(x_2)'_t = n_2x_1 + m_2x_2 + k_2x_3,$$
$$(x_3)'_t = n_3x_1 + m_3x_2 + k_3x_3,$$

using the formulae $x(t) = x_1/x_3$, $y(t) = x_2/x_3$.

23. $(Axy + Bx^2 + ay + bx + c)y'_x = kAxy + kBx^2 + my + k(ak + b - m)x + s.$

The substitution $y = z + kx$ leads to the Riccati equation with respect to $x = x(z)$:

$$[(m - ak)z + s - ck]x'_z = (Ak + B)x^2 + (Az + ak + b)x + az + c.$$

24. $(2Axy + Bx^2 + ay + bx + c)y'_x = Ay^2 + k(Ak + B)x^2 + aky + bkx + s.$

The substitution $y = z + kx$ leads to the Riccati equation with respect to $x = x(z)$:

$$(Az^2 + s - ck)x'_z = (2Ak + B)x^2 + (2Az + ak + b)x + az + c.$$

25. $(2Axy - Akx^2 + ay + bx + c)y'_x = Ay^2 + my + k(ak + b - m)x + s.$

The substitution $y = z + kx$ leads to the Riccati equation with respect to $x = x(z)$:

$$[Az^2 + (m - ak)z + s - ck]x'_z = Akx^2 + (2Az + ak + b)x + az + c.$$

26. $(2Axy + Bx^2 + ay - akx + b)y'_x = Ay^2 + k(Ak + B)x^2 + my - mkx + s.$

The substitution $y = z + kx$ leads to the Riccati equation with respect to $x = x(z)$:

$$[Az^2 + (m - ak)z + s - bk]x'_z = (2Ak + B)x^2 + 2Azx + az + c.$$

27. $(2Axy + Bx^2 + ay + bx + c)y'_x = Ay^2 + k(Ak + B)x^2 + by + ak^2x + s.$

The substitution $y = z + kx$ leads to the Riccati equation with respect to $x = x(z)$:

$$[Az^2 + (b - ak)z + s - ck]x'_z = (2Ak + B)x^2 + (2Az + ak + b)x + az + c.$$

28. $[Axy + Bx^2 + (m-1)Aay - (Abm + Ba)x]y'_x$
$= Ay^2 + Bxy - (Ab + Bam)y + (m-1)Bbx.$

This is a special case of equation 1.3.4.22.

Solution in the parametric form:

$$x = \frac{at + ACt^m}{t + C}, \qquad y = \frac{bt - BCt^m}{t + C}.$$

The solution can be presented in an implicit form as well:

$$C^m(Ay + Bx)^m + [A(b - y) + B(a - x)]^{m-1}(ay - bx) = 0.$$

29. $[(ax + c)y + (1 - n)x^2 + (2n - 1)x - n]y'_x = 2ay^2 + 2xy.$

The substitution $w = ay + x$ leads to an equation of the form 1.3.4.8:

$$2yww'_y = (1 - n)w^2 + [a(2n + 1)y + 2n - 1]w - a^2ny^2 - by - n,$$

where $b = (2n - 1)a - c$.

30. $[(x + c)y + (n + 1)x^2 - a(2n + 1)x + a^2n]y'_x = \dfrac{2n}{3n - 1}y^2 + 2xy.$

The transformation

$$z = \frac{3n - 1}{n - 1}\frac{1}{y}, \qquad w = \frac{3n - 1}{n - 1}\frac{x}{y} + \frac{n}{n - 1}$$

leads to an equation of the form 1.3.4.8:

$$2zww'_z = (1 - n)w^2 + [a(2n + 1)z + 2n - 1]w - a^2nz^2 - bz - n,$$

where $b = \dfrac{(3n - 1)c + an(2n + 1)}{n - 1}$.

31. $x(2axy + b)y'_x = -a(m + 3)xy^2 - b(m + 2)y + cx^m.$

The transformation

$$z = xy, \qquad w = -cx^{m+1} + a(m + 1)x^2y^2 + b(m + 1)xy$$

leads to the equation with separated variables: $ww'_z = (m + 1)^2(2az + b)(az^2 + bz)$.

32. $[(a_2x^2 + a_1x + a_0)y + b_2x^2 + b_1x + b_0]y'_x = c_2y^2 + c_1y + c_0.$

This is the Riccati equation with respect to $x = x(y)$.

The substitution $x = -\dfrac{c_2y^2 + c_1y + c_0}{a_2y + b_2}\dfrac{w'_y}{w}$ yields a second order linear equation:

$$f_2w''_{yy} - [(f_2)'_y + f_1f_2]w'_y + f_0f_2^2w = 0,$$

where $f_i = \dfrac{a_iy + b_i}{c_2y^2 + c_1y + c_0}$; $i = 1,\ 2,\ 3.$

33. $[(12a^2x^2 - 7ax + 1)y + 4cx^2 - 5bx]y'_x = -2x(3a^2y^2 + 2cy + 3b^2)$.

The substitution $w = x(3a^2y^2 + 2cy + 3b^2)$ leads to an equation of the form 1.3.3.3: $2ww'_y = (7ay + 5b)w - 3a^2y^3 - 2cy^2 - 3b^2y$.

34. $x[(m-1)(Ax+B)y + m(Dx^2 + Ex + F)]y'_x$
$= [A(1-n)x - Bn]y^2 + [D(2-n)x^2 + E(1-n)x - Fn]y$.

Solution: $Axy + Dx^2 + Ex + By + F = Cx^ny^m$.

35. $x^2(2axy + b)y'_x = -4ax^2y^2 - 3bxy + cx^2 + k$.

The transformation $z = xy$, $w = 2ax^2y^2 + 2bxy - cx^2 - k$ leads to an equation with separated variables: $ww'_z = 2(2az + b)(2az^2 + 2bz - k)$.

36. $(xy + ax^n + bx^2)y'_x = y^2 + cx^n + bxy$.

The transformation $t = y/x$, $z = x^{n-2}$ leads to a linear equation: $(c - at)z'_t = (n-2)(az + t + b)$.

37. $x(2ax^ny + b)y'_x = -a(3n + m)x^ny^2 - b(2n + m)y + Ax^m + cx^{-n}$.

The transformation

$$z = x^ny, \qquad w = -Ax^{n+m} + (n+m)(az^2 + bz) - c$$

leads to an equation with separated variables:

$$ww'_z = (n+m)^2(2az + b)\Big(az^2 + bz - \frac{c}{n+m}\Big).$$

38. $yy'_x = -ny^2 + a(2n+1)e^xy + by - a^2ne^{2x} - abe^x + c$.

Performing the transformation $x = \ln\xi$, $y = w + a\xi$, we obtain the Bernoulli equation: $(-nw^2 + bw + c)\xi'_w = w\xi + a\xi^2$.

1.3.5. Some Types of First and Second Order Equations Reducible to Abel Equations of the Second Kind

► *Notation: f, g, h, p, φ, ψ, Φ, F, and G are arbitrary functions of their arguments.*

1. Quasi-homogeneous equation

$$f(x^\nu y)x^{\nu+1}y'_x + g(x^\nu y) + Ax^\lambda = 0.$$

In the particular case when $\lambda = 0$ this equation is homogeneous. The transformation

$$z = x^\nu y, \quad v = Ax^\lambda + g(z) - \nu z f(z)$$

leads to the Abel equation:

$$vv'_z = [-(\lambda+\nu)f + g'_z - \nu z f'_z]v + \lambda f(g - \nu z f).$$

2. Quasi-homogeneous equation

$$f(x^\nu y)x^{\nu+1}y'_x + g(x^\nu y) + x^\lambda[h(x^\nu y)x^{\nu+1}y'_x + p(x^\nu y)] = 0.$$

The transformation $z = x^\nu y$, $\zeta = x^{-\lambda}$ leads to the Abel equation:

$$\{[g(z) - \nu z f(z)]\zeta + p(z) - \nu z h(z)\}\zeta'_z = \lambda f(z)\zeta^2 + \lambda h(z)\zeta.$$

3. Equations of the theory of chemical reactors and the combustion theory

$$y''_{xx} - ay'_x = f(y).$$

The substitution $w(y) = y'_x/a$ leads to the Abel equation:

$$ww'_y - w = a^{-2}f(y),$$

whose solvable cases are given in Subsection 1.3.1.

4. Equations of the theory of nonlinear oscillations

$$y''_{xx} + \varphi(y)y'_x + y = 0.$$

The substitution $z(y) = y'_x$ leads to the Abel equation:

$$zz'_y + \varphi(y)z + y = 0, \qquad (1)$$

which is reduced, with the aid of the substitution $\tau = \frac{1}{2}(a - y^2)$, to the following form:

$$zz'_\tau = g(\tau)z + 1, \qquad \text{where} \quad g(\tau) = \pm\frac{\varphi(\pm\sqrt{a - 2\tau})}{\sqrt{a - 2\tau}}. \qquad (2)$$

Concrete cases of equation (2) are outlined in Subsection 1.3.2.

5. Equations of the theory of nonlinear oscillations

$$y''_{xx} + \Phi(y'_x) + y = 0.$$

The transformation $z = y'_x$, $w = -y - \Phi(y'_x)$ leads to the Abel equation of the form (1):

$$ww'_z + \Phi'_z(z)w + z = 0.$$

6. Homogeneous equation with respect to the independent variable

$$x^2y''_{xx} = xg(y)y'_x + f(y).$$

The substitution $w(y) = xy'_x$ leads to the Abel equation:

$$ww'_y = [g(y) + 1]w + f(y).$$

7. Homogeneous equation in the extended sense

$$xy''_{xx} = g(yx^k)y'_x + x^{-k-1}f(yx^k).$$

The transformation $t = yx^k$, $u = x^k(xy'_x + ky)$ leads to the Abel equation:

$$uu'_t = [g(t) + 2k + 1]u + f(t) - ktg(t) - k(k+1)t.$$

To the Emden—Fowler equation, discussed in Section 2.3, correspond $g(t) = 0$, $f(t) = At^m$, $k = \frac{n+2}{m-1}$.

8. Homogeneous equation in the extended sense

$$y''_{xx} = x^\alpha y^\beta F\Big(\frac{x}{y}y'_x\Big) + yx^{-2}G\Big(\frac{x}{y}y'_x\Big).$$

The transformation

$$\eta = \frac{x}{y}y'_x, \quad v = x^{\alpha+2}y^{\beta-1}$$

leads to the Abel equation:

$$[F(\eta)v + G(\eta) + \eta - \eta^2]v'_\eta = [(\beta - 1)\eta + \alpha + 2]v.$$

To the generalized Emden—Fowler equation, discussed in Section 2.5, correspond $\alpha = n - l$, $\beta = m + l$, $F(\eta) = A\eta^l$, $G(\eta) = 0$.

9. Exponential-homogeneous equation

$$y''_{xx} = x^\alpha e^{\beta y} f(xy'_x) + x^{-2}g(xy'_x).$$

The substitution $\zeta = xy'_x$, $u = x^{\alpha+2}e^{\beta y}$ leads to the Abel equation:

$$[f(\zeta)u + g(\zeta) + \zeta]u'_\zeta = (\beta\zeta + \alpha + 2)u.$$

10. Exponential-homogeneous equation

$$y''_{xx} = e^{\alpha x} y^\beta f\Big(\frac{y'_x}{y}\Big) + yg\Big(\frac{y'_x}{y}\Big).$$

The substitution $\xi = y'_x/y$, $w = e^{\alpha x}y^{\beta-1}$ leads to the Abel equation:

$$[f(\xi)w + g(\xi) - \xi^2]w'_\xi = [(\beta - 1)\xi + \alpha]w.$$

1.4. Equations Containing Polynomial Functions of y

1.4.1. Abel Equations of the First Kind $y'_x = f_3(x)y^3 + f_2(x)y^2 + f_1(x)y + f_0(x)$

Preliminary comments.

1. If $y_0 = y_0(x)$ is a particular solution of the equation in question, the substitution $y - y_0 = 1/w$ reduces it to the Abel equation of the second kind:

$$ww'_x = -(3f_3y_0^2 + 2f_2y_0 + f_1)w^2 - (3f_3y_0 + f_2)w - f_3,$$

which is discussed in Section 1.3. For $f_0(x) \equiv 0$, we may choose $y_0 \equiv 0$ as a particular solution.

2. The transformation

$$\xi = \int f_3E^2\,dx, \quad u = \Big(y + \frac{f_2}{3f_3}\Big)E^{-1}, \qquad \text{where} \quad E = \exp\Big[\int\Big(f_1 - \frac{f_2^2}{3f_3}\Big)\,dx\Big],$$

brings the original equation to the normal form:

$$u'_\xi = u^3 + \Phi(\xi),$$

where

$$\Phi = \frac{1}{f_3E^3}\Big[f_0 + \frac{1}{3}\frac{d}{dx}\Big(\frac{f_2}{f_3}\Big) - \frac{f_1f_2}{3f_3} + \frac{2f_2^3}{27f_3^2}\Big].$$

1. $y'_x = ay^3 + bx^{-3/2}$.

This is a special case of equation 1.4.1.9 with $n = -1/2$.

2. $y'_x = -y^3 + 3a^2x^2y - 2a^3x^3 + a$.

The substitution $y = 1/w + ax$ leads to the Abel equation of the form 1.3.2.1: $ww'_x = 3axw + 1$.

3. $y'_x = -y^3 + (ax+b)y^2$.

The substitution $y = -1/w$ leads to the Abel equation of the form 1.3.2.1: $ww'_x = (ax+b)w + 1$.

4. $y'_x = -y^3 + (ax+b)^{-2}y^2$.

The substitution $y = -1/w$ leads to the Abel equation of the form 1.3.2.2: $ww'_x = (ax+b)^{-2}w + 1$.

5. $y'_x = -y^3 + (ax+b)^{-1/2}y^2$.

The substitution $y = -1/w$ leads to the Abel equation of the form 1.3.2.4: $ww'_x = (ax+b)^{-1/2}w + 1$.

6. $y'_x = ay^3 + 3abxy^2 - b - 2ab^3x^3$.

This is a special case of equation 1.4.1.10 with $n = 0$, $m = 1$.

7. $y'_x = axy^3 + by^2$.

The substitution $u = xy$ leads to an equation with separation of variables: $xu'_x = au^3 + bu^2 + u$.

8. $y'_x = axy^3 + 3abx^2y^2 - b - 2ab^3x^4$.

This is a special case of equation 1.4.1.10 with $n = m = 1$.

9. $y'_x = ax^{2n+1}y^3 + bx^{-n-2}$.

The substitution $w = yx^{n+1}$ leads to an equation with separation of variables: $xw'_x = aw^3 + (n+1)w + b$.

For $a = -\dfrac{n+1}{3A^2}$ and $b = \frac{2}{3}A(n+1)$, the solution in the parametric form is written as

$$x = \exp\Bigl(\frac{F}{n+1}\Bigr), \qquad y = -A\Bigl(1 + \frac{1}{\tau}\Bigr)\exp(-F),$$

where $F = \tau - \frac{1}{3}\ln|\tau + \frac{1}{3}| + C$.

10. $y'_x = ax^ny^3 + 3abx^{n+m}y^2 - bmx^{m-1} - 2ab^3x^{n+3m}$.

The substitution $w = y + bx^m$ leads to the Bernoulli equation: $w'_x = ax^nw^3 - 3ab^2x^{n+2m}w$.

11. $y'_x = ax^ny^3 + 3abx^{n+m}y^2 + cx^ky - 2ab^3x^{n+3m} + bcx^{m+k} - bmx^{m-1}$.

The substitution $w = y + bx^m$ leads to the Bernoulli equation: $w'_x = ax^nw^3 + (cx^k - 3ab^2x^{n+2m})w$.

12. $9y'_x = -x^m(ax^{1-m}+b)^{2\lambda+1}y^3 - x^{-2m}(9a+2+9bmx^{m-1})(ax^{1-m}+b)^{-\lambda-2}$.

With $\lambda = \dfrac{1}{3a(1-m)}$, the substitution

$$y = \Big(\frac{3}{w} + \frac{1}{ax+bx^m}\Big)\big(ax^{1-m}+b\big)^{-\lambda}$$

leads to the equation $ww'_x = w + ax + bx^m$ outlined in Subsection 1.3.1.

13. $xy'_x = ax^4y^3 + (bx^2-1)y + cx$.

The substitution $w = xy$ leads to an equation with separation of variables: $w'_x = x(aw^3 + bw + c)$.

14. $xy'_x = ay^3 + 3abx^ny^2 - bnx^n - 2ab^3x^{3n}$.

The substitution $w = y + bx^n$ leads to the Bernoulli equation: $w'_x = ax^{-1}w^3 - 3ab^2x^{2n-1}w$.

15. $xy'_x = ax^{2n+1}y^3 + (bx-n)y + cx^{1-n}$.

The substitution $w = yx^n$ leads to an equation with separated variables: $w'_x = aw^3 + bw + c$.

16. $xy'_x = ax^{n+2}y^3 + (bx^n-1)y + cx^{n-1}$.

The substitution $w = xy$ leads to an equation with separation of variables: $w'_x = x^{n-1}(aw^3 + bw + c)$.

17. $x^2y'_x = y^3 - 3a^2x^4y + 2a^3x^6 + 2ax^3$.

The transformation $x = \dfrac{1}{\xi}$, $y = \dfrac{1}{w} + ax^2$ leads to an equation of the form 1.3.2.2: $ww'_\xi = 3a\xi^{-2}w + 1$.

18. $y'_x = -(ax+bx^m)y^3 + y^2$.

The substitution $y = -1/w$ leads to the equation $ww'_x = w + ax + bx^m$ outlined in Subsection 1.3.1.

19. $y'_x = (Ax^2+Bx+C)^{-1/2}y^3 + y^2$.

The substitution $y = -1/w$ leads to the Abel equation of the form 1.3.1.63: $ww'_x = w - (Ax^2+Bx+C)^{-1/2}$.

20. $y'_x = -y^3 + ae^{\lambda x}y^2$.

The substitution $y = -1/w$ leads to the Abel equation of the form 1.3.2.7: $ww'_x = ae^{\lambda x}w + 1$.

21. $y'_x = -y^3 + 3a^2e^{2\lambda x}y - 2a^3e^{3\lambda x} + a\lambda e^{\lambda x}$.

The substitution $y = \dfrac{1}{w} + ae^{\lambda x}$ leads to the Abel equation of the form 1.3.2.7: $ww'_x = 3ae^{\lambda x}w + 1$.

22. $y'_x = -\frac{1}{3}\lambda^{-1}e^{2\lambda x}y^3 + \frac{2}{3}\lambda^2 e^{-\lambda x}.$

The solution in the parametric form is written as

$$x = \frac{F}{\lambda}, \quad y = -\lambda\Big(1 + \frac{1}{\tau}\Big)e^{-F}, \qquad \text{where} \quad F = \tau - \tfrac{1}{3}\ln|\tau + \tfrac{1}{3}| + C.$$

23. $y'_x = ae^{2\lambda x}y^3 + be^{\lambda x}y^2 + cy + de^{-\lambda x}.$

The substitution $y = we^{-\lambda x}$ leads to an equation with separated variables: $w'_x = aw^3 + bw^2 + (c+\lambda)w + d$.

24. $y'_x = ae^{\lambda x}y^3 + 3abe^{\lambda x}y^2 + cy - 2ab^3e^{\lambda x} + bc.$

The substitution $w = y+b$ yields the Bernoulli equation: $w'_x = ae^{\lambda x}w^3 + (c - 3ab^2e^{\lambda x})w$.

25. $y'_x = ae^{\lambda x}y^3 + 3abe^{(\lambda+\mu)x}y^2 - 2ab^3e^{(\lambda+3\mu)x} - b\mu e^{\mu x}.$

The substitution $w = y + be^{\mu x}$ leads to the Bernoulli equation: $w'_x = ae^{\lambda x}w^3 - 3ab^2e^{(\lambda+2\mu)x}w$.

26. $y'_x = ae^{\lambda x}y^3 + 3abe^{(\lambda+\mu)x}y^2 + 2ab^2e^{(\lambda+2\mu)x}y - b\mu e^{\mu x}.$

The substitution $w = y + be^{\mu x}$ leads to the Bernoulli equation: $w'_x = ae^{\lambda x}w^3 - ab^2e^{(\lambda+2\mu)x}w$.

27. $y'_x = ae^{\lambda x}y^3 + 3abe^{(\lambda+\mu)x}y^2 + \mu y - 2ab^3e^{(\lambda+3\mu)x}.$

The substitution $w = y + be^{\mu x}$ leads to the Bernoulli equation: $w'_x = ae^{\lambda x}w^3 + \big[\mu - 3ab^2e^{(\lambda+2\mu)x}\big]w$.

28. $$y'_x = ae^{\lambda x}y^3 + 3abe^{(\lambda+\mu)x}y^2 + [(3ab^2 + c)e^{(\lambda+2\mu)x} + s]y + b(ab^2+c)e^{(\lambda+3\mu)x} + b(s-\mu)e^{\mu x}.$$

The substitution $w = y + be^{\mu x}$ leads to the Bernoulli equation: $w'_x = ae^{\lambda x}w^3 + \big[ce^{(\lambda+2\mu)x} + s\big]w$.

29. $y'_x = [a + b\exp(2x/a)]y^3 + y^2.$

The substitution $y = -1/w$ leads to an equation of the form 1.3.1.8: $ww'_x = w - a - b\exp(2x/a)$.

30. $y'_x = -\frac{2}{3}ax^{-1}\exp(2ax^2)y^3 + (1 - \frac{4}{3}ax^2)\exp(-ax^2).$

The substitution $y = \Big(\dfrac{1}{2aw} + x\Big)\exp(-ax^2)$ leads to an equation of the form 1.3.1.16: $ww'_x = w + (6ax)^{-1}$.

31. $y'_x = -a\exp(2ax^3)y^3 + (1 - 2ax^3)\exp(-ax^3).$

The transformation $\xi = x^2$, $y = \Big(\dfrac{2}{3aw} + x\Big)\exp(-ax^3)$ leads to an equation of the form 1.3.1.32: $ww'_\xi = w \pm 2(9a)^{-1}\xi^{-1/2}$.

32. $y'_x = -ax^{-2}\exp(2ax^3)y^3 + 2x(1-ax^3)\exp(-ax^3).$

The substitution $y = \left(\dfrac{1}{3aw} + x^2\right)\exp(-ax^3)$ leads to an equation of the form 1.3.1.33: $ww'_x = w + (9a)^{-1}x^{-2}$.

33. $y'_x = ay^3 + b\cosh(\lambda x)y^2.$

The transformation $t = \dfrac{1}{y} + \dfrac{b}{\lambda}\sinh(\lambda x)$, $\xi = e^{\lambda x}$ leads to the Riccati equation: $2a\xi'_t = b\xi^2 - 2\lambda t\xi - b$.

34. $y'_x = ay^3 + b\sinh(\lambda x)y^2.$

The transformation $t = \dfrac{1}{y} + \dfrac{b}{\lambda}\cosh(\lambda x)$, $\xi = e^{\lambda x}$ leads to the Riccati equation: $2a\xi'_t = b\xi^2 - 2\lambda t\xi + b$.

35. $y'_x = -y^3 + 3a^2\cosh^2 x\, y - 2a^3\cosh^3 x + a\sinh x.$

The substitution $y = \dfrac{1}{w} + a\cosh x$ leads to the Abel equation of the form 1.3.2.9: $ww'_x = 3a\cosh x\, w + 1$.

36. $y'_x = -y^3 + 3a^2\sinh^2 x\, y - 2a^3\sinh^3 x + a\cosh x.$

The substitution $y = \dfrac{1}{w} + a\sinh x$ leads to the Abel equation of the form 1.3.2.10: $ww'_x = 3a\sinh x\, w + 1$.

37. $y'_x = -y^3 + a\cos(\omega x)y^2.$

The substitution $y = -1/w$ leads to the Abel equation of the form 1.3.2.11: $ww'_x = a\cos(\omega x)w + 1$.

38. $y'_x = -y^3 + a\sin(\omega x)y^2.$

The substitution $y = -1/w$ leads to the Abel equation of the form 1.3.2.12: $ww'_x = a\sin(\omega x)w + 1$.

39. $y'_x = -y^3 + 3a^2\cos^2(\lambda x)y + a\lambda\sin(\lambda x) + 2a^3\cos^3(\lambda x).$

The substitution $y = \dfrac{1}{w} - a\cos(\lambda x)$ leads to the Abel equation of the form 1.3.2.11: $ww'_x = -3a\cos(\lambda x)w + 1$.

40. $y'_x = -y^3 + 3a^2\sin^2(\lambda x)y + a\lambda\cos(\lambda x) - 2a^3\sin^3(\lambda x).$

The substitution $y = \dfrac{1}{w} + a\sin(\lambda x)$ leads to an equation of the form the Abel equation of the form 1.3.2.12: $ww'_x = 3a\sin(\lambda x)w + 1$.

41. $y'_x = afy^3 + \left(bfg^2 + \dfrac{g'_x}{g}\right)y + cfg^3, \qquad f = f(x), \quad g = g(x).$

The substitution $y = gw$ leads to an equation with separation of variables: $w'_x = fg^2(aw^3 + bw + c)$.

42. $y'_x = fy^3 + 3fhy^2 + (g + 3fh^2)y + fh^3 + gh - h'_x$,
where $f = f(x)$, $g = g(x)$, $h = h(x)$.

The substitution $w = y + h(x)$ leads to the Bernoulli equation: $w'_x = g(x)w + f(x)w^3$.

43. $y'_x = \dfrac{g'_x}{f^2(ag+b)^3}y^3 + \dfrac{f'_x}{f}y + fg'_x, \qquad f = f(x), \quad g = g(x).$

Solution: $\displaystyle\int \frac{du}{u^3 - au + 1} + C = \frac{1}{a}\ln|ag+b|, \qquad \text{where} \quad u = \frac{y}{f(ag+b)}.$

44. $y'_x = (y-f)(y-g)\left(y - \dfrac{af+bg}{a+b}\right)h + \dfrac{y-g}{f-g}f'_x + \dfrac{y-f}{g-f}g'_x$,
where $f = f(x)$, $g = g(x)$, $h = h(x)$.

Solution: $\displaystyle |y-f|^a|y-g|^b\left|y - \frac{af+bg}{a+b}\right|^{-a-b} = C\exp\left[\frac{ab}{a+b}\int (f-g)^2 h\,dx\right].$

1.4.2. Equations of the Form $(A_{22}y^2 + A_{12}xy + A_{11}x^2 + A_0)y'_x = B_{22}y^2 + B_{12}xy + B_{11}x^2 + B_0$

Preliminary comments.

1. For $A_{22} = 0$, this is the Abel equation (see Subsection 1.3.4). For $B_{11} = 0$, this is the Abel equation with respect to $x = x(y)$.

2. The transformation $z = y/x$, $\zeta = x^{-2}$, leads to the Abel equation of the second kind:

$$[(A_0z - B_0)\zeta + A_{22}z^3 + (A_{12} - B_{22})z^2 + (A_{11} - B_{12})z - B_{11}]\zeta'_z = 2A_0\zeta^2 + 2(A_{22}z^2 + A_{12}z + A_{11})\zeta.$$

3. The transformation $x = \bar{x} + \alpha$, $y = \bar{y} + \beta$, where α and β are the parameters which are determined by solving the second order algebraic system

$$A_{22}\beta^2 + A_{12}\alpha\beta + A_{11}\alpha^2 + A_0 = 0, \quad B_{22}\beta^2 + B_{12}\alpha\beta + B_{11}\alpha^2 + B_0 = 0,$$

leads to the equation

$$[A_{22}\bar{y}^2 + A_{12}\bar{x}\bar{y} + A_{11}\bar{x}^2 + (2A_{22}\beta + A_{12}\alpha)\bar{y} + (2A_{11}\alpha + A_{12}\beta)\bar{x})]\bar{y}'_{\bar{x}} = B_{22}\bar{y}^2 + B_{12}\bar{x}\bar{y} + B_{11}\bar{x}^2 + (2B_{22}\beta + B_{12}\alpha)\bar{y} + (2B_{11}\alpha + B_{12}\beta)\bar{x},$$

The transformation $\xi = \bar{y}/\bar{x}$, $w = 1/\bar{x}$ reduces this equation to the Abel equation of the second kind:

$$\{[a_2\xi^2 + (a_1 - b_2)\xi - b_1]w + A_{22}\xi^3 + (A_{12} - B_{22})\xi^2 + (A_{11} - B_{12})\xi - B_{11}\}w'_\xi = (a_2\xi + a_1)w^2 + (A_{22}\xi^2 + A_{12}\xi + A_{11})w.$$

where

$$a_2 = 2A_{22}\beta + A_{12}\alpha, \quad b_2 = 2B_{22}\beta + B_{12}\alpha, \quad a_1 = 2A_{11}\alpha + A_{12}\beta, \quad b_1 = 2B_{11}\alpha + B_{12}\beta.$$

4. The substitution $y = t + \varepsilon x$, where parameter ε is determined by solving the cubic equation

$$(A_{22}\varepsilon^2 + A_{12}\varepsilon + A_{11})\varepsilon - B_{22}\varepsilon^2 - B_{12}\varepsilon - B_{11} = 0,$$

leads to the Abel equation of the second kind with respect to $x = x(t)$:

$$[Qtx + (B_{22} - A_{22}\varepsilon)t^2 + B_0 - A_0\varepsilon]x'_t = (A_{22}\varepsilon^2 + A_{12}\varepsilon + A_{11})x^2 + (2A_{22}\varepsilon + A_{12})tx + A_{22}t^2 + A_0,$$

where $Q = 2B_{22}\varepsilon + B_{12} - \varepsilon(2A_{22}\varepsilon + A_{12})$.

1. $\boldsymbol{(Ay^2 + x^2)y'_x = -2xy + Bx^2 + a.}$

Solution: $Ay^3 - Bx^3 + 3(x^2y - ax) = C.$

2. $\boldsymbol{(Ay^2 + Bx^2 - a^2B)y'_x = Cy^2 + 2Bxy.}$

The transformation $x = w + a$, $y = \xi w$ leads to the linear equation:

$$(-A\xi^3 + C\xi^2 + B\xi)w'_\xi = (A\xi^2 + B)w + 2aB.$$

3. $\boldsymbol{(Ay^2 + Bxy + Cx^2)y'_x = Dy^2 + Exy + Fx^2.}$

Homogeneous equation. The substitution $z = y/x$ leads to an equation with separated variables:

$$xz'_x = (Az^2 + Bz + C)^{-1}[-Az^3 + (D - B)z^2 + (E - C)z + F].$$

4. $\boldsymbol{(Ay^2 - 2Akxy + Bkx^2)y'_x = -By^2 + 2Bkxy - Ak^3x^2 + a.}$

The substitution $y = z + kx$ leads to the Riccati equation with respect to $x = x(z)$:

$$[-(Ak + B)z^2 + a]x'_z = k(B - Ak)x^2 + Az^2.$$

5. $\boldsymbol{(Ay^2 + 2Bxy + Ak^2x^2)y'_x = By^2 + 2Ak^2xy + Bk^2x^2 + a.}$

The substitution $y = z + kx$ leads to the Riccati equation with respect to $x = x(z)$:

$$[(B - Ak)z^2 + a]x'_z = 2k(Ak + B)x^2 + 2(Ak + B)zx + Az^2.$$

6. $\boldsymbol{(Ay^2 + Bxy + Cx^2 + a)y'_x = Aky^2 + Bkxy + Ckx^2 + b.}$

The substitution $y = z + kx$ leads to the Riccati equation with respect to $x = x(z)$:

$$(b - ak)x'_z = (Ak^2 + Bk + C)x^2 + (2Ak + B)zx + Az^2 + a.$$

7. $\boldsymbol{(Ay^2 + 2Bxy + Dx^2 + a)y'_x = -By^2 - 2Dxy + Ex^2 + b.}$

Solution: $Ay^3 - Ex^3 + 3(Bxy^2 + Dx^2y + ay - bx) = C.$

8. $\boldsymbol{(Ay^2 - 2Axy + Bx^2 + A - B)y'_x = -Ay^2 + 2Bxy - Bx^2 + A - B.}$

This is a special case of equation 1.4.2.21 with $a = 1$, $b = 1$.

9. $(Ay^2 + 2Axy + Bx^2 + A - B)y'_x = Ay^2 + 2Bxy + Bx^2 - A + B.$

This is a special case of equation 1.4.2.21 with $a = 1$, $b = -1$.

10. $(Ay^2 - 4Axy + Bx^2 + 4A - B)y'_x = -2Ay^2 + 2Bxy - 2Bx^2 + 8A - 2B.$

This is a special case of equation 1.4.2.21 with $a = 1$, $b = 2$.

11. $(Ay^2 + 4Axy + Bx^2 + 4A - B)y'_x = 2Ay^2 + 2Bxy + 4Bx^2 - 8A + 2B.$

This is a special case of equation 1.4.2.21 with $a = 1$, $b = -2$.

12. $(Ay^2 - 6Axy + Bx^2 + 9A - B)y'_x = -3Ay^2 + 2Bxy - 3Bx^2 + 27A - 3B.$

This is a special case of equation 1.4.2.21 with $a = 1$, $b = 3$.

13. $(Ay^2 + 6Axy + Bx^2 + 9A - B)y'_x = 3Ay^2 + 2Bxy + 3Bx^2 - 27A + 3B.$

This is a special case of equation 1.4.2.21 with $a = 1$, $b = -3$.

14. $2(Ay^2 - Axy + Bx^2 + A - 4B)y'_x = -Ay^2 + 4Bxy - Bx^2 + A - 4B.$

This is a special case of equation 1.4.2.21 with $a = 2$, $b = 1$.

15. $2(Ay^2 + Axy + Bx^2 + A - 4B)y'_x = Ay^2 + 4Bxy + Bx^2 - A + 4B.$

This is a special case of equation 1.4.2.21 with $a = 2$, $b = -1$.

16. $(ay^2 - 2bxy + ax^2 + ab^2 - a^3)y'_x = -by^2 + 2axy - bx^2 + b^3 - a^2b.$

This is a special case of equation 1.4.2.21 with $A = 1$, $B = 1$.

17. $(ay^2 - 2bxy - ax^2 + ab^2 + a^3)y'_x = -by^2 - 2axy + bx^2 + b^3 + a^2b.$

This is a special case of equation 1.4.2.21 with $A = 1$, $B = -1$.

18. $(ay^2 - 2bxy + 2ax^2 + ab^2 - 2a^3)y'_x = -by^2 + 4axy - 2bx^2 + b^3 - 2a^2b.$

This is a special case of equation 1.4.2.21 with $A = 1$, $B = 2$.

19. $(ay^2 - 2bxy - 2ax^2 + ab^2 + 2a^3)y'_x = -by^2 - 4axy + 2bx^2 + b^3 + 2a^2b.$

This is a special case of equation 1.4.2.21 with $A = 1$, $B = -2$.

20. $(Ay^2 + Bxy + Cx^2 + a)y'_x$
$= Dy^2 + k(2Ak + B - 2D)xy + k(-Ak^2 + Dk + C)x^2 + b.$

The substitution $y = z + kx$ leads to the Riccati equation with respect to $x = x(z)$:

$$[(D - Ak)z^2 + b - ak]x'_z = (Ak^2 + Bk + C)x^2 + (2Ak + B)zx + Az^2 + a.$$

21. $(aAy^2 - 2bAxy + aBx^2 + ab^2A - a^3B)y'_x$
$= -bAy^2 + 2aBxy - bBx^2 + b^3A - a^2bB.$

The transformation $x = w + a$, $y = \xi w + b$ leads to the linear equation:

$$(-aA\xi^3 + bA\xi^2 + aB\xi - bB)w'_\xi = (aA\xi^2 - 2bA\xi + aB)w + 2a^2B - 2b^2A.$$

1.4.3. Equations of the Form $(A_{22}y^2 + A_{12}xy + A_{11}x^2 + A_2y + A_1x)y'_x = B_{22}y^2 + B_{12}xy + B_{11}x^2 + B_2y + B_1x$*

Preliminary comments.

1. For $A_{22} = 0$, this is the Abel equation (see Subsection 1.3.4). For $B_{11} = 0$ this is the Abel equation with respect to $x = x(y)$.

2. The transformation $\xi = y/x$, $w = 1/x$ leads to the Abel equation of the second kind:

$$\{[A_2\xi^2 + (A_1 - B_2)\xi - B_1]w + A_{22}\xi^3 + (A_{12} - B_{22})\xi^2 + (A_{11} - B_{12})\xi - B_{11}\}w'_\xi = (A_2\xi + A_1)w^2 + (A_{22}\xi^2 + A_{12}\xi + A_{11})w.$$

3. In Paragraph 3 of Subsection 1.4.4, another transformation is given which reduces the original equation to the Abel equation of the second kind.

4. Dynamical systems of the second order

$$\frac{dx}{dt} = P(x, y), \qquad \frac{dy}{dt} = Q(x, y) \tag{1}$$

which describe the behavior of the simplest Lagrangian and Hamiltonian systems in mechanics are often reduced to equations of the considered type when

$$\begin{aligned} P(x, y) &= f(x, y)(A_{22}y^2 + A_{12}xy + A_{11}x^2 + A_2y + A_1x), \\ Q(x, y) &= f(x, y)(B_{22}y^2 + B_{12}xy + B_{11}x^2 + B_2y + B_1x), \end{aligned} \tag{2}$$

where $f = f(x, y)$ is an arbitrary function.

In particular, dynamical systems (1) with functions (2) and $f \equiv 1$ are met with in analyzing complex equilibrium states. In this case, functions P and Q are substituted by their Taylor-series expansions in the vicinity of the equilibrium state $x = y = 0$ with the first and second order terms retained.

When obtained the solution of the ordinary differential equation

$$(A_{22}y^2 + A_{12}xy + A_{11}x^2 + A_2y + A_1x)y'_x = B_{22}y^2 + B_{12}xy + B_{11}x^2 + B_2y + B_1x$$

in the parametric form $x = x(u, C_1)$, $y = y(u, C_1)$, the solution of the system (1), (2) is determined by the formulae

$$x = x(u, C_1), \quad y = y(u, C_1), \quad t = \int \frac{x'_u\,du}{P\big(x(u, C_1),\, y(u, C_1)\big)} + C_2.$$

The latter relation defines the implicit dependence of parameter u on t: $u = u(t, C_1, C_2)$, and makes it possible to find, with the aid of two former formulae, the dependence of x and y on t.

1. $(y^2 - x^2 + ay)y'_x = y^2 - x^2 + ax.$

Solution in the parametric form:

$$x = at + C|t|^{-1}e^{4t}, \quad y = -at + C|t|^{-1}e^{4t}.$$

---

* This section was written with A.I. Zhurov

2. $(y^2 - x^2 + ay)y'_x = 2y^2 - 2xy + ay.$

Solution in the parametric form:

$$x = t + Ct^2 e^{a/t}, \quad y = Ct^2 e^{a/t}.$$

3. $(y^2 - x^2 + ay - ax)y'_x = y^2 - x^2 - ay + ax.$

Solution in the parametric form:

$$x = at + Ce^{2t}, \quad y = -at + Ce^{2t}.$$

4. $(y^2 - x^2 + ay + 2ax)y'_x = y^2 - x^2 + 2ay + ax.$

Solution in the parametric form:

$$x = -at + C|t|^3 e^{4t}, \quad y = at + C|t|^3 e^{4t}.$$

5. $(y^2 - x^2 + ay + 2ax)y'_x = 2xy - 2x^2 + ay + 2ax.$

Solution in the parametric form:

$$x = t + Ct^{-2} e^{-a/t}, \quad y = -2t + Ct^{-2} e^{-a/t}.$$

6. $(y^2 - x^2 + ay - 2ax)y'_x = 4y^2 - 6xy + 2x^2 + ay - 2ax.$

Solution in the parametric form:

$$x = \tfrac{1}{3}t + C|t|^{2/3} e^{a/t}, \quad y = \tfrac{2}{3}t + C|t|^{2/3} e^{a/t}.$$

7. $(y^2 - x^2 + ay + 3ax)y'_x = -y^2 + 4xy - 3x^2 + ay + 3ax.$

Solution in the parametric form:

$$x = \tfrac{1}{2}t + C|t|^{-1} e^{-a/t}, \quad y = -\tfrac{3}{2}t + C|t|^{-1} e^{-a/t}.$$

8. $(y^2 - xy + ay + ax)y'_x = xy - x^2 + ay + ax.$

Solution in the parametric form:

$$x = -t + C|t|^{-1} e^{a/t}, \quad y = t + C|t|^{-1} e^{a/t}.$$

9. $(y^2 - xy + ay + ax)y'_x = y^2 - xy + 2ay.$

Solution in the parametric form:

$$x = -at + Ct^2 e^t, \quad y = Ct^2 e^t.$$

10. $(y^2 - xy + ay - 2ax)y'_x = 3y^2 - 5xy + 2x^2 + ay - 2ax.$

Solution in the parametric form:

$$x = \tfrac{1}{2}t + C|t|^{1/2} e^{a/t}, \quad y = t + C|t|^{1/2} e^{a/t}.$$

11. $(y^2 + xy - 2x^2 + ay + ax)y'_x = y^2 + xy - 2x^2 + 2ax.$

Solution in the parametric form:

$$x = at + Ct^{-2}e^{9t}, \quad y = -2at + Ct^{-2}e^{9t}.$$

12. $(y^2 + xy - 2x^2 + ay + ax)y'_x = 2y^2 - xy - x^2 + ay + ax.$

Solution in the parametric form:

$$x = t + C|t|^3 e^{a/t}, \quad y = -t + C|t|^3 e^{a/t}.$$

13. $(y^2 + xy - 2x^2 + ay - ax)y'_x = y^2 + xy - 2x^2 - 2ay + 2ax.$

Solution in the parametric form:

$$x = at + Ce^{3t}, \quad y = -2at + Ce^{3t}.$$

14. $(y^2 + xy - 2x^2 + ay - 2ax)y'_x = 5y^2 - 7xy + 2x^2 + ay - 2ax.$

Solution in the parametric form:

$$x = \tfrac{1}{4}t + C|t|^{3/4}e^{a/t}, \quad y = \tfrac{1}{2}t + C|t|^{3/4}e^{a/t}.$$

15. $(y^2 - 2xy + x^2 + ay)y'_x = ay.$

Solution: $x = y + \dfrac{a}{C - \ln|y|}.$

16. $(y^2 - 2xy + x^2 + ay + ax)y'_x = -y^2 + 2xy - x^2 + ay + ax.$

Solution in the parametric form:

$$x = -\frac{a}{2\ln|t|} + Ct, \quad y = \frac{a}{2\ln|t|} + Ct.$$

17. $(y^2 - 2xy + x^2 + ay + 2ax)y'_x = -2(y^2 - 2xy + x^2) + ay + 2ax.$

Solution in the parametric form:

$$x = -\frac{a}{3\ln|t|} + Ct, \quad y = \frac{2a}{3\ln|t|} + Ct.$$

18. $(y^2 - 2xy + x^2 + ay - 2ax)y'_x = 2(y^2 - 2xy + x^2) + ay - 2ax.$

Solution in the parametric form:

$$x = \frac{a}{\ln|t|} + Ct, \quad y = \frac{2a}{\ln|t|} + Ct.$$

19. $(y^2 + 2xy + x^2 + ay + 2ax)y'_x = -y^2 - 2xy - x^2 + 2ay + ax.$

Solution in the parametric form:

$$x = C^2\left(t^{1/3} + \frac{4t^2}{5a}\right) + Ct, \quad y = -C^2\left(t^{1/3} + \frac{4t^2}{5a}\right) + Ct, \qquad a \neq 0.$$

20. $(y^2 + 2xy + x^2 + ay - ax)y'_x = -y^2 - 2xy - x^2 + ay - ax.$

Solution in the parametric form:

$$x = C^3\sqrt{1 - \frac{4t^3}{3a}} + C^2t, \quad y = -C^3\sqrt{1 - \frac{4t^3}{3a}} + C^2t, \qquad a \neq 0.$$

21. $(y^2 + 2xy + x^2 + ay - 2ax)y'_x = -y^2 - 2xy - x^2 - 2ay + ax.$

Solution in the parametric form:

$$x = C^2\left(t^3 + \frac{4t^2}{a}\right) + Ct, \quad y = -C^2\left(t^3 + \frac{4t^2}{a}\right) + Ct, \qquad a \neq 0.$$

22. $(y^2 + 2xy - 3x^2 + ay + ax)y'_x = 3y^2 - 2xy - 1x^2 + ay + ax.$

Solution in the parametric form:

$$x = \tfrac{1}{2}t + Ct^2e^{a/t}, \quad x = -\tfrac{1}{2}t + Ct^2e^{a/t}.$$

23. $(y^2 + 2xy - 3x^2 + ay + ax)y'_x = y^2 + 2xy - 3x^2 - ay + 3ax.$

Solution in the parametric form:

$$x = at + C|t|^{-1}e^{8t}, \quad y = -3at + C|t|^{-1}e^{8t}.$$

24. $(y^2 + 2xy - 3x^2 + ay + 2ax)y'_x = y^2 + 2xy - 3x^2 + 3ax.$

Solution in the parametric form:

$$x = at + C|t|^{-3}e^{16t}, \quad y = -3at + C|t|^{-3}e^{16t}.$$

25. $(y^2 - x^2 + ay + bx)y'_x = y^2 - x^2 + by + ax.$

Solution in the parametric form:

$$x = (a-b)t + C|t|^{-\frac{a+b}{a-b}}e^{4t}, \quad y = (b-a)t + C|t|^{-\frac{a+b}{a-b}}e^{4t}, \qquad a \neq b.$$

26. $(y^2 - xy + ay + bx)y'_x = y^2 - xy + (a+b)y.$

Solution in the parametric form:

$$x = -bt + C|t|^{\frac{a+b}{b}}e^{t}, \quad y = C|t|^{\frac{a+b}{b}}e^{t}, \qquad b \neq 0.$$

27. $(y^2 + xy - 2x^2 + ay + bx)y'_x = y^2 + xy - 2x^2 + (b-a)y + 2ax.$

Solution in the parametric form:

$$x = (2a-b)t + C|t|^{-\frac{a+b}{2a-b}}e^{9t}, \quad y = 2(b-2a)t + C|t|^{-\frac{a+b}{2a-b}}e^{9t}, \qquad b \neq 2a.$$

28. $(y^2 - 2xy + x^2 + ay - abx)y'_x = b(y^2 - 2xy + x^2) + ay - abx.$

Solution in the parametric form:

$$x = \frac{a}{b-1}\,\frac{1}{\ln|t|} + Ct, \quad y = \frac{ab}{b-1}\,\frac{1}{\ln|t|} + Ct, \qquad b \neq 1.$$

29. $(y^2 + 2xy - 3x^2 + ay + bx)y'_x = y^2 + 2xy - 3x^2 + (b - 2a)y + 3ax.$

Solution in the parametric form:

$$x = (3a - b)t + C|t|^{-\frac{a+b}{3a-b}} e^{16t}, \quad y = 3(b - 3a)t + C|t|^{-\frac{a+b}{3a-b}} e^{16t}, \qquad b \neq 3a.$$

30. $(y^2 - 3xy + 2x^2 + ay + bx)y'_x = y^2 - 3xy + 2x^2 + (3a + b)y - 2ax.$

Solution in the parametric form:

$$x = (2a + b)t + C|t|^{\frac{a+b}{2a+b}} e^{-t}, \quad y = 2(2a + b)t + C|t|^{\frac{a+b}{2a+b}} e^{-t}, \qquad b \neq -2a.$$

31. $(y^2 + 3xy - 4x^2 + ay + bx)y'_x = y^2 + 3xy - 4x^2 + (b - 3a)y + 4ax.$

Solution in the parametric form:

$$x = (4a - b)t + C|t|^{-\frac{a+b}{4a-b}} e^{25t}, \quad y = 4(b - 4a)t + C|t|^{-\frac{a+b}{4a-b}} e^{25t}, \qquad b \neq 4a.$$

32. $[y^2 + Axy - (A+1)x^2 + by - 2bx]y'_x = (A+4)y^2 - (A+6)xy + 2x^2 + by - 2bx.$

Solution in the parametric form:

$$x = \frac{t}{A+3} + C|t|^{\frac{A+2}{A+3}} e^{b/t}, \quad y = \frac{2t}{A+3} + C|t|^{\frac{A+2}{A+3}} e^{b/t}, \qquad A \neq -3.$$

33. $(y^2 - 2Axy + A^2x^2 + by - bx)y'_x = Ay^2 - 2A^2xy + A^3x^2 + by - bx.$

Solution in the parametric form:

$$x = C^3\sqrt{1 + \frac{2(A-1)}{3b}t^3} + C^2t, \quad y = AC^3\sqrt{1 + \frac{2(A-1)}{3b}t^3} + C^2t, \qquad b \neq 0.$$

34. $[y^2 - 2Axy + (2A-1)x^2 + by - Abx]y'_x = (2-A)y^2 - 2xy + Ax^2 + by - Abx.$

Solution in the parametric form:

$$x = \frac{t}{1-A} + Ct^2e^{b/t}, \quad y = \frac{At}{1-A} + Ct^2e^{b/t}, \qquad A \neq 1.$$

35. $(y^2 - 2Axy + A^2x^2 + ay + bx)y'_x = A(y^2 - 2Axy + A^2x^2) + (aA + a + b)y - aAx.$

Solution in the parametric form:

$$x = C^2\left[t^{\frac{aA+b}{a+b}} + \frac{(1-A)^2}{(2-A)a+b}t^2\right] + Ct, \quad y = AC^2\left[t^{\frac{aA+b}{a+b}} + \frac{(1-A)^2}{(2-A)a+b}t^2\right] + Ct,$$

where $a + b \neq 0$ and $(2 - A)a + b \neq 0$.

36. $[y^2 - (A+2)xy + (A+1)x^2 + by - Abx]y'_x = -Axy + Ax^2 + by - Abx.$

Solution in the parametric form:

$$x = \frac{t}{1-A} + C|t|^A e^{(A-1)b/t}, \quad y = \frac{At}{1-A} + C|t|^A e^{(A-1)b/t}, \qquad A \neq 1.$$

37. $[Ay^2 + xy - (A+1)x^2 + by + bx]y'_x = (A+1)y^2 - xy - Ax^2 + by + bx.$

Solution in the parametric form:

$$x = t + C|t|^{2A+1}e^{b/t}, \quad y = -t + C|t|^{2A+1}e^{b/t}.$$

38. $(Ay^2 + Bxy + Cx^2 + kx)y'_x = Dy^2 + Exy + Fx^2 + ky.$

The substitution $y = xz$ leads to a linear equation with respect to $x = x(z)$:

$$[-Az^3 + (D-B)z^2 + (E-C)z + F]x'_z = (Az^2 + Bz + C)x + k.$$

39. $(Ay^2 + Bxy + Cx^2 - \alpha By - \alpha Cx)y'_x = Dy^2 + Exy + \alpha(C-E)y.$

The transformation $x = w + \alpha$, $y = \xi w$ leads to a linear equation:

$$[-A\xi^3 + (D-B)\xi^2 + (E-C)\xi]w'_\xi = (A\xi^2 + B\xi + C)w + \alpha C.$$

40. $(Ay^2 + 2Bxy + Ak^2x^2 + ay + bx)y'_x = By^2 + 2Ak^2xy + Bk^2x^2 + by + ak^2x.$

This is a special case of equation 1.4.3.57 with $C = Ak^2$.

41. $(Ay^2 + 2Bxy + Ak^2x^2 + ay + bx)y'_x = By^2 + 2Ak^2xy + Bk^2x^2 + aky + bkx.$

This is a special case of equation 1.4.3.62 with $C = Ak^2$.

42. $(Ay^2 + 2Bxy + Ak^2x^2 + ay - akx)y'_x = By^2 + 2Ak^2xy + Bk^2x^2 + my - mkx.$

This is a special case of equation 1.4.3.61 with $C = Ak^2$.

43. $(Ay^2 + 2Bxy - Bkx^2 + ay + bx)y'_x = By^2 + 2Ak^2xy - Ak^3x^2 + by + ak^2x.$

This is a special case of equation 1.4.3.58 with $m = b$.

44. $(Ay^2 + 2Bxy - Bkx^2 + ay + bx)y'_x = By^2 + 2Ak^2xy - Ak^3x^2 + aky + bkx.$

This is a special case of equation 1.4.3.62 with $C = -Bk$.

45. $(Ay^2 + 2Bxy - Bkx^2 + ay - akx)y'_x = By^2 + 2Ak^2xy - Ak^3x^2 + my - mkx.$

This is a special case of equation 1.4.3.61 with $C = -Bk$.

46. $(Ay^2 + 2Akxy + Cx^2 + ay + bx)y'_x = Aky^2 + 2Ak^2xy + Ckx^2 + by + ak^2x.$

This is a special case of equation 1.4.3.57 with $B = Ak$.

47. $(Ay^2 + 2Akxy + Cx^2 + ay + bx)y'_x = Aky^2 + 2Ak^2xy + Ckx^2 + aky + bkx.$

This is a special case of equation 1.4.3.62 with $B = Ak$.

48. $(Ay^2 + 2Akxy + Cx^2 + ay - akx)y'_x = Aky^2 + 2Ak^2xy + Ckx^2 + my - mkx.$

This is a special case of equation 1.4.3.61 with $B = Ak$.

49. $(Ay^2 - 2Akxy + Bkx^2 + ay + bx)y'_x = -By^2 + 2Bkxy - Ak^3x^2 + by + ak^2x.$

This is a special case of equation 1.4.3.59 with $m = b$.

50. $(Ay^2 - 2Akxy + Bkx^2 + ay + bx)y'_x = -By^2 + 2Bkxy - Ak^3x^2 + aky + bkx.$

This is a special case of equation 1.4.3.59 with $m = ak$.

51. $[y^2 + 2Axy + A^2x^2 + (A-1)By - 2ABx]y'_x$
$= -A(y^2 + 2Axy + A^2x^2) - (A^2+1)By + A(A-1)Bx.$

Solution in the parametric form:

$$x = C^2\left[t^A + \frac{A+1}{(A-2)B}t^2\right] + Ct, \quad y = -AC^2\left[t^A + \frac{A+1}{(A-2)B}t^2\right] + Ct, \qquad A \neq 2,\ B \neq 0.$$

52. $[y^2 - 2Axy + A^2x^2 + (B-1)ky + (A-B)kx]y'_x$
$= A(y^2 - 2Axy + A^2x^2) + (AB-1)ky - A(B-1)kx.$

Solution in the parametric form:

$$x = C^2\left[t^B - \frac{A-1}{(B-2)k}t^2\right] + Ct, \quad y = AC^2\left[t^B - \frac{A-1}{(B-2)k}t^2\right] + Ct, \qquad B \neq 2,\ k \neq 0.$$

53. $[2y^2 - (A+3)xy + (A+1)x^2 + By - ABx]y'_x$
$= (A+1)y^2 - (3A+1)xy + 2Ax^2 + By - ABx.$

Solution in the parametric form:

$$x = \frac{t}{1-A} + C|t|^{-1}e^{-B/t}, \quad y = \frac{At}{1-A} + C|t|^{-1}e^{-B/t}, \qquad A \neq 1.$$

54. $[2y^2 - (3A+1)xy + (3A-1)x^2 + By - ABx]y'_x$
$= (3-A)y^2 - (A+3)xy + 2Ax^2 + By - ABx.$

Solution in the parametric form:

$$x = \frac{t}{1-A} + C|t|^3e^{B/t}, \quad y = \frac{At}{1-A} + C|t|^3e^{B/t}, \qquad A \neq 1.$$

55. $[A(y^2 - 2xy + x^2) - A(A-B)y + B(A-B)x]y'_x$
$= B(y^2 - 2xy + x^2) - A(A-B)y + B(A-B)x.$

Solution in the parametric form:

$$x = \frac{A}{\ln|t|} + Ct, \quad y = \frac{B}{\ln|t|} + Ct.$$

56. $(Ay^2 + Bxy + Cx^2 + ay + bx)y'_x = Aky^2 + Bkxy + Ckx^2 + ny + (ak + b - n)x.$

The substitution $y = z + kx$ leads to the Riccati equation with respect to $x = x(z)$:

$$(n - ak)zx'_z = (Ak^2 + Bk + C)x^2 + [(2Ak + B)z + ak + b]x + Az^2 + az.$$

57. $(Ay^2 + 2Bxy + Cx^2 + ay + bx)y'_x$
$= By^2 + 2Ak^2xy + k(-Ak^2 + Bk + C)x^2 + by + ak^2x.$

The substitution $y = z + kx$ leads to the Riccati equation with respect to $x = x(z)$:

$$[(B - Ak)z + b - ak]zx'_z = (Ak^2 + 2Bk + C)x^2 + [2(Ak + B)z + ak + b]x + Az^2 + az.$$

58. $(Ay^2 + 2Bxy - Bkx^2 + ay + bx)y'_x$
$= By^2 + 2Ak^2xy - Ak^3x^2 + my + k(ak + b - m)x.$

The substitution $y = z + kx$ leads to the Riccati equation with respect to $x = x(z)$:

$$[(B - Ak)z + m - ak]zx'_z = (Ak^2 + Bk)x^2 + [2(Ak + B)z + ak + b]x + Az^2 + az.$$

59. $(Ay^2 - 2Akxy + Bkx^2 + ay + bx)y'_x$
$= -By^2 + 2Bkxy - Ak^3x^2 + my + k(ak + b - m)x.$

The substitution $y = z + kx$ leads to the Riccati equation with respect to $x = x(z)$:

$$[-(Ak + B)z + m - ak]zx'_z = k(B - Ak)x^2 + (ak + b)x + Az^2 + az.$$

60. $(Ay^2 + 2Bxy + Ak^2x^2 + ay + bx)y'_x$
$= By^2 + 2Ak^2xy + Bk^2x^2 + my + k(ak + b - m)x.$

The substitution $y = z + kx$ leads to the Riccati equation with respect to $x = x(z)$:

$$[(B - Ak)z + m - ak]zx'_z = 2k(Ak + B)x^2 + [2(Ak + B)z + ak + b]x + Az^2 + az.$$

61. $(Ay^2 + 2Bxy + Cx^2 + ay - akx)y'_x$
$= By^2 + 2Ak^2xy + k(-Ak^2 + Bk + C)x^2 + my - mkx.$

The substitution $y = z + kx$ leads to the Riccati equation with respect to $x = x(z)$:

$$[(B - Ak)z^2 + m - ak]zx'_z = (Ak^2 + 2Bk + C)x^2 + 2(Ak + B)zx + Az^2 + az.$$

62. $(Ay^2 + 2Bxy + Cx^2 + ay + bx)y'_x$
$= By^2 + 2Ak^2xy + k(-Ak^2 + Bk + C)x^2 + aky + bkx.$

The substitution $y = z + kx$ leads to the Riccati equation with respect to $x = x(z)$:

$$(B - Ak)z^2x'_z = (Ak^2 + 2Bk + C)x^2 + [2(Ak + B)z + ak + b]x + Az^2 + az.$$

63. $\{(A - 1)y^2 + [2 - A(k + 1)]xy + (Ak - 1)x^2 + By - Bkx\}y'_x$
$= (A - k)y^2 + [2k - A(k + 1)]xy + (A - 1)kx^2 + By - Bkx.$

Solution in the parametric form:

$$x = \frac{t}{1 - k} + C|t|^A e^{B/t}, \quad y = \frac{kt}{1 - k} + C|t|^A e^{B/t}, \qquad k \neq 1.$$

64. $[A(\alpha y^2 + \beta xy + \gamma x^2) + (2\alpha - A^2\sigma)y + (\beta - AB\sigma)x]y'_x$
$+ B(\alpha y^2 + \beta xy + \gamma x^2) + (\beta - AB\sigma)y + (2\gamma - B^2\sigma)x = 0.$

Solution: $\alpha y^2 + \beta xy + \gamma x^2 - A\sigma y - B\sigma x + \sigma = C\exp(-Ay - Bx).$

65. $(A_{22}y^2 + A_{12}xy + A_{11}x^2 + A_2y + A_1x)y'_x$
$= B_{22}y^2 + k(2A_{22}k + A_{12} - 2B_{22})xy + k(-A_{22}k^2 + B_{22}k + A_{11})x^2$
$+ B_2y + k(A_2k + A_1 - B_2)x.$

The substitution $y = z + kx$ leads to the Riccati equation with respect to $x = x(z)$:

$$[(B_{22} - A_{22}k)z + B_2 - A_2k]zx'_z$$
$$= (A_{22}k^2 + A_{12}k + A_{11})x^2 + [(2A_{22}k + A_{12})z + A_2k + A_1]x + A_{22}z^2 + A_2z.$$

▶ *In equations 66–70, the following notation is used:*

$$\Delta = Ab - aB, \quad \delta = Ab + aB.$$

66. $(Aa^2y^2 - 2Aabxy + Ab^2x^2 - \Delta Aay + \Delta aBx)y'_x$
$= a^2By^2 - 2aBbxy + Bb^2x^2 - \Delta Aby + \Delta Bbx.$

Solution in the parametric form:

$$x = \frac{A}{\ln|t|} + aCt, \qquad y = \frac{B}{\ln|t|} + bCt.$$

67. $[kAa^2y^2 - k\delta axy + kaBbx^2 - m\Delta Aay + (maB - \Delta)\Delta x]y'_x$
$= kAaby^2 - k\delta bxy + kBb^2x^2 - (mAb + \Delta)\Delta y + m\Delta Bbx.$

Solution in the parametric form:

$$x = At + aC|t|^{m+1}e^{kt}, \qquad y = Bt + bC|t|^{m+1}e^{kt}.$$

68. $[mAa^2y^2 - a(m\delta - \Delta)xy + b(maB - \Delta)x^2 + k\Delta Aay - k\Delta aBx]y'_x$
$= a(mBb + \Delta)y^2 - b(m\delta + \Delta)xy + mBb^2x^2 + k\Delta Aby - k\Delta Bbx.$

Solution in the parametric form:

$$x = At + aC|t|^{m+1}e^{k/t}, \qquad y = Bt + bC|t|^{m+1}e^{k/t}.$$

69. $(kA^3y^2 - 2kA^2Bxy + kAB^2x^2 - 2\Delta a^2y + 2\Delta abx)y'_x$
$= kA^2By^2 - 2kAB^2xy + kB^3x^2 - 2\Delta aby + 2\Delta b^2x.$

Solution in the parametric form:

$$x = AC^3\sqrt{\tfrac{1}{3}kt^3 + 1} + aC^2t, \qquad y = BC^3\sqrt{\tfrac{1}{3}kt^3 + 1} + bC^2t.$$

70. $[kA^3y^2 - 2kA^2Bxy + kAB^2x^2 + m\Delta Aay - (mAb + \Delta)\Delta x]y'_x$
$= kA^2By^2 - 2kAB^2xy + kB^3x^2 + (maB - \Delta)\Delta y - m\Delta Bbx.$

Solution in the parametric form:

$$x = AC^2\Big(t^{m+1} + \frac{k}{m-1}t^2\Big) + aCt, \quad y = BC^2\Big(t^{m+1} + \frac{k}{m-1}t^2\Big) + bCt, \qquad m \neq 1.$$

1.4.4. Equations of the Form $(A_{22}y^2 + A_{12}xy + A_{11}x^2 + A_2y + A_1x + A_0)y'_x = B_{22}y^2 + B_{12}xy + B_{11}x^2 + B_2y + B_1x + B_0$

Preliminary comments.

1. With $A_{22} = 0$, this is the Abel equation (see Subsection 1.3.4). With $B_{11} = 0$, this is the Abel equation with respect to $x = x(y)$.

See Subsection 1.4.2 for the case $A_2 = A_1 = B_2 = B_1 = 0$.

See Subsection 1.4.3 for the case $A_0 = B_0 = 0$.

2. The transformation $x = \bar{x} + \alpha$, $y = \bar{y} + \beta$, where α and β are the parameters which are determined by solving the second order algebraic system

$$A_{22}\beta^2 + A_{12}\alpha\beta + A_{11}\alpha^2 + A_2\beta + A_1\alpha + A_0 = 0,$$
$$B_{22}\beta^2 + B_{12}\alpha\beta + B_{11}\alpha^2 + B_2\beta + B_1\alpha + B_0 = 0,$$

leads to the equation

$$(A_{22}\bar{y}^2 + A_{12}\bar{x}\bar{y} + A_{11}\bar{x}^2 + a_2\bar{y} + a_1\bar{x})\bar{y}'_{\bar{x}} = B_{22}\bar{y}^2 + B_{12}\bar{x}\bar{y} + B_{11}\bar{x}^2 + b_2\bar{y} + b_1\bar{x}, \qquad (1)$$

where

$$a_2 = 2A_{22}\beta + A_{12}\alpha + A_2, \qquad a_1 = 2A_{11}\alpha + A_{12}\beta + A_1,$$
$$b_2 = 2B_{22}\beta + B_{12}\alpha + B_2, \qquad b_1 = 2B_{11}\alpha + B_{12}\beta + B_1.$$

The transformation $\xi = \bar{y}/\bar{x}$, $w = 1/\bar{x}$ reduces equation (1) to the Abel equation of the second kind:

$$\{[a_2\xi^2 + (a_1 - b_2)\xi - b_1]w + A_{22}\xi^3 + (A_{12} - B_{22})\xi^2 + (A_{11} - B_{12})\xi - B_{11}\}w'_\xi$$
$$= (a_2\xi + a_1)w^2 + (A_{22}\xi^2 + A_{12}\xi + A_{11})w.$$

3. The substitution $y = z + \varepsilon x$, where parameter ε is determined by solving the cubic equation

$$(A_{22}\varepsilon^2 + A_{12}\varepsilon + A_{11})\varepsilon - B_{22}\varepsilon^2 - B_{12}\varepsilon - B_{11} = 0,$$

leads to the Abel equation of the second kind with respect to $x = x(z)$:

$$[(Qz + R)x + (B_{22} - A_{22}\varepsilon)z^2 + (B_2 - A_2\varepsilon)z + B_0 - A_0\varepsilon]x'_z$$
$$= (A_{22}\varepsilon^2 + A_{12}\varepsilon + A_{11})x^2 + [(2A_{22}\varepsilon + A_{12})z + A_2\varepsilon + A_1]x + A_{22}z^2 + A_2z + A_0,$$

where

$$Q = 2B_{22}\varepsilon + B_{12} - \varepsilon(2A_{22}\varepsilon + A_{12}), \quad R = B_2\varepsilon + B_1 - \varepsilon(A_2\varepsilon + A_1).$$

1. $\boldsymbol{(y + ax + b)^2 y'_x = (\alpha y + \beta x + \gamma)^2.}$

This is a special case of equation 1.7.1.6 with $f(z) = z^{-2}$.

2. $\boldsymbol{(Ay^2 + Bxy - \alpha By + kx - \alpha k)y'_x = Cy^2 + Dxy + (k - \alpha D)y.}$

The transformation $x = w + \alpha$, $y = w\xi$ leads to a linear equation with respect to $w = w(\xi)$: $[-A\xi^3 + (C - B)\xi^2 + D\xi]w'_\xi = (A\xi^2 + B\xi)w + k$.

3. $\boldsymbol{(Ay^2 + 2Axy + Bx^2 + A - B)y'_x = Ay^2 + 2Bxy + Dx^2 + 2(B - D)x + D - A.}$

The transformation $x = w + 1$, $y = \xi w - 1$ leads to a linear equation:

$$(-A\xi^3 - A\xi^2 + B\xi + D)w'_\xi = (A\xi^2 + 2A\xi + B)w + 2(B - A).$$

4. $\boldsymbol{(Ay^2 - 2Axy + Bx^2 + A - B)y'_x = -Ay^2 + 2Bxy + Cx^2 + 2(B + C)x + A + C.}$

The transformation $x = w - 1$, $y = \xi w - 1$ leads to a linear equation:

$$(-A\xi^3 + A\xi^2 + B\xi + C)w'_\xi = (A\xi^2 - 2A\xi + B)w + 2(A - B).$$

5. $(Ay^2 + 2Axy + Bx^2 + A - B)y'_x = Ay^2 + 2Bxy + Cx^2 + 2(C - B)x - A + C.$

The transformation $x = w - 1$, $y = \xi w + 1$ leads to a linear equation:

$$(-A\xi^3 - A\xi^2 + B\xi + C)w'_\xi = (A\xi^2 + 2A\xi + B)w + 2(A - B).$$

6. $(Ay^2 - 2Axy + Bx^2 + A - B)y'_x = -Ay^2 + 2Bxy + Cx^2 - 2(B + C)x + A + C.$

The transformation $x = w + 1$, $y = \xi w + 1$ leads to a linear equation:

$$(-A\xi^3 + A\xi^2 + B\xi + C)w'_\xi = (A\xi^2 - 2A\xi + B)w + 2(B - A).$$

7. $(Ay^2 - 2Axy + Bx^2 + A - B)y'_x$
$= Cy^2 + 2Bxy + Dx^2 - 2(A + C)y - 2(B + D)x + 2A + C + D.$

The transformation $x = w + 1$, $y = \xi w + 1$ leads to a linear equation:

$$[-A\xi^3 + (2A + C)\xi^2 + B\xi + D]w'_\xi = (A\xi^2 - 2A\xi + B)w + 2(B - A).$$

8. $(2Ay^2 - 2Axy + Bx^2 + 2A - 4B)y'_x$
$= -Ay^2 + 2Bxy + Dx^2 - 2(B + 2D)x + A + 4D.$

This is a special case of equation 1.4.4.34 with $\alpha = 2$, $\beta = 1$, $C = -A$.

9. $(Ay^2 + 4Axy + Bx^2 + 4A - B)y'_x = 2Ay^2 + 2Bxy + Cx^2 - 2(C - 2B)x + C - 8A.$

The transformation $x = w + 1$, $y = \xi w - 2$ leads to a linear equation:

$$(-A\xi^3 - 2A\xi^2 + B\xi + C)w'_\xi = (A\xi^2 + 4A\xi + B)w + 2B - 8A.$$

10. $(Ay^2 - 4Axy + Bx^2 + 4A - B)y'_x = -2Ay^2 + 2Bxy + Cx^2 - 2(2B + C)x + 8A + C.$

The transformation $x = w + 1$, $y = \xi w + 2$ leads to a linear equation:

$$(-A\xi^3 + 2A\xi^2 + B\xi + C)w'_\xi = (A\xi^2 - 4A\xi + B)w + 2B - 8A.$$

11. $(Ay^2 + 4Axy + Bx^2 + 4A - B)y'_x$
$= Cy^2 + 2Bxy + 2Bx^2 + 4(C - 2A)y + 2B + 4C - 16A.$

The transformation $x = w + 1$, $y = \xi w - 2$ leads to a linear equation:

$$[-A\xi^3 + (C - 4A)\xi^2 + B\xi + 2B]w'_\xi = (A\xi^2 + 4A\xi + B)w + 2B - 8A.$$

12. $(2Ay^2 + 2Axy + Bx^2 + 2A - 4B)y'_x = Ay^2 + 2Bxy + Dx^2 + 2(B - 2D)x + 4D - A.$

This is a special case of equation 1.4.4.34 with $\alpha = 2$, $\beta = -1$, $C = A$.

13. $(2Ay^2 + 2Axy - Bx^2 + 2A + 4B)y'_x = Ay^2 - 2Bxy - Dx^2 + 2(B - 2D)x - A - 4D.$

This is a special case of equation 1.4.4.34 with $\alpha = -2$, $\beta = 1$, $C = -A$.

14. $(Ay^2 + 2Bxy + Ak^2x^2 + ay + bx + m)y'_x$
$= By^2 + 2Ak^2xy + Bk^2x^2 + by + ak^2x + s.$

This is a special case of equation 1.4.4.27 with $C = Ak^2$.

15. $(Ay^2 + 2Bxy + Ak^2x^2 + ay + bx + m)y'_x$
$$= By^2 + 2Ak^2xy + Bk^2x^2 + aky + bkx + s.$$

This is a special case of equation 1.4.4.32 with $C = Ak^2$.

16. $(Ay^2 + 2Bxy + Ak^2x^2 + ay - akx + b)y'_x$
$$= By^2 + 2Ak^2xy + Bk^2x^2 + my - mkx + s.$$

This is a special case of equation 1.4.4.31 with $C = Ak^2$.

17. $(Ay^2+2Bxy-Bkx^2+ay+bx+c)y'_x = By^2+2Ak^2xy-Ak^3x^2+by+ak^2x+s.$

This is a special case of equation 1.4.4.28 with $m = b$.

18. $(Ay^2+2Bxy-Bkx^2+ay+bx+m)y'_x = By^2+2Ak^2xy-Ak^3x^2+aky+bkx+s.$

This is a special case of equation 1.4.4.32 with $C = -Bk$.

19. $(Ay^2 + 2Bxy - Bkx^2 + ay - akx + b)y'_x$
$$= By^2 + 2Ak^2xy - Ak^3x^2 + my - mkx + s.$$

This is a special case of equation 1.4.4.31 with $C = -Bk$.

20. $(Ay^2+2Akxy+Cx^2+ay+bx+m)y'_x = Aky^2+2Ak^2xy+Ckx^2+by+ak^2x+s.$

This is a special case of equation 1.4.4.27 with $B = Ak$.

21. $(Ay^2+2Akxy+Cx^2+ay+bx+m)y'_x = Aky^2+2Ak^2xy+Ckx^2+aky+bkx+s.$

This is a special case of equation 1.4.4.32 with $B = Ak$.

22. $(Ay^2 + 2Akxy + Cx^2 + ay - akx + b)y'_x$
$$= Aky^2 + 2Ak^2xy + Ckx^2 + my - mkx + s.$$

This is a special case of equation 1.4.4.31 with $B = Ak$.

23. $(Ay^2 - 2Akxy + Bkx^2 + ay + bx + c)y'_x$
$$= -By^2 + 2Bkxy - Ak^3x^2 + by + ak^2x + s.$$

This is a special case of equation 1.4.4.29 with $m = b$.

24. $(Ay^2 - 2Akxy + Bkx^2 + ay + bx + c)y'_x$
$$= -By^2 + 2Bkxy - Ak^3x^2 + aky + bkx + s.$$

This is a special case of equation 1.4.4.29 with $m = ak$.

25. $(Ay^2 + 2Bxy + Cx^2 - 2A\beta y + kx + A\beta^2)y'_x$
$$= By^2 + Exy + Fx^2 + ky - E\beta x - B\beta^2 - k\beta.$$

The substitution $w = y - \beta$ leads to an equation of the form 1.4.3.38:

$$(Aw^2 + 2Bxw + Cx^2 + \bar{k}x)w'_x = Bw^2 + Exw + Fx^2 + \bar{k}w,$$

where $\bar{k} = k + 2B\beta$.

26. $(Ay^2 + Bxy + Cx^2 + ay + bx + m)y'_x$
$$= Aky^2 + Bkxy + Ckx^2 + ny + k(ak + b - n)x + s.$$

The substitution $y = z + kx$ leads to the Riccati equation with respect to $x = x(z)$:

$$[(n - ak)z + s - mk]x'_z = (Ak^2 + Bk + C)x^2 + [(2Ak + B)z + ak + b]x + Az^2 + az + m.$$

27. $(Ay^2 + 2Bxy + Cx^2 + ay + bx + m)y'_x$
$$= By^2 + 2Ak^2xy + k(-Ak^2 + Bk + C)x^2 + by + ak^2x + s.$$

The substitution $y = z + kx$ leads to the Riccati equation with respect to $x = x(z)$:

$$[(B - Ak)z^2 + (b - ak)z + s - mk]x'_z = (Ak^2 + 2Bk + C)x^2 + [2(Ak + B)z + ak + b]x + Az^2 + az + m.$$

28. $(Ay^2 + 2Bxy - Bkx^2 + ay + bx + c)y'_x$
$$= By^2 + 2Ak^2xy - Ak^3x^2 + my + k(ak + b - m)x + s.$$

The substitution $y = z + kx$ leads to the Riccati equation with respect to $x = x(z)$:

$$[(B-Ak)z^2+(m-ak)z+s-ck]x'_z = (Ak^2+Bk)x^2+[2(Ak+B)z+ak+b]x+Az^2+az+c.$$

29. $(Ay^2 - 2Akxy + Bkx^2 + ay + bx + c)y'_x$
$$= -By^2 + 2Bkxy - Ak^3x^2 + my + k(ak + b - m)x + s.$$

The substitution $y = z + kx$ leads to the Riccati equation with respect to $x = x(z)$:

$$[-(Ak + B)z^2 + (m - ak)z + s - ck]x'_z = k(B - Ak)x^2 + (ak + b)x + Az^2 + az + c.$$

30. $(Ay^2 + 2Bxy + Ak^2x^2 + ay + bx + c)y'_x$
$$= By^2 + 2Ak^2xy + Bk^2x^2 + my + k(ak + b - m)x + s.$$

The substitution $y = z + kx$ leads to the Riccati equation with respect to $x = x(z)$:

$$[(B-Ak)z^2+(m-ak)z+s-ck]x'_z = 2k(Ak+B)x^2+[2(Ak+B)z+ak+b]x+Az^2+az+c.$$

31. $(Ay^2 + 2Bxy + Cx^2 + ay - akx + b)y'_x$
$$= By^2 + 2Ak^2xy + k(-Ak^2 + Bk + C)x^2 + my - mkx + s.$$

The substitution $y = z + kx$ leads to the Riccati equation with respect to $x = x(z)$:

$$[(B - Ak)z^2 + (m - ak)z + s - bk]x'_z = (Ak^2 + 2Bk + C)x^2 + 2(Ak + B)zx + Az^2 + az + b.$$

32. $(Ay^2 + 2Bxy + Cx^2 + ay + bx + m)y'_x$
$$= By^2 + 2Ak^2xy + k(-Ak^2 + Bk + C)x^2 + aky + bkx + s.$$

The substitution $y = z + kx$ leads to the Riccati equation with respect to $x = x(z)$:

$$[(B - Ak)z^2 + s - mk]x'_z = (Ak^2 + 2Bk + C)x^2 + [2(Ak + B)z + ak + b]x + Az^2 + az + m.$$

33. $[A(\alpha y^2 + \beta xy + \gamma x^2) + (A\delta + 2\alpha)y + (A\varepsilon + \beta)x + A\sigma + \delta]y'_x$
$$+ B(\alpha y^2 + \beta xy + \gamma x^2) + (B\delta + \beta)y + (B\varepsilon + 2\gamma)x + B\sigma + \varepsilon = 0.$$

Solution: $\alpha y^2 + \beta xy + \gamma x^2 + \delta y + \varepsilon x + \sigma = C \exp(-Ay - Bx).$

34. $(\alpha Ay^2 - 2\beta Axy + Bx^2 + \alpha\beta^2 A - \alpha^2 B)y'_x$
$= Cy^2 + 2Bxy + Dx^2 - 2\beta(\beta A + C)y - 2(\alpha D + \beta B)x + \alpha^2 D + \beta^2(2\beta A + C).$

The substitution $x = w + \alpha$, $y = \xi w + \beta$ leads to a linear equation:

$$[-\alpha A\xi^3 + (2\beta A + C)\xi^2 + B\xi + D]w'_\xi = (\alpha A\xi^2 - 2\beta A\xi + B)w + 2(\alpha B - \beta^2 A).$$

35. $(A_{22}y^2 + A_{12}xy + A_{11}x^2 + A_2 y + A_1 x + A_0)y'_x$
$= B_{22}y^2 + k(2A_{22}k + A_{12} - 2B_{22})xy + k(-A_{22}k^2 + B_{22}k + A_{11})x^2$
$+ B_2 y + k(A_2 k + A_1 - B_2)x + B_0.$

The substitution $y = z + kx$ leads to the Riccati equation with respect to $x = x(z)$:

$$[(B_{22} - A_{22}k)z^2 + (B_2 - A_2 k)z + B_0 - A_0 k]x'_z =$$
$$(A_{22}k^2 + A_{12}k + A_{11})x^2 + [(2A_{22}k + A_{12})z + A_2 k + A_1]x + A_{22}z^2 + A_2 z + A_0.$$

36. $(A_{22}y^2 + A_{12}xy + A_{11}x^2 + A_2 y + A_1 x + A_0)y'_x$
$= B_{22}y^2 + B_{12}xy + B_{11}x^2 + B_2 y + B_1 x + B_0.$

Here A_{ij}, B_{ij}, and A_1 are arbitrary parameters, and the other parameters are defined by the relations

$$A_2 = -A_{12}\alpha - 2A_{22}\beta,$$
$$A_0 = -A_{11}\alpha^2 + A_{22}\beta^2 - A_1\alpha,$$
$$B_2 = (2A_{11} - B_{12})\alpha + (A_{12} - 2B_{22})\beta + A_1,$$
$$B_1 = -2B_{11}\alpha - B_{12}\beta,$$
$$B_0 = B_{11}\alpha^2 + (B_{12} - 2A_{11})\alpha\beta + (B_{22} - A_{12})\beta^2 - A_1\beta$$

(α, β are arbitrary parameters).

The transformation $x = w + \alpha$, $y = \xi w + \beta$ leads to a linear equation:

$$[-A_{22}\xi^3 + (B_{22} - A_{12})\xi^2 + (B_{12} - A_{11})\xi + B_{11}]w'_\xi = (A_{22}\xi^2 + A_{12}\xi + A_{11})w + k,$$

where $k = 2A_{11}\alpha + A_{12}\beta + A_1$.

1.5. Nonlinear Equations of the Form $f(x,y)y'_x = g(x,y)$ Containing Arbitrary Parameters

1.5.1. Equations Containing Power Functions

1. $y'_x = A\sqrt{y} + Bx^{-1/2}.$

The substitution $w = (2/A)\sqrt{y}$ leads to the Abel equation of the form 1.3.1.32: $ww'_x = w + 2BA^{-2}x^{-1/2}$.

2. $y'_x = A\sqrt{y} + Bx^{-1}.$

Let $A = \pm 2a^{-1}\sqrt{b}$, $B = \mp 4b$ $(b > 0)$.

The solution in the parametric form is written as

$$x = af(\tau), \qquad y = b[2\tau \pm f(\tau)]^2,$$

where $f(\tau) = \exp(\mp\tau^2)\left[\int \exp(\mp\tau^2)\,d\tau + C\right]^{-1}$.

3. $y'_x = A\sqrt{y} + Bx^{-2}$.

The substitution $w = 2A^{-1}\sqrt{y}$ leads to the Abel equation of the form 1.3.1.33: $ww'_x = w + 2BA^{-2}x^{-2}$.

4. $y'_x = a\sqrt{y} + bx + cx^m$.

The substitution $w = 2a^{-1}\sqrt{y}$ leads to the Abel equation of the second kind: $ww'_x = w + 2a^{-2}(bx + cx^m)$, whose special cases are outlined in Subsection 1.3.1.

5. $y'_x = ay^n + bx^{\frac{n}{1-n}}$.

Solution:

$$\int \frac{dw}{aw^n + \frac{1}{1-n}w + b} = \ln|x| + C, \qquad \text{where} \quad w = yx^{\frac{1}{n-1}}.$$

6. $y'_x = Ay^s - Bx^k$.

The transformation $x = \left(w'_z\right)^{\frac{1}{k}}$, $y = \lambda\left(\frac{w}{z}\right)^{\frac{1}{s}}$, where $\lambda = \left(\frac{B}{A}\right)^{\frac{1}{s}}$, leads to the generalized Emden—Fowler equation:

$$w''_{zz} = -\frac{\lambda k}{sB} z^{-\frac{1}{s}} w^{\frac{1-s}{s}} \left(w'_z\right)^{\frac{k-1}{k}},$$

which is discussed in Section 2.5 (in the classification table, one should search for the equations satisfying the condition $n + m + 1 = 0$).

7. $y'_x = (ax + by + c)^n$.

This is a special case of equation 1.7.1.1 with $f(\xi) = \xi^n$.

8. $y'_x = ax^{m-n-nm}y^n + bx^m$.

Solution:

$$\int \frac{dw}{w^n - \lambda w + 1} + C = b\sqrt[n]{\frac{a}{b}}\ln|x|,$$

where $w = \sqrt[n]{\frac{a}{b}}\,yx^{-m-1}$, $\lambda = \frac{m+1}{b}\sqrt[n]{\frac{b}{a}}$.

9. $y'_x = ax^{n-1}y^{m+1} + bx^{nk-1}y^{mk+1}$.

This is the homogeneous equation in the extended sense of the form 1.7.1.3 with $f(\xi) = a\xi + b\xi^k$.

10. $y'_x = ax^k y\sqrt{y} + bx^m y + cx^s\sqrt{y}$.

This is a special case of equation 1.7.1.4 with $f(x) = ax^k$, $g(x) = bx^m$, $h(x) = cx^s$, $n = 1/2$.

11. $y'_x = ax^k y^{1+n} + bx^m y + cx^s y^{1-n}$.

This is a special case of equation 1.7.1.4 with $f(x) = ax^k$, $g(x) = bx^m$, $h(x) = cx^s$.

12. $y'_x = x^{n-1}y^{1-m}(ax^n + by^m)^k.$

This is a special case of equation 1.7.1.7 with $f(\xi) = \xi^k$.

13. $xy'_x = y + ax^{n-m}y^m + bx^{n-k}y^k.$

The substitution $y = xw$ leads to an equation with separation of variables: $w'_x = x^{n-2}(aw^m + bw^k)$.

14. $(ay^n + bx)y'_x = 1.$

Solution: $x = e^{by}\left(C + a\int y^n e^{-by}\,dy\right)$.

15. $x(xy^n + a)y'_x + by = 0.$

Solution: $nb - a = x\bigl(Cy^{a/b} + y^n\bigr)$.

16. $x(ay^m + m)y'_x = y\bigl[bx^{n(\lambda-1)}y^{m\lambda} - n\bigr].$

This is a special case of equation 1.7.1.15 with $f(\xi) = a\xi$, $g(\xi) = 1$, $h(\xi) = b\xi^\lambda$, $k = n$.

17. $(ax^n + bx^2 + cxy)y'_x = kx^n + bxy + cy^2.$

The transformation $t = y/x$, $z = x^{n-2}$ leads to a linear equation: $(k - at)z'_t = (n-2)(az + b + ct)$.

18. $(ay^n + bx^2 + cxy)y'_x = ky^n + bxy + cy^2.$

The transformation $t = y/x$, $z = x^{n-2}$ leads to a linear equation: $t^n(k - at)z'_t = (n-2)(at^n z + b + ct)$.

19. $(ax^n + by^n + x)y'_x = \alpha x^k y^{n-k} + \beta x^m y^{n-m} + y.$

The transformation $t = y/x$, $z = x^{n-1}$ leads to a linear equation:

$$(\alpha t^{n-k} + \beta t^{n-m} - bt^{n+1} - at)z'_t = (n-1)(bt^n + a)z + n - 1.$$

20. $(ax^n + by^n + Ax^2 + Bxy)y'_x = \alpha x^k y^{n-k} + \beta x^m y^{n-m} + Axy + By^2.$

The transformation $t = y/x$, $z = x^{n-2}$ leads to a linear equation:

$$(\alpha t^{n-k} + \beta t^{n-m} - bt^{n+1} - at)z'_t = (n-2)(bt^n + a)z + (n-2)(Bt + A).$$

21. $[(ax + by)^n + bx]y'_x = c(ax + by)^m - ax.$

This is a special case of equation 1.7.1.13 with $f(\xi) = \xi^n$, $g(\xi) = 1$, $h(\xi) = c\xi^m$.

22. $[(ax + by)^n + by]y'_x = c(ax + by)^m - ay.$

This is a special case of equation 1.7.1.14 with $f(\xi) = \xi^n$, $g(\xi) = 1$, $h(\xi) = c\xi^m$.

23. $(\alpha x + \beta y + \gamma)^n y'_x = (ax + by + c)^n.$

This is a special case of equation 1.7.1.6 with $f(\xi) = \xi^n$.

24. $(ax^n + by^m)y'_x = x^{n-1}y^{1-m}.$

This is a special case of equation 1.7.1.7 with $f(\xi) = 1/\xi$.

25. $(ay^m + bx^n + s)y'_x + \alpha x^k + bnx^{n-1}y + \beta = 0.$

Solution:

$$a\varphi(y) + \alpha\psi(x) + bx^n y + sy + \beta x = C,$$

where

$$\varphi(y) = \begin{cases} \dfrac{y^{m+1}}{m+1} & \text{if } m \neq -1, \\ \ln|y| & \text{if } m = -1, \end{cases} \qquad \psi(x) = \begin{cases} \dfrac{x^{k+1}}{k+1} & \text{if } k \neq -1, \\ \ln|x| & \text{if } k = -1. \end{cases}$$

26. $(ax^2y^n + bxy^m + cy^k)y'_x = \alpha y^p + \beta y^q + \gamma.$

This is the Riccati equation with respect to $x = x(y)$.

27. $(ax^ny^m + x)y'_x = bx^ky^{n+m-k} + y.$

The transformation $t = y/x$, $z = x^{n+m-1}$ leads to a linear equation: $t^m(bt^{n-k} - at)z'_t = (n+m-1)(at^m z + 1)$.

28. $x(ax^ny^m + \alpha)y'_x + y(bx^ny^m + \beta) = 0.$

Solution:

$$\frac{(y^a x^b)^A}{A} + \frac{(y^\alpha x^\beta)^B}{B} = C, \qquad \text{where} \quad A = \frac{m\beta - n\alpha}{a\beta - b\alpha}, \qquad B = \frac{mb - na}{a\beta - b\alpha}.$$

29. $x(anx^ky^{n+k} + s)y'_x + y(bmx^{m+k}y^k + s) = 0.$

Solution: $aky^n + bkx^m - s(xy)^{-k} = C$.

30. $(ax^ny^m + Ax^2 + Bxy)y'_x = bx^ky^{n+m-k} + Axy + By^2.$

The transformation $t = y/x$, $z = x^{n+m-2}$ leads to a linear equation: $t^m(bt^{n-k} - at)z'_t = (n+m-2)(at^m z + Bt + A)$.

31. $(amx^ny^{m-1} + by^k)y'_x + anx^{n-1}y^m + cx^s = 0.$

This is a special case of equation 1.7.1.18 with $f(y) = by^k$, $g(x) = cx^s$.

32. $(ax^ny^m + bxy^k)y'_x = \alpha y^s + \beta.$

This is the Bernoulli equation with respect to $x = x(y)$ (see Subsection 1.1.5).

33. $x(ax^{n-k}y^m + m)y'_x = y(bx^{\lambda n-k}y^{\lambda m} - n).$

This is a special case of equation 1.7.1.15 with $f(\xi) = a\xi$, $g(\xi) = 1$, $h(\xi) = b\xi^\lambda$.

34. $x(ax^ny^{m-k} + m)y'_x = y(bx^{\lambda n}y^{\lambda m-k} - n).$

This is a special case of equation 1.7.1.16 with $f(\xi) = a\xi$, $g(\xi) = 1$, $h(\xi) = b\xi^\lambda$.

35. $(ax^{n+1}y^{m-1} + bx^{nk+1}y^{mk-1})y'_x = cx^{ns}y^{ms}.$

This is a special case of equation 1.7.1.3 with $f(\xi) = c\xi^s(a\xi + b\xi^k)^{-1}$.

36. $(ax^n + by^m)^k y'_x = cx^{n-1}y^{1-m}.$

This is a special case of equation 1.7.1.7 with $f(\xi) = c\xi^{-k}$.

37. $xy'_x = y + a\sqrt{y^2 + bx^2}.$

Solution: $y + \sqrt{y^2 + bx^2} = Cx^{a+1}$.

38. $\left(e_1\dfrac{x+a}{r_1^3} + e_2\dfrac{x-a}{r_2^3}\right)y'_x - y\left(\dfrac{e_1}{r_1^3} + \dfrac{e_2}{r_2^3}\right) = 0,$

where $r_1^2 = (x+a)^2 + y^2, \quad r_2^2 = (x-a)^2 + y^2.$

This is the equation of force lines corresponding to the Coulomb law.

Solution: $e_1\dfrac{x+a}{r_1} + e_2\dfrac{x-a}{r_2} = C$.

1.5.2. Equations Containing Exponential Functions

1. $y'_x = ae^y + be^x.$

Solution in the parametric form:

$$x = \ln\tau, \qquad y = b\tau - \ln\left(C - a\int\frac{e^{b\tau}}{\tau}\,d\tau\right).$$

2. $y'_x = ae^y + bx^n.$

Solution in the parametric form:

with $n \neq -1$,

$$x = \tau^{\frac{1}{n+1}}, \qquad y = \frac{b}{n+1}\tau - \ln\left[C - \frac{a}{n+1}\int\tau^{-\frac{n}{n+1}}\exp\left(\frac{b\tau}{n+1}\right)d\tau\right];$$

with $n = -1$, $b \neq -1$,

$$x = e^\tau, \qquad y = -\ln\left(Ce^{-b\tau} - \frac{a}{b+1}e^\tau\right);$$

with $n = -1$, $b = -1$,

$$x = e^\tau, \qquad y = -\tau - \ln(C - a\tau).$$

3. $y'_x = ay^{-1} + be^x.$

Solution in the parametric form:

$$x = \ln(AE^{-1}) \mp \tau^2, \qquad y = B[2 \pm \exp(\mp\tau^2)E^{-1}],$$

where $a = \mp 2B^2$, $b = \pm A^{-1}B$, $E = \int\exp(\mp\tau^2)\,d\tau + C$.

4. $y'_x = Ae^{y+ax} - a.$

This is a special case of equation 1.7.1.2 with $f(\xi) = Ae^{\xi}$, $n = 1$, $b = 0$.

5. $y'_x = ae^{\nu x+\lambda y} + be^{\mu x}.$

This is a special case of equation 1.7.2.5 with $f(x) = ae^{\nu x}$, $g(x) = be^{\mu x}$.

6. $y'_x = ae^{\nu x+\lambda y} + bx^n.$

This is a special case of equation 1.7.2.5 with $f(x) = ae^{\nu x}$, $g(x) = bx^n$.

7. $y'_x = ax^n e^{\lambda y} + be^{\nu x}.$

This is a special case of equation 1.7.2.5 with $f(x) = ax^n$, $g(x) = be^{\nu x}$.

8. $y'_x = ax^n e^{\lambda y} + bx^m.$

This is a special case of equation 1.7.2.5 with $f(x) = ax^n$, $g(x) = bx^m$.

9. $y'_x = ae^{\nu x+\lambda y} + be^{\mu x-\lambda y}.$

This is a special case of equation 1.7.2.8 with $f(x) = ae^{\nu x}$, $g(x) = 0$, $h(x) = be^{\mu x}$.

10. $y'_x = ax^n e^{\lambda y} + bx^m e^{-\lambda y}.$

This is a special case of equation 1.7.2.8 with $f(x) = ax^n$, $g(x) = 0$, $h(x) = bx^m$.

11. $y'_x = (y + ae^{\lambda x})^n - a\lambda e^{\lambda x}.$

This is a special case of equation 1.7.2.10 with $f(\xi) = \xi^n$.

12. $y'_x = (ae^y + bx^{-k})^{1/k}.$

Solution in the parametric form:

$$x = \exp\Bigl\{\tau - \frac{1}{k}\bigl[f(\tau)+C\bigr]\Bigr\}, \quad y = f(\tau)+C, \qquad \text{where} \quad f(\tau) = \int \frac{k\,d\tau}{k(b+ae^{k\tau})^{-1/k}+1}.$$

13. $y'_x = (ay^k + be^x)^{1/k}.$

Solution in the parametric form:

$$x = f(\tau)+C, \quad y = \exp\Bigl\{\tau + \frac{1}{k}\bigl[f(\tau)+C\bigr]\Bigr\}, \qquad \text{where} \quad f(\tau) = \int \frac{k\,d\tau}{k(a+be^{-k\tau})^{1/k}-1}.$$

14. $y'_x = ax^{n-1}e^{\lambda ny} + bx^{m-1}e^{\lambda my}.$

This is a special case of equation 1.7.2.2 with $f(\xi) = a\xi^{n-1} + b\xi^{m-1}$.

15. $y'_x = ax^{n-1}e^{\alpha y} + bx^{nm-1}e^{\alpha my}.$

This is a special case of equation 1.7.2.4 with $f(\xi) = a\xi + b\xi^m$.

16. $y'_x = ae^{\lambda nx}y^{n+1} + be^{-\lambda x}$.

This is a special case of equation 1.7.2.1 with $f(\xi) = a\xi^{n+1} + b$.

17. $y'_x = ae^{\alpha x}y^{m+1} + be^{\alpha nx}y^{nm+1}$.

This is a special case of equation 1.7.2.3 with $f(\xi) = a\xi + b\xi^n$.

18. $y'_x = ae^{\lambda nx}y^{n+1} + be^{\lambda mx}y^{m+1}$.

This is a special case of equation 1.7.2.1 with $f(\xi) = a\xi^{n+1} + b\xi^{m+1}$.

19. $y'_x = ae^{2\alpha x-\beta y} + be^{\alpha x} + ce^{\alpha x-\beta y}$.

This is a special case of equation 1.7.2.9 with $f(\xi) = \xi + c$.

20. $y'_x = ax^n y^k + bx^n e^{\alpha x}y^{k+1} - \alpha y$.

This is a special case of equation 1.7.2.7 with $f(x) = x^n$, $g(\xi) = a + b\xi$, $m = 1$.

21. $y'_x = ae^{\lambda x}y^{1+n} + be^{\mu x}y + ce^{\nu x}y^{1-n}$.

This is a special case of equation 1.7.1.4 with $f(x) = ae^{\lambda x}$, $g(x) = be^{\mu x}$, $h(x) = ce^{\nu x}$.

22. $y'_x = ae^{\lambda x}y^{1+n} + be^{\mu x}y + cx^m y^{1-n}$.

This is a special case of equation 1.7.1.4 with $f(x) = ae^{\lambda x}$, $g(x) = be^{\mu x}$, $h(x) = cx^m$.

23. $y'_x = ax^k y^{1+n} + be^{\lambda x}y + cx^m y^{1-n}$.

This is a special case of equation 1.7.1.4 with $f(x) = ax^k$, $g(x) = be^{\lambda x}$, $h(x) = cx^m$.

24. $y'_x = ae^{\lambda x}y^{1+n} + bx^m y + ce^{\mu x}y^{1-n}$.

This is a special case of equation 1.7.1.4 with $f(x) = ae^{\lambda x}$, $g(x) = bx^m$, $h(x) = ce^{\mu x}$.

25. $y'_x = ae^{\lambda x}y^{1+n} + bx^m y + cx^k y^{1-n}$.

This is a special case of equation 1.7.1.4 with $f(x) = ae^{\lambda x}$, $g(x) = bx^m$, $h(x) = cx^k$.

26. $xy'_x = ax^{n+k}e^y + bx^{nm+k}e^{my} - n$.

This is a special case of equation 1.7.2.6 with $f(x) = x^{k-1}$, $g(\xi) = a\xi + b\xi^m$.

27. $(by + \lambda)y'_x = ce^{ax+by} - ay$.

This is a special case of equation 1.7.1.14 with $f(\xi) = \lambda$, $g(\xi) = 1$, $h(\xi) = ce^{\xi}$.

28. $xyy'_x = ax^n e^y - ny$.

This is a special case of equation 1.7.2.11 with $f(\xi) = a\xi$, $\alpha = 1$.

29. $xy^2y'_x = ax^n e^y - ny^2$.

This is a special case of equation 1.7.2.12 with $f(\xi) = a\xi$, $\alpha = 1$.

30. $(ae^y + be^x)y'_x = 1.$

Solution in the parametric form:

$$x = a\tau - \ln\left(C - b\int \frac{e^{a\tau}}{\tau}\,d\tau\right), \qquad y = \ln\tau.$$

31. $(ay^n + be^x)y'_x = 1.$

Solution in the parametric form:
with $n \neq -1$:

$$x = \frac{a}{n+1}\tau - \ln\left[C - \frac{b}{n+1}\int \tau^{-\frac{n}{n+1}}\exp\left(\frac{a\tau}{n+1}\right)d\tau\right], \qquad y = \tau^{\frac{1}{n+1}};$$

with $n = -1,\ a \neq -1$:

$$x = -\ln\left(Ce^{-a\tau} - \frac{b}{a+1}e^{\tau}\right), \qquad y = e^{\tau};$$

with $n = -1,\ a = -1$:

$$x = -\tau - \ln(C - b\tau), \qquad y = e^{\tau}.$$

32. $(ae^y + bx)y'_x = 1.$

Solution in the implicit form:

$$x = \begin{cases} Ce^{by} + \dfrac{a}{1-b}e^y & \text{if } b \neq 1, \\ e^y(C + ay) & \text{if } b = 1. \end{cases}$$

33. $(ae^y + bx^2)y'_x = 1.$

Solutions in the parametric form:

$$x = -\frac{1}{2b}\tau(\ln Z)'_\tau, \quad y = \ln\left(\frac{\tau^2}{4ab}\right), \qquad Z = C_1J_0(\tau) + C_2Y_0(\tau)$$

and

$$x = -\frac{1}{2b}\tau(\ln Z)'_\tau, \quad y = \ln\left(-\frac{\tau^2}{4ab}\right), \qquad Z = C_1I_0(\tau) + C_2K_0(\tau),$$

where J_0 and Y_0 are Bessel functions, I_0 and K_0 are modified Bessel functions.

34. $(ae^y + bx^{-1})y'_x = 1.$

Let $a = \pm A/B$, $b = \mp 2A^2$. The solution in the parametric form is written as

$$x = A\bigl[2\tau \pm \exp(\mp\tau^2)f(\tau)\bigr], \qquad y = \ln\bigl[Bf(\tau)\bigr] \mp \tau^2,$$

where $f(\tau) = \left[\int \exp(\mp\tau^2)\,d\tau + C\right]^{-1}$.

35. $(be^{\alpha y} + c)y'_x = e^{\alpha x + by} - ae^{\alpha y}.$

This is a special case of equation 1.7.2.13 with $f(\xi) = c$, $g(\xi) = 1$, $h(\xi) = e^{\xi}$.

36. $(ae^{\alpha x} + be^{\beta y})y'_x = e^{\alpha x - \beta y}$.

This is a special case of equation 1.7.2.9 with $f(\xi) = \xi^{-1}$.

37. $(e^{\alpha x + \gamma y} + a\beta)y'_x + be^{\nu x + \beta y} + a\alpha = 0$.

This is a special case of equation 1.7.2.15 with $f(y) = e^{\gamma y}$, $g(x) = be^{\nu x}$.

38. $(e^{ax+by} + bx)y'_x = ce^{ax+by} - ax$.

This is a special case of equation 1.7.1.13 with $f(\xi) = e^{\xi}$, $g(\xi) = 1$, $h(\xi) = ce^{\xi}$.

39. $(e^{ax+by} + by)y'_x = ce^{ax+by} - ay$.

This is a special case of equation 1.7.1.14 with $f(\xi) = e^{\xi}$, $g(\xi) = 1$, $h(\xi) = ce^{\xi}$.

40. $(ae^{\alpha x}y^m + b)y'_x = y$.

This is a special case of equation 1.7.2.3 with $f(\xi) = (a\xi + b)^{-1}$.

41. $(e^{\alpha x}y^n + a\beta)y'_x + be^{\nu x + \beta y} + a\alpha = 0$.

This is a special case of equation 1.7.2.15 with $f(y) = y^n$, $g(x) = be^{\nu x}$.

42. $(e^{\alpha x}y^n + a\beta)y'_x + bx^m e^{\beta y} + a\alpha = 0$.

This is a special case of equation 1.7.2.15 with $f(y) = y^n$, $g(x) = bx^m$.

43. $(e^{\alpha x}y^m + mx)y'_x = y(be^{\alpha n x}y^{nm} - \alpha x)$.

This is a special case of equation 1.7.2.17 with $f(\xi) = \xi$, $g(\xi) = 1$, $h(\xi) = b\xi^n$.

44. $x(x^n e^{\alpha y} + \alpha y)y'_x = bx^{nm}e^{\alpha m y} - ny$.

This is a special case of equation 1.7.2.16 with $f(\xi) = \xi$, $g(\xi) = 1$, $h(\xi) = b\xi^m$.

45. $(ax^n e^{\lambda y} + bxe^{\mu y})y'_x = e^{\nu y}$.

This is the Bernoulli equation with respect to $x = x(y)$ (see 1.1.5).

46. $(ax^n e^{\lambda y} + bxy^m)y'_x = e^{\mu y}$.

This is the Bernoulli equation with respect to $x = x(y)$.

47. $(ax^n y^m + bxe^{\lambda y})y'_x = y^k$.

This is the Bernoulli equation with respect to $x = x(y)$.

48. $(ax^n y^m + bxy^k)y'_x = e^{\lambda y}$.

This is the Bernoulli equation with respect to $x = x(y)$.

49. $(amx^n y^{m-1} + b)y'_x + anx^{n-1}y^m + ce^{\lambda x} = 0$.

This is a special case of equation 1.7.1.18 with $f(y) = b$, $g(x) = ce^{\lambda x}$.

50. $(amx^n y^{m-1} + be^{\lambda y})y'_x + anx^{n-1}y^m + c = 0.$

This is a special case of equation 1.7.1.18 with $f(y) = be^{\lambda y}$, $g(x) = c$.

51. $(amx^n y^{m-1} + by^k)y'_x + anx^{n-1}y^m + ce^{\lambda x} = 0.$

This is a special case of equation 1.7.1.18 with $f(y) = by^k$, $g(x) = ce^{\lambda x}$.

52. $[(ax+by)^n + be^{\alpha x}]y'_x = c(ax+by)^m - ae^{\alpha x}.$

This is a special case of equation 1.7.2.14 with $f(\xi) = \xi^n$, $g(x) = 1$, $h(\xi) = c\xi^m$.

53. $[(ax+by)^n + be^{\alpha y}]y'_x = c(ax+by)^m - ae^{\alpha y}.$

This is a special case of equation 1.7.2.13 with $f(\xi) = \xi^n$, $g(x) = 1$, $h(\xi) = c\xi^m$.

1.5.3. Equations Containing Hyperbolic Functions

1. $y'_x = a\cosh(\lambda y) + b\cosh(\nu x).$

This is a special case of equation 1.7.2.18 with $f(x) = 0$, $g(x) = a$, $h(x) = b\cosh(\nu x)$.

2. $y'_x = a\sinh(\lambda y) + b\sinh(\nu x).$

This is a special case of equation 1.7.2.18 with $f(x) = a$, $g(x) = 0$, $h(x) = b\sinh(\nu x)$.

3. $y'_x = ax^n\cosh(\lambda y) + bx^m.$

This is a special case of equation 1.7.2.18 with $f(x) = 0$, $g(x) = ax^n$, $h(x) = bx^m$.

4. $y'_x = ax^n\sinh(\lambda y) + bx^m.$

This is a special case of equation 1.7.2.18 with $f(x) = ax^n$, $g(x) = 0$, $h(x) = bx^m$.

5. $y'_x = ay^{1+n} + by + c\sinh(\lambda x)y^{1-n}.$

This is a special case of equation 1.7.1.4 with $f(x) = a$, $g(x) = b$, $h(x) = c\sinh(\lambda x)$.

6. $y'_x = ay^{1+n} + b\sinh(\lambda x)y + cy^{1-n}.$

This is a special case of equation 1.7.1.4 with $f(x) = a$, $g(x) = b\sinh(\lambda x)$, $h(x) = c$.

7. $y'_x = y\cosh x(ay^{nm}\sinh^{n-1}x + by^m).$

This is a special case of equation 1.7.2.20 with $f(\xi) = a\xi^n + b\xi$.

8. $y'_x = y\sinh x(ay^{nm}\cosh^{n-1}x + by^m).$

This is a special case of equation 1.7.2.22 with $f(\xi) = a\xi^n + b\xi$.

9. $xy'_x = (ax^n\cosh y + b)\coth y.$

This is a special case of equation 1.7.2.23 with $f(\xi) = a\xi + b$.

10. $xy'_x = (ax^n \sinh y + b) \tanh y$.

This is a special case of equation 1.7.2.21 with $f(\xi) = a\xi + b$.

11. $(ay^m \cosh x + b)y'_x = y^{m+1} \sinh x$.

This is a special case of equation 1.7.2.22 with $f(\xi) = \xi(a\xi + b)^{-1}$.

12. $(ay^m \sinh x + b)y'_x = y^{m+1} \cosh x$.

This is a special case of equation 1.7.2.20 with $f(\xi) = \xi(a\xi + b)^{-1}$.

13. $(ax^n + bx \cosh^m y)y'_x = y^k$.

This is the Bernoulli equation with respect to $x = x(y)$ (see Subsection 1.1.5).

14. $(ax^n + bx \tanh^m y)y'_x = y^k$.

This is the Bernoulli equation with respect to $x = x(y)$.

15. $(ax^n + bx \cosh^m y)y'_x = \cosh^k(\lambda y)$.

This is the Bernoulli equation with respect to $x = x(y)$.

16. $(ax^n + bx \tanh^m y)y'_x = \tanh^k(\lambda y)$.

This is the Bernoulli equation with respect to $x = x(y)$.

17. $(amx^n y^{m-1} + b)y'_x + anx^{n-1}y^m + c \sinh^k(\lambda x) = 0$.

This is a special case of equation 1.7.1.18 with $f(y) = b$, $g(x) = c \sinh^k(\lambda x)$.

18. $(amx^n y^{m-1} + b)y'_x + anx^{n-1}y^m + c \tanh^k(\lambda x) = 0$.

This is a special case of equation 1.7.1.18 with $f(y) = b$, $g(x) = c \tanh^k(\lambda x)$.

19. $(ax^n y^m + bx)y'_x = \cosh^k(\lambda y)$.

This is the Bernoulli equation with respect to $x = x(y)$.

20. $(ax^n y^m + bx)y'_x = \tanh^k(\lambda y)$.

This is the Bernoulli equation with respect to $x = x(y)$.

21. $(ax^n \cosh^m y + bx)y'_x = \sinh^k(\lambda y)$.

This is the Bernoulli equation with respect to $x = x(y)$.

22. $(ax^n \tanh^m y + bx)y'_x = y^k$.

This is the Bernoulli equation with respect to $x = x(y)$.

23. $(amx^n y^{m-1} + b \sinh^k y)y'_x + anx^{n-1}y^m + c = 0$.

This is a special case of equation 1.7.1.18 with $f(y) = b \sinh^k y$, $g(x) = c$.

24. $(amx^n y^{m-1} + b \tanh^k y)y'_x + anx^{n-1}y^m + c = 0$.

This is a special case of equation 1.7.1.18 with $f(y) = b \tanh^k y$, $g(x) = c$.

1.5.4. Equations Containing Logarithmic Functions

1. $y'_x = y(\alpha x + m \ln y + \beta)$.

This is a special case of equation 1.7.2.3 with $f(\xi) = \ln \xi + \beta$.

2. $y'_x = ax^{kn-1}y^{km+1}(n \ln x + m \ln y)$.

This is a special case of equation 1.7.1.3 with $f(\xi) = a\xi^k \ln \xi$.

3. $y'_x = ax^n y \ln^2 y + bx^m y \ln y + cx^k y$.

This is a special case of equation 1.7.3.1 with $f(x) = ax^n$, $g(x) = bx^m$, $h(x) = cx^k$.

4. $xy'_x = (\alpha y + n \ln x)^m + \beta$.

This is a special case of equation 1.7.2.4 with $f(\xi) = \ln^m \xi + \beta$.

5. $xy'_x = y(n \ln x + m \ln y)$.

This is a special case of equation 1.7.1.3 with $f(\xi) = \ln \xi$.

6. $mxy'_x = ax^s y^k (n \ln x + m \ln y) - ny$.

This is a special case of equation 1.7.1.5 with $f(x) = \dfrac{a}{m} x^{s-1}$, $g(\xi) = \ln \xi$.

7. $(x^a + b)y'_x = yx^{a-1} + c(\ln y - \ln x)$.

This is a special case of equation 1.7.1.12 with $f(\xi) = b$, $g(\xi) = c \ln \xi$, $h(\xi) = 1$.

8. $x(\alpha y + \beta)y'_x = n \ln x + (\alpha - n)y$.

This is a special case of equation 1.7.2.16 with $f(\xi) = \beta$, $g(\xi) = 1$, $h(\xi) = \ln \xi$.

9. $x(a + mx^k)y'_x = y(bn \ln x + bm \ln y - nx^k)$.

This is a special case of equation 1.7.1.15 with $f(\xi) = a$, $g(\xi) = 1$, $h(\xi) = b \ln \xi$.

10. $x(a + my^k)y'_x = y(bn \ln x + bm \ln y - ny^k)$.

This is a special case of equation 1.7.1.16 with $f(\xi) = a$, $g(\xi) = 1$, $h(\xi) = b \ln \xi$.

11. $(amx^n y^{m-1} + b)y'_x + anx^{n-1}y^m + c \ln^k(\lambda x) = 0$.

This is a special case of equation 1.7.1.18 with $f(y) = b$, $g(x) = c \ln^k(\lambda x)$.

12. $(a \ln y + bx)y'_x = 1$.

Solution:

$$x = e^{by}\left(\frac{a}{b} \int \frac{e^{-by}}{y}\, dy + C\right) - \frac{a}{b} \ln y.$$

13. $x(\ln y)y'_x = y(ax^{nk}y^k + bx^n y) - ny \ln y$.

This is a special case of equation 1.7.3.7 with $f(\xi) = a\xi^k + b\xi$, $m = 1$.

14. $x(a+m\ln y)y'_x = y(bx^n y^m - n\ln y + c)$.

This is a special case of equation 1.7.3.9 with $f(\xi)=a$, $g(\xi)=1$, $h(\xi)=b\xi+c$.

15. $(ax^n + bx\ln^m y)y'_x = \ln^k(\lambda y)$.

This is the Bernoulli equation with respect to $x=x(y)$.

16. $x(ax^n y^m + m\ln x)y'_x = y(bx^{nk}y^{mk} - n\ln x)$.

This is a special case of equation 1.7.3.10 with $f(\xi)=a\xi$, $g(\xi)=1$, $h(\xi)=b\xi^k$.

17. $x(ax^n y^m + m\ln y)y'_x = y(bx^{nk}y^{mk} - n\ln y)$.

This is a special case of equation 1.7.3.9 with $f(\xi)=a\xi$, $g(\xi)=1$, $h(\xi)=b\xi^k$.

18. $(amx^n y^{m-1} + b\ln^k y)y'_x + anx^{n-1}y^m + c = 0$.

This is a special case of equation 1.7.1.18 with $f(y)=b\ln^k y$, $g(x)=c$.

19. $(ax^n\ln^m y + bx)y'_x = \ln^k(\lambda y)$.

This is the Bernoulli equation with respect to $x=x(y)$.

20. $(ax^n\ln^m y + bx\ln^k y)y'_x = y^s$.

This is the Bernoulli equation with respect to $x=x(y)$.

1.5.5. Equations Containing Trigonometric Functions

1. $y'_x = \alpha\cos(ay) + \beta\cos(bx)$.

This is a special case of equation 1.7.4.11 with $f(x)=\alpha$, $g(x)=0$, $h(x)=\beta\cos(bx)$.

2. $y'_x = \sin(ax)\cos(by) + \cos(ax)\sin(by)$.

This is a special case of equation 1.7.1.1 with $f(\xi)=\sin\xi$, $c=0$.

3. $y'_x = a\tan(bxy)$.

The solution is given by the relation

$$\int_0^w \exp\bigl(\tfrac{1}{2}t^2\bigr)\cos\bigl(\sqrt{ab}\,xt\bigr)\,dt = C\exp\bigl(\tfrac{1}{2}abx^2\bigr), \qquad \text{where} \quad w = y\sqrt{\frac{b}{a}}.$$

4. $y'_x = bx^n\cos(ay) + cx^m$.

This is a special case of equation 1.7.4.11 with $f(x)=bx^n$, $g(x)=0$, $h(x)=cx^m$.

5. $y'_x = bx^n\sin(ay) + cx^m$.

This is a special case of equation 1.7.4.11 with $f(x)=0$, $g(x)=bx^n$, $h(x)=cx^m$.

6. $y'_x = y\cos x(ay^{nm}\sin^{n-1}x + by^m)$.

This is a special case of equation 1.7.4.4 with $f(\xi) = a\xi^n + b\xi$.

7. $y'_x = y\sin x(ay^{nm}\cos^{n-1}x + by^m)$.

This is a special case of equation 1.7.4.3 with $f(\xi) = a\xi^n + b\xi$.

8. $y'_x = a\dfrac{\sin^2 y}{\cos^2 x} + b\dfrac{\cos^2 y}{\sin^2 x}$.

This is a special case of equation 1.7.4.14 with $f(\xi) = a\xi + b\xi^{-1}$.

9. $y'_x = ay^{1+n} + by + c\sin(\lambda x)y^{1-n}$.

This is a special case of equation 1.7.1.4 with $f(x) = a$, $g(x) = b$, $h(x) = c\sin(\lambda x)$.

10. $y'_x = ay^{1+n} + b\sin(\lambda x)y + cy^{1-n}$.

This is a special case of equation 1.7.1.4 with $f(x) = a$, $g(x) = b\sin(\lambda x)$, $h(x) = c$.

11. $xy'_x + a\sin(bx + cy) = 0$.

The substitution $w = x\tan\dfrac{bx + cy}{2}$ leads to the Riccati equation of the form 1.2.2.22: $2xw'_x - bw^2 + 2(ac - 1)w - bx^2 = 0$.

12. $xy'_x = ax^2\tan(by) + y$.

The substitution $y = xw$ leads to an equation of the form 1.5.5.3: $w'_x = a\tan(bxw)$.

13. $xy'_x = ax^n\cos^2 y + b\cos y\sin y$.

This is a special case of equation 1.7.4.8 with $f(\xi) = \frac{1}{2}(a\xi + b)$.

14. $xy'_x = ax^n\sin^2 y + b\cos y\sin y$.

This is a special case of equation 1.7.4.7 with $f(\xi) = \frac{1}{2}(a\xi + b)$.

15. $xy'_x = ax^m\sin^k y\cos^{2-k} y - n\sin 2y$.

This is a special case of equation 1.7.4.18 with $f(x) = ax^{m-2nk-1}$, $g(\xi) = \xi^k$.

16. $(1 + \tan^2 y)y'_x = a\tan^{m+1} y + b\tan y + cx^n\tan^{1-m} y$.

This is a special case of equation 1.7.4.19 with $f(x) = a$, $g(x) = b$, $h(x) = cx^n$.

17. $(amx^ny^{m-1} + b)y'_x + anx^{n-1}y^m + c\sin^k(\lambda x) = 0$.

This is a special case of equation 1.7.1.18 with $f(y) = b$, $g(x) = c\sin^k(\lambda x)$.

18. $(amx^ny^{m-1} + b)y'_x + anx^{n-1}y^m + c\tan^k(\lambda x) = 0$.

This is a special case of equation 1.7.1.18 with $f(y) = b$, $g(x) = c\tan^k(\lambda x)$.

19. $(ax^n y^m + bx)y'_x = \cos^k(\lambda y)$.

This is the Bernoulli equation with respect to $x = x(y)$ (see 1.1.5).

20. $(ax^n y^m + bx)y'_x = \tan^k(\lambda y)$.

This is the Bernoulli equation with respect to $x = x(y)$.

21. $(ay^m \cos x + b)y'_x = y^{m+1} \sin x$.

This is a special case of equation 1.7.4.3 with $f(\xi) = \xi(a\xi + b)^{-1}$.

22. $(ay^m \sin x + b)y'_x = y^{m+1} \cos x$.

This is a special case of equation 1.7.4.4 with $f(\xi) = \xi(a\xi + b)^{-1}$.

23. $(ax^n + bx \cos^m y)y'_x = y^k$.

This is the Bernoulli equation with respect to $x = x(y)$.

24. $(ax^n + bx \cos^m y)y'_x = \cos^k(\lambda y)$.

This is the Bernoulli equation with respect to $x = x(y)$.

25. $(amx^n y^{m-1} + b \cos^k y)y'_x + anx^{n-1} y^m + c = 0$.

This is a special case of equation 1.7.1.18 with $f(y) = b \cos^k y$, $g(x) = c$.

26. $(ax^n \cos^m y + bx)y'_x = \cos^k(\lambda y)$.

This is the Bernoulli equation with respect to $x = x(y)$.

27. $(ax^n + bx \tan^m y)y'_x = y^k$.

This is the Bernoulli equation with respect to $x = x(y)$.

28. $(ax^n + bx \tan^m y)y'_x = \tan^k(\lambda y)$.

This is the Bernoulli equation with respect to $x = x(y)$.

29. $(amx^n y^{m-1} + b \tan^k y)y'_x + anx^{n-1} y^m + c = 0$.

This is a special case of equation 1.7.1.18 with $f(y) = b \tan^k y$, $g(x) = c$.

30. $(ax^n \tan^m y + bx)y'_x = \tan^k(\lambda y)$.

This is the Bernoulli equation with respect to $x = x(y)$.

1.5.6. Equations Containing Combinations of Exponential, Hyperbolic, Logarithmic, and Trigonometric Functions

1. $y'_x = ax^n e^{\lambda y} + b \ln^m x$.

This is a special case of equation 1.7.2.5 with $f(x) = ax^n$, $g(x) = b \ln^m x$.

2. $y'_x = a\ln^n(\nu x)e^{\lambda y} + bx^m.$

This is a special case of equation 1.7.2.5 with $f(x) = a\ln^n(\nu x)$, $g(x) = bx^m$.

3. $y'_x = ae^{\lambda y}(\lambda y + \ln x)^m.$

This is a special case of equation 1.7.2.2 with $f(\xi) = a\ln^m \xi$.

4. $y'_x = ae^{-\lambda x}(\lambda x + \ln y)^m.$

This is a special case of equation 1.7.2.1 with $f(\xi) = a\ln^m \xi$.

5. $y'_x = ay\ln^2 y + by\ln y + ce^{\lambda x}y.$

This is a special case of equation 1.7.3.1 with $f(x) = a$, $g(x) = b$, $h(x) = ce^{\lambda x}$.

6. $y'_x = ay\ln^2 y + be^{\lambda x}y\ln y + cy.$

This is a special case of equation 1.7.3.1 with $f(x) = a$, $g(x) = be^{\lambda x}$, $h(x) = c$.

7. $y'_x = ae^y \sin x + b\tan x.$

This is a special case of equation 1.7.5.6 with $f(\xi) = a\xi + b$.

8. $y'_x = (ae^x \sin y + b)\tan y.$

This is a special case of equation 1.7.5.4 with $f(\xi) = a\xi + b$.

9. $y'_x = ae^x \sin^2 y + be^{-x}\cos^2 y.$

This is a special case of equation 1.7.5.8 with $f(\xi) = \frac{1}{2}(a\xi + b/\xi)$.

10. $y'_x = a\cos^n(\mu x)e^{\lambda y} + bx^m.$

This is a special case of equation 1.7.2.5 with $f(x) = a\cos^n(\mu x)$, $g(x) = bx^m$.

11. $y'_x = ax^n e^{\lambda y} + b\cos^m(\mu x).$

This is a special case of equation 1.7.2.5 with $f(x) = ax^n$, $g(x) = b\cos^m(\mu x)$.

12. $y'_x = ax^n e^{\lambda y} + b\tan^m(\mu x).$

This is a special case of equation 1.7.2.5 with $f(x) = ax^n$, $g(x) = b\tan^m(\mu x)$.

13. $y'_x = a\tan^n(\mu x)e^{\lambda y} + bx^m.$

This is a special case of equation 1.7.2.5 with $f(x) = a\tan^n(\mu x)$, $g(x) = bx^m$.

14. $y'_x = Ae^{\lambda x}\cos(ay) + Be^{\mu x}\sin(ay) + Ae^{\lambda x}.$

The substitution $w = \tan(\frac{1}{2}ay)$ leads to a linear equation: $w'_x = aBe^{\mu x}w + aAe^{\lambda x}$.

15. $y'_x = a\sin(\mu x)\sinh(\lambda y) + b\cos(\mu x)\cosh(\lambda y).$

This is a special case of equation 1.7.2.18 with $f(x) = a\sin(\mu x)$, $g(x) = b\cos(\mu x)$, $h(x) = 0$.

16. $y'_x = ay\ln^2 y + by\ln y + c\sin^n(\lambda x)y.$

This is a special case of equation 1.7.3.1 with $f(x) = a$, $g(x) = b$, $h(x) = c\sin^n(\lambda x)$.

17. $(1+\tan^2 y)y'_x = a\tan^{1+m} y + b\tan y + ce^{\lambda x}\tan^{1-m} y.$

This is a special case of equation 1.7.4.19 with $f(x) = a$, $g(x) = b$, $h(x) = ce^{\lambda x}$.

18. $(ae^x\cos y + b)y'_x = \cot y.$

This is a special case of equation 1.7.5.5 with $f(\xi) = (a\xi + b)^{-1}$.

19. $(ae^x\sin y + b)y'_x = \tan y.$

This is a special case of equation 1.7.5.4 with $f(\xi) = (a\xi + b)^{-1}$.

20. $(ae^y\cos x + b)y'_x = \tan x.$

This is a special case of equation 1.7.5.6 with $f(\xi) = (a\xi + b)^{-1}$.

21. $(ae^y\sin x + b)y'_x = \cot x.$

This is a special case of equation 1.7.5.7 with $f(\xi) = (a\xi + b)^{-1}$.

22. $(e^{\alpha x}y^n + a\beta)y'_x + be^{\beta y}\ln^m x + a\alpha = 0.$

This is a special case of equation 1.7.2.15 with $f(y) = y^n$, $g(x) = b\ln^m x$.

23. $(e^{\alpha x}y^n + a\beta)y'_x + be^{\beta y}\cos^m x + a\alpha = 0.$

This is a special case of equation 1.7.2.15 with $f(y) = y^n$, $g(x) = b\cos^m x$.

24. $(e^{\alpha x}\cos^n y + a\beta)y'_x + be^{\beta y}\cos^m(\lambda x) + a\alpha = 0.$

This is a special case of equation 1.7.2.15 with $f(y) = \cos^n y$, $g(x) = b\cos^m(\lambda x)$.

1.6. Equations Not Solved for Derivative

1.6.1. Equations of the Second Degree in y'_x

1. $(y'_x)^2 = ay + bx^2.$

See equation 1.6.3.43.

2. $(y'_x)^2 = y + ax^2 + bx + c.$

The substitution $w = 2\sqrt{y + ax^2 + bx + c}$ leads to an equation of the form 1.3.1.2: $ww'_x - w = 4ax + 2b$.

3. $(y'_x)^2 = ay^3 + by + c.$

Solution: $x = C \pm \displaystyle\int \frac{dy}{\sqrt{ay^3 + by + c}}.$

4. $(y'_x)^2 = ay + b\sqrt{x}.$

See equation 1.6.3.26.

5. $(y'_x)^2 = ay + b\sqrt{x} + c, \qquad a \neq 0.$

The substitution $aw = 2\sqrt{ay + b\sqrt{x} + c}$ leads to the Abel equation of the form 1.3.1.32: $ww'_x - w = ba^{-2}x^{-1/2}$.

6. $(y'_x)^2 + ay'_x + by = 0.$

Solution in the parametric form:

$$bx = -2t - a\ln t + C, \qquad by = -t^2 - at.$$

7. $(y'_x)^2 + ayy'_x = bx + c.$

Differentiate the equation with respect to x, take y as the independent variable, and assume $\xi = y'_x$. As a result we obtain a linear equation with respect to $y = y(\xi)$:

$$(a\xi^2 - b)y'_\xi + a\xi y + 2\xi^2 = 0.$$

8. $(y'_x)^2 + axy'_x + by + cx^2 = 0.$

The transformation $x = e^t$, $y = x^2u$ leads to an autonomous equation:

$$\left(u'_t + 2u + \frac{a}{2}\right)^2 = \frac{a^2}{4} - c - bu.$$

Having extracted the root and carried over the terms $2u + \frac{1}{2}a$ from the left-hand side to the right-hand side, we obtain an equation of the form 1.1.2.

9. $y = xy'_x + ax^2 + b(y'_x)^2 + cy'_x + d, \qquad a \neq 0.$

Differentiating with respect to x and changing to new variables $t = y'_x$ and $w(t) = -2ax$, we arrive at the Abel equation of the form 1.3.1.2: $ww'_t - w = -4abt - 2ac$.

10. $(y'_x)^2 + (ax + b)y'_x - ay + c = 0, \qquad a \neq 0.$

Solutions:

$$y = (ax + b)C + aC^2 + ca^{-1} \qquad \text{and} \qquad 4ay = 4c - (ax + b)^2.$$

11. $(y'_x)^2 + (ay + bx)y'_x + abxy = 0.$

This equation can be factorized:

$$(y'_x + ay)(y'_x + bx) = 0.$$

Therefore, the solutions are

$$y = Ce^{-ax} \qquad \text{and} \qquad y = -\tfrac{1}{2}bx^2 + C.$$

12. $(y'_x)^2 + ax^2y'_x + bxy = 0.$

The transformation $z = \ln x$, $u = yx^{-3}$ leads to an equation not depending implicitly on z:

$$(u'_z)^2 + (a + 6u)u'_z + (3a + b + 9u)u = 0.$$

Rewriting the latter equation to solve for u'_z, we obtain an equation of the form 1.1.2.

13. $(y'_x)^2 = y + ax^{m+1} - \dfrac{m+1}{2(m+3)^2}x^2 + b.$

The substitution

$$w = 2\Big[y + ax^{m+1} - \frac{m+1}{2(m+3)^2}x^2 + b\Big]^{1/2}$$

leads to the Abel equation of the form 1.3.1.10:

$$ww'_x - w = -\frac{2(m+1)}{(m+3)^2}x + 2a(m+1)x^m.$$

14. $(y'_x)^2 = \lambda y + ax^2 + bx^{m+1} + c.$

With $\lambda \neq 0$, the substitution $\lambda w = 2(\lambda y + ax^2 + bx^{m+1} + c)^{1/2}$ leads to the Abel equation:

$$ww'_x - w = 4a\lambda^{-2}x + 2b\lambda^{-2}(m+1)x^m,$$

which is outlined in Subsection 1.3.1.

The special cases of the original equation are equations 1.6.1.1, 1.6.1.2, 1.6.1.4, 1.6.1.5, and 1.6.1.13.

15. $a(y'_x)^2 - yy'_x - x = 0.$

Solution in the parametric form:

$$x = \frac{t}{\sqrt{t^2+1}}\big[C + a\ln\big(t + \sqrt{t^2+1}\,\big)\big], \qquad y = at - \frac{x}{t}.$$

16. $x(y'_x)^2 = axy + b.$

See equation 1.6.3.32.

17. $x(y'_x)^2 = axy + bx + c, \qquad a \neq 0.$

The substitution $aw = 2\sqrt{ay + b + cx^{-1}}$ leads to the Abel equation of the form 1.3.1.33: $ww'_x - w = -2ca^{-2}x^{-2}$.

18. $x(y'_x)^2 - ayy'_x + b = 0.$

With $a \neq 1$, the solution in the parametric form is written as

$$x = Ct^k + \frac{b}{2a-1}t^2, \quad aty = xt^2 + b, \qquad \text{where} \quad k = \frac{1}{a-1}.$$

With $a = 1$, the solution is $\;C(y - Cx) = b.$

19. $x(y'_x)^2 + ayy'_x + bx = 0.$

With $a \neq -1$, the solution in the parametric form is written as

$$x = Ct|(a+1)t^2 + b|^{-\frac{a+2}{2(a+1)}}, \qquad y = -\frac{x}{at}(t^2 + b).$$

In addition, there is the solution $y = \pm x\sqrt{-\dfrac{b}{a+1}}$.

With $a = -1$, the solution in the parametric form is written as

$$x = Ct\exp\Big(-\frac{t^2}{2b}\Big), \qquad y = x\Big(t + \frac{b}{t}\Big).$$

20. $x(y'_x)^2 - yy'_x + ay = 0.$

Solution in the parametric form:

$$x = C(t-a)\exp(-t/a), \qquad y = Ct^2\exp(-t/a),$$

In addition, there is the solution $y = 0$.

21. $x(y'_x)^2 - yy'_x + ax^2y'_x + by'_x + c = 0, \qquad a \neq 0.$

Divide the both sides by y'_x and differentiate with respect to x. Changing to new variables $t = y'_x$ and $w(t) = -2ax$, we arrive at the Abel equation of the form 1.3.1.33: $ww'_t - w = act^{-2}$.

22. $y(y'_x)^2 + axy'_x + by = 0.$

Solution in the parametric form is defined by the relations

$$axt + y(b + t^2) = 0, \quad Cy(t^2 + a + b)^m = t^{b/(a+b)}, \qquad \text{where} \quad m = \frac{a+2b}{2(a+b)}.$$

In addition, there is the solution $y = \pm x\sqrt{-a-b}$ corresponding to the limit $C \to \infty$.

23. $x(y'_x)^2 + (a-y)y'_x + b = 0.$

Solutions: $C(Cx - y + a) + b = 0$ and $(y-a)^2 = 4bx$.

24. $ax(y'_x)^2 + (bx - ay + k)y'_x - by = 0.$

Solution: $y = Cx + \dfrac{kC}{aC + b}$.

In addition, there is the exceptional solution which may be written in the parametric form as

$$x = -\frac{bk}{(at+b)^2}, \qquad y = xt + \frac{kt}{at+b}.$$

25. $ax(y'_x)^2 - (ay + bx - a - b)y'_x + by = 0.$

Differentiating with respect to x and factorizing, we obtain

$$(2axy'_x - ay - bx + a + b)y''_{xx} = 0.$$

Equating both factors to zero and integrating, we arrive at the solutions:

$$y = Cx + \frac{C(a+b)}{aC - b} \quad \text{and} \quad (ay + bx - a - b)^2 - 4abxy = 0.$$

26. $x(y'_x)^2 + ayy'_x + bx^n y^m = 0.$

The substitution $x = e^t$ leads to an equation of the form 1.6.1.56: $(y'_t)^2 + ayy'_t + be^{(n+1)t}y^m = 0$.

27. $x^2(y'_x)^2 - (2xy + a)y'_x + y^2 = 0.$

Solutions:

$$y = aC^2x + aC \quad \text{and} \quad y = -\frac{a}{4x}.$$

28. $ax^2(y'_x)^2 - 2axyy'_x + y^2 - a(a-1)x^2 = 0.$

Solutions:

$$y \pm \sqrt{y^2 + ax^2} = Cx^{1+k}, \qquad \text{where} \quad k = \sqrt{(a-1)/a}.$$

29. $(a^2 - 1)x^2(y'_x)^2 + 2xyy'_x - y^2 + a^2x^2 = 0.$

Solution in the parametric form:

$$x = C(t^2+1)^{-1/2}\left(t + \sqrt{t^2+1}\right)^{-1/a}, \qquad y = xt + ax\sqrt{t^2+1}.$$

30. $x^2(y'_x)^2 + (ax^2y^3 + b)y'_x + aby^3 = 0.$

The equation can be factorized:

$$(y'_x + ay^3)(x^2y'_x + b) = 0.$$

Equating each of the factors to zero, we obtain the solutions:

$$y^{-2} = 2ax + C \quad \text{and} \quad y = b/x + C.$$

31. $axy(y'_x)^2 - (ay^2 + bx^2 + k)y'_x + bxy = 0.$

This differential equation presents an equation of curvature lines of a surface defined by the relation

$$Ax^2 + By^2 + Cz^2 = 1,$$

where

$$a = AB(C - B), \quad b = AB(A - C), \quad k = C(B - A).$$

Solutions: $(aC - b)y^2 = C(aC - b)x^2 - kC$ and $ay^2 = bx^2 \pm 2x\sqrt{-bk} - k.$

32. $y^2(y'_x)^2 = ax^2y^2 + b.$

See equation 1.6.3.34.

33. $y^2(y'_x)^2 = ax^{-2/5}y^2 + b.$

See equation 1.6.3.28.

34. $y^2(y'_x)^2 + 2axyy'_x + (1 - a)y^2 + ax^2 + (a - 1)b = 0.$

Solutions:

$$y^2 + ax^2 - b = (a-1)(x+C)^2 \quad \text{and} \quad y^2 + ax^2 - b = 0.$$

35. $(a-b)y^2(y'_x)^2 - 2bxyy'_x + ay^2 - bx^2 - ab = 0.$

Solutions:

$$x^2+y^2 = Cx + b - \frac{a-b}{4a}C^2, \quad \text{and} \quad (a-b)y^2 - bx^2 = (a-b)b.$$

36. $(x^2-a)(y'_x)^2 - 2xyy'_x - x^2 = 0.$

Solving for y, differentiating with respect to x, and setting $w(x) = y'_x$, we obtain a factorized equation:

$$(xw'_x - w)(x^2w^2 + x^2 - aw^2) = 0.$$

Equating each of the factors to zero, we arrive at the solutions:

$$y = \frac{1}{2C}(x^2 - a - C^2) \quad \text{and} \quad y^2 + x^2 = a \quad (y \neq 0).$$

37. $(x^2-a^2)(y'_x)^2 + 2xyy'_x + y^2 = 0.$

The equation can be factorized:

$$(xy'_x + ay'_x + y)(xy'_x - ay'_x + y) = 0.$$

Equating each of the factors to zero, we obtain the solutions:

$$(x+a)y = C \quad \text{and} \quad (x-a)y = C.$$

38. $(x^2+a)(y'_x)^2 - 2xyy'_x + y^2 + b = 0.$

Differentiating with respect to x, we obtain a factorized equation:

$$[(x^2+a)y'_x - xy]y''_{xx} = 0.$$

Therefore, the solutions of the original equation are

$$y = C_1x + C_2, \quad \text{where } aC_1^2 + C_2^2 + b = 0; \qquad bx^2 + ay^2 + ab = 0.$$

39. $(ay-x^2)(y'_x)^2 + 2xyy'_x - y^2 = 0.$

Solution: $(Cy+x)^2 = 4ay.$

40. $(ay^2+bx)(y'_x)^2 = 1.$

See equation 1.6.3.44.

41. $(ax^2+by)(y'_x)^2 = x^2y.$

See equation 1.6.3.46.

42. $(axy+b)(y'_x)^2 = y.$

See equation 1.6.3.33.

43. $(y^2 - a^2x^2)(y'_x)^2 + 2xyy'_x + (1 - a^2)x^2 = 0.$

Solution in the parametric form:

$$x = \frac{Ct}{\sqrt{t^2+1}}, \qquad y = aC - \frac{C}{\sqrt{t^2+1}}.$$

44. $(ay - bx)^2[a^2(y'_x)^2 + b^2] - k^2(ay'_x + b)^2 = 0.$

Solve the equation for $ay - bx$ and differentiate with respect to x. Setting $w(x) = y'_x$, we obtain a factorized equation with respect to $w(x)$:

$$(aw - b)[(a^2w^2 + b^2)^{3/2} \pm abkw'_x] = 0.$$

Equating each of the factors to zero and integrating, we arrive at the solutions:

$$(bx - C)^2 + (ay - C)^2 = k^2 \quad \text{and} \quad ay - bx = \pm k\sqrt{2}.$$

45. $x^3(y'_x)^2 + x^2yy'_x + a = 0.$

Solutions:

$$Cxy = C^2x + a \quad \text{and} \quad xy^2 = 4a.$$

46. $xy^2(y'_x)^2 = ay^2 + bx.$

See equation 1.6.3.45.

47. $(ax^2y^2 + b)(y'_x)^2 = x^2.$

See equation 1.6.3.35.

48. $(xy'_x + a)^2 - 2ay + x^2 = 0, \qquad a \neq 0.$

The substitution $2ay - x^2 = u^2$ leads to the equation $xuu'_x - a(u - a) + x^2 = 0$. Next assuming $u - a = xw(x)$, we obtain $(xw + a)w'_x + w^2 + 1 = 0$. Taking w as the independent variable, we arrive at a linear equation whose solution is

$$x = (w^2 + 1)^{-1/2}\left[C - a\ln\left(w + \sqrt{w^2+1}\right)\right].$$

49. $(xy'_x + y + 2ax)^2 = 4(xy + ax^2 + b).$

The substitution $u = xy + ax^2 + b$ leads to an equation of the form 1.1.2: $u'_x = \pm 2\sqrt{u}$.

50. $(a\sqrt{y} + bx)(y'_x)^2 = 1.$

See equation 1.6.3.27.

51. $(ax^2y^{3/5} + by)(y'_x)^2 = x^2y.$

See equation 1.6.3.29.

52. $(a_2x + b_2y + c_2)(y'_x)^2 + (a_1x + b_1y + c_1)y'_x + a_0x + b_0y + c_0 = 0.$

The Legendre transformation $x = u'_t$, $y = tu'_t - u$ $(y'_x = t)$ leads to a linear equation:

$$[f(t) + tg(t)]u'_t = g(t)u + h(t),$$

where

$$f(t) = a_2t^2 + a_1t + a_0, \quad g(t) = b_2t^2 + b_1t + b_0, \quad h(t) = -c_2t^2 - c_1t - c_0.$$

53. $(y'_x)^2 = ae^y + b.$

See equation 1.6.3.3 with $k = 2$.

54. $(y'_x)^2 = a + be^x.$

See equation 1.6.3.4 with $k = 2$.

55. $(y'_x)^2 = ay^2 + be^x.$

See equation 1.6.3.8 with $k = 2$.

56. $(y'_x)^2 + ayy'_x + be^{\lambda x}y^m = 0.$

With $m \neq 2$, solving for y'_x and performing the substitution $w = e^{\lambda x}y^{m-2}$, we arrive at an equation with separated variables of the form 1.1.2:

$$w'_x = \lambda w + \frac{2-m}{2}\left(a \pm \sqrt{a^2 - 4bw}\,\right)w.$$

With $m = 2$, solving the original equation for y'_x, we obtain an equation with separation of variables:

$$2y'_x = y\left(-a \pm \sqrt{a^2 - 4be^{\lambda x}}\,\right).$$

57. $x^2(y'_x)^2 = ax^2e^y + b.$

See equation 1.6.3.9 with $k = 2$.

58. $(ae^y + bx^2)(y'_x)^2 = 1.$

See equation 1.6.3.9 with $k = -2$.

59. $(ae^xy^2 + b)(y'_x)^2 = y^2.$

See equation 1.6.3.8 with $k = -2$.

60. $(y'_x)^2 = ay + b\ln x.$

See equation 1.6.3.13.

61. $(y'_x)^2 = \lambda y + a\ln x + b, \qquad \lambda \neq 0.$

The substitution $\lambda w = 2\sqrt{\lambda y + a\ln x + b}$ leads to the Abel equation of the form 1.3.1.16: $ww'_x - w = 2a\lambda^{-2}x^{-1}$.

62. $(y'_x)^2 - xyy'_x + y^2\ln(ay) = 0.$

Solutions:

$$ay = \exp(Cx - C^2) \quad \text{and} \quad ay = \exp(\tfrac{1}{4}x^2).$$

63. $(a\ln y + bx)(y'_x)^2 = 1.$

See equation 1.6.3.14.

1.6.2. Equations of the Third Degree in y'_x

1. $(y'_x)^3 + ax + by + c = 0.$

This is a special case of equation 1.8.1.14 with $f(w) = w^3$.

2. $a(y'_x)^3 + by'_x = x.$

This is a special case of equation 1.8.1.1 with $f(w) = aw^3 + bw$.

3. $a(y'_x)^3 + by'_x = y.$

This is a special case of equation 1.8.1.2 with $f(w) = aw^3 + bw$.

4. $a(y'_x)^3 + xy'_x = y.$

This is a special case of equation 1.8.1.6 with $f(w) = aw^3$.

5. $a(y'_x)^3 + bxy'_x = y.$

This is a special case of equation 1.8.1.7 with $f(w) = bw$, $g(x) = aw^3$.

6. $(y'_x)^3 - axy'_x + x^3 = 0, \qquad a \neq 0.$

Solution in the parametric form:

$$x = \frac{at}{t^3+1}, \qquad y = C + \frac{a^2}{6}\frac{4t^3+1}{(t^3+1)^2}.$$

7. $(y'_x)^3 - axyy'_x + 2ay^2 = 0.$

Differentiating with respect to x and eliminating y, we obtain a factorized equation with respect to $w(x) = y'_x$:

$$[2(w'_x)^2 - axw'_x + aw](9w - ax^2) = 0.$$

Equating each of the factors to zero and integrating, we find the solutions:

$$y = \frac{a}{4}C(x-C)^2 \qquad \text{and} \qquad y = \frac{a}{27}x^3.$$

8. $a(y'_x)^3 + b(y'_x)^2 = x.$

This is a special case of equation 1.8.1.1 with $f(w) = aw^3 + bw^2$.

9. $a(y'_x)^3 + b(y'_x)^2 = y.$

This is a special case of equation 1.8.1.2 with $f(w) = aw^3 + bw^2$.

10. $(y'_x)^3 + a(y'_x)^2 + by + abx + d = 0.$

Solution in the parametric form:

$$2bx = -3t^2 + 2at - 2a^2\ln(t+a) + C, \qquad by = -abx - t^3 - at^2 - d,$$

In addition, there is the solution $by = -abx - d$.

11. $a(y'_x)^3 + b(y'_x)^2 + cy'_x = y + d.$

Solution in the parametric form:

$$x = C + \tfrac{3}{2}at^2 + 2bt + c\ln|t|, \qquad y = at^3 + bt^2 + ct - d.$$

12. $a(y'_x)^3 + bx(y'_x)^2 = y.$

This is a special case of equation 1.8.1.7 with $f(w) = bw^2$, $g(x) = aw^3$.

13. $ax(y'_x)^3 + by'_x = y.$

This is a special case of equation 1.8.1.7 with $f(w) = aw^3$, $g(w) = bw$.

14. $ax(y'_x)^3 + b(y'_x)^2 = y.$

This is a special case of equation 1.8.1.7 with $f(w) = aw^3$, $g(w) = bw^2$.

15. $(ax + by + c)(y'_x)^3 = \alpha x + \beta y + \gamma.$

Dividing both sides by $ax + by + c$ and raising to the power 1/3, we finally arrive at an equation of the form 1.7.1.6 with $f(w) = w^{-1/3}$.

16. $ax^{3/2}(y'_x)^3 + 2xy'_x = y.$

Solution: $(y - aC^{3/2})^2 = 4Cx.$

17. $(x^2 - a^2)(y'_x)^3 + bx(x^2 - a^2)(y'_x)^2 + y'_x + bx = 0.$

The equation can be factorized:

$$(y'_x + bx)[(y'_x)^2(x^2 - a^2) + 1] = 0,$$

whence we find the solutions:

$$y = -\frac{1}{2}bx^2 + C \qquad \text{and} \qquad y = \pm\arcsin\frac{x}{a} + C.$$

18. $ax^n(y'_x)^3 + xy'_x = y.$

This is a special case of equation 1.8.1.8 with $f(w) = aw^3$.

19. $(xy'_x - y)^3 + ay + bx = 0.$

This is a special case of equation 1.8.1.11 with $f(w) = 1$, $g(w) = a$, $h(w) = b$, $n = 3$.

20. $(xy'_x - y)^3 + ayy'_x + bx = 0.$

This is a special case of equation 1.8.1.11 with $f(w) = 1$, $g(w) = aw$, $h(w) = b$, $n = 3$.

21. $(xy'_x - y)^3 + axy'_x + by = 0.$

This is a special case of equation 1.8.1.11 with $f(w) = 1$, $g(w) = b$, $h(w) = aw$, $n = 3$.

1.6.3. Equations of the Form $(y'_x)^k = f(y) + g(x)$

Preliminary comments.

1. In the general case, the equation

$$(y'_x)^k = f(y) + g(x) \tag{1}$$

is reduced, with the aid of the transformation

$$t = \int [g(x)]^{1/k}\,dx, \qquad u = \int [f(y)]^{-1/k}\,dy,$$

to the same form

$$(u'_t)^k = F(u) + G(t),$$

where functions $F = F(u)$ and $G = G(t)$ are defined parametrically by the following formulae:

$$F(u) = \frac{1}{f(y)}, \qquad u = \int [f(y)]^{-1/k}\,dy,$$
$$G(t) = \frac{1}{g(x)}, \qquad t = \int [g(x)]^{1/k}\,dx.$$

2. Taking y as the independent variable, we obtain from equation (1) an equation of the same class for $x = x(y)$:

$$(x'_y)^{-1/k} = g(x) + f(y).$$

3. The equation

$$y'_x = a\sqrt{y} + g(x) \qquad (k = 1,\ \ f = a\sqrt{y}\,)$$

is reduced, with the aid of the substitution $w(x) = 2a^{-1}\sqrt{y}$, to the Abel equation $ww'_x - w = 2a^{-2}g(x)$ which is outlined in Subsection 1.3.1.

4. The equation

$$y'_x = y^{-1} + g(x) \qquad (k = 1,\ \ f = y^{-1})$$

is an alternative form of writing the Abel equation $yy'_x = g(x)y + 1$ which is outlined in Subsection 1.3.2.

5. The equation

$$y'_x = ay^s + g(x) \qquad (k = 1,\ \ f = ay^s\,)$$

is reduced, with the aid of the substitution $aw = y - \int g(x)\,dx$ followed by raising both sides of the equation to the power of $1/s$, to an equation of the class in question:

$$(w'_x)^{1/s} = aw + \int g(x)\,dx.$$

6. The equation

$$(y'_x)^2 = ay + g(x) \qquad (k = 2,\ \ f = ay,\ \ a \neq 0)$$

is reduced, with the aid of the substitution $aw = 2\sqrt{ay + g(x)}$, to the Abel equation of the second kind:

$$ww'_x = w + \varphi(x), \qquad \text{where} \quad \varphi = 2a^{-2}g'_x(x),$$

which is outlined in Subsection 1.3.1.

7. The equation

$$(y'_x)^{1/2} = ay + g(x) \qquad (k = 1/2, \;\; f = ay)$$

is reduced, by squaring both sides and performing the substitution $z = ay + g(x)$, to the Riccati equation

$$z'_x = az^2 + g'_x.$$

For some specific functions $g = g(x)$, the solutions of the latter equation are given in Section 1.2.

8. The equation

$$(y'_x)^{1/2} = ay^{1/2} + g(x) \qquad (k = 1/2, \;\; f = ay^{1/2})$$

can be reduced, by squaring both sides and performing the substitution $y = \exp(a^2 x)\xi^2$, to the Abel equation of the second kind:

$$\xi\xi'_x = a\exp(-\tfrac{1}{2}a^2 x)g\xi + \tfrac{1}{2}\exp(-a^2 x)g^2$$

(see Subsection 1.3.3).

9. The equation

$$(y'_x)^{-1/2} = f(y) + ax \qquad (k = -1/2, \;\; g = ax)$$

can be reduced, by squaring both sides and performing the substitution $v = f(y) + ax$, to the Riccati equation:

$$v'_y = av^2 + f'_y.$$

For some specific functions $f = f(y)$, the solutions of the latter equation are given in Section 1.2.

10. For the sake of convenience, in Tables 1.5–1.9 are listed all the equations outlined in Subsection 1.6.3. Five classification tables are given below which classify the equations wherein functions f and g are of the same form. The rightmost columns of the tables present the numbers of equations where the corresponding solutions are given. After the tables follow the equations—they are combined into groups so that the solutions of the equations within each group are expressed in terms of the functions indicated before the groups as a notation list.

1. $\boldsymbol{(y'_x)^k = Ay^s + B.}$

Solution: $x = \displaystyle\int (Ay^s + B)^{-1/k}\,dy + C.$

2. $\boldsymbol{(y'_x)^k = A + Bx^r.}$

Solution: $y = \displaystyle\int (A + Bx^r)^{1/k}\,dx + C.$

3. $\boldsymbol{(y'_x)^k = Ae^y + B.}$

Solution: $x = \displaystyle\int (Ae^y + B)^{-1/k}\,dy + C.$

TABLE 1.5
Solvable equations of the form $(y'_x)^k = Ay^s + Bx^r$

k	s	r	Equation	k	s	r	Equation
arbitrary	arbitrary ($s \neq k$)	$\dfrac{ks}{k-s}$	1.6.3.7	-1	1	-1	1.6.3.23
arbitrary	arbitrary	0	1.6.3.1	-1	1	$1/2$	1.6.3.42
arbitrary ($k \neq -1,\,1$)	$\dfrac{k}{1-k}$	$-\dfrac{k}{1+k}$	1.6.3.6	$-1/2$	arbitrary ($s \neq -1,\,0$)	1	1.6.3.17
arbitrary	0	arbitrary	1.6.3.2	$-1/2$	-1	1	1.6.3.38
arbitrary	1	1	1.6.3.5	$1/2$	1	arbitrary ($r \neq -1,\,0$)	1.6.3.16
-2	-1	-2	1.6.3.46	$1/2$	1	-1	1.6.3.37
-2	-1	1	1.6.3.33	1	-1	-2	1.6.3.20
-2	$-2/5$	-2	1.6.3.29	1	-1	$-1/2$	1.6.3.39
-2	$1/2$	1	1.6.3.27	1	-1	1	1.6.3.22
-2	2	-2	1.6.3.35	1	$1/2$	-2	1.6.3.30
-2	2	1	1.6.3.44	1	$1/2$	-1	1.6.3.11
-1	arbitrary ($s \neq 0$)	1	1.6.3.10	1	$1/2$	$-1/2$	1.6.3.24
-1	arbitrary ($s \neq -2,\,0$)	2	1.6.3.15	1	$1/2$	1	1.6.3.41
-1	-2	-1	1.6.3.21	2	-2	-1	1.6.3.45
-1	-2	$1/2$	1.6.3.31	2	-2	$-2/5$	1.6.3.28
-1	-2	2	1.6.3.36	2	-2	2	1.6.3.34
-1	-1	$1/2$	1.6.3.12	2	1	-1	1.6.3.32
-1	$-1/2$	-1	1.6.3.40	2	1	$1/2$	1.6.3.26
-1	$-1/2$	$1/2$	1.6.3.25	2	1	2	1.6.3.43

4. $(y'_x)^k = A + Be^x$.

Solution: $y = \displaystyle\int (A + Be^x)^{1/k}\,dx + C.$

5. $(y'_x)^k = Ay + Bx$.

Solution in the parametric form:

$$x = \int (A\tau^{1/k} + B)^{-1}\,d\tau + C, \qquad y = \frac{1}{A}\Big[\tau - B\int (A\tau^{1/k} + B)^{-1}\,d\tau - BC\Big].$$

TABLE 1.6
Solvable equations of the form $(y'_x)^k = Ae^y + Bx^r$

k	r	Equation
arbitrary	$-k$	1.6.3.9
arbitrary	0	1.6.3.3
-1	-1	1.5.2.34
-1	1	1.5.2.32
-1	2	1.5.2.33
$-1/2$	1	1.6.3.19
1	arbitrary	1.5.2.2

TABLE 1.7
Solvable equations of the form $(y'_x)^k = Ay^s + Be^x$

k	s	Equation
arbitrary	k	1.6.3.8
arbitrary	0	1.6.3.4
-1	arbitrary	1.5.2.31
$1/2$	1	1.6.3.18
1	-1	1.5.2.3

TABLE 1.8
Solvable equations of the form $(y'_x)^k = Ae^y + Be^x$

k	Equation
-1	1.5.2.30
1	1.5.2.1

TABLE 1.9
Solvable equations containing logarithmic function

Form of equation	Equation
$(y'_x)^{-2} = A\ln y + Bx$	1.6.3.14
$(y'_x)^{-1} = A\ln y + Bx$	1.5.4.12
$(y'_x)^2 = Ay + B\ln x$	1.6.3.13

6. $(y'_x)^k = Ay^{\frac{k}{1-k}} + Bx^{-\frac{k}{1+k}}, \qquad |k| \neq 1.$

Solution in the parametric form:

$$x = a\left[\int (\gamma\tau^{1/k} + \beta)^{-1}\,d\tau + C\right]^{\frac{k+1}{k}}, \qquad y = b\left[\tau - \beta\int (\gamma\tau^{1/k} + \beta)^{-1}\,d\tau - \beta C\right]^{\frac{k-1}{k}},$$

where

$$A = a^{-\frac{k}{1+k}} b^{-\frac{k}{1-k}} \beta B, \quad B = a^{\frac{k}{1+k}}\left[\frac{b(k-1)}{a(k+1)}\gamma\right]^k.$$

7. $(y'_x)^k = Ay^s + Bx^{\frac{ks}{k-s}}, \qquad k \neq s.$

Solution in the parametric form:

$$x = C^{k-s}\exp\left\{\int\left[\frac{k-s}{s}\left(A + B\tau^{\frac{ks}{s-k}}\right)^{\frac{1}{k}} - \tau\right]^{-1} d\tau\right\},$$

$$y = C^k\left\{\tau\exp\int\left[\frac{k-s}{s}\left(A + B\tau^{\frac{ks}{s-k}}\right)^{\frac{1}{k}} - \tau\right]^{-1} d\tau\right\}^{\frac{k}{k-s}}.$$

8. $(y'_x)^k = Ay^k + Be^x.$

Solution in the parametric form:

$$x = \int \left[(A + Be^{-k\tau})^{1/k} - \frac{1}{k} \right]^{-1} d\tau + C,$$

$$y = \exp\left\{ \tau + \frac{1}{k} \int \left[(A + Be^{-k\tau})^{1/k} - \frac{1}{k} \right]^{-1} d\tau + \frac{C}{k} \right\}.$$

9. $(y'_x)^k = Ae^y + Bx^{-k}.$

Solution in the parametric form:

$$x = \exp\left\{ \tau - \frac{1}{k} \int \left[(B + Ae^{k\tau})^{-1/k} + \frac{1}{k} \right]^{-1} d\tau - \frac{C}{k} \right\},$$

$$y = \int \left[(B + Ae^{k\tau})^{-1/k} + \frac{1}{k} \right]^{-1} d\tau + C.$$

10. $(y'_x)^{-1} = Ay^s + Bx.$

Solution: $x = e^{By}\left(A \int y^s e^{-By}\, dy + C \right).$

▶ *In the solutions of equations 11–14, the following notation is used:*

$$F = \left[\int \exp(\mp\tau^2)\, d\tau + C \right]^{-1}.$$

11. $y'_x = Ay^{1/2} + Bx^{-1}.$

Solution in the parametric form:

$$x = aF\exp(\mp\tau^2), \qquad y = b[2\tau \pm F\exp(\mp\tau^2)]^2,$$

where $A = \pm 2a^{-1}b^{1/2}$, $B = \mp 4b$.

12. $(y'_x)^{-1} = Ay^{-1} + Bx^{1/2}.$

Solution in the parametric form:

$$x = a[2\tau \pm F\exp(\mp\tau^2)]^2, \qquad y = bF\exp(\mp\tau^2),$$

where $A = \mp 4a$, $B = \pm 2a^{1/2}b^{-1}$.

13. $(y'_x)^2 = Ay + B\ln x.$

Solution in the parametric form:

$$x = aF\exp(\mp\tau^2), \qquad y = b\left\{ [2\tau \pm F\exp(\mp\tau^2)]^2 \pm 4\ln(aF) - 4\tau^2 \right\},$$

where $A = 4a^{-2}b$, $B = \mp 4bA$.

14. $(y'_x)^{-2} = A\ln y + Bx.$

Solution in the parametric form:

$$x = a\left\{ [2\tau \pm F\exp(\mp\tau^2)]^2 \pm 4\ln(bF) - 4\tau^2 \right\}, \qquad y = bF\exp(\mp\tau^2),$$

where $A = \mp 4aB$, $B = 4ab^{-2}$.

▶ *In the solutions of equations 15–19, the following notation is used:*

$$Z = \begin{cases} C_1 J_\nu(\tau) + C_2 Y_\nu(\tau) & \text{for the upper sign,} \\ C_1 I_\nu(\tau) + C_2 K_\nu(\tau) & \text{for the lower sign,} \end{cases}$$

where J_ν and Y_ν are Bessel functions, I_ν and K_ν are modified Bessel functions.

Remark. *The solutions of equations 15–19 contain only the ratio $Z'_\tau/Z = (\ln Z)'_\tau$. Therefore, for the sake of symmetric appearance, two "arbitrary" constants C_1 and C_2 are indicated in the definition of function Z (instead, we may set, for instance, $C_1 = 1$ and $C_2 = C$).*

15. $(y'_x)^{-1} = Ay^s + Bx^2, \qquad s \neq -2,\ s \neq 0.$

Solution in the parametric form:

$$x = a\tau^{-2\nu}\big[\tau(\ln Z)'_\tau + \nu\big], \qquad y = b\tau^{2\nu},$$

where $\nu = \dfrac{1}{s+2}$, $A = \mp\dfrac{s+2}{2}ab^{-1-s}$, $B = -\dfrac{s+2}{2}a^{-1}b^{-1}$.

16. $(y'_x)^{1/2} = Ay + Bx^r, \qquad r \neq -1,\ r \neq 0.$

Solution in the parametric form:

$$x = a\tau^{2\nu}, \qquad y = b\tau^{2\nu}\Big[\tau(\ln Z)'_\tau + \nu \pm \frac{r+1}{2r}\tau^2\Big],$$

where $\nu = \dfrac{1}{r+1}$, $A = b^{-1}\Big[-\dfrac{(r+1)b}{2a}\Big]^{1/2}$, $B = \mp\dfrac{r+1}{2r}a^{-r}bA$.

17. $(y'_x)^{-1/2} = Ay^s + Bx, \qquad s \neq -1,\ s \neq 0.$

Solution in the parametric form:

$$x = a\tau^{2\nu}\Big[\tau(\ln Z)'_\tau + \nu \pm \frac{s+1}{2s}\tau^2\Big], \qquad y = b\tau^{2\nu},$$

where $\nu = \dfrac{1}{s+1}$, $A = \mp\dfrac{s+1}{2s}ab^{-s}B$, $B = a^{-1}\Big[-\dfrac{(s+1)a}{2b}\Big]^{1/2}$.

18. $(y'_x)^{1/2} = Ay + Be^x.$

Solution in the parametric form:

$$x = \ln(a\tau^2), \qquad y = b\Big[\tau(\ln Z)'_\tau \pm \frac{1}{2}\tau^2\Big],$$

where $\nu = 0$, $A = b^{-1}\Big(-\dfrac{b}{2}\Big)^{1/2}$, $B = \mp\dfrac{1}{2}a^{-1}bA$.

19. $(y'_x)^{-1/2} = Ae^y + Bx.$

Solution in the parametric form:

$$x = a\Big[\tau(\ln Z)'_\tau \pm \frac{1}{2}\tau^2\Big], \qquad y = \ln(b\tau^2),$$

where $\nu = 0$, $A = \mp\dfrac{1}{2}ab^{-1}B$, $B = a^{-1}\Big(-\dfrac{a}{2}\Big)^{1/2}$.

▶ *In the solutions of equations 20–35, the following notation is used:*

$$Z = \begin{cases} C_1 J_{1/3}(\tau) + C_2 Y_{1/3}(\tau) & \text{for the upper sign,} \\ C_1 I_{1/3}(\tau) + C_2 K_{1/3}(\tau) & \text{for the lower sign,} \end{cases}$$

where $J_{1/3}$ and $Y_{1/3}$ are Bessel functions, $I_{1/3}$ and $K_{1/3}$ are modified Bessel functions;

$$U_1 = \tau Z'_\tau + \tfrac{1}{3} Z, \qquad U_2 = U_1^2 \pm \tau^2 Z^2, \qquad U_3 = \pm\tfrac{2}{3}\tau^2 Z^3 - 2U_1 U_2.$$

Remark. *The solutions of equations 20–35 contain only the ratio $Z'_\tau / Z = (\ln Z)'_\tau$. Therefore, for the sake of symmetric appearance, two "arbitrary" constants C_1 and C_2 are indicated in the definition of function Z (instead, we may set, for instance, $C_1 = 1$ and $C_2 = C$).*

20. $y'_x = Ay^{-1} + Bx^{-2}$.

Solution in the parametric form:

$$x = a\tau^{-4/3} Z^{-2} U_2, \quad y = b\tau^{-2/3} Z^{-1} U_2^{-1} U_3, \qquad \text{where} \quad A = 2a^{-1}b^2, \quad B = \mp\tfrac{2}{3}ab.$$

21. $(y'_x)^{-1} = Ay^{-2} + Bx^{-1}$.

Solution in the parametric form:

$$x = a\tau^{-2/3} Z^{-1} U_2^{-1} U_3, \quad y = b\tau^{-4/3} Z^{-2} U_2, \qquad \text{where} \quad A = \mp\tfrac{2}{3}ab, \quad B = 2a^2 b^{-1}.$$

22. $y'_x = Ay^{-1} + Bx$.

Solution in the parametric form:

$$x = a\tau^{-2/3} Z^{-1} U_1, \quad y = b\tau^{-4/3} Z^{-2} U_2, \qquad \text{where} \quad A = \mp\tfrac{2}{3}a^{-1}b^2, \quad B = 2a^{-2}b.$$

23. $(y'_x)^{-1} = Ay + Bx^{-1}$.

Solution in the parametric form:

$$x = a\tau^{-4/3} Z^{-2} U_2, \quad y = b\tau^{-2/3} Z^{-1} U_1, \qquad \text{where} \quad A = 2a^{-2}b, \quad B = \mp\tfrac{2}{3}a^2 b^{-1}.$$

24. $y'_x = Ay^{1/2} + Bx^{-1/2}$.

Solution in the parametric form:

$$x = a\tau^{-4/3} Z^{-2} U_1^2, \quad y = b\tau^{-8/3} Z^{-4} U_2^2, \qquad \text{where} \quad A = 2a^{-1}b^{1/2}, \quad B = \mp\tfrac{2}{3}a^{-1/2}b.$$

25. $(y'_x)^{-1} = Ay^{-1/2} + Bx^{1/2}$.

Solution in the parametric form:

$$x = a\tau^{-8/3} Z^{-4} U_2^2, \quad y = b\tau^{-4/3} Z^{-2} U_1^2, \qquad \text{where} \quad A = \mp\tfrac{2}{3}ab^{-1/2}, \quad B = 2a^{1/2}b^{-1}.$$

26. $(y'_x)^2 = Ay + Bx^{1/2}$.

Solution in the parametric form:

$$x = a\tau^{-4/3} Z^{-2} U_1^2, \quad y = b\tau^{-8/3} Z^{-4} (U_2^2 \pm \tfrac{4}{3}\tau^2 Z^3 U_1),$$

where $A = 4a^{-2}b$, $B = \mp\tfrac{4}{3}a^{-1/2}bA$.

27. $(y'_x)^{-2} = Ay^{1/2} + Bx$.

Solution in the parametric form:

$$x = a\tau^{-8/3}Z^{-4}(U_2^2 \pm \tfrac{4}{3}\tau^2 Z^3 U_1), \quad y = b\tau^{-4/3}Z^{-2}U_1^2,$$

where $A = \mp\frac{4}{3}ab^{-1/2}B$, $B = 4ab^{-2}$.

28. $(y'_x)^2 = Ay^{-2} + Bx^{-2/5}$.

Solution in the parametric form:

$$x = a\tau^{-5/3}Z^{-5/2}U_1^{5/2}, \quad y = b\tau^{-4/3}Z^{-2}(U_2^2 \pm \tfrac{4}{3}\tau^2 Z^3 U_1)^{1/2},$$

where $A = \mp\frac{4}{3}a^{-2/5}b^2 B$, $B = \frac{16}{25}a^{-8/5}b^2$.

29. $(y'_x)^{-2} = Ay^{-2/5} + Bx^{-2}$.

Solution in the parametric form:

$$x = a\tau^{-4/3}Z^{-2}(U_2^2 \pm \tfrac{4}{3}\tau^2 Z^3 U_1)^{1/2}, \quad y = b\tau^{-5/3}Z^{-5/2}U_1^{5/2},$$

where $A = \frac{16}{25}a^2 b^{-8/5}$, $B = \mp\frac{4}{3}a^2 b^{-2/5}A$.

30. $y'_x = Ay^{1/2} + Bx^{-2}$.

Solution in the parametric form:

$$x = a\tau^{4/3}Z^2 U_2^{-1}, \quad y = b\tau^{-4/3}Z^{-2}U_2^{-2}U_3^2, \qquad \text{where} \quad A = \pm\tfrac{4}{3}a^{-1}b^{1/2}, \quad B = -4ab.$$

31. $(y'_x)^{-1} = Ay^{-2} + Bx^{1/2}$.

Solution in the parametric form:

$$x = a\tau^{-4/3}Z^{-2}U_2^{-2}U_3^2, \quad y = b\tau^{4/3}Z^2 U_2^{-1}, \qquad \text{where} \quad A = -4ab, \quad B = \pm\tfrac{4}{3}a^{1/2}b^{-1}.$$

32. $(y'_x)^2 = Ay + Bx^{-1}$.

Solution in the parametric form:

$$x = a\tau^{4/3}Z^2 U_2^{-1}, \quad y = b\tau^{-4/3}Z^{-2}U_2^{-2}(U_3^2 - 4U_2^3), \qquad \text{where } A = \tfrac{16}{9}a^{-2}b, \; B = 4abA.$$

33. $(y'_x)^{-2} = Ay^{-1} + Bx$.

Solution in the parametric form:

$$x = a\tau^{-4/3}Z^{-2}U_2^{-2}(U_3^2 - 4U_2^3), \quad y = b\tau^{4/3}Z^2 U_2^{-1}, \qquad \text{where } A = 4abB, \; B = \tfrac{16}{9}ab^{-2}.$$

34. $(y'_x)^2 = Ay^{-2} + Bx^2$.

Solution in the parametric form:

$$x = a\tau^{2/3}ZU_2^{-1/2}, \quad y = b\tau^{-2/3}Z^{-1}U_2^{-1}(U_3^2 - 4U_2^3)^{1/2},$$

where $A = 4a^2b^2B$, $B = \frac{16}{9}a^{-4}b^2$.

35. $(y'_x)^{-2} = Ay^2 + Bx^{-2}$.

Solution in the parametric form:

$$x = a\tau^{-2/3}Z^{-1}U_2^{-1}(U_3^2 - 4U_2^3)^{1/2}, \quad y = b\tau^{2/3}ZU_2^{-1/2},$$

where $A = \frac{16}{9}a^2b^{-4}$, $B = 4a^2b^2A$.

▶ *In the solutions of equations 36–46, the following notation is used:*

$$R = \begin{cases} C_1\tau^\nu + C_2\tau^{-\nu} & \textit{for the upper sign,} \\ C_1\sin(\nu\ln\tau) + C_2\cos(\nu\ln\tau) & \textit{for the lower sign,} \\ C_1\ln\tau + C_2 & \textit{for } \nu = 0. \end{cases}$$

$$Q = \begin{cases} (1+\nu)C_1\tau^\nu + (1-\nu)C_2\tau^{-\nu} & \textit{for the upper sign,} \\ (C_1 - \nu C_2)\sin(\nu\ln\tau) + (C_2 + \nu C_1)\cos(\nu\ln\tau) & \textit{for the lower sign,} \\ C_1\ln\tau + C_1 + C_2 & \textit{for } \nu = 0. \end{cases}$$

Remark. *The expressions for R and Q contains two "arbitrary" constants C_1 an C_2. One of them may be fixed to set it equal to any nonzero number (for example, we may set $C_2 = \pm 1$), while another constant remains arbitrary.*

36. $\boldsymbol{(y'_x)^{-1} = Ay^{-2} + Bx^2}$.

Solution in the parametric form:

$$x = a\tau^{-2}R^{-1}Q, \quad y = b\tau^2, \qquad \nu = \sqrt{|1 - 4AB|},$$

where $A = -\dfrac{1 \mp \nu^2}{2}ab$, $B = -\dfrac{1}{2}a^{-1}b^{-1}$.

37. $\boldsymbol{(y'_x)^{1/2} = Ay + Bx^{-1}}$.

Solution in the parametric form:

$$x = a\tau^2, \quad y = b\tau^{-2}\left(R^{-1}Q - \frac{1 \mp \nu^2}{2}\right),$$

where $A = b^{-1}\left(-\dfrac{b}{2a}\right)^{1/2}$, $B = \dfrac{1 \mp \nu^2}{2}abA$.

38. $\boldsymbol{(y'_x)^{-1/2} = Ay^{-1} + Bx}$.

Solution in the parametric form:

$$x = a\tau^{-2}\left(R^{-1}Q - \frac{1 \mp \nu^2}{2}\right), \quad y = b\tau^2,$$

where $A = \dfrac{1 \mp \nu^2}{2}abB$, $B = a^{-1}\left(-\dfrac{a}{2b}\right)^{1/2}$.

39. $\boldsymbol{y'_x = Ay^{-1} + Bx^{-1/2}}$.

Solution in the parametric form:

$$x = a\tau^2R^2, \quad y = b\tau Q, \qquad \text{where} \quad A = (-1 \pm \nu^2)\frac{b^2}{2a}, \quad B = a^{-1/2}b.$$

40. $\boldsymbol{(y'_x)^{-1} = Ay^{-1/2} + Bx^{-1}}$.

Solution in the parametric form:

$$x = a\tau Q, \quad y = b\tau^2R^2, \qquad \text{where} \quad A = ab^{-1/2}, \quad B = (-1 \pm \nu^2)\frac{b^2}{2a}.$$

41. $\boldsymbol{y'_x = Ay^{1/2} + Bx}$.

Solution in the parametric form:

$$x = a\tau R, \quad y = b\tau^2 Q^2, \qquad \text{where} \quad A = 2(-1 \pm \nu^2)a^{-1}b^{1/2}, \quad B = 4a^{-2}b.$$

42. $\boldsymbol{(y'_x)^{-1} = Ay + Bx^{1/2}}$.

Solution in the parametric form:

$$x = a\tau^2 Q^2, \quad y = b\tau R, \qquad \text{where} \quad A = 4ab^{-2}, \quad B = 2(-1 \pm \nu^2)a^{1/2}b^{-1}.$$

43. $\boldsymbol{(y'_x)^2 = Ay + Bx^2}$.

Solution in the parametric form:

$$x = a\tau R, \quad y = b\tau^2[Q^2 - (-1 \pm \nu^2)R^2], \qquad \text{where} \quad A = 16a^{-2}b, \quad B = (-1 \pm \nu^2)a^{-2}bA.$$

44. $\boldsymbol{(y'_x)^{-2} = Ay^2 + Bx}$.

Solution in the parametric form:

$$x = a\tau^2[Q^2 - (-1 \pm \nu^2)R^2], \quad y = b\tau R, \qquad \text{where} \quad A = (-1 \pm \nu^2)ab^{-2}B, \quad B = 16ab^{-2}.$$

45. $\boldsymbol{(y'_x)^2 = Ay^{-2} + Bx^{-1}}$.

Solution in the parametric form:

$$x = a\tau^2 R^2, \quad y = b\tau[Q^2 - (-1 \pm \nu^2)R^2]^{1/2},$$

where $A = (-1 \pm \nu^2)a^{-1}b^2B$, $B = a^{-1}b^2$.

46. $\boldsymbol{(y'_x)^{-2} = Ay^{-1} + Bx^{-2}}$.

Solution in the parametric form:

$$x = a\tau[Q^2 - (-1 \pm \nu^2)R^2]^{1/2}, \quad y = b\tau^2 R^2,$$

where $A = a^2b^{-1}$, $B = (-1 \pm \nu^2)a^2b^{-1}A$.

1.6.4. Other equations

1. $\boldsymbol{y = xy'_x + ax^2 + b\sqrt{y'_x} + c}, \qquad \boldsymbol{a \neq 0}$.

Differentiating the equation with respect to x and changing to new variables $t = y'_x$ and $w(t) = -2ax$, we arrive at the Abel equation of the form 1.3.1.32: $ww'_t - w = -abt^{-1/2}$.

2. $\boldsymbol{y = xy'_x + ax^2 + b(y'_x)^2 + c(y'_x)^{m+1} + d}, \qquad \boldsymbol{a \neq 0}$.

Differentiating the equation with respect to x and changing to new variables $t = y'_x$ and $w(t) = -2ax$, we arrive at the Abel equation:

$$ww'_t - w = -4abt - 2ac(m+1)t^m,$$

whose solvable case are outlined in Subsection 1.3.1.

3. $a(y'_x)^n + b(y'_x)^m = x.$

Solution in the parametric form:

with $n \neq -1$, $m \neq -1$,

$$x = at^n + bt^m, \qquad y = C + \frac{an}{n+1}t^{n+1} + \frac{bm}{m+1}t^{m+1};$$

with $n = -1$, $m \neq -1$,

$$x = \frac{a}{t} + bt^m, \qquad y = C + a\ln|t| + \frac{bm}{m+1}t^{m+1}.$$

4. $a(y'_x)^n + b(y'_x)^m = y.$

Solution in the parametric form:

with $n \neq 1$, $m \neq 1$,

$$x = C + \frac{an}{n-1}t^{n-1} + \frac{bm}{m-1}t^{m-1}, \qquad y = at^n + bt^m;$$

with $n = 1$, $m \neq 1$,

$$x = C + a\ln|t| + \frac{bm}{m-1}t^{m-1}, \qquad y = at + bt^m.$$

5. $y = xy'_x + a(y'_x)^n.$

Solution: $y = Cx + aC^n$. In addition, there is the solution $y = Ax^{\frac{n}{n-1}}$, where $aA^{n-1}n^n = -(n-1)^{n-1}$, $n \neq 1$.

6. $y = xy'_x + ax^n(y'_x)^m.$

This is a special case of equation 1.8.1.8 with $f(w) = aw^m$.

7. $y = ax^n(y'_x)^{2n} + 2xy'_x.$

This is a special case of equation 1.8.1.9 with $f(w) = aw^n$.

8. $y'_x = ax^n(xy'_x - y)^m.$

The Legendre transformation $x = w'_t$, $y = tw'_t - w$ $(y'_x = t)$ leads to the equation $t = aw^m(w'_t)^n$. By integrating, we obtain the solution in the parametric form:

with $m \neq -n$, $n \neq -1$,

$$x = \left(\frac{t}{a}\right)^{\frac{1}{n}} \left[\frac{m+n}{n+1}t\left(\frac{t}{a}\right)^{\frac{1}{n}} + C\right]^{-\frac{m}{m+n}},$$

$$y = \left[\frac{1-m}{1+n}t\left(\frac{t}{a}\right)^{\frac{1}{n}} - C\right]\left[\frac{m+n}{n+1}t\left(\frac{t}{a}\right)^{\frac{1}{n}} + C\right]^{-\frac{m}{m+n}};$$

with $m = -n$, $n \neq -1$,

$$x = C\left(\frac{t}{a}\right)^{\frac{1}{n}} \exp\left[\frac{n}{n+1}t\left(\frac{t}{a}\right)^{\frac{1}{n}}\right], \qquad y = C\left[t\left(\frac{t}{a}\right)^{\frac{1}{n}} - 1\right]\exp\left[\frac{n}{n+1}t\left(\frac{t}{a}\right)^{\frac{1}{n}}\right];$$

with $m \neq -n$, $n = -1$,

$$x = \frac{a}{t}\left[a(1-m)\ln|t| + C\right]^{\frac{m}{1-m}}, \qquad y = tx - \left[a(1-m)\ln|t| + C\right]^{\frac{1}{1-m}};$$

with $m = 1$, $n = -1$,

$$y = Cx^{\frac{a}{a-1}}.$$

9. $x = a\exp(\lambda y'_x) + b\exp(\mu y'_x)$.

This is a special case of equation 1.8.1.1 with $f(w) = a\exp(\lambda w) + b\exp(\mu w)$.

10. $y = a\exp(\lambda y'_x) + b\exp(\mu y'_x)$.

This is a special case of equation 1.8.1.2 with $f(w) = a\exp(\lambda w) + b\exp(\mu w)$.

11. $y = xy'_x + ax^n\exp(\lambda y'_x)$.

This is a special case of equation 1.8.1.8 with $f(w) = a\exp(\lambda w)$.

12. $y = ax\exp(\lambda y'_x) + b\exp(\mu y'_x)$.

This is a special case of equation 1.8.1.7 with $f(w) = a\exp(\lambda w)$, $g(w) = b\exp(\mu w)$.

13. $\ln y'_x + xy'_x + ay + b = 0$.

Solution in the parametric form:

with $a \neq 0$ and $\neq -1$,

$$x = \frac{1}{at} + Ct^{-\frac{1}{a+1}}, \qquad y = -\frac{1}{a}(xt + \ln t + b);$$

with $a = 0$,

$$x = -\frac{\ln t + b}{t}, \qquad y = C + (b-1)\ln t + \frac{1}{2}(\ln t)^2;$$

with $a = -1$,

$$y = Cx + \ln C + b \quad \text{and} \quad y = \ln(-1/x) + b - 1.$$

14. $y = xy'_x + ax^2 + b\ln y'_x + c, \qquad a \neq 0$.

Differentiating the equation with respect to x and changing to new variables $t = y'_x$ and $w(t) = -2ax$, we arrive at the Abel equation of the form 1.3.1.16: $ww'_t - w = -2abt^{-1}$.

15. $y = xy'_x + ax^n\ln^m(\lambda y'_x)$.

This is a special case of equation 1.8.1.8 with $f(w) = a\ln^m(\lambda w)$.

16. $y = xy'_x + ax^n\sin^m(ky'_x)$.

This is a special case of equation 1.8.1.8 with $f(w) = a\sin^m(kw)$.

17. $y = xy'_x + ax^n\cos^m(ky'_x)$.

This is a special case of equation 1.8.1.8 with $f(w) = a\cos^m(kw)$.

18. $y = xy'_x + ax^n\tan^m(ky'_x)$.

This is a special case of equation 1.8.1.8 with $f(w) = a\tan^m(kw)$.

1.7. Equations of the Form $F(x,y)y'_x = G(x,y)$ Containing Arbitrary Functions

Notation: f, g, and h are arbitrary composite functions whose argument, indicated after the function name, may depend on both x an y.

1.7.1. Equations Containing Power Functions

1. $y'_x = f(ax + by + c)$.

With $b = 0$, we have an equation of the form 1.1.1. With $b \neq 0$, the substitution $u(x) = ax + by + c$ leads to an equation of the form 1.1.2: $u'_x = bf(u)$.

2. $y'_x = f(y + ax^n + b) - anx^{n-1}$.

The substitution $u = y + ax^n + b$ leads to an equation of the form 1.1.2: $u'_x = f(u)$.

3. $y'_x = \dfrac{y}{x} f(x^n y^m)$.

Homogeneous equation in the extended sense.

The substitution $z = x^n y^m$ leads to an equation with separation of variables: $xz'_x = nz + mzf(z)$.

4. $y'_x = f(x)y^{1+n} + g(x)y + h(x)y^{1-n}$.

The substitution $w = y^n$ leads to the Riccati equation: $w'_x = nf(x)w^2 + ng(x)w + nh(x)$.

5. $y'_x = -\dfrac{n}{m}\dfrac{y}{x} + y^k f(x) g(x^n y^m)$.

The substitution $z = x^n y^m$ leads to an equation with separation of variables: $z'_x = mx^{\frac{n-nk}{m}} f(x) z^{\frac{k+m-1}{m}} g(z)$.

6. $y'_x = f\left(\dfrac{ax + by + c}{\alpha x + \beta y + \gamma}\right)$.

With $\Delta = a\beta - b\alpha \neq 0$, the transformation $x = u + \dfrac{b\gamma - c\beta}{\Delta}$, $y = v(u) + \dfrac{c\alpha - a\gamma}{\Delta}$ yields

$$v'_u = f\left(\frac{au + bv}{\alpha u + \beta v}\right).$$

Dividing both the numerator and denominator of the fraction on the right-hand side by u, we obtain a homogeneous equation of the form 1.1.6.

With $\Delta = 0$, $b \neq 0$, the substitution $v(x) = ax + by + c$ leads to an equation of the form 1.1.2:

$$v'_x = a + bf\left(\frac{bv}{\beta v + b\gamma - c\beta}\right).$$

With $\Delta = 0$, $\beta \neq 0$, the substitution $v(x) = \alpha x + \beta y + \gamma$ also leads to an equation of the form 1.1.2:

$$v'_x = \alpha + \beta f\left(\frac{bv + c\beta - b\gamma}{\beta v}\right).$$

7. $y'_x = x^{n-1}y^{1-m}f(ax^n + by^m)$.

The substitution $w = ax^n + by^m$ leads to an equation with separation of variables: $w'_x = x^{n-1}[an + bmf(w)]$.

8. $y^n y'_x + ax^n + g(x)f(y^{n+1} + ax^{n+1}) = 0$.

The substitution $w = y^{n+1} + ax^{n+1}$ leads to an equation with separation of variables: $w'_x + (n+1)g(x)f(w) = 0$.

9. $[x^n f(y) + xg(y)]y'_x = h(y)$.

This is the Bernoulli equation with respect to $x = x(y)$ (see 1.1.5).

10. $[x^2 + xf(y) + g(y)]y'_x = h(y)$.

This is the Riccati equation with respect to $x = x(y)$ (see Section 1.2).

11. $y'_x = [f(x)y + g(x)]\sqrt{(y-a)(y-b)}$.

The substitution $u^2 = (y-a)/(y-b)$ leads to the Riccati equation:

$$\pm 2u'_x = [bf(x) + g(x)]u^2 - af(x) - g(x).$$

12. $\left[f\left(\frac{y}{x}\right) + x^a h\left(\frac{y}{x}\right)\right]y'_x = g\left(\frac{y}{x}\right) + yx^{a-1}h\left(\frac{y}{x}\right)$.

The substitution $y = xt$ leads to the Bernoulli equation with respect to $x = x(t)$: $[g(t) - tf(t)]x'_t = f(t)x + h(t)x^{a+1}$.

13. $[f(ax+by) + bxg(ax+by)]y'_x = h(ax+by) - axg(ax+by)$.

The substitution $t = ax + by$ leads to a linear equation with respect to $x = x(t)$: $[af(t) + bh(t)]x'_t = bg(t)x + f(t)$.

14. $[f(ax+by) + byg(ax+by)]y'_x = h(ax+by) - ayg(ax+by)$.

The substitution $t = ax + by$ leads to a linear equation with respect to $y = y(t)$: $[af(t) + bh(t)]y'_t = -ag(t)y + h(t)$.

15. $x[f(x^n y^m) + mx^k g(x^n y^m)]y'_x = y[h(x^n y^m) - nx^k g(x^n y^m)]$.

The transformation $t = x^n y^m$, $z = x^{-k}$ leads to a linear equation: $t[nf(t) + mh(t)]z'_t = -kf(t)z - kmg(t)$.

16. $x[f(x^n y^m) + my^k g(x^n y^m)]y'_x = y[h(x^n y^m) - ny^k g(x^n y^m)]$.

The transformation $t = x^n y^m$, $z = y^{-k}$ leads to a linear equation: $t[nf(t) + mh(t)]z'_t = -kh(t)z + kng(t)$.

17. $x[sf(x^n y^m) - mg(x^k y^s)]y'_x = y[ng(x^k y^s) - kf(x^n y^m)]$.

The transformation $t = x^n y^m$, $w = x^k y^s$ leads to an equation with separated variables: $tf(t)w'_t = wg(w)$.

18. $[f(y) + amx^n y^{m-1}]y'_x + g(x) + anx^{n-1}y^m = 0.$

Solution: $\displaystyle\int f(y)\,dy + \int g(x)\,dx + ax^n y^m = C.$

19. $f(x,y)y'_x + g(x,y) = 0,$ where $\dfrac{\partial f}{\partial x} = \dfrac{\partial g}{\partial y}.$

Total differential equation.

Solution:

$$\int_{y_0}^{y} f(x_0,t)\,dt + \int_{x_0}^{x} g(t,y)\,dt = C,$$

where x_0 and y_0 are arbitrary numbers.

1.7.2. Equations Containing Exponential and Hyperbolic Functions

1. $y'_x = e^{-\lambda x} f(e^{\lambda x} y).$

The substitution $u = e^{\lambda x} y$ leads to an equation of the form 1.1.2: $u'_x = f(u) + \lambda u$.

2. $y'_x = e^{\lambda y} f(e^{\lambda y} x).$

The substitution $u = e^{\lambda y} x$ leads to an equation with separated variables: $xu'_x = \lambda u^2 f(u) + u$.

3. $y'_x = y f(e^{\alpha x} y^m).$

Exponential homogeneous equation.

The substitution $z = e^{\alpha x} y^m$ leads to an equation of the form 1.1.2: $z'_x = \alpha z + mzf(z)$.

4. $y'_x = \dfrac{1}{x} f(x^n e^{\alpha y}).$

Exponential homogeneous equation.

The substitution $z = x^n e^{\alpha y}$ leads to an equation with separated variables: $xz'_x = nz + \alpha z f(z)$.

5. $y'_x = f(x)e^{\lambda y} + g(x).$

The substitution $u = e^{-\lambda y}$ leads to a linear equation: $u'_x = -\lambda g(x)u - \lambda f(x)$.

6. $y'_x = -\dfrac{n}{x} + f(x)g(x^n e^y).$

The substitution $z = x^n e^y$ leads to an equation with separation of variables: $z'_x = f(x)zg(z)$.

7. $y'_x = -\dfrac{\alpha}{m} y + y^k f(x) g(e^{\alpha x} y^m).$

The substitution $z = e^{\alpha x} y^m$ leads to an equation with separation of variables:

$$z'_x = m \exp\Big[\frac{\alpha}{m}(1-k)x\Big] f(x) z^{\frac{k+m-1}{m}} g(z).$$

8. $y'_x = f(x)e^{\lambda y} + g(x) + h(x)e^{-\lambda y}$.

The substitution $u = e^{\lambda y}$ leads to the Riccati equation: $u'_x = \lambda f(x)u^2 + \lambda g(x)u + \lambda h(x)$.

9. $y'_x = e^{\alpha x - \beta y} f(ae^{\alpha x} + be^{\beta y})$.

The substitution $w = ae^{\alpha x} + be^{\beta y}$ leads to an equation with separated variables: $w'_x = e^{\alpha x}[a\alpha + b\beta f(w)]$.

10. $y'_x = f(y + ae^{\lambda x} + b) - a\lambda e^{\lambda x}$.

The substitution $w = y + ae^{\lambda x} + b$ leads to an equation of the form 1.1.2: $w'_x = f(w)$.

11. $y'_x = -\dfrac{n}{\alpha x} + \dfrac{f(x^n e^{\alpha y})}{xy}$.

The substitution $t = x^n e^{\alpha y}$ leads to a linear equation with respect to $y = y(t)$: $\alpha^2 t f(t) y'_t = -ny + \alpha f(t)$.

12. $y'_x = -\dfrac{n}{\alpha x} + \dfrac{f(x^n e^{\alpha y})}{xy^2}$.

The substitution $t = x^n e^{\alpha y}$ leads to the Riccati equation: $\alpha^2 t f(t) y'_t = -ny^2 + \alpha f(t)$.

13. $[f(ax+by) + be^{\alpha y} g(ax+by)]y'_x = h(ax+by) - ae^{\alpha y} g(ax+by)$.

The transformation $t = ax+by$, $z = e^{-\alpha y}$ leads to a linear equation: $[af(t) + bh(t)]z'_t = -\alpha h(t)z + \alpha a g(t)$.

14. $[f(ax+by) + be^{\alpha x} g(ax+by)]y'_x = h(ax+by) - ae^{\alpha x} g(ax+by)$.

The transformation $t = ax+by$, $z = e^{-\alpha x}$ leads to a linear equation: $[af(t) + bh(t)]z'_t = -\alpha f(t)z - \alpha b g(t)$.

15. $[e^{\alpha x} f(y) + a\beta]y'_x + e^{\beta y} g(x) + a\alpha = 0$.

Solution:

$$\int e^{-\beta y} f(y)\,dy + \int e^{-\alpha x} g(x)\,dx - ae^{-\alpha x - \beta y} = C.$$

16. $x[f(x^n e^{\alpha y}) + \alpha y g(x^n e^{\alpha y})]y'_x = h(x^n e^{\alpha y}) - nyg(x^n e^{\alpha y})$.

The substitution $t = x^n e^{\alpha y}$ leads to a linear equation with respect to $y = y(t)$: $t[nf(t) + \alpha h(t)]y'_t = -ng(t)y + h(t)$.

17. $[f(e^{\alpha x} y^m) + mxg(e^{\alpha x} y^m)]y'_x = y[h(e^{\alpha x} y^m) - \alpha x g(e^{\alpha x} y^m)]$.

The substitution $t = e^{\alpha x} y^m$ leads to a linear equation with respect to $x = x(t)$: $t[\alpha f(t) + mh(t)]x'_t = mg(t)x + f(t)$.

18. $y'_x = f(x)\sinh(\lambda y) + g(x)\cosh(\lambda y) + h(x)$.

The substitution $u = e^{\lambda y}$ leads to the Riccati equation: $2u'_x = \lambda(f+g)u^2 + 2\lambda hu + \lambda(g-f)$.

19. $y'_x = f(x)\sinh^2(\lambda y) + g(x)\cosh^2(\lambda y) + h(x)\sinh(2\lambda y) + s(x).$

The substitution $w = \tanh(\lambda y)$ leads to the Riccati equation: $w'_x = \lambda(f + s)w^2 + 2\lambda hw + \lambda(g - s)$.

20. $y'_x = y\coth x\, f(y^m \sinh x).$

The transformation $t = \sinh x$, $z = y^m$ leads to an equation of the form 1.7.1.3: $tz'_t = mzf(tz)$.

21. $y'_x = x^{-1}\tanh y\, f(x^n \sinh y).$

The transformation $t = x^n$, $z = \sinh y$ leads to an equation of the form 1.7.1.3: $ntz'_t = zf(tz)$.

22. $y'_x = y\tanh x\, f(y^m \cosh x).$

The substitution $t = \cosh x$ leads to an equation of the form 1.7.1.3: $ty'_t = yf(ty^m)$.

23. $y'_x = x^{-1}\coth y\, f(x^n \cosh y).$

The substitution $z = \cosh y$ leads to an equation of the form 1.7.1.3: $xz'_x = zf(x^n z)$.

1.7.3. Equations Containing Logarithmic Functions

1. $y'_x = f(x)y\ln^2 y + g(x)y\ln y + h(x)y.$

The substitution $u = \ln y$ leads to the Riccati equation: $u'_x = f(x)u^2 + g(x)u + h(x)$.

2. $y'_x = x^{-1}y^{m+1}f(y^m \ln x).$

The substitution $t = \ln x$ leads to an equation of the form 1.7.1.3: $y'_t = \dfrac{y}{t}[ty^m f(ty^m)]$.

3. $y'_x = x^{-n-1}yf(x^n \ln y).$

The substitution $z = \ln y$ leads to an equation of the form 1.7.1.3: $z'_x = \dfrac{z}{x}\dfrac{f(x^n z)}{x^n z}$.

4. $y'_x = x^{-1}e^y f(e^y \ln x).$

The substitution $t = \ln x$ leads to an equation of the form 1.7.2.4: $y'_t = \dfrac{1}{t}[te^y f(te^y)]$.

5. $y'_x = ye^{-x}f(e^x \ln y).$

The substitution $z = \ln y$ leads to an equation of the form 1.7.2.3: $z'_x = z\dfrac{f(e^x z)}{e^x z}$.

6. $y'_x = -nx^{-1}y\ln y + yf(x)g(x^n \ln y).$

The substitution $w(x) = x^n \ln y$ leads to an equation with separation of variables: $w'_x = x^n f(x)g(w)$.

7. $y'_x = -\dfrac{n}{m}\dfrac{y}{x} + \dfrac{yf(x^n y^m)}{x \ln y}$.

The transformation $t = x^n y^m$, $z = \ln y$ leads to a linear equation: $m^2 t f(t) z'_t = -nz + mf(t)$.

8. $y'_x = -\dfrac{n}{m}\dfrac{y}{x} + \dfrac{yf(x^n y^m)}{x (\ln y)^2}$.

The transformation $t = x^n y^m$, $z = \ln y$ leads to the Riccati equation: $m^2 t f(t) z'_t = -nz^2 + mf(t)$.

9. $x[f(x^n y^m) + m \ln y\, g(x^n y^m)]y'_x = y[h(x^n y^m) - n \ln y\, g(x^n y^m)]$.

The transformation $t = x^n y^m$, $z = \ln y$ leads to a linear equation: $t[nf(t) + mh(t)]z'_t = -ng(t)z + h(t)$.

10. $x[f(x^n y^m) + m \ln x\, g(x^n y^m)]y'_x = y[h(x^n y^m) - n \ln x\, g(x^n y^m)]$.

The transformation $t = x^n y^m$, $z = \ln x$ leads to a linear equation: $t[nf(t) + mh(t)]z'_t = mg(t)z + f(t)$.

1.7.4. Equations Containing Trigonometric Functions

1. $y'_x = y^{m+1} \sin x\, F(y^m \cos x)$.

This is an equation of the form 1.7.4.3 with $f(\xi) = \xi F(\xi)$.

2. $y'_x = y^{m+1} \cos x\, F(y^m \sin x)$.

This is an equation of the form 1.7.4.4 with $f(\xi) = \xi F(\xi)$.

3. $y'_x = y \tan x\, f(y^m \cos x)$.

The substitution $t = \cos x$ leads to an equation of the form 1.7.1.3: $y'_t = -\dfrac{y}{t} f(ty^m)$.

4. $y'_x = y \cot x\, f(y^m \sin x)$.

The substitution $t = \sin x$ leads to an equation of the form 1.7.1.3: $y'_t = \dfrac{y}{t} f(ty^m)$.

5. $y'_x = x^{-1} \tan y\, f(x^n \sin y)$.

The transformation $t = x^n$, $z = \sin y$ leads to an equation of the form 1.7.1.3: $ntz'_t = zf(tz)$.

6. $y'_x = x^{-1} \cot y\, f(x^n \cos y)$.

The transformation $t = x^n$, $z = \cos y$ leads to an equation of the form 1.7.1.3: $ntz'_t = -zf(tz)$.

7. $y'_x = x^{-1} \sin 2y\, f(x^n \tan y)$.

The transformation $t = x^n$, $z = \tan y$ leads to an equation of the form 1.7.1.3: $ntz'_t = 2zf(tz)$.

8. $y'_x = x^{-1} \sin 2y\, f(x^n \cot y)$.

The transformation $t = x^n$, $z = \cot y$ leads to an equation of the form 1.7.1.3: $ntz'_t = -2zf(tz)$.

9. $y'_x = \dfrac{y}{\sin 2x} f(y^m \tan x)$.

The substitution $t = \tan x$ leads to an equation of the form 1.7.1.3: $2ty'_t = yf(ty^m)$.

10. $y'_x = \dfrac{y}{\sin 2x} f(y^m \cot x)$.

The substitution $t \doteq \cot x$ leads to an equation of the form 1.7.1.3: $2ty'_t = -yf(ty^m)$.

11. $y'_x = f(x)\cos(ay) + g(x)\sin(ay) + h(x)$.

The substitution $u = \tan(ay/2)$ leads to the Riccati equation: $2u'_x = a(h - f)u^2 + 2agu + a(f + h)$.

12. $y'_x = f(x)\cos^2(ay) + g(x)\sin^2(ay) + h(x)\sin(2ay) + s(x)$.

The substitution $u = \tan(ay)$ leads to the Riccati equation: $u'_x = a(g+s)u^2 + 2ahu + a(f + s)$.

13. $y'_x = f(y + a\tan x) - a\tan^2 x$.

The substitution $u = y + a\tan x$ leads to an equation of the form 1.1.2: $u'_x = a + f(u)$.

14. $y'_x = \dfrac{\sin 2y}{\sin 2x} f(\tan x \tan y)$.

The transformation $t = \tan x$, $z = \tan y$ leads to an equation of the form 1.7.1.3: $tz'_t = zf(tz)$.

15. $y'_x = \cot x \tan y\, f(\sin x \sin y)$.

The transformation $t = \sin x$, $z = \sin y$ leads to an equation of the form 1.7.1.3: $tz'_t = zf(tz)$.

16. $y'_x = -\cot x \tan y + \dfrac{f(x)}{\cos y} g(\sin x \sin y)$.

The substitution $w(x) = \sin x \sin y$ leads to an equation with separated variables: $w'_x = \sin x\, f(x)g(w)$.

17. $y'_x = -\dfrac{\sin 2y}{\sin 2x} + \cos^2 y\, f(x)g(\tan x \tan y)$.

The substitution $w(x) = \tan x \tan y$ leads to an equation with separated variables: $w'_x = \tan x\, f(x)g(w)$.

18. $y'_x = -nx^{-1}\sin 2y + \cos^2 y\, f(x)g(x^{2n}\tan y)$.

The substitution $w(x) = x^{2n}\tan y$ leads to an equation with separated variables: $w'_x = x^{2n} f(x)g(w)$.

19. $(1 + \tan^2 y)y'_x = f(x)\tan^{m+1} y + g(x)\tan y + h(x)\tan^{1-m} y$.

The substitution $u = \tan^m y$ leads to the Riccati equation: $u'_x = mf(x)u^2 + mg(x)u + mh(x)$.

1.7.5. Equations Containing Combinations of Exponential, Logarithmic, and Trigonometric Functions

1. $y'_x = -\sin 2y + \cos^2 y\, f(x) g(e^{2x} \tan y)$.

The substitution $w(x) = e^{2x} \tan y$ leads to the equation with separated variables: $w'_x = e^{2x} f(x) g(w)$.

2. $y'_x = \dfrac{F(e^x \cos y)}{e^x \sin y}$.

This is an equation of the type 1.7.5.5 with $f(\xi) = F(\xi)/\xi$.

3. $y'_x = e^y \cos x\, F(e^y \sin x)$.

This is an equation of the type 1.7.5.7 with $f(\xi) = \xi F(\xi)$

4. $y'_x = \tan y\, f(e^x \sin y)$.

The substitution $z = \sin y$ leads to an equation of the form 1.7.2.3: $z'_x = z f(e^x z)$.

5. $y'_x = \cot y\, f(e^x \cos y)$.

The substitution $z = \cos y$ leads to an equation of the form 1.7.2.3: $z'_x = -z f(e^x z)$.

6. $y'_x = \tan x\, f(e^y \cos x)$.

The substitution $t = \cos x$ leads to an equation of the form 1.7.2.4: $t y'_t = -f(t e^y)$.

7. $y'_x = \cot x\, f(e^y \sin x)$.

The substitution $t = \sin x$ leads to an equation of the form 1.7.2.4: $t y'_t = f(t e^y)$.

8. $y'_x = \sin 2y\, f(e^x \tan y)$.

The substitution $z = \tan x$ leads to an equation of the form 1.7.2.3: $z'_x = 2z f(e^x z)$.

9. $y'_x = \sin 2y\, f(e^x \cot y)$.

The substitution $z = \cot x$ leads to an equation of the form 1.7.2.3: $z'_x = -2z f(e^x z)$.

10. $y'_x = \dfrac{F(e^x \sin y)}{e^x \cos y}$.

This is an equation of the form 1.7.5.4 with $f(\xi) = F(\xi)/\xi$.

11. $y'_x = e^y \sin x\, F(e^y \cos x)$.

This is an equation of the from 1.7.5.6 with $f(\xi) = \xi F(\xi)$.

12. $y'_x = \dfrac{f(e^y \tan x)}{\sin 2x}$.

The substitution $t = \tan x$ leads to an equation of the form 1.7.2.4: $2t y'_t = f(t e^y)$.

13. $y'_x = \dfrac{f(e^y \cot x)}{\sin 2x}$.

The substitution $t = \cot x$ leads to an equation of the form 1.7.2.4: $2ty'_t = -f(te^y)$.

14. $y'_x = e^{-\lambda x} f(\lambda x + \ln y)$.

The substitution $u = \lambda x + \ln y$ leads to an equation of the form 1.1.2: $u'_x = e^{-u} f(u) + \lambda$.

15. $y'_x = e^{\lambda y} f(\lambda y + \ln x)$.

The substitution $u = \lambda y + \ln x$ leads to the equation with separated variables: $xu'_x = \lambda e^u f(u) + 1$.

1.8. Equations of the Form $F(x, y, y'_x) = 0$ Not Solved for the Derivative and Containing Arbitrary Functions

1.8.1. Some Equations

1. $x = f(y'_x)$.

The solution in the parametric form:

$$x = f(t), \qquad y = \int t f'_t(t)\,dt + C.$$

2. $y = f(y'_x)$.

The solution in the parametric form:

$$x = \int f'_t(t)\frac{dt}{t} + C, \qquad y = f(t).$$

3. $x^n y^m = f\left(\dfrac{xy'_x}{y}\right)$.

This is an equation of the form 1.8.2.3. Change to a new variable $w(x) = xy'_x/y$; divide both sides of the equation by $x^n y^m$ and differentiate with respect to x. As a result we arrive at an equation with separation of variables: $xf'_w(w)w'_x = (mw + n)f(w)$.

The solution is written in the parametric form:

$$\ln|x| = \int \frac{f'_w(w)\,dw}{(mw + n)f(w)} + C, \qquad x^n y^m = f(w).$$

In addition, there are solutions $y = A_k x^{-n/m}$, where A_k are roots of the equation $A_k^m - f(-n/m) = 0$.

4. $x^n e^{\alpha y} = f(xy'_x)$.

The substitution $y = \ln u$ leads to an equation of the form 1.8.1.3: $x^n u^\alpha = f(xu'_x/u)$.

5. $e^{\alpha x} y^n = f(y'_x/y)$.

The substitution $x = \ln t$ leads to an equation of the form 1.8.1.3: $t^\alpha y^n = f(ty'_t/y)$.

6. $y = xy'_x + f(y'_x)$.

The Clerot equation.

Solution: $y = Cx + f(C)$.

In addition, there is a particular solution which may be written in the parametric form as

$$x = -f'_t(t), \qquad y = -tf'_t(t) + f(t).$$

7. $y = xf(y'_x) + g(y'_x)$.

The Lagrange—d'Alembert equation.

With $f(t) = t$, see equation 1.8.1.6. Having differentiated with respect to x, we arrive at a linear equation with respect to $x = x(t)$, where $t = y'_x$:

$$[t - f(t)]x'_t = f'_t(t)x + g'_t(t).$$

8. $y = x^n f(y'_x) + xy'_x$.

Differentiating with respect to x and denoting $t = y'_x$, we obtain the Bernoulli equation for $x = x(t)$:

$$x'_t - \frac{1}{n}\frac{f'_t(t)}{f(t)}x - \frac{1}{nf(t)}x^{2-n} = 0.$$

9. $y = f\big(x(y'_x)^2\big) + 2xy'_x$.

Solution: $[y - f(C)]^2 = 4Cx$.

10. $y = f\big(x(y'_x)^n\big) + \dfrac{n}{n-1}xy'_x$.

Solution: $y = f(C^n) + \dfrac{nC}{n-1}x^{\frac{n-1}{n}}$.

11. $(xy'_x - y)^n f(y'_x) + yg(y'_x) + xh(y'_x) = 0$.

With the aid the Legendre transformation $x = u'_t$, $y = tu'_t - u$ ($y'_x = t$), we obtain the Bernoulli equation:

$$[tg(t) + h(t)]u'_t = g(t)u + f(t)u^n.$$

12. $(y'_x)^2 + [f(x) + g(x)]y'_x + f(x)g(x) = 0$.

The equation can be factorized: $[y'_x + f(x)][y'_x + g(x)] = 0$, i.e., it falls into two simpler equations $y'_x + f(x) = 0$ and $y'_x + g(x) = 0$. Therefore, the solutions are

$$y + \int f(x)\,dx = C \quad \text{and} \quad y + \int g(x)\,dx = C.$$

13. $(y'_x)^2 + 2fyy'_x + gy^2 = (g - f^2)\exp\left(-2\displaystyle\int_a^x f\,dx\right)$, $\qquad f = f(x),\ g = g(x)$.

Solution:

$$y = \begin{cases} \exp\left(-\displaystyle\int_a^x f\,dx\right)\sin\left(\displaystyle\int_a^x \sqrt{g - f^2}\,dx + C\right) & \text{if } g > f^2, \\ C\exp\left(-\displaystyle\int_a^x f\,dx\right) & \text{if } g \equiv f^2, \\ \exp\left(-\displaystyle\int_a^x f\,dx\right)\cosh\left(\displaystyle\int_a^x \sqrt{f^2 - g}\,dx + C\right) & \text{if } g < f^2. \end{cases}$$

14. $f(y'_x) + ax + by + s = 0.$

Solution in the parametric form:

$$x = C - \int \frac{f'_t(t)\,dt}{a + bt}, \qquad by = -ax - s - f(t).$$

In addition, there is a particular solution $y = \alpha x + \beta$, where α and β are determined by solving the system of two algebraic equations $a + b\alpha = 0$ and $f(\alpha) + b\beta + s = 0$.

15. $f(yy'_x + x) = y^2[(y'_x)^2 + 1].$

Setting $u(x) = yy'_x + x$ and differentiating with respect to x, we obtain

$$u'_x[f'_u(u) - 2u + 2x] = 0. \tag{1}$$

Equating the first factor to zero, after integrating we find $y^2 = -(x - C)^2 + B$. Substituting the latter into the original equation, we have $B = f(C)$. As a result we obtain the solution: $y^2 = f(C) - (x - C)^2$.

There is also an exceptional solution that corresponds to equating the second factor of (1) to zero. The solution in the parametric form is written as

$$x = u - \tfrac{1}{2} f'_u(u), \qquad y^2 = f(u) - \tfrac{1}{4}[f'_u(u)]^2.$$

16. $y = a(y'_x)^2 + f(x - 2ay'_x).$

This is a special case of equation 1.8.1.18 with $n = 2$.

Solution:

$$y = f(C) + \frac{1}{4a}(x - C)^2.$$

In addition, there is the following solution written in the parametric form:

$$x = t + 2af'_t(t), \qquad y = f(t) + a\bigl[f'_t(t)\bigr]^2.$$

17. $y = 2a(y'_x)^3 + f\bigl(x - 3a(y'_x)^2\bigr).$

This is a special case of equation 1.8.1.18 with $n = 3$.

Solution:

$$y = f(C) + 2a\Bigl(\frac{x - C}{3a}\Bigr)^{3/2}.$$

In addition, there is the following solution written in the parametric form:

$$x = t + 3a\bigl[f'_t(t)\bigr]^2, \qquad y = f(t) + 2a\bigl[f'_t(t)\bigr]^3.$$

18. $y = a(n - 1)(y'_x)^n + f\bigl(x - an(y'_x)^{n-1}\bigr).$

Differentiating with respect to x, we obtain a factorized equation:

$$\bigl[1 - an(n - 1)(y'_x)^{n-2} y''_{xx}\bigr]\bigl[y'_x - f'_t(t)\bigr] = 0, \tag{1}$$

where $t = x - an(y'_x)^{n-1}$. Equate the first factor to zero and integrate the obtained equation. Substituting the expression obtained into the original equation, we find the solution:

$$y = f(C) + a(n - 1)\Bigl(\frac{x - C}{an}\Bigr)^{\frac{n}{n-1}}.$$

Equating the second factor in (1) to zero, we have another solution which can be written in the parametric form as

$$x = t + an\bigl[f'_t(t)\bigr]^{n-1}, \qquad y = f(t) + a(n - 1)\bigl[f'_t(t)\bigr]^n.$$

19. $f(x^2 + y^2)\sqrt{(y'_x)^2 + 1} = xy'_x - y.$

Setting $x = r(t)\cos t$, $y = r(t)\sin t$ and integrating, we obtain the solution:

$$t = \int \frac{f(r^2)\,dr}{r\sqrt{r^2 - f^2(r^2)}} + C.$$

20. $\Phi\big(f_x + f_y y'_x,\ f - x(f_x + f_y y'_x)\big) = 0, \qquad f = f(x,y),\ f_x = \frac{\partial f}{\partial x},\ f_y = \frac{\partial f}{\partial y}.$

Differentiating with respect to x, we obtain

$$(f_x + f_y y'_x)'_x(\Phi_u - x\Phi_v) = 0,$$

where $\Phi_u = \dfrac{\partial \Phi}{\partial u}$ and $\Phi_v = \dfrac{\partial \Phi}{\partial v}$ are partial derivatives of function $\Phi(u,v)$. Equating the first factor to zero, we find the solution:

$$f(x,y) = Cx + A, \qquad \text{where} \quad \Phi(C, A) = 0.$$

It remains to be checked whether the equation $\Phi_u - x\Phi_v = 0$ possesses any solutions and which of them satisfy the original equation.

1.8.2. Some Transformations

1. $x = f(y, y'_x).$

Substituting $t = y'_x$ and differentiating both sides of the equation with respect to x, we obtain an equation with respect to $y = y(t)$:

$$[1 - t f_y(y,\ t)]y'_t = t f_t(y,\ t), \qquad \text{where} \quad f_t = \frac{\partial f}{\partial t}, \quad f_y = \frac{\partial f}{\partial y}.$$

If $y = y(t)$ is the solution of the latter equation, the solution of the original equation may be presented in the parametric form as

$$x = f(y(t),\ t), \qquad y = y(t).$$

2. $y = f(x, y'_x).$

Differentiating with respect to x and setting $t = y'_x$, we obtain an equation with respect to $x = x(t)$:

$$[t - f_x(x,\ t)]x'_t = f_t(x,\ t), \qquad \text{where} \quad f_t = \frac{\partial f}{\partial t}, \quad f_x = \frac{\partial f}{\partial x}.$$

If $x = x(t)$ is the solution of the latter equation, the solution of the original equation may be presented in the parametric form as

$$x = x(t), \qquad y = f\big(x(t),\ t\big).$$

3. $\displaystyle x^n y^m = f\left(x^k y^s, \frac{xy'_x}{y}\right).$

Set $z = x^k y^s$ and $w = \dfrac{xy'_x}{y}$. Divide both sides of the equation by $x^n y^m$ and differentiate with respect to x. As a result we arrive at the following equation with respect to $w = w(z)$:

$$z(sw+k)(f_z + f_w w'_z) = (mw+n)f, \qquad \text{where} \quad f = f(z,\, w),$$

which is usually is simpler than the original equation, since it is readily solved for the derivative. If $w = w(z)$ is the solution of the equation obtained, the solution of the original equation is written in the parametric form as

$$x^k y^s = z, \qquad x^n y^m = f\bigl(z,\, w(z)\bigr).$$

4. $x^n e^{\alpha y} = f(x^m e^{\beta y},\, xy'_x).$

The substitution $y = \ln u$ leads to an equation of the form 1.8.2.3:

$$x^n u^\alpha = f\left(x^m u^\beta, \frac{xu'_x}{u}\right).$$

5. $e^{\alpha x} y^n = f(e^{\beta x} y^m,\, y'_x/y).$

The substitution $x = \ln t$ leads to an equation of the form 1.8.2.3:

$$t^\alpha y^n = f\left(t^\beta y^m, \frac{ty'_t}{y}\right).$$

6. $f(x,\, xy'_x - y,\, y'_x) = 0.$

The Legendre transformation $x = u'_t$, $y = tu'_t - u$ $(y'_x = t)$, where $u = u(t)$, leads to the equation $f(u'_t,\, u,\, t) = 0$. The inverse transformation: $t = y'_x$, $u = xy'_x - y$, $u'_t = x$.

7. $(y'_x)^2 = \lambda y + f(x).$

With $\lambda \neq 0$, the substitution $\lambda w = 2\sqrt{\lambda y + f(x)}$ leads to the Abel equation of the second kind:

$$ww'_x = w + \varphi(x), \qquad \text{where} \quad \varphi = 2\lambda^{-2} f'_x(x),$$

which is outlined in Subsection 1.3.1 for specific functions φ.

8. $y = xy'_x + ax^2 + f(y'_x), \qquad a \neq 0.$

Differentiating the equation with respect to x and changing to new variables $t = y'_x$ and $w = -2ax$, we arrive at the Abel equation of the second kind:

$$ww'_t = w + \varphi(t), \qquad \text{where} \quad \varphi = -2af'_t(t),$$

which is outlined in Subsection 1.3.1 for specific functions φ.

Chapter 2
Second Order Differential Equations

2.1. Linear Equations

2.1.1. Preliminary Comments

1. A homogeneous linear equation of the second order has the form

$$f_2(x)y''_{xx} + f_1(x)y'_x + f_0(x)y = 0. \qquad (1)$$

Let $y_0 = y_0(x)$ be a nontrivial particular solution ($y_0 \not\equiv 0$) of this equation. Then the general solution of equation (1) can be found from the formula

$$y = y_0\Big(C_1 + C_2\int \frac{e^{-F}}{y_0^2}\,dx\Big), \qquad \text{where} \quad F = \int \frac{f_1}{f_2}\,dx. \qquad (2)$$

For specific equations described below in 2.1.2–2.1.8, often only particular solutions are given, while the general solutions can be obtained with formula (2).

2. The substitution $u = y'_x/y$ brings equation (1) to the Riccati equation

$$f_2(x)u'_x + f_2(x)u^2 + f_1(x)u + f_0(x) = 0$$

which is discussed in Section 1.2.

3. Assuming

$$y = u(x)\exp\Big(-\frac{1}{2}\int \frac{f_1}{f_2}\,dx\Big) \qquad (3)$$

yields from equation (1) the canonical (or normal) form

$$u''_{xx} + f(x)u = 0, \qquad \text{where} \quad f = \frac{f_0}{f_2} - \frac{1}{4}\Big(\frac{f_1}{f_2}\Big)^2 - \frac{1}{2}\Big(\frac{f_1}{f_2}\Big)'_x. \qquad (4)$$

Substitution (3) is a special case of the more general transformation (φ is an arbitrary function):

$$x = \varphi(\xi), \quad y(x) = u(\xi)\sqrt{|\varphi'_\xi(\xi)|}\,\exp\Big(-\frac{1}{2}\int \frac{f_1(\varphi)}{f_2(\varphi)}\,d\varphi\Big),$$

which reduces the original equation to the canonical form.

4. A nonhomogeneous linear equation of the second order has the form

$$f_2(x)y''_{xx} + f_1(x)y'_x + f_0(x)y = g(x). \tag{5}$$

Let $y_1 = y_1(x)$ and $y_2 = y_2(x)$ be two nontrivial linearly-independent ($y_1/y_2 \not\equiv \text{const}$) solutions of the corresponding homogeneous equation with $g \equiv 0$. Then the general solution of equation (5) can be found from the formula

$$y = C_1 y_1 + C_2 y_2 + y_2 \int y_1 \frac{g}{f_2}\frac{dx}{W} - y_1 \int y_2 \frac{g}{f_2}\frac{dx}{W}, \tag{6}$$

where $W = y_1(y_2)'_x - y_2(y_1)'_x$.

Given a nontrivial particular solution $y_1 = y_1(x)$ of the homogeneous equation with $g \equiv 0$, formula (6) can be used for the construction of the general solution of equation (5) with the second solution $y_2 = y_2(x)$ taken in the form

$$y_2 = y_1 \int \frac{e^{-F}}{y_1^2}\,dx, \qquad \text{where} \quad F = \int \frac{f_1}{f_2}\,dx, \quad W = e^{-F}, \tag{7}$$

In Subsections 2.1.2–2.1.8, mainly homogeneous equations are given; the corresponding nonhomogeneous equations may be solved by means of the formulae (6) and (7).

2.1.2. Equations Containing Power Functions

1. $\boldsymbol{y''_{xx} + ay = 0.}$

Solution:

$$y = \begin{cases} C_1 \sinh(x\sqrt{|a|}\,) + C_2 \cosh(x\sqrt{|a|}\,) & \text{if } a < 0, \\ C_1 + C_2 x & \text{if } a = 0, \\ C_1 \sin(x\sqrt{a}\,) + C_2 \cos(x\sqrt{a}\,) & \text{if } a > 0. \end{cases}$$

2. $\boldsymbol{y''_{xx} - (ax + b)y = 0, \qquad a \neq 0.}$

The substitution $\xi = a^{-2/3}(ax + b)$ leads to the Airy equation

$$y''_{\xi\xi} - \xi y = 0, \tag{1}$$

which is often met with in various applications. The solution of equation (1) can be written as

$$y = C_1 \operatorname{Ai}(\xi) + C_2 \operatorname{Bi}(\xi),$$

where $\operatorname{Ai}(\xi)$ and $\operatorname{Bi}(\xi)$ are the Airy functions of the first and second kind, respectively.

The Airy functions admits the following integral representation:

$$\operatorname{Ai}(\xi) = \frac{1}{\pi}\int_0^\infty \cos\Bigl(\frac{1}{3}t^3 + \xi t\Bigr)\,dt,$$

$$\operatorname{Bi}(\xi) = \frac{1}{\pi}\int_0^\infty \Bigl[\exp\Bigl(-\frac{1}{3}t^3 + \xi t\Bigr) + \sin\Bigl(\frac{1}{3}t^3 + \xi t\Bigr)\Bigr]\,dt.$$

The Airy functions can be expressed in terms of the Bessel functions and the modified Bessel functions of the order 1/3 with the formulae

$$\mathrm{Ai}(\xi) = \frac{1}{3}\sqrt{\xi}\,\big[I_{-1/3}(z) - I_{1/3}(z)\big] = \frac{1}{\pi}\sqrt{\frac{\xi}{3}}\,K_{1/3}(z),$$
$$\mathrm{Bi}(\xi) = \sqrt{\frac{\xi}{3}}\,\big[I_{-1/3}(z) + I_{1/3}(z)\big],$$
$$\mathrm{Ai}(-\xi) = \frac{1}{3}\sqrt{\xi}\,\big[J_{-1/3}(z) + J_{1/3}(z)\big],$$
$$\mathrm{Bi}(-\xi) = \sqrt{\frac{\xi}{3}}\,\big[J_{-1/3}(z) - J_{1/3}(z)\big],$$

where $z = \frac{2}{3}\xi^{3/2}$.

For large values of ξ, the leading terms of the asymptotic expansions of the Airy functions are

$$\mathrm{Ai}(\xi) = \frac{1}{2\sqrt{\pi}}\xi^{-1/4}\exp(-z),$$
$$\mathrm{Ai}(-\xi) = \frac{1}{\sqrt{\pi}}\xi^{-1/4}\sin\Big(z + \frac{\pi}{4}\Big),$$
$$\mathrm{Bi}(\xi) = \frac{1}{\sqrt{\pi}}\xi^{-1/4}\exp(z),$$
$$\mathrm{Bi}(-\xi) = \frac{1}{\sqrt{\pi}}\xi^{-1/4}\cos\Big(z + \frac{\pi}{4}\Big).$$

The Airy equation (1) is a special case of the equation 2.1.2.7 with $n = 1$.

3. $\boldsymbol{y''_{xx} - (a^2x^2 + a)y = 0.}$

Particular solution: $y_0 = \exp\Big(\dfrac{ax^2}{2}\Big)$.

4. $\boldsymbol{y''_{xx} - (ax^2 + b)y = 0.}$

The transformation $z = x^2\sqrt{a}$, $u = e^{z/2}y$ leads to the degenerate hypergeometric equation 2.1.2.65:

$$zu''_{zz} + \Big(\frac{1}{2} - z\Big)u'_z - \frac{1}{4}\Big(\frac{b}{\sqrt{a}} + 1\Big)u = 0.$$

5. $\boldsymbol{y''_{xx} + a^3x(2 - ax)y = 0.}$

Particular solution: $y_0 = \exp\Big(-\dfrac{a^2x}{2} + ax\Big)$.

6. $\boldsymbol{y''_{xx} - (ax^2 + bx + c)y = 0.}$

The substitution $\xi = x + \dfrac{b}{2a}$ leads to an equation of the form 2.1.2.4:

$$y''_{\xi\xi} - \Big(a\xi^2 + c - \frac{b^2}{4a}\Big)y = 0.$$

7. $y''_{xx} - ax^n y = 0.$

1°. For $n = -2$, this is the Euler equation 2.1.2.118, while for $n = -4$, this is the equation 2.1.2.198 (in both cases the solution is expressed in terms of elementary function).

2°. Assume $2/(n+2) = 2m+1$, where m is an integer. Then the solution is

$$y = x(x^{1-2q}D)^{m+1}\left[C_1 \exp\left(\frac{\sqrt{a}}{q}x^q\right) + C_2 \exp\left(-\frac{\sqrt{a}}{q}x^q\right)\right] \quad \text{if } m \geq 0,$$

$$y = x(x^{1-2q}D)^{-m}\left[C_1 \exp\left(\frac{\sqrt{a}}{q}x^q\right) + C_2 \exp\left(-\frac{\sqrt{a}}{q}x^q\right)\right] \quad \text{if } m < 0,$$

where $D = \dfrac{d}{dx}$, $q = \dfrac{n+2}{2} = \dfrac{1}{2m+1}$.

3°. For any n, the solution is expressed in terms of Bessel functions and modified Bessel functions of the first or second kind (see 2.1.2.121 and 2.1.2.122):

$$y = \sqrt{x}\left[C_1 J_{\frac{1}{2q}}\left(\frac{\sqrt{-a}}{q}x^q\right) + C_2 Y_{\frac{1}{2q}}\left(\frac{\sqrt{-a}}{q}x^q\right)\right], \quad \text{if } a < 0,$$

$$y = \sqrt{x}\left[C_1 I_{\frac{1}{2q}}\left(\frac{\sqrt{a}}{q}x^q\right) + C_2 K_{\frac{1}{2q}}\left(\frac{\sqrt{a}}{q}x^q\right)\right], \quad \text{if } a > 0,$$

where $q = \frac{1}{2}(n+2)$.

8. $y''_{xx} - a(ax^{2n} + nx^{n-1})y = 0.$

Particular solution: $y_0 = \exp\left(\dfrac{a}{n+1}x^{n+1}\right)$.

9. $y''_{xx} - ax^{n-2}(ax^n + n + 1)y = 0.$

Particular solution: $y_0 = x\exp\left(\dfrac{ax^n}{n}\right)$.

10. $y''_{xx} + (ax^{2n} + bx^{n-1})y = 0.$

The substitution $\xi = x^{n+1}$ leads to an equation of the form 2.1.2.103:

$$(n+1)^2 \xi y''_{\xi\xi} + n(n+1)y'_\xi + (a\xi + b)y = 0.$$

11. $y''_{xx} + ay'_x + by = 0.$

The equation of damping oscillation.

1°. Solution with $\lambda^2 = a^2 - 4b > 0$:

$$y = C_1 \exp\left(\frac{\lambda - a}{2}x\right) + C_2 \exp\left(\frac{-\lambda - a}{2}x\right).$$

2°. Solution with $\lambda^2 = 4b - a^2 > 0$:

$$y = \exp\left(-\frac{ax}{2}\right)\left(C_1 \sin\frac{\lambda x}{2} + C_2 \cos\frac{\lambda x}{2}\right).$$

3°. Solution with $a^2 = 4b$:

$$y = \exp\left(-\frac{ax}{2}\right)(C_1 x + C_2).$$

12. $y''_{xx} + ay'_x + (bx + c)y = 0.$

This is a special case of equation 2.1.2.103.

13. $y''_{xx} + ay'_x - (bx^2 + c)y = 0.$

The substitution $y = u\exp\left(\frac{1}{2}x^2\sqrt{b}\right)$ leads to an equation of the form 2.1.2.103: $u''_{xx} + (2\sqrt{b}\,x + a)u'_x + (a\sqrt{b}\,x - c + \sqrt{b}\,)u = 0$.

14. $y''_{xx} + ay'_x + b(-bx^2 + ax + 1)y = 0.$

Particular solution: $y_0 = \exp\left(-\dfrac{bx^2}{2}\right)$.

15. $y''_{xx} + ay'_x + bx(-bx^3 + ax + 2)y = 0.$

Particular solution: $y_0 = \exp\left(-\dfrac{bx^3}{3}\right)$.

16. $y''_{xx} + ay'_x + b(-bx^{2n} + ax^n + nx^{n-1})y = 0.$

Particular solution: $y_0 = \exp\left(-\dfrac{b}{n+1}x^{n+1}\right)$.

17. $y''_{xx} + ay'_x + b(-bx^{2n} - ax^n + nx^{n-1})y = 0.$

Particular solution: $y_0 = \exp\left(-\dfrac{b}{n+1}x^{n+1} - ax\right)$.

18. $y''_{xx} + xy'_x + (n + 1)y = 0, \qquad n = 1,\ 2,\ 3,\ \ldots$

Solution: $y = \dfrac{d^n}{dx^n}\left\{\exp\left(-\dfrac{x^2}{2}\right)\left[C_1 + C_2\displaystyle\int \exp\left(\dfrac{x^2}{2}\right)dx\right]\right\}$.

19. $y''_{xx} - 2xy'_x + 2ny = 0, \qquad n = 1,\ 2,\ 3,\ \ldots$

Solution: $y = \exp(x^2)\dfrac{d^n}{dx^n}\left\{\exp(-x^2)\left[C_1 + C_2\displaystyle\int \exp(x^2)\,dx\right]\right\}$.

20. $y''_{xx} + 2axy'_x + (bx^4 + a^2x^2 + cx + a)y = 0.$

This is a special case of equation 2.1.2.46 with $n = 1$, $m = 2$.

21. $y''_{xx} + (ax + b)y'_x + ay = 0.$

Particular solution: $y_0 = \exp\left(-\dfrac{a}{2}x^2 - bx\right)$.

22. $y''_{xx} + (ax + b)y'_x - ay = 0.$

Particular solution: $y_0 = ax + b$.

23. $y''_{xx} + (ax + b)y'_x + c(ax + b - c)y = 0.$

Particular solution: $y_0 = e^{-cx}$.

24. $y''_{xx} + (ax + 2b)y'_x + (abx - a + b^2)y = 0.$

Particular solution: $y_0 = xe^{-bx}$.

25. $y''_{xx} + (ax + b)y'_x + (cx + d)y = 0.$

This is a special case of equation 2.1.2.103.

26. $y''_{xx} + (ax + b)y'_x + c[(a - c)x^2 + bx + 1]y = 0.$

Particular solution: $y_0 = \exp\Bigl(-\dfrac{cx^2}{2}\Bigr)$.

27. $y''_{xx} + 2(ax + b)y'_x + (a^2x^2 + 2abx + c)y = 0.$

The substitution $u = y\exp\bigl(\frac{1}{2}ax^2 + bx\bigr)$ leads to an equation of the form 2.1.2.1: $u''_{xx} + (c - a - b^2)u = 0$.

28. $y''_{xx} + (ax + b)y'_x + (\alpha x^2 + \beta x + \gamma)y = 0.$

Assuming $y = u\exp(sx^2)$, where s is a root of the quadratic equation $4s^2 + 2as + \alpha = 0$, yields an equation of the form 2.1.2.103:

$$u''_{xx} + [(a + 4s)x + b]u'_x + [(\beta + 2bs)x + \gamma + 2s]u = 0.$$

29. $y''_{xx} + (ax + b)y'_x + c(-cx^{2n} + ax^{n+1} + bx^n + nx^{n-1})y = 0.$

Particular solution: $y_0 = \exp\Bigl(-\dfrac{c}{n+1}x^{n+1}\Bigr)$.

30. $y''_{xx} + a(x^2 - b^2)y'_x - a(x + b)y = 0.$

Particular solution: $y_0 = x - b$.

31. $y''_{xx} + (ax^2 + b)y'_x + c(ax^2 + b - c)y = 0.$

Particular solution: $y_0 = e^{-cx}$.

32. $y''_{xx} + (ax^2 + 2b)y'_x + (abx^2 - ax + b^2)y = 0.$

Particular solution: $y_0 = xe^{-bx}$.

33. $y''_{xx} + (2x^2 + a)y'_x + (x^4 + ax^2 + 2x + b)y = 0.$

The substitution $u = y\exp\bigl(\frac{1}{3}x^3\bigr)$ leads to a constant coefficient equation: $u''_{xx} + au'_x + bu = 0$.

34. $y''_{xx} + (ax^2 + bx)y'_x + (3ax + 2b)y = 0.$

Particular solution: $y_0 = x\exp\bigl(-\frac{1}{3}ax^3 - \frac{1}{2}bx^2\bigr)$.

35. $y''_{xx} + (abx^2 + bx + 2a)y'_x + a^2(bx^2 + 1)y = 0.$

Particular solution: $y_0 = (ax + 1)e^{-ax}$.

36. $y''_{xx} + (ax^2 + bx + c)y'_x + x(abx^2 + bc + 2a)y = 0.$

Particular solution: $y_0 = \exp\Bigl(-\frac{a}{3}x^3 - cx\Bigr).$

37. $y''_{xx} + (ax^2 + bx + c)y'_x + (abx^3 + acx^2 + b)y = 0.$

Particular solution: $y_0 = \exp\Bigl(-\frac{b}{2}x^2 - cx\Bigr).$

38. $y''_{xx} + (ax^3 + 2b)y'_x + (abx^3 - ax^2 + b^2)y = 0.$

Particular solution: $y_0 = xe^{-bx}.$

39. $y''_{xx} + (ax^3 + bx)y'_x + 2(2ax^2 + b)y = 0.$

Particular solution: $y_0 = x\exp\Bigl(-\frac{a}{4}x^4 - \frac{b}{2}x^2\Bigr).$

40. $y''_{xx} + (abx^3 + bx^2 + 2a)y'_x + a^2(bx^3 + 1)y = 0.$

Particular solution: $y_0 = (ax + 1)e^{-ax}.$

41. $y''_{xx} + ax^n y'_x = 0.$

This equation is encountered in the theory of diffusion boundary layer.

Solution: $y = C_1 + C_2 \int \exp\Bigl(-\frac{ax^{n+1}}{n+1}\Bigr)\,dx.$

42. $y''_{xx} + ax^n y'_x + bx^{n-1}y = 0.$

For $n = -1$, we obtain the Euler equation 2.1.2.118. For $n \neq -1$, the substitution $z = x^{n+1}$ leads to an equation of the form 2.1.2.103:

$$(n+1)^2 z y''_{zz} + (n+1)(az + n)y'_z + by = 0.$$

43. $y''_{xx} + 2ax^n y'_x + a(ax^{2n} + nx^{n-1})y = 0.$

Particular solution: $y_0 = x\exp\Bigl(-\frac{a}{n+1}x^{n+1}\Bigr).$

44. $y''_{xx} + ax^n y'_x + (bx^{2n} + cx^{n-1})y = 0.$

The substitution $\xi = x^{n+1}$ leads to an equation of the form 2.1.2.103:

$$(n+1)^2 \xi y''_{\xi\xi} + (n+1)(a\xi + n)y'_\xi + (b\xi + c)y = 0.$$

45. $y''_{xx} + ax^n y'_x - b(ax^{n+m} + bx^{2m} + mx^{m-1})y = 0.$

Particular solution: $y_0 = \exp\Bigl(\frac{b}{m+1}x^{m+1}\Bigr).$

46. $y''_{xx} + 2ax^n y'_x + (a^2x^{2n} + bx^{2m} + anx^{n-1} + cx^{m-1})y = 0.$

The substitution $w = y\exp\Bigl(\frac{a}{n+1}x^{n+1}\Bigr)$ leads to an equation of the form 2.1.2.10: $w''_{xx} + (bx^{2m} + cx^{m-1})w = 0.$

47. $y''_{xx} + (ax^n + b)y'_x + c(ax^n + b - c)y = 0.$

Particular solution: $y_0 = e^{-cx}$.

48. $y''_{xx} + (ax^n + 2b)y'_x + (abx^n - ax^{n-1} + b^2)y = 0.$

Particular solution: $y_0 = xe^{-bx}$.

49. $y''_{xx} + (abx^n + bx^{n-1} + 2a)y'_x + a^2(bx^n + 1)y = 0.$

Particular solution: $y_0 = (ax + 1)e^{-ax}$.

50. $y''_{xx} + (abx^n + 2bx^{n-1} - a^2x)y'_x + a(abx^n + bx^{n-1} - a^2x)y = 0.$

Particular solution: $y_0 = (ax + 2)e^{-ax}$.

51. $y''_{xx} + x^n[ax^2 + (ac + b)x + bc]y'_x - x^n(ax + b)y = 0.$

Particular solution: $y_0 = x + c$.

52. $y''_{xx} + (ax^n + bx^m)y'_x - (ax^{n-1} + bx^{m-1})y = 0.$

Particular solution: $y_0 = x$.

53. $y''_{xx} + (ax^n + bx^m)y'_x + (anx^{n-1} + bmx^{m-1})y = 0.$

Integrating, we obtain a first order linear equation: $y'_x + (ax^n + bx^m)y = C$.

54. $y''_{xx} + (ax^n + bx^m)y'_x + [a(n + 1)x^{n-1} + b(m + 1)x^{m-1}]y = 0.$

Particular solution: $y_0 = x\exp\left(-\frac{a}{n+1}x^{n+1} - \frac{b}{m+1}x^{m+1}\right)$.

55. $y''_{xx} + (ax^n + bx^m)y'_x + c(ax^n + bx^m - c)y = 0.$

Particular solution: $y_0 = e^{-cx}$.

56. $y''_{xx} + (ax^n + bx^m)y'_x + [abx^{m+n} + b(m + 1)x^{m-1} - ax^{n-1}]y = 0.$

Particular solution: $y_0 = x\exp\left(-\frac{b}{m+1}x^{m+1}\right)$.

57. $y''_{xx} + (ax^n + bx^m + c)y'_x + (abx^{m+n} + bcx^m + anx^{n-1})y = 0.$

Particular solution: $y_0 = \exp\left(-\frac{a}{n+1}x^{n+1} - cx\right)$.

58. $xy''_{xx} + \frac{1}{2}y'_x + ay = 0.$

Solution:

$$y = \begin{cases} C_1 \cos\sqrt{4ax} + C_2 \sin\sqrt{4ax} & \text{if } ax > 0, \\ C_1 \cosh\sqrt{4|ax|} + C_2 \sinh\sqrt{4|ax|} & \text{if } ax < 0. \end{cases}$$

59. $xy''_{xx} + ay'_x + (bx + c)y = 0.$

This is a special case of equation 2.1.2.103.

60. $xy''_{xx} + ny'_x + bx^{1-2n}y = 0.$

For $n = 1$, this is the Euler equation 2.1.2.118. For $n \neq 1$, the solution is

$$y = \begin{cases} C_1 \sin\left(\dfrac{\sqrt{b}}{n-1}x^{1-n}\right) + C_2 \cos\left(\dfrac{\sqrt{b}}{n-1}x^{1-n}\right) & \text{if } b > 0, \\ C_1 \exp\left(\dfrac{\sqrt{-b}}{n-1}x^{1-n}\right) + C_2 \exp\left(\dfrac{-\sqrt{-b}}{n-1}x^{1-n}\right) & \text{if } b < 0. \end{cases}$$

61. $xy''_{xx} + (1-3n)y'_x - a^2n^2x^{2n-1}y = 0.$

Particular solution: $y_0 = (ax^n + 1)\exp(-ax^n)$.

62. $xy''_{xx} + ay'_x + bx^n y = 0.$

If $n = -1$ and $b = 0$, we have the Euler equation 2.1.2.118. If $n \neq -1$ and $b \neq 0$, the solution is expressed in terms of Bessel functions:

$$y = x^{\frac{1-a}{2}}\left[C_1 J_\nu\left(\frac{2\sqrt{b}}{n+1}x^{\frac{n+1}{2}}\right) + C_2 Y_\nu\left(\frac{2\sqrt{b}}{n+1}x^{\frac{n+1}{2}}\right)\right], \qquad \nu = \frac{|1-a|}{n+1}.$$

63. $xy''_{xx} + ay'_x + bx^n(-bx^{n+1} + a + n)y = 0.$

Particular solution: $y_0 = \exp\left(-\dfrac{b}{n+1}x^{n+1}\right)$.

64. $xy''_{xx} + axy'_x + ay = 0.$

Particular solution: $y_0 = xe^{-ax}$.

65. $xy''_{xx} + (b-x)y'_x - ay = 0.$

The degenerate hypergeometric equation.

If $b \neq 0, -1, -2, -3, \ldots$, Kummer's series is a particular solution:

$$\Phi(a,b;x) = 1 + \sum_{k=1}^{\infty} \frac{(a)_k}{(b)_k}\frac{x^k}{k!},$$

where $(a)_k = a(a+1)\ldots(a+k-1)$, $(a)_0 = 1$. If $b > a > 0$, this solution may be written in terms of the definite integral

$$\Phi(a,b;x) = \frac{\Gamma(b)}{\Gamma(a)\,\Gamma(b-a)}\int_0^1 e^{xt}t^{a-1}(1-t)^{b-a-1}\,dt,$$

where $\Gamma(z) = \int_0^\infty e^{-t}t^{z-1}\,dt$ is the gamma-function.

If b is not an integer, then the general solution has the form

$$y = C_1\Phi(a,b;x) + C_2x^{1-b}\Phi(a-b+1,\ 2-b;\ x).$$

TABLE 2.1
Special cases of Kummer's function $\Phi(a, b; x)$.

a	b	z	Φ	Conventional notation
a	a	x	e^x	
1	2	$2x$	$\frac{1}{x}e^x \sinh x$	
a	$a+1$	$-x$	$ax^{-a}\gamma(a, x)$	Incomplete gamma-function $\gamma(a, x) = \int_0^x e^{-t}t^{a-1}\,dt$
$\frac{1}{2}$	$\frac{3}{2}$	$-x^2$	$\frac{\sqrt{\pi}}{2}\operatorname{erf} x$	Error function $\operatorname{erf} x = \frac{2}{\sqrt{\pi}}\int_0^x \exp(-t^2)\,dt$
$-n$	$\frac{1}{2}$	$\frac{x^2}{2}$	$\frac{n!}{(2n)!}\left(-\frac{1}{2}\right)^{-n} H_{2n}(x)$	Hermite polynomials $H_n = (-1)^n e^{x^2}\frac{d^n}{dx^n}(e^{-x^2})$, $n = 0, 1, 2, 3, \ldots$
$-n$	$\frac{3}{2}$	$\frac{x^2}{2}$	$\frac{n!}{(2n+1)!}\left(-\frac{1}{2}\right)^{-n} H_{2n+1}(x)$	
$-n$	b	x	$\frac{n!}{(b)_n}L_n^{(b-1)}(x)$	Laguerre polynomials $L_n^{(\alpha)}(x) = \frac{e^x x^{-\alpha}}{n!}\frac{d^n}{dx^n}(e^{-x}x^{n+\alpha})$, $\alpha = b-1$, $(b)_n = b(b+1)\ldots(b+n-1)$
$\nu+\frac{1}{2}$	$2\nu+1$	$2x$	$\Gamma(1+\nu)e^x\left(\frac{x}{2}\right)^{-\nu} I_\nu(x)$	Modified Bessel functions $I_\nu(x)$
$n+1$	$2n+2$	$2x$	$\Gamma\left(n+\frac{3}{2}\right)e^x\left(\frac{x}{2}\right)^{-n-\frac{1}{2}} I_{n+\frac{1}{2}}(x)$	

In Table 2.1 are given some special cases where Φ is expressed in terms of simpler functions.

Function Φ possesses the properties

$$\Phi(a, b; x) = e^x\Phi(b-a,\ b;\ -x); \qquad \frac{d^n}{dx^n}\Phi(a, b; x) = \frac{(a)_n}{(b)_n}\Phi(a+n,\ b+n;\ x).$$

The following asymptotic relations hold:

$$\Phi(a, b; x) \to \frac{\Gamma(b)}{\Gamma(a)}e^x x^{a-b}\left[1 + O\left(\frac{1}{|x|}\right)\right], \qquad \text{as } x \to +\infty,$$

$$\Phi(a, b; x) \to \frac{\Gamma(b)}{\Gamma(b-a)}(-x)^{-a}\left[1 + O\left(\frac{1}{|x|}\right)\right], \qquad \text{as } x \to -\infty.$$

The following function is a solution of the degenerate hypergeometric equation:

$$\Psi(a, b; x) = \frac{\Gamma(1-b)}{\Gamma(a-b+1)}\Phi(a, b; x) + \frac{\Gamma(b-1)}{\Gamma(a)}x^{1-b}\Phi(a-b+1,\ 2-b;\ x).$$

Calculate the limit as $b \to n$ (n is an integer) to obtain

$$\Psi(a,n;x) = \frac{(-1)^{n-1}}{n!\,\Gamma(a-n)}\biggl\{\Phi(a,n{+}1;x)\ln x + \sum_{r=0}^{\infty}\frac{(a)_r}{(n+1)_r}\bigl[\psi(a+r)-\psi(1+r)-\psi(1+n+r)\bigr]\frac{x^r}{r!}\biggr\} + \frac{(n-1)!}{\Gamma(a)}\sum_{r=0}^{n-1}\frac{(a-n)_r}{(1-n)_r}\frac{x^{r-n}}{r!},$$

where $n = 0,\ 1,\ 2,\ \dots$ (the last sum is omitted for $n=0$), $\psi(z) = [\ln\Gamma(z)]'_z$ is the logarithmic derivative of the gamma-function:

$$\psi(1) = -\gamma, \quad \psi(n) = -\gamma + \sum_{k=1}^{n-1} k^{-1},$$

$\gamma = 0.5572\dots$ is the Euler constant.

If b is a negative number, then function Ψ may be presented with the formula

$$\Psi(a,b;x) = x^{1-b}\Psi(a-b+1,\ 2-b;\ x)$$

which holds for any value of x.

For $b \neq 0,\ -1,\ -2,\ -3,\ \dots$, the general solution of the degenerate hypergeometric equation may be written in the form

$$y = C_1\Phi(a,b;x) + C_2\Psi(a,b;x),$$

while for $b = 0,\ -1,\ -2,\ -3,\ \dots$, it may be written as

$$y = x^{1-b}\bigl[C_1\Phi(a-b+1,\ 2-b;\ x) + C_2\Psi(a-b+1,\ 2-b;\ x)\bigr].$$

The functions Φ and Ψ are described in the books by Abramowitz & Stegun (1964) and Bateman & Erdélyi (1953, vol. 1) in more detail (see also Supplement 2).

66. $\boldsymbol{xy''_{xx} + (ax+b)y'_x + c[(a-c)x+b]y = 0.}$

Particular solution: $y_0 = e^{-cx}$.

67. $\boldsymbol{xy''_{xx} + (2ax+b)y'_x + a(ax+b)y = 0.}$

Solution: $y = e^{-ax}(C_1 + C_2x^{1-b})$.

68. $\boldsymbol{xy''_{xx} + [(a+b)x+n+m]y'_x + (abx+an+bm)y = 0,}$
where n and m are positive integers; $a \neq b$ or $n \neq m$.

Solution:

$$y = C_1e^{-ax}\frac{d^{m-1}}{dx^{m-1}}x^{-n}e^{(a-b)x} + C_2e^{-bx}\frac{d^{n-1}}{dx^{n-1}}x^{-m}e^{(b-a)x}.$$

69. $\boldsymbol{xy''_{xx} + (ax+b)y'_x + (cx+d)y = 0.}$

This is a special case of equation 2.1.2.103.

70. $xy''_{xx} - (ax+1)y'_x - bx^2(bx+a)y = 0.$

Particular solution: $y_0 = \exp\Bigl(-\dfrac{bx^2}{2}\Bigr).$

71. $xy''_{xx} - (2ax+1)y'_x + (bx^3 + a^2x + a)y = 0.$

Solution: $y = e^{ax}\Bigl[C_1\sin\Bigl(\dfrac{x^2}{2}\sqrt{b}\Bigr) + C_2\cos\Bigl(\dfrac{x^2}{2}\sqrt{b}\Bigr)\Bigr].$

72. $xy''_{xx} + (ax+b)y'_x + cx(-cx^2 + ax + b + 1) = 0.$

Particular solution: $y_0 = \exp\Bigl(-\dfrac{cx^2}{2}\Bigr).$

73. $xy''_{xx} - (2ax^2+1)y'_x + bx^3y = 0.$

Solution: $y = C_1\exp\bigl[\frac{1}{2}\bigl(a+\sqrt{a^2-b}\bigr)x^2\bigr] + C_2\exp\bigl[\frac{1}{2}\bigl(a-\sqrt{a^2-b}\bigr)x^2\bigr].$

74. $xy''_{xx} + (abx^2 + b - 5)y'_x + 2a^2(b-2)x^3y = 0.$

Particular solution: $y_0 = (ax^2+1)\exp(-ax^2).$

75. $xy''_{xx} + (ax^2+bx)y'_x - [acx^2 + (a+bc+c^2)x + b + 2c]y = 0.$

Particular solution: $y_0 = xe^{cx}.$

76. $xy''_{xx} + (ax^2+bx+2)y'_x + by = 0.$

Particular solution: $y_0 = a + \dfrac{b}{x}.$

77. $xy''_{xx} + (ax^2+bx+c)y'_x + (2ax+b)y = 0.$

This equation can be integrated to obtain the first order linear equation: $xy'_x + (ax^2+bx+c-1)y = C.$

78. $xy''_{xx} + (ax^2+bx+c)y'_x + (c-1)(ax+b)y = 0.$

Particular solution: $y_0 = x^{1-c}.$

79. $xy''_{xx} + (ax^2+bx+c)y'_x + (Ax^2+Bx+C)y = 0.$

1°. Let $A = ak$, $B = k(b-k)$, $C = ck$, where k is an arbitrary number.
Particular solution: $y_0 = e^{-kx}.$

2°. Let $A = a(b+k)$, $B = a(c+1) - k(b+k)$, $C = -ck$.
Particular solution: $y_0 = \exp\Bigl(-\dfrac{ax^2}{2} + kx\Bigr).$

3°. Let $A = a(b+k)$, $B = 2a - bk - k^2$, $C = b(c-1) + k(c-2)$.
Particular solution: $y_0 = x^{1-c}\exp\Bigl(-\dfrac{ax^2}{2} + kx\Bigr).$

4°. Let $A = -ak$, $B = a(c-1) - k(b+k)$, $C = b(c-1) + k(c-2)$.
Particular solution: $y_0 = x^{1-c}e^{kx}.$

80. $xy''_{xx} + (ax^2 + bx + 2)y'_x + (cx^2 + dx + b)y = 0.$

The substitution $u = xy$ leads to an equation of the form 2.1.2.103: $u''_{xx} + (ax + b)u'_x + (cx + d - a)u = 0$.

81. $xy''_{xx} + (ax^3 + b)y'_x + a(b-1)x^2 y = 0.$

Particular solution: $y_0 = x^{1-b}$.

82. $xy''_{xx} + x(ax^2 + b)y'_x + (3ax^2 + b)y = 0.$

Particular solution: $y_0 = x\exp\Bigl(-\dfrac{ax^3}{3} - bx\Bigr)$.

83. $xy''_{xx} + (ax^3 + bx^2 + 2)y'_x + bxy = 0.$

Particular solution: $y_0 = a + \dfrac{b}{x}$.

84. $xy''_{xx} + (abx^3 + bx^2 + ax - 1)y'_x + a^2bx^3 y = 0.$

Particular solution: $y_0 = (ax+1)e^{-ax}$.

85. $xy''_{xx} + (ax^3 + bx^2 + cx + d)y'_x + (d-1)(ax^2 + bx + c)y = 0.$

Particular solution: $y_0 = x^{1-d}$.

86. $xy''_{xx} + ax^n y'_x + (abx^n - ax^{n-1} - b^2x + 2b)y = 0.$

Particular solution: $y_0 = xe^{-bx}$.

87. $xy''_{xx} + (ax^n + 2)y'_x + ax^{n-1}y = 0.$

Particular solution: $y_0 = x^{-1}$.

88. $xy''_{xx} + (x^n + 1 - n)y'_x + bx^{2n-1}y = 0.$

For $b \neq \frac{1}{4}$, the general solution has the form

$$y = C_1\exp\Bigl(\frac{\beta_1}{n}x^n\Bigr) + C_2\exp\Bigl(\frac{\beta_2}{n}x^n\Bigr),$$

where β_1 and β_2 are the roots of the quadratic equation $\beta^2 + \beta + b = 0$.

For $b = \frac{1}{4}$, the solution is

$$y = (C_1 + C_2x^n)\exp\Bigl(-\frac{x^n}{2n}\Bigr).$$

89. $xy''_{xx} + (ax^n + b)y'_x + anx^{n-1}y = 0.$

Particular solution: $y_0 = x^{1-b}\exp\Bigl(-\dfrac{ax^n}{n}\Bigr)$.

90. $xy''_{xx} + (ax^n + b)y'_x + a(b-1)x^{n-1}y = 0.$

Particular solution: $y_0 = x^{1-b}$.

91. $xy''_{xx} + (ax^n + b)y'_x + a(b + n - 1)x^{n-1}y = 0.$

Particular solution: $y_0 = \exp\Bigl(-\dfrac{ax^n}{n}\Bigr).$

92. $xy''_{xx} + (ax^n + b)y'_x + c(ax^n - cx + b)y = 0.$

Particular solution: $y_0 = e^{-cx}.$

93. $xy''_{xx} + (abx^n + b - 3n + 1)y'_x + a^2n(b - n)x^{2n-1}y = 0.$

Particular solution: $y_0 = (ax^n + 1)\exp(-ax^n).$

94. $xy''_{xx} + (ax^n + b)y'_x + (cx^{2n-1} + dx^{n-1})y = 0.$

This is a special case of equation 2.1.2.141 with $\gamma = 0$.

95. $xy''_{xx} + (ax^n + bx^{n-1} + 2)y'_x + bx^{n-2}y = 0.$

Particular solution: $y_0 = a + \dfrac{b}{x}.$

96. $xy''_{xx} + (ax^n + bx)y'_x + (abx^n + anx^{n-1} - b)y = 0.$

Particular solution: $y_0 = x\exp\Bigl(-\dfrac{ax^n}{n}\Bigr).$

97. $xy''_{xx} + (abx^n + bx^{n-1} + ax - 1)y'_x + a^2bx^ny = 0.$

Particular solution: $y_0 = (ax + 1)e^{-ax}.$

98. $xy''_{xx} + (ax^n + bx^m + c)y'_x + (c - 1)(ax^{n-1} + bx^{m-1})y = 0.$

Particular solution: $y_0 = x^{1-c}.$

99. $xy''_{xx} + (abx^{n+m} + anx^n + bx^m + 1 - 2n)y'_x + a^2bnx^{2n+m-1}y = 0.$

Particular solution: $y_0 = (ax^n + 1)\exp(-ax^n).$

100. $(x + a)y''_{xx} + (bx + c)y'_x + by = 0.$

Particular solution: $y_0 = \exp\Bigl(-\displaystyle\int \frac{bx + c - 1}{x + a}\,dx\Bigr).$

101. $(a_1x + a_0)y''_{xx} + (b_1x + b_0)y'_x - mb_1y = 0.$

If $m = 1, 2, 3, \ldots$, a polynomial of order m in x is a particular solution of the equation, which may be presented as

$$y_0 = \sum_{k=0}^{m}\Bigl(-\frac{1}{b_1}\Bigr)^k \bigl\{x^m I x^{-m-1}[(a_1x + a_0)D^2 + b_0D]\bigr\}^k x^m,$$

where $D = \dfrac{d}{dx}$, $Ix^\nu = \dfrac{x^{\nu+1}}{\nu + 1}$ with $\nu \neq -1$.

TABLE 2.2
Solutions of equation 2.1.2.103 for different values of the determining parameters

Solution: $y = e^{hx} z(\xi)$, where $\xi = \dfrac{x-\mu}{\lambda}$					
Constraints	h	λ	μ	z	Parameters
$a_2 \neq 0$, $D \neq 0$	$\dfrac{D-a_1}{2a_2}$	$-\dfrac{a_2}{A(h)}$	$-\dfrac{b_2}{a_2}$	$\mathcal{J}(a,b;\xi)$	$a = B(h)/A(h)$, $b = (a_2 b_1 - a_1 b_2) a_2^{-2}$
$a_2 = 0$, $a_1 \neq 0$	$-\dfrac{a_0}{a_1}$	1	$-\dfrac{2b_2 h + b_1}{a_1}$	$\mathcal{J}\bigl(a, \frac{1}{2}; k\xi^2\bigr)$	$a = B(h)/(2a_1)$, $k = -a_1/(2b_2)$
$a_2 \neq 0$, $a_1^2 = 4a_0 a_2$	$-\dfrac{a_1}{2a_2}$	a_2	$-\dfrac{b_2}{a_2}$	$\xi^\alpha Z_{2\alpha}\bigl(\beta\sqrt{\xi}\,\bigr)$	$\alpha = \dfrac{1}{2} - \dfrac{2b_2 h + b_1}{2a_2}$, $\beta = 2\sqrt{B(h)}$
$a_2 = a_1 = 0$, $a_0 \neq 0$	$-\dfrac{b_1}{2b_2}$	1	$\dfrac{4b_0 b_2 - b_1^2}{4a_0 b_2}$	$\xi^{1/2} Z_{1/3}\bigl(k\xi^{3/2}\bigr)$	$k = \dfrac{2}{3}\Bigl(\dfrac{a_0}{b_2}\Bigr)^{1/2}$
Notation: $D^2 = a_1^2 - 4a_0 a_2$, $A(h) = 2a_2 h + a_1$, $B(h) = b_2 h^2 + b_1 h + b_0$					

102. $(ax+b)y''_{xx} + s(cx+d)y'_x - s^2[(a+c)x + b + d]y = 0.$

Particular solution: $y_0 = e^{sx}$.

103. $(a_2 x + b_2)y''_{xx} + (a_1 x + b_1)y'_x + (a_0 x + b_0)y = 0.$

Let $\mathcal{J}(a,b;x)$ be an arbitrary solution of the degenerate hypergeometric equation $xy''_{xx} + (b-x)y'_x - ay = 0$ (see 2.1.2.65), and $Z_\nu(x)$ be an arbitrary solution of the Bessel equation (see 2.1.2.121). The results of solving the original equation are presented in Table 2.2.

104. $(x+\lambda)y''_{xx} + (ax^n + bx^m + c)y'_x + (anx^{n-1} + bmx^{m-1})y = 0.$

Particular solution: $y_0 = \exp\Bigl(-\displaystyle\int \frac{ax^n + bx^m + c - 1}{x+\lambda}\,dx\Bigr).$

105. $x^2 y''_{xx} + ay = 0.$

This is a special case of equation 2.1.2.118. The substitution $x = e^t$ leads to a constant coefficient equation: $y''_{tt} - y'_t + ay = 0$.

106. $x^2 y''_{xx} + (ax+b)y = 0.$

This is a special case of equation 2.1.2.127.

107. $x^2 y''_{xx} + [a^2 x^2 - n(n+1)]y = 0, \qquad n = 0,\ 1,\ 2,\ \ldots$

Solution: $yx^{n+1} = (x^3 D)^n \Bigl(\dfrac{C_1 e^{ax} + C_2 e^{-ax}}{x^{2n-1}}\Bigr)$, where $D = \dfrac{d}{dx}$.

108. $x^2y''_{xx} - [a^2x^2 + n(n+1)]y = 0, \qquad n = 0,\ 1,\ 2,\ \ldots$

Solution: $yx^{n+1} = (x^3D)^n\Big(\dfrac{C_1\cos ax + C_2\sin ax}{x^{2n-1}}\Big)$, where $D = \dfrac{d}{dx}$.

109. $x^2y''_{xx} - (a^2x^2 + 2abx + b^2 - b)y = 0.$

Particular solution: $y_0 = x^b e^{ax}$.

110. $x^2y''_{xx} + (ax^2 + bx + c)y = 0.$

The substitution $y = x^\lambda u$, where λ is a root of the quadratic equation $\lambda^2 - \lambda + c = 0$, leads to an equation of the form 2.1.2.103: $xu''_{xx} + 2\lambda u'_x + (ax + b)u = 0$.

For $a = -\frac{1}{4}$, $b = k$, $c = \frac{1}{4} - m^2$, the original equation is referred to as Whittaker's equation.

111. $x^2y''_{xx} - (ax^3 + \frac{5}{16})y = 0.$

Particular solution: $y_0 = x^{-1/4}\exp\Big(\dfrac{2}{3}\sqrt{a}\,x^{3/2}\Big)$.

112. $x^2y''_{xx} - [a^2x^4 + a(2b-1)x^2 + b(b+1)]y = 0.$

Particular solution: $y_0 = x^{-b}\exp\Big(-\dfrac{ax^2}{2}\Big)$.

113. $x^2y''_{xx} + (ax^n + b)y = 0.$

This is a special case of equation 2.1.2.127.

114. $x^2y''_{xx} - [a^2x^{2n} + a(2b + n - 1)x^n + b(b-1)]y = 0.$

Particular solution: $y_0 = x^b\exp\Big(\dfrac{a}{n}x^n\Big)$.

115. $x^2y''_{xx} + (ax^{2n} + bx^n + c)y = 0.$

This is a special case of equation 2.1.2.141.

116. $x^2y''_{xx} + \Big(ax^{3n} + bx^{2n} + \dfrac{1-n^2}{4}\Big)y = 0.$

The transformation $\xi = ax^n + b$, $w = yx^{\frac{n-1}{2}}$ leads to an equation of the form 2.1.2.7: $w''_{\xi\xi} + (an)^{-2}\xi w = 0$.

117. $x^2y''_{xx} + \Big[ax^{2n}(bx^n + c)^m + \dfrac{1-n^2}{4}\Big]y = 0.$

The transformation $\xi = bx^n + c$, $w = yx^{\frac{n-1}{2}}$ leads to an equation of the form 2.1.2.7: $w''_{\xi\xi} + a(bn)^{-2}\xi^m w = 0$.

118. $x^2y''_{xx} + axy'_x + by = 0.$

The Euler equation.

Solution:

$$y = \begin{cases} |x|^{\frac{1-a}{2}}\left(C_1|x|^{\mu} + C_2|x|^{-\mu}\right) & \text{if } (1-a)^2 > 4b, \\ |x|^{\frac{1-a}{2}}\left(C_1 + C_2\ln|x|\right) & \text{if } (1-a)^2 = 4b, \\ |x|^{\frac{1-a}{2}}\left[C_1\sin(\mu\ln|x|) + C_2\cos(\mu\ln|x|)\right] & \text{if } (1-a)^2 < 4b, \end{cases}$$

where $\mu = \frac{1}{2}|(1-a)^2 - 4b|^{1/2}$.

119. $x^2y''_{xx} + xy'_x + [x^2 - (n+\frac{1}{2})^2]y = 0, \qquad n = 0,\ 1,\ 2,\ \ldots$

This is a special case of equation 2.1.2.121.

Solution: $y = x^{n+1/2}\left[\left(\dfrac{1}{x}\dfrac{d}{dx}\right)^n\left(C_1\dfrac{\sin x}{x} + C_2\dfrac{\cos x}{x}\right)\right].$

120. $x^2y''_{xx} + xy'_x - [x^2 + (n+\frac{1}{2})^2]y = 0, \qquad n = 0,\ 1,\ 2,\ \ldots$

This is a special case of equation 2.1.2.122.

Solution: $y = x^{n+1/2}\left[\left(\dfrac{1}{x}\dfrac{d}{dx}\right)^n\left(C_1\dfrac{e^x}{x} + C_2\dfrac{e^{-x}}{x}\right)\right].$

121. $x^2y''_{xx} + xy'_x + (x^2 - \nu^2)y = 0.$

The Bessel equation.

1°. Let ν be an arbitrary noninteger. Then the general solution is

$$y = C_1J_\nu(x) + C_2Y_\nu(x), \tag{1}$$

where J_ν and Y_ν are Bessel functions of the first and second kind:

$$J_\nu(x) = \sum_{k=0}^{\infty}\frac{(-1)^k(x/2)^{\nu+2k}}{k!\,\Gamma(\nu+k+1)}, \qquad Y_\nu(x) = \frac{J_\nu(x)\cos\pi\nu - J_{-\nu}(x)}{\sin\pi\nu}. \tag{2}$$

Solution (1) is denoted by $y = Z_\nu(x)$ which is referred to as the cylindrical function.

The cylindrical functions possess the following properties:

$$\begin{aligned} 2\nu Z_\nu(x) &= x[Z_{\nu-1}(x) + Z_{\nu+1}(x)], \\ \frac{d}{dx}[x^\nu Z_\nu(x)] &= x^\nu Z_{\nu-1}(x), \\ \frac{d}{dx}[x^{-\nu} Z_\nu(x)] &= -x^{-\nu} Z_{\nu+1}(x). \end{aligned}$$

Functions J_ν and Y_ν may be presented in terms of definite integrals (with $x > 0$):

$$\begin{aligned} \pi J_\nu(x) &= \int_0^\pi \cos(x\sin\theta - \nu\theta)\,d\theta - \sin\pi\nu\int_0^\infty \exp(-x\sinh t - \nu t)\,dt, \\ \pi Y_\nu(x) &= \int_0^\pi \sin(x\sin\theta - \nu\theta)\,d\theta - \int_0^\infty (e^{\nu t} + e^{-\nu t}\cos\pi\nu)e^{-x\sinh t}\,dt. \end{aligned}$$

2°. In the case $\nu = n + \frac{1}{2}$, where $n = 0, 1, 2, \ldots$, the Bessel functions are expressed in terms of elementary functions:

$$J_{n+\frac{1}{2}}(x) = \sqrt{\frac{2}{\pi}}\, x^{n+\frac{1}{2}} \left(-\frac{1}{x}\frac{d}{dx}\right)^n \frac{\sin x}{x},$$

$$J_{-n-\frac{1}{2}}(x) = \sqrt{\frac{2}{\pi}}\, x^{n+\frac{1}{2}} \left(\frac{1}{x}\frac{d}{dx}\right)^n \frac{\cos x}{x},$$

$$Y_{n+\frac{1}{2}}(x) = (-1)^{n+1} J_{-n-\frac{1}{2}}(x).$$

3°. Let $\nu = n$ be an arbitrary integer. The following relations hold:

$$J_{-n}(x) = (-1)^n J_n(x), \quad Y_{-n}(x) = (-1)^n Y_n(x).$$

The solution is given by formula (1), wherein function $J_n(x)$ is obtained by substituting $\nu = n$ into formula (2), while function $Y_n(x)$ is found by taking the limit as $\nu \to n$ and for nonnegative n becomes

$$Y_n(x) = \frac{2}{\pi} J_n(x) \ln\frac{x}{2} - \frac{1}{\pi}\sum_{k=0}^{n-1} \frac{(n-k-1)!}{k!}\left(\frac{2}{x}\right)^{n-2k}$$
$$- \frac{1}{\pi}\sum_{k=0}^{\infty} (-1)^k \left(\frac{x}{2}\right)^{n+2k} \frac{\psi(k+1) + \psi(n+k+1)}{k!\,(n+k)!},$$

where $\psi(1) = -\gamma$, $\psi(n) = -\gamma + \sum_{k=1}^{n-1} k^{-1}$, $\gamma = 0.5572\ldots$ is the Euler constant, $\psi(x) = [\ln\Gamma(x)]'_x$ is the logarithmic derivative of the gamma-function.

For nonnegative integer n and large x, we may write

$$\sqrt{\pi x}\, J_{2n}(x) = (-1)^n(\cos x + \sin x) + O(x^{-2}),$$

$$\sqrt{\pi x}\, J_{2n+1}(x) = (-1)^{n+1}(\cos x - \sin x) + O(x^{-2}).$$

Function J_n may be presented in terms of the definite integral

$$J_n(x) = \frac{1}{\pi}\int_0^\pi \cos(x\sin t - nt)\,dt; \qquad n = 0, 1, 2, \ldots$$

The Bessel functions are described in the books by Abramowitz & Stegun (1964) and Bateman & Erdélyi (1953, vol. 2) in more detail (see also Supplement 2).

122. $x^2 y''_{xx} + x y'_x - (x^2 + \nu^2) y = 0.$

The modified Bessel equation.

It can be reduced to the equation 2.1.2.121 by means of the substitution $x = i\bar{x}$.
Solution:

$$y = C_1 I_\nu(x) + C_2 K_\nu(x),$$

where I_ν and K_ν are modified Bessel functions:

$$I_\nu(x) = \sum_{k=0}^{\infty} \frac{(x/2)^{2k+\nu}}{k!\,\Gamma(\nu+k+1)}, \qquad K_\nu(x) = \frac{\pi}{2}\frac{I_{-\nu} - I_\nu}{\sin \pi\nu}.$$

$I_\nu(x)$ may be expressed in terms of Bessel function:

$$I_\nu(x) = e^{-\pi\nu i/2} J_\nu(xe^{\pi i/2}), \quad i^2 = -1.$$

The case $\nu = n + \frac{1}{2}$, where $n = 0, 1, 2, \ldots$, is given in 2.1.2.120.
If $\nu = n$ is a nonnegative integer, we have

$$K_n(x) = (-1)^{n+1} I_n(x) \ln \frac{x}{2} + \frac{1}{2} \sum_{m=0}^{n-1} (-1)^m \left(\frac{x}{2}\right)^{2m-n} \frac{(n-m-1)!}{m!}$$
$$+ \frac{1}{2}(-1)^n \sum_{m=0}^{\infty} \left(\frac{x}{2}\right)^{n+2m} \frac{\psi(n+m+1)+\psi(m+1)}{m!\,(n+m)!}; \quad n = 0, 1, 2, \ldots,$$

where $\psi(z)$ is the logarithmic derivative of the gamma-function (see 2.1.2.121); for $n = 0$, the first sum is omitted.

As $x \to +\infty$, the leading terms of the asymptotic expansion are

$$I_\nu(x) \simeq \frac{e^x}{\sqrt{2\pi x}}, \quad K_\nu(x) \simeq \frac{\sqrt{\pi}}{\sqrt{2x}} e^{-x}.$$

The modified Bessel functions are described in the books by Abramowitz & Stegun (1964) and Bateman & Erdélyi (1953, vol. 2) in more detail (see also Supplement 2).

123. $\boldsymbol{x^2 y''_{xx} + 2xy'_x - (a^2x^2 + 2)y = 0.}$

Solution: $x^2 y = C_1(ax - 1)e^{ax} + C_2(ax + 1)e^{-ax}$.

124. $\boldsymbol{x^2 y''_{xx} - 2axy'_x + [b^2x^2 + a(a+1)]y = 0.}$

Solution: $y = x^a(C_1 \sin bx + C_2 \cos bx)$.

125. $\boldsymbol{x^2 y''_{xx} - 2axy'_x + [-b^2x^2 + a(a+1)]y = 0.}$

Solution: $y = x^a(C_1 e^{bx} + C_2 e^{-bx})$.

126. $\boldsymbol{x^2 y''_{xx} + \lambda xy'_x + (ax^2 + bx + c)y = 0.}$

The substitution $y = x^k u$, where k is a root of the quadratic equation $k^2 + (\lambda - 1)k + c = 0$, leads to an equation of the form 2.1.2.103:

$$xu''_{xx} + (\lambda + 2k)u'_x + (ax + b)u = 0.$$

127. $\boldsymbol{x^2 y''_{xx} + axy'_x + (bx^n + c)y = 0,}$ $n \neq 0$.

The case $b = 0$ corresponds to the Euler equation 2.1.2.118.

For $b \neq 0$, the solution is

$$y = x^{\frac{1-a}{2}} \left[C_1 J_\nu\left(\frac{2}{n}\sqrt{b}\, x^{\frac{n}{2}}\right) + C_2 Y_\nu\left(\frac{2}{n}\sqrt{b}\, x^{\frac{n}{2}}\right) \right],$$

where $\nu = \frac{1}{n}\sqrt{(1-a)^2 - 4c}$, J_ν and Y_ν are Bessel functions of the first and second kind.

128. $x^2 y''_{xx} + axy'_x + x^n(bx^n + c)y = 0.$

The substitution $\xi = x^n$ leads to an equation of the form 2.1.2.103:

$$n^2 \xi y''_{\xi\xi} + n(n-1+a)y'_\xi + (b\xi + c)y = 0.$$

129. $x^2 y''_{xx} + (ax + b)y'_x + cy = 0.$

The transformation $x = \xi^{-1}$, $y = \xi^k e^\xi w$, where k is a root of the quadratic equation $k^2 + (1-a)k + c = 0$, leads to the equation of the form 2.1.2.103:

$$\xi w''_{\xi\xi} + [(2-b)\xi + 2k + 2 - a]w'_\xi + [(1-b)\xi + 2k + 2 - a - bk]w = 0.$$

130. $x^2 y''_{xx} + ax^2 y'_x + (bx^2 + cx + d)y = 0.$

The substitution $y = u\exp\bigl(-\frac{1}{2}ax\bigr)$ leads to an equation of the form 2.1.2.110:

$$x^2 u''_{xx} + [(\tfrac{1}{4}a^2 + b)x^2 + cx + d]u = 0.$$

131. $x^2 y''_{xx} + (ax^2 + b)y'_x + c[(a-c)x^2 + b]y = 0.$

Particular solution: $y_0 = e^{-cx}$.

132. $x^2 y''_{xx} + (ax^2 + bx)y'_x - by = 0.$

Particular solution: $y_0 = x^{-b} e^{-ax}$.

133. $x^2 y''_{xx} + (ax^2 + bx)y'_x + [k(a-k)x^2 + (an + bk - 2kn)x + n(b-n-1)]y = 0.$

Particular solution: $y_0 = x^{-n} e^{-kx}$.

134. $a_2 x^2 y''_{xx} + (a_1 x^2 + b_1 x)y'_x + (a_0 x^2 + b_0 x + c_0)y = 0.$

The substitution $y = x^k w$, where k is a root of the quadratic equation $a_2 k^2 + (b_1 - a_2)k + c_0 = 0$, leads to the equation of the form 2.1.2.103:

$$a_2 x w''_{xx} + (a_1 x + 2a_2 k + b_1)w'_x + (a_0 x + a_1 k + b_0)w = 0.$$

135. $x^2 y''_{xx} + [ax^2 + (ab-1)x + b]y'_x + a^2 bxy = 0.$

Particular solution: $y_0 = (ax + 1)e^{-ax}$.

136. $x^2 y''_{xx} - 2x(x^2 - a)y'_x + \{2nx^2 + [(-1)^n - 1]a\}y = 0.$

For $n = 0, 1, 2, \ldots$, particular solutions are polynomials: $y_0 = P_n(x)$, where $P_0 = 1$, $P_1 = x$, $P_2 = 2x^2 - 1 - 2a$, $P_3 = 2x^3 - (3 + 2a)x$, ... The polynomials contain only even powers of x for even n and only odd powers of x for odd n.

137. $x^2 y''_{xx} + x(ax^2 + bx + c)y'_x + (Ax^3 + Bx^2 + Cx + D)y = 0.$

1°. The substitution $y = x^k w$, where k is a root of the quadratic equation $k^2 + (c-1)k + D = 0$, leads to an equation of the form 2.1.2.79 (see also 2.1.2.75–2.1.2.78):

$$x w''_{xx} + (ax^2 + bx + c + 2k)w'_x + [Ax^2 + (B + ak)x + C + bk]y = 0.$$

2°. Let s and r be arbitrary parameters.

For $A = ar$, $B = as + br - r^2$, $C = bs + cr - 2rs$, $D = s(c - s - 1)$, a particular solution is $y_0 = x^{-s} e^{-rx}$.

For $A = a(b-r)$, $B = a(c-s+1) + r(b-r)$, $C = bs + cr - 2rs$, $D = s(c-s-1)$, a particular solution is $y_0 = x^{-s}\exp(-\frac{1}{2}ax^2 - rx)$.

138. $x^2y''_{xx} + ax^ny'_x - (abx^n + acx^{n-1} + b^2x^2 + 2bcx + c^2 - c)y = 0.$

Particular solution: $y_0 = x^c e^{bx}$.

139. $x^2y''_{xx} + ax^ny'_x + (abx^{n+2m} - b^2x^{4m+2} + amx^{n-1} - m^2 - m)y = 0.$

Particular solution: $y_0 = x^{-m}\exp\Bigl(-\dfrac{b}{2m+1}x^{2m+1}\Bigr)$.

140. $x^2y''_{xx} + x(ax^n + b)y'_x + b(ax^n - 1)y = 0.$

Particular solution: $y_0 = x^{-b}$.

141. $x^2y''_{xx} + x(ax^n + b)y'_x + (\alpha x^{2n} + \beta x^n + \gamma)y = 0.$

The transformation $\xi = x^n$, $w = y\xi^{-k}$, where k is a root of the quadratic equation $n^2k^2 + n(b-1)k + \gamma = 0$, leads to an equation of the form 2.1.2.103:

$$n^2\xi w''_{\xi\xi} + [na\xi + 2kn^2 + n(n-1+b)]w'_\xi + (\alpha\xi + kna + \beta)w = 0.$$

142. $x^2y''_{xx} + x(2ax^n + b)y'_x + [a^2x^{2n} + a(b+n-1)x^n + \alpha x^{2m} + \beta x^m + \gamma]y = 0.$

The substitution $w = y\exp\Bigl(\dfrac{a}{n}x^n\Bigr)$ leads to the equation of the form 2.1.2.141:

$$x^2w''_{xx} + bxw'_x + (\alpha x^{2m} + \beta x^m + \gamma)w = 0.$$

143. $x^2y''_{xx} + (ax^{n+2} + bx^2 + c)y'_x + (anx^{n+1} + acx^n + bc)y = 0.$

Particular solution: $y_0 = \exp\Bigl(-\dfrac{a}{n+1}x^{n+1} - bx\Bigr)$.

144. $(1 - x^2)y''_{xx} + n(n-1)y = 0, \qquad n = 0,\ 1,\ 2,\ \ldots$

This equation is encountered in hydrodynamics when describing axially symmetric Stokes flows.

For $n \ge 2$, the solution is

$$y = C_1\mathcal{J}_n(x) + C_2\mathcal{H}_n(x),$$

where $\mathcal{J}_n$ and $\mathcal{H}_n$ are the Gegenbauer functions which may be presented in terms of the Legendre functions of the first and second kind (see 2.1.2.147) as follows:

$$\mathcal{J}_n(x) = \frac{P_{n-2}(x) - P_n(x)}{2n-1}, \quad \mathcal{H}_n(x) = \frac{Q_{n-2}(x) - Q_n(x)}{2n-1}.$$

For $n = 0$ and $n = 1$, the solution is $y = C_1 + C_2x$.

145. $(x^2 - a^2)y''_{xx} + by'_x - 6y = 0.$

Particular solution: $y_0 = (4x - b)|x + a|^{\frac{2a+b}{2a}}|x - a|^{\frac{2a-b}{2a}}$.

146. $(x^2-1)y''_{xx}+xy'_x+ay=0.$

1°. For $a=k^2>0$, the solution is

$$y=\begin{cases} C_1\cos(k\,\mathrm{Arcosh}\,|x|)+C_2\sin(k\,\mathrm{Arcosh}\,|x|) & \text{if } |x|>1,\\ C_1\exp(k\arccos x)+C_2\exp(-k\arccos x) & \text{if } |x|<1,\end{cases}$$

where $\mathrm{Arcosh}\,x=\ln\bigl(x+\sqrt{x^2-1}\bigr)$.

2°. For $a=-k^2<0$, the solution is

$$y=\begin{cases} C_1\exp(k\,\mathrm{Arcosh}\,|x|)+C_2\exp(-k\,\mathrm{Arcosh}\,|x|) & \text{if } |x|>1,\\ C_1\cos(k\arccos x)+C_2\sin(k\arccos x) & \text{if } |x|<1.\end{cases}$$

3°. For $a=-n^2$, where n is a nonnegative integer, particular solutions are the Tchebycheff polynomials $T_n(x)=2^{1-n}\cos(n\arccos x)$.

147. $(1-x^2)y''_{xx}-2xy'_x+n(n+1)y=0, \qquad n=0,\ 1,\ 2,\ \ldots$

The Legendre equation.

The solution is

$$y=C_1P_n(x)+C_2Q_n(x),$$

where the Legendre polynomials $P_n(x)$ and the Legendre functions of the second kind $Q_n(x)$ are given by the formulae

$$P_n(x)=\frac{1}{n!\,2^n}\frac{d^n}{dx^n}(x^2-1)^n,$$

$$Q_n(x)=\frac{1}{2}P_n(x)\ln\frac{1+x}{1-x}-\sum_{m=1}^{n}\frac{1}{m}P_{m-1}(x)P_{n-m}(x).$$

Functions $P_n=P_n(x)$ can be conveniently calculated by the recurrence relations

$$P_0=1,\quad P_1=x,\quad P_2=\frac{1}{2}(3x^2-1),\quad \ldots,\quad P_{n+1}=\frac{2n+1}{n+1}xP_n-\frac{n}{n+1}P_{n-1}.$$

Three leading functions $Q_n=Q_n(x)$ are

$$Q_0=\frac{1}{2}\ln\frac{1+x}{1-x},\quad Q_1=\frac{x}{2}\ln\frac{1+x}{1-x}-1,\quad Q_2=\frac{3x^2-1}{4}\ln\frac{1+x}{1-x}-\frac{3}{2}x.$$

All n zeros of the polynomial $P_n(x)$ are real and lie on the interval $-1<x<+1$; functions $P_n(x)$ form an orthogonal system on the closed interval $-1\le x\le +1$, with the relations taking place

$$\int_{-1}^{+1}P_n(x)P_m(x)\,dx=\begin{cases}0 & \text{if } n\ne m,\\ \dfrac{2}{2n+1} & \text{if } n=m.\end{cases}$$

148. $(1-x^2)y''_{xx}-2xy'_x+\nu(\nu+1)y=0.$

The Legendre equation; ν is an arbitrary number.

The case $\nu=n$ where n is a nonnegative integer is considered in 2.1.2.147.

The substitution $z=x^2$ leads to the hypergeometric equation. Therefore, with $|x|<1$ the solution can be written as

$$y=C_1F\Bigl(-\frac{\nu}{2},\ \frac{1+\nu}{2},\ \frac{1}{2};\ x\Bigr)+C_2xF\Bigl(\frac{1-\nu}{2},\ 1+\frac{\nu}{2},\ \frac{3}{2};\ x\Bigr),$$

where F is the hypergeometric series (see 2.1.2.158).

The Legendre equation is discussed in the books by Abramowitz & Stegun (1964), Bateman & Erdélyi (1953, vol. 1), and Kamke (1976) in more detail (see also Supplement 2).

149. $(x^2 - 1)y''_{xx} + 2(n + 1)xy'_x - (\nu + n + 1)(\nu - n)y = 0, \qquad n = 1,\ 2,\ 3,\ \ldots$

Solution: $y = \dfrac{d^n}{dx^n} y_\nu(x)$, where y_ν is the general solution of the Legendre equation 2.1.2.148.

150. $(x^2 - 1)y''_{xx} - 2(n - 1)xy'_x - (\nu - n + 1)(\nu + n)y = 0, \qquad n = 1,\ 2,\ 3,\ \ldots$

Solution: $y = |x^2 - 1|^n \dfrac{d^n}{dx^n} y_\nu(x)$, where y_ν is the general solution of the Legendre equation 2.1.2.148.

151. $(ax^2 + b)y''_{xx} + axy'_x + cy = 0.$

The substitution $z = \displaystyle\int \frac{dx}{\sqrt{ax^2 + b}}$ leads to a constant coefficient equation: $y''_{zz} + cy = 0$.

152. $(x^2 + a)y''_{xx} + 2bxy'_x + 2(b - 1)y = 0.$

Particular solution: $y_0 = (x^2 + a)^{1-b}$.

153. $(ax^2 + b)y''_{xx} + (2n + 1)axy'_x + cy = 0, \qquad n = 1,\ 2,\ 3,\ \ldots$

This equation can be obtained by n-fold differentiation of the equation of the form 2.1.2.151:

$$(ax^2 + b)u''_{xx} + axu'_x + (c - an^2)u = 0.$$

Solution: $y = \dfrac{d^n u}{dx^n}$.

154. $(1 - x^2)y''_{xx} - xy'_x + (2ax^2 + b)y = 0.$

This is an algebraic form of the Mathieu equation.

The substitution $x = \cos z$ leads to the Mathieu equation 2.1.6.4: $y''_{zz} + (a + b + a\cos 2z)y = 0$.

155. $(1 - x^2)y''_{xx} + (ax + b)y'_x + cy = 0.$

The substitution $2z = 1 + x$ leads to the hypergeometric equation 2.1.2.158:

$$z(1 - z)y''_{zz} + \left[az + \tfrac{1}{2}(b - a)\right]y'_z + cy = 0.$$

156. $(ax^2 + b)y''_{xx} + (cx^2 + d)y'_x + \lambda[(c - a\lambda)x^2 + d - b\lambda]y = 0.$

Particular solution: $y_0 = e^{-\lambda x}$.

157. $(ax^2 + b)y''_{xx} + [\lambda(c + a)x^2 + (c - a)x + 2b\lambda]y'_x + \lambda^2(cx^2 + b)y = 0.$

Particular solution: $y_0 = (\lambda x + 1)e^{-\lambda x}$.

158. $x(x-1)y''_{xx}+[(\alpha+\beta+1)x-\gamma]y'_x+\alpha\beta y=0.$

The Gauss hypergeometric equation.

For $\gamma\neq 0,\ -1,\ -2,\ -3,\ \dots$, a solution can be expressed in terms of the hypergeometric series:

$$F(\alpha,\beta,\gamma;x)=1+\sum_{k=1}^{\infty}\frac{(\alpha)_k(\beta)_k}{(\gamma)_k}\frac{x^k}{k!},\qquad (\alpha)_k=\alpha(\alpha+1)\dots(\alpha+k-1),$$

which is a fortiori convergent for $|x|<1$.

For $\gamma>\beta>0$, this solution can be expressed in terms of the definite integral

$$F(\alpha,\beta,\gamma;x)=\frac{\Gamma(\gamma)}{\Gamma(\beta)\,\Gamma(\gamma-\beta)}\int_0^1 t^{\beta-1}(1-t)^{\gamma-\beta-1}(1-tx)^{-\alpha}\,dt,$$

where $\Gamma(\beta)$ is the gamma-function.

If γ is not an integer, the general solution of the hypergeometric equation has the form

$$y=C_1F(\alpha,\beta,\gamma;x)+C_2x^{1-\gamma}F(\alpha-\gamma+1,\ \beta-\gamma+1,\ 2-\gamma;\ x).$$

In degenerate cases $\gamma=0,\ -1,\ -2,\ -3,\ \dots$, a particular solution of the hypergeometric equation corresponds to $C_1=0$, $C_2=1$. If γ is a positive integer, another particular solution corresponds to $C_1=1$, $C_2=0$. In both these cases, the general solution can be constructed by means of the formula given in 2.1.1.

Table 2.3 represents some special cases where F is expressed in terms of elementary functions.

In Table 2.4 are given the general solutions of the hypergeometric equation for some values of the determining parameters.

Function F possesses the following properties:

$$\begin{aligned}
F(\alpha,\beta,\gamma;x)&=F(\beta,\alpha,\gamma;x),\\
F(\alpha,\beta,\gamma;x)&=(1-x)^{\gamma-\alpha-\beta}F(\gamma-\alpha,\ \gamma-\beta,\ \gamma;\ x),\\
F(\alpha,\beta,\gamma;x)&=(1-x)^{-\alpha}F(\alpha,\ \gamma-\beta,\ \gamma;\ \frac{x}{x-1}),\\
\frac{d^n}{dx^n}F(\alpha,\beta,\gamma;x)&=\frac{(\alpha)_n(\beta)_n}{(\gamma)_n}F(\alpha+n,\ \beta+n,\ \gamma+n;\ x).
\end{aligned}$$

The hypergeometric functions are discussed in the books by Abramowitz & Stegun (1964) and Bateman & Erdélyi (1953, vol. 1) in more detail.

159. $x(x+a)y''_{xx}+(bx+c)y'_x+dy=0.$

The substitution $x=-az$ leads to the hypergeometric equation 2.1.2.158:

$$z(1-z)y''_{zz}+[(c/a)-bz]y'_z-dy=0.$$

160. $2x(x-1)y''_{xx}+(2x-1)y'_x+(ax+b)y=0.$

The substitution $x=\cos^2\xi$ leads to the Mathieu equation 2.1.6.4:

$$y''_{\xi\xi}-(a+2b+a\cos 2\xi)y=0.$$

TABLE 2.3
Some special cases in which the hypergeometric function $F(\alpha,\beta,\gamma;z)$ is expressed in terms of elementary functions

α	β	γ	z	F
$-n$	β	γ	x	$\sum_{k=0}^{n}\frac{(-n)_k(\beta)_k}{(\gamma)_k}\frac{x^k}{k!}$, where $n=1,2,3,\ldots$
$-n$	β	$-n-m$	x	$\sum_{k=0}^{n}\frac{(-n)_k(\beta)_k}{(-n-m)_k}\frac{x^k}{k!}$, where $n=1,2,3,\ldots$
α	β	β	x	$(1-x)^{-\alpha}$
α	$\alpha+\frac{1}{2}$	$\frac{1}{2}$	x^2	$\frac{1}{2}\left[(1+x)^{-2\alpha}+(1-x)^{-2\alpha}\right]$
α	$\alpha+\frac{1}{2}$	$\frac{3}{2}$	x^2	$\frac{(1+x)^{1-2\alpha}-(1-x)^{1-2\alpha}}{2x(1-2\alpha)}$
α	$-\alpha$	$\frac{1}{2}$	$-x^2$	$\frac{1}{2}\left[\left(\sqrt{1+x^2}+x\right)^{2\alpha}+\left(\sqrt{1+x^2}-x\right)^{2\alpha}\right]$
α	$1-\alpha$	$\frac{1}{2}$	$-x^2$	$\frac{\left(\sqrt{1+x^2}+x\right)^{2\alpha-1}+\left(\sqrt{1+x^2}-x\right)^{2\alpha-1}}{2\sqrt{1+x^2}}$
α	$\alpha-\frac{1}{2}$	$2\alpha-1$	x	$2^{2\alpha-2}\left(1+\sqrt{1-x}\right)^{2-2\alpha}$
α	$1-\alpha$	$\frac{3}{2}$	$\sin^2 x$	$\frac{\sin[(2\alpha-1)x]}{(\alpha-1)\sin(2x)}$
α	$2-\alpha$	$\frac{3}{2}$	$\sin^2 x$	$\frac{\sin[(2\alpha-2)x]}{(\alpha-1)\sin(2x)}$
α	$1-\alpha$	$\frac{1}{2}$	$\sin^2 x$	$\frac{\cos[(2\alpha-1)x]}{\cos x}$
α	$\alpha+1$	$\frac{1}{2}\alpha$	x	$(1+x)(1-x)^{-\alpha-1}$
α	$\alpha+\frac{1}{2}$	$2\alpha+1$	x	$\left(\frac{1+\sqrt{1-x}}{2}\right)^{-2\alpha}$
α	$\alpha+\frac{1}{2}$	2α	x	$\frac{1}{\sqrt{1-x}}\left(\frac{1+\sqrt{1-x}}{2}\right)^{1-2\alpha}$
$\frac{1}{2}$	$\frac{1}{2}$	$\frac{3}{2}$	x^2	$\frac{1}{x}\arcsin x$
$\frac{1}{2}$	1	$\frac{3}{2}$	$-x^2$	$\frac{1}{x}\arctan x$
1	1	2	$-x$	$\frac{1}{x}\ln(x+1)$
$\frac{1}{2}$	1	$\frac{3}{2}$	x^2	$\frac{1}{2x}\ln\frac{1+x}{1-x}$
$n+1$	$n+m+1$	$n+m+l+2$	x	$\frac{(-1)^m(n+m+l+1)!}{n!\,l!\,(n+m)!\,(m+l)!}\frac{d^{n+m}}{dx^{n+m}}\left\{(1-x)^{m+l}\frac{d^lF}{dx^l}\right\}$, $F=-\frac{\ln(1-x)}{x}$, $n,m,l=0,1,2,3,\ldots$

TABLE 2.4
General solutions of the hypergeometric equation for some values of the determining parameters

α	β	γ	Solution: $y = y(x)$
0	β	γ	$C_1 + C_2 \int \lvert x\rvert^{-\gamma}\lvert x-1\rvert^{\gamma-\beta-1}\,dx$
α	$\alpha+\frac{1}{2}$	$2\alpha+1$	$C_1\bigl(1+\sqrt{1-x}\bigr)^{-2\alpha} + C_2 x^{-2\alpha}\bigl(1+\sqrt{1-x}\bigr)^{2\alpha}$
α	$\alpha-\frac{1}{2}$	$\frac{1}{2}$	$C_1\bigl(1+\sqrt{x}\bigr)^{1-2\alpha} + C_2\bigl(1-\sqrt{x}\bigr)^{1-2\alpha}$
α	$\alpha+\frac{1}{2}$	$\frac{3}{2}$	$\frac{1}{\sqrt{x}}\Bigl[C_1\bigl(1+\sqrt{x}\bigr)^{1-2\alpha} + C_2\bigl(1-\sqrt{x}\bigr)^{1-2\alpha}\Bigr]$
1	β	γ	$\lvert x\rvert^{1-\gamma}\lvert x-1\rvert^{\gamma-\beta-1}\Bigl(C_1 + C_2 \int \lvert x\rvert^{\gamma-2}\lvert x-1\rvert^{\beta-\gamma}\,dx\Bigr)$
α	β	α	$\lvert x-1\rvert^{-\beta}\Bigl(C_1 + C_2 \int \lvert x\rvert^{-\alpha}\lvert x-1\rvert^{\beta-1}\,dx\Bigr)$
α	β	$\alpha+1$	$\lvert x\rvert^{-\alpha}\Bigl(C_1 + C_2 \int \lvert x\rvert^{\alpha-1}\lvert x-1\rvert^{-\beta}\,dx\Bigr)$

161. $(x^2 + 2ax + b)y''_{xx} + (x + a)y'_x - m^2 y = 0.$

Solution: $y = C_1\bigl(x + a + \sqrt{x^2 + 2ax + b}\bigr)^m + C_2\bigl(x + a + \sqrt{x^2 + 2ax + b}\bigr)^{-m}.$

162. $(ax^2 + bx + c)y''_{xx} + (dx + k)y'_x + (d - 2a)y = 0.$

Integrating yields a first order linear equation:

$$(ax^2 + bx + c)y'_x + [(d - 2a)x + k - b]y = C.$$

163. $(ax^2 + bx + c)y''_{xx} + (kx + d)y'_x - ky = 0.$

Particular solution: $y_0 = kx + d.$

164. $(ax^2 + 2bx + c)y''_{xx} + (ax + b)y'_x + dy = 0.$

The substitution $\xi = \displaystyle\int \frac{dx}{\sqrt{ax^2 + 2bx + c}}$ leads to a constant coefficient equation: $y''_{\xi\xi} + dy = 0.$

165. $(ax^2 + 2bx + c)y''_{xx} + 3(ax + b)y'_x + dy = 0.$

The substitution $u = y\sqrt{|ax^2 + 2bx + c|}$ leads to an equation of the form 2.1.2.164:

$$(ax^2 + 2bx + c)u''_{xx} + (ax + b)u'_x + (d - a)u = 0.$$

166. $(a_2x^2 + b_2x + c_2)y''_{xx} + (b_1x + c_1)y'_x + c_0y = 0.$

Let λ_1 and λ_2 be the roots of the quadratic equation $a_2\lambda^2 + b_2\lambda + c_2 = 0$.

1°. For $\lambda_1 \neq \lambda_2$, the substitution $z = \dfrac{x - \lambda_1}{\lambda_2 - \lambda_1}$ leads to the hypergeometric equation 2.1.2.158:

$$z(1 - z)y''_{zz} - (Ax + B)y'_z - Cy = 0,$$

where $A = \dfrac{b_1}{a_2}$, $B = \dfrac{b_1\lambda_1 + c_1}{a_2(\lambda_2 - \lambda_1)}$, $C = \dfrac{c_0}{a_2}$.

2°. For $\lambda_1 = \lambda_2 = \lambda$, the transformation $x = \lambda + \xi^{-1}$, $y = \xi^k u$, where k is a root of the quadratic equation $a_2k^2 + (a_2 - b_1)k + c_0 = 0$, leads to an equation of the form 2.1.2.103:

$$a_2\xi u''_{\xi\xi} - [(c_1 + \lambda b_1)\xi + b_1 - 2a_2(k + 1)]u'_\xi - k(c_1 + \lambda b_1)u = 0.$$

3°. Let $c_0 = -a_2n(n-1) - b_1n$, where n is a positive integer. Then, amongst solutions there exists a polynomial of the degree $\leq n$.

167. $(ax^2 + bx + c)y''_{xx} - (x^2 - k^2)y'_x + (x + k)y = 0.$

Particular solution: $y_0 = x - k$.

168. $(ax^2 + bx + c)y''_{xx} + (x^3 + k^3)y'_x - (x^2 - kx + k^2)y = 0.$

Particular solution: $y_0 = x + k$.

169. $x^3y''_{xx} + (ax + b)y = 0.$

This is a special case of equation 2.1.2.127 with $n = -1$.

170. $x^3y''_{xx} + (ax^2 + b)y'_x + cxy = 0.$

The substitution $x = 1/z$ leads to an equation of the form 2.1.2.141:

$$z^2y''_{zz} + z(2 - a - bz)y'_z + cy = 0.$$

171. $x^3y''_{xx} + (ax^2 + bx)y'_x + by = 0.$

Particular solution: $y_0 = a - 2 + \dfrac{b}{x}$.

172. $x^3y''_{xx} + (ax^2 + bx)y'_x + cy = 0.$

The substitution $x = 1/z$ leads to an equation of the form 2.1.2.103:

$$zy''_{zz} + (2 - a - bz)y'_z + cy = 0.$$

173. $x^3y''_{xx} + (ax^2 + bx)y'_x + (cx + d)y = 0.$

The substitution $y = x^k u$, where $k = -d/b$,leads to the equation of the form 2.1.2.129:

$$x^2u''_{xx} + [(a + 2k)x + b]u'_x + [k(a + k - 1) + c]u = 0.$$

If $c = 0$ and $d = b(a - 2)$, a particular solution is $y_0 = e^{b/x}$.

174. $x^3y''_{xx} + (ax^3 - x^2 + abx + b)y'_x + a^2bxy = 0.$

Particular solution: $y_0 = (ax+1)e^{-ax}$.

175. $x^3y''_{xx} + x(ax^n + b)y'_x - (ax^n - abx^{n-1} + b)y = 0.$

Particular solution: $y_0 = x\exp\left(\dfrac{b}{x}\right)$.

176. $x(ax^2 + b)y''_{xx} + 2(ax^2 + b)y'_x - 2axy = 0.$

Particular solution: $y_0 = ax + \dfrac{b}{x}$.

177. $x(x^2 + a)y''_{xx} + (bx^2 + c)y'_x + sxy = 0.$

The substitution $az = -x^2$ leads to the hypergeometric equation 2.1.2.158:

$$z(1-z)y''_{zz} + \frac{1}{2}\left[1 + \frac{c}{a} - (1+b)z\right]y'_z - \frac{1}{4}sy = 0.$$

178. $x^2(ax + b)y''_{xx} + [cx^2 + (2b + a\lambda)x + b\lambda]y'_x + \lambda(c - 2a)y = 0.$

Particular solution: $y_0 = \exp\left(\dfrac{\lambda}{x}\right)$.

179. $x^2(ax + b)y''_{xx} - 2x(ax + 2b)y'_x + 2(ax + 3b)y = 0.$

Solution: $y = \dfrac{C_1x^2 + C_2x^3}{ax + b}$.

180. $x^2(ax + b)y''_{xx} + [a(2 - n - m)x^2 - b(n + m)x]y'_x + [am(n - 1)x + bn(m + 1)]y = 0.$

Solution: $y = \dfrac{C_1x^n + C_2x^{m+1}}{ax + b}$.

181. $x^2(x + a_2)y''_{xx} + x(b_1x + a_1)y'_x + (b_0x + a_0)y = 0.$

The substitution $y = x^k u$, where k is a root of the quadratic equation $a_2k^2 + k(a_1 - a_2) + a_0 = 0$ leads to an equation of the form 2.1.2.159:

$$x(x + a_2)u''_{xx} + [(2k + b_1)x + 2ka_2 + a_1]u'_x + [k^2 + k(b_1 - 1) + b_0]u = 0.$$

182. $(ax^3 + bx^2 + cx)y''_{xx} + (\alpha x^2 + \beta x + 2c)y'_x + (\beta - 2b)y = 0.$

Particular solution: $y_0 = 2a - \alpha + \dfrac{2b - \beta}{x}$.

183. $(ax^3 + bx^2 + cx)y''_{xx} + (\alpha x^2 + \beta x + 2c)y'_x - (\alpha x + 2b - \beta)y = 0.$

Particular solution:

$$y_0 = \alpha x + 2(\beta - b) + \frac{\lambda}{x}, \qquad \text{where} \quad \lambda = \frac{c\alpha + (b-\beta)(2b-\beta)}{\alpha - a}.$$

184. $(ax^3 + bx^2 + cx)y''_{xx} + [-2ax^2 - (b+1)x + k]y'_x + 2(ax+1)y = 0.$

Particular solution: $y_0 = (ak + b - 1)x^2 + (c+k)(2x - k).$

185. $(ax^3 + bx^2 + cx)y''_{xx} + (nx^2 + mx + k)y'_x + (k-1)[(n - ak)x + m - bk]y = 0.$

Particular solution: $y_0 = x^{1-k}.$

186. $(ax^3 + bx^2 + cx)y''_{xx} + [(m-a)x^2 + (2cm - 1)x - c]y'_x + (-2mx + 1)y = 0.$

Particular solution: $y_0 = (a+m)x^2 + (2b + 4cm - 1)(x + c).$

187. $(ax^3 + bx^2 + cx)y''_{xx} + (nx^2 + mx + k)y'_x + [-2(a+n)x + 1]y = 0.$

With the constraint

$$2(2a+n)(c+k) + (2b + 2m + 1)[m + 1 + 2k(a+n)] = 0,$$

a particular solution has the form $y_0 = (2a+n)x^2 + (2b + 2m + 1)(x - k).$

188. $(ax^3 + x^2 + b)y''_{xx} + a^2x(x^2 - b)y'_x - a^3bxy = 0.$

Particular solution: $y_0 = (ax + 2)e^{-ax}.$

189. $2x(ax^2 + bx + c)y''_{xx} + (ax^2 - c)y'_x + \lambda x^2 y = 0.$

The substitution $\xi = \int \left(\dfrac{x}{ax^2 + bx + c}\right)^{1/2} dx$ leads to a constant coefficient equation: $2y''_{\xi\xi} + \lambda y = 0.$

190. $x(ax^2 + bx + 1)y''_{xx} + (\alpha x^2 + \beta x + \gamma)y'_x + (nx + m)y = 0.$

The substitution $y = x^{1-\gamma}w$ leads to an equation of the similar form:

$$x(ax^2 + bx + 1)w''_{xx} + [(\alpha + 2a - 2a\gamma)x^2 + (\beta + 2b - 2b\gamma)x + 2 - \gamma]w'_x + \\ \{[n + (1-\gamma)(\alpha - a\gamma)]x + m + (1-\gamma)(\beta - b\gamma)\}w = 0.$$

191. $x(x-1)(x-a)y''_{xx} +$
$\{(\alpha + \beta + 1)x^2 - [\alpha + \beta + 1 + a(\gamma + \delta) - \delta]x + a\gamma\}y'_x + (\alpha\beta x - q)y = 0.$

Heun's equation.

For $|a| \geq 1$ and $\gamma \neq 0, -1, -2, -3, \ldots$, a solution can be represented as the power series

$$F(a, q; \alpha, \beta, \gamma, \delta, x) = \sum_{n=0}^{\infty} c_n x^n,$$

wherein the coefficients are determined by the recurrence formulae

$$c_0 = 1, \quad a\gamma c_1 = q,$$
$$a(n+1)(\gamma + n)c_{n+1} = \Big[a(\gamma + \delta + n - 1) + \alpha + \beta - \delta + n + \frac{q}{n}\Big]nc_n \\ - \big[(n-1)(n-2) + (n-1)(\alpha + \beta + 1) + \alpha\beta\big]c_{n-1}.$$

The series is a fortiori convergent for $|x| \leq 1$.

TABLE 2.5
Some transformations retaining the form of Heun's equation

No	New variables	Parameters of transformed equation for $w=w(\xi)$					
1*	$\xi=x,\ w=y$	a	q	α	β	γ	δ
2	$\xi=1-x,\ w=y$	$1-a$	$\alpha\beta-q$	α	β	δ	γ
3	$\xi=x,\ w=\lvert x\rvert^{\gamma-1}y$	a	q_1	$\alpha-\gamma+1$	$\beta-\gamma+1$	$2-\gamma$	δ
4	$\xi=\frac{x}{a},\ w=\lvert x\rvert^{\gamma-1}y$	$\frac{1}{a}$	$\frac{q_2}{a}$	$\alpha-\gamma+1$	$\beta-\gamma+1$	$2-\gamma$	$\alpha+\beta-\gamma-\delta+1$
5	$\xi=\frac{1}{x},\ w=\lvert x\rvert^{\alpha}y$	$\frac{1}{a}$	q_3	α	$\alpha-\gamma+1$	$\alpha-\beta+1$	δ
6	$\xi=\frac{x}{a},\ w=y$	$\frac{1}{a}$	q	α	β	γ	$\alpha+\beta-\gamma-\delta+1$
7	$\xi=1-\frac{x}{a},\ w=y$	$1-\frac{1}{a}$	q	α	β	$\alpha+\beta-\gamma-\delta+1$	γ
8	$\xi=\frac{x}{a},\ w=\lvert x\rvert^{\alpha}y$	a	q	$\alpha-\gamma+1$	$\alpha+\gamma-1$	$\alpha-\beta+1$	$\alpha+\beta-\gamma-\delta+1$
9	$\xi=\frac{x-1}{x},\ w=\lvert x\rvert^{\alpha}y$	$1-\frac{1}{a}$	q	α	$\alpha-\gamma+1$	δ	$\alpha-\beta+1$
10	$\xi=\frac{a(x-1)}{x(a-1)},\ w=\lvert x\rvert^{\alpha}y$	$\frac{a}{a-1}$	q	α	$\alpha-\gamma+1$	δ	$\alpha+\beta-\gamma-\delta+1$
11	$\xi=\frac{x}{x-1},\ w=\lvert x-1\rvert^{\alpha}y$	$\frac{a}{a-1}$	q	α	$\alpha-\delta+1$	γ	$\alpha-\beta+1$
12	$\xi=\frac{x(a-1)}{a(x-1)},\ w=\lvert x-1\rvert^{\alpha}y$	$1-\frac{1}{a}$	q	α	$\alpha-\delta+1$	γ	$\alpha+\beta-\gamma-\delta+1$
Notation: $q_1=q+(\alpha-\gamma+1)(\beta-\gamma+1)-\alpha\beta+\delta(\gamma-1)$, $q_2=q_1+a\delta(1-\gamma)$, $q_3=qa^{-1}+\alpha(\alpha-\gamma+1)+\alpha a^{-1}(\delta-\beta)-a\delta$.							

* This row corresponds to the original equation, while the others refer to the transformed equation for $w=w(\xi)$

If γ is not an integer, the general solution of Heun's equation can be presented as follows:

$$y = C_1 F(a,q;\alpha,\beta,\gamma,\delta,x) + C_2|x|^{1-\gamma}F(a,\ q_1;\ \alpha-\gamma+1,\ \beta-\gamma+1,\ 2-\gamma,\ \delta,\ x),$$

where $q_1 = q + (\alpha-\gamma+1)(\beta-\gamma+1) - \alpha\beta + \delta(\gamma-1)$.

In Table 2.5 are listed some transformations retaining the form of Heun's equation. (Whenever at least one of the indicated equations is integrable by quadrature with some values of parameters, all the other equations are also integrable for those values of the parameters.)

For Heun's equation, see also the book by Bateman & Erdélyi (1955, vol. 3).

192. $(ax^3 + bx^2 + cx + d)y''_{xx} - (x^2 - \lambda^2)y'_x + (x+\lambda)y = 0.$

Particular solution: $y_0 = x - \lambda$.

193. $2(ax^3 + bx^2 + cx + d)y''_{xx} + (3ax^2 + 2bx + c)y'_x + \lambda y = 0.$

The substitution $\xi = \int \frac{dx}{\sqrt{ax^3 + bx^2 + cx + d}}$ leads to a constant coefficient equation: $2y''_{\xi\xi} + \lambda y = 0$.

194. $2(ax^3 + bx^2 + cx + d)y''_{xx} + 3(3ax^2 + 2bx + c)y'_x + (6ax + 2b + \lambda)y = 0.$

This equation is obtained by differentiating the equation 2.1.2.193.

195. $(ax^3 + bx^2 + cx + d)y''_{xx} + [\alpha x^2 + (\alpha\gamma + \beta)x + \beta\gamma]y'_x - (\alpha x + \beta)y = 0.$

Particular solution: $y_0 = x + \gamma$.

196. $(ax^3 + bx^2 + cx + d)y''_{xx} + (x^3 + \lambda^3)y'_x - (x^2 - \lambda x + \lambda^2)y = 0.$

Particular solution: $y_0 = x + \lambda$.

197. $2x(ax^2 + bx + c)y''_{xx} + [a(2-k)x^2 + b(1-k)x - ck]y'_x + \lambda x^{k+1}y = 0.$

The substitution $\xi = \int \frac{x^{k/2}\,dx}{\sqrt{ax^2 + bx + c}}$ leads to a constant coefficient equation: $2y''_{\xi\xi} + \lambda y = 0$.

198. $x^4 y''_{xx} + ay = 0.$

The transformation $z = 1/x$, $u = y/x$ leads to a constant coefficient equation: $u''_{zz} + au = 0$.

199. $x^4 y''_{xx} + (ax^2 + bx + c)y = 0.$

The transformation $z = 1/x$, $u = y/x$ leads to an equation of the form 2.1.2.110: $z^2 u''_{zz} + (cz^2 + bz + a)u = 0$.

200. $x^4 y''_{xx} - (a + b)x^2 y'_x + [(a + b)x + ab]y = 0.$

Solution:

$$y = \begin{cases} C_1 x e^{-a/x} + C_2 x e^{-b/x} & \text{if } a \neq b, \\ (C_1 x + C_2)e^{-a/x} & \text{if } a = b. \end{cases}$$

201. $x^4 y''_{xx} + 2x^2(x + a)y'_x + by = 0.$

The substitution $z = 1/x$ leads to a constant coefficient equation: $y''_{zz} - 2ay'_z + by = 0$.

202. $x^4 y''_{xx} + ax^n y'_x - (ax^{n-1} + abx^{n-2} + b^2)y = 0.$

Particular solution: $y_0 = xe^{-b/x}$.

203. $x^2(x - a)^2 y''_{xx} + by = 0.$

Solution: $C_1|x|^m|x-a|^{1-m} + C_2|x|^{1-m}|x-a|^m$, where m is a root of the quadratic equation $m(m-1)a^2 = -b$.

204. $x^2(x-a)^2 y''_{xx} + by = cx^2(x-a)^2.$

Solution:

$$y = |x|^m |x-a|^{1-m}\left(C_1 + \frac{c}{a(2m-1)}\int |x|^{1-m}|x-a|^m\,dx\right) + |x|^{1-m}|x-a|^m\left(C_2 - \frac{c}{a(2m-1)}\int |x|^m|x-a|^{1-m}\,dx\right),$$

where m is a root of the quadratic equation $m(m-1)a^2 = -b$.

205. $ax^2(x-1)^2 y''_{xx} + (bx^2 + cx + d)y = 0.$

Find roots p and q of the quadratic equations

$$ap(p-1) + d = 0, \qquad aq(q-1) + b + c + d = 0.$$

Then, the substitution $y = x^p(x-1)^q w$ leads to the hypergeometric equation 2.1.2.158:

$$ax(x-1)w''_{xx} + 2a[(p+q)x - p]w'_x + (2apq - c - 2d)w = 0.$$

206. $x^2(x^2+a)y''_{xx} + (bx^2+c)xy'_x + dy = 0.$

The substitution $\xi = x^2$ leads to an equation of the form 2.1.2.181:

$$4\xi^2(\xi+a)y''_{\xi\xi} + 2\xi[(b+1)\xi + a + c]y'_\xi + dy = 0.$$

207. $(x^2+1)^2 y''_{xx} + ay = 0.$

Solution:

$$\frac{y}{\sqrt{x^2+1}} = \begin{cases} C_1\cos(\beta\arctan x) + C_2\sin(\beta\arctan x) & \text{if } a+1=\beta^2>0,\\ C_1\cosh(\beta\arctan x) + C_2\sinh(\beta\arctan x) & \text{if } a+1=-\beta^2<0,\\ C_1 + C_2\arctan x & \text{if } a=-1.\end{cases}$$

208. $(x^2-1)^2 y''_{xx} + ay = 0.$

Solution:

$$y = \begin{cases} \sqrt{|x^2-1|}\left[C_1\cos\left(\beta\ln\left|\frac{x+1}{x-1}\right|\right) + C_2\sin\left(\beta\ln\left|\frac{x+1}{x-1}\right|\right)\right] & \text{if } a-1=4\beta^2>0,\\ (x+1)\left[C_1\left|\frac{x+1}{x-1}\right|^{(2\beta-1)/2} + C_2\left|\frac{x+1}{x-1}\right|^{-(2\beta+1)/2}\right] & \text{if } a-1=-4\beta^2<0,\\ \sqrt{|x^2-1|}\left(C_1 + C_2\ln\left|\frac{x+1}{x-1}\right|\right) & \text{if } a=1.\end{cases}$$

209. $(x^2 \pm a^2)^2 y''_{xx} + b^2 y = 0.$

This is the bending equation of a double-walled compressed bar with a parabolic cross-section.

For the upper sign (constricted bar), the solution is as follows:

$$y = \sqrt{x^2+a^2}\,(C_1\cos u + C_2\sin u), \qquad u = \frac{\sqrt{a^2+b^2}}{a}\arctan\left(\frac{x}{a}\right).$$

For the lower sign (bar with salients), the solution is

$$y = \sqrt{a^2-x^2}\,(C_1\cos u + C_2\sin u), \qquad u = \frac{\sqrt{b^2-a^2}}{2a}\ln\frac{a+x}{a-x}; \quad |x| < a.$$

210. $(ax^2 + b)^2 y''_{xx} + 2ax(ax^2 + b)y'_x + cy = 0.$

The substitution $\xi = \int \frac{dx}{ax^2 + b}$ leads to a constant coefficient equation: $y''_{\xi\xi} + cy = 0$.

211. $(x^2 - 1)^2 y''_{xx} + 2x(x^2 - 1)y'_x - [\nu(\nu + 1)(x^2 - 1) + n^2]y = 0,$
where ν is an arbitrary number, n is a nonnegative integer.

1°. With $n = 0$, this equation coincides with the Legendre equation 2.1.2.148. Denote its general solution by $y_\nu(x)$.

2°. With $n = 1, 2, 3, \ldots$, the general solution of the original equation is given by the formula

$$y = |x^2 - 1|^{n/2} \frac{d^n}{dx^n} y_\nu(x).$$

212. $(1 - x^2)^2 y''_{xx} - 2x(1 - x^2)y'_x + [\nu(\nu + 1)(1 - x^2) - \mu^2]y = 0,$
where ν and μ are arbitrary numbers.

The Legendre equation.

The transformation $x = 1 - 2\xi$, $y = |x^2 - 1|^{\mu/2} w$ leads to the hypergeometric equation 2.1.2.158:

$$\xi(\xi - 1)w''_{\xi\xi} + (\mu + 1)(1 - 2\xi)w'_\xi + (\nu - \mu)(\nu + \mu + 1)w = 0$$

with parameters $\alpha = \mu - \nu$, $\beta = \mu + \nu + 1$, $\gamma = \mu + 1$.

In particular, the original equation is integrable by quadrature with $\nu = \mu$ or $\nu = -\mu - 1$.

See Supplement 2 for more detail on the Legendre equation.

213. $a(x^2 - 1)^2 y''_{xx} + bx(x^2 - 1)y'_x + (cx^2 + dx + e)y = 0.$

The transformation

$$\xi = \tfrac{1}{2}(x + 1), \qquad w = |x + 1|^{-p}|x - 1|^{-q} y,$$

where p and q are parameters which are determined by solving the second order algebraic system

$$4aq(q - 1) + 2bq + c + d + e = 0, \qquad (p - q)[2a(p + q - 1) + b] = d$$

leads to the hypergeometric equation 2.1.2.158 in $w = w(\xi)$.

214. $(ax^2 + b)^2 y''_{xx} + (2ax + c)(ax^2 + b)y'_x + ky = 0.$

The substitution $\xi = \int \frac{dx}{ax^2 + b}$ leads to a constant coefficient equation: $y''_{\xi\xi} + cy'_\xi + ky = 0$.

215. $(ax^2 + b)^2 y''_{xx} + (ax^2 + b)(cx^2 + d)y'_x + 2(bc - ad)xy = 0.$

Particular solution: $y_0 = \exp\left(-\int \frac{cx^2 + d}{ax^2 + b}\, dx\right)$.

216. $(x^2 + a)^2 y''_{xx} + bx^n (x^2 + a) y'_x - (bx^{n+1} + a) y = 0.$

Particular solution: $y_0 = \sqrt{x^2 + a}$.

217. $(x^2 + a)^2 y''_{xx} + bx^n (x^2 + a) y'_x - m[bx^{n+1} + (m-1)x^2 + a] y = 0.$

Particular solution: $y_0 = (x^2 + a)^{m/2}$.

218. $(x - a)^2 (x - b)^2 y''_{xx} - cy = 0, \quad a \neq b.$

The transformation $\xi = \ln \dfrac{x-a}{x-b}$, $y = (x-b)\eta$ leads to a constant coefficient equation: $(a-b)^2(\eta''_{\xi\xi} - \eta'_\xi) - c\eta = 0$. Therefore, the solution is as follows:

$$y = C_1 |x-a|^{(1+\lambda)/2} |x-b|^{(1-\lambda)/2} + C_2 |x-a|^{(1-\lambda)/2} |x-b|^{(1+\lambda)/2},$$

where $\lambda^2 = 4c(a-b)^{-2} + 1 \neq 0$.

219. $(x - a)^2 (x - b)^2 y''_{xx} + (x - a)(x - b)(2x + \lambda) y'_x + \mu y = 0.$

Let k_1 and k_2 be the roots of the quadratic equation

$$(a-b)^2 k^2 + (a-b)(a+b+\lambda)k + \mu = 0.$$

1°. With $k_1 \neq k_2$, the solution is

$$y = C_1 \left| \frac{x-a}{x-b} \right|^{k_1} + C_2 \left| \frac{x-a}{x-b} \right|^{k_2}.$$

2°. With $k_1 = k_2 = k$, the solution is

$$y = \left| \frac{x-a}{x-b} \right|^{k} \left(C_1 + C_2 \ln \left| \frac{x-a}{x-b} \right| \right).$$

220. $(ax^2 + bx + c)^2 y''_{xx} + Ay = 0.$

The transformation

$$\xi = \int \frac{dx}{ax^2 + bx + c}, \qquad w = \frac{y}{\sqrt{ax^2 + bx + c}}$$

leads to a constant coefficient equation: $w''_{\xi\xi} + (A + ac - \frac{1}{4}b^2) w = 0$.

221. $(ax^2 + bx + c)^2 y''_{xx} + (2ax + k)(ax^2 + bx + c) y'_x + my = 0.$

The substitution $\xi = \displaystyle\int \frac{dx}{ax^2 + bx + c}$ leads to a constant coefficient equation: $y''_{\xi\xi} + (k - b) y'_\xi + my = 0$.

222. $x^6 y''_{xx} - x^5 y'_x + ay = 0.$

The transformation $\xi = x^{-2}$, $w = yx^{-2}$ leads to a constant coefficient equation: $4w''_{\xi\xi} + aw = 0$.

223. $x^6 y''_{xx} + (3x^2 + a)x^3 y'_x + by = 0.$

The substitution $\xi = x^{-2}$ leads to a constant coefficient equation: $4y''_{\xi\xi} - 2ay'_\xi + by = 0$.

224. $$y''_{xx} + y'_x \sum_{n=1}^{3} (1 - \alpha_n - \beta_n) \frac{b_n}{b_n x - a_n} - \frac{y}{(b_1 x - a_1)(b_2 x - a_2)(b_3 x - a_3)} \sum_{n=1}^{3} \alpha_n \beta_n \frac{\Delta_n \Delta_{n-1}}{b_n x - a_n} = 0,$$

where $\sum_{n=1}^{3} (\alpha_n + \beta_n) = 1$, $|a_n| + |b_n| > 0$, $\Delta_n = a_n b_{n+1} - a_{n+1} b_n \neq 0$, $a_{n+3} = a_n$, and $b_{n+3} = b_n$.

Denote this equation by

$$\left\{ \begin{array}{ccc|ccc|c} a_1 & a_2 & a_3 & \alpha_1 & \alpha_2 & \alpha_3 & x \\ b_1 & b_2 & b_3 & \beta_1 & \beta_2 & \beta_3 & y \end{array} \right\} = 0. \tag{1}$$

With $a_1 = b_2 = 0$, $a_3 = b_3 = 1$, $\alpha_1 = \alpha_3 = 0$, $\alpha_2 = \alpha$, $\beta_1 = 1 - \gamma$, $\beta_2 = \beta$, and $\beta_3 = \gamma - \alpha - \beta$, equation (1) transforms into the hypergeometric equation 2.1.2.158.

The transformation

$$\xi = \frac{Ax + B}{Cx + D}, \qquad w = \frac{|b_1 x - a_1|^r |b_3 x - a_3|^s}{|b_2 x - a_2|^{r+s}} y, \tag{2}$$

where $AD - BC \neq 0$, brings the original equation into an equation of the similar form:

$$\left\{ \begin{array}{ccc|ccc|c} A_1 & A_2 & A_3 & \alpha_1 + r & \alpha_2 - r - s & \alpha_3 + s & \xi \\ B_1 & B_2 & B_3 & \beta_1 + r & \beta_2 - r - s & \beta_3 + s & w \end{array} \right\} = 0, \tag{3}$$

where $A_n = Aa_n + Bb_n$, $B_n = Ca_n + Db_n$.

In (2), assume $r = -\alpha_1$, $s = -\alpha_3$, $A = b_1/\Delta_3$, $B = -a_1/\Delta_3$, $C = -b_2/\Delta_2$, and $D = a_2/\Delta_2$ to obtain the hypergeometric equation (3).

225. $x^n y''_{xx} + c(ax + b)^{n-4} y = 0.$

The transformation $\xi = \dfrac{x}{ax + b}$, $w = \dfrac{y}{ax + b}$ leads to an equation of the form 2.1.2.7: $w''_{\xi\xi} + cb^{-2}\xi^{-n} w = 0$.

226. $x^n y''_{xx} + axy'_x - (b^2 x^n + 2bx^{n-1} + abx + a)y = 0.$

Particular solution: $y_0 = xe^{bx}$.

227. $x^n y''_{xx} + (ax + b)y'_x - ay = 0.$

Particular solution: $y_0 = ax + b$.

228. $x^n y''_{xx} + (ax^{n-1} + bx)y'_x + (a - 1)by = 0.$

Particular solution: $y_0 = x^{1-a}$.

229. $x^n y''_{xx} + (2x^{n-1} + ax^2 + bx)y'_x + by = 0.$

Particular solution: $y_0 = a + \dfrac{b}{x}$.

230. $x^n y''_{xx} + (ax^n + b)y'_x + c[(a-c)x^n + b]y = 0.$

Particular solution: $y_0 = e^{-cx}$.

231. $x^n y''_{xx} + (ax^n - x^{n-1} + abx + b)y'_x + a^2bxy = 0.$

Particular solution: $y_0 = (ax+1)e^{-ax}$.

232. $x^n y''_{xx} + (ax^{n+m} + 1)y'_x + ax^m(1 + mx^{n-1})y = 0.$

Particular solution: $y_0 = \exp\Big(-\dfrac{a}{m+1}x^{m+1}\Big)$.

233. $(ax^n + b)y''_{xx} + (cx^n + d)y'_x + \lambda[(c - a\lambda)x^n + d - b\lambda]y = 0.$

Particular solution: $y_0 = e^{-\lambda x}$.

234. $(ax^n + bx + c)y''_{xx} = an(n-1)x^{n-2}y.$

Particular solution: $y_0 = ax^n + bx + c$.

235. $x(x^n + 1)y''_{xx} + [(a-b)x^n + a - n]y'_x + b(1-a)x^{n-1}y = 0.$

Particular solution: $y_0 = (x^n + 1)^{b/n}$.

236. $x(x^{2n} + a)y''_{xx} + (x^{2n} + a - an)y'_x - b^2x^{2n-1}y = 0.$

Solution: $y = C_1\big(x^n + \sqrt{x^{2n} + a}\,\big)^{b/n} + C_2\big(x^n + \sqrt{x^{2n} + a}\,\big)^{-b/n}$.

237. $x^2(a^2x^{2n} - 1)y''_{xx} + x[a^2(n+1)x^{2n} + n - 1]y'_x - \nu(\nu+1)a^2n^2x^{2n}y = 0.$

Solution: $y = y_\nu(ax^n)$, where $y_\nu(x)$ is the general solution of the Legendre equation 2.1.2.148.

238. $x^2(ax^n - 1)y''_{xx} + x(apx^n + q)y'_x + (arx^n + s)y = 0.$

Find the roots A_1, A_2 and B_1, B_2 of the quadratic equations

$$A^2 - (q+1)A - s = 0, \quad B^2 - (p-1)B + r = 0$$

and define parameters c, α, β, and γ by the relations

$$c = A_1, \quad \alpha = \frac{A_1 + B_1}{n}, \quad \beta = \frac{A_1 + B_2}{n}, \quad \gamma = \frac{A_1 - A_2}{n} + 1.$$

Then the solution of the original equation has the form

$$y = x^c w(ax^n),$$

where $w(\xi)$ is the general solution of the hypergeometric equation 2.1.2.158:

$$\xi(\xi - 1)w''_{\xi\xi} + [(\alpha + \beta + 1)\xi - \gamma]w'_\xi + \alpha\beta w = 0.$$

239. $(x^n+a)^2y''_{xx}-bx^{n-2}[(b-1)x^n+a(n-1)]y=0.$

Particular solution: $y_0=(x^n+a)^{b/n}$.

240. $(ax^n+b)^2y''_{xx}+(ax^n+b)(cx^n+d)y'_x+n(bc-ad)x^{n-1}y=0.$

Particular solution: $y_0=\exp\Bigl(-\int\dfrac{cx^n+d}{ax^n+b}\,dx\Bigr)$.

241. $(x^n+a)^2y''_{xx}+bx^m(x^n+a)y'_x-x^{n-2}(bx^{m+1}+an-a)y=0.$

Particular solution: $y_0=(x^n+a)^{1/n}$.

242. $(ax^n+b)^2y''_{xx}+cx^m(ax^n+b)y'_x+(cx^m-anx^{n-1}-1)y=0.$

Particular solution: $y_0=\exp\Bigl(-\int\dfrac{dx}{ax^n+b}\Bigr)$.

243. $x^2(ax^n+b)^2y''_{xx}+(n+1)x(a^2x^{2n}-b^2)y'_x+cy=0.$

The substitution $\xi=\dfrac{1}{nb}\ln\Bigl(\dfrac{ax^n}{ax^n+b}\Bigr)$ leads to a constant coefficient equation: $y''_{\xi\xi}-b(n+2)y'_\xi+cy=0$.

244. $(ax^{n+1}+bx^n+c)y''_{xx}+(\alpha x^n+\beta x^{n-1}+\gamma)y'_x$
$+[n(\alpha-a-an)x^{n-1}+(n-1)(\beta-bn)x^{n-2}]y=0.$

Particular solution: $y_0=\exp\biggl[\int\dfrac{(an+a-\alpha)x^n+(bn-\beta)x^{n-1}-\gamma}{ax^{n+1}+bx^n+c}\,dx\biggr]$.

245. $(ax^n+bx^m+c)y''_{xx}+(\lambda-x)y'_x+y=0.$

Particular solution: $y_0=x-\lambda$.

246. $(ax^n+bx^m+c)y''_{xx}+(\lambda^2-x^2)y'_x+(x+\lambda)y=0.$

Particular solution: $y_0=x-\lambda$.

247. $2(ax^n+bx^m+c)y''_{xx}+(anx^{n-1}+bmx^{m-1})y'_x+dy=0.$

The substitution $\xi=\displaystyle\int\frac{dx}{\sqrt{ax^n+bx^m+c}}$ leads to a constant coefficient equation: $2y''_{\xi\xi}+dy=0$.

248. $(ax^n+b)^{m+1}y''_{xx}+(ax^n+b)y'_x-anmx^{n-1}y=0.$

Particular solution: $y_0=\exp\biggl[-\int\dfrac{dx}{(ax^n+b)^m}\biggr]$.

249. $xP_ny''_{xx}+[2P_n+(ax^2+bx)Q_{n-2}]y'_x+bQ_{n-2}y=0,$
where $P_n=P_n(x)$ and $Q_{n-2}=Q_{n-2}(x)$ are arbitrary polynomials of the degrees n and $n-2$, respectively.

Particular solution: $y_0=a+\dfrac{b}{x}$.

2.1.3. Equations Containing Exponential Functions

1. $y''_{xx} + ae^{\lambda x}y = 0, \qquad \lambda \neq 0.$

Solution:

$$y = C_1 J_0\Big(\frac{2\sqrt{a}}{\lambda}e^{\lambda x/2}\Big) + C_2 Y_0\Big(\frac{2\sqrt{a}}{\lambda}e^{\lambda x/2}\Big),$$

where J_0 and Y_0 are Bessel functions.

2. $y''_{xx} + (ae^{x} - b)y = 0.$

Solution:

$$y = C_1 J_{2\sqrt{b}}(2\sqrt{a}\,e^{x/2}) + C_2 Y_{2\sqrt{b}}(2\sqrt{a}\,e^{x/2}),$$

where J_ν and Y_ν are Bessel functions.

3. $y''_{xx} + a(\lambda e^{\lambda x} - ae^{2\lambda x})y = 0.$

Particular solution: $y_0 = \exp\Big(-\frac{a}{\lambda}e^{\lambda x}\Big).$

4. $y''_{xx} - [a^2e^{2x} + a(2b+1)e^{x} + b^2]y = 0.$

Particular solution: $y_0 = \exp(ae^{x} + bx).$

5. $y''_{xx} - (ae^{2\lambda x} + be^{\lambda x} + c)y = 0.$

The transformation $z = e^{\lambda x}$, $w = z^{-k}y$, where $k = \sqrt{c}/\lambda$, leads to an equation of the form 2.1.2.103:

$$\lambda^2 z w''_{zz} + \lambda^2(2k+1)w'_z - (az+b)w = 0.$$

6. $y''_{xx} + (ae^{4\lambda x} + be^{3\lambda x} + ce^{2\lambda x} - \frac{1}{4}\lambda^2)y = 0.$

The transformation $\xi = e^{\lambda x}$, $w = ye^{\lambda x/2}$ leads to an equation of the form 2.1.2.6:

$$w''_{\xi\xi} + \lambda^{-2}(a\xi^2 + b\xi + c)w = 0.$$

7. $y''_{xx} + [ae^{2\lambda x}(be^{\lambda x} + c)^n - \frac{1}{4}\lambda^2]y = 0.$

The transformation $\xi = be^{\lambda x} + c$, $w = ye^{\lambda x/2}$ leads to an equation of the form 2.1.2.7: $w''_{\xi\xi} + a(b\lambda)^{-2}\xi^n w = 0.$

8. $y''_{xx} + ay'_x + be^{2ax}y = 0.$

The transformation $\xi = e^{ax}$, $u = ye^{ax}$ leads to a constant coefficient linear equation: $u''_{\xi\xi} + ba^{-2}u = 0.$

9. $y''_{xx} - ay'_x + be^{2ax}y = 0.$

The substitution $\xi = e^{ax}$ leads to a constant coefficient equation: $y''_{\xi\xi} + ba^{-2}y = 0.$

10. $y''_{xx} + ay'_x + (be^{\lambda x} + c)y = 0.$

Solution:

$$y = e^{-ax/2}\left[C_1 J_\nu\left(\frac{2\sqrt{b}}{\lambda}e^{\lambda x/2}\right) + C_2 Y_\nu\left(\frac{2\sqrt{b}}{\lambda}e^{\lambda x/2}\right)\right], \qquad \nu = \frac{\sqrt{a^2 - 4c}}{\lambda},$$

where J_ν and Y_ν are Bessel functions.

11. $y''_{xx} - y'_x + \left(ae^{3\lambda x} + be^{2\lambda x} + \dfrac{1 - \lambda^2}{4}\right)y = 0.$

The substitution $z = e^x$ leads to an equation of the form 2.1.2.116:

$$z^2 y''_{zz} + \left(az^{3\lambda} + bz^{2\lambda} + \frac{1 - \lambda^2}{4}\right)y = 0.$$

12. $y''_{xx} - y'_x + \left[ae^{2\lambda x}(be^{\lambda x} + c)^n + \dfrac{1 - \lambda^2}{4}\right]y = 0.$

The substitution $z = e^x$ leads to an equation of the form 2.1.2.117:

$$z^2 y''_{zz} + \left[az^{2\lambda}(bz^\lambda + c)^n + \frac{1 - \lambda^2}{4}\right]y = 0.$$

13. $y''_{xx} + 2ae^{\lambda x}y'_x + ae^{\lambda x}(ae^{\lambda x} + \lambda)y = 0.$

Solution: $y = \exp\left(-\dfrac{a}{\lambda}e^{\lambda x}\right)(C_1 + C_2 x).$

14. $y''_{xx} + (a + b)e^{\lambda x}y'_x + ae^{\lambda x}(be^{\lambda x} + \lambda)y = 0.$

Particular solution: $y_0 = \exp\left(-\dfrac{a}{\lambda}e^{\lambda x}\right).$

15. $y''_{xx} + ae^{\lambda x}y'_x - be^{\mu x}(ae^{\lambda x} + be^{\mu x} + \mu)y = 0.$

Particular solution: $y_0 = \exp\left(\dfrac{b}{\mu}e^{\mu x}\right).$

16. $y''_{xx} + 2ke^{\mu x}y'_x + (ae^{2\lambda x} + be^{\lambda x} + k^2e^{2\mu x} + k\mu e^{\mu x} + c)y = 0.$

The substitution $w = y\exp\left(\dfrac{k}{\mu}e^{\mu x}\right)$ leads to an equation of the form 2.1.3.5: $w''_{xx} + (ae^{2\lambda x} + be^{\lambda x} + c)w = 0$.

17. $y''_{xx} + ae^{\lambda x}y'_x + b(ax^ne^{\lambda x} - bx^{2n} + nx^{n-1})y = 0.$

Particular solution: $y_0 = \exp\left(-\dfrac{b}{n+1}x^{n+1}\right).$

18. $y''_{xx} + 2ae^{\lambda x}y'_x + (a^2e^{2\lambda x} + a\lambda e^{\lambda x} + bx^{2n} + cx^{n-1})y = 0.$

The substitution $w = y\exp\left(\dfrac{a}{\lambda}e^{\lambda x}\right)$ leads to an equation of the form 2.1.2.10: $w''_{xx} + (bx^{2n} + cx^{n-1})w = 0$.

19. $y''_{xx} - (a + 2be^{ax})y'_x + b^2 e^{2ax} y = 0.$

Particular solution: $y_0 = \exp\left(\frac{b}{a} e^{ax}\right).$

20. $y''_{xx} + (ae^{2\lambda x} + \lambda)y'_x - a\lambda e^{2\lambda x} y = 0.$

Particular solution: $y_0 = ae^{\lambda x} + \lambda e^{-\lambda x}.$

21. $y''_{xx} + (ae^{\lambda x} - \lambda)y'_x + be^{2\lambda x} y = 0.$

The substitution $\xi = e^{\lambda x}$ leads to a constant coefficient linear equation: $\lambda^2 y''_{\xi\xi} + a\lambda y'_\xi + by = 0.$

22. $y''_{xx} + (ae^{\lambda x} + b)y'_x + c(ae^{\lambda x} + b - c)y = 0.$

Particular solution: $y_0 = e^{-cx}.$

23. $y''_{xx} + (a + be^{2\lambda x})y'_x + \lambda(a - \lambda - be^{2\lambda x})y = 0.$

Particular solution: $y_0 = be^{\lambda x} + ae^{-\lambda x}.$

24. $y''_{xx} + (abe^{\lambda x} + b - 3\lambda)y'_x + a^2\lambda(b - \lambda)e^{2\lambda x} y = 0.$

Particular solution: $y_0 = (ae^{\lambda x} + 1)\exp(-ae^{\lambda x}).$

25. $y''_{xx} + (2ae^{\lambda x} - \lambda)y'_x + (a^2 e^{2\lambda x} + be^{\mu x})y = 0.$

This is a special case of equation 2.1.3.30.

26. $y''_{xx} + (2ae^{\lambda x} + b)y'_x + [a^2 e^{2\lambda x} + a(b + \lambda)e^{\lambda x} + c]y = 0.$

The substitution $w = y \exp\left(\frac{a}{\lambda} e^{\lambda x}\right)$ leads to a constant coefficient linear equation: $w''_{xx} + bw'_x + cw = 0.$

27. $y''_{xx} + (ae^{\lambda x} + 2b - \lambda)y'_x + (ce^{2\lambda x} + abe^{\lambda x} + b^2 - b\lambda)y = 0.$

The transformation $\xi = \frac{1}{\lambda} e^{\lambda x}$, $w = e^{bx} y$ leads to a constant coefficient linear equation: $w''_{\xi\xi} + aw'_\xi + cw = 0.$

28. $y''_{xx} + (ae^{x} + b)y'_x + [c(a - c)e^{2x} + (ak + bc + c - 2ck)e^{x} + k(b - k)]y = 0.$

Particular solution: $y_0 = \exp(-ce^x - kx).$

29. $y''_{xx} + (ae^{\lambda x} + b)y'_x + (\alpha e^{2\lambda x} + \beta e^{\lambda x} + \gamma)y = 0.$

The substitution $\xi = e^x$ leads to an equation of the form 2.1.2.141:

$$\xi^2 y''_{\xi\xi} + (a\xi^\lambda + b + 1)\xi y'_\xi + (\alpha\xi^{2\lambda} + \beta\xi^\lambda + \gamma)y = 0.$$

30. $y''_{xx} + (2ae^{\lambda x} - \lambda)y'_x + (a^2 e^{2\lambda x} + be^{2\mu x} + ce^{\mu x} + k)y = 0.$

The substitution $w = y \exp\left(\frac{a}{\lambda} e^{\lambda x} - \frac{\lambda x}{2}\right)$ leads to an equation of the form 2.1.3.5:

$$w''_{xx} + \left(be^{2\mu x} + ce^{\mu x} + k - \tfrac{1}{4}\lambda^2\right)w = 0.$$

31. $y''_{xx} + (2ae^{\lambda x} + b - \lambda)y'_x + (a^2e^{2\lambda x} + abe^{\lambda x} + ce^{2\mu x} + de^{\mu x} + k)y = 0.$

The substitution $w = y\exp\Big(\frac{a}{\lambda}e^{\lambda x} + \frac{b-\lambda}{2}x\Big)$ leads to an equation of the form 2.1.3.5:

$$w''_{xx} + \big[ce^{2\mu x} + de^{\mu x} + k - \tfrac{1}{4}(b-\lambda)^2\big]w = 0.$$

32. $y''_{xx} + (ae^{\lambda x} + be^{\mu x})y'_x + ae^{\lambda x}(be^{\mu x} + \lambda)y = 0.$

Particular solution: $y_0 = \exp\Big(-\frac{a}{\lambda}e^{\lambda x}\Big).$

33. $y''_{xx} + e^{\lambda x}(ae^{2\mu x} + b)y'_x + \mu[e^{\lambda x}(b - ae^{2\mu x}) - \mu]y = 0.$

Particular solution: $y_0 = ae^{\mu x} + be^{-\mu x}.$

34. $y''_{xx} + (ae^{\lambda x} + be^{\mu x} + c)y'_x + (a\lambda e^{\lambda x} + b\mu e^{\mu x})y = 0.$

Particular solution: $y_0 = \exp\Big(-\frac{a}{\lambda}e^{\lambda x} - \frac{b}{\mu}e^{\mu x} - cx\Big).$

35. $y''_{xx} + (ae^{\lambda x} + be^{\mu x} + c)y'_x + [abe^{(\lambda+\mu)x} + ace^{\lambda x} + b\mu e^{\mu x}]y = 0.$

Particular solution: $y_0 = \exp\Big(-\frac{b}{\mu}e^{\mu x} - cx\Big).$

36. $y''_{xx} + (ae^{\lambda x} + 2be^{\mu x} - \lambda)y'_x +$
$[abe^{(\lambda+\mu)x} + ce^{2\lambda x} + b^2e^{2\mu x} + b(\mu - \lambda)e^{\mu x}]y = 0.$

1°. With $\lambda = 0$, the equation transforms into 2.1.3.26, and with $\mu = 0$ it transforms into 2.1.3.27.

2°. With $\lambda\mu \neq 0$, the transformation $\xi = \frac{1}{\lambda}e^{\lambda x}$, $w = y\exp\Big(\frac{b}{\mu}e^{\mu x}\Big)$ leads to a constant coefficient equation: $w''_{\xi\xi} + aw'_\xi + cw = 0.$

37. $y''_{xx} + [abe^{(\lambda+\mu)x} + a\lambda e^{\lambda x} + be^{\mu x} - 2\lambda]y'_x + a^2b\lambda e^{(2\lambda+\mu)x}y = 0.$

Particular solution: $y_0 = (ae^{\lambda x} + 1)\exp(-ae^{\lambda x}).$

38. $y''_{xx} + (ax + b)e^{\lambda x}y'_x - ae^{\lambda x}y = 0.$

Particular solution: $y_0 = ax + b.$

39. $y''_{xx} + (axe^{\lambda x} + 2b)y'_x + (abxe^{\lambda x} - ae^{\lambda x} + b^2)y = 0.$

Particular solution: $y_0 = xe^{-bx}.$

40. $y''_{xx} + x(ae^{\lambda x} + be^{\mu x})y'_x - (ae^{\lambda x} + be^{\mu x})y = 0.$

Particular solution: $y_0 = x.$

41. $y''_{xx} + (ax^n + be^{\lambda x})y'_x + (abx^ne^{\lambda x} + anx^{n-1})y = 0.$

Particular solution: $y_0 = \exp\Big(-\frac{a}{n+1}x^{n+1}\Big).$

42. $y''_{xx} + a\exp(bx^n)y'_x + c[a\exp(bx^n) - c]y = 0.$

Particular solution: $y_0 = e^{-cx}$.

43. $y''_{xx} + (ax+b)\exp(\lambda x^n)y'_x - a\exp(\lambda x^n)y = 0.$

Particular solution: $y_0 = ax + b$.

44. $y''_{xx} + ax^n\exp(bx^m)y'_x - ax^{n-1}\exp(bx^m)y = 0.$

Particular solution: $y_0 = x$.

45. $xy''_{xx} - (2ax^2+1)y'_x + 4bx^3\exp(2\lambda x^2)y = 0.$

Solution:

$$y = \exp\left(\frac{ax^2}{2}\right)\left[C_1 J_{\frac{a}{2\lambda}}\left(\frac{\sqrt{b}}{\lambda}e^{\lambda x^2}\right) + C_2 Y_{\frac{a}{2\lambda}}\left(\frac{\sqrt{b}}{\lambda}e^{\lambda x^2}\right)\right],$$

where J_ν and Y_ν are Bessel functions.

46. $xy''_{xx} + axe^{\lambda x}y'_x + ae^{\lambda x}(1+\lambda x)y = 0.$

Particular solution: $y_0 = x\exp\left(-\frac{a}{\lambda}e^{\lambda x}\right)$.

47. $xy''_{xx} + axe^{\lambda x}y'_x - [a(bx+1)e^{\lambda x} + b(bx+2)]y = 0.$

Particular solution: $y_0 = xe^{bx}$.

48. $xy''_{xx} + (axe^{\lambda x} + b)y'_x + a(b-1)e^{\lambda x}y = 0.$

Particular solution: $y_0 = x^{1-b}$.

49. $xy''_{xx} + [a(bx+1)e^{\lambda x} + bx - 1]y'_x + ab^2xe^{\lambda x}y = 0.$

Particular solution: $y_0 = (bx+1)e^{-bx}$.

50. $xy''_{xx} + [(ax^2+bx)e^{\lambda x} + 2]y'_x + be^{\lambda x}y = 0.$

Particular solution: $y_0 = a + \frac{b}{x}$.

51. $xy''_{xx} + (ax^n + be^{\lambda x})y'_x + ax^{n-1}(be^{\lambda x} + n - 1)y = 0.$

Particular solution: $y_0 = \exp\left(-\frac{a}{n}x^n\right)$.

52. $xy''_{xx} + (axe^{\lambda x} + bx^n)y'_x + [a(bx^n - 1)e^{\lambda x} + bnx^{n-1}]y = 0.$

Particular solution: $y_0 = x\exp\left(-\frac{b}{n}x^n\right)$.

53. $xy''_{xx} + [(ax^n+1)e^{\lambda x} + anx^n + 1 - 2n]y'_x + a^2nx^{2n-1}e^{\lambda x}y = 0.$

Particular solution: $y_0 = (ax^n+1)\exp(-ax^n)$.

54. $xy''_{xx} + (ae^{\lambda x} + be^{\mu x})y'_x + (a\lambda e^{\lambda x} + b\mu e^{\mu x})y = 0.$

Integrate to obtain a first order linear equation: $xy'_x + (ae^{\lambda x} + be^{\mu x} - 1)y = C$.

55. $xy''_{xx} + [ax^n \exp(bx^m) + c]y'_x + a(c-1)x^{n-1}\exp(bx^m)y = 0.$

Particular solution: $y_0 = x^{1-c}$.

56. $(x+a)y''_{xx} + (be^{\lambda x} + c)y'_x + b\lambda e^{\lambda x}y = 0.$

Particular solution: $y_0 = \exp\biggl(\int \frac{1-c-be^{\lambda x}}{x+a}\,dx\biggr)$.

57. $4x^2y''_{xx} + [ax^{2n}\exp(bx^n) + 1 - n^2]y = 0.$

The transformation $\xi = bx^n$, $w = yx^{\frac{n-1}{2}}$ leads to an equation of the form 2.1.3.1: $4w''_{\xi\xi} + a(bn)^{-2}e^{\xi}w = 0$.

58. $x^2y''_{xx} + 2axy'_x + [(b^2e^{2cx} - \nu^2)c^2x^2 + a(a-1)]y = 0.$

Solution: $y = x^{-a}[C_1J_\nu(be^{cx}) + C_2Y_\nu(be^{cx})]$, where J_ν and Y_ν are Bessel functions.

59. $x^2y''_{xx} + axe^{\lambda x}y'_x + b(ae^{\lambda x} - b - 1)y = 0.$

Particular solution: $y_0 = x^{-b}$.

60. $x^2y''_{xx} + x(ae^{\lambda x} + 2b)y'_x + [a(cx+b)e^{\lambda x} - c^2x^2 + b(b-1)]y = 0.$

Particular solution: $y_0 = x^{-b}e^{-cx}$.

61. $x^4y''_{xx} + (e^{2/x} - \nu^2)y = 0.$

Solution: $y = x[C_1J_\nu(e^{1/x}) + C_2Y_\nu(e^{1/x})]$, where J_ν and Y_ν are Bessel functions.

62. $x^4y''_{xx} + \left[a\exp\left(\frac{2\lambda}{x}\right) + b\exp\left(\frac{\lambda}{x}\right) + c\right]y = 0.$

The transformation $\xi = 1/x$, $w = y/x$ leads to an equation of the form 2.1.3.5: $w''_{\xi\xi} + (ae^{2\lambda\xi} + be^{\lambda\xi} + c)w = 0$.

63. $x^4y''_{xx} + ax^2e^{\lambda x}y'_x + [a(b-x)e^{\lambda x} - b^2]y = 0.$

Particular solution: $y_0 = x\exp(b/x)$.

64. $(x^2+a)^2y''_{xx} + be^{\lambda x}(x^2+a)y'_x - (bxe^{\lambda x} + a)y = 0.$

Particular solution: $y_0 = \sqrt{x^2+a}$.

65. $(x^n+a)^2y''_{xx} + b(x^n+a)e^{\lambda x}y'_x - x^{n-2}(bxe^{\lambda x} + an - a)y = 0.$

Particular solution: $y_0 = (x^n+a)^{1/n}$.

66. $(ax^n+b)^2y''_{xx}+c(ax^n+b)e^{\lambda x}y'_x+(ce^{\lambda x}-anx^{n-1}-1)y=0.$

Particular solution: $y_0=\exp\Big(-\int\frac{dx}{ax^n+b}\Big).$

67. $(a^2e^{2\lambda x}+b)y''_{xx}-b\lambda y'_x-a^2\lambda^2\nu^2e^{2\lambda x}y=0.$

Solution: $y=C_1\big(ae^{\lambda x}+\sqrt{a^2e^{2\lambda x}+b}\big)^{\nu}+C_2\big(ae^{\lambda x}+\sqrt{a^2e^{2\lambda x}+b}\big)^{-\nu}.$

68. $2(ae^{\lambda x}+b)y''_{xx}+a\lambda e^{\lambda x}y'_x+cy=0.$

The substitution $\xi=\int\frac{dx}{\sqrt{ae^{\lambda x}+b}}$ leads to a constant coefficient linear equation: $2y''_{\xi\xi}+cy=0.$

69. $(ae^{\lambda x}+b)y''_{xx}+(ce^{\lambda x}+d)y'_x+k[(c-ak)e^{\lambda x}+d-bk]y=0.$

Particular solution: $y_0=e^{-kx}.$

70. $(ae^{\lambda x}+b)y''_{xx}+(ce^{\lambda x}+d)y'_x+(ne^{\lambda x}+m)y=0.$

For $a=0$, this is an equation of the form 2.1.3.29. For $a\neq 0$, the transformation $\xi=ae^{\lambda x}$, $w=y\xi^{-k}$, where k is a root of the quadratic equation $b\lambda^2k^2+d\lambda k+m=0$, leads to an equation of the form 2.1.2.159:

$$a\lambda^2\xi(\xi+b)w''_{\xi\xi}+\lambda[(2ak\lambda+a\lambda+c)\xi+a(2bk\lambda+b\lambda+d)]w'_{\xi}+(ak^2\lambda^2+ck\lambda+n)w=0.$$

71. $(e^x+k)y''_{xx}+(ae^{\lambda x}+be^{\mu x}+c)y'_x+(a\lambda e^{\lambda x}+b\mu e^{\mu x}-e^x)y=0.$

Integrating yields a first order linear equation:

$$(e^x+k)y'_x+(ae^{\lambda x}+be^{\mu x}-e^x+c)y=C.$$

72. $(ae^{\lambda x}+b)^2y''_{xx}+ce^{\lambda x}(\lambda b-ce^{\lambda x})y=0.$

Particular solution: $y_0=(ae^{\lambda x}+b)^k$, where $k=-\frac{c}{a\lambda}.$

73. $(ae^{\lambda x}+b)^2y''_{xx}+\sigma(ae^{\lambda x}+b)y'_x+ce^{\lambda x}(\sigma+\lambda b-ce^{\lambda x})y=0.$

Particular solution: $y_0=(ae^{\lambda x}+b)^k$, where $k=-\frac{c}{a\lambda}.$

74. $(ae^{\lambda x}+b)^2y''_{xx}+(a\lambda e^{\lambda x}+c)(ae^{\lambda x}+b)y'_x+my=0.$

The substitution $\xi=\int\frac{dx}{ae^{\lambda x}+b}$ leads to the constant coefficient linear equation: $y''_{\xi\xi}+cy'_{\xi}+my=0.$

75. $(ae^{\lambda x}+b)^2y''_{xx}+ke^{\mu x}(ae^{\lambda x}+b)y'_x+ce^{\lambda x}(ke^{\mu x}-ce^{\lambda x}+\lambda b)y=0.$

Particular solution: $y_0=(ae^{\lambda x}+b)^k$, where $k=-\frac{c}{a\lambda}.$

76. $4(ae^{\lambda x}+b)^n y''_{xx}+[ke^{2\lambda x}(ce^{\lambda x}+d)^{n-4}-\lambda^2(ae^{\lambda x}+b)^n]y=0.$

The transformation

$$\xi=\frac{ae^{\lambda x}+b}{ce^{\lambda x}+d},\qquad w=\frac{ye^{\lambda x/2}}{ce^{\lambda x}+d}$$

leads to an equation of the form 2.1.2.7: $4w''_{\xi\xi}+k(\Delta\lambda)^{-2}\xi^{-n}w=0$, where $\Delta=ad-bc$.

77. $(ae^{\lambda x}+bx+c)y''_{xx}-a\lambda^2e^{\lambda x}y=0.$

Particular solution: $y_0=ae^{\lambda x}+bx+c$.

78. $[(ax+b)e^{\lambda x}+c]y''_{xx}-c\lambda^2y=0.$

Particular solution: $y_0=ce^{-\lambda x}+ax+b$.

2.1.4. Equations Containing Hyperbolic Functions

1. $y''_{xx}-(a-2q\cosh 2x)y=0.$

The modified Mathieu equation.

The substitution $x=i\xi$ leads to the Mathieu equation 2.1.6.4:

$$y''_{\xi\xi}+(a-2q\cos 2\xi)y=0.$$

For eigenvalues $a=a_n(q)$ and $a=b_n(q)$, the correponding solutions of the modified Mathieu equation are

$$\mathrm{Ce}_{2n+p}(x,q)=\mathrm{ce}_{2n+p}(ix,q)=\sum_{k=0}^{\infty}A_{2k+p}^{2n+p}\cosh[(2k+p)x],$$

$$\mathrm{Se}_{2n+p}(x,q)=-i\,\mathrm{se}_{2n+p}(ix,q)=\sum_{k=0}^{\infty}B_{2k+p}^{2n+p}\sinh[(2k+p)x],$$

where p may be equal to 0 and 1, and coefficients A_{2k+p}^{2n+p} and B_{2k+p}^{2n+p} are indicated in 2.1.6.4.

The modified Mathieu equation is discussed in the books by Abramowitz & Stegun (1964) and Bateman & Erdélyi (1955, vol. 3) in more detail.

2. $y''_{xx}+(a\cosh^2 x+b)y=0.$

Utilize the formula $\cosh 2x=2\cosh^2 x-1$ to obtain an equation of the form 2.1.4.1:

$$y''_{xx}+\left(\frac{a}{2}+b+\frac{a}{2}\cosh 2x\right)y=0.$$

3. $y''_{xx}-a[a\cosh^2(bx)+b\sinh(bx)]y=0.$

Particular solution: $y_0=\exp\left[\frac{a}{b}\sinh(bx)\right]$.

4. $y''_{xx}-a[a\sinh^2(bx)+b\cosh(bx)]y=0.$

Particular solution: $y_0=\exp\left[\frac{a}{b}\cosh(bx)\right]$.

5. $\boldsymbol{y''_{xx} + (a\cosh^2 x + b\sinh^2 x + c)y = 0.}$

Utilize the formulae $2\sinh^2 x = \cosh(2x) - 1$ and $2\cosh^2 x = \cosh(2x) + 1$ to yield an equation of the form 2.1.4.1:

$$y''_{xx} + \left[\frac{a-b}{2} + c + \frac{a+b}{2}\cosh(2x)\right]y = 0.$$

6. $\boldsymbol{y''_{xx} + [a\tanh(\lambda x) + b]y = 0.}$

The transformation

$$\xi = \frac{1-\tanh(\lambda x)}{1+\tanh(\lambda x)}, \qquad w = y\xi^{-k/\lambda},$$

where k is a root of the quadratic equation $4k^2 + b - a = 0$, leads to an equation of the form 2.1.2.159:

$$4\lambda^2\xi(\xi+1)w''_{\xi\xi} + 4\lambda(2k+\lambda)(\xi+1)w'_{\xi} + (4k^2+a+b)w = 0.$$

7. $\boldsymbol{y''_{xx} - 4a^2\tanh^2(3ax)y = 0.}$

Particular solution: $y_0 = \sinh(3ax)[\cosh(3ax)]^{-1/3}$.

8. $\boldsymbol{y''_{xx} + [a\lambda - a(a+\lambda)\tanh^2(\lambda x)]y = 0.}$

Particular solution: $y_0 = [\cosh(\lambda x)]^{-a/\lambda}$.

9. $\boldsymbol{y''_{xx} + [3a\lambda - \lambda^2 - a(a+\lambda)\tanh^2(\lambda x)]y = 0.}$

Particular solution: $y_0 = \sinh(\lambda x)[\cosh(\lambda x)]^{-a/\lambda}$.

10. $\boldsymbol{y''_{xx} + [a\coth(\lambda x) + b]y = 0.}$

The transformation

$$\xi = \frac{1-\tanh(\lambda x)}{1+\tanh(\lambda x)}, \qquad w = y\xi^{-k/\lambda},$$

where k is a root of the quadratic equation $4k^2 + b - a = 0$, leads to an equation of the form 2.1.2.159:

$$4\lambda^2\xi(\xi-1)w''_{\xi\xi} + 4\lambda(2k+\lambda)(\xi-1)w'_{\xi} + (4k^2+a+b)w = 0.$$

11. $\boldsymbol{y''_{xx} - 4a^2\coth^2(3ax)y = 0.}$

Particular solution: $y_0 = \cosh(3ax)[\sinh(3ax)]^{-1/3}$.

12. $\boldsymbol{y''_{xx} + [a\lambda - a(a+\lambda)\coth^2(\lambda x)]y = 0.}$

Particular solution: $y_0 = [\sinh(\lambda x)]^{-a/\lambda}$.

13. $\boldsymbol{y''_{xx} + [3a\lambda - \lambda^2 - a(a+\lambda)\coth^2(\lambda x)]y = 0.}$

Particular solution: $y_0 = \cosh(\lambda x)[\sinh(\lambda x)]^{-a/\lambda}$.

14. $y''_{xx} + ay'_x - \lambda[\lambda + a\tanh(\lambda x)]y = 0.$

Particular solution: $y_0 = \cosh(\lambda x)$.

15. $y''_{xx} + a\sinh(\lambda x)y'_x + b[a\sinh(\lambda x) - b]y = 0.$

Particular solution: $y_0 = e^{-bx}$.

16. $y''_{xx} + a\sinh(\lambda x)y'_x - \lambda[\lambda + \cosh(\lambda x)]y = 0.$

Particular solution: $y_0 = \sinh(\lambda x)$.

17. $y''_{xx} + a\cosh(\lambda x)y'_x - \lambda[\lambda + \sinh(\lambda x)]y = 0.$

Particular solution: $y_0 = \cosh(\lambda x)$.

18. $y''_{xx} + 2\tanh x\, y'_x + ay = 0.$

Solution:

$$y\cosh x = \begin{cases} C_1\cos(bx) + C_2\sin(bx) & \text{if } a-1=b^2>0, \\ C_1\cosh(bx) + C_2\sinh(bx) & \text{if } a-1=-b^2<0. \end{cases}$$

19. $y''_{xx} + a\tanh(\lambda x)y'_x + b[a\tanh(\lambda x) - b]y = 0.$

Particular solution: $y_0 = e^{-bx}$.

20. $y''_{xx} - \lambda\tanh(\lambda x)y'_x - a^2\cosh^2(\lambda x)y = 0.$

Solution: $y = C_1\exp\Bigl[\frac{a}{\lambda}\sinh(\lambda x)\Bigr] + C_2\exp\Bigl[-\frac{a}{\lambda}\sinh(\lambda x)\Bigr]$.

21. $y''_{xx} - \tanh x\, y'_x + a^2\coth^2 x\,(\sinh x)^{2m-2}y = 0.$

Solution:

$$y = \sqrt{\sinh x}\left[C_1 J_{\frac{1}{2m}}\Bigl(\frac{a}{m}\sinh^m x\Bigr) + C_2 Y_{\frac{1}{2m}}\Bigl(\frac{a}{m}\sinh^m x\Bigr)\right],$$

where J_ν and Y_ν are Bessel functions.

22. $y''_{xx} + 2\tanh x\, y'_x + (ax^2 + bx + c)y = 0.$

The substitution $u = y\cosh x$ leads to an equation of the form 2.1.2.6:

$$u''_{xx} + (ax^2 + bx + c - 1)u = 0.$$

23. $y''_{xx} + 2\tanh x\, y'_x + (ax^n + 1)y = 0.$

The substitution $u = y\cosh x$ leads to an equation of the form 2.1.2.7: $u''_{xx} + ax^n u = 0$.

24. $y''_{xx} + 2\tanh x\, y'_x + (ax^{2n} + bx^{n-1} + 1)y = 0.$

The substitution $u = y\cosh x$ leads to an equation of the form 2.1.2.10:

$$u''_{xx} + (ax^{2n} + bx^{n-1})u = 0.$$

25. $y''_{xx} + 2n \coth x \, y'_x + (n^2 - a^2)y = 0, \qquad n = 1,\ 2,\ 3,\ \ldots$

Solution: $y = \left(\dfrac{1}{\sinh x}\dfrac{d}{dx}\right)^n (C_1 e^{ax} + C_2 e^{-ax}).$

26. $y''_{xx} + (2\tanh x + a)y'_x + (a\tanh x + b)y = 0.$

The substitution $u = y\cosh x$ leads to a constant coefficient linear equation: $u''_{xx} + au'_x + (b-1)u = 0$.

27. $y''_{xx} + a\tanh^n(\lambda x)y'_x - \lambda[\lambda + a\tanh^{n+1}(\lambda x)]y = 0.$

Particular solution: $y_0 = \cosh(\lambda x)$.

28. $y''_{xx} + (ax+b)\sinh^n(\lambda x)y'_x - a\sinh^n(\lambda x)y = 0.$

Particular solution: $y_0 = ax + b$.

29. $y''_{xx} + (ax+b)\tanh^n(\lambda x)y'_x - a\tanh^n(\lambda x)y = 0.$

Particular solution: $y_0 = ax + b$.

30. $y''_{xx} + ax^n\cosh^m(\lambda x)y'_x - ax^{n-1}\cosh^m(\lambda x)y = 0.$

Particular solution: $y_0 = x$.

31. $y''_{xx} + ax^n\tanh^m(\lambda x)y'_x - ax^{n-1}\tanh^m(\lambda x)y = 0.$

Particular solution: $y_0 = x$.

32. $xy''_{xx} + ax\cosh^n(\lambda x)y'_x - [a(bx+1)\cosh^n(\lambda x) + b(bx+2)]y = 0.$

Particular solution: $y_0 = xe^{bx}$.

33. $xy''_{xx} + ax\tanh^n(\lambda x)y'_x - [a(bx+1)\tanh^n(\lambda x) + b(bx+2)]y = 0.$

Particular solution: $y_0 = xe^{bx}$.

34. $xy''_{xx} + [ax\cosh^n(\lambda x) + b]y'_x + a(b-1)\cosh^n(\lambda x)y = 0.$

Particular solution: $y_0 = x^{1-b}$.

35. $xy''_{xx} + [ax\tanh^n(\lambda x) + b]y'_x + a(b-1)\tanh^n(\lambda x)y = 0.$

Particular solution: $y_0 = x^{1-b}$.

36. $xy''_{xx} + [(ax^2+bx)\cosh^n(\lambda x) + 2]y'_x + b\cosh^n(\lambda x)y = 0.$

Particular solution: $y_0 = a + \dfrac{b}{x}$.

37. $xy''_{xx} + [(ax^2+bx)\tanh^n(\lambda x) + 2]y'_x + b\tanh^n(\lambda x)y = 0.$

Particular solution: $y_0 = a + \dfrac{b}{x}$.

38. $xy''_{xx} + (a\sinh^n x + bx^{m+1})y'_x + bx^m(a\sinh^n x + m)y = 0.$

Particular solution: $y_0 = \exp\Bigl(-\dfrac{b}{m+1}x^{m+1}\Bigr).$

39. $xy''_{xx} + (a\tanh^n x + bx^{m+1})y'_x + bx^m(a\tanh^n x + m)y = 0.$

Particular solution: $y_0 = \exp\Bigl(-\dfrac{b}{m+1}x^{m+1}\Bigr).$

40. $x^2y''_{xx} + ax\cosh^n(\lambda x)y'_x + b[a\cosh^n(\lambda x) - b - 1]y = 0.$

Particular solution: $y_0 = x^{-b}.$

41. $x^2y''_{xx} + ax\tanh^n(\lambda x)y'_x + b[a\tanh^n(\lambda x) - b - 1]y = 0.$

Particular solution: $y_0 = x^{-b}.$

42. $(a\sinh x + b)y''_{xx} + (c\sinh x + d)y'_x + \lambda[(c - a\lambda)\sinh x + d - b\lambda]y = 0.$

Particular solution: $y_0 = e^{-\lambda x}.$

43. $(a\tanh x + b)y''_{xx} + (c\tanh x + d)y'_x + \lambda[(c - a\lambda)\tanh x + d - b\lambda]y = 0.$

Particular solution: $y_0 = e^{-\lambda x}.$

44. $[a\tanh(\lambda x) + b]y''_{xx} + [c\tanh(\lambda x) + d]y'_x + [n\tanh(\lambda x) + m]y = 0.$

The transformation

$$\xi = \frac{1 + \tanh(\lambda x)}{1 - \tanh(\lambda x)}, \qquad w = y\xi^{-k/\lambda},$$

where k is a root of the quadratic equation $4(a - b)k^2 + 2(c - d)k + n - m = 0$, leads to an equation of the form 2.1.2.159:

$$4\lambda^2\xi[(a+b)\xi + b - a]w''_{\xi\xi} + 2\lambda\{[2(2k+\lambda)(a+b) + c + d]\xi + 2(2k+\lambda)(b-a) + d - c\}w'_\xi + [4(a+b)k^2 + 2(c+d)k + n + m]w = 0.$$

45. $[a + b\coth(\lambda x)]y''_{xx} + [c + d\coth(\lambda x)]y'_x + [n + m\coth(\lambda x)]y = 0.$

Multiply this equation by $\tanh(\lambda x)$ to obtain the equation 2.1.4.44.

46. $\cosh^2(ax)y''_{xx} - by = 0.$

The substitution $ax = \ln\sqrt{\dfrac{\xi}{1-\xi}}$ $(0 < \xi < 1)$ leads to the hypergeometric equation 2.1.2.158:

$$\xi(\xi - 1)y''_{\xi\xi} + (2\xi - 1)y'_\xi + a^{-2}by = 0.$$

47. $\sinh^2(ax)y''_{xx} - by = 0.$

The substitution $ax = \pm\ln\dfrac{\xi}{\sqrt{\xi^2 + 1}}$ $(\xi > 0)$ leads to the equation 2.1.2.177:

$$\xi(\xi^2 + 1)y''_{\xi\xi} + (3\xi^2 + 1)y'_\xi - 4a^{-2}b\xi y = 0.$$

48. $y''_{xx}\sinh^2 x - [a^2\sinh^2 x + n(n-1)]y = 0, \qquad a \neq 0; \quad n = 1,\ 2,\ 3,\ \ldots$

Solution: $y = \sinh^n x\left(\dfrac{1}{\sinh x}\dfrac{d}{dx}\right)^n (C_1 e^{ax} + C_2 e^{-ax})$.

49. $\sinh^n(\lambda x) y''_{xx} + [a\cosh^{n-4}(\lambda x) - \lambda^2 \sinh^n(\lambda x)]y = 0.$

The transformation $\xi = \tanh(\lambda x)$, $w = \dfrac{y}{\cosh(\lambda x)}$ leads to an equation of the form 2.1.2.7: $w''_{\xi\xi} + a\lambda^{-2}\xi^{-n} w = 0$.

50. $\cosh^n(\lambda x) y''_{xx} + [a\sinh^{n-4}(\lambda x) - \lambda^2 \cosh^n(\lambda x)]y = 0.$

The transformation $\xi = \coth(\lambda x)$, $w = \dfrac{y}{\sinh(\lambda x)}$ leads to an equation of the form 2.1.2.7: $w''_{\xi\xi} + a\lambda^{-2}\xi^{-n} w = 0$.

51. $[a\sinh(\lambda x) + bx + c]y''_{xx} - a\lambda^2 \sinh(\lambda x)y = 0.$

Particular solution: $y_0 = a\sinh(\lambda x) + bx + c$.

52. $[a\cosh(\lambda x) + bx + c]y''_{xx} - a\lambda^2 \cosh(\lambda x)y = 0.$

Particular solution: $y_0 = a\cosh(\lambda x) + bx + c$.

2.1.5. Equations Containing Logarithmic Functions

1. $y''_{xx} - (a^2x^2\ln^2 x + a\ln x + a)y = 0.$

Particular solution: $y_0 = e^{-ax^2/4}x^{ax^2/2}$.

2. $y''_{xx} - (a^2x^{2n}\ln^2 x + anx^{n-1}\ln x + ax^{n-1})y = 0.$

Particular solution: $y_0 = e^{-F}x^{(n+1)F}$, where $F = \dfrac{ax^{n+1}}{(n+1)^2}$.

3. $y''_{xx} + a\ln^n(bx)y'_x + c[a\ln^n(bx) - c]y = 0.$

Particular solution: $y_0 = e^{-cx}$.

4. $y''_{xx} + [a\ln^n(bx) + c]y'_x + ac\ln^n(bx)y = 0.$

Particular solution: $y_0 = e^{-cx}$.

5. $y''_{xx} + (ax+b)\ln^n(cx)y'_x - a\ln^n(cx)y = 0.$

Particular solution: $y_0 = ax + b$.

6. $y''_{xx} + ax^n\ln^m(bx)y'_x - ax^{n-1}\ln^m(bx)y = 0.$

Particular solution: $y_0 = x$.

7. $xy''_{xx} - (a^2x\ln^2 x + a)y = 0.$

Particular solution: $y_0 = e^{-ax}x^{ax}$.

8. $xy''_{xx} - [a^2 x \ln^{2n}(bx) + an \ln^{n-1}(bx)]y = 0.$

Particular solution: $y_0 = \exp\left[a \int \ln^n(bx)\,dx\right].$

9. $xy''_{xx} + ax \ln x\, y'_x + a(\ln x + 1)y = 0.$

Particular solution: $y_0 = e^{ax} x^{1-ax}.$

10. $xy''_{xx} + (ax \ln x + b)y'_x + (ab \ln x + a)y = 0.$

Particular solution: $y_0 = e^{ax} x^{-ax}.$

11. $xy''_{xx} + (2ax \ln x + 1)y'_x + (a^2 x \ln^2 x + a \ln x + a)y = 0.$

Solution: $y = e^{ax} x^{-ax}(C_1 + C_2 \ln x).$

12. $xy''_{xx} + \ln x(ax + b)y'_x + a(b \ln^2 x + 1)y = 0.$

Particular solution: $y_0 = e^{ax} x^{-ax}.$

13. $xy''_{xx} + ax \ln^n(bx)y'_x + an \ln^{n-1}(bx)y = 0.$

Particular solution: $y_0 = \exp\left[-a \int \ln^n(bx)\,dx\right].$

14. $xy''_{xx} + ax \ln^n x y'_x + (a \ln^n x + an \ln^{n-1} x)y = 0.$

Particular solution: $y_0 = x \exp\left(-a \int \ln^n x\,dx\right).$

15. $xy''_{xx} + (ax^n \ln x + 1)y'_x - ax^{n-1}\, y = 0.$

Particular solution: $y_0 = \ln x.$

16. $xy''_{xx} + (ax \ln^n x + 1)y'_x - a \ln^{n-1} x\, y = 0.$

Particular solution: $y_0 = \ln x.$

17. $xy''_{xx} + (ax \ln^n x + b)y'_x + a(b-1) \ln^n x\, y = 0.$

Particular solution: $y_0 = x^{1-b}.$

18. $xy''_{xx} + [(ax^2 + bx) \ln^n(cx) + 2]y'_x + b \ln^n(cx)y = 0.$

Particular solution: $y_0 = a + \dfrac{b}{x}.$

19. $xy''_{xx} + (ax^n + b \ln^m x)y'_x + ax^{n-1}(b \ln^m x + n - 1)y = 0.$

Particular solution: $y_0 = \exp\left(-\dfrac{a}{n} x^n\right).$

20. $xy''_{xx} + (ax^n + bx \ln^m x)y'_x + [b(ax^n - 1) \ln^m x + anx^{n-1}]y = 0.$

Particular solution: $y_0 = x \exp\left(-\dfrac{a}{n} x^n\right).$

21. $x^2y''_{xx} + (a\ln x + b)y = 0.$

The transformation $\xi = a\ln x + b - \frac{1}{4}$, $w = yx^{-1/2}$ leads to an equation of the form 2.1.2.7: $w''_{\xi\xi} + a^{-2}\xi w = 0$.

22. $x^2y''_{xx} + (a\ln^2 x + b\ln x + c)y = 0.$

The transformation $\xi = \ln x$, $w = yx^{-1/2}$ leads to an equation of the form 2.1.2.6: $w''_{\xi\xi} + (a\xi^2 + b\xi + c - \frac{1}{4})w = 0$.

23. $x^2y''_{xx} + [a(b\ln x + c)^n + \frac{1}{4}]y = 0.$

The transformation $\xi = b\ln x + c$, $w = yx^{-1/2}$ leads to an equation of the form 2.1.2.7: $w''_{\xi\xi} + ab^{-2}\xi^n w = 0$.

24. $x^2y''_{xx} + xy'_x + a\ln^n(bx)y = 0.$

Solution:

$$y = \sqrt{\ln(bx)}\left[C_1 J_{\frac{1}{2m}}\left(\frac{\sqrt{a}}{m}\ln^m(bx)\right) + C_2 Y_{\frac{1}{2m}}\left(\frac{\sqrt{a}}{m}\ln^m(bx)\right)\right],$$

where $m = \frac{1}{2}(n+2)$; J_ν and Y_ν are Bessel functions.

25. $x^2y''_{xx} + xy'_x + (a\ln^{2n} x + b\ln^{n-1} x)y = 0.$

The substitution $\xi = \ln x$ leads to an equation of the form 2.1.2.10: $y''_{\xi\xi} + (a\xi^{2n} + b\xi^{n-1})y = 0$.

26. $x^2y''_{xx} + x(2a\ln x + 1)y'_x + (x^2 + a^2\ln^2 x + b)y = 0.$

The substitution $y = w\exp(-\frac{1}{2}a\ln^2 x)$ leads to the Bessel equation 2.1.2.121: $x^2w''_{\xi\xi} + xw'_\xi + (x^2 + b - a)w = 0$.

27. $x^2y''_{xx} + x(2\ln x + a + 1)y'_x + (\ln^2 x + a\ln x + b)y = 0.$

The transformation $\xi = \ln x$, $w = y\exp(\frac{1}{2}\ln^2 x)$ leads to a constant coefficient equation: $w''_{\xi\xi} + aw'_\xi + (b-1)w = 0$.

28. $x^2y''_{xx} + x(2\ln x + a)y'_x + [\ln^2 x + (a-1)\ln x + bx^n + c]y = 0.$

The substitution $w = y\exp(\frac{1}{2}\ln^2 x)$ leads to an equation of the form 2.1.2.127:

$$x^2w''_{xx} + axw'_x + (bx^n + c - 1)w = 0.$$

29. $x^2y''_{xx} + ax\ln^n(bx)y'_x + c[a\ln^n(bx) - c - 1]y = 0.$

Particular solution: $y_0 = x^{-c}$.

30. $x^2y''_{xx} + x(ax^n + b\ln x)y'_x + b(ax^n\ln x - \ln x + 1)y = 0.$

Particular solution: $y_0 = \exp\left(-\frac{b}{2}\ln^2 x\right)$.

31. $x(x+a)y''_{xx} + x(b\ln x + c)y'_x + by = 0.$

Particular solution: $y_0 = \exp\left(-\int \frac{b\ln x + c - 1}{x+a}\,dx\right).$

32. $x^4 y''_{xx} + ax^2 \ln^n(bx)y'_x + [a(c-x)\ln^n(bx) - c^2]y = 0.$

Particular solution: $y_0 = x\exp\left(\frac{c}{x}\right).$

33. $(a\ln x + b)y''_{xx} + (c\ln x + d)y'_x + \lambda[(c - a\lambda)\ln x + d - b\lambda]y = 0.$

Particular solution: $y_0 = e^{-\lambda x}.$

34. $x\ln x\, y''_{xx} - ny'_x - a^2 x(\ln x)^{2n+1}y = 0.$

Solution: $y = C_1 e^{aF} + C_2 e^{-aF}$, where $F = \int \ln^n x\,dx.$

35. $x\ln(ax)y''_{xx} - [n\ln(ax) + m]y'_x - b^2 x^{2n+1}\ln^{2m+1}(ax)y = 0.$

Solution: $y = C_1 e^{bF} + C_2 e^{-bF}$, where $F = \int x^n \ln^m(ax)\,dx.$

36. $x^2 \ln(ax)y''_{xx} + y = 0.$

Solution: $y = C_1 \ln(ax) + C_2 \ln(ax) \int [\ln(ax)]^{-2}\,dx.$

37. $x\ln^2 x\, y''_{xx} + (ax+1)\ln x\, y'_x + bxy = 0.$

The substitution $\xi = \int \frac{dx}{\ln x}$ leads to a constant coefficient linear equation: $y''_{\xi\xi} + ay'_\xi + by = 0.$

38. $x(ax\ln x + bx + c)y''_{xx} - ay = 0.$

Particular solution: $y_0 = ax\ln x + bx + c.$

39. $x^2(a\ln x + bx + c)y''_{xx} + ay = 0.$

Particular solution: $y_0 = a\ln x + bx + c.$

40. $\ln^n(ax)y''_{xx} + (b^2 - x^2)y'_x + (x+b)y = 0.$

Particular solution: $y_0 = x - b.$

2.1.6. Equations Containing Trigonometric Functions

1. $y''_{xx} + a^2 y = b\sin(\lambda x).$

Solution:

$$y = \begin{cases} C_1\sin(ax) + C_2\cos(ax) + \dfrac{b}{a^2-\lambda^2}\sin(\lambda x) & \text{if } a \neq \lambda, \\ C_1\sin(ax) + C_2\cos(ax) - \dfrac{b}{2a}x\cos(ax) & \text{if } a = \lambda. \end{cases}$$

TABLE 2.6
The general solution of the Mathieu equation expressed in terms of subsidiary periodical functions $\varphi_1(x)$ and $\varphi_2(x)$.

Constraint	General solution $y=y(x)$	Period of φ_1 and φ_2	Index
$y_1(\pi)>1$	$C_1e^{2\mu x}\varphi_1(x)+C_2e^{-2\mu x}\varphi_2(x)$	π	μ is a real number
$y_1(\pi)<-1$	$C_1e^{2\rho x}\varphi_1(x)+C_2e^{-2\rho x}\varphi_2(x)$	2π	$\mu=\rho+\frac{1}{2}i,\quad i^2=-1$ ρ is the real part of μ
$\lvert y_1(\pi)\rvert<1$	$(C_1\cos\nu x+C_2\sin\nu x)\varphi_1(x)$ $+(C_1\cos\nu x-C_2\sin\nu x)\varphi_2(x)$	π	$\mu=i\nu$ is a pure imaginary number, $\cos(2\pi\nu)=y_1(\pi)$
$y_1(\pi)=\pm1$	$C_1\varphi_1(x)+C_2x\varphi_2(x)$	π	$\mu=0$

2. $\boldsymbol{y''_{xx}+a^2y=b\cos(\lambda x)}$.

Solution:

$$y=\begin{cases} C_1\sin(ax)+C_2\cos(ax)+\dfrac{b}{a^2-\lambda^2}\cos(\lambda x) & \text{if } a\neq\lambda,\\[2mm] C_1\sin(ax)+C_2\cos(ax)+\dfrac{b}{2a}x\sin(ax) & \text{if } a=\lambda.\end{cases}$$

3. $\boldsymbol{y''_{xx}+[a\sin(\lambda x)+b]y=0}$.

The substitution $\lambda x=2\xi+\frac{\pi}{2}$ leads to the Mathieu equation 2.1.6.4:

$$y''_{\xi\xi}+(4a\lambda^{-2}\cos 2\xi+4b\lambda^{-2})y=0.$$

4. $\boldsymbol{y''_{xx}+(a-2q\cos 2x)y=0}$.

The Mathieu equation.

1°. Given numbers a and q, there exists the general solution $y(x)$ and a characteristic index μ such that

$$y(x+\pi)=e^{2\pi\mu}y(x).$$

For small values of q, the approximate value of μ can be found from the equation

$$\cosh(\pi\mu)=1+2\sin^2\left(\tfrac{1}{2}\pi\sqrt{a}\right)+\frac{\pi q^2}{(1-a)\sqrt{a}}\sin\left(\pi\sqrt{a}\right)+O(q^4).$$

If $y_1(x)$ is the solution of the Mathieu equation, which satisfies the initial conditions $y_1(0)=1$ and $y_1'(0)=0$, the characteristic index can be determined from the relation

$$\cosh(2\pi\mu)=y_1(\pi).$$

The solution $y_1(x)$ and hence μ can be determined with any degree of accuracy by means of numerical and approximate methods.

The general solution differs depending on the value of $y_1(\pi)$ and can be expressed in terms of two subsidiary periodical functions $\varphi_1(x)$ and $\varphi_2(x)$ (see Table 2.6).

TABLE 2.6a

Periodical solutions of the Mathieu equation $\mathrm{ce}_n = \mathrm{ce}_n(x,q)$ and $\mathrm{se}_n = \mathrm{se}_n(x,q)$ (for odd n, functions ce_n and se_n are 2π-periodical, and for even n, they are π-periodical); definite eigenvalues $a = a_n(q)$ and $b = b_n(q)$ correspond to each value of parameter q.

Mathieu functions	Recurrence relations for coefficients	Normalization conditions
$\mathrm{ce}_{2n} = \sum\limits_{m=0}^{\infty} A_{2m}^{2n} \cos 2mx$	$qA_2^{2n} = a_{2n}A_0^{2n}$; $qA_4^{2n} = (a_{2n}-4)A_2^{2n} - 2qA_0^{2n}$; $qA_{2m+2}^{2n} = (a_{2n}-4m^2)A_{2m}^{2n} - qA_{2m-2}^{2n}$, $m \geq 2$	$(A_0^{2n})^2 + \sum\limits_{m=0}^{\infty}(A_{2m}^{2n})^2 = \begin{cases} 2 & \text{if } n=0 \\ 1 & \text{if } n\geq 1 \end{cases}$
$\mathrm{ce}_{2n+1} = \sum\limits_{m=0}^{\infty} A_{2m+1}^{2n+1} \cos(2m+1)x$	$qA_3^{2n+1} = (a_{2n+1}-1-q)A_1^{2n+1}$; $qA_{2m+3}^{2n+1} = [a_{2n+1}-(2m+1)^2]A_{2m+1}^{2n+1} - qA_{2m-1}^{2n+1}$, $m \geq 1$	$\sum\limits_{m=0}^{\infty}(A_{2m+1}^{2n+1})^2 = 1$
$\mathrm{se}_{2n} = \sum\limits_{m=0}^{\infty} B_{2m}^{2n} \sin 2mx$, $\mathrm{se}_0 = 0$	$qB_4^{2n} = (b_{2n}-4)B_2^{2n}$; $qB_{2m+2}^{2n} = (b_{2n}-4m^2)B_{2m}^{2n} - qB_{2m-2}^{2n}$, $m \geq 2$	$\sum\limits_{m=0}^{\infty}(B_{2m}^{2n})^2 = 1$
$\mathrm{se}_{2n+1} = \sum\limits_{m=0}^{\infty} B_{2m+1}^{2n+1} \sin(2m+1)x$	$qB_3^{2n+1} = (b_{2n+1}-1-q)B_1^{2n+1}$; $qB_{2m+3}^{2n+1} = [b_{2n+1}-(2m+1)^2]B_{2m+1}^{2n+1} - qB_{2m-1}^{2n+1}$, $m \geq 1$	$\sum\limits_{m=0}^{\infty}(B_{2m+1}^{2n+1})^2 = 1$

2°. In applications, of major interest are periodical solutions of the Mathieu equation which exist for definite values of parameters a and q (those values of a are referred to as eigenvalues). The most important solutions are listed in Table 2.6a.

The Mathieu functions possess the following properties:

$$\mathrm{ce}_{2n}(x,\,-q) = (-1)^n\,\mathrm{ce}_{2n}\Big(\frac{\pi}{2}-x,\;q\Big),\qquad \mathrm{ce}_{2n+1}(x,\,-q) = (-1)^n\,\mathrm{se}_{2n+1}\Big(\frac{\pi}{2}-x,\;q\Big),$$
$$\mathrm{se}_{2n}(x,\,-q) = (-1)^{n-1}\,\mathrm{se}_{2n}\Big(\frac{\pi}{2}-x,\;q\Big),\quad \mathrm{se}_{2n+1}(x,\,-q) = (-1)^n\,\mathrm{ce}_{2n+1}\Big(\frac{\pi}{2}-x,\;q\Big).$$

Selecting sufficiently large number m and omitting the term with the maximum number in the recurrence relations (indicated in Table 2.6a), we can obtain approximate relations for eigenvalues a_n (or b_n) with respect to parameter q. Then, equating the determinant of the corresponding homogeneous linear system of equations for coefficients A_m^n (or B_m^n) to zero, we obtain an algebraic equation for finding $a_n(q)$ (or $b_n(q)$).

For fixed real $q \neq 0$, eigenvalues a_n and b_n are all real and different, while

$$\begin{aligned} &\text{if}\quad q>0\quad \text{then}\quad a_0 < b_1 < a_1 < b_2 < a_2 < \cdots \\ &\text{if}\quad q<0\quad \text{then}\quad a_0 < a_1 < b_1 < b_2 < a_2 < a_3 < b_3 < b_4 < \cdots \end{aligned}$$

The eigenvalues possess the properties

$$a_{2n}(-q) = a_{2n}(q),\quad b_{2n}(-q) = b_{2n}(q),\quad a_{2n+1}(-q) = b_{2n+1}(q).$$

The solution of the Mathieu equation corresponding to eigenvalue a_n (or b_n) has n zeros on the interval $0 \le x < \pi$ (q is a real number).

Listed below are two leading terms of asymptotic expansions of the Mathieu functions $\mathrm{ce}_n(x,q)$ and $\mathrm{se}_n(x,q)$, as well as of the corresponding eigenvalues $a_n(q)$ and $b_n(q)$, as $q \to 0$:

$$\mathrm{ce}_0 = \frac{1}{\sqrt{2}}\Bigl(1 - \frac{q}{2}\cos 2x\Bigr), \quad a_0 = -\frac{q^2}{2} + \frac{7q^4}{128};$$

$$\mathrm{ce}_1 = \cos x - \frac{q}{8}\cos 3x, \quad a_1 = 1 + q;$$

$$\mathrm{ce}_2 = \cos 2x + \frac{q}{4}\Bigl(1 - \frac{\cos 4x}{3}\Bigr), \quad a_2 = 4 + \frac{5q^2}{12};$$

$$\mathrm{ce}_n = \cos nx + \frac{q}{4}\biggl[\frac{\cos(n+2)x}{n+1} - \frac{\cos(n-2)x}{n-1}\biggr], \quad a_n = n^2 + \frac{q^2}{2(n^2-1)} \quad (n \ge 3);$$

$$\mathrm{se}_1 = \sin x - \frac{q}{8}\sin 3x, \quad b_1 = 1 - q;$$

$$\mathrm{se}_2 = \sin 2x - q\frac{\sin 4x}{12}, \quad b_2 = 4 - \frac{q^2}{12};$$

$$\mathrm{se}_n = \sin nx - \frac{q}{4}\biggl[\frac{\sin(n+2)x}{n+1} - \frac{\sin(n-2)x}{n-1}\biggr], \quad b_n = n^2 + \frac{q^2}{2(n^2-1)} \quad (n \ge 3).$$

The Mathieu functions are discussed in the books by Abramowitz & Stegun (1964) and Bateman & Erdélyi (1955, vol. 3) in more detail.

5. $\boldsymbol{y''_{xx} + (a\sin^2 x + b)y = 0.}$

Utilizing the formula $2\sin^2 x = 1 - \cos 2x$, we obtain the Mathieu equation 2.1.6.4: $y''_{xx} + (\frac{1}{2}a + b - \frac{1}{2}a\cos 2x)y = 0$.

6. $\boldsymbol{y''_{xx} + (a\cos^2 x + b)y = 0.}$

Utilizing the formula $2\cos^2 x = 1 + \cos 2x$, we obtain the Mathieu equation 2.1.6.4: $y''_{xx} + (\frac{1}{2}a + b + \frac{1}{2}a\cos 2x)y = 0$.

7. $\boldsymbol{y''_{xx} - a[a\sin^2(bx) + b\cos(bx)]y = 0.}$

Particular solution: $y_0 = \exp\Bigl[-\frac{a}{b}\cos(bx)\Bigr]$.

8. $\boldsymbol{y''_{xx} - a[a\cos^2(bx) + b\sin(bx)]y = 0.}$

Particular solution: $y_0 = \exp\Bigl[-\frac{a}{b}\sin(bx)\Bigr]$.

9. $\boldsymbol{y''_{xx} + a[\lambda + (\lambda - a)\tan^2(\lambda x)]y = 0.}$

Particular solution: $y_0 = [\cos(\lambda x)]^{a/\lambda}$.

10. $\boldsymbol{y''_{xx} + (a\tan^2 x + b)y = 0.}$

The transformation $\xi = \sin^2 x$, $w = y\cos^m x$, where m is a root of the quadratic equation $m^2 + m + a = 0$ leads to the hypergeometric equation 2.1.2.158:

$$\xi(\xi-1)w''_{\xi\xi} + [(1-m)\xi - \tfrac{1}{2}]w'_\xi - \tfrac{1}{4}(m+b)w = 0.$$

11. $y''_{xx} + a[\lambda + (\lambda - a)\cot^2(\lambda x)]y = 0.$

Particular solution: $y_0 = [\sin(\lambda x)]^{a/\lambda}.$

12. $y''_{xx} + (a\cot^2 x + b)y = 0.$

The substitution $x = \xi + \frac{\pi}{2}$ leads to the equation 2.1.6.10: $y''_{\xi\xi} + (a\tan^2\xi + b)y = 0.$

13. $y''_{xx} + a\sin(bx)y'_x + c[ax^n\sin(bx) - cx^{2n} + nx^{n-1}]y = 0.$

Particular solution: $y_0 = \exp\Bigl(-\dfrac{c}{n+1}x^{n+1}\Bigr).$

14. $y''_{xx} + (a\sin x + b)y'_x + a(b\sin x + \cos x)y = 0.$

Particular solution: $y_0 = \exp(a\cos x).$

15. $y''_{xx} + a\sin^n(bx)y'_x + c[a\sin^n(bx) - c]y = 0.$

Particular solution: $y_0 = e^{-cx}.$

16. $y''_{xx} + [a\sin^n(bx) + c]y'_x + ac\sin^n(bx)y = 0.$

Particular solution: $y_0 = e^{-cx}.$

17. $y''_{xx} + (a\sin^n x + b\sin x)y'_x + b(a\sin^{n+1} x + \cos x)y = 0.$

Particular solution: $y_0 = \exp(b\cos x).$

18. $y''_{xx} + (ax + b)\sin^n(cx)y'_x - a\sin^n(cx)y = 0.$

Particular solution: $y_0 = ax + b.$

19. $y''_{xx} + ax^n\sin^m(bx)y'_x - ax^{n-1}\sin^m(bx)y = 0.$

Particular solution: $y_0 = x.$

20. $y''_{xx} + ax^n\sin^m(bx)y'_x + c[ax^{n+k}\sin^m(bx) - cx^{2k} + kx^{k-1}]y = 0.$

Particular solution: $y_0 = \exp\Bigl(-\dfrac{c}{k+1}x^{k+1}\Bigr).$

21. $y''_{xx} + (a\cos x + b)y'_x + a(b\cos x - \sin x)y = 0.$

Particular solution: $y_0 = \exp(-a\sin x).$

22. $y''_{xx} + a\cos^n(bx)y'_x + c[a\cos^n(bx) - c]y = 0.$

Particular solution: $y_0 = e^{-cx}.$

23. $y''_{xx} + [a\cos^n(bx) + c]y'_x + ac\cos^n(bx)y = 0.$

Particular solution: $y_0 = e^{-cx}.$

24. $y''_{xx} + (a\cos^n x + b\cos x)y'_x + b(a\cos^{n+1} x - \sin x)y = 0.$

Particular solution: $y_0 = \exp(-b\sin x).$

25. $y''_{xx} + (ax+b)\cos^n(cx)y'_x - a\cos^n(cx)y = 0.$

Particular solution: $y_0 = ax + b$.

26. $y''_{xx} + ax^n\cos^m(bx)y'_x - ax^{n-1}\cos^m(bx)y = 0.$

Particular solution: $y_0 = x$.

27. $y''_{xx} + ax^n\cos^m(bx)y'_x + c[ax^{n+k}\cos^m(bx) - cx^{2k} + kx^{k-1}]y = 0.$

Particular solution: $y_0 = \exp\Bigl(-\dfrac{c}{k+1}x^{k+1}\Bigr)$.

28. $y''_{xx} + (a-\lambda)\tan(\lambda x)y'_x + a\lambda y = 0.$

Particular solution: $y_0 = [\cos(\lambda x)]^{a/\lambda}$.

29. $y''_{xx} + a\tan x\, y'_x + by = 0.$

1°. The substitution $\xi = \sin x$ leads to an equation of the form 2.1.2.155:

$$(\xi^2-1)y''_{\xi\xi} + (1-a)\xi y'_\xi - by = 0.$$

2°. Solution with $a = -2$:

$$y\cos x = \begin{cases} C_1\sin(kx) + C_2\cos(kx) & \text{if } b+1 = k^2 > 0, \\ C_1\sinh(kx) + C_2\cosh(kx) & \text{if } b+1 = -k^2 < 0. \end{cases}$$

3°. Solution with $a = 2$, $b = 3$:

$$y = C_1\cos^3 x + C_2\sin x(1 + 2\cos^2 x).$$

30. $y''_{xx} + n\tan x\, y'_x + a^2(\cos x)^{2n}y = 0.$

Solution: $y = C_1\sin(au) + C_2\cos(au)$, where $u = \int\cos^n x\,dx$.

31. $y''_{xx} + \tan x\, y'_x + a^2\cos^2 x(\sin x)^{2n-2}y = 0.$

Solution:

$$y = \sqrt{\sin x}\left[C_1 J_{\frac{1}{2n}}\Bigl(\frac{a}{n}\sin^n x\Bigr) + C_2 Y_{\frac{1}{2n}}\Bigl(\frac{a}{n}\sin^n x\Bigr)\right],$$

where J_ν and Y_ν are Bessel functions.

32. $y''_{xx} + a\tan x\, y'_x + (b\tan^2 x + c)y = 0.$

This is a special case of equation 2.1.6.55.

33. $y''_{xx} + \tan x\, y'_x - a(a-1)\cot^2 x\, y = 0.$

Solution:

$$y = \begin{cases} C_1|\sin x|^a + C_2|\sin x|^{1-a} & \text{if } a \neq \frac{1}{2}, \\ \sqrt{|\sin x|}\,(C_1 + C_2\ln|\sin x|) & \text{if } a = \frac{1}{2}. \end{cases}$$

34. $y''_{xx} - 2\lambda\tan(\lambda x)y'_x + (ax^2 + bx + c)y = 0.$

The substitution $u = y\cos(\lambda x)$ leads to an equation of the form 2.1.2.6:

$$u''_{xx} + (ax^2 + bx + c + \lambda^2)u = 0.$$

35. $y''_{xx} - 2\lambda\tan(\lambda x)y'_x + (ax^{2n} + bx^{n-1} - \lambda^2)y = 0.$

The substitution $u = y\cos(\lambda x)$ leads to an equation of the form 2.1.2.10:

$$u''_{xx} + (ax^{2n} + bx^{n-1})u = 0.$$

36. $y''_{xx} + a\tan^n(bx)y'_x + c[a\tan^n(bx) - c]y = 0.$

Particular solution: $y_0 = e^{-cx}$.

37. $y''_{xx} + a\tan^n(\lambda x)y'_x + b[a\tan^{n+1}(\lambda x) + (\lambda - b)\tan^2(\lambda x) + \lambda]y = 0.$

Particular solution: $y_0 = [\cos(\lambda x)]^{b/\lambda}$.

38. $y''_{xx} + a\tan^n x\, y'_x + (a\tan^{n+1} x - a\tan^{n-1} x + 4)y = 0.$

Particular solution: $y_0 = \sin x\cos x$.

39. $y''_{xx} + [a\tan^n(bx) + c]y'_x + ac\tan^n(bx)y = 0.$

Particular solution: $y_0 = e^{-cx}$.

40. $y''_{xx} + \tan x\,(a\tan^n x + b - 1)y'_x + (ab\tan^{n+2} x - a\tan^n x + 2b + 2)y = 0.$

Particular solution: $y_0 = \sin x\cos^b x$.

41. $y''_{xx} + (ax + b)\tan^n(cx)y'_x - a\tan^n(cx)y = 0.$

Particular solution: $y_0 = ax + b$.

42. $y''_{xx} + ax^n\tan^m(bx)y'_x - ax^{n-1}\tan^m(bx)y = 0.$

Particular solution: $y_0 = x$.

43. $y''_{xx} + ax^n\tan^m(bx)y'_x + c[ax^{n+k}\tan^m(bx) - cx^{2k} + kx^{k-1}]y = 0.$

Particular solution: $y_0 = \exp\Bigl(-\dfrac{c}{k+1}x^{k+1}\Bigr)$.

44. $y''_{xx} + \cot x\, y'_x + \nu(\nu + 1)y = 0.$

The substitution $\xi = \cos x$ leads to the Legendre equation 2.1.2.148:

$$(\xi^2 - 1)y''_{\xi\xi} + 2\xi y'_\xi - \nu(\nu + 1)y = 0.$$

45. $y''_{xx} + 2a\cot(ax)y'_x + (b^2 - a^2)y = 0.$

Particular solution: $y_0 = \dfrac{\cos(bx)}{\sin(ax)}$.

46. $y''_{xx} + (\lambda - a)\cot(\lambda x)y'_x + a\lambda y = 0.$

Particular solution: $y_0 = [\sin(\lambda x)]^{a/\lambda}$.

47. $y''_{xx} + a\cot(\lambda x)y'_x + by = 0.$

The substitution $\xi = \lambda x + \frac{\pi}{2}$ leads to an equation of the form 2.1.6.29:

$$y''_{\xi\xi} - a\lambda^{-1}\tan\xi\, y'_\xi + b\lambda^{-2}y = 0.$$

48. $y''_{xx} - 2a\cot(2ax)y'_x - b^2\sin^2(2ax)y = 0.$

Solution: $y = C_1\exp\Big[\frac{b}{a}\sin^2(ax)\Big] + C_2\exp\Big[-\frac{b}{a}\sin^2(ax)\Big].$

49. $y''_{xx} - n\cot x\, y'_x + a^2(\sin x)^{2n}y = 0.$

Solution: $y = C_1\sin(au) + C_2\cos(au)$, where $u = \int \sin^n x\,dx$.

50. $y''_{xx} - 2\cot(2x)y'_x + a\tan^2 x\, y = 0.$

The substitution $\xi = \cos x$ leads to the Euler equation 2.1.2.118: $\xi^2 y''_{\xi\xi} - \xi y'_\xi + ay = 0$.

51. $y''_{xx} + a\cot(\lambda x)y'_x + b[\lambda + (\lambda - a - b)\cot^2(\lambda x)]y = 0.$

Particular solution: $y_0 = [\sin(\lambda x)]^{b/\lambda}$.

52. $y''_{xx} + a\cot x\, y'_x + (b\cot^2 x + c)y = 0.$

This is a special case of equation 2.1.6.55.

53. $y''_{xx} + 2\lambda\cot(\lambda x)y'_x + (ax^2 + bx + c)y = 0.$

The substitution $u = y\sin(\lambda x)$ leads to 2.1.2.6: $u''_{xx} + (ax^2 + bx + c + \lambda^2)u = 0$.

54. $y''_{xx} + 2\lambda\cot(\lambda x)y'_x + (ax^{2n} + bx^{n-1} - \lambda^2)y = 0.$

The substitution $u = y\sin(\lambda x)$ leads to 2.1.2.10: $u''_{xx} + (ax^{2n} + bx^{n-1})u = 0$.

55. $y''_{xx} + (a\tan x + b\cot x)y'_x + (\alpha\tan^2 x + \beta\cot^2 x + \gamma)y = 0.$

The transformation $\xi = \sin^2 x$, $y = w\sin^n x\cos^m x$, where n and m are roots of the quadratic equations

$$n^2 + (b-1)n + \beta = 0, \qquad m^2 - (a+1)m + \alpha = 0,$$

leads to the hypergeometric equation 2.1.2.158:

$$4\xi(\xi-1)w''_{\xi\xi} + 2[(2n+2m+2+b-a)\xi - 2n - b - 1]w'_\xi + (2nm+n+m+bm-an-\gamma)w = 0.$$

56. $y''_{xx} + a\cot^n(bx)y'_x + c[a\cot^n(bx) - c]y = 0.$

Particular solution: $y_0 = e^{-cx}$.

57. $y''_{xx} + [a\cot^n(bx) + c]y'_x + ac\cot^n(bx)y = 0.$

Particular solution: $y_0 = e^{-cx}$.

58. $y''_{xx} + (ax + b)\cot^n(cx)y'_x - a\cot^n(cx)y = 0.$

Particular solution: $y_0 = ax + b$.

59. $y''_{xx} + ax^n\cot^m(bx)y'_x - ax^{n-1}\cot^m(bx)y = 0.$

Particular solution: $y_0 = x$.

60. $y''_{xx} + ax^n\cot^m(bx)y'_x + c[ax^{n+k}\cot^m(bx) - cx^{2k} + kx^{k-1}]y = 0.$

Particular solution: $y_0 = \exp\Bigl(-\dfrac{c}{k+1}x^{k+1}\Bigr)$.

61. $xy''_{xx} + [(ax^2 + bx)\sin^n(cx) + 2]y'_x + b\sin^n(cx)y = 0.$

Particular solution: $y_0 = a + \dfrac{b}{x}$.

62. $xy''_{xx} + (ax^{n+1} + b\sin^m x)y'_x + ax^n(b\sin^m x + n)y = 0.$

Particular solution: $y_0 = \exp\Bigl(-\dfrac{a}{n+1}x^{n+1}\Bigr)$.

63. $xy''_{xx} + (ax^n + bx\sin^m x)y'_x + [b(ax^n - 1)\sin^m x + anx^{n-1}]y = 0.$

Particular solution: $y_0 = x\exp\Bigl(-\dfrac{a}{n}x^n\Bigr)$.

64. $xy''_{xx} + ax^n\sin^m(bx)y'_x - [a(cx + 1)x^{n-1}\sin^m(bx) + c^2x + 2c]y = 0.$

Particular solution: $y_0 = xe^{cx}$.

65. $xy''_{xx} + [ax^n\sin^m(bx) + c]y'_x + a(c - 1)x^{n-1}\sin^m(bx)y = 0.$

Particular solution: $y_0 = x^{1-c}$.

66. $xy''_{xx} + ax\cos^n(bx)y'_x - [a(cx + 1)\cos^n(bx) + c^2x + 2c]y = 0.$

Particular solution: $y_0 = xe^{cx}$.

67. $xy''_{xx} + [(ax^2 + bx)\cos^n(cx) + 2]y'_x + b\cos^n(cx)y = 0.$

Particular solution: $y_0 = a + \dfrac{b}{x}$.

68. $xy''_{xx} + (ax^{n+1} + b\cos^m x)y'_x + ax^n(b\cos^m x + n)y = 0.$

Particular solution: $y_0 = \exp\Bigl(-\dfrac{a}{n+1}x^{n+1}\Bigr)$.

69. $xy''_{xx} + [ax^n\cos^m(bx) + c]y'_x + a(c - 1)x^{n-1}\cos^m(bx)y = 0.$

Particular solution: $y_0 = x^{1-c}$.

70. $xy''_{xx} + (ax^n + bx\cos^m x)y'_x + [b(ax^n - 1)\cos^m x + anx^{n-1}]y = 0.$

Particular solution: $y_0 = x\exp\Bigl(-\dfrac{a}{n}x^n\Bigr).$

71. $xy''_{xx} - 2\lambda x\tan(\lambda x)y'_x + (ax+b)y = 0.$

The substitution $u = y\cos(\lambda x)$ leads to an equation of the form 2.1.2.59:

$$xu''_{xx} + [(a+\lambda^2)x + b]u = 0.$$

72. $xy''_{xx} + ax\tan^n(bx)y'_x - [a(cx+1)\tan^n(bx) + c^2x + 2c]y = 0.$

Particular solution: $y_0 = xe^{cx}.$

73. $xy''_{xx} + [(ax^2+bx)\tan^n(cx) + 2]y'_x + b\tan^n(cx)y = 0.$

Particular solution: $y_0 = a + \dfrac{b}{x}.$

74. $xy''_{xx} + (ax^{n+1} + b\tan^m x)y'_x + ax^n(b\tan^m x + n)y = 0.$

Particular solution: $y_0 = \exp\Bigl(-\dfrac{a}{n+1}x^{n+1}\Bigr).$

75. $xy''_{xx} + (ax^n + bx\tan^m x)y'_x + [b(ax^n - 1)\tan^m x + anx^{n-1}]y = 0.$

Particular solution: $y_0 = x\exp\Bigl(-\dfrac{a}{n}x^n\Bigr).$

76. $xy''_{xx} + [ax^n\tan^m(bx) + c]y'_x + a(c-1)x^{n-1}\tan^m(bx)y = 0.$

Particular solution: $y_0 = x^{1-c}.$

77. $xy''_{xx} + 2\lambda x\cot(\lambda x)y'_x + (ax+b)y = 0.$

The substitution $u = y\sin(\lambda x)$ leads to an equation of the form 2.1.2.59:

$$xu''_{xx} + [(a+\lambda^2)x + b]u = 0.$$

78. $xy''_{xx} + ax\cot^n(bx)y'_x - [a(cx+1)\cot^n(bx) + c^2x + 2c]y = 0.$

Particular solution: $y_0 = xe^{cx}.$

79. $xy''_{xx} + (ax^{n+1} + b\cot^m x)y'_x + ax^n(b\cot^m x + n)y = 0.$

Particular solution: $y_0 = \exp\Bigl(-\dfrac{a}{n+1}x^{n+1}\Bigr).$

80. $xy''_{xx} + (ax^n + bx\cot^m x)y'_x + [b(ax^n - 1)\cot^m x + anx^{n-1}]y = 0.$

Particular solution: $y_0 = x\exp\Bigl(-\dfrac{a}{n}x^n\Bigr).$

81. $xy''_{xx} + [ax^n\cot^m(bx) + c]y'_x + a(c-1)x^{n-1}\cot^m(bx)y = 0.$

Particular solution: $y_0 = x^{1-c}.$

82. $x^2 y''_{xx} + x(a\sin^n x + 1)y'_x + b(a\sin^n x - b)y = 0.$

Particular solution: $y_0 = x^{-b}$.

83. $x^2 y''_{xx} + x(a\sin^n x + b)y'_x + b(a\sin^n x - 1)y = 0.$

Particular solution: $y_0 = x^{-b}$.

84. $x^2 y''_{xx} + ax^n \sin^m(bx) y'_x + c[ax^{n-1}\sin^m(bx) - c - 1]y = 0.$

Particular solution: $y_0 = x^{-c}$.

85. $x^2 y''_{xx} + x(a\cos^n x + 1)y'_x + b(a\cos^n x - b)y = 0.$

Particular solution: $y_0 = x^{-b}$.

86. $x^2 y''_{xx} + x(a\cos^n x + b)y'_x + b(a\cos^n x - 1)y = 0.$

Particular solution: $y_0 = x^{-b}$.

87. $x^2 y''_{xx} + ax^n \cos^m(bx) y'_x + c[ax^{n-1}\cos^m(bx) - c - 1]y = 0.$

Particular solution: $y_0 = x^{-c}$.

88. $x^2 y''_{xx} - 2\lambda x^2 \tan(\lambda x) y'_x + (ax^2 + bx + c)y = 0.$

The substitution $u = y\cos(\lambda x)$ leads to an equation of the form 2.1.2.110:

$$x^2 u''_{xx} + [(a+\lambda^2)x^2 + bx + c]u = 0.$$

89. $x^2 y''_{xx} + x(1 - 2x\tan x)y'_x - (x\tan x + \nu^2)y = 0.$

Solution: $y\cos x = C_1 J_\nu(x) + C_2 Y_\nu(x)$, where J_ν and Y_ν are Bessel functions.

90. $x^2 y''_{xx} - x(2x\tan x + k)y'_x + (ax^2 + bx + c + kx\tan x)y = 0.$

The substitution $u = y\cos x$ leads to an equation of the form 2.1.2.126:

$$x^2 u''_{xx} - kxu'_x + [(a+1)x^2 + bx + c]u = 0.$$

91. $x^2 y''_{xx} + x(a\tan^n x + 1)y'_x + b(a\tan^n x - b)y = 0.$

Particular solution: $y_0 = x^{-b}$.

92. $x^2 y''_{xx} + x(a\tan^n x + b)y'_x + b(a\tan^n x - 1)y = 0.$

Particular solution: $y_0 = x^{-b}$.

93. $x^2 y''_{xx} + ax^n \tan^m(bx) y'_x + c[ax^{n-1}\tan^m(bx) - c - 1]y = 0.$

Particular solution: $y_0 = x^{-c}$.

94. $x^2 y''_{xx} + 2\lambda x^2 \cot(\lambda x) y'_x + (ax^2 + bx + c)y = 0.$

The substitution $u = y\sin(\lambda x)$ leads to an equation of the form 2.1.2.110:

$$x^2 u''_{xx} + [(a+\lambda^2)x^2 + bx + c]u = 0.$$

95. $x^2y''_{xx} + x(2x\cot x + k)y'_x + (ax^2 + bx + c + kx\cot x)y = 0.$

The substitution $u = y\sin x$ leads to an equation of the form 2.1.2.126:

$$x^2u''_{xx} + kxu'_x + [(a+1)x^2 + bx + c]u = 0.$$

96. $x^2y''_{xx} + x(a\cot^n x + 1)y'_x + b(a\cot^n x - b)y = 0.$

Particular solution: $y_0 = x^{-b}$.

97. $x^2y''_{xx} + x(a\cot^n x + b)y'_x + b(a\cot^n x - 1)y = 0.$

Particular solution: $y_0 = x^{-b}$.

98. $x^2y''_{xx} + ax^n\cot^m(bx)y'_x + c[ax^{n-1}\cot^m(bx) - c - 1]y = 0.$

Particular solution: $y_0 = x^{-c}$.

99. $x^4y''_{xx} + \left[a\sin\left(\frac{\lambda}{x}\right) + b\right]y = 0.$

The transformation $\xi = 1/x$, $w = y/x$ leads to an equation of the form 2.1.6.3:

$$w''_{\xi\xi} + [a\sin(\lambda\xi) + b]w = 0.$$

100. $x^4y''_{xx} + ax^2\sin^n(bx)y'_x + [a(c - x)\sin^n(bx) - c^2]y = 0.$

Particular solution: $y_0 = x\exp\left(\frac{c}{x}\right)$.

101. $x^4y''_{xx} + ax^2\cos^n(bx)y'_x + [a(c - x)\cos^n(bx) - c^2]y = 0.$

Particular solution: $y_0 = x\exp\left(\frac{c}{x}\right)$.

102. $x^4y''_{xx} + ax^2\tan^n(bx)y'_x + [a(c - x)\tan^n(bx) - c^2]y = 0.$

Particular solution: $y_0 = x\exp\left(\frac{c}{x}\right)$.

103. $x^4y''_{xx} + ax^2\cot^n(bx)y'_x + [a(c - x)\cot^n(bx) - c^2]y = 0.$

Particular solution: $y_0 = x\exp\left(\frac{c}{x}\right)$.

104. $\sin(2x)y''_{xx} - y'_x + 2a^2\sin^2 x\, y = 0.$

Solution: $y = C_1\sin(au) + C_2\cos(au)$, where $u = \int\sqrt{\tan x}\,dx$.

105. $\sin(2x)y''_{xx} - 2ny'_x + 2a^2\sin^2 x\,(\tan x)^{2n-1}y = 0.$

Solution: $y = C_1\sin(au) + C_2\cos(au)$, where $u = \int\tan^n x\,dx$.

106. $\sin x\, y''_{xx} + \cos x\, y'_x + \nu(\nu + 1)\sin x\, y = 0.$

The substitution $\xi = \cos x$ leads to the Legendre equation 2.1.2.148:

$$(1 - \xi^2)y''_{\xi\xi} - 2\xi\, y'_\xi + \nu(\nu + 1)y = 0.$$

107. $\sin x\, y''_{xx} + (2n+1)\cos x\, y'_x + (\nu - n)(\nu + n + 1)\sin x\, y = 0,$
where ν is an arbitrary number, n is a positive integer.

The substitution $\xi = \cos x$ leads to an equation of the form 2.1.2.149:

$$(\xi^2 - 1)y''_{\xi\xi} + 2(n+1)\xi\, y'_\xi + (n-\nu)(\nu+n+1)y = 0.$$

108. $\sin^2 x\, y''_{xx} + ay = 0.$

This is a special case of equation 2.1.6.110.

109. $\sin^2 x\, y''_{xx} - [a\sin^2 x + n(n-1)]y = 0, \qquad n = 1,\ 2,\ 3,\ \ldots$

Solution: $y = \sin^n x \left(\dfrac{1}{\sin x}\dfrac{d}{dx}\right)^n \left(C_1 e^{x\sqrt{a}} + C_2 e^{-x\sqrt{a}}\right).$

110. $\sin^2 x\, y''_{xx} + (a\sin^2 x + b)y = 0.$

Set $x = 2\xi$. Utilizing the trigonometric formulae $\sin 2\xi = 2\sin\xi\cos\xi$ and $b = b(\sin^2\xi + \cos^2\xi)^2$ and dividing both sides of the equation by $\sin^2 x$, we arrive at an equation of the form 2.1.6.55:

$$y''_{\xi\xi} + (b\tan^2\xi + b\cot^2\xi + 4a + 2b)y = 0.$$

111. $\sin^2 x\, y''_{xx} - \{[(a^2b^2 - (a+1)^2]\sin^2 x + a(a+1)b\sin 2x + a(a-1)\}y = 0.$

Particular solution: $y_0 = e^{abx}\sin^a x\,(\cos x + b\sin x).$

112. $\sin^2 x\, y''_{xx} + \sin x\cos x\, y'_x + [\nu(\nu+1)\sin^2 x - n^2]y = 0,$
where ν is an arbitrary number, n is a nonnegative integer.

The transformation $\xi = \cos x$, $y = w\sin^n x$ leads to an equation of the form 2.1.2.149:

$$(\xi^2 - 1)w''_{\xi\xi} + 2(n+1)\xi w'_\xi + (n-\nu)(n+\nu+1)w = 0.$$

113. $\sin^2 x\, y''_{xx} + \sin x(a\cos x + b)y'_x + (\alpha\cos^2 x + \beta\cos x + \gamma)y = 0.$

Set $x = 2\xi$. Utilizing the trigonometric formulae

$$\sin(2\xi) = 2\sin\xi\cos\xi, \quad \cos(2\xi) = \cos^2\xi - \sin^2\xi, \quad b = b(\sin^2\xi + \cos^2\xi),$$
$$\beta = \beta(\sin^2\xi + \cos^2\xi), \quad \gamma = \gamma(\sin^2\xi + \cos^2\xi)^2,$$

and dividing all the terms by $\sin^2 x$, we arrive at an equation of the form 2.1.6.55:

$$y''_{\xi\xi} + [(b-a)\tan\xi + (b+a)\cot\xi]y'_\xi + [(\alpha-\beta+\gamma)\tan^2\xi + (\alpha+\beta+\gamma)\cot^2\xi + 2\gamma - 2\alpha]y = 0.$$

114. $\cos^2 x\, y''_{xx} - [a\cos^2 x + n(n-1)]y = 0, \qquad n = 1,\ 2,\ 3,\ \ldots$

Solution: $y = \cos^n x \left(\dfrac{1}{\cos x}\dfrac{d}{dx}\right)^n \left(C_1 e^{x\sqrt{a}} + C_2 e^{-x\sqrt{a}}\right).$

115. $\cos^2 x\, y''_{xx} + (a\cos^2 x + b)y = 0.$

The substitution $x = \xi + \frac{\pi}{2}$ leads to 2.1.6.110: $\sin^2\xi\, y''_{\xi\xi} + (a\sin^2\xi + b)y = 0.$

116. $\cos^2 x\, y''_{xx} + a\sin(2x)y'_x + [b\cos(2x) + c]y = 0.$

Dividing the equation by $\cos^2 x$ and utilizing the formulae

$$\sin(2x) = 2\sin x\cos x, \quad \cos(2x) = \cos^2 x - \sin^2 x, \quad c = c(\sin^2 x + \cos^2 x),$$

we obtain an equation of the form 2.1.6.55:

$$y''_{xx} + 2a\tan x\, y'_x + [(c-b)\tan^2 x + b + c]y = 0.$$

117. $\cos^2(ax)y''_{xx} + (n-1)a\sin(2ax)y'_x + na^2[(n-1)\sin^2(ax) + \cos^2(ax)]y = 0.$

Particular solution: $y_0 = \cos^n(ax)$.

118. $\cos^2 x\, y''_{xx} + \cos x\,(a\sin x + b)y'_x + (\alpha\sin^2 x + \beta\sin x + \gamma)y = 0.$

The substitution $x = \xi + \frac{\pi}{2}$ leads to an equation of the form 2.1.6.113: $\sin^2\xi y''_{\xi\xi} - \sin\xi\,(a\cos\xi + b)y'_\xi + (\alpha\cos^2\xi + \beta\cos\xi + \gamma)y = 0$.

119. $\sin x\cos^2 x\, y''_{xx} + \cos x\,(a\sin^2 x + b)y'_x + c\sin x\, y = 0.$

1°. Dividing the equation by $\sin x\cos^2 x$ and assuming $b = b(\sin^2 x + \cos^2 x)$, $c = c(\sin^2 x + \cos^2 x)$, we obtain the equation 2.1.6.55:

$$y''_{xx} + [(a+b)\tan x + b\cot x]y'_x + c(\tan^2 x + 1)y = 0.$$

2°. Particular solutions:

$$\begin{aligned} &y_0 = \cos^a x && \text{for } c = a(b+1),\\ &y_0 = \tan^{1-b} x && \text{for } c = (a+2)(b-1),\\ &y_0 = \sin^{1-b} x\cos^{a+b-1} x && \text{for } c = 2(a+b-1). \end{aligned}$$

120. $\sin x\cos^2 x\, y''_{xx} + \cos x(a\sin^2 x - 1)y'_x + b\sin^3 x\, y = 0.$

Solution: $y = C_1(\cos x)^{k_1} + C_2(\cos x)^{k_2}$, where k_1 and k_2 are the roots of the quadratic equation $k^2 - ak + b = 0$.

121. $\sin^2 x\cos^2 x\, y''_{xx} + (a\sin^2 x + b\cos^2 x + c\sin^2 x\cos^2 x)y = 0.$

Dividing the equation by $\sin^2 x\cos^2 x$ and assuming $a = a(\sin^2 x + \cos^2 x)$, $b = b(\sin^2 x + \cos^2 x)$, we arrive at the equation 2.1.6.55:

$$y''_{xx} + (a\tan^2 x + b\cot^2 x + a + b + c)y = 0.$$

122. $[a\sin(\lambda x) + bx + c]y''_{xx} + a\lambda^2\sin(\lambda x)y = 0.$

Particular solution: $y_0 = a\sin(\lambda x) + bx + c$.

123. $[a\cos(\lambda x) + bx + c]y''_{xx} + a\lambda^2\cos(\lambda x)y = 0.$

Particular solution: $y_0 = a\cos(\lambda x) + bx + c$.

124. $\sin^n(ax)y''_{xx} + (x^2 - b^2)y'_x - (x+b)y = 0.$

Particular solution: $y_0 = x - b$.

125. $\sin^n(\lambda x)y''_{xx} + [\lambda^2 \sin^n(\lambda x) + a\cos^{n-4}(\lambda x)]y = 0.$

The transformation $\xi = \tan(\lambda x)$, $w = \dfrac{y}{\cos(\lambda x)}$ leads to an equation of the form 2.1.2.7: $w''_{\xi\xi} + a\lambda^{-2}\xi^{-n}w = 0$.

126. $(a\sin^n x + b)y''_{xx} + (c\sin^n x + d)y'_x + \lambda[(c - a\lambda)\sin^n x + d - b\lambda]y = 0.$

Particular solution: $y_0 = e^{-\lambda x}$.

127. $\cos^n(\lambda x)y''_{xx} + [\lambda^2 \cos^n(\lambda x) + a\sin^{n-4}(\lambda x)]y = 0.$

The substitution $\lambda x = \frac{\pi}{2} - \lambda\xi$ leads to an equation of the form 2.1.6.125.

128. $\cos^n(ax)y''_{xx} + (x^2 - b^2)y'_x - (x + b)y = 0.$

Particular solution: $y_0 = x - b$.

129. $(a\cos^n x + b)y''_{xx} + (c\cos^n x + d)y'_x + \lambda[(c - a\lambda)\cos^n x + d - b\lambda]y = 0.$

Particular solution: $y_0 = e^{-\lambda x}$.

130. $(a\tan^n x + b)y''_{xx} + (cx + d)y'_x - cy = 0.$

Particular solution: $y_0 = cx + d$.

131. $(a\tan^n x + b)y''_{xx} + (c\tan^n x + d)y'_x + \lambda[(c - a\lambda)\tan^n x + d - b\lambda]y = 0.$

Particular solution: $y_0 = e^{-\lambda x}$.

132. $(a\cot^n x + b)y''_{xx} + (c\cot^n x + d)y'_x + \lambda[(c - a\lambda)\cot^n x + d - b\lambda]y = 0.$

Particular solution: $y_0 = e^{-\lambda x}$.

2.1.7. Equations Containing Inverse Trigonometric Functions

1. $y''_{xx} + (ax + b + c\arcsin x)y'_x + [c(ax + b)\arcsin x + a]y = 0.$

Particular solution: $y_0 = \exp(-\frac{1}{2}ax^2 - bx)$.

2. $y''_{xx} + b(\arcsin x)^n y'_x + c[b(\arcsin x)^n - c]y = 0.$

Particular solution: $y_0 = e^{-cx}$.

3. $y''_{xx} + b(\arcsin x)^n y'_x + a[bx^m(\arcsin x)^n - ax^{2m} + mx^{m-1}]y = 0.$

Particular solution: $y_0 = \exp\Big(-\dfrac{a}{m+1}x^{m+1}\Big)$.

4. $y''_{xx} + (ax + b)(\arcsin x)^n y'_x - a(\arcsin x)^n y = 0.$

Particular solution: $y_0 = ax + b$.

5. $y''_{xx} + ax^n(\arcsin x)^m y'_x - ax^{n-1}(\arcsin x)^m y = 0.$

Particular solution: $y_0 = x.$

6. $y''_{xx} + (ax + b + c\arccos x)y'_x + [c(ax + b)\arccos x + a]y = 0.$

Particular solution: $y_0 = \exp(-\frac{1}{2}ax^2 - bx).$

7. $y''_{xx} + b(\arccos x)^n y'_x + c[b(\arccos x)^n - c]y = 0.$

Particular solution: $y_0 = e^{-cx}.$

8. $y''_{xx} + b(\arccos x)^n y'_x + a[bx^m(\arccos x)^n - ax^{2m} + mx^{m-1}]y = 0.$

Particular solution: $y_0 = \exp\Bigl(-\frac{a}{m+1}x^{m+1}\Bigr).$

9. $y''_{xx} + (ax + b)(\arccos x)^n y'_x - a(\arccos x)^n y = 0.$

Particular solution: $y_0 = ax + b.$

10. $y''_{xx} + ax^n(\arccos x)^m y'_x - ax^{n-1}(\arccos x)^m y = 0.$

Particular solution: $y_0 = x.$

11. $y''_{xx} + (ax + b + c\arctan x)y'_x + [c(ax + b)\arctan x + a]y = 0.$

Particular solution: $y_0 = \exp(-\frac{1}{2}ax^2 - bx).$

12. $y''_{xx} + b(\arctan x)^n y'_x + c[b(\arctan x)^n - c]y = 0.$

Particular solution: $y_0 = e^{-cx}.$

13. $y''_{xx} + b(\arctan x)^n y'_x + a[bx^m(\arctan x)^n - ax^{2m} + mx^{m-1}]y = 0.$

Particular solution: $y_0 = \exp\Bigl(-\frac{a}{m+1}x^{m+1}\Bigr).$

14. $y''_{xx} + (ax + b)(\arctan x)^n y'_x - a(\arctan x)^n y = 0.$

Particular solution: $y_0 = ax + b.$

15. $y''_{xx} + ax^n(\arctan x)^m y'_x - ax^{n-1}(\arctan x)^m y = 0.$

Particular solution: $y_0 = x.$

16. $y''_{xx} + (ax + b + c\,\mathrm{arccot}\, x)y'_x + [c(ax + b)\,\mathrm{arccot}\, x + a]y = 0.$

Particular solution: $y_0 = \exp(-\frac{1}{2}ax^2 - bx).$

17. $y''_{xx} + b(\mathrm{arccot}\, x)^n y'_x + c[b(\mathrm{arccot}\, x)^n - c]y = 0.$

Particular solution: $y_0 = e^{-cx}.$

18. $y''_{xx} + b(\mathrm{arccot}\, x)^n y'_x + a[bx^m(\mathrm{arccot}\, x)^n - ax^{2m} + mx^{m-1}]y = 0.$

Particular solution: $y_0 = \exp\Bigl(-\frac{a}{m+1}x^{m+1}\Bigr).$

19. $y''_{xx} + (ax+b)(\operatorname{arccot} x)^n y'_x - a(\operatorname{arccot} x)^n y = 0.$

Particular solution: $y_0 = ax + b.$

20. $y''_{xx} + ax^n(\operatorname{arccot} x)^m y'_x - ax^{n-1}(\operatorname{arccot} x)^m y = 0.$

Particular solution: $y_0 = x.$

21. $xy''_{xx} + ax \arcsin x\, y'_x - [a(bx+1)\arcsin x + b(bx+2)]y = 0.$

Particular solution: $y_0 = xe^{bx}.$

22. $xy''_{xx} + [a(bx+1)\arcsin x + bx - 1]y'_x + ab^2x \arcsin x\, y = 0.$

Particular solution: $y_0 = (bx+1)e^{-bx}.$

23. $xy''_{xx} + [(ax^2+bx)\arcsin x + 2]y'_x + b\arcsin x\, y = 0.$

Particular solution: $y_0 = a + \dfrac{b}{x}.$

24. $xy''_{xx} + [ax(\arcsin x)^n + b]y'_x + a(b-1)(\arcsin x)^n y = 0.$

Particular solution: $y_0 = x^{1-b}.$

25. $xy''_{xx} + (ax^{n+1} + b\arcsin x)y'_x + ax^n(b\arcsin x + n)y = 0.$

Particular solution: $y_0 = \exp\Bigl(-\dfrac{a}{n+1}x^{n+1}\Bigr).$

26. $xy''_{xx} + (ax^n + bx\arcsin x)y'_x + [b(ax^n - 1)\arcsin x + anx^{n-1}]y = 0.$

Particular solution: $y_0 = x\exp\Bigl(-\dfrac{a}{n}x^n\Bigr).$

27. $x^2y''_{xx} + bx\arcsin x\, y'_x + a(b\arcsin x - a - 1)y = 0.$

Particular solution: $y_0 = x^{-a}.$

28. $x^2y''_{xx} + x(b\arcsin x + 2)y'_x + [b(ax+1)\arcsin x - a^2x^2]y = 0.$

Particular solution: $y_0 = \dfrac{1}{x}e^{-ax}.$

29. $xy''_{xx} + ax\arccos x\, y'_x - [a(bx+1)\arccos x + b(bx+2)]y = 0.$

Particular solution: $y_0 = xe^{bx}.$

30. $xy''_{xx} + [a(bx+1)\arccos x + bx - 1]y'_x + ab^2x\arccos x\, y = 0.$

Particular solution: $y_0 = (bx+1)e^{-bx}.$

31. $xy''_{xx} + [(ax^2+bx)\arccos x + 2]y'_x + b\arccos x\, y = 0.$

Particular solution: $y_0 = a + \dfrac{b}{x}.$

32. $xy''_{xx} + [ax(\arccos x)^n + b]y'_x + a(b-1)(\arccos x)^n y = 0.$

Particular solution: $y_0 = x^{1-b}$.

33. $xy''_{xx} + (ax^{n+1} + b\arccos x)y'_x + ax^n(b\arccos x + n)y = 0.$

Particular solution: $y_0 = \exp\Bigl(-\frac{a}{n+1}x^{n+1}\Bigr)$.

34. $xy''_{xx} + (ax^n + bx\arccos x)y'_x + [b(ax^n - 1)\arccos x + anx^{n-1}]y = 0.$

Particular solution: $y_0 = x\exp\Bigl(-\frac{a}{n}x^n\Bigr)$.

35. $x^2y''_{xx} + bx\arccos x\, y'_x + a(b\arccos x - a - 1)y = 0.$

Particular solution: $y_0 = x^{-a}$.

36. $x^2y''_{xx} + x(b\arccos x + 2)y'_x + [b(ax+1)\arccos x - a^2x^2]y = 0.$

Particular solution: $y_0 = \frac{1}{x}e^{-ax}$.

37. $xy''_{xx} + ax\arctan x\, y'_x - [a(bx+1)\arctan x + b(bx+2)]y = 0.$

Particular solution: $y_0 = xe^{bx}$.

38. $xy''_{xx} + [a(bx+1)\arctan x + bx - 1]y'_x + ab^2x\arctan x\, y = 0.$

Particular solution: $y_0 = (bx+1)e^{-bx}$.

39. $xy''_{xx} + [(ax^2 + bx)\arctan x + 2]y'_x + b\arctan x\, y = 0.$

Particular solution: $y_0 = a + \frac{b}{x}$.

40. $xy''_{xx} + [ax(\arctan x)^n + b]y'_x + a(b-1)(\arctan x)^n y = 0.$

Particular solution: $y_0 = x^{1-b}$.

41. $xy''_{xx} + (ax^{n+1} + b\arctan x)y'_x + ax^n(b\arctan x + n)y = 0.$

Particular solution: $y_0 = \exp\Bigl(-\frac{a}{n+1}x^{n+1}\Bigr)$.

42. $xy''_{xx} + (ax^n + bx\arctan x)y'_x + [b(ax^n - 1)\arctan x + anx^{n-1}]y = 0.$

Particular solution: $y_0 = x\exp\Bigl(-\frac{a}{n}x^n\Bigr)$.

43. $xy''_{xx} + ax(\arctan^n x + b)y'_x - a(\arctan^n x + b)y = 0.$

Particular solution: $y_0 = x$.

44. $xy''_{xx} + b\arctan^n x\, y'_x + a(b\arctan^n x - ax)y = 0.$

Particular solution: $y_0 = e^{-ax}$.

45. $xy''_{xx} + a(\arctan^n x + bx)y'_x + ab\arctan^n x\,y = 0.$

Particular solution: $y_0 = e^{-bx}$.

46. $xy''_{xx} + b\arctan^n x\,y'_x + ax(b\arctan^n x - ax^2 + 1)y = 0.$

Particular solution: $y_0 = \exp(-frac12ax^2)$.

47. $x^2y''_{xx} + bx\arctan x\,y'_x + a(b\arctan x - a - 1)y = 0.$

Particular solution: $y_0 = x^{-a}$.

48. $x^2y''_{xx} + x(b\arctan x + 2)y'_x + [b(ax+1)\arctan x - a^2x^2]y = 0.$

Particular solution: $y_0 = \dfrac{1}{x}e^{-ax}$.

49. $x^2y''_{xx} + ax(\arctan^n x + b)y'_x - a(\arctan^n x + b)y = 0.$

Particular solution: $y_0 = x$.

50. $x^2y''_{xx} + b\arctan^n x\,y'_x + a(b\arctan^n x - ax^2)y = 0.$

Particular solution: $y_0 = e^{-ax}$.

51. $x^2y''_{xx} + a(\arctan^n x + bx^2)y'_x + ab\arctan^n x\,y = 0.$

Particular solution: $y_0 = e^{-bx}$.

52. $x^2y''_{xx} + x[(ax+b)\arctan^n x + 2]y'_x + b\arctan^n x\,y = 0.$

Particular solution: $y_0 = a + \dfrac{b}{x}$.

53. $xy''_{xx} + ax\,\mathrm{arccot}\,x\,y'_x - [a(bx+1)\,\mathrm{arccot}\,x + b(bx+2)]y = 0.$

Particular solution: $y_0 = xe^{bx}$.

54. $xy''_{xx} + [a(bx+1)\,\mathrm{arccot}\,x + bx - 1]y'_x + ab^2x\,\mathrm{arccot}\,x\,y = 0.$

Particular solution: $y_0 = (bx+1)e^{-bx}$.

55. $xy''_{xx} + [(ax^2+bx)\,\mathrm{arccot}\,x + 2]y'_x + b\,\mathrm{arccot}\,x\,y = 0.$

Particular solution: $y_0 = a + \dfrac{b}{x}$.

56. $xy''_{xx} + [ax(\mathrm{arccot}\,x)^n + b]y'_x + a(b-1)(\mathrm{arccot}\,x)^n y = 0.$

Particular solution: $y_0 = x^{1-b}$.

57. $xy''_{xx} + (ax^{n+1} + b\,\mathrm{arccot}\,x)y'_x + ax^n(b\,\mathrm{arccot}\,x + n)y = 0.$

Particular solution: $y_0 = \exp\Big(-\dfrac{a}{n+1}x^{n+1}\Big)$.

58. $xy''_{xx} + (ax^n + bx \operatorname{arccot} x)y'_x + [b(ax^n - 1)\operatorname{arccot} x + anx^{n-1}]y = 0.$

Particular solution: $y_0 = x \exp\left(-\frac{a}{n}x^n\right)$.

59. $x^2y''_{xx} + bx \operatorname{arccot} x\, y'_x + a(b \operatorname{arccot} x - a - 1)y = 0.$

Particular solution: $y_0 = x^{-a}$.

60. $x^2y''_{xx} + x(b \operatorname{arccot} x + 2)y'_x + [b(ax + 1)\operatorname{arccot} x - a^2x^2]y = 0.$

Particular solution: $y_0 = \frac{1}{x}e^{-ax}$.

61. $(ax^2 + b)y''_{xx} + c(ax^2 + b)(\arcsin x)^n y'_x - 2a[cx(\arcsin x)^n + 1]y = 0.$

Particular solution: $y_0 = ax^2 + b$.

62. $(ax^2 + b)y''_{xx} + c(ax^2 + b)(\arccos x)^n y'_x - 2a[cx(\arccos x)^n + 1]y = 0.$

Particular solution: $y_0 = ax^2 + b$.

63. $(x^2 + 1)y''_{xx} - [a^2(x^2 + 1)(\arctan x)^2 + a]y = 0.$

Particular solution: $y_0 = (x^2 + 1)^{-a/2} \exp(ax \arctan x)$.

64. $(ax^2 + b)y''_{xx} + c(ax^2 + b)(\arctan x)^n y'_x - 2a[cx(\arctan x)^n + 1]y = 0.$

Particular solution: $y_0 = ax^2 + b$.

65. $(ax^2 + b)y''_{xx} + c(ax^2 + b)(\operatorname{arccot} x)^n y'_x - 2a[cx(\operatorname{arccot} x)^n + 1]y = 0.$

Particular solution: $y_0 = ax^2 + b$.

66. $x^4y''_{xx} + ax^2 \arcsin x\, y'_x + [a(b - x)\arcsin x - b^2]y = 0.$

Particular solution: $y_0 = x \exp\left(\frac{b}{x}\right)$.

67. $x^4y''_{xx} + ax^2 \arccos x\, y'_x + [a(b - x)\arccos x - b^2]y = 0.$

Particular solution: $y_0 = x \exp\left(\frac{b}{x}\right)$.

68. $x^4y''_{xx} + ax^2 \arctan x\, y'_x + [a(b - x)\arctan x - b^2]y = 0.$

Particular solution: $y_0 = x \exp\left(\frac{b}{x}\right)$.

69. $x^4y''_{xx} + ax^2 \operatorname{arccot} x\, y'_x + [a(b - x)\operatorname{arccot} x - b^2]y = 0.$

Particular solution: $y_0 = x \exp\left(\frac{b}{x}\right)$.

70. $(x^2 + 1)^2y''_{xx} + [a(\arctan x)^2 + b \arctan x + c]y = 0.$

The transformation $\xi = \arctan x$, $w = \dfrac{y}{\sqrt{x^2 + 1}}$ leads to an equation of the form 2.1.2.6: $w''_{\xi\xi} + (a\xi^2 + b\xi + c + 1)w = 0$.

71. $(x^2+1)^2 y''_{xx} + [b(\arctan x)^n - 1]y = 0.$

The transformation $\xi = \arctan x$, $w = \dfrac{y}{\sqrt{x^2+1}}$ leads to an equation of the form 2.1.2.7: $w''_{\xi\xi} + b\xi^n w = 0$.

72. $(x^2+1)^2 y''_{xx} + [a(\operatorname{arccot} x)^2 + b \operatorname{arccot} x + c]y = 0.$

The transformation $\xi = \operatorname{arccot} x$, $w = \dfrac{y}{\sqrt{x^2+1}}$ leads to an equation of the form 2.1.2.6: $w''_{\xi\xi} + (a\xi^2 + b\xi + c + 1)w = 0$.

73. $(x^2+1)^2 y''_{xx} + [b(\operatorname{arccot} x)^n - 1]y = 0.$

The transformation $\xi = \operatorname{arccot} x$, $w = \dfrac{y}{\sqrt{x^2+1}}$ leads to an equation of the form 2.1.2.7: $w''_{\xi\xi} + b\xi^n w = 0$.

74. $(ax^2+b)^2 y''_{xx} + (cx+d)(\arcsin x)^n y'_x - c(\arcsin x)^n y = 0.$

Particular solution: $y_0 = cx + d$.

75. $(x^2+a)^2 y''_{xx} + b(x^2+a)(\arcsin x)^n y'_x - [bx(\arcsin x)^n + a]y = 0.$

Particular solution: $y_0 = \sqrt{x^2+a}$.

76. $(ax^2+b)^2 y''_{xx} + c(ax^2+b)(\arcsin x)^n y'_x + [c(\arcsin x)^n - 2ax - 1]y = 0.$

Particular solution: $y_0 = \exp\Bigl(-\displaystyle\int \frac{dx}{ax^2+b}\Bigr)$.

77. $(ax^2+b)^2 y''_{xx} + (cx+d)(\arccos x)^n y'_x - c(\arccos x)^n y = 0.$

Particular solution: $y_0 = cx + d$.

78. $(x^2+a)^2 y''_{xx} + b(x^2+a)(\arccos x)^n y'_x - [bx(\arccos x)^n + a]y = 0.$

Particular solution: $y_0 = \sqrt{x^2+a}$.

79. $(ax^2+b)^2 y''_{xx} + c(ax^2+b)(\arccos x)^n y'_x + [c(\arccos x)^n - 2ax - 1]y = 0.$

Particular solution: $y_0 = \exp\Bigl(-\displaystyle\int \frac{dx}{ax^2+b}\Bigr)$.

80. $(ax^2+b)^2 y''_{xx} + (cx+d)(\arctan x)^n y'_x - c(\arctan x)^n y = 0.$

Particular solution: $y_0 = cx + d$.

81. $(x^2+a)^2 y''_{xx} + b(x^2+a)(\arctan x)^n y'_x - [bx(\arctan x)^n + a]y = 0.$

Particular solution: $y_0 = \sqrt{x^2+a}$.

82. $(ax^2+b)^2 y''_{xx} + c(ax^2+b)(\arctan x)^n y'_x + [c(\arctan x)^n - 2ax - 1]y = 0.$

Particular solution: $y_0 = \exp\Bigl(-\displaystyle\int \frac{dx}{ax^2+b}\Bigr)$.

83. $(ax^2+b)^2y''_{xx}+(cx+d)(\operatorname{arccot} x)^n y'_x - c(\operatorname{arccot} x)^n y = 0.$

Particular solution: $y_0 = cx + d.$

84. $(x^2+a)^2y''_{xx}+b(x^2+a)(\operatorname{arccot} x)^n y'_x - [bx(\operatorname{arccot} x)^n + a]y = 0.$

Particular solution: $y_0 = \sqrt{x^2+a}.$

85. $(ax^2+b)^2y''_{xx}+c(ax^2+b)(\operatorname{arccot} x)^n y'_x + [c(\operatorname{arccot} x)^n - 2ax - 1]y = 0.$

Particular solution: $y_0 = \exp\Bigl(-\int \frac{dx}{ax^2+b}\Bigr).$

2.1.8. Equations Containing Combinations of Exponential, Logarithmic, Trigonometric, and Other Functions

1. $y''_{xx}+ae^{\lambda x}y'_x+b[b+ae^{\lambda x}\tan(bx)]y=0.$

Particular solution: $y_0 = \cos(bx).$

2. $y''_{xx}+ae^{\lambda x}y'_x+b[b-ae^{\lambda x}\cot(bx)]y=0.$

Particular solution: $y_0 = \sin(bx).$

3. $y''_{xx}+a\cosh^n(\lambda x)y'_x+b[b+a\cosh^n(\lambda x)\tan(bx)]y=0.$

Particular solution: $y_0 = \cos(bx).$

4. $y''_{xx}+a\cosh^n(\lambda x)y'_x+b[b-a\cosh^n(\lambda x)\cot(bx)]y=0.$

Particular solution: $y_0 = \sin(bx).$

5. $y''_{xx}+a\cosh^n(kx)y'_x+be^{\lambda x}[a\cosh^n(kx)-be^{\lambda x}+\lambda]y=0.$

Particular solution: $y_0 = \exp\Bigl(-\frac{b}{\lambda}e^{\lambda x}\Bigr).$

6. $y''_{xx}+a\sinh^n(\lambda x)y'_x+b[b+a\sinh^n(\lambda x)\tan(bx)]y=0.$

Particular solution: $y_0 = \cos(bx).$

7. $y''_{xx}+a\sinh^n(\lambda x)y'_x+b[b-a\sinh^n(\lambda x)\cot(bx)]y=0.$

Particular solution: $y_0 = \sin(bx).$

8. $y''_{xx}+a\sinh^n(kx)y'_x+be^{\lambda x}[a\sinh^n(kx)-be^{\lambda x}+\lambda]y=0.$

Particular solution: $y_0 = \exp\Bigl(-\frac{b}{\lambda}e^{\lambda x}\Bigr).$

9. $y''_{xx}+a\tanh^n(\lambda x)y'_x+b[b+a\tanh^n(\lambda x)\tan(bx)]y=0.$

Particular solution: $y_0 = \cos(bx).$

10. $y''_{xx} + a\tanh^n(\lambda x)y'_x + b[b - a\tanh^n(\lambda x)\cot(bx)]y = 0.$

Particular solution: $y_0 = \sin(bx)$.

11. $y''_{xx} + a\tanh^n(kx)y'_x + be^{\lambda x}[a\tanh^n(kx) - be^{\lambda x} + \lambda]y = 0.$

Particular solution: $y_0 = \exp\left(-\frac{b}{\lambda}e^{\lambda x}\right)$.

12. $y''_{xx} + a\coth^n(\lambda x)y'_x + b[b + a\coth^n(\lambda x)\tan(bx)]y = 0.$

Particular solution: $y_0 = \cos(bx)$.

13. $y''_{xx} + a\coth^n(\lambda x)y'_x + b[b - a\coth^n(\lambda x)\cot(bx)]y = 0.$

Particular solution: $y_0 = \sin(bx)$.

14. $y''_{xx} + a\coth^n(kx)y'_x + be^{\lambda x}[a\coth^n(kx) - be^{\lambda x} + \lambda]y = 0.$

Particular solution: $y_0 = \exp\left(-\frac{b}{\lambda}e^{\lambda x}\right)$.

15. $y''_{xx} + a\ln^n(\lambda x)y'_x + b[b + a\ln^n(\lambda x)\tan(bx)]y = 0.$

Particular solution: $y_0 = \cos(bx)$.

16. $y''_{xx} + a\ln^n(\lambda x)y'_x + b[b - a\ln^n(\lambda x)\cot(bx)]y = 0.$

Particular solution: $y_0 = \sin(bx)$.

17. $y''_{xx} + a\ln^n(kx)y'_x + be^{\lambda x}[a\ln^n(kx) - be^{\lambda x} + \lambda]y = 0.$

Particular solution: $y_0 = \exp\left(-\frac{b}{\lambda}e^{\lambda x}\right)$.

18. $y''_{xx} + a\cos^n(kx)y'_x + be^{\lambda x}[a\cos^n(kx) - be^{\lambda x} + \lambda]y = 0.$

Particular solution: $y_0 = \exp\left(-\frac{b}{\lambda}e^{\lambda x}\right)$.

19. $y''_{xx} + a\sin^n(kx)y'_x + be^{\lambda x}[a\sin^n(kx) - be^{\lambda x} + \lambda]y = 0.$

Particular solution: $y_0 = \exp\left(-\frac{b}{\lambda}e^{\lambda x}\right)$.

20. $y''_{xx} + a\tan^n(kx)y'_x + be^{\lambda x}[a\tan^n(kx) - be^{\lambda x} + \lambda]y = 0.$

Particular solution: $y_0 = \exp\left(-\frac{b}{\lambda}e^{\lambda x}\right)$.

21. $y''_{xx} + a\cot^n(kx)y'_x + be^{\lambda x}[a\cot^n(kx) - be^{\lambda x} + \lambda]y = 0.$

Particular solution: $y_0 = \exp\left(-\frac{b}{\lambda}e^{\lambda x}\right)$.

22. $y''_{xx} + (ae^{\lambda x} + b\ln^n x)y'_x + ae^{\lambda x}(b\ln^n x + \lambda)y = 0.$

Particular solution: $y_0 = \exp\Bigl(-\dfrac{a}{\lambda}e^{\lambda x}\Bigr).$

23. $y''_{xx} + (ae^{\lambda x} + b\cos x)y'_x + b(ae^{\lambda x}\cos x - \sin x)y = 0.$

Particular solution: $y_0 = \exp(-b\sin x).$

24. $y''_{xx} + (ae^{\lambda x} + b\cos^n x)y'_x + ae^{\lambda x}(b\cos^n x + \lambda)y = 0.$

Particular solution: $y_0 = \exp\Bigl(-\dfrac{a}{\lambda}e^{\lambda x}\Bigr).$

25. $y''_{xx} + (ae^{\lambda x} + b\cos^n x)y'_x + b\cos^{n-1} x\,(ae^{\lambda x}\cos x - n\sin x)y = 0.$

Particular solution: $y_0 = \exp\Bigl(-b\displaystyle\int \cos^n x\,dx\Bigr).$

26. $y''_{xx} + (ae^{\lambda x} + b\sin x)y'_x + b(ae^{\lambda x}\sin x + \cos x)y = 0.$

Particular solution: $y_0 = \exp(b\cos x).$

27. $y''_{xx} + (ae^{\lambda x} + b\sin^n x)y'_x + ae^{\lambda x}(b\sin^n x + \lambda)y = 0.$

Particular solution: $y_0 = \exp\Bigl(-\dfrac{a}{\lambda}e^{\lambda x}\Bigr).$

28. $y''_{xx} + (ae^{\lambda x} + b\sin^n x)y'_x + b\sin^{n-1} x\,(ae^{\lambda x}\sin x + n\cos x)y = 0.$

Particular solution: $y_0 = \exp\Bigl(-b\displaystyle\int \sin^n x\,dx\Bigr).$

29. $y''_{xx} + (ae^{\lambda x} + b\tan x)y'_x + (b+1)(ae^{\lambda x}\tan x + 1)y = 0.$

Particular solution: $y_0 = \cos^{b+1} x.$

30. $y''_{xx} + (ae^{\lambda x} + b\tan^n x)y'_x + ae^{\lambda x}(b\tan^n x + \lambda)y = 0.$

Particular solution: $y_0 = \exp\Bigl(-\dfrac{a}{\lambda}e^{\lambda x}\Bigr).$

31. $y''_{xx} + (ae^{\lambda x} + b\cot x)y'_x + (b-1)(ae^{\lambda x}\cot x - 1)y = 0.$

Particular solution: $y_0 = \sin^{1-b} x.$

32. $y''_{xx} + (ae^{\lambda x} + b\cot^n x)y'_x + ae^{\lambda x}(b\cot^n x + \lambda)y = 0.$

Particular solution: $y_0 = \exp\Bigl(-\dfrac{a}{\lambda}e^{\lambda x}\Bigr).$

33. $y''_{xx} + (a\cosh^n x + b\cos x)y'_x + b(a\cosh^n x\cos x - \sin x)y = 0.$

Particular solution: $y_0 = \exp(-b\sin x).$

34. $y''_{xx} + (a\cosh^n x + b\cos^m x)y'_x + b\cos^{m-1} x\,(a\cosh^n x\cos x - m\sin x)y = 0.$

Particular solution: $y_0 = \exp\Bigl(-b\int \cos^m x\,dx\Bigr).$

35. $y''_{xx} + (a\cosh^n x + b\sin x)y'_x + b(a\cosh^n x\sin x + \cos x)y = 0.$

Particular solution: $y_0 = \exp(b\cos x).$

36. $y''_{xx} + (a\cosh^n x + b\sin^m x)y'_x + b\sin^{m-1} x\,(a\cosh^n x\sin x + m\cos x)y = 0.$

Particular solution: $y_0 = \exp\Bigl(-b\int \sin^m x\,dx\Bigr).$

37. $y''_{xx} + (a\cosh^n x + b\tan x)y'_x + (b+1)(a\cosh^n x\tan x + 1)y = 0.$

Particular solution: $y_0 = \cos^{b+1} x.$

38. $y''_{xx} + (a\cosh^n x + b\cot x)y'_x + (b-1)(a\cosh^n x\cot x - 1)y = 0.$

Particular solution: $y_0 = \sin^{1-b} x.$

39. $y''_{xx} + (a\sinh^n x + b\cos x)y'_x + b(a\sinh^n x\cos x - \sin x)y = 0.$

Particular solution: $y_0 = \exp(-b\sin x).$

40. $y''_{xx} + (a\sinh^n x + b\cos^m x)y'_x + b\cos^{m-1} x\,(a\sinh^n x\cos x - m\sin x)y = 0.$

Particular solution: $y_0 = \exp\Bigl(-b\int \cos^m x\,dx\Bigr).$

41. $y''_{xx} + (a\sinh^n x + b\sin x)y'_x + b(a\sinh^n x\sin x + \cos x)y = 0.$

Particular solution: $y_0 = \exp(b\cos x).$

42. $y''_{xx} + (a\sinh^n x + b\sin^m x)y'_x + b\sin^{m-1} x\,(a\sinh^n x\sin x + m\cos x)y = 0.$

Particular solution: $y_0 = \exp\Bigl(-b\int \sin^m x\,dx\Bigr).$

43. $y''_{xx} + (a\sinh^n x + b\tan x)y'_x + (b+1)(a\sinh^n x\tan x + 1)y = 0.$

Particular solution: $y_0 = \cos^{b+1} x.$

44. $y''_{xx} + (a\sinh^n x + b\cot x)y'_x + (b-1)(a\sinh^n x\cot x - 1)y = 0.$

Particular solution: $y_0 = \sin^{1-b} x.$

45. $y''_{xx} + (a\tanh^n x + b\cos x)y'_x + b(a\tanh^n x\cos x - \sin x)y = 0.$

Particular solution: $y_0 = \exp(-b\sin x).$

46. $y''_{xx} + (a\tanh^n x + b\cos^m x)y'_x + b\cos^{m-1} x\,(a\tanh^n x\cos x - m\sin x)y = 0.$

Particular solution: $y_0 = \exp\Bigl(-b\int \cos^m x\,dx\Bigr).$

47. $y''_{xx} + (a\tanh^n x + b\sin x)y'_x + b(a\tanh^n x\sin x + \cos x)y = 0.$

Particular solution: $y_0 = \exp(b\cos x).$

48. $y''_{xx} + (a\tanh^n x + b\sin^m x)y'_x + b\sin^{m-1} x\,(a\tanh^n x\sin x + m\cos x)y = 0.$

Particular solution: $y_0 = \exp\Bigl(-b\int \sin^m x\,dx\Bigr).$

49. $y''_{xx} + (a\tanh^n x + b\tan x)y'_x + (b+1)(a\tanh^n x\tan x + 1)y = 0.$

Particular solution: $y_0 = \cos^{b+1} x.$

50. $y''_{xx} + (a\tanh^n x + b\cot x)y'_x + (b-1)(a\tanh^n x\cot x - 1)y = 0.$

Particular solution: $y_0 = \sin^{1-b} x.$

51. $y''_{xx} + (a\coth^n x + b\cos x)y'_x + b(a\coth^n x\cos x - \sin x)y = 0.$

Particular solution: $y_0 = \exp(-b\sin x).$

52. $y''_{xx} + (a\coth^n x + b\cos^m x)y'_x + b\cos^{m-1} x\,(a\coth^n x\cos x - m\sin x)y = 0.$

Particular solution: $y_0 = \exp\Bigl(-b\int \cos^m x\,dx\Bigr).$

53. $y''_{xx} + (a\coth^n x + b\sin x)y'_x + b(a\coth^n x\sin x + \cos x)y = 0.$

Particular solution: $y_0 = \exp(b\cos x).$

54. $y''_{xx} + (a\coth^n x + b\sin^m x)y'_x + b\sin^{m-1} x\,(a\coth^n x\sin x + m\cos x)y = 0.$

Particular solution: $y_0 = \exp\Bigl(-b\int \sin^m x\,dx\Bigr).$

55. $y''_{xx} + (a\coth^n x + b\tan x)y'_x + (b+1)(a\coth^n x\tan x + 1)y = 0.$

Particular solution: $y_0 = \cos^{b+1} x.$

56. $y''_{xx} + (a\coth^n x + b\cot x)y'_x + (b-1)(a\coth^n x\cot x - 1)y = 0.$

Particular solution: $y_0 = \sin^{1-b} x.$

57. $y''_{xx} + (a\ln^n x + b\cos x)y'_x + b(a\ln^n x\cos x - \sin x)y = 0.$

Particular solution: $y_0 = \exp(-b\sin x).$

58. $y''_{xx} + (a\ln^n x + b\cos^m x)y'_x + b\cos^{m-1} x\,(a\ln^n x\cos x - m\sin x)y = 0.$

Particular solution: $y_0 = \exp\Bigl(-b\int \cos^m x\,dx\Bigr).$

59. $y''_{xx} + (a\ln^n x + b\sin x)y'_x + b(a\ln^n x\sin x + \cos x)y = 0.$

Particular solution: $y_0 = \exp(b\cos x).$

60. $y''_{xx} + (a\ln^n x + b\sin^m x)y'_x + b\sin^{m-1} x\,(a\ln^n x\sin x + m\cos x)y = 0.$

Particular solution: $y_0 = \exp\Bigl(-b\int \sin^m x\,dx\Bigr).$

61. $y''_{xx} + (a\ln^n x + b\tan x)y'_x + (b+1)(a\ln^n x\tan x + 1)y = 0.$

Particular solution: $y_0 = \cos^{b+1} x.$

62. $y''_{xx} + (a\ln^n x + b\cot x)y'_x + (b-1)(a\ln^n x\cot x - 1)y = 0.$

Particular solution: $y_0 = \sin^{1-b} x.$

63. $y''_{xx} + ae^{\lambda x}\cos(bx)y'_x + b[b + ae^{\lambda x}\sin(bx)]y = 0.$

Particular solution: $y_0 = \cos(bx).$

64. $y''_{xx} + ae^{\lambda x}\sin(bx)y'_x + b[b - ae^{\lambda x}\cos(bx)]y = 0.$

Particular solution: $y_0 = \sin(bx).$

65. $y''_{xx} + a\cosh(bx)\ln^n(\lambda x)y'_x - b[b + a\sinh(bx)\ln^n(\lambda x)]y = 0.$

Particular solution: $y_0 = \cosh(bx).$

66. $y''_{xx} + a\cosh(bx)\cos^n(\lambda x)y'_x - b[b + a\sinh(bx)\cos^n(\lambda x)]y = 0.$

Particular solution: $y_0 = \cosh(bx).$

67. $y''_{xx} + a\cosh^n(\lambda x)\cos(bx)y'_x + b[b + a\cosh^n(\lambda x)\sin(bx)]y = 0.$

Particular solution: $y_0 = \cos(bx).$

68. $y''_{xx} + a\cosh(bx)\sin^n(\lambda x)y'_x - b[b + a\sinh(bx)\sin^n(\lambda x)]y = 0.$

Particular solution: $y_0 = \cosh(bx).$

69. $y''_{xx} + a\cosh^n(\lambda x)\sin(bx)y'_x + b[b - a\cosh^n(\lambda x)\cos(bx)]y = 0.$

Particular solution: $y_0 = \sin(bx).$

70. $y''_{xx} + a\cosh(bx)\tan^n(\lambda x)y'_x - b[b + a\sinh(bx)\tan^n(\lambda x)]y = 0.$

Particular solution: $y_0 = \cosh(bx).$

71. $y''_{xx} + a\cosh(bx)\cot^n(\lambda x)y'_x - b[b + a\sinh(bx)\cot^n(\lambda x)]y = 0.$

Particular solution: $y_0 = \cosh(bx).$

72. $y''_{xx} + a\sinh(bx)\ln^n(kx)y'_x - b[b + a\cosh(bx)\ln^n(kx)]y = 0.$

Particular solution: $y_0 = \sinh(bx).$

73. $y''_{xx} + a\sinh(bx)\cos^n(kx)y'_x - b[b + a\cosh(bx)\cos^n(kx)]y = 0.$

Particular solution: $y_0 = \sinh(bx).$

74. $y''_{xx} + a\sinh^n(\lambda x)\cos(bx)y'_x + b[b + a\sinh^n(\lambda x)\sin(bx)]y = 0.$

Particular solution: $y_0 = \cos(bx)$.

75. $y''_{xx} + a\sinh(bx)\sin^n(kx)y'_x - b[b + a\cosh(bx)\sin^n(kx)]y = 0.$

Particular solution: $y_0 = \sinh(bx)$.

76. $y''_{xx} + a\sinh^n(\lambda x)\sin(bx)y'_x + b[b - a\sinh^n(\lambda x)\cos(bx)]y = 0.$

Particular solution: $y_0 = \sin(bx)$.

77. $y''_{xx} + a\sinh(bx)\tan^n(kx)y'_x - b[b + a\cosh(bx)\tan^n(kx)]y = 0.$

Particular solution: $y_0 = \sinh(bx)$.

78. $y''_{xx} + a\sinh(bx)\cot^n(\lambda x)y'_x - b[b + a\cosh(bx)\cot^n(\lambda x)]y = 0.$

Particular solution: $y_0 = \sinh(bx)$.

79. $y''_{xx} + a\tanh^n(\lambda x)\cos(bx)y'_x + b[b + a\tanh^n(\lambda x)\sin(bx)]y = 0.$

Particular solution: $y_0 = \cos(bx)$.

80. $y''_{xx} + a\tanh^n(\lambda x)\sin(bx)y'_x + b[b - a\tanh^n(\lambda x)\cos(bx)]y = 0.$

Particular solution: $y_0 = \sin(bx)$.

81. $y''_{xx} + a\coth^n(\lambda x)\cos(bx)y'_x + b[b + a\coth^n(\lambda x)\sin(bx)]y = 0.$

Particular solution: $y_0 = \cos(bx)$.

82. $y''_{xx} + a\coth^n(\lambda x)\sin(bx)y'_x + b[b - a\coth^n(\lambda x)\cos(bx)]y = 0.$

Particular solution: $y_0 = \sin(bx)$.

83. $y''_{xx} + a\ln^n(\lambda x)\cos(bx)y'_x + b[b + a\ln^n(\lambda x)\sin(bx)]y = 0.$

Particular solution: $y_0 = \cos(bx)$.

84. $y''_{xx} + a\ln^n(\lambda x)\sin(bx)y'_x + b[b - a\ln^n(\lambda x)\cos(bx)]y = 0.$

Particular solution: $y_0 = \sin(bx)$.

85. $y''_{xx} + (a + be^{2\lambda x})\ln^n(kx)y'_x + \lambda[(a - be^{2\lambda x})\ln^n(kx) - \lambda]y = 0.$

Particular solution: $y_0 = be^{\lambda x} + ae^{-\lambda x}$.

86. $y''_{xx} + (a + be^{2\lambda x})\cos^n(kx)y'_x + \lambda[(a - be^{2\lambda x})\cos^n(kx) - \lambda]y = 0.$

Particular solution: $y_0 = be^{\lambda x} + ae^{-\lambda x}$.

87. $y''_{xx} + (a + be^{2\lambda x})\sin^n(kx)y'_x + \lambda[(a - be^{2\lambda x})\sin^n(kx) - \lambda]y = 0.$

Particular solution: $y_0 = be^{\lambda x} + ae^{-\lambda x}$.

88. $y''_{xx} + (a + be^{2\lambda x})\tan^n(kx)y'_x + \lambda[(a - be^{2\lambda x})\tan^n(kx) - \lambda]y = 0.$

Particular solution: $y_0 = be^{\lambda x} + ae^{-\lambda x}$.

89. $y''_{xx} + (a + be^{2\lambda x})\cot^n(kx)y'_x + \lambda[(a - be^{2\lambda x})\cot^n(kx) - \lambda]y = 0.$

Particular solution: $y_0 = be^{\lambda x} + ae^{-\lambda x}$.

90. $y''_{xx} + (a\,\mathrm{sn}^2\,x + b)y = 0.$

The Lamé equation in the form of Jacobi, $\mathrm{sn}\,x$ is the Jacobi elliptic function. See the books by Bateman & Erdélyi (1955, vol. 3) and Kamke (1976) for more information on this equation.

91. $y''_{xx} + [A\wp(x) + B]y = 0.$

The Lamé equation in the form of Weierstrass, $\wp$ is the Weierstrass function. See the books by Bateman & Erdélyi (1955, vol. 3) and Kamke (1976) for more information on this equation.

92. $xy''_{xx} + (ax\ln x + be^{\lambda x})y'_x + a(be^{\lambda x}\ln x + 1)y = 0.$

Particular solution: $y_0 = e^{ax}x^{-ax}$.

93. $xy''_{xx} + (1 - axe^{\lambda x}\ln x)y'_x + ae^{\lambda x}y = 0.$

Particular solution: $y_0 = \ln x$.

94. $xy''_{xx} + (ax\ln x + b\cosh^n x)y'_x + a(b\cosh^n x\ln x + 1)y = 0.$

Particular solution: $y_0 = e^{ax}x^{-ax}$.

95. $xy''_{xx} + (1 - ax\cosh^n x\ln x)y'_x + a\cosh^n x\,y = 0.$

Particular solution: $y_0 = \ln x$.

96. $xy''_{xx} + (ax\ln x + b\sinh^n x)y'_x + a(b\sinh^n x\ln x + 1)y = 0.$

Particular solution: $y_0 = e^{ax}x^{-ax}$.

97. $xy''_{xx} + (1 - ax\sinh^n x\ln x)y'_x + a\sinh^n x\,y = 0.$

Particular solution: $y_0 = \ln x$.

98. $xy''_{xx} + (ax\ln x + b\tanh^n x)y'_x + a(b\tanh^n x\ln x + 1)y = 0.$

Particular solution: $y_0 = e^{ax}x^{-ax}$.

99. $xy''_{xx} + (1 - ax\tanh^n x\ln x)y'_x + a\tanh^n x\,y = 0.$

Particular solution: $y_0 = \ln x$.

100. $xy''_{xx} + (ax\ln x + b\coth^n x)y'_x + a(b\coth^n x\ln x + 1)y = 0.$

Particular solution: $y_0 = e^{ax}x^{-ax}$.

101. $xy''_{xx} + (1 - ax\coth^n x \ln x)y'_x + a\coth^n x\, y = 0.$

Particular solution: $y_0 = \ln x.$

102. $xy''_{xx} + (ax\ln x + b\cos^n x)y'_x + a(b\cos^n x \ln x + 1)y = 0.$

Particular solution: $y_0 = e^{ax}x^{-ax}.$

103. $xy''_{xx} + (1 - ax\cos^n x \ln x)y'_x + a\cos^n x\, y = 0.$

Particular solution: $y_0 = \ln x.$

104. $xy''_{xx} + (ax\ln x + b\sin^n x)y'_x + a(b\sin^n x \ln x + 1)y = 0.$

Particular solution: $y_0 = e^{ax}x^{-ax}.$

105. $xy''_{xx} + (1 - ax\sin^n x \ln x)y'_x + a\sin^n x\, y = 0.$

Particular solution: $y_0 = \ln x.$

106. $xy''_{xx} + (ax\ln x + b\tan^n x)y'_x + a(b\tan^n x \ln x + 1)y = 0.$

Particular solution: $y_0 = e^{ax}x^{-ax}.$

107. $xy''_{xx} + (1 - ax\tan^n x \ln x)y'_x + a\tan^n x\, y = 0.$

Particular solution: $y_0 = \ln x.$

108. $xy''_{xx} + (ax\ln x + b\cot^n x)y'_x + a(b\cot^n x \ln x + 1)y = 0.$

Particular solution: $y_0 = e^{ax}x^{-ax}.$

109. $xy''_{xx} + (1 - ax\cot^n x \ln x)y'_x + a\cot^n x\, y = 0.$

Particular solution: $y_0 = \ln x.$

110. $x^2y''_{xx} + x(a\ln x + be^{\lambda x})y'_x + a(be^{\lambda x}\ln x - \ln x + 1)y = 0.$

Particular solution: $y_0 = \exp(-\tfrac{1}{2}a\ln^2 x).$

111. $x^2y''_{xx} + x(a\ln x + b\cosh^n x)y'_x + a(b\cosh^n x \ln x - \ln x + 1)y = 0.$

Particular solution: $y_0 = \exp(-\tfrac{1}{2}a\ln^2 x).$

112. $x^2y''_{xx} + x(a\ln x + b\sinh^n x)y'_x + a(b\sinh^n x \ln x - \ln x + 1)y = 0.$

Particular solution: $y_0 = \exp(-\tfrac{1}{2}a\ln^2 x).$

113. $x^2y''_{xx} + x(a\ln x + b\tanh^n x)y'_x + a(b\tanh^n x \ln x - \ln x + 1)y = 0.$

Particular solution: $y_0 = \exp(-\tfrac{1}{2}a\ln^2 x).$

114. $x^2y''_{xx} + x(a\ln x + b\coth^n x)y'_x + a(b\coth^n x \ln x - \ln x + 1)y = 0.$

Particular solution: $y_0 = \exp(-\tfrac{1}{2}a\ln^2 x).$

115. $x^2 y''_{xx} + x(a \ln x + b \cos^n x) y'_x + a(b \cos^n x \ln x - \ln x + 1) y = 0.$

Particular solution: $y_0 = \exp(-\frac{1}{2} a \ln^2 x)$.

116. $x^2 y''_{xx} + x(a \ln x + b \sin^n x) y'_x + a(b \sin^n x \ln x - \ln x + 1) y = 0.$

Particular solution: $y_0 = \exp(-\frac{1}{2} a \ln^2 x)$.

117. $x^2 y''_{xx} + x(a \ln x + b \tan^n x) y'_x + a(b \tan^n x \ln x - \ln x + 1) y = 0.$

Particular solution: $y_0 = \exp(-\frac{1}{2} a \ln^2 x)$.

118. $x^2 y''_{xx} + x(a \ln x + b \cot^n x) y'_x + a(b \cot^n x \ln x - \ln x + 1) y = 0.$

Particular solution: $y_0 = \exp(-\frac{1}{2} a \ln^2 x)$.

119. $\sin^2 x \, y''_{xx} + \sin x \, (a + b e^{\lambda x}) y'_x + a(b e^{\lambda x} - \cos x) y = 0.$

Particular solution: $y_0 = \left(\cot \frac{x}{2}\right)^a$.

120. $\sin^2 x \, y''_{xx} + \sin x \, (a + b \cosh^n x) y'_x + a(b \cosh^n x - \cos x) y = 0.$

Particular solution: $y_0 = \left(\cot \frac{x}{2}\right)^a$.

121. $\sin^2 x \, y''_{xx} + \sin x \, (a + b \sinh^n x) y'_x + a(b \sinh^n x - \cos x) y = 0.$

Particular solution: $y_0 = \left(\cot \frac{x}{2}\right)^a$.

122. $\sin^2 x \, y''_{xx} + \sin x \, (a + b \tanh^n x) y'_x + a(b \tanh^n x - \cos x) y = 0.$

Particular solution: $y_0 = \left(\cot \frac{x}{2}\right)^a$.

123. $\sin^2 x \, y''_{xx} + \sin x \, (a + b \coth^n x) y'_x + a(b \coth^n x - \cos x) y = 0.$

Particular solution: $y_0 = \left(\cot \frac{x}{2}\right)^a$.

124. $\sin^2 x \, y''_{xx} + \sin x \, (a + b \ln^n x) y'_x + a(b \ln^n x - \cos x) y = 0.$

Particular solution: $y_0 = \left(\cot \frac{x}{2}\right)^a$.

125. $\cos^2 x \, y''_{xx} + \cos x \, (a + b e^{\lambda x}) y'_x + a(b e^{\lambda x} + \sin x) y = 0.$

Particular solution: $y_0 = \left[\cot\left(\frac{x}{2} + \frac{\pi}{4}\right)\right]^a$.

126. $\cos^2 x \, y''_{xx} + \cos x \, (a + b \cosh^n x) y'_x + a(b \cosh^n x + \sin x) y = 0.$

Particular solution: $y_0 = \left[\cot\left(\frac{x}{2} + \frac{\pi}{4}\right)\right]^a$.

127. $\cos^2 x\, y''_{xx} + \cos x\,(a + b\sinh^n x)y'_x + a(b\sinh^n x + \sin x)y = 0.$

Particular solution: $y_0 = \left[\cot\left(\frac{x}{2} + \frac{\pi}{4}\right)\right]^a.$

128. $\cos^2 x\, y''_{xx} + \cos x\,(a + b\tanh^n x)y'_x + a(b\tanh^n x + \sin x)y = 0.$

Particular solution: $y_0 = \left[\cot\left(\frac{x}{2} + \frac{\pi}{4}\right)\right]^a.$

129. $\cos^2 x\, y''_{xx} + \cos x\,(a + b\coth^n x)y'_x + a(b\coth^n x + \sin x)y = 0.$

Particular solution: $y_0 = \left[\cot\left(\frac{x}{2} + \frac{\pi}{4}\right)\right]^a.$

130. $\cos^2 x\, y''_{xx} + \cos x\,(a + b\ln^n x)y'_x + a(b\ln^n x + \sin x)y = 0.$

Particular solution: $y_0 = \left[\cot\left(\frac{x}{2} + \frac{\pi}{4}\right)\right]^a.$

2.1.9. Equations Containing Arbitrary Functions

Notation: $f = f(x)$ and $g = g(x)$ are arbitrary functions; a, b, c, d, n, m, k, λ, α, β, and γ are arbitrary parameters.

1. $y''_{xx} + ay = f.$

Solution:

$$y = \begin{cases} C_1\cos(kx) + C_2\sin(kx) + \dfrac{1}{k}\displaystyle\int_{x_0}^{x} f(\xi)\sin[k(x-\xi)]\,d\xi & \text{if } a = k^2 > 0, \\ C_1\cosh(kx) + C_2\sinh(kx) + \dfrac{1}{k}\displaystyle\int_{x_0}^{x} f(\xi)\sinh[k(x-\xi)]\,d\xi & \text{if } a = -k^2 < 0, \end{cases}$$

where x_0 is an arbitrary number.

2. $y''_{xx} + ay'_x + by = f.$

The substitution $y = w\exp(-\frac{1}{2}ax)$ leads to an equation of the form 2.1.9.1:

$$w''_{xx} + (b - \tfrac{1}{4}a^2)w = f\exp(\tfrac{1}{2}ax).$$

3. $y''_{xx} + fy'_x = g.$

Solution: $y = C_1 + \displaystyle\int e^{-F}\Big(C_2 + \int e^F g\,dx\Big)\,dx$, where $F = \displaystyle\int f\,dx.$

4. $y''_{xx} + xfy'_x - fy = 0.$

Particular solution: $y_0 = x.$

5. $y''_{xx} + fy'_x + a(f - a)y = 0.$

Particular solution: $y_0 = e^{-ax}.$

6. $\boldsymbol{y''_{xx} + fy'_x + a(x^n f - ax^{2n} + nx^{n-1})y = 0.}$

Particular solution: $y_0 = \exp\Bigl(-\dfrac{a}{n+1}x^{n+1}\Bigr)$.

7. $\boldsymbol{y''_{xx} + (f + ax^n + b)y'_x + [(ax^n + b)f + anx^{n-1}]y = 0.}$

Particular solution: $y_0 = \exp\Bigl(-\dfrac{a}{n+1}x^{n+1} - bx\Bigr)$.

8. $\boldsymbol{xy''_{xx} + xfy'_x - [(ax+1)f + a(ax+2)]y = 0.}$

Particular solution: $y_0 = xe^{ax}$.

9. $\boldsymbol{xy''_{xx} + (xf + a)y'_x + (a-1)fy = 0.}$

Particular solution: $y_0 = x^{1-a}$.

10. $\boldsymbol{xy''_{xx} + [(ax+1)f + ax - 1]y'_x + a^2 xfy = 0.}$

Particular solution: $y_0 = (ax+1)e^{-ax}$.

11. $\boldsymbol{xy''_{xx} + [(ax^2 + bx)f + 2]y'_x + bfy = 0.}$

Particular solution: $y_0 = a + \dfrac{b}{x}$.

12. $\boldsymbol{xy''_{xx} + (f + ax^{n+1})y'_x + ax^n(f+n)y = 0.}$

Particular solution: $y_0 = \exp\Bigl(-\dfrac{a}{n+1}x^{n+1}\Bigr)$.

13. $\boldsymbol{xy''_{xx} + (xf + ax^n)y'_x + [(ax^n - 1)f + anx^{n-1}]y = 0.}$

Particular solution: $y_0 = x\exp\Bigl(-\dfrac{a}{n}x^n\Bigr)$.

14. $\boldsymbol{xy''_{xx} + [(ax^n + 1)f + anx^n + 1 - 2n]y'_x + a^2 nx^{2n-1}fy = 0.}$

Particular solution: $y_0 = (ax^n + 1)\exp(-ax^n)$.

15. $\boldsymbol{x^2 y''_{xx} + \alpha xy'_x + \beta y = f.}$

The nonhomogeneous Euler equation.

The substitution $x = e^t$ leads to an equation of the form 2.1.9.2:

$$y''_{tt} + (\alpha - 1)y'_t + \beta y = f(e^t).$$

16. $\boldsymbol{x^2 y''_{xx} + xy'_x + (x^2 - \nu^2)y = f.}$

The nonhomogeneous Bessel equation.

The general solution is expressed in terms of Bessel functions:

$$y = C_1 J_\nu(x) + C_2 Y_\nu(x) + \frac{\pi}{2}Y_\nu(x)\int xJ_\nu(x)f(x)\,dx - \frac{\pi}{2}J_\nu(x)\int xY_\nu(x)f(x)\,dx.$$

17. $x^2 y''_{xx} + xfy'_x + a(f - a - 1)y = 0.$

Particular solution: $y_0 = x^{-a}.$

18. $x^2 y''_{xx} + x(f + 2a)y'_x + [(bx + a)f - b^2x^2 + a(a - 1)]y = 0.$

Particular solution: $y_0 = x^{-a}e^{-bx}.$

19. $x^2 y''_{xx} + xfy'_x + [(ax^{2n+1} + n)f - a^2x^{4n+2} - n^2 - n]y = 0.$

Particular solution: $y_0 = x^{-n}\exp\Bigl(-\frac{a}{2n+1}x^{2n+1}\Bigr).$

20. $(ax^2 + bx + c)y''_{xx} + (x + k)fy'_x - fy = 0.$

Particular solution: $y_0 = x + k.$

21. $x^4 y''_{xx} + x^2 fy'_x + [(\lambda - x)f - \lambda^2]y = 0.$

Particular solution: $y_0 = x\exp\Bigl(\frac{\lambda}{x}\Bigr).$

22. $x^2(ax^2 + b)y''_{xx} + x(ax^2 + b)fy'_x - [(ax^2 - b)f + 2b]y = 0.$

Particular solution: $y_0 = ax + \frac{b}{x}.$

23. $(x^2 + a)^2 y''_{xx} + (x^2 + a)fy'_x - (xf + a)y = 0.$

Particular solution: $y_0 = \sqrt{x^2 + a}.$

24. $(x^2 + a)^2 y''_{xx} + (x^2 + a)fy'_x - m[xf + (m - 1)x^2 + a]y = 0.$

Particular solution: $y_0 = (x^2 + a)^{m/2}.$

25. $(ax^n + b)y''_{xx} + (ax^n + b)fy'_x - anx^{n-2}(xf + n - 1)y = 0.$

Particular solution: $y_0 = ax^n + b.$

26. $(ax^n + bx)y''_{xx} + (ax^n + bx)fy'_x - [(anx^{n-1} + b)f + an(n - 1)x^{n-2}]y = 0.$

Particular solution: $y_0 = ax^n + bx.$

27. $(x^n + a)^2 y''_{xx} + (x^n + a)fy'_x - x^{n-2}(xf + an - a)y = 0.$

Particular solution: $y_0 = (x^n + a)^{1/n}.$

28. $(ax^n + b)^2 y''_{xx} + (ax^n + b)fy'_x + (f - anx^{n-1} - 1)y = 0.$

Particular solution: $y_0 = \exp\Bigl(-\int \frac{dx}{ax^n + b}\Bigr).$

29. $f(x)y''_{xx} + [ax^2 + (ac + b)x + bc]y'_x - (ax + b)y = 0.$

Particular solution: $y_0 = x + c.$

30. $y''_{xx} - (f^2 + f'_x)y = 0.$

Particular solution: $y_0 = \exp\Bigl(\int f\,dx\Bigr).$

31. $y''_{xx} + fy'_x - [a(a+1)f^2 + af'_x]y = 0.$

Particular solution: $y_0 = \exp\Bigl(a\int f\,dx\Bigr).$

32. $y''_{xx} + 2fy'_x + (f^2 + f'_x)y = 0.$

Solution: $y = (C_2x + C_1)\exp\Bigl(-\int f\,dx\Bigr).$

33. $y''_{xx} + (1-a)fy'_x - a(f^2 + f'_x)y = 0.$

Particular solution: $y_0 = \exp\Bigl(a\int f\,dx\Bigr).$

34. $y''_{xx} + fy'_x + (fg - g^2 + g'_x)y = 0.$

Particular solution: $y_0 = \exp\Bigl(-\int g\,dx\Bigr).$

35. $y''_{xx} + 2fy'_x + (f^2 + f'_x + a)y = 0.$

The substitution $u = y\exp\Bigl(\int f\,dx\Bigr)$ leads to a constant coefficient linear equation: $u''_{xx} + au = 0.$

36. $y''_{xx} + 2fy'_x + (f^2 + f'_x + ax^{2n} + bx^{n-1})y = 0.$

The substitution $w = y\exp\Bigl(\int f\,dx\Bigr)$ leads to an equation of the form 2.1.2.10: $w''_{xx} + a(x^{2n} + bx^{n-1})w = 0.$

37. $y''_{xx} + (2f + a)y'_x + (f^2 + af + f'_x + b)y = 0.$

The substitution $u = y\exp\Bigl(\int f\,dx\Bigr)$ leads to a constant coefficient linear equation: $u''_{xx} + au'_x + bu = 0.$

38. $y''_{xx} + (f + g)y'_x + (fg + f'_x)y = 0.$

Particular solution: $y_0 = \exp\Bigl(-\int f\,dx\Bigr).$

39. $xy''_{xx} + xfy'_x + (f + xf'_x)y = 0.$

Particular solution: $y_0 = x\exp\Bigl(-\int f\,dx\Bigr).$

40. $xy''_{xx} + (xf + a)y'_x + (af + xf'_x)y = 0.$

Particular solution: $y_0 = \exp\Bigl(-\int f\,dx\Bigr).$

41. $(x+a)y''_{xx}+(f+b)y'_x+f'_x y=0.$

Particular solution: $y_0=\exp\Bigl(\int\frac{1-b-f}{x+a}\,dx\Bigr).$

42. $x^2y''_{xx}+x(2f+1)y'_x+(f^2+xf'_x+x^2-a)y=0.$

The substitution $y=w\exp\bigl(-\int\frac{f}{x}\,dx\bigr)$ leads to the Bessel equation 2.1.2.121:

$$x^2w''_{xx}+xw'_x+(x^2-a)w=0.$$

43. $x^2y''_{xx}+x(2f+a)y'_x+[f^2+(a-1)f+xf'_x+bx^n+c]y=0.$

The substitution $w=y\exp\Bigl(\int\frac{f}{x}\,dx\Bigr)$ leads to an equation of the form 2.1.2.127:

$$x^2w''_{xx}+axw'_x+(bx^n+c)w=0.$$

44. $x^2y''_{xx}+x(2f+ax^n+b)y'_x+[f^2+(ax^n+b-1)f+xf'_x+\alpha x^{2n}+\beta x^n+\gamma]y=0.$

The substitution $w=y\exp\Bigl(\int\frac{f}{x}\,dx\Bigr)$ leads to an equation of the form 2.1.2.141:

$$x^2w''_{xx}+(ax^n+b)xw'_x+(\alpha x^{2n}+\beta x^n+\gamma)w=0.$$

45. $2fy''_{xx}+f'_xy'_x+ay=0.$

The substitution $\xi=\int\frac{dx}{\sqrt{f}}$ leads to a constant coefficient equation: $2y''_{\xi\xi}+ay=0.$

46. $fy''_{xx}-f'_xy'_x-af^3y=0.$

Solution: $y=C_1e^u+C_2e^{-u}$, where $u=\sqrt{a}\int f\,dx.$

47. $fy''_{xx}-(f'_x+af^2)y'_x+bf^3y=0.$

Solution:

$$y=C_1\exp\Bigl(\lambda_1\int f\,dx\Bigr)+C_2\exp\Bigl(\lambda_2\int f\,dx\Bigr),$$

where λ_1 and λ_2 are the roots of the quadratic equation $\lambda^2-a\lambda+b=0.$

48. $fy''_{xx}-(f'_x+af)y'_x-bf^2(a+bf)y=0.$

Particular solution: $y_0=\exp\Bigl(-b\int f\,dx\Bigr).$

49. $fy''_{xx}-(f'_x+2af)y'_x+(af'_x+a^2f-b^2f^3)y=0.$

Particular solution: $y_0=e^{ax}\exp\Bigl(b\int f\,dx\Bigr).$

50. $f^2 y''_{xx} + f(f'_x + a)y'_x + by = 0.$

The substitution $\xi = \int \frac{dx}{f}$ leads to a constant coefficient linear equation: $y''_{\xi\xi} + ay'_\xi + by = 0$.

51. $f^2 y''_{xx} + f(f'_x + 2g + a)y'_x + (fg'_x + g^2 + ag + b)y = 0.$

The transformation $\xi = \int \frac{dx}{f}$, $u = y\exp\Bigl(\int \frac{g}{f}\,dx\Bigr)$ leads to a constant coefficient equation: $u''_{\xi\xi} + au'_\xi + bu = 0$.

52. $fgy''_{xx} - (af'_x g + bfg'_x)y'_x - \lambda f^{2a+1} g^{2b+1} y = 0.$

Solution: $y = C_1 e^u + C_2 e^{-u}$, where $u = \sqrt{\lambda}\int f^a g^b\,dx$.

53. $fy''_{xx} - f''_{xx} y = 0.$

Solution: $y = C_1 f + C_2 f \int f^{-2}\,dx$.

54. $4f^2 y''_{xx} - [2ff''_{xx} - (f'_x)^2 + a]y = 0.$

1°. Solution with $a = 0$:

$$y = \sqrt{f}\left(C_1 + C_2 \int \frac{dx}{f}\right).$$

2°. Solution with $a > 0$:

$$y = \sqrt{f}\,(C_1 e^{\varphi} + C_2 e^{-\varphi}), \qquad \text{where} \quad \varphi = \frac{\sqrt{a}}{2}\int \frac{dx}{f}.$$

3°. Solution with $a < 0$:

$$y = \sqrt{f}\,(C_1 \cos\varphi + C_2 \sin\varphi), \qquad \text{where} \quad \varphi = \frac{\sqrt{-a}}{2}\int \frac{dx}{f}.$$

55. $y''_{xx} - \dfrac{f''_{xx}}{f'_x} y'_x + a^2 (f'_x)^2 f^{2n-2} y = 0.$

Solution:

$$y = \sqrt{f}\left[C_1 J_{\frac{1}{2n}}\Bigl(\frac{a}{n} f^n\Bigr) + C_2 Y_{\frac{1}{2n}}\Bigl(\frac{a}{n} f^n\Bigr)\right],$$

where J_m and Y_m are Bessel functions.

56. $y''_{xx} + \left(\dfrac{ff'_x}{f^2 + a} - \dfrac{f''_{xx}}{f'_x}\right) y'_x - \dfrac{b^2 (f'_x)^2}{f^2 + a} y = 0.$

Solution: $y = C_1\bigl(f + \sqrt{f^2 + a}\bigr)^b + C_2\bigl(f + \sqrt{f^2 + a}\bigr)^{-b}$.

57. $y''_{xx} + fy'_x + ae^{\lambda x}(f - ae^{\lambda x} + \lambda)y = 0.$

Particular solution: $y_0 = \exp\Bigl(-\frac{a}{\lambda} e^{\lambda x}\Bigr)$.

58. $y''_{xx} + (f + ae^{\lambda x})y'_x + ae^{\lambda x}(f + \lambda)y = 0.$

Particular solution: $y_0 = \exp\left(-\frac{a}{\lambda}e^{\lambda x}\right).$

59. $y''_{xx} + (a + be^{2\lambda x})fy'_x + \lambda[(a - be^{2\lambda x})f - \lambda]y = 0.$

Particular solution: $y_0 = be^{\lambda x} + ae^{-\lambda x}.$

60. $y''_{xx} + 2fy'_x + (f^2 + f'_x + ae^{2\lambda x} + be^{\lambda x} + c)y = 0.$

The substitution $u = y\exp\left(\int f\,dx\right)$ leads to an equation of the form 2.1.3.5: $u''_{xx} + (ae^{2\lambda x} + be^{\lambda x} + c)u = 0.$

61. $y''_{xx} + f\sinh(ax)y'_x - a[a + f\cosh(ax)]y = 0.$

Particular solution: $y_0 = \sinh(ax).$

62. $y''_{xx} + f\cosh(ax)y'_x - a[a + f\sinh(ax)]y = 0.$

Particular solution: $y_0 = \cosh(ax).$

63. $y''_{xx} - f'_x y'_x + a^2e^{2f}y = 0.$

Solution: $y = C_1\sin\left(a\int e^f\,dx\right) + C_2\cos\left(a\int e^f\,dx\right).$

64. $y''_{xx} - f'_x y'_x - a^2e^{2f}y = 0.$

Solution: $y = C_1\exp\left(a\int e^f\,dx\right) + C_2\exp\left(-a\int e^f\,dx\right).$

65. $(ae^{\lambda x} + b)^2y''_{xx} + (ae^{\lambda x} + b)fy'_x + ce^{\lambda x}(f - ce^{\lambda x} + \lambda b)y = 0.$

Particular solution: $y_0 = \left(ae^{\lambda x} + b\right)^{-\frac{c}{a\lambda}}.$

66. $xy''_{xx} + (1 - fx\ln x)y'_x + fy = 0.$

Particular solution: $y_0 = \ln x.$

67. $xy''_{xx} + (f + ax\ln x)y'_x + a(f\ln x + 1)y = 0.$

Particular solution: $y_0 = e^{ax}x^{-ax}.$

68. $x^2y''_{xx} + 2x(\ln x + a)fy'_x + [\frac{1}{4} - (\ln x + a + 2)f]y = 0.$

Particular solution: $y_0 = \sqrt{x}\,(\ln x + a).$

69. $x^2y''_{xx} + x(f + a\ln x)y'_x + a(f\ln x - \ln x + 1)y = 0.$

Particular solution: $y_0 = \exp(-\frac{1}{2}a\ln^2 x).$

70. $y''_{xx} + f\sin(ax)y'_x + a[a - f\cos(ax)]y = 0.$

Particular solution: $y_0 = \sin(ax).$

71. $y''_{xx} + f\cos(ax)y'_x + a[a + f\sin(ax)]y = 0.$

Particular solution: $y_0 = \cos(ax)$.

72. $y''_{xx} + fy'_x + a[\lambda + f\tan(\lambda x) + (\lambda - a)\tan^2(\lambda x)]y = 0.$

Particular solution: $y_0 = [\cos(\lambda x)]^{a/\lambda}$.

73. $y''_{xx} + fy'_x + a[\lambda - f\cot(\lambda x) + (\lambda - a)\cot^2(\lambda x)]y = 0.$

Particular solution: $y_0 = [\sin(\lambda x)]^{a/\lambda}$.

74. $y''_{xx} + (f + a\sin x)y'_x + a(f\sin x + \cos x)y = 0.$

Particular solution: $y_0 = \exp(a\cos x)$.

75. $y''_{xx} + (f + a\cos x)y'_x + a(f\cos x - \sin x)y = 0.$

Particular solution: $y_0 = \exp(-a\sin x)$.

76. $y''_{xx} + (f + a\tan x)y'_x + (a+1)(f\tan x + 1)y = 0.$

Particular solution: $y_0 = \cos^{a+1} x$.

77. $y''_{xx} + (f + a\cot x)y'_x + (a-1)(f\cot x - 1)y = 0.$

Particular solution: $y_0 = \sin^{1-a} x$.

78. $y''_{xx} + \tan x\,(f + a - 1)y'_x + [(a\tan^2 x - 1)f + 2a + 2]y = 0.$

Particular solution: $y_0 = \sin x\cos^a x$.

79. $y''_{xx} + (f + a\cos^n x)y'_x + a\cos^{n-1} x\,(f\cos x - n\sin x)y = 0.$

Particular solution: $y_0 = \exp\Bigl(-a\int\cos^n x\,dx\Bigr)$.

80. $y''_{xx} + (f + a\sin^n x)y'_x + a\sin^{n-1} x\,(f\sin x + n\cos x)y = 0.$

Particular solution: $y_0 = \exp\Bigl(-a\int\sin^n x\,dx\Bigr)$.

81. $\sin^2 x\,y''_{xx} + \sin x\,(f + a)y'_x + a(f - \cos x)y = 0.$

Particular solution: $y_0 = \Bigl(\cot\dfrac{x}{2}\Bigr)^a$.

82. $\cos^2 x\,y''_{xx} + \cos x\,(a + f)y'_x + a(f + \sin x)y = 0.$

Particular solution: $y_0 = \Bigl[\cot\Bigl(\dfrac{x}{2} + \dfrac{\pi}{4}\Bigr)\Bigr]^a$.

2.1.10. Some Transformations

Notation: f, g, and h are arbitrary composite functions of their argument which is written in parentheses following the name of a function (the argument is a function of x).

1. $y''_{xx} + x^{-4} f\left(\frac{1}{x}\right) y = 0.$

The transformation $\xi = \frac{1}{x}$, $w = \frac{y}{x}$ leads to the equation $w''_{\xi\xi} + f(\xi)w = 0$.

2. $y''_{xx} + (cx+d)^{-4} f\left(\frac{ax+b}{cx+d}\right) y = 0.$

The transformation $\xi = \frac{ax+b}{cx+d}$, $w = \frac{y}{cx+d}$ leads to a simpler equation: $w''_{\xi\xi} + \Delta^{-2} f(\xi)w = 0$, where $\Delta = ad - bc$.

3. $x^2 y''_{xx} + \left[\frac{1-n^2}{4} + x^{2n} f(ax^n + b)\right] y = 0.$

The transformation $\xi = ax^n + b$, $w = yx^{\frac{n-1}{2}}$ leads to a simpler equation: $w''_{\xi\xi} + (an)^{-2} f(\xi)w = 0$.

4. $x^2 y''_{xx} + x(xf + a)y'_x + (xg + b)y = 0, \qquad f = f(x), g = g(x).$

The substitution $y = x^k w$, where k is a root of the quadratic equation $k^2 + (a-1)k + b = 0$, leads to the equation $xw''_{xx} + (xf + a + 2k)w'_x + (g + kf)w = 0$.

5. $xP_n(x)y''_{xx} + Q_n(x)y'_x + R_{n-1}(x)y = 0,$

where $P_n(x) = \sum_{m=0}^{n} a_m x^m$, $Q_n(x) = \sum_{m=0}^{n} b_m x^m$, $R_{n-1}(x) = \sum_{m=0}^{n-1} c_m x^m$.

The substitution $y = x^k w$, where $k = 1 - \frac{b_0}{a_0}$, leads to an equation of the similar form:

$$xP_n(x)w''_{xx} + [Q_n(x) + 2kP_n(x)]w'_x + [R_{n-1}(x) + F_{n-1}(x)]w = 0,$$

where $F_{n-1}(x) = \frac{k}{x}[Q_n(x) + (k-1)P_n(x)]$.

6. $x(x-1)P_{n-1}(x)y''_{xx} + Q_n(x)y'_x + R_{n-1}(x)y = 0,$

where $P_{n-1}(x) = \sum_{m=0}^{n-1} a_m x^m$, $Q_n(x) = \sum_{m=0}^{n} b_m x^m$, $R_{n-1}(x) = \sum_{m=0}^{n-1} c_m x^m$.

The transformation $\xi = \frac{x}{x-1}$, $w = |x-1|^{-k} y$, where k is a root of the quadratic equation $a_{n-1}k^2 + (b_n - a_{n-1})k + c_{n-1} = 0$, leads to an equation of the similar form:

$$\xi(\xi-1)\widehat{P}_{n-1}(\xi)w''_{\xi\xi} + [2(1-k)\xi\widehat{P}_{n-1}(\xi) - \widehat{Q}_n(\xi)]w'_\xi + [k(k-1)\widehat{P}_{n-1}(\xi) + F_{n-1}(\xi)]w = 0,$$

where

$$\widehat{P}_{n-1}(\xi) = \sum_{m=0}^{n-1} a_m \xi^m (\xi-1)^{n-m-1}, \quad \widehat{Q}_n(\xi) = \sum_{m=0}^{n} b_m \xi^m (\xi-1)^{n-m},$$

$$\widehat{R}_{n-1}(\xi) = \sum_{m=0}^{n-1} c_m \xi^m (\xi-1)^{n-m-1}, \quad F_{n-1}(\xi) = \frac{\widehat{R}_{n-1}(\xi) + k\widehat{Q}_n(\xi) + k(k-1)\widehat{P}_{n-1}(\xi)}{\xi - 1}.$$

7. $\boldsymbol{y''_{xx} + \left[e^{2\lambda x} f(ae^{\lambda x} + b) - \frac{1}{4}\lambda^2\right] y = 0.}$

The transformation $\xi = ae^{\lambda x} + b$, $w = ye^{\lambda x/2}$ leads to a simpler equation: $w''_{\xi\xi} + (a\lambda)^{-2} f(\xi) w = 0$.

8. $\boldsymbol{y''_{xx} + f(e^{\lambda x}) y'_x + g(e^{\lambda x}) y = 0.}$

The substitution $z = e^{\lambda x}$ leads to the equation $\lambda^2 z^2 y''_{zz} + \lambda z[f(z) + \lambda] y'_z + g(z) y = 0$.

9. $\boldsymbol{y''_{xx} + \left[-\lambda^2 + \sinh^{-4}(\lambda x) f\big(\coth(\lambda x)\big)\right] y = 0.}$

The transformation $\xi = \coth(\lambda x)$, $w = \dfrac{y}{\sinh(\lambda x)}$ leads to a simpler equation: $w''_{\xi\xi} + \lambda^{-2} f(\xi) w = 0$.

10. $\boldsymbol{y''_{xx} + \left[-\lambda^2 + \cosh^{-4}(\lambda x) f\big(\tanh(\lambda x)\big)\right] y = 0.}$

The transformation $\xi = \tanh(\lambda x)$, $w = \dfrac{y}{\cosh(\lambda x)}$ leads to a simpler equation: $w''_{\xi\xi} + \lambda^{-2} f(\xi) w = 0$.

11. $\boldsymbol{y''_{xx} + \left[-\frac{1}{4}\lambda^2 + \frac{e^{2\lambda x}}{(ce^{\lambda x} + d)^4} f\left(\frac{ae^{\lambda x} + b}{ce^{\lambda x} + d}\right)\right] y = 0.}$

The transformation $\xi = \dfrac{ae^{\lambda x} + b}{ce^{\lambda x} + d}$, $w = \dfrac{ye^{\lambda x/2}}{ce^{\lambda x} + d}$ leads to a simpler equation: $w''_{\xi\xi} + (\Delta\lambda)^{-2} f(\xi) w = 0$, where $\Delta = ad - bc$.

12. $\boldsymbol{f y''_{xx} + (2f \tanh x + g) y'_x + (g \tanh x + h) y = 0, \qquad f = f(x),\ g = g(x),\ h = h(x).}$

The substitution $u = y \cosh x$ leads to the simpler equation: $f u''_{xx} + g u'_x + (h - f) u = 0$.

13. $\boldsymbol{f y''_{xx} + (2f \coth x + g) y'_x + (g \coth x + h) y = 0, \qquad f = f(x),\ g = g(x),\ h = h(x).}$

The substitution $u = y \sinh x$ leads to a simpler equation $f u''_{xx} + g u'_x + (h - f) u = 0$.

14. $\boldsymbol{x^2 y''_{xx} + [f(a \ln x + b) + \frac{1}{4}] y = 0.}$

The transformation $\xi = a \ln x + b$, $w = yx^{-1/2}$ leads to a simpler equation $w''_{\xi\xi} + a^{-2} f(\xi) w = 0$.

15. $\boldsymbol{(x^2 - 1)^2 y''_{xx} + f\left(\ln \frac{ax - a}{x + 1}\right) y = 0.}$

The transformation $\xi = \ln \dfrac{ax - a}{x + 1}$, $w = \dfrac{y}{\sqrt{|x^2 - 1|}}$ leads to a simpler equation: $4w''_{\xi\xi} + [f(\xi) - 1] w = 0$.

16. $\boldsymbol{x^2 f(\ln x) y''_{xx} + x g(\ln x) y'_x + h(\ln x) y = 0.}$

The substitution $\xi = \ln x$ leads to the equation $f(\xi) y''_{\xi\xi} + [g(\xi) - f(\xi)] y'_\xi + h(\xi) y = 0$.

17. $y''_{xx} + [\lambda^2 + \sin^{-4}(\lambda x) f(\cot(\lambda x))]y = 0.$

The transformation $\xi = \cot(\lambda x)$, $w = \dfrac{y}{\sin(\lambda x)}$ leads to a simpler equation: $w''_{\xi\xi} + \lambda^{-2} f(\xi) w = 0$.

18. $y''_{xx} + \big[\lambda^2 + \cos^{-4}(\lambda x) f\big(\tan(\lambda x)\big)\big]y = 0.$

The transformation $\xi = \tan(\lambda x)$, $w = \dfrac{y}{\cos(\lambda x)}$ leads to a simpler equation: $w''_{\xi\xi} + \lambda^{-2} f(\xi) w = 0$.

19. $y''_{xx} + \left[\lambda^2 + \dfrac{1}{\sin^4(\lambda x + b)} f\left(\dfrac{\sin(\lambda x + a)}{\sin(\lambda x + b)}\right)\right] y = 0.$

The transformation $\xi = \dfrac{\sin(\lambda x + a)}{\sin(\lambda x + b)}$, $w = \dfrac{y}{\sin(\lambda x + b)}$ leads to a simpler equation: $w''_{\xi\xi} + [\lambda \sin(b - a)]^{-2} f(\xi) w = 0$.

20. $f y''_{xx} + (g - 2f \tan x) y'_x + (h - g \tan x) y = 0, \qquad f = f(x),\ g = g(x),\ h = h(x).$

The substitution $u = y \cos x$ leads to a simpler equation: $f u''_{xx} + g u'_x + (f + h) u = 0$.

21. $f y''_{xx} + (g + 2f \cot x) y'_x + (h + g \cot x) y = 0, \qquad f = f(x),\ g = g(x),\ h = h(x).$

The substitution $u = y \sin x$ leads to a simpler equation: $f u''_{xx} + g u'_x + (f + h) u = 0$.

22. $(x^2 + 1)^2 y''_{xx} + f(\arctan x + b) y = 0.$

The transformation $\xi = \arctan x + b$, $w = \dfrac{y}{\sqrt{x^2 + 1}}$ leads to a simpler equation: $w''_{\xi\xi} + [f(\xi) + 1] w = 0$.

23. $(x^2 + 1)^2 y''_{xx} + f(\operatorname{arccot} x + b) y = 0.$

The transformation $\xi = \operatorname{arccot} x + b$, $w = \dfrac{y}{\sqrt{x^2 + 1}}$ leads to a simpler equation: $w''_{\xi\xi} + [f(\xi) + 1] w = 0$.

24. $y''_{xx} + f(x) y = 0.$

The transformation $x = \varphi(\xi)$, $y = w\sqrt{|\varphi'_\xi|}$ leads to an equation of the similar form

$$w''_{\xi\xi} + \left[\frac{1}{2}\frac{\varphi'''_{\xi\xi\xi}}{\varphi'_\xi} - \frac{3}{4}\left(\frac{\varphi''_{\xi\xi}}{\varphi'_\xi}\right)^2 + (\varphi'_\xi)^2 f(\varphi)\right] w = 0.$$

2.1.11. Asymptotic Solutions

This subsection presents asymptotic solutions, as $\varepsilon \to 0$ ($\varepsilon > 0$), of some second-order linear ordinary differential equations containing arbitrary functions (sufficiently smooth), with the independent variable being a real number.

1. Consider the equation

$$\varepsilon^2 y''_{xx} - f(x) y = 0 \tag{1}$$

on a closed interval $a \le x \le b$.

Case 1. With the condition $f \neq 0$, the leading terms of the asymptotic expansions of the fundamental system of solutions, as $\varepsilon \to 0$, are given by the formulae

$$y_1 = f^{-1/4} \exp\biggl(-\frac{1}{\varepsilon}\int \sqrt{f}\,dx\biggr), \qquad y_2 = f^{-1/4} \exp\biggl(\frac{1}{\varepsilon}\int \sqrt{f}\,dx\biggr) \qquad \text{if } f > 0,$$

$$y_1 = (-f)^{-1/4} \cos\biggl(\frac{1}{\varepsilon}\int \sqrt{-f}\,dx\biggr), \quad y_2 = (-f)^{-1/4} \sin\biggl(\frac{1}{\varepsilon}\int \sqrt{-f}\,dx\biggr) \quad \text{if } f < 0.$$

Case 2. Discuss the asymptotic solution of equation (1) in the vicinity of the point $x = x_0$ where function $f(x)$ vanishes (such a point is referred to as a transition point). We assume that function f can be presented in the form

$$f(x) = (x_0 - x)\psi(x), \qquad \text{where} \quad \psi(x) > 0.$$

In this case, the asymptotic solution, as $\varepsilon \to 0$ ($\varepsilon > 0$), is described by three different formulae:

with $x > x_0$,

$$y = [-f(x)]^{-1/4}\biggl\{C_1 \cos\biggl[\frac{1}{\varepsilon}\int_{x_0}^{x} \sqrt{-f(x)}\,dx + \frac{\pi}{4}\biggr] + C_2 \sin\biggl[\frac{1}{\varepsilon}\int_{x_0}^{x} \sqrt{-f(x)}\,dx + \frac{\pi}{4}\biggr]\biggr\};$$

with $x < x_0$,

$$y = [f(x)]^{-1/4}\biggl\{C_1 \exp\biggl[\frac{1}{\varepsilon}\int_{x}^{x_0} \sqrt{f(x)}\,dx\biggr] + \frac{1}{2}C_2 \exp\biggl[-\frac{1}{\varepsilon}\int_{x}^{x_0} \sqrt{f(x)}\,dx\biggr]\biggr\};$$

in the vicinity of the transition point $x = x_0$,

$$y = \widetilde{C}_1 \operatorname{Bi}(z) + \widetilde{C}_2 \operatorname{Ai}(z), \qquad z = \varepsilon^{-2/3}[\psi(x_0)]^{1/3}(x_0 - x),$$

where $\operatorname{Ai}(z)$ and $\operatorname{Bi}(z)$ are the Airy function of the first and second kind, respectively (see equation 2.1.2.2).

Constants C_1, C_2 and $\widetilde{C}_1$, $\widetilde{C}_2$ in the above asymptotic solutions are related by

$$C_1 = \frac{1}{\sqrt{\pi}}[\varepsilon\psi(x_0)]^{1/6}\widetilde{C}_1, \quad C_2 = \frac{1}{\sqrt{\pi}}[\varepsilon\psi(x_0)]^{1/6}\widetilde{C}_2.$$

2. The two-term asymptotic expansions of the solution of equation (1) with $f > 0$, as $\varepsilon \to 0$, on a closed interval $a \le x \le b$ has the form

$$\begin{aligned} y_1 &= f^{-1/4} \exp\biggl(-\frac{1}{\varepsilon}\int_{x_0}^{x} \sqrt{f}\,dx\biggr)\biggl\{1 - \varepsilon\int_{x_0}^{x}\biggl[\frac{1}{8}\frac{f''_{xx}}{f^{3/2}} - \frac{5}{32}\frac{(f'_x)^2}{f^{5/2}}\biggr]dx + O(\varepsilon^2)\biggr\},\\ y_2 &= f^{-1/4} \exp\biggl(\frac{1}{\varepsilon}\int_{x_0}^{x} \sqrt{f}\,dx\biggr)\biggl\{1 + \varepsilon\int_{x_0}^{x}\biggl[\frac{1}{8}\frac{f''_{xx}}{f^{3/2}} - \frac{5}{32}\frac{(f'_x)^2}{f^{5/2}}\biggr]dx + O(\varepsilon^2)\biggr\}, \end{aligned} \qquad (2)$$

where x_0 is an arbitrary number satisfying the inequality $a \le x_0 \le b$.

The asymptotic expansions of the fundamental system of solutions of equation (1) with $f < 0$ are derived by separating the real and imaginary parts in either formula (2).

3. Consider the equation

$$\varepsilon^2 y''_{xx} - x^{m-2} f(x) y = 0 \tag{3}$$

on a closed interval $a \le x \le b$, where $a < 0$ and $b > 0$, under the condition that m is a positive integer and $f(x) \ne 0$. In this case, the leading term of the asymptotic solution, as $\varepsilon \to 0$, in the vicinity of the point $x = 0$ is expressed in terms of a simpler model equation which results from substituting function $f(x)$ in equation (3) by constant $f(0)$ (the solution of the model equation is expressed in terms of the Bessel functions of the order $1/m$, see equation 2.1.2.7).

We specify below the formulae by which the leading terms of the asymptotic expansions of the fundamental system of solutions of equation (3) with $a < x < 0$ and $0 < x < b$ are related (excluding a small vicinity of the point $x = 0$). Three different cases can be extracted:

1°. Let m be an even integer, and $f(x) > 0$. Then,

$$y_1 = \begin{cases} [f(x)]^{-1/4} \exp\left[\dfrac{1}{\varepsilon}\displaystyle\int_0^x \sqrt{f(x)}\,dx\right] & \text{if } x < 0, \\ k^{-1}[f(x)]^{-1/4} \exp\left[\dfrac{1}{\varepsilon}\displaystyle\int_0^x \sqrt{f(x)}\,dx\right] & \text{if } x > 0; \end{cases}$$

$$y_2 = \begin{cases} [f(x)]^{-1/4} \exp\left[-\dfrac{1}{\varepsilon}\displaystyle\int_0^x \sqrt{f(x)}\,dx\right] & \text{if } x < 0, \\ k[f(x)]^{-1/4} \exp\left[-\dfrac{1}{\varepsilon}\displaystyle\int_0^x \sqrt{f(x)}\,dx\right] & \text{if } x > 0, \end{cases}$$

where $f = f(x)$, $k = \sin\left(\dfrac{\pi}{m}\right)$.

2°. Let m be an even integer, and $f(x) < 0$. Then,

$$y_1 = \begin{cases} |f(x)|^{-1/4} \cos\left[-\dfrac{1}{\varepsilon}\displaystyle\int_0^x \sqrt{|f(x)|}\,dx + \dfrac{\pi}{4}\right] & \text{if } x < 0, \\ k^{-1}|f(x)|^{-1/4} \cos\left[\dfrac{1}{\varepsilon}\displaystyle\int_0^x \sqrt{|f(x)|}\,dx - \dfrac{\pi}{4}\right] & \text{if } x > 0; \end{cases}$$

$$y_2 = \begin{cases} |f(x)|^{-1/4} \cos\left[-\dfrac{1}{\varepsilon}\displaystyle\int_0^x \sqrt{|f(x)|}\,dx - \dfrac{\pi}{4}\right] & \text{if } x < 0, \\ k|f(x)|^{-1/4} \cos\left[\dfrac{1}{\varepsilon}\displaystyle\int_0^x \sqrt{|f(x)|}\,dx + \dfrac{\pi}{4}\right] & \text{if } x > 0, \end{cases}$$

where $f = f(x)$, $k = \tan\left(\dfrac{\pi}{2m}\right)$.

3°. Let m be an odd integer. Then,

$$y_1 = \begin{cases} |f(x)|^{-1/4} \cos\left[-\dfrac{1}{\varepsilon}\displaystyle\int_0^x \sqrt{|f(x)|}\,dx + \dfrac{\pi}{4}\right] & \text{if } x < 0, \\ \dfrac{1}{2}k^{-1}[f(x)]^{-1/4} \exp\left[\dfrac{1}{\varepsilon}\displaystyle\int_0^x \sqrt{f(x)}\,dx\right] & \text{if } x > 0; \end{cases}$$

$$y_2 = \begin{cases} |f(x)|^{-1/4} \cos\left[-\dfrac{1}{\varepsilon}\displaystyle\int_0^x \sqrt{|f(x)|}\,dx - \dfrac{\pi}{4}\right] & \text{if } x < 0, \\ k[f(x)]^{-1/4} \exp\left[-\dfrac{1}{\varepsilon}\displaystyle\int_0^x \sqrt{f(x)}\,dx\right] & \text{if } x > 0, \end{cases}$$

where $f = f(x)$, $k = \sin\left(\dfrac{\pi}{2m}\right)$.

4. Consider an equation of the form

$$\varepsilon^2 y''_{xx} - f(x,\varepsilon)y = 0 \tag{4}$$

on a closed interval $a \le x \le b$ under the condition that $f \neq 0$. Assume that the asymptotic relation holds

$$f(x,\varepsilon) = \sum_{k=0}^{\infty} f_k(x)\varepsilon^k, \qquad \varepsilon \to 0.$$

Then, the leading terms of the asymptotic expansions of the fundamental system of solutions of equation (4) are given by the formulae

$$y_1 = f_0^{-1/4}(x) \exp\left[-\frac{1}{\varepsilon}\int \sqrt{f_0(x)}\,dx + \frac{1}{2}\int \frac{f_1(x)}{\sqrt{f_0(x)}}\,dx\right]\bigl[1 + O(\varepsilon)\bigr],$$
$$y_2 = f_0^{-1/4}(x) \exp\left[\frac{1}{\varepsilon}\int \sqrt{f_0(x)}\,dx + \frac{1}{2}\int \frac{f_1(x)}{\sqrt{f_0(x)}}\,dx\right]\bigl[1 + O(\varepsilon)\bigr].$$

5. Consider an equation of the form

$$\varepsilon y''_{xx} + g(x)y'_x + f(x)y = 0$$

on a closed interval $0 \le x \le 1$. With $g(x) > 0$, the asymptotic solution of this equation, satisfying the boundary conditions $y(0) = C_1$ and $y(1) = C_2$, can be presented in the following form:

$$y = (C_1 - kC_2)\exp\bigl[-\varepsilon^{-1}g(0)x\bigr] + C_2 \exp\left[\int_x^1 \frac{f(x)}{g(x)}\,dx\right] + O(\varepsilon),$$

where $k = \exp\left[\displaystyle\int_0^1 \frac{f(x)}{g(x)}\,dx\right]$.

6. Now let us take a look at an equation of the form

$$\varepsilon^2 y''_{xx} + \varepsilon g(x)y'_x + f(x)y = 0 \tag{5}$$

on a closed interval $a \le x \le b$. Assume

$$D(x) \equiv [g(x)]^2 - 4f(x) \neq 0.$$

Then, the leading terms of the asymptotic expansions of the fundamental system of solutions of equation (5), as $\varepsilon \to 0$, are described by the formulae

$$y_1 = |D(x)|^{-1/4} \exp\left[-\frac{1}{2\varepsilon}\int \sqrt{D(x)}\,dx - \frac{1}{2}\int \frac{g'_x(x)}{\sqrt{D(x)}}\,dx\right]\bigl[1 + O(\varepsilon)\bigr],$$
$$y_2 = |D(x)|^{-1/4} \exp\left[\frac{1}{2\varepsilon}\int \sqrt{D(x)}\,dx - \frac{1}{2}\int \frac{g'_x(x)}{\sqrt{D(x)}}\,dx\right]\bigl[1 + O(\varepsilon)\bigr].$$

7. The more general equation

$$\varepsilon^2 y''_{xx} + \varepsilon g(x,\varepsilon) y'_x + f(x,\varepsilon) y = 0. \tag{6}$$

is reducible, with the aid of the substitution $y = w \exp\Bigl(-\frac{1}{2\varepsilon}\int g\,dx\Bigr)$, to an equation of the form (3):

$$\varepsilon^2 w''_{xx} + \bigl(f - \tfrac{1}{4}g^2 - \tfrac{1}{2}\varepsilon g'_x\bigr) w = 0,$$

to which the asymptotic formulae given above in Paragraph 4 are applicable.

2.1.12. Series Solutions

Let us consider a homogeneous linear differential equation of the general form

$$y''_{xx} + f(x) y'_x + g(x) y = 0. \tag{1}$$

Case 1. Assume that the functions $f(x)$ and $g(x)$ are representable, in the vicinity of the point $x = x_0$, in the form

$$f(x) = \sum_{n=0}^{\infty} A_n (x - x_0)^n, \qquad g(x) = \sum_{n=0}^{\infty} B_n (x - x_0)^n, \tag{2}$$

in the region $|x - x_0| < R$, where R stands for the minimum radius of convergence of the two series in (2). In this case, the point $x = x_0$ is referred to as an ordinary point, and equation (1) possesses two linearly independent solutions of the form

$$y_1(x) = \sum_{n=0}^{\infty} a_n (x - x_0)^n, \qquad y_2(x) = \sum_{n=0}^{\infty} b_n (x - x_0)^n. \tag{3}$$

Coefficients a_n and b_n are determined by substituting the series (2) into equation (1) followed by extracting the coefficients with respect to identical powers of $(x - x_0)$.

Case 2. Assume that the functions $f(x)$ and $g(x)$ are representable, in the vicinity of the point $x = x_0$, in the form

$$f(x) = \sum_{n=-1}^{\infty} A_n (x - x_0)^n, \qquad g(x) = \sum_{n=-2}^{\infty} B_n (x - x_0)^n, \tag{4}$$

in the region $|x - x_0| < R$. In this case, the point $x = x_0$ is referred to as a regular singular point. Let λ_1 and λ_2 be the roots of the quadratic equation

$$\lambda_1^2 + (A_{-1} - 1)\lambda + B_{-2} = 0.$$

There are three cases, depending on the values of the exponents of the singularity.

1) If $\lambda_1 \neq \lambda_2$ and $\lambda_1 - \lambda_2$ is not equal to an integer, equation (1) has two linearly independent solutions

$$\begin{aligned} y_1(x) &= |x - x_0|^{\lambda_1} \Bigl[1 + \sum_{n=1}^{\infty} a_n (x - x_0)^n\Bigr], \\ y_2(x) &= |x - x_0|^{\lambda_2} \Bigl[1 + \sum_{n=1}^{\infty} b_n (x - x_0)^n\Bigr]. \end{aligned} \tag{5}$$

It is clear that in the general case, coefficients a_n and b_n of the series (4) and (5) will be different.

2) If $\lambda_1 = \lambda_2 = \lambda$, equation (1) possesses two linearly independent solutions in the forms

$$y_1(x) = |x - x_0|^\lambda \left[1 + \sum_{n=1}^{\infty} a_n (x - x_0)^n\right],$$

$$y_2(x) = y_1(x) \ln|x - x_0| + |x - x_0|^\lambda \sum_{n=0}^{\infty} b_n (x - x_0)^n.$$

3) If $\lambda_1 = \lambda_2 + N$, where N is an integer greater than 0, equation (1) has two linearly independent solutions in the forms

$$y_1(x) = |x - x_0|^{\lambda_1} \left[1 + \sum_{n=1}^{\infty} a_n (x - x_0)^n\right],$$

$$y_2(x) = k y_1(x) \ln|x - x_0| + |x - x_0|^{\lambda_2} \sum_{n=0}^{\infty} b_n (x - x_0)^n,$$

where k may be equal to zero.

To construct the solution in each of the three cases, the following procedure is to be done: Substitute the above expressions for y_1 and y_2 into the original equation (1) and equate the coefficients of $(x - x_0)^n$ and $(x - x_0)^n \ln|x - x_0|$ for different values of n to obtain recurrence relations for the unknown coefficients. From these recurrence relations the solution sought can be found.

2.2. Autonomous Equations $y''_{xx} = F(y, y'_x)$

Preliminary Comments. Equations of this form are often encountered in different areas of mechanics, applied mathematics, physics, and chemical engineering science.

1. The substitution $w(y) = y'_x$ leads to the first order equation

$$w'_y = w^{-1} F(y,\ w) \tag{1}$$

(see Chapter 1).

2. The solution of the original autonomous equation can be represented in the implicit form

$$x = \int \frac{dy}{w(y,\ C_1)} + C_2, \tag{2}$$

where $w = w(y,\ C_1)$ is the solution of the first order equation (1).

3. The solution of the original autonomous equation can be written in the parametric form

$$x = \int \frac{y'_\tau(\tau,\ C_1)}{w(\tau,\ C_1)}\, d\tau + C_2, \qquad y = y(\tau,\ C_1), \tag{3}$$

where $y = y(\tau,\ C_1)$, $w = w(\tau,\ C_1)$ is the parametric form of the solution of the first order equation (1).

Formula (2) is a special case of formula (3) with $y = \tau$.

2.2.1. Equations of the Form $y''_{xx} - y'_x = f(y)$

Preliminary Comments. Equations of this form are encountered in the theory of combustion and the theory of chemical reactors.

1. The substitution $w(y) = y'_x$ leads to the Abel equation $ww'_y - w = f(y)$ which is considered in Subsection 1.3.1 for some concrete functions f.

2. The solution of the original autonomous equation can be written in the parametric form (3), where $y = y(\tau,\ C_1)$, $w = w(\tau,\ C_1)$ is the parametric form of the solution to the Abel equation of the second kind $ww'_y - w = f(y)$.

1. $\quad y''_{xx} - y'_x = -\dfrac{2(m+1)}{(m+3)^2}y \pm \dfrac{m+1}{2a^2}y^m, \qquad m \neq \pm 1, \quad m \neq -3.$

Solution in the parametric form:

$$x = \frac{m+3}{m-1}\ln\left[aC_1^{1-m}\frac{m-1}{m+3}\int\frac{d\tau}{\sqrt{1\pm\tau^{m+1}}} + C_2\right],$$

$$y = C_1^2\tau\left[aC_1^{1-m}\frac{m-1}{m+3}\int\frac{d\tau}{\sqrt{1\pm\tau^{m+1}}} + C_2\right]^{\frac{2}{m-1}}.$$

2. $\quad y''_{xx} - y'_x = \pm 2a^2y^{-1}.$

Solution in the parametric form:

$$x = -\ln\left[C_1\int\exp(\pm\tau^2)\,d\tau + C_2\right], \quad y = aC_1\exp(\pm\tau^2)\left[C_1\int\exp(\pm\tau^2)\,d\tau + C_2\right]^{-1}.$$

3. $\quad y''_{xx} - y'_x = -\frac{2}{9}y + \frac{16}{9}a^{3/2}y^{-1/2}.$

Solution in the parametric form:

$$x = -3\ln\bigl\{C_1\exp(-\tau)\bigl[\exp(3\tau) + C_2\sin\bigl(\sqrt{3}\,\tau\bigr)\bigr]\bigr\},$$

$$y = a\exp(2\tau)\frac{\bigl[2\exp(3\tau) - C_2\sin\bigl(\sqrt{3}\,\tau\bigr) + \sqrt{3}\,C_2\cos\bigl(\sqrt{3}\,\tau\bigr)\bigr]^2}{\bigl[\exp(3\tau) + C_2\sin\bigl(\sqrt{3}\,\tau\bigr)\bigr]^2}.$$

4. $\quad y''_{xx} - y'_x = -\frac{9}{100}y \pm \frac{9}{100}a^{8/3}y^{-5/3}.$

Solution in the parametric form:

$$x = -\tfrac{5}{4}\ln\bigl[\pm(\tau^4 - 6\tau^2 + 4C_1\tau - 3)\bigr] + C_2,$$

$$y = a(\tau^3 - 3\tau + C_1)^{3/2}\bigl[\pm(\tau^4 - 6\tau^2 + 4C_1\tau - 3)\bigr]^{-9/8}.$$

5. $\quad y''_{xx} - y'_x = -\frac{3}{16}y - \frac{3}{64}a^{8/3}y^{-5/3}.$

Solution in the parametric form:

$$x = C_1 - 2\ln[\sin\tau\cosh(\tau + C_2) + \cos\tau\sinh(\tau + C_2)], \qquad y = a[\tan\tau + \tanh(\tau + C_2)]^{-3/2}.$$

► *In the solutions of equations 6–9, the following notation is used:*

$$Z = \begin{cases} C_1 J_\nu(\tau) + Y_\nu(\tau) & \text{for the upper sign,} \\ C_1 I_\nu(\tau) + K_\nu(\tau) & \text{for the lower sign,} \end{cases}$$

where J_ν and Y_ν are Bessel functions, I_ν and K_ν are modified Bessel functions.

6. $\boldsymbol{y''_{xx} - y'_x = Ay^{-1/2}}$.

Solution in the parametric form:

$$x = -2\int \tau^{-1} Z^{-1}(\tau Z'_\tau + \tfrac{1}{3}Z)\,d\tau + C_2, \qquad y = a\tau^{-4/3} Z^{-2}[(\tau Z'_\tau + \tfrac{1}{3}Z)^2 \pm \tau^2 Z^2],$$

where $\nu = \frac{1}{3}$, $A = \mp\frac{1}{3}a^{3/2}$.

7. $\boldsymbol{y''_{xx} - y'_x = Ay^{-2}}$.

Solution in the parametric form:

$$x = \mp\tfrac{2}{3}\int \tau Z^2\big[(\tau Z'_\tau + \tfrac{1}{3}Z)^2 \pm \tau^2 Z^2\big]^{-1}\,d\tau + C_2,$$
$$y = 2a\tau^{4/3} Z^2\big[(\tau Z'_\tau + \tfrac{1}{3}Z)^2 \pm \tau^2 Z^2\big]^{-1},$$

where $\nu = \frac{1}{3}$, $A = -36a^3$.

8. $\boldsymbol{y''_{xx} - y'_x = 2A^2 - Ay^{1/2}}$.

Solution in the parametric form:

$$x = \pm 2\int \tau^{-1}(Z'_\tau)^{-1}(\tau Z \pm 2Z'_\tau)\,d\tau + C_2, \quad y = a(Z'_\tau)^{-2}(\tau Z \pm 2Z'_\tau)^2,$$

where $\nu = 0$, $A = a^{1/2}$.

9. $\boldsymbol{y''_{xx} - y'_x = Ay^{-1/2} + 2B^2 + By^{1/2}}$.

Solution in the parametric form:

$$x = -2\int \tau^{-1} Z^{-1}(\tau Z'_\tau - Z)\,d\tau + C_2, \quad y = B^2 Z^{-2}(\tau Z'_\tau - Z)^2,$$

where $A = (1 - \nu^2)B^3$.

► *In the solutions of equations 10–14, the following notation is used: the function $\wp = \wp(\tau)$ is defined in the implicit form*

$$\tau = \int \frac{d\wp}{\sqrt{\pm(4\wp^3 - 1)}} - C_1.$$

The upper sign in this formula corresponds to the classical elliptic Weierstrass function $\wp = \wp(\tau + C_1,\ 0,\ 1)$.

10. $y''_{xx} - y'_x = Ay^2 - \frac{9}{625}A^{-1}$.

Solution in the parametric form:

$$x = 5\ln\tau + C_2, \qquad y = 5a(\tau^2\wp \mp \tfrac{1}{2}), \qquad \text{where} \quad A = \pm\tfrac{6}{125}a^{-1}.$$

11. $y''_{xx} - y'_x = Ay^2 - \frac{6}{25}y$.

Solution in the parametric form:

$$x = 5\ln\tau + C_2, \qquad y = 5a\tau^2\wp, \qquad \text{where} \quad A = \pm\tfrac{6}{125}a^{-1}.$$

12. $y''_{xx} - y'_x = Ay^2 + \frac{6}{25}y$.

Solution in the parametric form:

$$x = 5\ln\tau + C_2, \qquad y = 5a(\tau^2\wp \mp 1), \qquad \text{where} \quad A = \pm\tfrac{6}{125}a^{-1}.$$

13. $y''_{xx} - y'_x = 12y + Ay^{-5/2}$.

Solution in the parametric form:

$$x = \mp\tfrac{2}{7}\int \wp^{-1}\bigl(f \pm 2\tau\wp^2\bigr)^{-1}\,d\tau + C_2, \quad y = a\wp^{-6/7}\bigl(f \pm 2\tau\wp^2\bigr)^{-4/7},$$

where $f = \sqrt{\pm(4\wp^3 - 1)}$, $A = \mp 147a^{7/2}$.

14. $y''_{xx} - y'_x = \frac{63}{4}y + Ay^{-5/3}$.

Solution in the parametric form:

$$x = -\tfrac{3}{4}\int (f \pm 2\tau\wp^2)(\tau f + 2\wp)^{-1}\,d\tau + C_2, \quad y = 2a\bigl(f \pm 2\tau\wp^2\bigr)^{3/2}(\tau f + 2\wp)^{-9/8},$$

where $f = \sqrt{\pm(4\wp^3 - 1)}$, $A = -\frac{128}{3}a^2(2a)^{2/3}$.

► *In the solutions of equations 15–18, the following notation is used:*

$$I = \int \frac{\tau\,d\tau}{\sqrt{\pm(4\tau^3 - 1)}} + C_1$$

is the incomplete elliptic integral of the second kind in the Weierstrass form,

$$R = \sqrt{\pm(4\tau^3 - 1)}, \quad I_1 = 2\tau I \mp R, \quad I_2 = \tau^{-1}(2\tau RI \mp R^2 - 1).$$

15. $y''_{xx} - y'_x = Ay^{1/2} - \frac{12}{49}y$.

Solution in the parametric form:

$$x = -7\int \tau R^{-1}I^{-1}\,d\tau + C_2, \quad y = 7a\tau^2 I^{-4}, \qquad \text{where} \quad A = \pm\tfrac{12}{49}(7a)^{1/2}.$$

16. $y''_{xx} - y'_x = 6y + Ay^{-4}$.

Solution in the parametric form:

$$x = -\tfrac{1}{5}\int \tau^{-1}R^{-1}I_1^{-1}\,d\tau + C_2, \quad y = a\tau^{-3/5}I_1^{-2/5}, \qquad \text{where} \quad A = \mp 150a^5.$$

17. $y''_{xx} - y'_x = 20y + Ay^{-1/2}$.

Solution in the parametric form:

$$x = \tfrac{1}{3}\int R^{-1}I_1^{-1}I_2\,d\tau + C_2, \quad y = aI_1^{-4/3}I_2^2, \qquad \text{where} \quad A = \pm 108a^{3/2}.$$

18. $y''_{xx} - y'_x = \frac{15}{4}y + Ay^{-7}$.

Solution in the parametric form:

$$x = \int R^{-1}I_1\left(4\tau I_1^2 \mp I_2^2\right)^{-1}d\tau + C_2, \quad y = aI_1^{1/2}\left(4\tau I_1^2 \mp I_2^2\right)^{-3/8}, \qquad \text{where} \quad A = \pm\tfrac{3}{4}a^8.$$

19. $y''_{xx} - y'_x = Ay + By^{-1} - B^2y^{-3}$.

The substitution $w(y) = y'_x$ leads to the Abel equation of the form 1.3.1.5:

$$ww'_y - w = Ay + By^{-1} - B^2y^{-3}.$$

20. $y''_{xx} - y'_x = -\frac{3}{16}y + Ay^{-1/3} + By^{-5/3}$.

The substitution $w(y) = y'_x$ leads to the Abel equation of the form 1.3.1.61:

$$ww'_y - w = -\tfrac{3}{16}y + Ay^{-1/3} + By^{-5/3}.$$

21. $y''_{xx} - y'_x = -\frac{5}{36}y + Ay^{-3/5} + By^{-7/5}$.

The substitution $w(y) = y'_x$ leads to the Abel equation of the form 1.3.1.62:

$$ww'_y - w = -\tfrac{5}{36}y + Ay^{-3/5} + By^{-7/5}.$$

22. $y''_{xx} - y'_x = \frac{4}{9}y + 2Ay^2 + 2A^2y^3$.

The substitution $w(y) = y'_x$ leads to the Abel equation of the form 1.3.1.14:

$$ww'_y - w = \tfrac{4}{9}y + 2Ay^2 + 2A^2y^3.$$

23. $y''_{xx} - y'_x = Ay^{k-1} - kBy^k + kB^2y^{2k-1}$.

The substitution $w(y) = y'_x$ leads to the Abel equation of the form 1.3.1.6:

$$ww'_y - w = Ay^{k-1} - kBy^k + kB^2y^{2k-1}.$$

24. $y''_{xx} - y'_x = \pm \dfrac{2a^2}{\sqrt{y^2 \pm 8a^2}}$.

Solution in the parametric form:

$$x = \mp \int E^{-1} F^{-1} (F^2 \pm 2E^2)\, d\tau + C_2, \qquad y = \pm a E^{-1} F^{-1} (F^2 \mp 2E^2),$$

where $E = \int \exp(\mp\tau^2)\, d\tau + C_1$, $F = 2\tau E \pm \exp(\mp\tau^2)$.

25. $y''_{xx} - y'_x = A + B\exp(-2y/A)$.

The substitution $w(y) = y'_x$ leads to the Abel equation of the form 1.3.1.8:

$$ww'_y - w = A + B\exp(-2y/A).$$

26. $y''_{xx} - y'_x = a^2\lambda e^{2\lambda y} - a(b\lambda + 1)e^{\lambda y} + b$.

The substitution $w(y) = y'_x$ leads to the Abel equation of the form 1.3.1.73:

$$ww'_y - w = a^2\lambda e^{2\lambda y} - a(b\lambda + 1)e^{\lambda y} + b.$$

27. $y''_{xx} - y'_x = a^2\lambda e^{2\lambda y} + a\lambda y e^{\lambda y} + b e^{\lambda y}$.

The substitution $w(y) = y'_x$ leads to the Abel equation of the form 1.3.1.74:

$$ww'_y - w = a^2\lambda e^{2\lambda y} + a\lambda y e^{\lambda y} + b e^{\lambda y}.$$

2.2.2. Equations of the Form $y''_{xx} + f(y)y'_x + y = 0$

Preliminary Comments. Equation of this form are often encountered in the theory of nonlinear oscillations.

1. The transformation

$$z = -\tfrac{1}{2}y^2 + a, \qquad w = y'_x$$

leads to the Abel equation

$$ww'_z = g(z)w + 1, \qquad \text{where} \quad g(z) = f(y)/y, \quad y = \pm\sqrt{2(a-z)},$$

whose special cases are outlined in Subsection 1.3.2.

2. For oscillatory systems with a weak nonlinearity

$$y''_{xx} + \varepsilon F(y) y'_x + y = 0,$$

two leading terms of the asymptotic solution, as $\varepsilon \to 0$, are described by the formula

$$y = A\cos(x + B),$$

where functions $A = A(\xi)$ and $B = B(\xi)$ depend on the slow variable $\xi = \varepsilon x$; they are determined from the autonomous system of the first order equations

$$A'_\xi = -\frac{A}{2\pi}\int_0^{2\pi} F(A\cos\varphi)\sin^2\varphi\, d\varphi, \quad B'_\xi = -\frac{1}{2\pi}\int_0^{2\pi} F(A\cos\varphi)\sin\varphi\cos\varphi\, d\varphi.$$

The right-hand sides of these equations depend only on function A. The system is solved consecutively starting from the first equation.

1. $\boldsymbol{y''_{xx} + ayy'_x + y = 0.}$

Solution in the parametric form:

$$x = -A\int \tau^{-1}(C_1 + 2A^2\ln|\tau| - 2A\tau)^{-1/2}\,d\tau + C_2, \quad y = (C_1 + 2A^2\ln|\tau| - 2A\tau)^{1/2},$$

where $A = a^{-1}$.

2. $\boldsymbol{y''_{xx} - \varepsilon(1 - y^2)y'_x + y = 0.}$

Van der Pol oscillator

Solution, as $\varepsilon \to 0$:

$$y = a\cos(x - \theta) - \frac{1}{32}\varepsilon a^3\sin[3(x - \theta)] + O(\varepsilon^2),$$

where

$$a^2 = \frac{4}{1 + (4C_1^{-2} - 1)e^{-\varepsilon x}}, \qquad \theta = \frac{1}{8}\varepsilon\ln a - \frac{7}{64}\varepsilon a^2 + \frac{1}{16}\varepsilon^2 x + C_2.$$

In applications, x plays the role of time, C_1 is the initial oscillation amplitude, and C_2 is the initial phase with $\varepsilon = 0$.

As $\varepsilon \to +\infty$, the periodic solution of the Van der Pol equation consists of intervals with fast and slow vibrations and describes damping oscillations with the period $T = (3 - 2\ln 2)\varepsilon + O(\varepsilon^{-1/3})$.

3. $\boldsymbol{y''_{xx} + y(ay^2 + b)y'_x + y = 0.}$

The transformation $z = -\frac{1}{2}y^2$, $w = y'_x$ leads to the Abel equation of the form 1.3.2.1: $ww'_z = (-2az + b)w + 1$.

Solution in the parametric form with $a < 0$:

$$x = \mp\frac{2}{3}k\int \tau^{-1/3}\left[\pm\frac{4}{3}k^2\tau^{-2/3}Z^{-1}\left(\tau Z'_\tau + \frac{1}{3}Z\right) - \frac{b}{a}\right]^{-1/2} d\tau + C_2,$$

$$y = \left[\pm\frac{4}{3}k^2\tau^{-2/3}Z^{-1}\left(\tau Z'_\tau + \frac{1}{3}Z\right) - \frac{b}{a}\right]^{1/2}, \qquad a = -\frac{9}{4}k^{-3},$$

where

$$Z = \begin{cases} C_1 J_{1/3}(\tau) + C_2 Y_{1/3}(\tau) & \text{for the upper sign,} \\ C_1 I_{1/3}(\tau) + C_2 K_{1/3}(\tau) & \text{for the lower sign,} \end{cases}$$

$J_{1/3}$ and $Y_{1/3}$ are Bessel functions, $I_{1/3}$ and $K_{1/3}$ are modified Bessel functions.

4. $\boldsymbol{y''_{xx} + y(ay^2 + b)^{-2}y'_x + y = 0.}$

The transformation $z = -\frac{1}{2}y^2$, $w = y'_x$ leads to the Abel equation of the form 1.3.2.2: $ww'_z = (-2az + b)^{-2}w + 1$.

5. $\boldsymbol{y''_{xx} + y(ay^2+b)^{-1/2}y'_x + y = 0.}$

The transformation $z = -\frac{1}{2}y^2$, $w = y'_x$ leads to the Abel equation of the form 1.3.2.4: $ww'_z = (-2az+b)^{-1/2}w + 1$.

Solution in the parametric form:

$$x = -aC_1 \int \left(aC_1^2E^2 - \frac{b}{a}\right)^{-1/2} \frac{E\,d\tau}{\tau^2-\tau+a} + C_2, \qquad y = \left(aC_1^2E^2 - \frac{b}{a}\right)^{1/2},$$

where $E = \exp\left(-\int \frac{\tau\,d\tau}{\tau^2-\tau+a}\right)$.

6. $\boldsymbol{y''_{xx} - y\left(2a + \frac{1}{ay^2+b}\right)y'_x + y = 0.}$

The transformation $z = -\frac{1}{2}y^2 - \frac{b}{2a}$, $w = y'_x$ leads to the Abel equation of the form 1.3.2.3:

$$ww'_z = \left(A - \frac{1}{Az}\right)w + 1, \qquad \text{where} \quad A = -2a.$$

7. $\boldsymbol{y''_{xx} + ay\exp(\lambda y^2)y'_x + y = 0.}$

The transformation $z = -\frac{1}{2}y^2$, $w = y'_x$ leads to the Abel equation of the form 1.3.2.7: $ww'_z = a\exp(-2\lambda z)w + 1$.

8. $\boldsymbol{y''_{xx} + y[a\exp(\lambda y^2) + b\exp(-\lambda y^2)]y'_x + y = 0.}$

The transformation $z = -\frac{1}{2}y^2$, $w = y'_x$ leads to the Abel equation of the form 1.3.2.8: $ww'_z = [b\exp(2\lambda z) + a\exp(-2\lambda z)]w + 1$.

9. $\boldsymbol{y''_{xx} + 2ay\exp[a(b-y^2)]y'_x + y = 0.}$

Solution in the parametric form:

$$x = \mp 2k \int \left(b - 4k^2\tau^2 - \ln|kE_{\mp}^{-1}|\right)^{-1/2} d\tau + C_2, \quad y = \left(b - 4k^2\tau^2 - \ln|kE_{\mp}^{-1}|\right)^{1/2},$$

where $a = \mp\frac{1}{4}k^{-2}$, $E_{\mp} = \int \exp(\mp\tau^2)\,d\tau + C_1$.

10. $\boldsymbol{y''_{xx} + Ay\cosh(\lambda y^2)y'_x + y = 0.}$

This is a special case of equation 2.2.2.8 with $a = b = \frac{1}{2}A$.

11. $\boldsymbol{y''_{xx} + Ay\sinh(\lambda y^2)y'_x + y = 0.}$

This is a special case of equation 2.2.2.8 with $a = -b = \frac{1}{2}A$.

12. $\boldsymbol{y''_{xx} + 2Ayy'_x\sqrt{\sinh^2[A(B-y^2)] + 2A^{-1}} + y = 0.}$

Solution in the parametric form:

$$x = 2a\int (F^2 + 2E^2)G^{-1}Q^{-1}\,d\tau + C_2, \quad y = Q; \qquad A = \tfrac{1}{4}a^{-2},$$

where

$$E = \int \exp(-\tau^2)\,d\tau + C_1, \quad F = 2\tau E + \exp(-\tau^2), \quad G = \sqrt{F^2 - 2E^2 + 8E^2F^2},$$

$$Q = \sqrt{B - 4a^2\,\mathrm{Arsinh}[aE^{-1}F^{-1}(F^2-2E^2)]}, \quad \mathrm{Arsinh}\,z = \ln\left(z + \sqrt{z^2+1}\,\right).$$

13. $y''_{xx} - 2Ayy'_x\sqrt{\cosh^2[A(y^2 - B)] - 2A^{-1}} + y = 0.$

Solution in the parametric form:

$$x = 2a \int (F^2 - 2E^2)G^{-1}Q^{-1}\,d\tau + C_2, \quad y = Q; \qquad A = \tfrac{1}{4}a^{-2},$$

where

$$E = \int \exp(\tau^2)\,d\tau + C_1, \quad F = 2\tau E - \exp(\tau^2), \quad G = \sqrt{F^2 + 2E^2 - 8E^2F^2},$$

$$Q = \sqrt{B + 4a^2 \operatorname{Arcosh}[aE^{-1}F^{-1}(F^2 + 2E^2)]}, \quad \operatorname{Arcosh} z = \pm\ln\bigl(z + \sqrt{z^2 - 1}\,\bigr).$$

14. $y''_{xx} + Ay\cos(\omega y^2)y'_x + y = 0.$

The transformation $z = -\frac{1}{2}y^2$, $w = y'_x$ leads to the Abel equation of the form 1.3.2.11: $ww'_z = A\cos(2\omega z)w + 1$.

15. $y''_{xx} + Ay\sin(\omega y^2)y'_x + y = 0.$

The transformation $z = -\frac{1}{2}y^2$, $w = y'_x$ leads to the Abel equation of the form 1.3.2.12: $ww'_z = -A\sin(2\omega z)w + 1$.

2.2.3. Lienard Equations $y''_{xx} + f(y)y'_x + g(y) = 0$

1. $y''_{xx} + y + ay^3 = 0.$

Duffing equation.

Solution:

$$x = \pm \int \bigl(C_1 - y^2 - \tfrac{1}{2}ay^4\bigr)^{-1/2}\,dy + C_2.$$

The period of oscillations with the amplitude C is expressed in terms of the complete elliptic integral of the first kind:

$$T = \frac{4}{\sqrt{1 + aC^2}}K\Bigl(\frac{aC^2}{2 + 2aC^2}\Bigr), \qquad \text{where} \quad K(m) = \int_0^{\pi/2} \frac{dt}{\sqrt{1 - m\sin^2 t}}.$$

The asymptotic solution, as $a \to 0$, has the form

$$y = \widetilde{C}_1\cos[(1 + \tfrac{3}{8}a\widetilde{C}_1^2)x + \widetilde{C}_2] + \tfrac{1}{32}a\widetilde{C}_1^3\cos[3(1 + \tfrac{3}{8}a\widetilde{C}_1^2)x + 3\widetilde{C}_2] + O(a^2),$$

where $\widetilde{C}_1$ and $\widetilde{C}_2$ are arbitrary constants. The corresponding asymptotics for the period of oscillations with the amplitude C is described by the formula

$$T = 2\pi(1 - \tfrac{3}{8}aC^2) + O(a^2).$$

2. $y''_{xx} = (ay + 3b)y'_x + cy^3 - aby^2 - 2b^2y.$

The substitution $w(y) = y'_x$ leads to the Abel equation of the form 1.3.3.1:

$$ww'_y = (ay + 3b)w + cy^3 - aby^2 - 2b^2y.$$

3. $y''_{xx} = (3ay + b)y'_x - a^2y^3 - aby^2 + cy.$

The substitution $w(y) = y'_x$ leads to the Abel equation of the form 1.3.3.2:

$$ww'_y = (3ay + b)w - a^2y^3 - aby^2 + cy.$$

4. $2y''_{xx} = (7ay + 5b)y'_x - 3a^2y^3 - 2cy^2 - 3b^2y.$

The substitution $w(y) = y'_x$ leads to the Abel equation of the form 1.3.3.3:

$$2ww'_y = (7ay + 5b)w - 3a^2y^3 - 2cy^2 - 3b^2y.$$

5. $y''_{xx} = y^{n-1}[(1 + 2n)y + an]y'_x - ny^{2n}(y + a).$

The substitution $w(y) = y'_x$ leads to the Abel equation of the form 1.3.3.8:

$$ww'_y = y^{n-1}[(1 + 2n)y + an]w - ny^{2n}(y + a).$$

6. $y''_{xx} = a(y - nb)y^{n-1}y'_x + c[y^2 - (2n + 1)by + n(n + 1)b^2]y^{2n-1}.$

The substitution $w(y) = y'_x$ leads to the Abel equation of the form 1.3.3.9:

$$ww'_y = a(y - nb)y^{n-1}w + c[y^2 - (2n + 1)by + n(n + 1)b^2]y^{2n-1}.$$

7. $y''_{xx} = [a(2n + k)y^k + b]y^{n-1}y'_x + (-a^2ny^{2k} - aby^k + c)y^{2n-1}.$

The substitution $w(y) = y'_x$ leads to the Abel equation of the form 1.3.3.10:

$$ww'_y = [a(2n + k)y^k + b]y^{n-1}w + (-a^2ny^{2k} - aby^k + c)y^{2n-1}.$$

8. $y''_{xx} = [a(2m + k)y^{2k} + b(2m - k)]y^{m-k-1}y'_x - (a^2my^{4k} + cy^{2k} + b^2m)y^{2m-2k-1}.$

The substitution $w(y) = y'_x$ leads to the Abel equation of the form 1.3.3.11:

$$ww'_y = [a(2m + k)y^{2k} + b(2m - k)]y^{m-k-1}w - (a^2my^{4k} + cy^{2k} + b^2m)y^{2m-2k-1}.$$

9. $y''_{xx} = ae^{\lambda y}y'_x + be^{\lambda y}.$

Solution in the parametric form:

$$x = -\frac{A}{\lambda}\int \tau^{-1}\bigl(C_1 + A^2\ln|\tau| - A\tau\bigr)^{-1}\,d\tau + C_2, \qquad y = \frac{1}{\lambda}\ln\Bigl[-\frac{\lambda}{b}\bigl(C_1 + A^2\ln|\tau| - A\tau\bigr)\Bigr],$$

where $A = b/a$.

10. $y''_{xx} = (ae^y + b)y'_x + ce^{2y} - abe^y - b^2.$

The substitution $w(y) = y'_x$ leads to the Abel equation of the form 1.3.3.15:

$$ww'_y = (ae^y + b)w + ce^{2y} - abe^y - b^2.$$

11. $\boldsymbol{y''_{xx} = [a(2\mu + \lambda)e^{\lambda y} + b]e^{\mu y}y'_x + (-a^2\mu e^{2\lambda y} - abe^{\lambda y} + c)e^{2\mu y}}$.

The substitution $w(y) = y'_x$ leads to the Abel equation of the form 1.3.3.16:

$$ww'_y = [a(2\mu + \lambda)e^{\lambda y} + b]e^{\mu y}w + (-a^2\mu e^{2\lambda y} - abe^{\lambda y} + c)e^{2\mu y}.$$

12. $\boldsymbol{y''_{xx} = (ae^{\lambda y} + b)y'_x + c[a^2e^{2\lambda y} + ab(\lambda y + 1)e^{\lambda y} + b^2\lambda y]}$.

The substitution $w(y) = y'_x$ leads to the Abel equation of the form 1.3.3.17:

$$ww'_y = (ae^{\lambda y} + b)w + c[a^2e^{2\lambda y} + ab(\lambda y + 1)e^{\lambda y} + b^2\lambda y].$$

13. $\boldsymbol{y''_{xx} = e^{\lambda y}(2a\lambda y + a + b)y'_x - e^{2\lambda y}(a^2\lambda y^2 + aby + c)}$.

The substitution $w(y) = y'_x$ leads to the Abel equation of the form 1.3.3.18:

$$ww'_y = e^{\lambda y}(2a\lambda y + a + b)w - e^{2\lambda y}(a^2\lambda y^2 + aby + c).$$

14. $\boldsymbol{y''_{xx} = e^{ay}(2ay^2 + 2y + b)y'_x + e^{2ay}(-ay^4 - by^2 + c)}$.

The substitution $w(y) = y'_x$ leads to the Abel equation of the form 1.3.3.19:

$$ww'_y = e^{ay}(2ay^2 + 2y + b)w + e^{2ay}(-ay^4 - by^2 + c).$$

15. $\boldsymbol{y''_{xx} = (a\cosh y + b)y'_x - ab\sinh y + c}$.

The substitution $w(y) = y'_x$ leads to the Abel equation of the form 1.3.3.20:

$$ww'_y = (a\cosh y + b)w - ab\sinh y + c.$$

16. $\boldsymbol{y''_{xx} = (a\sinh y + b)y'_x - ab\cosh y + c}$.

The substitution $w(y) = y'_x$ leads to the Abel equation of the form 1.3.3.21:

$$ww'_y = (a\sinh y + b)w - ab\cosh y + c.$$

17. $\boldsymbol{y''_{xx} + a\sin y = 0}$.

This is the equation of oscillations of the mathematical pendulum.

Solution:

$$x = \pm\int (2a\cos y + C_1)^{-1/2}\,dy + C_2.$$

With $a > 0$ and initial conditions $y(0) = C > 0$, $y'_x(0) = 0$, where variable x takes the role of the time, and y is the angle deviation from the equilibrium state, the oscillations of the mathematical pendulum are described by the formula

$$\sin\frac{y}{2} = m\,\mathrm{sn}\big(\sqrt{a}\,x\big), \qquad m = \sin\frac{C}{2},$$

where $\mathrm{sn} = \mathrm{sn}(z)$ is the Jacobi elliptic function defined implicitly as follows:

$$\mathrm{sn}(z) = \sin\beta, \qquad z = \int_0^\beta \frac{d\beta}{\sqrt{1 - m^2\sin^2\beta}}$$

(parameter β is to be eliminated from these relations).

The period of oscillations of the mathematical pendulum is expressed in terms of the complete elliptic integral of the second kind:

$$T = \frac{4}{\sqrt{a}}K(m), \qquad K(m) = \int_0^{\pi/2} \frac{d\beta}{\sqrt{1 - m^2\sin^2\beta}}.$$

At small amplitudes, as $C \to 0$, the following asymptotic formula holds for the period:

$$T = \frac{2\pi}{\sqrt{a}}\left(1 + \frac{1}{16}C^2\right) + O(C^4).$$

18. $y''_{xx} + a\sin(\omega y)y'_x + b\sin(\omega y) = 0.$

Solution in the parametric form:

$$x = -A \int \tau^{-1}\left(b^2 - \omega^2 F^2\right)^{-1/2} d\tau + C_2, \qquad y = \frac{1}{\omega}\arccos\left(\frac{\omega}{b}F\right),$$

where $A = b/a$, $F = A\tau - A^2 \ln|\tau| + C_1$.

19. $y''_{xx} + a\cos(\omega y)y'_x + b\cos(\omega y) = 0.$

The substitution $\omega y = \omega u + \frac{\pi}{2}$ leads to an equation of the form 2.2.3.18:

$$u''_{xx} - a\sin(\omega u)u'_x - b\sin(\omega u) = 0.$$

2.2.4. Rayleigh Equations $y''_{xx} + f(y'_x) + g(y) = 0$

Preliminary Comments. Equations of this form are encountered in the theory of nonlinear oscillations.

1. Let us discuss the particular case $g(y) = y$ which corresponds to the equation

$$y''_{xx} + f(y'_x) + y = 0. \qquad (1)$$

Differentiating equation (1) with respect to x and substituting $z(x) = y'_x$, we obtain the equation of nonlinear oscillations

$$z''_{xx} + \Phi(z)z'_x + z = 0, \qquad \text{where} \quad \Phi(z) = f'_z(z), \qquad (2)$$

which is considered in Subsection 2.2.2.

The solution of equation (1) can be written in the parametric form:

$$x = x(\tau,\, C_1,\, C_2), \qquad y = -f(z) - \frac{z'_\tau}{x'_\tau},$$

where $x = x(\tau,\, C_1,\, C_2)$, $z = z(\tau,\, C_1,\, C_2)$ is the parametric form of the solution of equation (2).

2. The transformation

$$\xi = -\tfrac{1}{2}\left(y'_x\right)^2 + a, \qquad w = -y - f(y'_x),$$

reduces equation (1) to the Abel equation

$$ww'_\xi = H(\xi)w + 1, \qquad \text{where} \quad H(\xi) = z^{-1}\Phi(z), \quad z = \pm\sqrt{2(a-\xi)}, \qquad (3)$$

where function $\Phi = \Phi(z)$ is defined above in equation (2). Concrete equations of the form (3) are outlined in Subsection 1.3.2.

3. The equation of the special form

$$y''_{xx} + a\left(y'_x\right)^2 + g(y) = 0 \qquad (4)$$

is reduced, with the aid of the substitution $w(y) = (y'_x)^2$, to a first order linear equation $w'_y + 2aw + 2g(y) = 0$. Therefore, the solution of the equation (4) can be written in the implicit form

$$x = C_2 \pm \int \left[C_1 e^{-2ay} - G(y)\right]^{-1/2} dy, \qquad \text{where} \quad G(y) = 2e^{-2ay} \int e^{2ay} g(y)\, dy.$$

4. The equation of the special form

$$y''_{xx} + a(y'_x)^4 + b(y'_x)^2 + g(y) = 0 \tag{5}$$

is reduced, with the aid of the substitution $w(y) = (y'_x)^2$, to the Riccati equation $w'_y + 2aw^2 + 2bw + 2g(y) = 0$ which is outlined in Section 1.2.

5. For the oscillatory systems with a weak nonlinearity

$$y''_{xx} + \varepsilon F(y'_x) + y = 0,$$

two leading terms of the asymptotic solution, as $\varepsilon \to 0$, are described by the formula

$$y = A\cos(x + B),$$

where functions $A = A(\xi)$ and $B = B(\xi)$ depend on the slow variable $\xi = \varepsilon x$ and are defined from the autonomous system of the first order equations

$$A'_\xi = \frac{1}{2\pi} \int_0^{2\pi} F(-A\sin\varphi)\sin\varphi\, d\varphi, \quad AB'_\xi = \frac{1}{2\pi} \int_0^{2\pi} F(-A\sin\varphi)\cos\varphi\, d\varphi.$$

The right-hand sides of these equations depend only on function A. The system is solved consecutively starting from the first equation.

1. $\boldsymbol{y''_{xx} + a(y'_x)^2 + by = 0.}$

This equation describes small oscillations when the drag force is proportional to the speed squared.

Solution in the implicit form:

$$x = C_2 \pm a \int \left[C_1 a^2 e^{-2ay} + b\left(\tfrac{1}{2} - ay\right)\right]^{-1/2} dy.$$

2. $\boldsymbol{y''_{xx} + \varepsilon\left[\tfrac{1}{3}(y'_x)^3 - y'_x\right] + y = 0.}$

Van der Pol equation.

Differentiating the equation with respect to x and passing on to a new variable $w(x) = y'_x$, we arrive at an equation of the form 2.2.2.2: $w''_{xx} - \varepsilon(1 - w^2)w'_x + w = 0$.

Solution, as $\varepsilon \to 0$:

$$y = \frac{2C_1}{\sqrt{1 - C_2 e^{-\varepsilon x}}} \cos x + \frac{2\sqrt{1 - C_1^2}}{\sqrt{1 - C_2 e^{-\varepsilon x}}} \sin x + O(\varepsilon^2).$$

3. $y''_{xx} + a(y'_x)^4 + b(y'_x)^2 + y = 0.$

The transformation $\xi = -\frac{1}{2}(y'_x)^2$, $w = -y - a(y'_x)^4 - b(y'_x)^2$ leads to the Abel equation of the form 1.3.2.1: $ww'_\xi = (-8a\xi + 2b)w + 1$.

4. $y''_{xx} + (y'_x)^2\left[a(y'_x)^2 + b\right]^{-1} + y = 0.$

The transformation $\xi = -\frac{1}{2}(y'_x)^2$, $w = -y - (y'_x)^2[a(y'_x)^2 + b]^{-1}$ leads to the Abel equation of the form 1.3.2.2: $ww'_\xi = 2b(b - 2a\xi)^{-2}w + 1$.

5. $y''_{xx} + A\exp[\lambda(y'_x)^2] + B + y = 0.$

Differentiating the equation with respect to x and passing on to a new variable $w(x) = y'_x$, we arrive at an equation of the form 2.2.2.7:

$$w''_{xx} + 2A\lambda w\exp(\lambda w^2)w'_x + w = 0.$$

6. $y''_{xx} + A\cosh[\lambda(y'_x)^2] + B + y = 0.$

Differentiating the equation with respect to x and passing on to a new variable $w(x) = y'_x$, we arrive at an equation of the form 2.2.2.11:

$$w''_{xx} + 2A\lambda w\sinh(\lambda w^2)w'_x + w = 0.$$

7. $y''_{xx} + A\sinh[\lambda(y'_x)^2] + B + y = 0.$

Differentiating the equation with respect to x and passing on to a new variable $w(x) = y'_x$, we arrive at an equation of the form 2.2.2.10:

$$w''_{xx} + 2A\lambda w\cosh(\lambda w^2)w'_x + w = 0.$$

8. $y''_{xx} + a(y'_x)^2 + b\sin y = 0.$

This equation describes the oscillations of the mathematical pendulum when the drag force is proportional to the speed squared.

Solution in the implicit form:

$$x = C_2 \pm \int \left[C_1 e^{-2ay} + \frac{2b}{4a^2+1}(\cos y - 2a\sin y)\right]^{-1/2} dy.$$

9. $y''_{xx} + A\cos[\lambda(y'_x)^2] + B + y = 0.$

Differentiating the equation with respect to x and passing on to a new variable $w(x) = y'_x$, we arrive at an equation of the form 2.2.2.15:

$$w''_{xx} - 2A\lambda w\sin(\lambda w^2)w'_x + w = 0.$$

10. $y''_{xx} + A\sin[\lambda(y'_x)^2] + B + y = 0.$

Differentiating the equation with respect to x and passing on to a new variable $w(x) = y'_x$, we arrive at an equation of the form 2.2.2.14:

$$w''_{xx} + 2A\lambda w\cos(\lambda w^2)w'_x + w = 0.$$

2.3. Emden—Fowler Equation $y''_{xx} = Ax^n y^m$

2.3.1. Exact Solutions

Preliminary comments. The value of the insignificant parameter A is in many cases defined in the form of a function of two (one) auxiliary coefficients a and b:

$$A = \varphi(a,\ b) \tag{1}$$

and the corresponding solutions are represented in the parametric form

$$x = f_1(\tau,\ C_1,\ C_2, a), \qquad y = f_2(\tau,\ C_1,\ C_2, b), \tag{2}$$

where τ is a parameter, C_1 and C_2 are arbitrary constants, f_1 and f_2 are some functions.

Having fixed the auxiliary coefficient sign $a > 0$ (or $b > 0$) in (1), the coefficient b should be expressed in terms of both A and a with the help of

$$b = \psi(A,\ a).$$

Substituting this formula into (2), we obtain a solution of the equation under consideration (where the concrete numerical value of the coefficient a may be chosen arbitrarily). The case $a < 0$ (or $b < 0$), which may lead to the branch of the solution or to a different domain of determining the variables x and y in (2), should be considered in a similar manner.

One may also use a different approach by setting one of the auxiliary coefficients (e.g., a) equal to ± 1 in (1) and (2); then the other coefficients will be identically expressed in terms of A by means of (1).

Table 2.7 represents all solvable Emden—Fowler equations whose solutions are outlined in Subsection 2.3.1. The one-parameter families (in the space of parameters n, m) and isolated points are represented in a consecutive fashion. Equations are arranged in accordance with the growth of m and the growth of n (for identical m). The number of the equation sought is indicated in the last column in this table.

1. $\boldsymbol{y''_{xx} = Ax^n}$.

Solution:

$$y = \begin{cases} \dfrac{Ax^{n+2}}{(n+1)(n+2)} + C_1 x + C_2 & \text{if } n \neq -1;\ -2, \\ -A \ln|x| + C_1 x + C_2 & \text{if } n = -2, \\ A \displaystyle\int \ln|x|\, dx + C_1 x + C_2 & \text{if } n = -1. \end{cases}$$

2. $\boldsymbol{y''_{xx} = Ay^m}$.

Solution:

$$x = \begin{cases} \pm \displaystyle\int \Bigl(\frac{2A}{m+1} y^{m+1} + C_1\Bigr)^{-1/2} dy + C_2 & \text{if } m \neq -1, \\ \pm \displaystyle\int (2A \ln|y| + C_1)^{-1/2}\, dy + C_2 & \text{if } m = -1. \end{cases}$$

TABLE 2.7
Solvable Cases of the Emden—Fowler Equation $y''_{xx} = Ax^n y^m$

No	m	n	Equation
	One-parameter families		
1	arbitrary	0	2.3.1.2
2	arbitrary	$-m-3$	2.3.1.3
3	arbitrary	$-\frac{1}{2}(m+3)$	2.3.1.4
4	0	arbitrary	2.3.1.1
5	1	arbitrary	2.3.1.5
	Isolated points		
6	-7	1	2.3.1.15
7	-7	3	2.3.1.16
8	$-5/2$	$-1/2$	2.3.1.22
9	-2	-2	2.3.1.28
10	-2	1	2.3.1.27
11	$-5/3$	$-10/3$	2.3.1.10
12	$-5/3$	$-7/3$	2.3.1.8
13	$-5/3$	$-5/6$	2.3.1.23
14	$-5/3$	$-1/2$	2.3.1.24
15	$-5/3$	1	2.3.1.7
16	$-5/3$	2	2.3.1.9
17	$-7/5$	$-13/5$	2.3.1.14
18	$-7/5$	1	2.3.1.13
19	$-1/2$	$-7/2$	2.3.1.12
20	$-1/2$	$-5/2$	2.3.1.6
21	$-1/2$	-2	2.3.1.26
22	$-1/2$	$-4/3$	2.3.1.17
23	$-1/2$	$-7/6$	2.3.1.18
24	$-1/2$	$-1/2$	2.3.1.25
25	$-1/2$	1	2.3.1.11
26	2	-5	2.3.1.19
27	2	$-20/7$	2.3.1.21
28	2	$-15/7$	2.3.1.20

Special cases:

1°. In the case $m = -1/2$, the solution can be written in the parametric form

$$x = aC_1^3(\tau^3 - 3\tau + C_2), \quad y = bC_1^4(\tau^2 - 1)^2, \qquad \text{where} \quad A = \pm\frac{4}{9}a^{-2}b^{3/2}.$$

2°. In the case $m = -4$, the solution can be written in the parametric form

$$x = aC_1^5\tau^{-1}\left(2\tau\int\frac{\tau\,d\tau}{R} + C_2\tau \mp R\right), \quad y = bC_1^2\tau^{-1},$$

where $R = \sqrt{\pm(4\tau^3 - 1)}$, $A = \mp 6a^{-2}b^5$.

3°. In the case $m = 2$, the solution can be written in the parametric form

$$x = aC_1^{-1}\tau, \quad y = bC_1^2\wp, \qquad \text{where} \quad A = \pm 6a^{-2}b^{-1},$$

the function $\wp$ of the parameter τ is defined in the implicit form:

$$\tau = \int\frac{d\wp}{\sqrt{\pm(4\wp^3 - 1)}} - C_2.$$

The upper sign in this formula corresponds to the classical elliptic Weierstrass function $\wp = \wp(\tau + C_2,\ 0,\ 1)$.

4°. In the case $m = -5/2$, the solution can be written in the parametric form

$$x = aC_1^7\wp^{-2}\left[\sqrt{\pm(4\wp^3 - 1)} \pm 2\tau\wp^2\right], \quad y = bC_1^4\wp^{-2}, \qquad \text{where} \quad A = \mp 3a^{-2}b^{7/2}.$$

The function $\wp$ of the parameter τ is defined in the previous case.

3. $y''_{xx} = Ax^{-m-3}y^m$.

1°. Solution in the parametric form with $m \neq -1$:

$$x = aC_1^{m-1}\left[\int (1 \pm \tau^{m+1})^{-1/2}\,d\tau + C_2\right]^{-1}, \quad y = bC_1^{m+1}\tau\left[\int (1 \pm \tau^{m+1})^{-1/2}\,d\tau + C_2\right]^{-1},$$

where $A = \pm\dfrac{m+1}{2}a^{m+1}b^{1-m}$.

2°. Solution in the parametric form with $m = -1$:

$$x = C_1\left[\int \exp(\mp\tau^2)\,d\tau + C_2\right]^{-1}, \quad y = b\exp(\mp\tau^2)\left[\int \exp(\mp\tau^2)\,d\tau + C_2\right]^{-1},$$

where $A = \mp 2b^2$.

4. $y''_{xx} = Ax^{-\frac{m+3}{2}}y^m$.

1°. Solution in the parametric form with $m \neq -1$:

$$x = aC_2^2 \exp\left[\int \left(\frac{2}{m+1}\tau^{m+1} + \frac{1}{4}\tau^2 + C_1\right)^{-1/2} d\tau\right],$$
$$y = bC_2\tau \exp\left[\frac{1}{2}\int \left(\frac{2}{m+1}\tau^{m+1} + \frac{1}{4}\tau^2 + C_1\right)^{-1/2} d\tau\right],$$

where $A = \left(\dfrac{a}{b^2}\right)^{\frac{m-1}{2}}$.

2°. Solution in the parametric form with $m = -1$:

$$x = aC_2^2 \exp\left[\int \left(2\ln|\tau| + \frac{1}{4}\tau^2 + C_1\right)^{-1/2} d\tau\right],$$
$$y = bC_2\tau \exp\left[\frac{1}{2}\int \left(2\ln|\tau| + \frac{1}{4}\tau^2 + C_1\right)^{-1/2} d\tau\right],$$

where $A = b^2/a$.

5. $y''_{xx} = Ax^n y$.

For $n \neq -2$, see equation 2.1.2.7. For $n = -2$, see equation 2.1.2.118.

6. $y''_{xx} = Ax^{-5/2}y^{-1/2}$.

Solution in the parametric form:

$$x = aC_1^{-3}(\tau^3 - 3\tau + C_2)^{-1}, \qquad y = bC_1(\tau^2 - 1)^2(\tau^3 - 3\tau + C_2)^{-1},$$

where $A = \pm\frac{4}{9}a^{1/2}b^{3/2}$.

7. $y''_{xx} = Axy^{-5/3}$.

Solution in the parametric form:

$$x = \pm aC_1^8(\tau^4 - 6\tau^2 + 4C_2\tau - 3), \qquad y = bC_1^9(\tau^3 - 3\tau + C_2)^{3/2},$$

where $A = \pm\frac{9}{64}a^{-3}b^{8/3}$.

8. $y''_{xx} = Ax^{-7/3}y^{-5/3}$.

Solution in the parametric form:

$$x = \pm \frac{aC_1^{-8}}{\tau^4 - 6\tau^2 + 4C_2\tau - 3}, \qquad y = \pm \frac{bC_1(\tau^3 - 3\tau + C_2)^{3/2}}{\tau^4 - 6\tau^2 + 4C_2\tau - 3},$$

where $A = \pm\frac{9}{64}a^{1/3}b^{8/3}$.

9. $y''_{xx} = Ax^2y^{-5/3}$.

1°. Solution in the parametric form with $A < 0$:

$$x = aC_1^2 \cos\tau \cosh(\tau + C_2)[\tan\tau + \tanh(\tau + C_2)], \quad y = bC_1^3[\cos\tau \cosh(\tau + C_2)]^{3/2},$$

where $A = -\frac{3}{16}a^{-4}b^{8/3}$.

2°. Solution in the parametric form with $A > 0$:

$$x = aC_1^2[\sinh\tau + \cos(\tau + C_2)], \qquad y = bC_1^3[\cosh\tau - \sin(\tau + C_2)]^{3/2},$$

where $A = \frac{3}{4}a^{-4}b^{8/3}$.

10. $y''_{xx} = Ax^{-10/3}y^{-5/3}$.

1°. Solution in the parametric form with $A < 0$:

$$x = aC_1^{-2}[\cos\tau \cosh(\tau + C_2)]^{-1}[\tan\tau + \tanh(\tau + C_2)]^{-1},$$
$$y = bC_1[\cos\tau \cosh(\tau + C_2)]^{1/2}[\tan\tau + \tanh(\tau + C_2)]^{-1},$$

where $A = -\frac{3}{16}a^{4/3}b^{8/3}$.

2°. Solution in the parametric form with $A > 0$:

$$x = aC_1^{-2}[\sinh\tau + \cos(\tau + C_2)]^{-1},$$
$$y = bC_1[\cosh\tau - \sin(\tau + C_2)]^{3/2}[\sinh\tau + \cos(\tau + C_2)]^{-1},$$

where $A = \frac{3}{4}a^{4/3}b^{8/3}$.

11. $y''_{xx} = Axy^{-1/2}$.

Solution in the parametric form:

$$x = aC_1 \exp(-\tau)\bigl[\exp(3\tau) + C_2 \sin\bigl(\sqrt{3}\,\tau\bigr)\bigr],$$
$$y = bC_1^2 \exp(-2\tau)\bigl[2\exp(3\tau) - C_2 \sin\bigl(\sqrt{3}\,\tau\bigr) + \sqrt{3}C_2 \cos\bigl(\sqrt{3}\,\tau\bigr)\bigr]^2,$$

where $A = 16a^{-3}b^{3/2}$.

▶ *In the solutions of equations 12–14, the following notation is used:*

$$S_1 = \exp(3\tau) + C_2 \sin(\sqrt{3}\,\tau), \qquad S_2 = 2\exp(3\tau) - C_2 \sin(\sqrt{3}\,\tau) + \sqrt{3}\,C_2 \cos(\sqrt{3}\,\tau),$$
$$S_3 = 2S_1(S_2)'_\tau - (S_1)'_\tau S_2 - S_1 S_2.$$

12. $y''_{xx} = Ax^{-7/2}y^{-1/2}$.

Solution in the parametric form:

$$x = aC_1^{-1}\exp(\tau)S_1^{-1}, \quad y = bC_1\exp(-\tau)S_1^{-1}S_2^2, \qquad \text{where} \quad A = 16(ab)^{3/2}.$$

13. $y''_{xx} = Axy^{-7/5}$.

Solution in the parametric form:

$$x = aC_1^4\exp(-2\tau)S_3, \quad y = bC_1^5\exp\left(-\tfrac{5}{2}\tau\right)S_1^{5/2}, \qquad \text{where} \quad A = \tfrac{5}{1024}a^{-3}b^{12/5}.$$

14. $y''_{xx} = Ax^{-13/5}y^{-7/5}$.

Solution in the parametric form:

$$x = aC_1^{-4}\exp(2\tau)S_3^{-1}, \quad y = bC_1\exp\left(-\tfrac{1}{2}\tau\right)S_1^{5/2}S_3^{-1}, \quad \text{where} \quad A = \tfrac{5}{1024}a^{3/5}b^{12/5}.$$

▶ *In the solutions of equations 15–18, the following notation is used:*

$$R = \sqrt{\pm(4\tau^3 - 1)}, \qquad f = 2\tau I(\tau) + C_2\tau \mp R,$$

where $I(\tau) = \int \tau R^{-1}\,d\tau$ *is the incomplete elliptic integral of the second kind in the form of Weierstrass.*

15. $y''_{xx} = Axy^{-7}$.

Solution in the parametric form:

$$x = aC_1^8[4\tau f^2 \mp \tau^{-2}(fR-1)^2], \quad y = bC_1^3 f^{1/2}, \qquad \text{where} \quad A = \pm\tfrac{3}{64}a^{-3}b^8.$$

16. $y''_{xx} = Ax^3y^{-7}$.

Solution in the parametric form:

$$x = aC_1^8[4\tau f^2 \mp \tau^{-2}(fR-1)^2]^{-1}, \qquad y = bC_1^5 f^{1/2}[4\tau f^2 \mp \tau^{-2}(fR-1)^2]^{-1},$$

where $A = \pm\frac{3}{64}a^{-5}b^8$.

17. $y''_{xx} = Ax^{-4/3}y^{-1/2}$.

Solution in the parametric form:

$$x = aC_1^9 f^3, \quad y = bC_1^4\tau^{-2}(fR-1)^2, \qquad \text{where} \quad A = \pm\tfrac{4}{3}a^{-2/3}b^{3/2}.$$

18. $y''_{xx} = Ax^{-7/6}y^{-1/2}$.

Solution in the parametric form:

$$x = aC_1^9 f^{-3}, \quad y = bC_1^5\tau^{-2}f^{-3}(fR-1)^2, \qquad \text{where} \quad A = \pm\tfrac{4}{3}a^{-5/6}b^{3/2}.$$

▶ *In the solutions of equations 19–24, the function $\wp$ of the parameter τ is defined in the implicit form*

$$\tau = \int \frac{d\wp}{\sqrt{\pm(4\wp^3 - 1)}} - C_2.$$

The upper sign in this formula corresponds to the classical elliptic Weierstrass function $\wp = \wp(\tau + C_2,\ 0,\ 1)$.

19. $y''_{xx} = Ax^{-5}y^2.$

Solution in the parametric form:

$$x = aC_1\tau^{-1}, \quad y = bC_1^3\tau^{-1}\wp, \qquad \text{where} \quad A = \pm 6a^3b^{-1}.$$

20. $y''_{xx} = Ax^{-15/7}y^2.$

Solution in the parametric form:

$$x = aC_1^7\tau^7, \quad y = bC_1\tau(\tau^2\wp \mp 1), \qquad \text{where} \quad A = \pm\tfrac{6}{49}a^{1/7}b^{-1}.$$

21. $y''_{xx} = Ax^{-20/7}y^2.$

Solution in the parametric form:

$$x = aC_1^7\tau^{-7}, \quad y = bC_1^6\tau^{-6}(\tau^2\wp \mp 1), \qquad \text{where} \quad A = \pm\tfrac{6}{49}a^{6/7}b^{-1}.$$

22. $y''_{xx} = Ax^{-1/2}y^{-5/2}.$

Solution in the parametric form:

$$x = aC_1^7\wp^2\left[\sqrt{\pm(4\wp^3 - 1)} \pm 2\tau\wp^2\right]^{-1}, \qquad y = bC_1^3\left[\sqrt{\pm(4\wp^3 - 1)} \pm 2\tau\wp^2\right]^{-1},$$

where $A = \mp 3a^{-3/2}b^{7/2}$.

23. $y''_{xx} = Ax^{-5/6}y^{-5/3}.$

Solution in the parametric form:

$$x = \frac{aC_1^{16}}{\left[\tau\sqrt{\pm(4\wp^3 - 1)} + 2\wp\right]^2}, \qquad y = \frac{bC_1^7\left[\sqrt{\pm(4\wp^3 - 1)} \pm 2\tau\wp^2\right]^{3/2}}{\left[\tau\sqrt{\pm(4\wp^3 - 1)} + 2\wp\right]^2},$$

where $A = -\frac{1}{6}a^{-7/6}b^{8/3}$.

24. $y''_{xx} = Ax^{-1/2}y^{-5/3}.$

Solution in the parametric form:

$$x = aC_1^{16}\left[\tau\sqrt{\pm(4\wp^3 - 1)} + 2\wp\right]^2, \qquad y = bC_1^9\left[\sqrt{\pm(4\wp^3 - 1)} \pm 2\tau\wp^2\right]^{3/2},$$

where $A = -\frac{1}{6}a^{-3/2}b^{8/3}$.

▶ *In the solutions of equations 25–28, the following notation is used:*

$$Z = \begin{cases} C_1 J_{1/3}(\tau) + C_2 Y_{1/3}(\tau) & \textit{for the upper sign,} \\ C_1 I_{1/3}(\tau) + C_2 K_{1/3}(\tau) & \textit{for the lower sign,} \end{cases}$$

where $J_{1/3}$ and $Y_{1/3}$ are Bessel functions, $I_{1/3}$ and $K_{1/3}$ are modified Bessel functions.

25. $y''_{xx} = Ax^{-1/2} y^{-1/2}$.

Solution in the parametric form:

$$x = a\tau^{2/3} Z^2, \quad y = b\tau^{-2/3}\left(\tau Z'_\tau + \frac{1}{3}Z\right)^2, \qquad \text{where} \quad A = \mp \frac{b}{3a}\left(\mp \frac{b}{a}\right)^{1/2}.$$

26. $y''_{xx} = Ax^{-2} y^{-1/2}$.

Solution in the parametric form:

$$x = a\tau^{-2/3} Z^{-2}, \quad y = b\tau^{-4/3} Z^{-2}\left(\tau Z'_\tau + \frac{1}{3}Z\right)^2, \qquad \text{where} \quad A = \mp \frac{1}{3} b^{3/2}.$$

27. $y''_{xx} = Axy^{-2}$.

Solution in the parametric form:

$$x = a\tau^{-2/3}\left[\left(\tau Z'_\tau + \frac{1}{3}Z\right)^2 \pm \tau^2 Z^2\right], \quad y = b\tau^{2/3} Z^2, \qquad \text{where} \quad A = -\frac{9}{2}\left(\frac{b}{a}\right)^3.$$

28. $y''_{xx} = Ax^{-2} y^{-2}$.

Solution in the parametric form:

$$x = \tau^{2/3}\left[\left(\tau Z'_\tau + \frac{1}{3}Z\right)^2 \pm \tau^2 Z^2\right]^{-1}, \qquad y = b\tau^{4/3} Z^2\left[\left(\tau Z'_\tau + \frac{1}{3}Z\right)^2 \pm \tau^2 Z^2\right]^{-1},$$

where $A = -\frac{9}{2} b^3$.

2.3.2. First Integrals (Conservation Laws)

In this subsection, first integrals of the form

$$\sum_{\alpha=0}^{k} f_\alpha(x,y)(y'_x)^\alpha = C, \qquad \text{where} \quad k = 2,\ 3,\ 4,\ 5,$$

for the Emden—Fowler equation $y''_{xx} = Ax^n y^m$ are given.

First integrals with $k = 2$

1. For $n = 0$ and arbitrary m $(m \neq -1)$,

$$(y'_x)^2 - \frac{2A}{m+1} y^{m+1} = C.$$

2. For $n = -\dfrac{m+3}{2}$ and arbitrary m $(m \neq -1)$,

$$x(y'_x)^2 - yy'_x - \frac{2A}{m+1} x^{-\frac{m+1}{2}} y^{m+1} = C.$$

3. For $n = -m - 3$ and arbitrary m $(m \neq -1)$,

$$x^2(y'_x)^2 - 2xyy'_x + y^2 - \frac{2A}{m+1} x^{-m-1} y^{m+1} = C.$$

4. For $n = -\frac{20}{7}$, $m = 2$,

$$\tfrac{343}{24} Ax^{8/7}(y'_x)^2 - \left(\tfrac{49}{3} Ax^{1/7} y - x\right)y'_x - \tfrac{343}{36} A^2 x^{-12/7} y^3 + \tfrac{7}{6} Ax^{-6/7} y^2 - y = C.$$

5. For $n = -\frac{15}{7}$, $m = 2$,

$$\tfrac{343}{24} Ax^{6/7}(y'_x)^2 - \left(\tfrac{49}{4} Ax^{-1/7} y + 1\right)y'_x - \tfrac{343}{36} A^2 x^{-9/7} y^3 - \tfrac{7}{8} Ax^{-8/7} y^2 = C.$$

First integrals with $k = 3$

1. For $n = 0$, $m = -\frac{1}{2}$,

$$(y'_x)^3 - 6Ay^{1/2} y'_x + 6A^2 x = C,$$
$$x(y'_x)^3 - y(y'_x)^2 - 6Axy^{1/2} y'_x + \tfrac{16}{3} Ay^{3/2} + 3A^2 x^2 = C.$$

2. For $n = 1$, $m = -\frac{1}{2}$,

$$(y'_x)^3 - 6Axy^{1/2} y'_x + 4Ay^{3/2} + 2A^2 x^3 = C.$$

3. For $n = -\frac{4}{3}$, $m = -\frac{1}{2}$,

$$x(y'_x)^3 - y(y'_x)^2 - 6Ax^{-1/3} y^{1/2} y'_x - 9A^2 x^{-2/3} = C.$$

4. For $n = -\frac{5}{2}$, $m = -\frac{1}{2}$,

$$x^2(y'_x)^3 - 2xy(y'_x)^2 + (y^2 - 6Ax^{-1/2} y^{1/2})y'_x + \tfrac{2}{3} Ax^{-3/2} y^{3/2} - 3A^2 x^{-2} = C,$$
$$x^3(y'_x)^3 - 3x^2 y(y'_x)^2 + 3(xy^2 - 2Ax^{1/2} y^{1/2})y'_x - y^3 + 6Ax^{-1/2} y^{3/2} - 6A^2 x^{-1} = C.$$

5. For $n = -\frac{7}{6}$, $m = -\frac{1}{2}$,

$$x^2(y'_x)^3 - 2xy(y'_x)^2 + (y^2 - 6Ax^{5/6} y^{1/2})y'_x + 6Ax^{-1/6} y^{3/2} + 9A^2 x^{2/3} = C.$$

6. For $n = -\frac{7}{2}$, $m = -\frac{1}{2}$,

$$x^3(y'_x)^3 - 3x^2 y(y'_x)^2 + 3(xy^2 - 2Ax^{-1/2} y^{1/2})y'_x - y^3 + 2Ax^{-3/2} y^{3/2} - 2A^2 x^{-3} = C.$$

First integrals with $k = 4$

1. For $n = 1$, $m = -\frac{5}{3}$,

$$(y'_x)^4 + 6Axy^{-2/3}(y'_x)^2 - 18Ay^{1/3}y'_x + 9A^2x^2y^{-4/3} = C,$$
$$x(y'_x)^4 - y(y'_x)^3 + 6Ax^2y^{-2/3}(y'_x)^2 - 27Axy^{1/3}y'_x + \tfrac{81}{4}Ay^{4/3} + 9A^2x^3y^{-4/3} = C.$$

2. For $n = 2$, $m = -\frac{5}{3}$,

$$(y'_x)^4 + 6Ax^2y^{-2/3}(y'_x)^2 - 36Axy^{1/3}y'_x + 9A^2x^4y^{-4/3} = C.$$

3. For $n = 0$, $m = -\frac{5}{3}$,

$$x(y'_x)^4 - y(y'_x)^3 + 6Axy^{-2/3}(y'_x)^2 - 9Ay^{1/3}y'_x + 9A^2xy^{-4/3} = C,$$
$$x^2(y'_x)^4 - 2xy(y'_x)^3 + (y^2 + 6Ax^2y^{-2/3})(y'_x)^2 - 18Axy^{1/3}y'_x + 12Ay^{4/3} + 9A^2x^2y^{-4/3} = C.$$

4. For $n = -\frac{1}{2}$, $m = -\frac{5}{3}$,

$$x(y'_x)^4 - y(y'_x)^3 + 6Ax^{1/2}y^{-2/3}(y'_x)^2 + 9A^2y^{-4/3} = C.$$

5. For $n = -\frac{4}{3}$, $m = -\frac{5}{3}$,

$$x^2(y'_x)^4 - 2xy(y'_x)^3 + (y^2 + 6Ax^{2/3}y^{-2/3})(y'_x)^2 + 6Ax^{-1/3}y^{1/3}y'_x + 9A^2x^{-2/3}y^{-4/3} = C,$$
$$x^3(y'_x)^4 - 3x^2y(y'_x)^3 + 3x(y^2 + 2Ax^{2/3}y^{-2/3})(y'_x)^2$$
$$- (y^3 + 3Ax^{2/3}y^{1/3})y'_x - 3Ax^{-1/3}y^{4/3} + 9A^2x^{1/3}y^{-4/3} = C.$$

6. For $n = -\frac{7}{3}$, $m = -\frac{5}{3}$,

$$x^3(y'_x)^4 - 3x^2y(y'_x)^3 + 3x(y^2 + 2Ax^{-1/3}y^{-2/3})(y'_x)^2$$
$$- (y^3 - 15Ax^{-1/3}y^{1/3})y'_x - \tfrac{3}{4}Ax^{-4/3}y^{4/3} + 9A^2x^{-5/3}y^{-4/3} = C,$$
$$x^4(y'_x)^4 - 4x^3y(y'_x)^3 + 6x^2(y^2 + Ax^{-1/3}y^{-2/3})(y'_x)^2$$
$$-2x(2y^3 - 3Ax^{-1/3}y^{1/3})y'_x + y^4 - 12Ax^{-1/3}y^{4/3} + 9A^2x^{-2/3}y^{-4/3} = C.$$

7. For $n = -\frac{5}{6}$, $m = -\frac{5}{3}$,

$$x^3(y'_x)^4 - 3x^2y(y'_x)^3 + 3x(y^2 + 2Ax^{7/6}y^{-2/3})(y'_x)^2$$
$$- (y^3 + 12Ax^{7/6}y^{1/3})y'_x + 6Ax^{1/6}y^{4/3} + 9A^2x^{4/3}y^{-4/3} = C.$$

8. For $n = -\frac{10}{3}$, $m = -\frac{5}{3}$,

$$x^4(y'_x)^4 - 4x^3y(y'_x)^3 + 6x^2(y^2 + Ax^{-4/3}y^{-2/3})(y'_x)^2$$
$$- 4x(y^3 - 6Ax^{-4/3}y^{1/3})y'_x + y^4 - 30Ax^{-4/3}y^{4/3} + 9A^2x^{-8/3}y^{-4/3} = C.$$

9. For $n = 1$, $m = -7$,

$$x(y'_x)^4 - y(y'_x)^3 + \tfrac{2}{3}Ax^2y^{-6}(y'_x)^2 - \tfrac{1}{3}Axy^{-5}y'_x - \tfrac{1}{12}Ay^{-4} + \tfrac{1}{9}A^2x^3y^{-12} = C.$$

10. For $n = 3$, $m = -7$,

$$x^3(y'_x)^4 - 3x^2y(y'_x)^3 + 3x(y^2 + \tfrac{2}{9}Ax^5y^{-6})(y'_x)^2$$
$$- y(y^2 + Ax^5y^{-6})y'_x + \tfrac{1}{4}Ax^4y^{-4} + \tfrac{1}{9}A^2x^9y^{-12} = C.$$

It should be noted that in the case $k = 4$ we omitted the first integrals of the form

$$\alpha F^2 + \beta F + \gamma = C,$$

where function $F = F(x, y, y'_x)$ is the left-hand side of the above integrals for $k = 2$, and α, β, and γ are some constants.

First integrals with $k = 5$

1. For $n = 0$, $m = -\frac{2}{3}$,

$$(y'_x)^5 - 15Ay^{1/3}(y'_x)^3 + \tfrac{135}{2}A^2y^{2/3}y'_x - \tfrac{135}{2}A^3x = C.$$

2. For $n = -\frac{7}{3}$, $m = -\frac{2}{3}$,

$$\begin{aligned} x^5(y'_x)^5 - 5x^4y(y'_x)^4 + 5x^3y(2y - 3Ax^{-1/3}y^{-2/3})(y'_x)^3 \\ - 5x^2y^2(2y - 9Ax^{-1/3}y^{-2/3})(y'_x)^2 + 5x(y^4 - 9Ax^{-1/3}y^{7/3} + \tfrac{27}{2}A^2x^{-2/3}y^{2/3})y'_x \\ + 15A(x^{-1/3}y^{10/3} - \tfrac{9}{2}Ax^{-2/3}y^{5/3} - \tfrac{9}{2}A^2x^{-1}) = C. \end{aligned}$$

2.3.3. Some Formulas and Transformations

1. With $m \neq 1$, the Emden—Fowler equation has the particular solution

$$y = \lambda x^{\frac{n+2}{1-m}}, \qquad \text{where} \quad \lambda = \left[\frac{(n+2)(n+m+1)}{A(m-1)^2}\right]^{\frac{1}{m-1}}.$$

2. The transformation $y = w/t$, $x = 1/t$ leads to the Emden—Fowler equation with the independent variable to a different power:

$$w''_{tt} = At^{-n-m-3}w^m.$$

3. Some more complicated transformations leading to the Emden—Fowler equation are outlined in Subsection 2.5.3 (see Fig. 3).

4. With $m \neq 1$ and $m \neq -2n - 3$, the transformation

$$\xi = \frac{2n+m+3}{m-1}x^{\frac{n+2}{m-1}}y, \qquad u = x^{\frac{n+2}{m-1}}\left(xy'_x + \frac{n+2}{m-1}y\right)$$

leads to the Abel equation:

$$uu'_\xi - u = -\frac{(n+2)(n+m+1)}{(2n+m+3)^2}\xi + A\left(\frac{m-1}{2n+m+3}\right)^2\xi^m,$$

whose special cases are given in Subsection 1.3.1.

5. Some more complicated transformations leading to other Abel equations are outlined in Subsection 2.5.3.

2.4. Equations of the Form $y''_{xx} = A_1x^{n_1}y^{m_1} + A_2x^{n_2}y^{m_2}$

See Section 2.3 for the special cases $A_1 = 0$ and $A_2 = 0$.

2.4.1. Classification Table

Table 2.8 represents all solvable equations of the form $y''_{xx} = A_1x^{n_1}y^{m_1} + A_2x^{n_2}y^{m_2}$ whose solutions are outlined in Subsection 2.4.2. The two-parameter families (in the space of parameters m_1, m_2, n_1, and n_2), the one-parameter families, and isolated points are represented in a consecutive fashion. Equations are arranged in accordance with the growth of m_1, the growth of m_2 (for identical m_1), the growth of n_1 (for identical m_1 and m_2), and the growth of n_2 (for identical m_1, m_2, and n_1). The number of the equation sought is indicated in the last column in this table.

TABLE 2.8

Solvable equations of the form $y''_{xx} = A_1 x^{n_1} y^{m_1} + A_2 x^{n_2} y^{m_2}$

No	m_1	m_2	n_1	n_2	A_1	A_2	Equation
1	arbitrary	arbitrary	0	0	arbitrary	arbitrary	2.4.2.1
2	arbitrary	arbitrary	$-m_1-3$	$-m_2-3$	arbitrary	arbitrary	2.4.2.2
3	arbitrary	arbitrary	$-\frac{1}{2}(m_1+3)$	$-\frac{1}{2}(m_2+3)$	arbitrary	arbitrary	2.4.2.3
4	arbitrary	0	0	0	arbitrary	arbitrary	2.4.2.19
5	arbitrary	0	$-m_1-3$	-3	arbitrary	arbitrary	2.4.2.20
6	1	arbitrary	-2	-2	$-\dfrac{2(m_2+1)}{(m_2+3)^2}$	arbitrary	2.4.2.4
7	1	arbitrary	-2	$-m_2-1$	$-\dfrac{2(m_2+1)}{(m_2+3)^2}$	arbitrary	2.4.2.5
8	1	-3	arbitrary ($n_1 \neq -2$)	0	arbitrary	arbitrary	2.4.2.83
9	-7	-7	4	3	arbitrary	arbitrary	2.4.2.39
10	-5	-5	2	0	arbitrary	arbitrary	2.4.2.16
11	-3	-7	0	1	arbitrary	arbitrary	2.4.2.42
12	-3	-7	0	3	arbitrary	arbitrary	2.4.2.43
13	-3	-4	0	0	arbitrary	arbitrary	2.4.2.17
14	-3	-4	0	1	arbitrary	arbitrary	2.4.2.18
15	-2	-3	-2	0	arbitrary	arbitrary	2.4.2.88
16	-2	-3	1	0	arbitrary	arbitrary	2.4.2.87
17	-2	-2	-1	-2	arbitrary	arbitrary	2.4.2.28
18	$-\frac{5}{3}$	$-\frac{5}{3}$	$-\frac{7}{3}$	$-\frac{10}{3}$	arbitrary	arbitrary	2.4.2.48
19	$-\frac{5}{3}$	$-\frac{5}{3}$	$-\frac{4}{3}$	$-\frac{10}{3}$	arbitrary	arbitrary	2.4.2.49
20	$-\frac{5}{3}$	$-\frac{5}{3}$	$-\frac{4}{3}$	$-\frac{7}{3}$	arbitrary	arbitrary	2.4.2.24
21	$-\frac{5}{3}$	$-\frac{5}{3}$	$-\frac{2}{3}$	$-\frac{4}{3}$	arbitrary	arbitrary	2.4.2.90
22	$-\frac{5}{3}$	$-\frac{5}{3}$	0	$-\frac{2}{3}$	arbitrary	arbitrary	2.4.2.89
23	$-\frac{5}{3}$	$-\frac{5}{3}$	2	0	arbitrary	arbitrary	2.4.2.47
24	$-\frac{5}{3}$	$-\frac{5}{3}$	2	1	arbitrary	arbitrary	2.4.2.46
25	$-\frac{3}{2}$	-2	$-\frac{3}{2}$	-2	arbitrary	arbitrary	2.4.2.81
26	$-\frac{3}{2}$	-2	0	1	arbitrary	arbitrary	2.4.2.80
27	$-\frac{7}{5}$	$-\frac{7}{5}$	$-\frac{8}{5}$	$-\frac{13}{5}$	arbitrary	arbitrary	2.4.2.25
28	$-\frac{4}{3}$	$-\frac{5}{3}$	$-\frac{5}{3}$	$-\frac{7}{3}$	arbitrary	arbitrary	2.4.2.102
29	$-\frac{4}{3}$	$-\frac{5}{3}$	0	1	arbitrary	arbitrary	2.4.2.101

TABLE 2.8 *Continued*
Solvable equations of the form $y''_{xx} = A_1 x^{n_1} y^{m_1} + A_2 x^{n_2} y^{m_2}$

No	m_1	m_2	n_1	n_2	A_1	A_2	Equation
30	$-\frac{3}{5}$	$-\frac{7}{5}$	$-\frac{12}{5}$	$-\frac{13}{5}$	arbitrary	arbitrary	2.4.2.53
31	$-\frac{3}{5}$	$-\frac{7}{5}$	0	1	arbitrary	arbitrary	2.4.2.52
32	$-\frac{1}{2}$	$-\frac{1}{2}$	$-\frac{5}{2}$	$-\frac{7}{2}$	arbitrary	arbitrary	2.4.2.23
33	$-\frac{1}{3}$	$-\frac{5}{3}$	$-\frac{8}{3}$	$-\frac{10}{3}$	arbitrary	arbitrary	2.4.2.55
34	$-\frac{1}{3}$	$-\frac{5}{3}$	$-\frac{8}{3}$	$-\frac{7}{3}$	arbitrary	arbitrary	2.4.2.59
35	$-\frac{1}{3}$	$-\frac{5}{3}$	$-\frac{8}{3}$	$-\frac{4}{3}$	arbitrary	arbitrary	2.4.2.57
36	$-\frac{1}{3}$	$-\frac{5}{3}$	0	0	arbitrary	arbitrary	2.4.2.56
37	$-\frac{1}{3}$	$-\frac{5}{3}$	0	1	arbitrary	arbitrary	2.4.2.58
38	$-\frac{1}{3}$	$-\frac{5}{3}$	0	2	arbitrary	arbitrary	2.4.2.54
39	0	-2	-3	-2	arbitrary	arbitrary	2.4.2.108
40	0	-2	0	1	arbitrary	arbitrary	2.4.2.107
41	0	-1	-3	-2	arbitrary	arbitrary	2.4.2.22
42	0	-1	0	0	arbitrary	arbitrary	2.4.2.21
43	0	$-\frac{2}{3}$	-3	$-\frac{7}{3}$	arbitrary	arbitrary	2.4.2.73
44	0	$-\frac{2}{3}$	0	0	arbitrary	arbitrary	2.4.2.72
45	0	$-\frac{1}{2}$	-4	$-\frac{5}{2}$	arbitrary	arbitrary	2.4.2.96
46	0	$-\frac{1}{2}$	-3	$-\frac{7}{2}$	arbitrary	arbitrary	2.4.2.51
47	0	$-\frac{1}{2}$	-3	$-\frac{5}{2}$	arbitrary	arbitrary	2.4.2.45
48	0	$-\frac{1}{2}$	-3	-2	arbitrary	arbitrary	2.4.2.106
49	0	$-\frac{1}{2}$	-3	$-\frac{1}{2}$	arbitrary	arbitrary	2.4.2.85
50	0	$-\frac{1}{2}$	$-\frac{5}{3}$	$-\frac{7}{6}$	arbitrary	arbitrary	2.4.2.41
51	0	$-\frac{1}{2}$	$-\frac{3}{2}$	$-\frac{5}{2}$	arbitrary	arbitrary	2.4.2.100
52	0	$-\frac{1}{2}$	$-\frac{3}{2}$	-2	arbitrary	arbitrary	2.4.2.79
53	0	$-\frac{1}{2}$	$-\frac{3}{2}$	$-\frac{1}{2}$	arbitrary	arbitrary	2.4.2.78
54	0	$-\frac{1}{2}$	$-\frac{3}{2}$	0	arbitrary	arbitrary	2.4.2.99
55	0	$-\frac{1}{2}$	$-\frac{4}{3}$	$-\frac{4}{3}$	arbitrary	arbitrary	2.4.2.40
56	0	$-\frac{1}{2}$	0	-2	arbitrary	arbitrary	2.4.2.86
57	0	$-\frac{1}{2}$	0	$-\frac{1}{2}$	arbitrary	arbitrary	2.4.2.105
58	0	$-\frac{1}{2}$	0	0	arbitrary	arbitrary	2.4.2.44
59	0	$-\frac{1}{2}$	0	1	arbitrary	arbitrary	2.4.2.50

TABLE 2.8 *Continued*

Solvable equations of the form $y''_{xx} = A_1 x^{n_1} y^{m_1} + A_2 x^{n_2} y^{m_2}$

No	m_1	m_2	n_1	n_2	A_1	A_2	Equation
60	0	$-\frac{1}{2}$	1	0	arbitrary	arbitrary	2.4.2.95
61	$\frac{1}{3}$	$-\frac{5}{3}$	$-\frac{10}{3}$	$-\frac{7}{3}$	arbitrary	arbitrary	2.4.2.98
62	$\frac{1}{3}$	$-\frac{5}{3}$	0	1	arbitrary	arbitrary	2.4.2.97
63	1	-7	-2	-2	$\frac{15}{4}$	arbitrary	2.4.2.35
64	1	-7	-2	6	$\frac{15}{4}$	arbitrary	2.4.2.36
65	1	-4	-2	-2	6	arbitrary	2.4.2.31
66	1	-4	-2	3	6	arbitrary	2.4.2.32
67	1	-3	-5	0	arbitrary	arbitrary	2.4.2.84
68	1	-3	1	0	arbitrary	arbitrary	2.4.2.82
69	1	$-\frac{5}{2}$	-2	-2	12	arbitrary	2.4.2.64
70	1	$-\frac{5}{2}$	-2	$\frac{3}{2}$	12	arbitrary	2.4.2.65
71	1	-2	-2	-2	2	arbitrary	2.4.2.6
72	1	-2	-2	1	2	arbitrary	2.4.2.7
73	1	$-\frac{5}{3}$	-2	-2	$-\frac{3}{16}$	arbitrary	2.4.2.26
74	1	$-\frac{5}{3}$	-2	-2	$-\frac{9}{100}$	arbitrary	2.4.2.10
75	1	$-\frac{5}{3}$	-2	-2	$\frac{3}{4}$	arbitrary	2.4.2.12
76	1	$-\frac{5}{3}$	-2	-2	$\frac{63}{4}$	arbitrary	2.4.2.66
77	1	$-\frac{5}{3}$	-2	$\frac{2}{3}$	$-\frac{3}{16}$	arbitrary	2.4.2.27
78	1	$-\frac{5}{3}$	-2	$\frac{2}{3}$	$-\frac{9}{100}$	arbitrary	2.4.2.11
79	1	$-\frac{5}{3}$	-2	$\frac{2}{3}$	$\frac{3}{4}$	arbitrary	2.4.2.13
80	1	$-\frac{5}{3}$	-2	$\frac{2}{3}$	$\frac{63}{4}$	arbitrary	2.4.2.67
81	1	$-\frac{7}{5}$	-2	-2	$-\frac{5}{36}$	arbitrary	2.4.2.29
82	1	$-\frac{7}{5}$	-2	$\frac{2}{5}$	$-\frac{5}{36}$	arbitrary	2.4.2.30
83	1	$-\frac{1}{2}$	-2	-2	$-\frac{2}{9}$	arbitrary	2.4.2.14
84	1	$-\frac{1}{2}$	-2	-2	$-\frac{4}{25}$	arbitrary	2.4.2.8
85	1	$-\frac{1}{2}$	-2	-2	20	arbitrary	2.4.2.33
86	1	$-\frac{1}{2}$	-2	$-\frac{1}{2}$	$-\frac{2}{9}$	arbitrary	2.4.2.15
87	1	$-\frac{1}{2}$	-2	$-\frac{1}{2}$	$-\frac{4}{25}$	arbitrary	2.4.2.9
88	1	$-\frac{1}{2}$	-2	$-\frac{1}{2}$	20	arbitrary	2.4.2.34
89	1	0	-5	-3	arbitrary	arbitrary	2.4.2.77

TABLE 2.8 *Continued*
Solvable equations of the form $y''_{xx} = A_1 x^{n_1} y^{m_1} + A_2 x^{n_2} y^{m_2}$

No	m_1	m_2	n_1	n_2	A_1	A_2	Equation
90	1	0	1	0	arbitrary	arbitrary	2.4.2.76
91	1	$\frac{1}{2}$	-2	-2	$-\frac{12}{49}$	arbitrary	2.4.2.37
92	1	$\frac{1}{2}$	-2	$-\frac{3}{2}$	$-\frac{12}{49}$	arbitrary	2.4.2.38
93	2	0	-5	-4	arbitrary	arbitrary	2.4.2.92
94	2	0	-5	-3	arbitrary	arbitrary	2.4.2.69
95	2	0	$-\frac{20}{7}$	$-\frac{13}{7}$	arbitrary	arbitrary	2.4.2.94
96	2	0	$-\frac{20}{7}$	$-\frac{12}{7}$	arbitrary	arbitrary	2.4.2.71
97	2	0	$-\frac{15}{7}$	$-\frac{9}{7}$	arbitrary	arbitrary	2.4.2.70
98	2	0	$-\frac{15}{7}$	$-\frac{8}{7}$	arbitrary	arbitrary	2.4.2.93
99	2	0	0	0	arbitrary	arbitrary	2.4.2.68
100	2	0	0	1	arbitrary	arbitrary	2.4.2.91
101	2	1	-3	-2	arbitrary	$-\frac{6}{25}$	2.4.2.61
102	2	1	-3	-2	arbitrary	$\frac{6}{25}$	2.4.2.63
103	2	1	-2	-2	arbitrary	$-\frac{6}{25}$	2.4.2.60
104	2	1	-2	-2	arbitrary	$\frac{6}{25}$	2.4.2.62
105	3	1	-6	-5	arbitrary	arbitrary	2.4.2.104
106	3	1	0	1	arbitrary	arbitrary	2.4.2.103
107	3	2	$-\frac{18}{5}$	$-\frac{14}{5}$	arbitrary	arbitrary	2.4.2.74
108	3	2	$-\frac{12}{5}$	$-\frac{11}{5}$	arbitrary	arbitrary	2.4.2.75

2.4.2. Exact Solutions

1. $y''_{xx} = A_1 y^{m_1} + A_2 y^{m_2}, \qquad m_1 \neq -1,\ m_2 \neq -1.$

1°. Solution in the parametric form:

$$x = a \int \left(C_1 + \tau^{m_1+1} \pm \tau^{m_2+1}\right)^{-1/2} d\tau + C_2, \qquad y = b\tau,$$

where $A_1 = \frac{1}{2} a^{-2} b^{1-m_1}(m_1+1)$, $A_2 = \pm\frac{1}{2} a^{-2} b^{1-m_2}(m_2+1)$.

2°. Solution in the parametric form:

$$x = a \int \left(C_1 - \tau^{m_1+1} \pm \tau^{m_2+1}\right)^{-1/2} d\tau + C_2, \qquad y = b\tau,$$

where $A_1 = -\frac{1}{2} a^{-2} b^{1-m_1}(m_1+1)$, $A_2 = \pm\frac{1}{2} a^{-2} b^{1-m_2}(m_2+1)$.

2. $y''_{xx} = A_1x^{-m_1-3}y^{m_1} + A_2x^{-m_2-3}y^{m_2}, \qquad m_1 \neq -1,\ m_2 \neq -1.$

1°. Solution in the parametric form:

$$x = a\left[\int (C_1 + \tau^{m_1+1} \pm \tau^{m_2+1})^{-1/2}\,d\tau + C_2\right]^{-1},$$
$$y = b\tau\left[\int (C_1 + \tau^{m_1+1} \pm \tau^{m_2+1})^{-1/2}\,d\tau + C_2\right]^{-1},$$

where $A_1 = \frac{1}{2}a^{1+m_1}b^{1-m_1}(m_1+1)$, $A_2 = \pm\frac{1}{2}a^{1+m_2}b^{1-m_2}(m_2+1)$.

2°. Solution in the parametric form:

$$x = a\left[\int (C_1 - \tau^{m_1+1} \pm \tau^{m_2+1})^{-1/2}\,d\tau + C_2\right]^{-1},$$
$$y = b\tau\left[\int (C_1 - \tau^{m_1+1} \pm \tau^{m_2+1})^{-1/2}\,d\tau + C_2\right]^{-1},$$

where $A_1 = -\frac{1}{2}a^{1+m_1}b^{1-m_1}(m_1+1)$, $A_2 = \pm\frac{1}{2}a^{1+m_2}b^{1-m_2}(m_2+1)$.

3. $y''_{xx} = A_1x^{-\frac{m_1+3}{2}}y^{m_1} + A_2x^{-\frac{m_2+3}{2}}y^{m_2}.$

1°. Solution in the parametric form with $m_1 \neq -1$ and $m_2 \neq -1$:

$$x = C_1^2\exp\left[\int\left(C_2 + \frac{1}{4}\tau^2 + \frac{2A_1}{m_1+1}\tau^{m_1+1} + \frac{2A_2}{m_2+1}\tau^{m_2+1}\right)^{-1/2}d\tau\right],$$
$$y = C_1\tau\exp\left[\frac{1}{2}\int\left(C_2 + \frac{1}{4}\tau^2 + \frac{2A_1}{m_1+1}\tau^{m_1+1} + \frac{2A_2}{m_2+1}\tau^{m_2+1}\right)^{-1/2}d\tau\right].$$

2°. Solution in the parametric form with $m_1 \neq -1$ and $m_2 = -1$:

$$x = C_1^2\exp\left[\int\left(C_2 + \frac{1}{4}\tau^2 + \frac{2A_1}{m_1+1}\tau^{m_1+1} + 2A_2\ln|\tau|\right)^{-1/2}d\tau\right],$$
$$y = C_1\tau\exp\left[\frac{1}{2}\int\left(C_2 + \frac{1}{4}\tau^2 + \frac{2A_1}{m_1+1}\tau^{m_1+1} + 2A_2\ln|\tau|\right)^{-1/2}d\tau\right].$$

4. $y''_{xx} = -\dfrac{2(m+1)}{(m+3)^2}x^{-2}y + Ax^{-2}y^m, \qquad m \neq -3,\ m \neq -1.$

Solution in the parametric form:

$$x = C_1\left[\int (1 \pm \tau^{m+1})^{-1/2}\,d\tau + C_2\right]^{\frac{m+3}{m-1}}, \quad y = b\tau\left[\int (1 \pm \tau^{m+1})^{-1/2}\,d\tau + C_2\right]^{\frac{2}{m-1}},$$

where $A = \pm\dfrac{(m+1)(m-1)^2}{2(m+3)^2}b^{1-m}$.

5. $y''_{xx} = -\dfrac{2(m+1)}{(m+3)^2}x^{-2}y + Ax^{-m-1}y^m, \qquad m \neq -3,\ m \neq -1.$

Solution in the parametric form:

$$x = C_1\left[\int (1\pm\tau^{m+1})^{-1/2}\,d\tau + C_2\right]^{-\frac{m+3}{m-1}}, \quad y = bC_1\tau\left[\int (1\pm\tau^{m+1})^{-1/2}\,d\tau + C_2\right]^{-\frac{m+1}{m-1}},$$

where $A = \pm\dfrac{(m+1)(m-1)^2}{2(m+3)^2}b^{1-m}$.

6. $y''_{xx} = 2x^{-2}y + Ax^{-2}y^{-2}.$

Solution in the parametric form:

$$x = C_1\left[\sqrt{\tau(\tau+1)} - \ln\left(\sqrt{\tau} + \sqrt{\tau+1}\right) + C_2\right]^{-1/3},$$
$$y = b\tau\left[\sqrt{\tau(\tau+1)} - \ln\left(\sqrt{\tau} + \sqrt{\tau+1}\right) + C_2\right]^{-2/3},$$

where $A = -\frac{9}{2}b^3$.

7. $y''_{xx} = 2x^{-2}y + Axy^{-2}.$

Solution in the parametric form:

$$x = C_1\left[\sqrt{\tau(\tau+1)} - \ln\left(\sqrt{\tau} + \sqrt{\tau+1}\right) + C_2\right]^{1/3},$$
$$y = bC_1\tau\left[\sqrt{\tau(\tau+1)} - \ln\left(\sqrt{\tau} + \sqrt{\tau+1}\right) + C_2\right]^{-1/3},$$

where $A = -\frac{9}{2}b^3$.

8. $y''_{xx} = -\frac{4}{25}x^{-2}y + Ax^{-2}y^{-1/2}.$

Solution in the parametric form:

$$x = C_1(\tau^3 - 3\tau + C_2)^{-5/3}, \quad y = b(\tau^2-1)^2(\tau^3 - 3\tau + C_2)^{-4/3},$$

where $A = \pm\frac{4}{25}b^{3/2}$.

9. $y''_{xx} = -\frac{4}{25}x^{-2}y + Ax^{-1/2}y^{-1/2}.$

Solution in the parametric form:

$$x = C_1(\tau^3 - 3\tau + C_2)^{5/3}, \quad y = bC_1(\tau^2-1)^2(\tau^3 - 3\tau + C_2)^{1/3},$$

where $A = \pm\frac{4}{25}b^{3/2}$.

10. $y''_{xx} = -\frac{9}{100}x^{-2}y + Ax^{-2}y^{-5/3}.$

Solution in the parametric form:

$$x = C_1\left[\pm(\tau^4 - 6\tau^2 + 4C_2\tau - 3)\right]^{-5/4},$$
$$y = b(\tau^3 - 3\tau + C_2)^{3/2}\left[\pm(\tau^4 - 6\tau^2 + 4C_2\tau - 3)\right]^{-9/8},$$

where $A = \pm\frac{9}{100}b^{8/3}$.

11. $y''_{xx} = -\frac{9}{100}x^{-2}y + Ax^{2/3}y^{-5/3}$.

Solution in the parametric form:

$$x = C_1[\pm(\tau^4 - 6\tau^2 + 4C_2\tau - 3)]^{5/4}, \quad y = bC_1(\tau^3 - 3\tau + C_2)^{3/2}[\pm(\tau^4 - 6\tau^2 + 4C_2\tau - 3)]^{1/8},$$

where $A = \pm\frac{9}{100}b^{8/3}$.

12. $y''_{xx} = \frac{3}{4}x^{-2}y + Ax^{-2}y^{-5/3}$.

Solution in the parametric form:

$$x = C_1(\tau^3 \pm 3\tau + C_2)^{-1/2}, \quad y = b(\tau^2 \pm 1)^{3/2}(\tau^3 \pm 3\tau + C_2)^{-3/4},$$

where $A = \pm\frac{4}{3}b^{8/3}$.

13. $y''_{xx} = \frac{3}{4}x^{-2}y + Ax^{2/3}y^{-5/3}$.

Solution in the parametric form:

$$x = C_1(\tau^3 \pm 3\tau + C_2)^{1/2}, \quad y = bC_1(\tau^2 \pm 1)^{3/2}(\tau^3 \pm 3\tau + C_2)^{-1/4},$$

where $A = \pm\frac{4}{3}b^{8/3}$.

14. $y''_{xx} = -\frac{2}{9}x^{-2}y + Ax^{-2}y^{-1/2}$.

Solution in the parametric form:

$$x = C_1(C_1e^{2k\tau} + C_2e^{-k\tau}\sin\omega)^{-3}, \quad \omega = \sqrt{3}\,k\tau,$$
$$y = bk^2(C_1e^{2k\tau} + C_2e^{-k\tau}\sin\omega)^{-2}\big[2C_1e^{2k\tau} + C_2e^{-k\tau}(\sqrt{3}\cos\omega - \sin\omega)\big]^2,$$

where $A = \frac{16}{9}bk^3$.

15. $y''_{xx} = -\frac{2}{9}x^{-2}y + Ax^{-1/2}y^{-1/2}$.

Solution in the parametric form:

$$x = C_1(C_1e^{2k\tau} + C_2e^{-k\tau}\sin\omega)^3, \quad \omega = \sqrt{3}\,k\tau,$$
$$y = bk^2C_1(C_1e^{2k\tau} + C_2e^{-k\tau}\sin\omega)\big[2C_1e^{2k\tau} + C_2e^{-k\tau}(\sqrt{3}\cos\omega - \sin\omega)\big]^2,$$

where $A = \frac{16}{9}bk^3$.

16. $y''_{xx} = A_1x^2y^{-5} + A_2y^{-5}$.

Solution in the parametric form:

$$x = \left(\frac{A_2}{A_1}\right)^{1/2}\tan\left[\int\left(C_1 - \frac{1}{2A_1A_2}\tau^{-4} - \tau^2\right)^{-1/2}d\tau + C_2\right],$$
$$y = A_2^{1/2}\tau\left\{\cos\left[\int\left(C_1 - \frac{1}{2A_1A_2}\tau^{-4} - \tau^2\right)^{-1/2}d\tau + C_2\right]\right\}^{-1}.$$

17. $y''_{xx} = A_1 y^{-3} + A_2 y^{-4}$.

1°. Solution in the parametric form:

$$x = a\left[\int \left(C_1 + \tau^{-3} \pm \tau^{-2}\right)^{-1/2} d\tau + C_2\right], \qquad y = b\tau,$$

where $A_1 = \mp a^{-2} b^4$, $A_2 = -\frac{3}{2} a^{-2} b^5$.

2°. Solution in the parametric form:

$$x = a\left[\int \left(C_1 - \tau^{-3} \pm \tau^{-2}\right)^{-1/2} d\tau + C_2\right], \qquad y = b\tau,$$

where $A_1 = \mp a^{-2} b^4$, $A_2 = \frac{3}{2} a^{-2} b^5$.

18. $y''_{xx} = A_1 y^{-3} + A_2 x y^{-4}$.

1°. Solution in the parametric form:

$$x = a\left[\int (C_1 + \tau^{-3} \pm \tau^{-2})^{-1/2} d\tau + C_2\right]^{-1}, \quad y = b\tau\left[\int (C_1 + \tau^{-3} \pm \tau^{-2})^{-1/2} d\tau + C_2\right]^{-1},$$

where $A_1 = \mp a^{-2} b^4$, $A_2 = -\frac{3}{2} a^{-3} b^5$.

2°. Solution in the parametric form:

$$x = a\left[\int (C_1 - \tau^{-3} \pm \tau^{-2})^{-1/2} d\tau + C_2\right]^{-1}, \quad y = b\tau\left[\int (C_1 - \tau^{-3} \pm \tau^{-2})^{-1/2} d\tau + C_2\right]^{-1},$$

where $A_1 = \mp a^{-2} b^4$, $A_2 = \frac{3}{2} a^{-3} b^5$.

19. $y''_{xx} = A_1 y^m + A_2, \qquad m \neq -1$.

1°. Solution in the parametric form:

$$x = a\left[\int \left(C_1 + \tau^{m+1} \pm \tau\right)^{-1/2} d\tau + C_2\right], \qquad y = b\tau,$$

where $A_1 = \frac{1}{2} a^{-2} b^{1-m}(m+1)$, $A_2 = \pm\frac{1}{2} a^{-2} b$.

2°. Solution in the parametric form:

$$x = a\left[\int \left(C_1 - \tau^{m+1} \pm \tau\right)^{-1/2} d\tau + C_2\right], \qquad y = b\tau,$$

where $A_1 = -\frac{1}{2} a^{-2} b^{1-m}(m+1)$, $A_2 = \pm\frac{1}{2} a^{-2} b$.

3°. See equation 2.4.2.21 for the case $m = -1$.

20. $y''_{xx} = A_1 x^{-m-3} y^m + A_2 x^{-3}, \qquad m \neq -1.$

1°. Solution in the parametric form:

$$x = a\left[\int \left(C_1 + \tau^{m+1} \pm \tau\right)^{-1/2} d\tau + C_2\right]^{-1},$$

$$y = b\tau\left[\int \left(C_1 + \tau^{m+1} \pm \tau\right)^{-1/2} d\tau + C_2\right]^{-1},$$

where $A_1 = \frac{1}{2} a^{1+m} b^{1-m}(m+1)$, $A_2 = \pm\frac{1}{2}ab$.

2°. Solution in the parametric form:

$$x = a\left[\int \left(C_1 - \tau^{m+1} \pm \tau\right)^{-1/2} d\tau + C_2\right]^{-1},$$

$$y = b\tau\left[\int \left(C_1 - \tau^{m+1} \pm \tau\right)^{-1/2} d\tau + C_2\right]^{-1},$$

where $A_1 = -\frac{1}{2} a^{1+m} b^{1-m}(m+1)$, $A_2 = \pm\frac{1}{2}ab$.

21. $y''_{xx} = A_1 + A_2 y^{-1}.$

Solution: $x = \displaystyle\int (C_1 + 2A_1 y + 2A_2 \ln|y|)^{-1/2} dy + C_2.$

22. $y''_{xx} = A_1 x^{-3} + A_2 x^{-2} y^{-1}.$

Solution in the parametric form:

$$x = \left[\int (C_1 + 2A_1\tau + 2A_2 \ln|\tau|)^{-1/2} d\tau + C_2\right]^{-1},$$

$$y = \tau\left[\int (C_1 + 2A_1\tau + 2A_2 \ln|\tau|)^{-1/2} d\tau + C_2\right]^{-1}.$$

23. $y''_{xx} = A_1 x^{-5/2} y^{-1/2} + A_2 x^{-7/2} y^{-1/2}.$

Solution in the parametric form:

$$x = \frac{1}{F}, \qquad y = \frac{k^2}{F}\left\{2C_1 e^{2k\tau} + C_2 e^{-k\tau}\left[\sqrt{3}\cos(\omega\tau) - \sin(\omega\tau)\right]\right\}^2,$$

where $F = C_1 e^{2k\tau} + C_2 e^{-k\tau}\sin(\omega\tau) - \dfrac{A_1}{A_2}$, $A_2 = 16k^3$, $\omega = k\sqrt{3}$.

24. $y''_{xx} = A_1 x^{-4/3} y^{-5/3} + A_2 x^{-7/3} y^{-5/3}.$

Solution in the parametric form:

$$x = \left(\tfrac{1}{36}A_2\tau^4 + C_1\tau^3 + C_2\tau + C_3\right)^{-1},$$

$$y = \left(\tfrac{1}{9}A_2\tau^3 + 3C_1\tau^2 + C_2\right)^{3/2}\left(\tfrac{1}{36}A_2\tau^4 + C_1\tau^3 + C_2\tau + C_3\right)^{-1},$$

where $9C_1C_2 = A_1 + A_2C_3$.

25. $y''_{xx} = A_1 x^{-8/5} y^{-7/5} + A_2 x^{-13/5} y^{-7/5}$.

Solution in the parametric form:

$$x = \left(aC_1^4 F - \frac{A_1}{A_2}\right)^{-1}, \qquad y = bC_1^5 S^{5/2} \left(aC_1^4 F - \frac{A_1}{A_2}\right)^{-1},$$

where $S = C_1 e^{2k\tau} + C_2 e^{-k\tau} \sin(\sqrt{3}\,k\tau)$, $F = (S'_\tau)^2 - 2SS''_{\tau\tau}$, $A_2 = -\dfrac{5}{1024} a^{-3} b^{12/5} k^{-6}$.

26. $y''_{xx} = -\frac{3}{16} x^{-2} y + A x^{-2} y^{-5/3}$.

1°. Solution in the parametric form with $A < 0$:

$$x = C_1[\cosh(\tau + C_2)\cos\tau]^{-2}[\tanh(\tau + C_2) + \tan\tau]^{-2},$$
$$y = b[\tanh(\tau + C_2) + \tan\tau]^{-3/2},$$

where $A = -\frac{3}{64} b^{8/3}$.

2°. Solution in the parametric form with $A > 0$:

$$x = C_1[\sinh\tau + \cos(\tau + C_2)]^{-2},$$
$$y = b[\cosh\tau - \sin(\tau + C_2)]^{3/2}[\sinh\tau + \cos(\tau + C_2)]^{-3/2},$$

where $A = \frac{3}{16} b^{8/3}$.

27. $y''_{xx} = -\frac{3}{16} x^{-2} y + A x^{2/3} y^{-5/3}$.

1°. Solution in the parametric form with $A < 0$:

$$x = C_1[\cosh(\tau + C_2)\cos\tau]^2[\tanh(\tau + C_2) + \tan\tau]^2,$$
$$y = bC_1[\cosh(\tau + C_2)\cos\tau]^2[\tanh(\tau + C_2) + \tan\tau]^{1/2},$$

where $A = -\frac{3}{64} b^{8/3}$.

2°. Solution in the parametric form with $A > 0$:

$$x = C_1[\sinh\tau + \cos(\tau + C_2)]^2,$$
$$y = bC_1[\cosh\tau - \sin(\tau + C_2)]^{3/2}[\sinh\tau + \cos(\tau + C_2)]^{1/2},$$

where $A = \frac{3}{16} b^{8/3}$.

28. $y''_{xx} = A_1 x^{-1} y^{-2} + A_2 x^{-2} y^{-2}$.

Solution in the parametric form:

$$x = \left\{aC_1\tau^{-2/3}\left[(\tau Z'_\tau + \tfrac{1}{3}Z)^2 \pm \tau^2 Z^2\right] - \frac{A_1}{A_2}\right\}^{-1},$$
$$y = bC_1\tau^{2/3} Z^2 \left\{aC_1\tau^{-2/3}\left[(\tau Z'_\tau + \tfrac{1}{3}Z)^2 \pm \tau^2 Z^2\right] - \frac{A_1}{A_2}\right\}^{-1},$$

where

$$Z = \begin{cases} C_1 J_{1/3}(\tau) + C_2 Y_{1/3}(\tau) & \text{for the upper sign,} \\ C_1 I_{1/3}(\tau) + C_2 K_{1/3}(\tau) & \text{for the lower sign,} \end{cases}$$

$J_{1/3}$ and $Y_{1/3}$ are Bessel functions, $I_{1/3}$ and $K_{1/3}$ are modified Bessel functions; $A_2 = -\frac{9}{2} a^{-3} b^3$.

▶ *In the solutions of equations 29–30, the following notation is used:*

$$S_1 = C_1 e^{2k\tau} + C_2 e^{-k\tau} \sin\bigl(\sqrt{3}\,k\tau\bigr),$$
$$S_2 = 2kC_1 e^{2k\tau} + kC_2 e^{-k\tau} \bigl[\sqrt{3}\cos\bigl(\sqrt{3}\,k\tau\bigr) - \sin\bigl(\sqrt{3}\,k\tau\bigr)\bigr],$$
$$S_3 = S_2^2 - 2S_1 (S_2)'_\tau.$$

29. $y''_{xx} = -\frac{5}{36} x^{-2} y + A x^{-2} y^{-7/5}$.

Solution in the parametric form:

$$x = C_1 S_3^{-3/2}, \qquad y = b S_1^{5/2} S_3^{-5/4},$$

where $A = -\frac{5}{2304} b^{12/5} k^{-6}$.

30. $y''_{xx} = -\frac{5}{36} x^{-2} y + A x^{2/5} y^{-7/5}$.

Solution in the parametric form:

$$x = C_1 S_3^{3/2}, \qquad y = bC_1 S_1^{5/2} S_3^{1/4},$$

where $A = -\frac{5}{2304} b^{12/5} k^{-6}$.

▶ *In the solutions of equations 31–39, the following notation is used:*

$$R = \sqrt{\pm(4\tau^3 - 1)}, \quad I = \int \tau R^{-1}\,d\tau, \quad F_1 = 2\tau I + C_2\tau \mp R, \quad F_2 = \tau^{-1}(RF_1 - 1),$$

where $I = I(\tau)$ is the incomplete elliptic integral of the second kind in the Weierstrass form.

31. $y''_{xx} = 6x^{-2} y + A x^{-2} y^{-4}$.

Solution in the parametric form:

$$x = C_1 \tau^{-1/5} F_1^{1/5}, \qquad y = b\tau^{-3/5} F_1^{-2/5},$$

where $A = \mp 150 b^5$.

32. $y''_{xx} = 6x^{-2} y + A x^{3} y^{-4}$.

Solution in the parametric form:

$$x = C_1 \tau^{1/5} F_1^{-1/5}, \qquad y = bC_1 \tau^{-2/5} F_1^{-3/5},$$

where $A = \mp 150 b^5$.

33. $y''_{xx} = 20x^{-2} y + A x^{-2} y^{-1/2}$.

Solution in the parametric form:

$$x = C_1 F_1^{1/3}, \qquad y = b F_1^{-4/3} F_2^2,$$

where $A = \pm 108 b^{3/2}$.

34. $y''_{xx} = 20x^{-2}y + Ax^{-1/2}y^{-1/2}$.

Solution in the parametric form:

$$x = C_1 F_1^{-1/3}, \qquad y = bC_1 F_1^{-5/3} F_2^2,$$

where $A = \pm 108 b^{3/2}$.

35. $y''_{xx} = \frac{15}{4}x^{-2}y + Ax^{-2}y^{-7}$.

Solution in the parametric form:

$$x = C_1 (4\tau F_1^2 \mp F_2^2)^{1/4}, \qquad y = bF_1^{1/2} (4\tau F_1^2 \mp F_2^2)^{-3/8},$$

where $A = \pm\frac{3}{4}b^8$.

36. $y''_{xx} = \frac{15}{4}x^{-2}y + Ax^6y^{-7}$.

Solution in the parametric form:

$$x = C_1 (4\tau F_1^2 \mp F_2^2)^{-1/4}, \qquad y = bC_1 F_1^{1/2} (4\tau F_1^2 \mp F_2^2)^{-5/8},$$

where $A = \pm\frac{3}{4}b^8$.

37. $y''_{xx} = -\frac{12}{49}x^{-2}y + Ax^{-2}y^{1/2}$.

Solution in the parametric form:

$$x = C_1 (I + C_2)^{-7}, \qquad y = b\tau^2 (I + C_2)^{-4},$$

where $A = \pm\frac{12}{49}b^{1/2}$.

38. $y''_{xx} = -\frac{12}{49}x^{-2}y + Ax^{-3/2}y^{1/2}$.

Solution in the parametric form:

$$x = C_1 (I + C_2)^7, \qquad y = bC_1\tau^2 (I + C_2)^3,$$

where $A = \pm\frac{12}{49}b^{1/2}$.

39. $y''_{xx} = A_1x^4y^{-7} + A_2x^3y^{-7}$.

Solution in the parametric form:

$$x = \left[aC_1^8(4\tau F_1^2 \mp F_2^2) - \frac{A_1}{A_2}\right]^{-1}, \qquad y = bC_1^3F_1^{1/2}\left[aC_1^8(4\tau F_1^2 \mp F_2^2) - \frac{A_1}{A_2}\right]^{-1},$$

where $A_2 = \pm\frac{3}{64}a^{-3}b^8$.

▶ *In the solutions of equations 40–43, the following notation is used:*

$$R_1 = (C_1 + \tau^{-3} \pm \tau^{-2})^{1/2}, \qquad R_2 = (C_1 - \tau^{-3} \pm \tau^{-2})^{1/2},$$
$$E_1 = \int R_1^{-1}\,d\tau + C_2, \qquad E_2 = \int R_2^{-1}\,d\tau + C_2,$$
$$F_1 = \tau - R_1E_1, \qquad F_2 = \tau - R_2E_2,$$
$$H_1 = 3\tau^3F_1^2 + 3(1 \pm \tau)E_1^2, \qquad H_2 = 3\tau^3F_2^2 + 3(-1 \pm \tau)E_2^2.$$

40. $y''_{xx} = A_1x^{-4/3} + A_2x^{-4/3}y^{-1/2}.$

Solutions in the parametric form:

$$x = a\tau^{-3}E_k^3, \qquad y = bF_k^2,$$

where $A_1 = \mp\frac{2}{9}a^{-2/3}b$, $A_2 = \frac{1}{3}a^{-2/3}b^{3/2}(-1)^k$; $k = 1$ and $k = 2$.

41. $y''_{xx} = A_1x^{-5/3} + A_2x^{-7/6}y^{-1/2}.$

Solutions in the parametric form:

$$x = a\tau^3E_k^{-3}, \qquad y = b\tau^3E_k^{-3}F_k^2,$$

where $A_1 = \pm\frac{2}{9}a^{-1/3}b$, $A_2 = \frac{1}{3}a^{-5/6}b^{3/2}(-1)^{k+1}$; $k = 1$ and $k = 2$.

42. $y''_{xx} = A_1y^{-3} + A_2xy^{-7}.$

Solutions in the parametric form:

$$x = a\tau^{-3}H_k, \qquad y = b\tau^{-1/2}E_k^{1/2},$$

where $A_1 = \mp\frac{1}{36}a^{-2}b^4$, $A_2 = -\frac{1}{36}a^{-3}b^8$; $k = 1$ and $k = 2$.

43. $y''_{xx} = A_1y^{-3} + A_2x^3y^{-7}.$

Solutions in the parametric form:

$$x = a\tau^3H_k^{-1}, \qquad y = b\tau^{5/2}E_k^{1/2}H_k^{-1},$$

where $A_1 = \pm\frac{1}{36}a^{-2}b^4$, $A_2 = -\frac{1}{36}a^{-5}b^8$; $k = 1$ and $k = 2$.

▶ *In the solutions of equations 44–45, the following notation is used:*

$$f_1 = \begin{cases} C_1e^{k\tau} + C_2e^{-k\tau} - \dfrac{A_2}{A_1}\tau & \text{if } A_1 > 0, \\ C_1\sin(k\tau) + C_2\cos(k\tau) - \dfrac{A_2}{A_1}\tau & \text{if } A_1 < 0, \end{cases}$$

$$f_2 = \begin{cases} k(C_1e^{k\tau} - C_2e^{-k\tau}) - \dfrac{A_2}{A_1} & \text{if } A_1 > 0, \\ k[C_1\cos(k\tau) - C_2\sin(k\tau)] - \dfrac{A_2}{A_1} & \text{if } A_1 < 0, \end{cases}$$

where $k = \sqrt{\tfrac{1}{2}|A_1|}$.

44. $y''_{xx} = A_1 + A_2 y^{-1/2}.$

Solution in the parametric form:

$$x = f_1, \qquad y = f_2^2.$$

45. $y''_{xx} = A_1 x^{-3} + A_2 x^{-5/2} y^{-1/2}.$

Solution in the parametric form:

$$x = f_1^{-1}, \qquad y = f_1^{-1} f_2^2. \quad .$$

▶ *In the solutions of equations 46–47, the following notation is used:*

For $A_1 > 0$,

$$T_1 = C_1 e^{k\tau} + C_2 e^{-k\tau} + C_3 \sin(k\tau), \quad k = (\tfrac{4}{3} A_1)^{1/4},$$
$$T_2 = k(C_1 e^{k\tau} - C_2 e^{-k\tau}) + k C_3 \cos(k\tau);$$

For $A_1 < 0$,

$$T_1 = e^{s\tau}[C_1 \sin(s\tau) + C_2 \cos(s\tau)] + C_3 e^{-s\tau} \sin(s\tau), \quad s = (-\tfrac{1}{3} A_1)^{1/4},$$
$$T_2 = s e^{s\tau}[(C_1 - C_2)\sin(s\tau) + (C_1 + C_2)\cos(s\tau)] - s C_3 e^{-s\tau}[\sin(s\tau) - \cos(s\tau)].$$

46. $y''_{xx} = A_1 x^2 y^{-5/3} + A_2 x y^{-5/3}.$

Solution in the parametric form:

$$x = T_1 - \frac{A_2}{2A_1}, \qquad y = T_2^{3/2},$$

where arbitrary constants C_1, C_2, and C_3 are related by

$$4C_1C_2 + C_3^2 = \tfrac{1}{4} A_1^{-2} A_2^2 \quad \text{if } A_1 > 0,$$
$$C_1 C_3 = \tfrac{1}{16} A_1^{-2} A_2^2 \qquad \text{if } A_1 < 0.$$

47. $y''_{xx} = A_1 x^2 y^{-5/3} + A_2 y^{-5/3}.$

Solution in the parametric form:

$$x = T_1, \qquad y = T_2^{3/2},$$

where arbitrary constants C_1, C_2, and C_3 are related by

$$4C_1C_2 + C_3^2 = -\tfrac{1}{2} A_1^{-1} A_2 \quad \text{if } A_1 > 0,$$
$$C_1 C_3 = -\tfrac{1}{4} A_1^{-1} A_2 \qquad \text{if } A_1 < 0.$$

▶ *In the solutions of equations 48–49, the following notation is used:*

For $A_2 > 0$,

$$T_1 = C_1 e^{k\tau} + C_2 e^{-k\tau} + C_3 \sin(k\tau), \quad k = \left(\tfrac{4}{3}A_2\right)^{1/4},$$
$$T_2 = k(C_1 e^{k\tau} - C_2 e^{-k\tau}) + kC_3 \cos(k\tau);$$

For $A_2 < 0$,

$$T_1 = e^{s\tau}[C_1 \sin(s\tau) + C_2 \cos(s\tau)] + C_3 e^{-s\tau} \sin(s\tau), \quad s = \left(-\tfrac{1}{3}A_2\right)^{1/4},$$
$$T_2 = se^{s\tau}[(C_1 - C_2)\sin(s\tau) + (C_1 + C_2)\cos(s\tau)] - sC_3 e^{-s\tau}[\sin(s\tau) - \cos(s\tau)].$$

48. $y''_{xx} = A_1 x^{-7/3} y^{-5/3} + A_2 x^{-10/3} y^{-5/3}$.

Solution in the parametric form:

$$x = \left(T_1 - \frac{A_1}{2A_2}\right)^{-1}, \qquad y = T_2^{3/2}\left(T_1 - \frac{A_1}{2A_2}\right)^{-1},$$

where arbitrary constants C_1, C_2, and C_3 are related by

$$4C_1C_2 + C_3^2 = \tfrac{1}{4}A_1^2 A_2^{-2} \quad \text{if } A_2 > 0,$$
$$C_1C_3 = \tfrac{1}{16}A_1^2 A_2^{-2} \quad \text{if } A_2 < 0.$$

49. $y''_{xx} = A_1 x^{-4/3} y^{-5/3} + A_2 x^{-10/3} y^{-5/3}$.

Solution in the parametric form:

$$x = T_1^{-1}, \qquad y = T_1^{-1} T_2^{3/2},$$

where arbitrary constants C_1, C_2, and C_3 are related by

$$4C_1C_2 + C_3^2 = -\tfrac{1}{2}A_1 A_2^{-1} \quad \text{if } A_2 > 0,$$
$$C_1C_3 = -\tfrac{1}{4}A_1 A_2^{-1} \quad \text{if } A_2 < 0.$$

▶ *In the solutions of equations 50–53, the following notation is used:*

$$R_1 = C_1\tau^{k_1} + C_2\tau^{k_2} + C_3\tau^{k_3},$$
$$R_2 = (C_1 + C_2\tau)e^{k\tau} + C_3 e^{\omega\tau},$$
$$R_3 = C_1 e^{k\tau} + e^{s\tau}(C_2 \sin\omega\tau + C_3\cos\omega\tau),$$
$$Q_1 = C_1k_1\tau^{k_1} + C_2k_2\tau^{k_2} + C_3k_3\tau^{k_3},$$
$$Q_2 = (kC_1 + C_2 + kC_2\tau)e^{k\tau} + \omega C_3 e^{\omega\tau},$$
$$Q_3 = kC_1 e^{k\tau} + e^{s\tau}[(sC_2 - \omega C_3)\sin\omega\tau + (sC_3 + \omega C_2)\cos\omega\tau],$$
$$S_1 = \tau(Q_1)'_\tau, \quad S_2 = (Q_2)'_\tau, \quad S_3 = (Q_3)'_\tau,$$

where k_1, k_2, *and* k_3 *(real numbers) or* k *and* $s \pm i\omega$ *(one real and two complex numbers) are the roots of the cubic equation* $\lambda^3 - \frac{1}{2}B_2\lambda - \frac{1}{2}B_1 = 0$. *Subscripts of* *functions R_m, Q_m, and S_m ($m = 1, 2, 3$) are selected depending on the sign of the following expression:*

$$2B_2^3 - 27B_1^2 \quad \begin{cases} > 0 & \textit{subscript 1,} \\ = 0 & \textit{subscript 2,} \\ < 0 & \textit{subscript 3;} \end{cases}$$

if $2B_2^3 = 27B_1^2$ (subscript 2), then

$$k = (\tfrac{1}{6}B_2)^{1/2}, \quad \omega = -2(\tfrac{1}{6}B_2)^{1/2} \qquad \textit{if} \quad B_1 < 0,$$
$$k = -(\tfrac{1}{6}B_2)^{1/2}, \quad \omega = 2(\tfrac{1}{6}B_2)^{1/2} \qquad \textit{if} \quad B_1 > 0.$$

Remark. *The expressions for R_m, Q_m contain three constants C_1, C_2, and C_3. One of them may be arbitrarily fixed to set it equal to any nonzero number (for example, we may set $C_3 = \pm 1$), and the other constants may be arbitrary.*

50. $y''_{xx} = A_1 + A_2 x y^{-1/2}$.

Solution in the parametric form:

$$x = R_m, \quad y = Q_m^2, \qquad A_1 = B_2, \quad A_2 = B_1.$$

51. $y''_{xx} = A_1 x^{-3} + A_2 x^{-7/2} y^{-1/2}$.

Solution in the parametric form:

$$x = R_m^{-1}, \quad y = R_m^{-1} Q_m^2, \qquad A_1 = B_2, \quad A_2 = B_1.$$

52. $y''_{xx} = A_1 y^{-3/5} + A_2 x y^{-7/5}$.

Solution in the parametric form:

$$x = a(2Q_m^2 - 4R_m S_m + B_2 R_m^2), \quad y = bR_m^{5/2},$$

where $A_1 = -ab^{-4/5} A_2 B_2$, $A_2 = -\frac{5}{32} a^{-3} b^{12/5} B_1^{-2}$.

53. $y''_{xx} = A_1 x^{-12/5} y^{-3/5} + A_2 x^{-13/5} y^{-7/5}$.

Solution in the parametric form:

$$x = a\bigl(2Q_m^2 - 4R_m S_m + B_2 R_m^2\bigr)^{-1}, \quad y = bR_m^{5/2}\bigl(2Q_m^2 - 4R_m S_m + B_2 R_m^2\bigr)^{-1},$$

where $A_1 = \frac{5}{32} a^{2/5} b^{8/5} B_1^{-2} B_2$, $A_2 = -\frac{5}{32} a^{3/5} b^{12/5} B_1^{-2}$.

▶ *In the solutions of equations 54–55, the following notation is used:*

1°. *For $A_2 > 0$, $A_1 \neq 0$:*

$$T_1 = C_1 e^{k\tau} + C_2 e^{-k\tau} + C_3 \sin \omega\tau,$$
$$T_2 = k(C_1 e^{k\tau} - C_2 e^{-k\tau}) + \omega C_3 \cos \omega\tau,$$

where $k = \{\frac{2}{3}[(A_1^2 + 3A_2)^{1/2} + A_1]\}^{1/2}$, $\omega = \{\frac{2}{3}[(A_1^2 + 3A_2)^{1/2} - A_1]\}^{1/2}$; *arbitrary constants C_1, C_2, and C_3 are related by the constraint* $4k^2 C_1 C_2 + \omega^2 C_3^2 = 0$.

2°. *For* $-A_1^2 < 3A_2 < 0$, $A_1 > 0$:

$$T_1 = C_1\tau^{k_1} + C_2\tau^{-k_1} + C_3\tau^{k_2} + C_4\tau^{-k_2},$$
$$T_2 = k_1(C_1\tau^{k_1} - C_2\tau^{-k_1}) + k_2(C_3\tau^{k_2} - C_4\tau^{-k_2}),$$

where $k_1 = \{\frac{2}{3}[A_1 + (A_1^2 + 3A_2)^{1/2}]\}^{1/2}$, $k_2 = \{\frac{2}{3}[A_1 - (A_1^2 + 3A_2)^{1/2}]\}^{1/2}$; *arbitrary constants* C_1, C_2, *and* C_3 *are related by the constraint* $(C_1C_2 + C_3C_4)(A_1^2 + 3A_2)^{1/2} + (C_1C_2 - C_3C_4)A_1 = 0$.

3°. *For* $-A_1^2 < 3A_2 < 0$, $A_1 < 0$:

$$T_1 = C_1 \sin\omega_1\tau + C_2\cos\omega_1\tau + C_3\sin\omega_2\tau,$$
$$T_2 = \omega_1(C_1\cos\omega_1\tau - C_2\sin\omega_1\tau) + \omega_2 C_3\cos\omega_2\tau,$$

where $\omega_1 = \{-\frac{2}{3}[A_1 + (A_1^2 + 3A_2)^{1/2}]\}^{1/2}$, $\omega_2 = \{-\frac{2}{3}[A_1 - (A_1^2 + 3A_2)^{1/2}]\}^{1/2}$; *arbitrary constants* C_1, C_2, *and* C_3 *are related by* $\omega_1^2(C_1^2 + C_2^2) - \omega_2^2 C_3^2 = 0$.

4°. *For* $A_1^2 + 3A_2 = 0$, $A_1 > 0$:

$$T_1 = (C_1 + C_2\tau)e^{k\tau} + (C_3 + C_4\tau)e^{-k\tau},$$
$$T_2 = (kC_1 + C_2 + kC_2\tau)e^{k\tau} - (kC_3 - C_4 + kC_4\tau)e^{-k\tau},$$

where $k = (\frac{2}{3}A_1)^{1/2}$; *arbitrary constants* C_1, C_2, *and* C_3 *are related by the constraint* $(C_1C_4 - C_2C_3)(\frac{2}{3}A_1)^{1/2} + 2C_2C_4 = 0$.

5°. *For* $A_1^2 + 3A_2 = 0$, $A_1 < 0$:

$$T_1 = (C_1 + C_2\tau)\sin\omega\tau + C_3\tau\cos\omega\tau,$$
$$T_2 = (\omega C_1 + C_3 + \omega C_2\tau)\cos\omega\tau + (C_2 - \omega C_3\tau)\sin\omega\tau,$$

where $\omega = (-\frac{2}{3}A_1)^{1/2}$; *arbitrary constants* C_1, C_2, *and* C_3 *are related by the constraint* $C_1C_3(-\frac{2}{3}A_1)^{1/2} + C_2^2 + C_3^2 = 0$.

6°. *For* $3A_2 < -A_1^2$:

$$T_1 = e^{k\tau}(C_1\sin\omega\tau + C_2\cos\omega\tau) + C_3e^{-k\tau}\sin\omega\tau,$$
$$T_2 = e^{k\tau}[(kC_2 + \omega C_1)\cos\omega\tau + (kC_1 - \omega C_2)\sin\omega\tau)] + C_3e^{-k\tau}(\omega\cos\omega\tau - k\sin\omega\tau),$$

where $k = \{\frac{1}{3}[A_1 + (-3A_2)^{1/2}]\}^{1/2}$, $\omega = \{\frac{1}{3}[-A_1 + (-3A_2)^{1/2}]\}^{1/2}$; *arbitrary constants* C_1, C_2, *and* C_3 *are related by* $C_2A_1 + C_1(-A_1^2 - 3A_2)^{1/2} = 0$.

54. $\boldsymbol{y''_{xx} = A_1 y^{-1/3} + A_2 x^2 y^{-5/3}}$.

Solution in the parametric form:

$$x = T_1, \qquad y = T_2^{3/2}.$$

55. $\boldsymbol{y''_{xx} = A_1 x^{-8/3} y^{-1/3} + A_2 x^{-10/3} y^{-5/3}}$.

Solution in the parametric form:

$$x = T_1^{-1}, \qquad y = T_1^{-1}T_2^{3/2}.$$

▶ *In the solutions of equations 56–59, the following notation is used:*

$$T_1 = \begin{cases} C_1 e^{\omega\tau} + C_2 e^{-\omega\tau} + C_3\tau & \text{if } A_1 > 0, \\ C_1 \sin\omega\tau + C_2\cos\omega\tau + C_3\tau & \text{if } A_1 < 0; \end{cases}$$

$$T_2 = \begin{cases} \omega(C_1 e^{\omega\tau} - C_2 e^{-\omega\tau}) + C_3 & \text{if } A_1 > 0, \\ \omega(C_1 \cos\omega\tau - C_2\sin\omega\tau) + C_3 & \text{if } A_1 < 0; \end{cases}$$

where $\omega = |\frac{4}{3}A_1|^{1/2}$.

56. $y''_{xx} = A_1 y^{-1/3} + A_2 y^{-5/3}$.

Solution in the parametric form:

$$x = T_1, \qquad y = T_2^{3/2},$$

where arbitrary constants C_1, C_2, and C_3 are related by

$$\begin{aligned} &3(A_1C_3^2 + A_2) + 16A_1^2 C_1C_2 = 0 && \text{if } A_1 > 0, \\ &3(A_1C_3^2 + A_2) + 4A_1^2(C_1^2 + C_2^2) = 0 && \text{if } A_1 < 0. \end{aligned}$$

57. $y''_{xx} = A_1 x^{-8/3} y^{-1/3} + A_2 x^{-4/3} y^{-5/3}$.

Solution in the parametric form:

$$x = T_1^{-1}, \qquad y = T_1^{-1} T_2^{3/2},$$

where arbitrary constants C_1, C_2, and C_3 are related by

$$\begin{aligned} &3(A_1C_3^2 + A_2) + 16A_1^2 C_1C_2 = 0 && \text{if } A_1 > 0, \\ &3(A_1C_3^2 + A_2) + 4A_1^2(C_1^2 + C_2^2) = 0 && \text{if } A_1 < 0. \end{aligned}$$

58. $y''_{xx} = A_1 y^{-1/3} + A_2 x y^{-5/3}$.

Solution in the parametric form:

$$x = T_1 - \frac{A_2}{4A_1}\tau^2, \qquad y = \left(T_2 - \frac{A_2}{2A_1}\tau\right)^{3/2},$$

where arbitrary constants C_1, C_2, and C_3 are related by

$$\begin{aligned} &3A_1C_3^2 + 16A_1^2 C_1C_2 + \tfrac{9}{16}A_1^{-2}A_2^2 = 0 && \text{if } A_1 > 0, \\ &3A_1C_3^2 + 4A_1^2(C_1^2 + C_2^2) + \tfrac{9}{16}A_1^{-2}A_2^2 = 0 && \text{if } A_1 < 0. \end{aligned}$$

59. $y''_{xx} = A_1 x^{-8/3} y^{-1/3} + A_2 x^{-7/3} y^{-5/3}$.

Solution in the parametric form:

$$x = \left(T_1 - \frac{A_2}{4A_1}\tau^2\right)^{-1}, \qquad y = \left(T_1 - \frac{A_2}{4A_1}\tau^2\right)^{-1}\left(T_2 - \frac{A_2}{2A_1}\tau\right)^{3/2},$$

where arbitrary constants C_1, C_2, and C_3 are related by

$$\begin{aligned} &3A_1C_3^2 + 16A_1^2 C_1C_2 + \tfrac{9}{16}A_1^{-2}A_2^2 = 0 && \text{if } A_1 > 0, \\ &3A_1C_3^2 + 4A_1^2(C_1^2 + C_2^2) + \tfrac{9}{16}A_1^{-2}A_2^2 = 0 && \text{if } A_1 < 0. \end{aligned}$$

► *In the solutions of equations 60–67, the following notation is used:*

$$f = \sqrt{\pm(4\wp^3 - 1)}, \qquad \tau = \int \frac{d\wp}{\sqrt{\pm(4\wp^3 - 1)}} - C_2.$$

Function $\wp = \wp(\tau)$ is defined implicitly in terms of the above elliptic integral of the first kind. For the upper sign, function $\wp$ coincides with the classical elliptic Weierstrass function $\wp = \wp(\tau + C_2,\ 0,\ 1)$. In the solution given below, one can take $\wp$ as the parameter instead of τ and use the explicit dependence $\tau = \tau(\wp)$.

60. $y''_{xx} = Ax^{-2}y^2 - \frac{6}{25}x^{-2}y.$

Solution in the parametric form:

$$x = C_1\tau^5, \quad y = b\tau^2\wp, \qquad \text{where} \quad A = \pm\tfrac{6}{25}b^{-1}.$$

61. $y''_{xx} = Ax^{-3}y^2 - \frac{6}{25}x^{-2}y.$

Solution in the parametric form:

$$x = C_1\tau^{-5}, \quad y = bC_1\tau^{-3}\wp, \qquad \text{where} \quad A = \pm\tfrac{6}{25}b^{-1}.$$

62. $y''_{xx} = Ax^{-2}y^2 + \frac{6}{25}x^{-2}y.$

Solution in the parametric form:

$$x = C_1\tau^5, \quad y = b(\tau^2\wp \mp 1), \qquad \text{where} \quad A = \pm\tfrac{6}{25}b^{-1}.$$

63. $y''_{xx} = Ax^{-3}y^2 + \frac{6}{25}x^{-2}y.$

Solution in the parametric form:

$$x = C_1\tau^{-5}, \quad y = bC_1\tau^{-5}(\tau^2\wp \mp 1), \qquad \text{where} \quad A = \pm\tfrac{6}{25}b^{-1}.$$

64. $y''_{xx} = 12x^{-2}y + Ax^{-2}y^{-5/2}.$

Solution in the parametric form:

$$x = C_1\wp^{2/7}(f \pm 2\tau\wp^2)^{-1/7}, \qquad y = b\wp^{-6/7}(f \pm 2\tau\wp^2)^{-4/7},$$

where $A = \mp 147b^{7/2}$.

65. $y''_{xx} = 12x^{-2}y + Ax^{3/2}y^{-5/2}.$

Solution in the parametric form:

$$x = C_1\wp^{-2/7}(f \pm 2\tau\wp^2)^{1/7}, \qquad y = bC_1\wp^{-8/7}(f \pm 2\tau\wp^2)^{-3/7},$$

where $A = \mp 147b^{7/2}$.

66. $y''_{xx} = \frac{63}{4}x^{-2}y + Ax^{-2}y^{-5/3}.$

Solution in the parametric form:

$$x = C_1(\tau f + 2\wp)^{-1/4}, \qquad y = b(\tau f + 2\wp)^{-9/8}(f \pm 2\tau\wp^2)^{3/2},$$

where $A = -\frac{32}{3}b^{8/3}$.

67. $y''_{xx} = \frac{63}{4}x^{-2}y + Ax^{2/3}y^{-5/3}.$

Solution in the parametric form:

$$x = C_1(\tau f + 2\wp)^{1/4}, \qquad y = bC_1(\tau f + 2\wp)^{-7/8}(f \pm 2\tau\wp^2)^{3/2},$$

where $A = -\frac{32}{3}b^{8/3}$.

▶ *In the solutions of equations 68–73, the following notation is used:*

$$f_1 = \sqrt{\pm 4\wp_1^3 - 2\wp_1 - C_2}, \qquad \tau = \int \frac{d\wp_1}{\sqrt{\pm 4\wp_1^3 - 2\wp_1 - C_2}} - C_1,$$

and

$$f_2 = \sqrt{\pm 4\wp_2^3 + 2\wp_2 - C_2}, \qquad \tau = \int \frac{d\wp_2}{\sqrt{\pm 4\wp_2^3 + 2\wp_2 - C_2}} - C_1,$$

where functions $\wp_1 = \wp_1(\tau)$ and $\wp_2 = \wp_2(\tau)$ are the inverse functions for the above elliptic integrals. For the upper signs, they are the classical Weierstrass functions $\wp_1 = \wp(\tau + C_1,\ 2,\ C_2)$ and $\wp_2 = \wp(\tau + C_1,\ -2,\ C_2)$.

68. $\boldsymbol{y''_{xx} = A_1 y^2 + A_2}$.

Solutions in the parametric form:

$$x = a\tau, \qquad y = b\wp_k,$$

where $A_1 = \pm 6a^{-2}b^{-1}$, $A_2 = a^{-2}b(-1)^k$, $k = 1$ and $k = 2$.

69. $\boldsymbol{y''_{xx} = A_1 x^{-5} y^2 + A_2 x^{-3}}$.

Solutions in the parametric form:

$$x = a\tau^{-1}, \qquad y = b\tau^{-1}\wp_k,$$

where $A_1 = \pm 6a^3b^{-1}$, $A_2 = ab(-1)^k$, $k = 1$ and $k = 2$.

70. $\boldsymbol{y''_{xx} = A_1 x^{-15/7} y^2 + A_2 x^{-9/7}}$.

Solutions in the parametric form:

$$x = a\tau^7, \qquad y = b\tau(\tau^2\wp_k \mp 1),$$

where $A_1 = \pm\frac{6}{49}a^{1/7}b^{-1}$, $A_2 = \frac{1}{49}a^{-5/7}b(-1)^k$, $k = 1$ and $k = 2$.

71. $\boldsymbol{y''_{xx} = A_1 x^{-20/7} y^2 + A_2 x^{-12/7}}$.

Solutions in the parametric form:

$$x = a\tau^{-7}, \qquad y = b\tau^{-6}(\tau^2\wp_k \mp 1),$$

where $A_1 = \pm\frac{6}{49}a^{6/7}b^{-1}$, $A_2 = \frac{1}{49}a^{-2/7}b(-1)^k$, $k = 1$ and $k = 2$.

72. $\boldsymbol{y''_{xx} = A_1 + A_2 y^{-2/3}}$.

Solutions in the parametric form:

$$x = a[f_k - (-1)^k\tau], \qquad y = b\wp_k^3,$$

where $A_1 = \pm\frac{1}{2}a^{-2}b$, $A_2 = \frac{1}{12}a^{-2}b^{5/3}(-1)^k$, $k = 1$ and $k = 2$.

73. $\boldsymbol{y''_{xx} = A_1 x^{-3} + A_2 x^{-7/3} y^{-2/3}}$.

Solutions in the parametric form:

$$x = a[f_k - (-1)^k\tau]^{-1}, \qquad y = b\wp_k^3[f_k - (-1)^k\tau]^{-1},$$

where $A_1 = \pm\frac{1}{2}ab$, $A_2 = \frac{1}{12}a^{1/3}b^{5/3}(-1)^k$, $k = 1$ and $k = 2$.

▶ *In the solutions of equations 74–75, the following notation is used:*

$$E = \int (1 \pm \tau^4)^{-1/2}\, d\tau + C_2, \qquad k^2 = \pm 1.$$

Function E can be expressed in terms of elliptic integrals or lemniscate functions.

74. $y''_{xx} = A_1 x^{-18/5} y^3 + A_2 x^{-14/5} y^2.$

Solutions in the parametric form:

$$x = aC_1^5 E^{-5}, \qquad y = bC_1^4 E^{-4}(\tau E - k),$$

where $A_1 = \pm\frac{2}{25} a^{8/5} b^{-2}$, $A_2 = \pm\frac{6}{25} a^{4/5} b^{-1} k$.

75. $y''_{xx} = A_1 x^{-12/5} y^3 + A_2 x^{-11/5} y^2.$

Solutions in the parametric form:

$$x = aC_1^5 E^5, \qquad y = bC_1 E(\tau E - k),$$

where $A_1 = \pm\frac{2}{25} a^{2/5} b^{-2}$, $A_2 = \pm\frac{6}{25} a^{1/5} b^{-1} k$.

▶ *In the solutions of equations 76–81, the following notation is used:*

$$f = \begin{cases} J_{1/3}(\tau) & \text{for the upper sign (Bessel function),} \\ I_{1/3}(\tau) & \text{for the lower sign (modified Bessel function),} \end{cases}$$

$$g = \begin{cases} Y_{1/3}(\tau) & \text{for the upper sign (Bessel function),} \\ K_{1/3}(\tau) & \text{for the lower sign (modified Bessel function),} \end{cases}$$

$$H = C_1 f + C_2 g + \beta\omega\left(g \int f\, d\tau - f \int g\, d\tau\right), \qquad \omega = \begin{cases} \frac{1}{2}\pi & \text{for the upper sign,} \\ -1 & \text{for the lower sign.} \end{cases}$$

76. $y''_{xx} = A_1 x y + A_2.$

Solutions in the parametric form:

$$x = a\tau^{2/3}, \qquad y = \tau^{1/3} H,$$

where $A_1 = \mp\frac{9}{4} a^{-3}$, $A_2 = \frac{9}{4} a^{-2} \beta$.

77. $y''_{xx} = A_1 x^{-5} y + A_2 x^{-3}.$

Solutions in the parametric form:

$$x = a\tau^{-2/3}, \qquad y = \tau^{-1/3} H,$$

where $A_1 = \mp\frac{9}{4} a^3$, $A_2 = \frac{9}{4} a\beta$.

78. $y''_{xx} = A_1 x^{-3/2} + A_2 x^{-1/2} y^{-1/2}.$

Solutions in the parametric form:

$$x = a\tau^{2/3} H^2, \qquad y = b\tau^{-2/3}\big(\tau H'_\tau + \tfrac{1}{3}H\big)^2,$$

where $A_1 = -\frac{1}{2}a^{-1/2}b\beta$, $A_2 = \mp\frac{1}{3}a^{-3/2}b^{3/2}$.

79. $y''_{xx} = A_1 x^{-3/2} + A_2 x^{-2} y^{-1/2}.$

Solutions in the parametric form:

$$x = a\tau^{-2/3} H^{-2}, \qquad y = b\tau^{-4/3} H^{-2}\big(\tau H'_\tau + \tfrac{1}{3}H\big)^2,$$

where $A_1 = -\frac{1}{2}a^{-1/2}b\beta$, $A_2 = \mp\frac{1}{3}b^{3/2}$.

80. $y''_{xx} = A_1 y^{-3/2} + A_2 x y^{-2}.$

Solutions in the parametric form:

$$x = a\tau^{-2/3}\big[\mp\tau^2 H^2 + 2\beta\tau H - \big(\tau H'_\tau + \tfrac{1}{3}H\big)^2\big], \qquad y = b\tau^{2/3} H^2,$$

where $A_1 = -ab^{-1/2}\beta A_2$, $A_2 = \frac{9}{2}a^{-3}b^3$.

81. $y''_{xx} = A_1 x^{-3/2} y^{-3/2} + A_2 x^{-2} y^{-2}.$

Solutions in the parametric form:

$$x = a\tau^{2/3}\big[\mp\tau^2 H^2 + 2\beta\tau H - \big(\tau H'_\tau + \tfrac{1}{3}H\big)^2\big]^{-1},$$
$$y = b\tau^{4/3} H^2\big[\mp\tau^2 H^2 + 2\beta\tau H - \big(\tau H'_\tau + \tfrac{1}{3}H\big)^2\big]^{-1},$$

where $A_1 = -\frac{9}{2}a^{-1/2}b^{5/2}\beta$, $A_2 = \frac{9}{2}b^3$.

► *In the solutions of equations 82–88, the following notation is used:*

$$U_\nu = \begin{cases} C_1 J_\nu(\tau) & \textit{for the upper sign (Bessel function)}, \\ C_1 I_\nu(\tau) & \textit{for the lower sign (modified Bessel function)}, \end{cases}$$

$$V_\nu = \begin{cases} C_2 Y_\nu(\tau) & \textit{for the upper sign (Bessel function)}, \\ C_2 K_\nu(\tau) & \textit{for the lower sign (modified Bessel function)}, \end{cases}$$

$$Z_\nu = \alpha_1 U_\nu + \alpha_2 V_\nu, \qquad X_\nu = \beta_1 U_\nu + \beta_2 V_\nu,$$

$$N = \begin{cases} Z_\nu X_\nu & \textit{for } \Delta = -(\alpha_1\beta_2 - \alpha_2\beta_1)^2; \\ \alpha U_\nu^2 + \beta U_\nu V_\nu + \gamma V_\nu^2 & \textit{for } \Delta = 4\alpha\gamma - \beta^2, \end{cases}$$

$$F_\nu = \tau Z'_\nu + \nu Z_\nu, \qquad G_\nu = \tau X'_\nu + \nu X_\nu,$$

$$N_1 = \begin{cases} Z_\nu G_\nu + X_\nu F_\nu & \textit{for } \Delta = -(\alpha_1\beta_2 - \alpha_2\beta_1)^2, \\ \tau N' + 2\nu N & \textit{for } \Delta = 4\alpha\gamma - \beta^2, \end{cases}$$

$$N_2 = N_1^2 \pm 4\tau^2 N^2 + \omega^2\Delta, \qquad \omega = \begin{cases} \frac{2}{\pi} & \textit{for the upper sign}, \\ -1 & \textit{for the lower sign}. \end{cases}$$

Prime denotes differentiation with respect to τ.

82. $y''_{xx} = A_1 x y + A_2 y^{-3}.$

Solutions in the parametric form:

$$x = a\tau^{2/3}, \qquad y = b\tau^{1/3} N^{1/2},$$

where $\nu = \frac{1}{3}$, $A_1 = \mp\frac{9}{4}a^{-3}$, $A_2 = \frac{9}{16}a^{-2}b^4\omega^2\Delta$.

83. $y''_{xx} = A_1 x^n y + A_2 y^{-3}, \qquad n \neq -2.$

Solutions in the parametric form:

$$x = a\tau^{2\nu}, \qquad y = b\tau^{\nu} N^{1/2},$$

where $\nu = \dfrac{1}{n+2}$, $A_1 = \mp\dfrac{1}{4\nu^2}a^{-n-2}$, $A_2 = \dfrac{1}{16\nu^2}a^{-2}b^4\omega^2\Delta$.

84. $y''_{xx} = A_1 x^{-5} y + A_2 y^{-3}.$

Solutions in the parametric form:

$$x = a\tau^{-2/3}, \qquad y = b\tau^{-1/3} N^{1/2},$$

where $\nu = \frac{1}{3}$, $A_1 = \mp\frac{9}{4}a^3$, $A_2 = \frac{9}{16}a^{-2}b^4\omega^2\Delta$.

85. $y''_{xx} = A_1 x^{-3} + A_2 x^{-1/2} y^{-1/2}.$

Solutions in the parametric form:

$$x = a\tau^{2/3} N, \qquad y = b\tau^{-2/3} N^{-1} N_1^2,$$

where $\nu = \frac{1}{3}$, $A_1 = -2ab\omega^2\Delta$, $A_2 = \mp\frac{8}{3}a^{-3/2}b^{3/2}$.

86. $y''_{xx} = A_1 + A_2 x^{-2} y^{-1/2}.$

Solutions in the parametric form:

$$x = a\tau^{-2/3} N^{-1}, \qquad y = b\tau^{-4/3} N^{-2} N_1^2,$$

where $\nu = \frac{1}{3}$, $A_1 = -2a^{-2}b\omega^2\Delta$, $A_2 = \mp\frac{8}{3}b^{3/2}$.

87. $y''_{xx} = A_1 x y^{-2} + A_2 y^{-3}.$

Solutions in the parametric form:

$$x = a\tau^{-2/3} N^{-1} N_2, \qquad y = b\tau^{2/3} N,$$

where $\nu = \frac{1}{3}$, $A_1 = -\frac{9}{128}a^{-3}b^3$, $A_2 = \frac{9}{64}a^{-2}b^4\omega^2\Delta$.

88. $y''_{xx} = A_1 x^{-2} y^{-2} + A_2 y^{-3}.$

Solutions in the parametric form:

$$x = a\tau^{2/3} N N_2^{-1}, \qquad y = b\tau^{4/3} N^2 N_2^{-1},$$

where $\nu = \frac{1}{3}$, $A_1 = -\frac{9}{128}b^3$, $A_2 = \frac{9}{64}a^{-2}b^4\omega^2\Delta$.

▶ *In the solutions of equations 89–90, the following notation is used:*

$$\Delta = C_2^2 - 2C_1, \qquad R = (36\Delta + 54B\tau - 2\tau^3)^{1/2}, \qquad z = 3\int \tau^{-1}R^{-1}\,d\tau,$$

$$W(z) = \begin{cases} \dfrac{\sqrt{-\Delta}}{C_1}\tan\bigl(\pm\sqrt{-\Delta}\,z\bigr) + \dfrac{C_2}{C_1} & \text{if } \Delta < 0; \\ \dfrac{\sqrt{\Delta}}{C_1}\tanh\bigl(\mp\sqrt{\Delta}\,z\bigr) + \dfrac{C_2}{C_1} & \text{if } \Delta > 0; \\ \mp\dfrac{1}{C_1 z} - \sqrt{\dfrac{2}{|C_1|}} & \text{if } \Delta = 0,\ C_2 < 0; \\ \mp\dfrac{1}{C_1 z} + \sqrt{\dfrac{2}{|C_1|}} & \text{if } \Delta = 0,\ C_2 > 0. \end{cases}$$

89. $y''_{xx} = A_1 y^{-5/3} + A_2 x^{-2/3} y^{-5/3}$.

Solutions in the parametric form:

$$x = a\tau^{-3/2}(C_1W^2 - 2C_2W + 2)^{3/2},$$
$$y = b\tau^{-9/4}(C_1W^2 - 2C_2W + 2)^{3/4}(6C_1W - 6C_2 \mp R)^{3/2},$$

where $A_1 = 24a^{-2}b^{8/3}C_1$, $A_2 = -36a^{-4/3}b^{8/3}B$.

90. $y''_{xx} = A_1 x^{-2/3} y^{-5/3} + A_2 x^{-4/3} y^{-5/3}$.

Solutions in the parametric form:

$$x = a\tau^{3/2}(C_1W^2 - 2C_2W + 2)^{-3/2},$$
$$y = b\tau^{-3/4}(C_1W^2 - 2C_2W + 2)^{-3/4}(6C_1W - 6C_2 \mp R)^{3/2},$$

where $A_1 = -36a^{-4/3}b^{8/3}B$, $A_2 = 24a^{-2/3}b^{8/3}C_1$.

▶ *In the solutions of equations 91–102, the following notation is used: functions P_1 and P_2 are the general solution of four modifications of the first Painlevé equation:*

$$P_1'' = \pm 6P_1^2 + \tau, \qquad P_2'' = \pm 6P_2^2 - \tau$$

(in the case of the upper sign, the equation for P_1 is the canonical form of the first Painlevé equation, see Subsection 2.8.2). In addition,

$$\begin{aligned} Q_1 &= \pm 6P_1^2 + \tau, & Q_2 &= \pm 6P_2^2 - \tau, \\ T_1 &= \tau^2 P_1 \mp 1, & T_2 &= \tau^2 P_2 \mp 1, \\ U_1 &= (P_1')^2 - 2P_1Q_1 \pm 8P_1^3, & U_2 &= (P_2')^2 - 2P_2Q_2 \pm 8P_2^3, \\ V_1 &= P_1'Q_1' + P_1' - Q_1^2, & V_2 &= P_2'Q_2' - P_2' - Q_2^2, \end{aligned}$$

where primes denote differentiation with respect to τ.

91. $y''_{xx} = A_1y^2 + A_2x.$

Solutions in the parametric form:

$$x = a\tau, \qquad y = bP_k,$$

where $A_1 = \pm 6a^{-2}b^{-1}$, $A_2 = a^{-3}b(-1)^{k+1}$; $k = 1$ and $k = 2$.

92. $y''_{xx} = A_1x^{-5}y^2 + A_2x^{-4}.$

Solutions in the parametric form:

$$x = a\tau^{-1}, \qquad y = b\tau^{-1}P_k,$$

where $A_1 = \pm 6a^3b^{-1}$, $A_2 = a^2b(-1)^{k+1}$; $k = 1$ and $k = 2$.

93. $y''_{xx} = A_1x^{-15/7}y^2 + A_2x^{-8/7}.$

Solutions in the parametric form:

$$x = a\tau^7, \qquad y = b\tau T_k,$$

where $A_1 = \pm\frac{6}{49}a^{1/7}b^{-1}$, $A_2 = \frac{1}{49}a^{-6/7}b(-1)^{k+1}$; $k = 1$ and $k = 2$.

94. $y''_{xx} = A_1x^{-20/7}y^2 + A_2x^{-13/7}.$

Solutions in the parametric form:

$$x = a\tau^{-7}, \qquad y = b\tau^{-6}T_k,$$

where $A_1 = \pm\frac{6}{49}a^{6/7}b^{-1}$, $A_2 = \frac{1}{49}a^{-1/7}b(-1)^{k+1}$; $k = 1$ and $k = 2$.

95. $y''_{xx} = A_1x + A_2y^{-1/2}.$

Solutions in the parametric form:

$$x = aP_k, \qquad y = b(P'_k)^2,$$

where $A_1 = \pm 24a^{-3}b$, $A_2 = 2a^{-2}b^{3/2}(-1)^{k+1}$; $k = 1$ and $k = 2$.

96. $y''_{xx} = A_1x^{-4} + A_2x^{-5/2}y^{-1/2}.$

Solutions in the parametric form:

$$x = aP_k^{-1}, \qquad y = bP_k^{-1}(P'_k)^2,$$

where $A_1 = \pm 24a^2b$, $A_2 = 2a^{1/2}b^{3/2}(-1)^{k+1}$; $k = 1$ and $k = 2$.

97. $y''_{xx} = A_1y^{1/3} + A_2xy^{-5/3}.$

Solutions in the parametric form:

$$x = aU_k, \qquad y = bP_k^{3/2},$$

where $A_1 = \mp 8ab^{-2}A_2$, $A_2 = -\frac{3}{16}a^{-3}b^{8/3}$; $k = 1$ and $k = 2$.

98. $y''_{xx} = A_1 x^{-10/3} y^{1/3} + A_2 x^{-7/3} y^{-5/3}$.

Solutions in the parametric form:

$$x = aU_k^{-1}, \qquad y = bP_k^{3/2} U_k^{-1},$$

where $A_1 = \mp 8ab^{-2} A_2$, $A_2 = -\frac{3}{16} a^{1/3} b^{8/3}$; $k = 1$ and $k = 2$.

99. $y''_{xx} = A_1 x^{-3/2} + A_2 y^{-1/2}$.

Solutions in the parametric form:

$$x = a(P_k')^2, \qquad y = bQ_k^2,$$

where $A_1 = \frac{1}{2} a^{-1/2} b(-1)^k$, $A_2 = \pm 6a^{-2} b^{3/2}$; $k = 1$ and $k = 2$.

100. $y''_{xx} = A_1 x^{-3/2} + A_2 x^{-5/2} y^{-1/2}$.

Solutions in the parametric form:

$$x = a(P_k')^{-2}, \qquad y = b(P_k')^{-2} Q_k^2,$$

where $A_1 = \frac{1}{2} a^{-1/2} b(-1)^k$, $A_2 = \pm 6a^{1/2} b^{3/2}$; $k = 1$ and $k = 2$.

101. $y''_{xx} = A_1 y^{-4/3} + A_2 x y^{-5/3}$.

Solutions in the parametric form:

$$x = aV_k, \qquad y = b(P_k')^3,$$

where $A_1 = ab^{-1/3} A_2 (-1)^k$, $A_2 = \frac{1}{36} a^{-3} b^{8/3}$; $k = 1$ and $k = 2$.

102. $y''_{xx} = A_1 x^{-5/3} y^{-4/3} + A_2 x^{-7/3} y^{-5/3}$.

Solutions in the parametric form:

$$x = aV_k^{-1}, \qquad y = b(P_k')^3 V_k^{-1},$$

where $A_1 = \frac{1}{36} a^{-1/3} b^{7/3} (-1)^k$, $A_2 = \frac{1}{36} a^{1/3} b^{8/3}$; $k = 1$ and $k = 2$.

► *In the solutions of equations 103–108, the following notation is used: functions P_1 and P_2 are the general solution of four modifications of the second Painlevé equation (with parameter $a = 0$):*

$$P_1'' = \tau P_1 \pm 2P_1^3, \qquad P_2'' = -\tau P_2 \pm 2P_2^3.$$

where primes denote differentiation with respect to τ. In the case of the upper sign, the equation for P_1 is the canonical form of the second Painlevé equation (with parameter $a = 0$, see Subsection 2.8.2).

103. $y''_{xx} = A_1 y^3 + A_2 xy.$

Solutions in the parametric form:

$$x = a\tau, \qquad y = bP_k,$$

where $A_1 = \pm 2a^{-2}b^{-2}$, $A_2 = a^3(-1)^{k+1}$; $k = 1$ and $k = 2$.

104. $y''_{xx} = A_1 x^{-6} y^3 + A_2 x^{-5} y.$

Solutions in the parametric form:

$$x = a\tau^{-1}, \qquad y = b\tau^{-1}P_k,$$

where $A_1 = \pm 2a^4 b^{-2}$, $A_2 = a^3(-1)^{k+1}$; $k = 1$ and $k = 2$.

105. $y''_{xx} = A_1 + A_2 x^{-1/2} y^{-1/2}.$

Solutions in the parametric form:

$$x = aP_k^2, \quad y = b(P'_k)^2, \qquad P'_k = (P_k)'_\tau,$$

where $A_1 = \pm 2a^{-2}b$, $A_2 = \frac{1}{2}a^{-3/2}b^{3/2}(-1)^{k+1}$; $k = 1$ and $k = 2$.

106. $y''_{xx} = A_1 x^{-3} + A_2 x^{-2} y^{-1/2}.$

Solutions in the parametric form:

$$x = aP_k^{-2}, \quad y = bP_k^{-2}(P'_k)^2, \qquad P'_k = (P_k)'_\tau,$$

where $A_1 = \pm 2ab$, $A_2 = \frac{1}{2}b^{3/2}(-1)^{k+1}$; $k = 1$ and $k = 2$.

107. $y''_{xx} = A_1 + A_2 xy^{-2}.$

Solutions in the parametric form:

$$x = a[\tau P_k^2 \pm P_k^4 - (P'_k)^2], \quad y = bP_k^2, \qquad P'_k = (P_k)'_\tau,$$

where $A_1 = \mp 2a^{-2}b(-1)^k$, $A_2 = 2a^{-3}b^3(-1)^{k+1}$; $k = 1$ and $k = 2$.

108. $y''_{xx} = A_1 x^{-3} + A_2 x^{-2} y^{-2}.$

Solutions in the parametric form:

$$x = a\big[\tau P_k^2 \pm P_k^4 - (P'_k)^2\big]^{-1}, \quad y = bP_k^2\big[\tau P_k^2 \pm P_k^4 - (P'_k)^2\big]^{-1},$$

where $A_1 = \pm 2ab$, $A_2 = 2b^3$; $k = 1$ and $k = 2$.

2.5. Generalized Emden—Fowler Equation $y''_{xx} = Ax^n y^m (y'_x)^l$

2.5.1. Classification Table

The case $l = 0$ corresponding to the classical Emden—Fowler equation is outlined in Section 2.3. In this section, the case $l \neq 0$ is considered.

Table 2.9 represents all solvable equations of the form $y''_{xx} = Ax^n y^m (y'_x)^l$ whose solutions are outlined in Subsection 2.5.2. The two-parameter families (in the space of parameters n, m, and l), one-parameter families, and isolated points are represented in a consecutive fashion. Equations are arranged in accordance with the growth of l, the growth of m (for identical l), and the growth of n (for identical m and l). The number of the equation sought is indicated in the last column in this table.

TABLE 2.9

Solvable cases of the generalized Emden—Fowler equation $y''_{xx} = Ax^n y^m (y'_x)^l$

No	l	m	n	Equation
Two-parameter families				
1	arbitrary	arbitrary	0	2.5.2.1
2	arbitrary	0	arbitrary	2.5.2.2
3	$\dfrac{2n+m+3}{n+m+2}$	arbitrary $(m \neq -1)$	arbitrary $(n \neq -1)$	2.5.2.3
One-parameter families				
4	arbitrary $(l \neq 1, 2)$	-1	-1	2.5.2.6
5	arbitrary $(l \neq \frac{3}{2})$	$-\frac{1}{2}$	$-\frac{1}{2}$	2.5.2.97
6	$\dfrac{3m+5}{2m+3}$	arbitrary $(m \neq -\frac{3}{2})$	$-\frac{1}{2}$	2.5.2.13
7	$\dfrac{3m+5}{2m+3}$	arbitrary $(m \neq -\frac{3}{2})$	1	2.5.2.10
8	$\dfrac{3n+4}{2n+3}$	$-\frac{1}{2}$	arbitrary $(n \neq -\frac{3}{2})$	2.5.2.11
9	$\dfrac{3n+4}{2n+3}$	1	arbitrary $(n \neq -\frac{3}{2})$	2.5.2.12
10	$\dfrac{3n+4}{2n+3}$	$-n-3$	arbitrary $(n \neq -\frac{3}{2})$	2.5.2.107
11	1	arbitrary $(m \neq -1, 0)$	-1	2.5.2.5
12	2	-1	arbitrary $(n \neq -1, 0)$	2.5.2.4
13	3	arbitrary $(m \neq -2)$	1	2.5.2.96
14	3	$-n-3$	arbitrary	2.5.2.9
Isolated points				
15	$\frac{1}{2}$	$-\frac{1}{2}$	$-\frac{5}{2}$	2.5.2.33
16	$\frac{1}{2}$	1	$-\frac{15}{8}$	2.5.2.83
17	$\frac{1}{2}$	1	$-\frac{20}{13}$	2.5.2.86
18	$\frac{1}{2}$	1	$-\frac{5}{4}$	2.5.2.80
19	$\frac{1}{2}$	1	0	2.5.2.78
20	$\frac{2}{3}$	$-\frac{1}{2}$	$-\frac{7}{6}$	2.5.2.53
21	$\frac{4}{5}$	$-\frac{5}{2}$	$-\frac{1}{2}$	2.5.2.76

TABLE 2.9 *Continued*

Solvable cases of the generalized Emden—Fowler equation $y''_{xx} = Ax^n y^m (y'_x)^l$

No	l	m	n	Equation
22	1	-2	1	2.5.2.14
23	1	-1	-1	2.5.2.8
24	$\frac{8}{7}$	1	$-\frac{3}{4}$	2.5.2.54
25	$\frac{8}{7}$	1	$-\frac{1}{2}$	2.5.2.52
26	$\frac{6}{5}$	$-\frac{1}{2}$	$-\frac{2}{3}$	2.5.2.68
27	$\frac{5}{4}$	1	$-\frac{1}{2}$	2.5.2.58
28	$\frac{5}{4}$	1	0	2.5.2.56
29	$\frac{9}{7}$	$-\frac{13}{8}$	1	2.5.2.39
30	$\frac{9}{7}$	$-\frac{1}{2}$	1	2.5.2.38
31	$\frac{13}{10}$	$-\frac{1}{2}$	$-\frac{5}{2}$	2.5.2.47
32	$\frac{27}{20}$	$-\frac{1}{2}$	$-\frac{2}{3}$	2.5.2.72
33	$\frac{18}{13}$	$-\frac{1}{2}$	$-\frac{7}{2}$	2.5.2.40
34	$\frac{7}{5}$	$-\frac{7}{4}$	1	2.5.2.18
35	$\frac{7}{5}$	$-\frac{10}{7}$	1	2.5.2.46
36	$\frac{7}{5}$	$-\frac{2}{3}$	1	2.5.2.32
37	$\frac{7}{5}$	$-\frac{1}{2}$	1	2.5.2.17
38	$\frac{7}{5}$	1	0	2.5.2.89
39	$\frac{7}{5}$	1	1	2.5.2.91
40	$\frac{7}{5}$	5	1	2.5.2.75
41	$\frac{10}{7}$	$-\frac{1}{2}$	$-\frac{5}{2}$	2.5.2.19
42	$\frac{22}{15}$	$-\frac{1}{2}$	$-\frac{2}{3}$	2.5.2.70
43	$\frac{3}{2}$	-2	$-\frac{1}{2}$	2.5.2.106
44	$\frac{3}{2}$	-2	1	2.5.2.99
45	$\frac{3}{2}$	$-\frac{1}{2}$	-2	2.5.2.100
46	$\frac{3}{2}$	$-\frac{1}{2}$	$-\frac{1}{2}$	2.5.2.29
47	$\frac{3}{2}$	$-\frac{1}{2}$	1	2.5.2.98
48	$\frac{3}{2}$	1	-2	2.5.2.105
49	$\frac{3}{2}$	1	$-\frac{1}{2}$	2.5.2.103
50	$\frac{23}{15}$	$-\frac{2}{3}$	$-\frac{1}{2}$	2.5.2.84
51	$\frac{11}{7}$	$-\frac{5}{2}$	$-\frac{1}{2}$	2.5.2.27

TABLE 2.9 *Continued*

Solvable cases of the generalized Emden—Fowler equation $y''_{xx} = Ax^n y^m (y'_x)^l$

No	l	m	n	Equation
52	$\frac{8}{5}$	0	1	2.5.2.73
53	$\frac{8}{5}$	1	$-\frac{7}{4}$	2.5.2.26
54	$\frac{8}{5}$	1	$-\frac{10}{7}$	2.5.2.48
55	$\frac{8}{5}$	1	$-\frac{2}{3}$	2.5.2.35
56	$\frac{8}{5}$	1	$-\frac{1}{2}$	2.5.2.24
57	$\frac{8}{5}$	1	1	2.5.2.74
58	$\frac{8}{5}$	1	5	2.5.2.94
59	$\frac{21}{13}$	$-\frac{7}{2}$	$-\frac{1}{2}$	2.5.2.45
60	$\frac{33}{20}$	$-\frac{2}{3}$	$-\frac{1}{2}$	2.5.2.87
61	$\frac{17}{10}$	$-\frac{5}{2}$	$-\frac{1}{2}$	2.5.2.49
62	$\frac{12}{7}$	1	$-\frac{13}{8}$	2.5.2.44
63	$\frac{12}{7}$	1	$-\frac{1}{2}$	2.5.2.42
64	$\frac{7}{4}$	$-\frac{1}{2}$	1	2.5.2.51
65	$\frac{7}{4}$	0	1	2.5.2.50
66	$\frac{9}{5}$	$-\frac{2}{3}$	$-\frac{1}{2}$	2.5.2.81
67	$\frac{13}{7}$	$-\frac{3}{4}$	1	2.5.2.65
68	$\frac{13}{7}$	$-\frac{1}{2}$	1	2.5.2.61
69	2	-1	-1	2.5.2.7
70	2	1	-2	2.5.2.16
71	$\frac{11}{5}$	$-\frac{1}{2}$	$-\frac{5}{2}$	2.5.2.95
72	$\frac{7}{3}$	$-\frac{7}{6}$	$-\frac{1}{2}$	2.5.2.64
73	$\frac{5}{2}$	$-\frac{5}{2}$	$-\frac{1}{2}$	2.5.2.36
74	$\frac{5}{2}$	$-\frac{15}{8}$	1	2.5.2.69
75	$\frac{5}{2}$	$-\frac{20}{13}$	1	2.5.2.71
76	$\frac{5}{2}$	$-\frac{5}{4}$	1	2.5.2.67
77	$\frac{5}{2}$	0	1	2.5.2.66
78	3	-5	2	2.5.2.79
79	3	$-\frac{7}{2}$	$-\frac{1}{2}$	2.5.2.31
80	3	$-\frac{10}{3}$	$-\frac{5}{3}$	2.5.2.37
81	3	$-\frac{20}{7}$	2	2.5.2.85

TABLE 2.9 *Continued*

Solvable cases of the generalized Emden—Fowler equation $y''_{xx} = Ax^n y^m (y'_x)^l$

No	l	m	n	Equation
82	3	$-\frac{5}{2}$	$-\frac{1}{2}$	2.5.2.22
83	3	$-\frac{13}{5}$	$-\frac{7}{5}$	2.5.2.43
84	3	$-\frac{7}{3}$	$-\frac{5}{3}$	2.5.2.25
85	3	$-\frac{15}{7}$	2	2.5.2.82
86	3	-2	-2	2.5.2.104
87	3	-2	-1	2.5.2.15
88	3	-2	$-\frac{1}{2}$	2.5.2.101
89	3	-2	1	2.5.2.28
90	3	$-\frac{4}{3}$	$-\frac{1}{2}$	2.5.2.59
91	3	$-\frac{7}{6}$	$-\frac{1}{2}$	2.5.2.60
92	3	$-\frac{5}{6}$	$-\frac{5}{3}$	2.5.2.93
93	3	$-\frac{1}{2}$	$-\frac{5}{2}$	2.5.2.90
94	3	$-\frac{1}{2}$	$-\frac{5}{3}$	2.5.2.92
95	3	0	-4	2.5.2.55
96	3	0	$-\frac{5}{2}$	2.5.2.88
97	3	0	$-\frac{1}{2}$	2.5.2.20
98	3	0	2	2.5.2.77
99	3	1	-7	2.5.2.62
100	3	1	-4	2.5.2.57
101	3	1	-2	2.5.2.102
102	3	1	$-\frac{5}{3}$	2.5.2.23
103	3	1	$-\frac{7}{5}$	2.5.2.41
104	3	1	$-\frac{1}{2}$	2.5.2.30
105	3	1	0	2.5.2.21
106	3	2	$-\frac{5}{3}$	2.5.2.34
107	3	3	-7	2.5.2.63

2.5.2. Exact Solutions

1. $y''_{xx} = Ay^m (y'_x)^l$.

1°. Solution in the parametric form with $m \neq -1$, $l \neq 2$:

$$x = aC_1^{1-m-l} \int (1 \pm \tau^{m+1})^{\frac{1}{l-2}} \, d\tau + C_2, \qquad y = bC_1^{2-l} \tau,$$

where $A = \pm \dfrac{m+1}{2-l} a^{l-2} b^{1-m-l}$.

2°. Solution in the parametric form with $m = -1$, $l \neq 2$:

$$x = aC_1 \int \tau^{\frac{l}{l-2}} \exp(\mp\tau^2) \, d\tau + C_2, \qquad y = bC_1 \exp(\mp\tau^2),$$

where $A = \mp \dfrac{4b^2}{a^2(2-l)} \left(\mp \dfrac{a}{2b}\right)^l$.

3°. Solution in the parametric form with $m \neq -1$, $l = 2$:

$$x = C_1 \int \tau^{\frac{1-m}{1+m}} \exp(\mp\tau^2) \, d\tau + C_2, \qquad y = b\tau^{\frac{2}{m+1}},$$

where $A = \pm(m+1)b^{-1-m}$.

4°. Solution with $m = -1$, $l = 2$:

$$y = \begin{cases} (C_1 x + C_2)^{\frac{1}{1-A}} & \text{if } A \neq 1, \\ C_2 \exp(C_1 x) & \text{if } A = 1. \end{cases}$$

2. $y''_{xx} = Ax^n (y'_x)^l$.

1°. Solution in the parametric form with $n \neq -1$, $l \neq 1$:

$$x = aC_1^{1-l} \tau, \qquad y = bC_1^{2+n-l} \int (1 \pm \tau^{n+1})^{\frac{1}{1-l}} \, d\tau + C_2,$$

where $A = \pm \dfrac{n+1}{1-l} a^{l-n-2} b^{1-l}$.

2°. Solution in the parametric form with $n = -1$, $l \neq 1$:

$$x = aC_1 \exp(\mp\tau^2), \qquad y = bC_1 \int \tau^{\frac{3-l}{1-l}} \exp(\mp\tau^2) \, d\tau + C_2,$$

where $A = \mp \dfrac{4a^2}{b^2(1-l)} \left(\mp \dfrac{b}{2a}\right)^{3-l}$.

3°. Solution in the parametric form with $n \neq -1$, $l = 1$:

$$x = a\tau^{\frac{2}{n+1}}, \qquad y = C_1 \int \tau^{\frac{1-n}{1+n}} \exp(\mp\tau^2) \, d\tau + C_2,$$

where $A = \mp(n+1)a^{1-n}$.

4°. Solution with $n = -1$, $l = 1$:

$$y = \begin{cases} C_1 x^{A+1} + C_2 & \text{if } A \neq 1, \\ C_1 \ln|x| + C_2 & \text{if } A = 1. \end{cases}$$

3. $y''_{xx} = Ax^n y^m (y'_x)^{\frac{2n+m+3}{n+m+2}}$.

Solution in the parametric form with $n \neq -1$, $m \neq -1$:

$$x = \exp\left[\int \frac{d\tau}{f(\tau)} + C_2\right], \qquad y = \tau \exp\left[-\frac{n+1}{m+1}\int \frac{d\tau}{f(\tau)} - \frac{n+1}{m+1}C_2\right],$$

where function $f = f(\tau)$ is defined implicitly by the formula

$$[f + (\sigma - 1)\tau](f + \sigma\tau)^{\frac{\sigma}{1-\sigma}} = C_1 + \frac{A\tau^{m+2}}{n+m+2}, \qquad \sigma = -\frac{n+1}{m+1}.$$

See equation 2.5.2.5 for the case $n = -1$. See equation 2.5.2.4 for the case $m = -1$.

4. $y''_{xx} = Ax^n y^{-1} (y'_x)^2$.

Solution in the parametric form with $n \neq -1$, $n \neq 0$:

$$x = a\tau^{\frac{1}{n}}, \qquad y = \pm\exp\left[\int \tau^{\frac{1-n}{n}} \left(\frac{n}{n+1}\tau^{\frac{n+1}{n}} + n\tau^{\frac{1}{n}} + C_1\right)^{-1} d\tau + C_2\right],$$

where $A = -a^{-n}$.

See equation 2.5.2.7 for the case $n = -1$. See equation 2.5.2.1 for the case $n = 0$.

5. $y''_{xx} = Ax^{-1} y^m y'_x$.

Solution in the parametric form with $m \neq -1$, $m \neq 0$:

$$x = \pm\exp\left[\int \tau^{\frac{1-m}{m}} \left(\frac{m}{m+1}\tau^{\frac{m+1}{m}} + m\tau^{\frac{1}{m}} + C_1\right)^{-1} d\tau + C_2\right], \qquad y = (A\tau)^{\frac{1}{m}}.$$

See equation 2.5.2.8 for the case $m = -1$. See equation 2.5.2.2 for the case $m = 0$.

6. $y''_{xx} = Ax^{-1} y^{-1} (y'_x)^l$.

Solution in the parametric form with $l \neq 1$, $l \neq 2$:

$$x = \pm\left(f - \frac{\tau}{\lambda}\right)^{-1} \exp\left[\frac{1}{\lambda}\int \frac{d\tau}{f(\tau)} + C_2\right], \quad y = \pm\exp\left[\int \frac{d\tau}{f(\tau)} + \lambda C_2\right], \qquad \lambda = \frac{l-1}{l-2},$$

where the function $f = f(\tau)$ is defined implicitly by the formula

$$\ln\left(\frac{f}{\tau} - \frac{1}{\lambda}\right) - \frac{\tau}{\lambda f - \tau} = \pm\frac{A}{\lambda}\tau^{\lambda} - \ln\tau + C_1, \qquad \lambda = \frac{l-1}{l-2}.$$

See equation 2.5.2.7 for the case $l = 2$. See equation 2.5.2.8 for the case $l = 1$.

7. $y''_{xx} = Ax^{-1} y^{-1} (y'_x)^2$.

Solution in the parametric form:

$$x = \pm e^{\tau}, \qquad y = C_2(\mp A\tau + e^{\tau} + C_1)\exp\left[\pm A\int (\mp A\tau + e^{\tau} + C_1)^{-1}\, d\tau\right].$$

8. $y''_{xx} = Ax^{-1}y^{-1}y'_x$.

Solution in the parametric form:

$$x = C_2(\pm A\tau + e^\tau + C_1)\exp\left[\mp A\int(\pm A\tau + e^\tau + C_1)^{-1}\,d\tau\right], \qquad y = \pm e^\tau.$$

9. $y''_{xx} = Ax^n y^{-n-3}(y'_x)^3$.

Solution in the parametric form with $n \neq -1$:

$$x = aC_1^{n+1}\tau\left[\int(1\pm\tau^{n+1})^{-1/2}\,d\tau + C_2\right]^{-1}, \quad y = bC_1^{n-1}\left[\int(1\pm\tau^{n+1})^{-1/2}\,d\tau + C_2\right]^{-1},$$

where $A = \mp\dfrac{n+1}{2}a^{1-n}b^{n+1}$.

See equation 2.5.2.15 for the case $n = -1$.

10. $y''_{xx} = Axy^m(y'_x)^{\frac{3m+5}{2m+3}}$.

Solution in the parametric form with $m \neq -3/2$:

$$x = aC_1^{-2}\left\{(1\pm\tau^{\mu+1})^{1/2}\left[\int(1\pm\tau^{\mu+1})^{-1/2}\,d\tau + C_2\right] - \tau\right\},$$

$$y = bC_1^{(\mu+1)(\mu-2)}\left[\int(1\pm\tau^{\mu+1})^{-1/2}\,d\tau + C_2\right]^{\mu+2},$$

where $\mu = -\dfrac{2m+3}{m+1}$, $\quad A = -\dfrac{\mu b^{\frac{1}{\mu+2}}}{(\mu+2)a}\left[\pm\dfrac{(\mu+1)a}{2(\mu+2)b}\right]^{\frac{1}{\mu}}$.

11. $y''_{xx} = Ax^n y^{-\frac{1}{2}}(y'_x)^{\frac{3n+4}{2n+3}}$.

Solution in the parametric form with $n \neq -3/2$:

$$x = aC_1^{(\mu+1)^2}\tau^{\mu+1}\left[\int(1\pm\tau^{\mu+1})^{-1/2}\,d\tau + C_2\right]^{-\mu-1},$$

$$y = bC_1^4\left\{(1\pm\tau^{\mu+1})^{1/2}\left[\int(1\pm\tau^{\mu+1})^{-1/2}\,d\tau + C_2\right] - \tau\right\}^2,$$

where $\mu = -\dfrac{n}{n+1}$, $\quad A = \dfrac{\mu+3}{a(\mu+1)}a^{\frac{\mu}{\mu+1}}b^{\frac{1}{2}}\left(\pm\dfrac{a}{b}\right)^{\frac{1}{\mu+3}}$.

12. $y''_{xx} = Ax^n y(y'_x)^{\frac{3n+4}{2n+3}}$.

Solution in the parametric form with $n \neq -3/2$:

$$x = aC_1^{(\mu-1)(\mu+2)}\left[\int(1\pm\tau^{\mu+1})^{-1/2}\,d\tau + C_2\right]^{\mu+2},$$

$$y = bC_1^{-2}\left\{(1\pm\tau^{\mu+1})^{1/2}\left[\int(1\pm\tau^{\mu+1})^{-1/2}\,d\tau + C_2\right] - \tau\right\},$$

where $\mu = -\dfrac{2n+3}{n+1}$, $\quad A = \dfrac{\mu a^{\frac{1}{\mu+2}}}{(\mu+2)b}\left[\pm\dfrac{(\mu+1)b}{2(\mu+2)a}\right]^{\frac{1}{\mu}}$.

13. $y''_{xx} = Ax^{-\frac{1}{2}} y^m (y'_x)^{\frac{3m+5}{2m+3}}$.

Solution in the parametric form with $m \neq -3/2$:

$$x = aC_1^4 \left\{ (1 \pm \tau^{\mu+1})^{1/2} \left[\int (1 \pm \tau^{\mu+1})^{-1/2} \, d\tau + C_2 \right] - \tau \right\}^2,$$

$$y = bC_1^{(\mu+1)^2} \tau^{\mu+1} \left[\int (1 \pm \tau^{\mu+1})^{-1/2} \, d\tau + C_2 \right]^{-\mu-1},$$

where $\mu = -\dfrac{m}{m+1}$, $A = -\dfrac{\mu+3}{b(\mu+1)} a^{\frac{1}{2}} b^{\frac{\mu}{\mu+1}} \left(\pm \dfrac{b}{a} \right)^{\frac{1}{\mu+3}}$.

14. $y''_{xx} = Axy^{-2} y'_x$.

Solution in the parametric form:

$$x = aC_1 \left\{ 2\tau \left[\int \exp(\mp\tau^2) \, d\tau + C_2 \right] \pm \exp(\mp\tau^2) \right\}, \qquad y = bC_1 \left[\int \exp(\mp\tau^2) \, d\tau + C_2 \right],$$

where $A = \mp \frac{1}{2} a^{-2} b^2$.

15. $y''_{xx} = Ax^{-1} y^{-2} (y'_x)^3$.

Solution in the parametric form:

$$x = a \exp(\mp\tau^2) \left[\int \exp(\mp\tau^2) \, d\tau + C_2 \right]^{-1}, \qquad y = C_1 \left[\int \exp(\mp\tau^2) \, d\tau + C_2 \right]^{-1},$$

where $A = \pm 2a^2$.

16. $y''_{xx} = Ax^{-2} y (y'_x)^2$.

Solution in the parametric form:

$$x = aC_1 \left[\int \exp(\mp\tau^2) \, d\tau + C_2 \right], \qquad y = bC_1 \left\{ 2\tau \left[\int \exp(\mp\tau^2) \, d\tau + C_2 \right] \pm \exp(\mp\tau^2) \right\},$$

where $A = \pm a^2 b^{-2}$.

17. $y''_{xx} = Axy^{-1/2} (y'_x)^{7/5}$.

Solution in the parametric form:

$$x = \pm aC_1 (\tau^2 - 1)(\tau^3 - 3\tau + C_2)^{-1/2}, \qquad y = bC_1^{16} (\tau^4 - 6\tau^2 + 4C_2\tau - 3)^2,$$

where $A = \pm 15 a^{-2} b^{1/2} \left(\dfrac{a}{16b} \right)^{2/5}$.

18. $y''_{xx} = Axy^{-7/4} (y'_x)^{7/5}$.

Solution in the parametric form:

$$x = \pm aC_1^{27} (\tau^3 - 3\tau + C_2)^{-1/2} (\tau^6 - 15\tau^4 + 20C_2\tau^3 - 45\tau^2 + 12C_2\tau + 27 - 8C_2^2),$$

$$y = bC_1^{32} (\tau^4 - 6\tau^2 + 4C_2\tau - 3)^{4/3},$$

where $A = \pm \dfrac{5}{12} a^{-2} b^{7/4} \left(\dfrac{a}{9b} \right)^{2/5}$.

19. $y''_{xx} = Ax^{-5/2}y^{-1/2}(y'_x)^{10/7}$.

Solution in the parametric form:

$$x = aC_1^{-1}(\tau^3 - 3\tau + C_2)^{-1}(\tau^4 - 6\tau^2 + 4C_2\tau - 3)^{2/3},$$
$$y = bC_1^{27}(\tau^3 - 3\tau + C_2)^{-1}(\tau^6 - 15\tau^4 + 20C_2\tau^3 - 45\tau^2 + 12C_2\tau + 27 - 8C_2^2)^2,$$

where $A = 28a(ab)^{1/2}\left(\dfrac{a}{27b}\right)^{3/7}$.

20. $y''_{xx} = Ax^{-1/2}(y'_x)^3$.

Solution in the parametric form:

$$x = aC_1^4(\tau^2 - 1)^2, \qquad y = bC_1^3(\tau^3 - 3\tau + C_2),$$

where $A = \pm\frac{4}{9}a^{3/2}b^{-2}$.

21. $y''_{xx} = Ay(y'_x)^3$.

Solution in the parametric form:

$$x = aC_1^3(\tau^3 - 3\tau + C_2), \quad y = bC_1\tau, \qquad \text{where} \quad A = -6ab^{-3}.$$

22. $y''_{xx} = Ax^{-1/2}y^{-5/2}(y'_x)^3$.

Solution in the parametric form:

$$x = aC_1(\tau^2 - 1)^2(\tau^3 - 3\tau + C_2)^{-1}, \qquad y = bC_1^{-3}(\tau^3 - 3\tau + C_2)^{-1},$$

where $A = \mp\frac{4}{9}a^{3/2}b^{1/2}$.

23. $y''_{xx} = Ax^{-5/3}y(y'_x)^3$.

Solution in the parametric form:

$$x = aC_1^9(\tau^3 - 3\tau + C_2)^{3/2}, \qquad y = \pm bC_1^8(\tau^4 - 6\tau^2 + 4C_2\tau - 3),$$

where $A = \mp\frac{9}{64}a^{8/3}b^{-3}$.

24. $y''_{xx} = Ax^{-1/2}y(y'_x)^{8/5}$.

Solution in the parametric form:

$$x = aC_1^{16}(\tau^4 - 6\tau^2 + 4C_2\tau - 3)^2, \qquad y = \pm bC_1(\tau^2 - 1)(\tau^3 - 3\tau + C_2)^{-1/2},$$

where $A = \mp 15a^{1/2}b^{-2}\left(\dfrac{b}{16a}\right)^{2/5}$.

25. $y''_{xx} = Ax^{-5/3}y^{-7/3}(y'_x)^3$.

Solution in the parametric form:

$$x = \pm aC_1(\tau^3 - 3\tau + C_2)^{3/2}(\tau^4 - 6\tau^2 + 4C_2\tau - 3)^{-1}, \qquad y = \pm bC_1^{-8}(\tau^4 - 6\tau^2 + 4C_2\tau - 3)^{-1},$$

where $A = \mp\frac{9}{64}a^{8/3}b^{1/3}$.

26. $y''_{xx} = Ax^{-7/4} y (y'_x)^{8/5}$.

Solution in the parametric form:

$$x = aC_1^{32}(\tau^4 - 6\tau^2 + 4C_2\tau - 3)^{4/3},$$
$$y = \pm bC_1^{27}(\tau^3 - 3\tau + C_2)^{-1/2}(\tau^6 - 15\tau^4 + 20C_2\tau^3 - 45\tau^2 + 12C_2\tau + 27 - 8C_2^2),$$

where $A = \mp\frac{5}{12}a^{7/4}b^{-2}\left(\frac{b}{9a}\right)^{2/5}$.

27. $y''_{xx} = Ax^{-1/2} y^{-5/2} (y'_x)^{11/7}$.

Solution in the parametric form:

$$x = aC_1^{27}(\tau^3 - 3\tau + C_2)^{-1}(\tau^6 - 15\tau^4 + 20C_2\tau^3 - 45\tau^2 + 12C_2\tau + 27 - 8C_2^2)^2,$$
$$y = bC_1^{-1}(\tau^3 - 3\tau + C_2)^{-1}(\tau^4 - 6\tau^2 + 4C_2\tau - 3)^{2/3},$$

where $A = -28b(ab)^{1/2}\left(\frac{b}{27a}\right)^{3/7}$.

28. $y''_{xx} = Axy^{-2}(y'_x)^3$.

1°. Solution in the parametric form with $A < \frac{1}{4}$:

$$x = \tau(C_1\tau^\nu + C_2\tau^{-\nu}), \quad y = \tau^2, \qquad \text{where} \quad \nu = \sqrt{1 - 4A}.$$

2°. Solution in the parametric form with $A = \frac{1}{4}$:

$$x = \tau(C_1 \ln|\tau| + C_2), \quad y = \tau^2.$$

3°. Solution in the parametric form with $A > \frac{1}{4}$:

$$x = \tau C_1 \sin(\nu \ln \tau + C_2), \quad y = \tau^2, \qquad \text{where} \quad \nu = \sqrt{4A - 1}.$$

29. $y''_{xx} = Ax^{-1/2} y^{-1/2} (y'_x)^{3/2}$.

Solution in the parametric form:

$$x = \pm\tau^2(C_1\tau^\nu + C_2\tau^{-\nu})^2, \qquad y = \tfrac{1}{4}\tau^{-2}[(1+\nu)C_1\tau^\nu + (1-\nu)C_2\tau^{-\nu}]^2,$$

where $A = \mp k^2$, $\nu = k^{-2}(k^4 + 4)^{1/2}$.

30. $y''_{xx} = Ax^{-1/2} y (y'_x)^3$.

Solution in the parametric form:

$$x = aC_1^2 \exp(-2\tau)\left[2\exp(3\tau) - C_2 \sin(\sqrt{3}\,\tau) + \sqrt{3}\,C_2 \cos(\sqrt{3}\,\tau)\right]^2,$$
$$y = bC_1 \exp(-\tau)\left[\exp(3\tau) + C_2 \sin(\sqrt{3}\,\tau)\right],$$

where $A = -16a^{3/2}b^{-3}$.

31. $y''_{xx} = Ax^{-1/2}y^{-7/2}(y'_x)^3$.

Solution in the parametric form:

$$x = \frac{aC_1 \exp(-\tau)\left[2\exp(3\tau) - C_2\sin\left(\sqrt{3}\,\tau\right) + \sqrt{3}\,C_2\cos\left(\sqrt{3}\,\tau\right)\right]^2}{\exp(3\tau) + C_2\sin\left(\sqrt{3}\,\tau\right)},$$

$$y = \frac{bC_1^{-1}\exp(\tau)}{\exp(3\tau) + C_2\sin\left(\sqrt{3}\,\tau\right)},$$

where $A = -16(ab)^{3/2}$.

32. $y''_{xx} = Axy^{-2/3}(y'_x)^{7/5}$.

1°. Solution in the parametric form with $A < 0$:

$$x = aC_1[\cosh(\tau + C_2)\cos\tau]^{1/2}[\tanh(\tau + C_2) - \tan\tau],$$
$$y = bC_1^6\cosh^3(\tau + C_2)\cos^3\tau\,[\tanh(\tau + C_2) + \tan\tau]^3,$$

where $A = -5a^{-2}b^{2/3}\left(\dfrac{a}{12b}\right)^{2/5}$.

2°. Solution in the parametric form with $A > 0$:

$$x = aC_1[\cosh\tau - \sin(\tau + C_2)]^{-1/2}[\sinh\tau - \cos(\tau + C_2)],$$
$$y = bC_1^6[\sinh\tau + \cos(\tau + C_2)]^3,$$

where $A = 5a^{-2}b^{2/3}\left(\dfrac{a}{6b}\right)^{2/5}$.

33. $y''_{xx} = Ax^{-5/2}y^{-1/2}(y'_x)^{1/2}$.

Solution in the parametric form:

$$x = aC_1^{-1}[\cosh(\tau + C_2)\cos\tau]^{-1},$$
$$y = bC_1\cosh(\tau + C_2)\cos\tau\,[\tanh(\tau + C_2) - \tan\tau]^2,$$

where $A = -4ab$.

34. $y''_{xx} = Ax^{-5/3}y^2(y'_x)^3$.

1°. Solution in the parametric form with $A > 0$:

$$x = aC_1^3[\cosh(\tau + C_2)\cos\tau]^{3/2},$$
$$y = bC_1^2\cosh(\tau + C_2)\cos\tau\,[\tanh(\tau + C_2) + \tan\tau],$$

where $A = \frac{3}{16}a^{8/3}b^{-4}$.

2°. Solution in the parametric form with $A < 0$:

$$x = aC_1^3[\cosh\tau - \sin(\tau + C_2)]^{3/2},\; y = bC_1^2[\sinh\tau + \cos(\tau + C_2)],$$

where $A = -\frac{3}{4}a^{8/3}b^{-4}$.

35. $y''_{xx} = Ax^{-2/3} y (y'_x)^{8/5}$.

1°. Solution in the parametric form with $A > 0$:

$$x = aC_1^6 \cosh^3(\tau + C_2)\cos^3\tau[\tanh(\tau + C_2) + \tan\tau]^3,$$
$$y = bC_1[\cosh(\tau + C_2)\cos\tau]^{1/2}[\tanh(\tau + C_2) - \tan\tau],$$

where $A = 5a^{2/3}b^{-2}\left(\dfrac{b}{12a}\right)^{2/5}$.

2°. Solution in the parametric form with $A < 0$:

$$x = aC_1^6[\sinh\tau + \cos(\tau + C_2)]^3,$$
$$y = bC_1[\cosh\tau - \sin(\tau + C_2)]^{-1/2}[\sinh\tau - \cos(\tau + C_2)],$$

where $A = -5a^{2/3}b^{-2}\left(\dfrac{b}{6a}\right)^{2/5}$.

36. $y''_{xx} = Ax^{-1/2} y^{-5/2} (y'_x)^{5/2}$.

Solution in the parametric form:

$$x = aC_1 \cosh(\tau + C_2)\cos\tau\,[\tanh(\tau + C_2) - \tan\tau]^2,$$
$$y = bC_1^{-1}[\cosh(\tau + C_2)\cos\tau]^{-1},$$

where $A = 4ab$.

37. $y''_{xx} = Ax^{-5/3} y^{-10/3} (y'_x)^3$.

1°. Solution in the parametric form with $A > 0$:

$$x = aC_1[\cosh(\tau + C_2)\cos\tau]^{1/2}[\tanh(\tau + C_2) + \tan\tau]^{-1},$$
$$y = bC_1^{-2}[\cosh(\tau + C_2)\cos\tau]^{-1}[\tanh(\tau + C_2) + \tan\tau]^{-1},$$

where $A = \frac{3}{16}a^{8/3}b^{4/3}$.

2°. Solution in the parametric form with $A < 0$:

$$x = aC_1[\cosh\tau - \sin(\tau + C_2)]^{3/2}[\sinh\tau + \cos(\tau + C_2)]^{-1},$$
$$y = bC_1^{-2}[\sinh\tau + \cos(\tau + C_2)]^{-1},$$

where $A = -\frac{3}{4}a^{8/3}b^{4/3}$.

► *In the solutions of equations 38–45, the following notation is used:*

$$E = \exp(3\tau), \quad S_1 = E + C_2\sin\left(\sqrt{3}\,\tau\right), \quad S_2 = 2E - C_2\sin\left(\sqrt{3}\,\tau\right) + \sqrt{3}\,C_2\cos\left(\sqrt{3}\,\tau\right),$$
$$S_3 = 2S_1(S_2)'_\tau - (S_1)'_\tau S_2 - S_1S_2, \quad S_4 = 2S_1(S_3)'_\tau - 5(S_1)'_\tau S_3 + S_1S_3.$$

38. $y''_{xx} = Axy^{-1/2}(y'_x)^{9/7}$.

Solution in the parametric form:

$$x = aC_1E^{-1/6}S_1^{-1/2}S_2, \quad y = bC_1^8E^{-4/3}S_3^2,$$

where $A = 7a^{-2}b^{1/2}\left(\dfrac{a}{64b}\right)^{2/7}$.

39. $y''_{xx} = Axy^{-13/8}(y'_x)^{9/7}$.

Solution in the parametric form:

$$x = aC_1^{25}E^{-5/6}S_1^{-1/2}S_4, \qquad y = bC_1^{32}E^{-16/15}S_3^{8/5},$$

where $A = 7a^{-2}b^{13/8}\left(\dfrac{25a}{256b}\right)^{2/7}$.

40. $y''_{xx} = Ax^{-7/2}y^{-1/2}(y'_x)^{18/13}$.

Solution in the parametric form:

$$x = aC_1^{-1}E^{1/15}S_1^{-1}S_3^{2/5}, \qquad y = bC_1^{25}E^{-5/3}S_1^{-1}S_4^2,$$

where $A = -208a^{5/2}b^{1/2}\left(\dfrac{a}{25b}\right)^{5/13}$.

41. $y''_{xx} = Ax^{-7/5}y(y'_x)^3$.

Solution in the parametric form:

$$x = aC_1^5E^{-5/6}S_1^{5/2}, \qquad y = bC_1^4E^{-2/3}S_3,$$

where $A = -\frac{5}{1024}a^{12/5}b^{-3}$.

42. $y''_{xx} = Ax^{-1/2}y(y'_x)^{12/7}$.

Solution in the parametric form:

$$x = aC_1^8E^{-4/3}S_3^2, \qquad y = bC_1E^{-1/6}S_1^{-1/2}S_2,$$

where $A = -7a^{1/2}b^{-2}\left(\dfrac{b}{64a}\right)^{2/7}$.

43. $y''_{xx} = Ax^{-7/5}y^{-13/5}(y'_x)^3$.

Solution in the parametric form:

$$x = aC_1E^{-1/6}S_1^{5/2}S_3^{-1}, \qquad y = bC_1^{-4}E^{2/3}S_3^{-1},$$

where $A = -\frac{5}{1024}a^{12/5}b^{3/5}$.

44. $y''_{xx} = Ax^{-13/8}y(y'_x)^{12/7}$.

Solution in the parametric form:

$$x = aC_1^{32}E^{-16/15}S_3^{8/5}, \qquad y = bC_1^{25}E^{-5/6}S_1^{-1/2}S_4,$$

where $A = -7a^{13/8}b^{-2}\left(\dfrac{25b}{256a}\right)^{2/7}$.

45. $y''_{xx} = Ax^{-1/2}y^{-7/2}(y'_x)^{21/13}$.

Solution in the parametric form:

$$x = aC_1^{25}E^{-5/3}S_1^{-1}S_4^2, \qquad y = bC_1^{-1}E^{1/15}S_1^{-1}S_3^{2/5},$$

where $A = 208a^{1/2}b^{5/2}\left(\dfrac{b}{25a}\right)^{5/13}$.

▶ *In the solutions of equations 46–49, the following notation is used:*

$$T_1 = \cosh(\tau + C_2)\cos\tau, \qquad \theta_1 = \cosh\tau - \sin(\tau + C_2),$$
$$T_2 = \tanh(\tau + C_2) + \tan\tau, \qquad \theta_2 = \sinh\tau + \cos(\tau + C_2),$$
$$T_3 = \tanh(\tau + C_2) - \tan\tau, \qquad \theta_3 = \sinh\tau - \cos(\tau + C_2),$$
$$T_4 = 3T_2T_3 - 4; \qquad \theta_4 = 3\theta_2\theta_3 - 2\theta_1^2.$$

46. $y''_{xx} = Axy^{-10/7}(y'_x)^{7/5}$.

1°. Solution in the parametric form with $A < 0$:

$$x = aC_1^9 T_1^{3/2} T_4, \quad y = bC_1^{14} T_1^{7/3} T_2^{7/3}, \qquad \text{where} \quad A = -\frac{5}{9}a^{-2}b^{10/7}\Big(\frac{9a}{28b}\Big)^{2/5}.$$

2°. Solution in the parametric form with $A > 0$:

$$x = aC_1^9 \theta_1^{-1/2}\theta_4, \quad y = bC_1^{14}\theta_2^{7/3}, \qquad \text{where} \quad A = \frac{5}{9}a^{-2}b^{10/7}\Big(\frac{9a}{14b}\Big)^{2/5}.$$

47. $y''_{xx} = Ax^{-5/2}y^{-1/2}(y'_x)^{13/10}$.

Solution in the parametric form:

$$x = aC_1^{-1} T_1^{-1/3} T_2^{2/3}, \quad y = bC_1^9 T_1^3 T_4^2, \qquad \text{where} \quad A = -20a(ab)^{1/2}\Big(\frac{a}{9b}\Big)^{3/10}.$$

48. $y''_{xx} = Ax^{-10/7}y(y'_x)^{8/5}$.

1°. Solution in the parametric form with $A > 0$:

$$x = aC_1^{14} T_1^{7/3} T_2^{7/3}, \quad y = bC_1^9 T_1^{3/2} T_4, \qquad \text{where} \quad A = \frac{5}{9}a^{10/7}b^{-2}\Big(\frac{9b}{28a}\Big)^{2/5}.$$

2°. Solution in the parametric form with $A < 0$:

$$x = aC_1^{14}\theta_2^{7/3}, \quad y = bC_1^9\theta_1^{-1/2}\theta_4, \qquad \text{where} \quad A = -\frac{5}{9}a^{10/7}b^{-2}\Big(\frac{9b}{14a}\Big)^{2/5}.$$

49. $y''_{xx} = Ax^{-1/2}y^{-5/2}(y'_x)^{17/10}$.

Solution in the parametric form:

$$x = aC_1^9 T_1^3 T_4^2, \quad y = bC_1^{-1} T_1^{-1/3} T_2^{2/3}, \qquad \text{where} \quad A = 20b(ab)^{1/2}\Big(\frac{b}{9a}\Big)^{3/10}.$$

▶ *In the solutions of equations 50–65, the following notation is used:*

$$R = \sqrt{\pm(4\tau^3 - 1)}, \quad F_1 = 2\tau I(\tau) + C_2\tau \mp R, \quad F_2 = \tau^{-1}(RF_1 - 1), \quad F_3 = 4\tau F_1^2 \mp F_2^2,$$

where $I(\tau) = \displaystyle\int \frac{\tau\,d\tau}{R}$ *is the incomplete elliptic integral of the second kind in the Weierstrass form.*

50. $y''_{xx} = Ax(y'_x)^{7/4}$.

Solution in the parametric form:

$$x = aC_1^{-3}R, \quad y = bC_1^5\tau^{-1}F_1, \qquad \text{where} \quad A = \mp\frac{2}{3}a^{-2}(\mp\frac{6a}{b})^{3/4}.$$

51. $y''_{xx} = Axy^{-1/2}(y'_x)^{7/4}$.

Solution in the parametric form:

$$x = aC_1^{-1}F_2, \quad y = bC_1^5\tau^2F_1^{-2}, \qquad \text{where} \quad A = \mp\frac{2}{3}a^{-2}b^{1/2}\left(\pm\frac{3a}{b}\right)^{3/4}.$$

52. $y''_{xx} = Ax^{-1/2}y(y'_x)^{8/7}$.

Solution in the parametric form:

$$x = aC_1^{-16}F_3^2, \quad y = bC_1^5F_1^{-3/2}F_2, \qquad \text{where} \quad A = \mp\frac{7}{16}a^{-1/2}b^{-1}\left(\frac{16a}{3b}\right)^{1/7}.$$

53. $y''_{xx} = Ax^{-7/6}y^{-1/2}(y'_x)^{2/3}$.

Solution in the parametric form:

$$x = aC_1^5F_1^{-3}F_3^6, \quad y = bC_1F_1^{-3}(F_2F_3 - 8F_1^2)^2, \qquad \text{where} \quad A = \mp 4a^{-5/6}b^{3/2}\left(\frac{a}{b}\right)^{2/3}.$$

54. $y''_{xx} = Ax^{-3/4}y(y'_x)^{8/7}$.

Solution in the parametric form:

$$x = aC_1^{-32}F_3^{-4}, \qquad y = bC_1^3F_1^{-3/2}(F_2F_3 - 8F_1^2),$$

where $A = \mp\frac{7}{32}a^{-1/4}b^{-1}\left(\frac{32a}{3b}\right)^{1/7}$.

55. $y''_{xx} = Ax^{-4}(y'_x)^3$.

Solution in the parametric form:

$$x = aC_1^2\tau^{-1}, \quad y = bC_1^5\tau^{-1}F_1, \qquad \text{where} \quad A = \pm 6a^5b^{-2}.$$

56. $y''_{xx} = Ay(y'_x)^{5/4}$.

Solution in the parametric form:

$$x = aC_1^5\tau^{-1}F_1, \quad y = bC_1^{-3}R, \qquad \text{where} \quad A = \pm\frac{2}{3}b^{-2}\left(\mp\frac{6b}{a}\right)^{3/4}.$$

57. $y''_{xx} = Ax^{-4}y(y'_x)^3$.

Solution in the parametric form:

$$x = aC_1^3F_1^{-1}, \quad y = bC_1^5\tau F_1^{-1}, \qquad \text{where} \quad A = \pm 6a^5b^{-3}.$$

58. $y''_{xx} = Ax^{-1/2}y(y'_x)^{5/4}$.

Solution in the parametric form:

$$x = aC_1^5\tau^2F_1^{-2}, \quad y = bC_1^{-1}F_2, \qquad \text{where} \quad A = \pm\frac{2}{3}a^{1/2}b^{-2}\Big(\pm\frac{3b}{a}\Big)^{3/4}.$$

59. $y''_{xx} = Ax^{-1/2}y^{-4/3}(y'_x)^3$.

Solution in the parametric form:

$$x = aC_1^4F_2^2, \quad y = bC_1^9F_1^3, \qquad \text{where} \quad A = \mp\frac{4}{3}a^{3/2}b^{-2/3}.$$

60. $y''_{xx} = Ax^{-1/2}y^{-7/6}(y'_x)^3$.

Solution in the parametric form:

$$x = aC_1^5F_1^{-3}F_2^2, \quad y = bC_1^9F_1^{-3}, \qquad \text{where} \quad A = \mp\frac{4}{3}a^{3/2}b^{-5/6}.$$

61. $y''_{xx} = Axy^{-1/2}(y'_x)^{13/7}$.

Solution in the parametric form:

$$x = aC_1^5F_1^{-3/2}F_2, \quad y = bC_1^{-16}F_3^2, \qquad \text{where} \quad A = \pm\frac{7}{16}a^{-1}b^{-1/2}\Big(\frac{16b}{3a}\Big)^{1/7}.$$

62. $y''_{xx} = Ax^{-7}y(y'_x)^3$.

Solution in the parametric form:

$$x = aC_1^3F_1^{1/2}, \quad y = bC_1^8F_3, \qquad \text{where} \quad A = \mp\frac{3}{64}a^8b^{-3}.$$

63. $y''_{xx} = Ax^{-7}y^3(y'_x)^3$.

Solution in the parametric form:

$$x = aC_1^5F_1^{1/2}F_3^{-1}, \quad y = bC_1^8F_3^{-1}, \qquad \text{where} \quad A = \mp\frac{3}{64}a^8b^{-5}.$$

64. $y''_{xx} = Ax^{-1/2}y^{-7/6}(y'_x)^{7/3}$.

Solution in the parametric form:

$$x = aC_1F_1^{-3}(F_2F_3 - 8F_1^2)^2, \quad y = bC_1^5F_1^{-3}F_3^6, \qquad \text{where} \quad A = \pm 4a^{3/2}b^{-5/6}\Big(\frac{b}{a}\Big)^{2/3}.$$

65. $y''_{xx} = Axy^{-3/4}(y'_x)^{13/7}$.

Solution in the parametric form:

$$x = aC_1^3F_1^{-3/2}(F_2F_3 - 8F_1^2), \quad y = bC_1^{-32}F_3^{-4}, \qquad \text{where} \quad A = \pm\frac{7}{32}a^{-1}b^{-1/4}\Big(\frac{32b}{3a}\Big)^{1/7}.$$

▶ *In the solutions of equations 66–95, the following notation is used:*

$$\tau = \int \frac{d\wp}{\sqrt{\pm(4\wp^3 - 1)}} - C_2, \qquad f = \sqrt{\pm(4\wp^3 - 1)}.$$

Function $\wp = \wp(\tau)$ is defined implicitly. The upper sign in the formulae corresponds to the classical elliptic Weierstrass function $\wp = \wp(\tau + C_2,\ 0,\ 1)$. The solutions given below are written in the parametric form. One can assume as the parameter either τ, hence $\wp = \wp(\tau)$, or $\wp$, hence $\tau = \tau(\wp)$.

66. $y''_{xx} = Ax(y'_x)^{5/2}$.

Solution in the parametric form:

$$x = aC_1^{-3} f, \quad y = bC_1\tau, \qquad \text{where} \quad A = \mp\frac{2}{b}\Bigl(\pm\frac{6}{ab}\Bigr)^{1/2}.$$

67. $y''_{xx} = Axy^{-5/4}(y'_x)^{5/2}$.

Solution in the parametric form:

$$x = aC_1^{-1}(\tau f - \wp), \quad y = bC_1^2\tau^4, \qquad \text{where} \quad A = -\frac{1}{2}a^{-1}b^{1/4}\Bigl(\pm\frac{3a}{2b}\Bigr)^{1/2}.$$

68. $y''_{xx} = Ax^{-2/3}y^{-1/2}(y'_x)^{6/5}$.

Solution in the parametric form:

$$x = aC_1^9\tau^{-3}\wp^3, \quad y = bC_1^4(\tau f - \wp)^2, \qquad \text{where} \quad A = \pm\frac{5}{3}a^{-1/3}b^{1/2}\Bigl(\frac{a}{4b}\Bigr)^{1/5}.$$

69. $y''_{xx} = Axy^{-15/8}(y'_x)^{5/2}$.

Solution in the parametric form:

$$x = aC_1^3\tau^{-6}(\tau^3 f + 3\tau^2\wp \mp 1), \quad y = bC_1^4\tau^{-8}, \qquad \text{where} \quad A = \frac{1}{8}a^{-1}b^{7/8}\Bigl(\mp\frac{3a}{b}\Bigr)^{1/2}.$$

70. $y''_{xx} = Ax^{-2/3}y^{-1/2}(y'_x)^{22/15}$.

Solution in the parametric form:

$$x = aC_1^{-1}\tau^3(\tau^2\wp \mp 1)^3, \quad y = bC_1^4\tau^{-12}(\tau^3 f + 3\tau^2\wp \mp 1)^2,$$

where $A = -5a^{-1/3}b^{1/2}\Bigl(\pm\dfrac{a}{4b}\Bigr)^{7/15}$.

71. $y''_{xx} = Axy^{-20/13}(y'_x)^{5/2}$.

Solution in the parametric form:

$$x = aC_1\tau(\tau^3 f - 4\tau^2\wp \pm 6), \quad y = bC_1^{13}\tau^{13}, \qquad \text{where} \quad A = \mp\frac{2}{13}a^{-1}b^{7/13}\Bigl(\pm\frac{6a}{13b}\Bigr)^{1/2}.$$

72. $y''_{xx} = Ax^{-2/3}y^{-1/2}(y'_x)^{27/20}$.

Solution in the parametric form:

$$x = aC_1^{-9}\tau^{-18}(\tau^2\wp \mp 1)^3, \quad y = bC_1\tau^2(\tau^3 f - 4\tau^2\wp \pm 6)^2,$$

where $A = \dfrac{20}{3}a^{-1/3}b^{1/2}\Big(\pm\dfrac{a}{4b}\Big)^{7/20}$.

73. $y''_{xx} = Ax(y'_x)^{8/5}$.

Solution in the parametric form:

$$x = aC_1^{-3}f, \quad y = bC_1^7\wp^{-2}(f \pm 2\tau\wp^2), \qquad \text{where} \quad A = \mp\frac{5}{6}a^{-2}\Big(\frac{3a}{b}\Big)^{3/5}.$$

74. $y''_{xx} = Axy(y'_x)^{8/5}$.

Solution in the parametric form:

$$x = aC_1^{-8}(\tau f + 2\wp), \quad y = bC_1^7\wp(f \pm 2\tau\wp^2)^{-1/2},$$

where $A = \dfrac{10}{3}a^{-2}b^{-1}\Big(\dfrac{3a}{b}\Big)^{3/5}$.

75. $y''_{xx} = Axy^5(y'_x)^{7/5}$.

Solution in the parametric form:

$$x = aC_1^{-27}(\tau^2\wp \mp 1)(f \pm 2\tau\wp^2)^{-1/2}, \quad y = bC_1^8(\tau f + 2\wp)^{-1/3},$$

where $A = -10a^{-2}b^{-5}\Big(\dfrac{a}{b}\Big)^{2/5}$.

76. $y''_{xx} = Ax^{-1/2}y^{-5/2}(y'_x)^{4/5}$.

Solution in the parametric form:

$$x = aC_1^{27}(\tau^2\wp \mp 1)^2(f \pm 2\tau\wp^2)^{-1}, \quad y = bC_1^7(f \pm 2\tau\wp^2)^{-1}(\tau f + 2\wp)^{4/3},$$

where $A = -5a^{-3/2}b^{7/2}\Big(\dfrac{a}{2b}\Big)^{4/5}$.

77. $y''_{xx} = Ax^2(y'_x)^3$.

Solution in the parametric form:

$$x = aC_1^2\wp, \quad y = bC_1^{-1}\tau, \qquad \text{where} \quad A = \mp 6a^{-1}b^{-2}.$$

78. $y''_{xx} = Ay(y'_x)^{1/2}$.

Solution in the parametric form:

$$x = aC_1\tau, \quad y = bC_1^{-3}f, \qquad \text{where} \quad A = \pm\frac{2}{a}\Big(\pm\frac{6}{ab}\Big)^{1/2}.$$

79. $y''_{xx} = Ax^2y^{-5}(y'_x)^3$.

Solution in the parametric form:

$$x = aC_1^3\tau^{-1}\wp, \quad y = bC_1\tau^{-1}, \qquad \text{where} \quad A = \mp 6a^{-1}b^3.$$

80. $y''_{xx} = Ax^{-5/4}y(y'_x)^{1/2}$.

Solution in the parametric form:

$$x = aC_1^2\tau^4, \quad y = bC_1^{-1}(\tau f - \wp), \qquad \text{where} \quad A = \frac{1}{2}a^{1/4}b^{-1}\Big(\pm\frac{3b}{2a}\Big)^{1/2}.$$

81. $y''_{xx} = Ax^{-1/2}y^{-2/3}(y'_x)^{9/5}$.

Solution in the parametric form:

$$x = aC_1^4(\tau f - \wp)^2, \quad y = bC_1^9\tau^{-3}\wp^3, \qquad \text{where} \quad A = \mp\frac{5}{3}a^{1/2}b^{-1/3}\Big(\frac{b}{4a}\Big)^{1/5}.$$

82. $y''_{xx} = Ax^2y^{-15/7}(y'_x)^3$.

Solution in the parametric form:

$$x = aC_1\tau(\tau^2\wp \mp 1), \quad y = bC_1^7\tau^7, \qquad \text{where} \quad A = \mp\frac{6}{49}a^{-1}b^{1/7}.$$

83. $y''_{xx} = Ax^{-15/8}y(y'_x)^{1/2}$.

Solution in the parametric form:

$$x = aC_1^4\tau^{-8}, \quad y = bC_1^3\tau^{-6}(\tau^3 f + 3\tau^2\wp \mp 1), \qquad \text{where} \quad A = -\frac{1}{8}a^{7/8}b^{-1}\Big(\mp\frac{3b}{a}\Big)^{1/2}.$$

84. $y''_{xx} = Ax^{-1/2}y^{-2/3}(y'_x)^{23/15}$.

Solution in the parametric form:

$$x = aC_1^4\tau^{-12}(\tau^3 f + 3\tau^2\wp \mp 1)^2, \quad y = bC_1^{-1}\tau^3(\tau^2\wp \mp 1)^3,$$

where $A = 5a^{1/2}b^{-1/3}\Big(\pm\dfrac{b}{4a}\Big)^{7/15}$.

85. $y''_{xx} = Ax^2y^{-20/7}(y'_x)^3$.

Solution in the parametric form:

$$x = aC_1^6\tau^{-6}(\tau^2\wp \mp 1), \quad y = bC_1^7\tau^{-7}, \qquad \text{where} \quad A = \mp\frac{6}{49}a^{-1}b^{6/7}.$$

86. $y''_{xx} = Ax^{-20/13}y(y'_x)^{1/2}$.

Solution in the parametric form:

$$x = aC_1^{13}\tau^{13}, \quad y = bC_1\tau(\tau^3 f - 4\tau^2\wp \pm 6), \qquad \text{where} \quad A = \pm\frac{2}{13}a^{7/13}b^{-1}\Big(\pm\frac{6b}{13a}\Big)^{1/2}.$$

87. $y''_{xx} = Ax^{-1/2} y^{-2/3} (y'_x)^{33/20}$.

Solution in the parametric form:

$$x = aC_1\tau^2(\tau^3 f - 4\tau^2\wp \pm 6)^2, \quad y = bC_1^{-9}\tau^{-18}(\tau^2\wp \mp 1)^3,$$

where $A = -\dfrac{20}{3} a^{1/2} b^{-1/3} \Big(\pm\dfrac{b}{4a}\Big)^{7/20}$.

88. $y''_{xx} = Ax^{-5/2} (y'_x)^3$.

Solution in the parametric form:

$$x = aC_1^4\wp^{-2}, \quad y = bC_1^7\wp^{-2}(f \pm 2\tau\wp^2), \qquad \text{where} \quad A = \pm 3a^{7/2}b^{-2}.$$

89. $y''_{xx} = Ay (y'_x)^{7/5}$.

Solution in the parametric form:

$$x = aC_1^7\wp^{-2}(f \pm 2\tau\wp^2), \quad y = bC_1^{-3} f, \qquad \text{where} \quad A = \pm\frac{5}{6} b^{-2} \Big(\frac{3b}{a}\Big)^{3/5}.$$

90. $y''_{xx} = Ax^{-5/2} y^{-1/2} (y'_x)^3$.

Solution in the parametric form:

$$x = aC_1^3(f \pm 2\tau\wp^2)^{-1}, \quad y = bC_1^7\wp^2(f \pm 2\tau\wp^2)^{-1}, \qquad \text{where} \quad A = \pm 3a^{7/2}b^{-3/2}.$$

91. $y''_{xx} = Axy (y'_x)^{7/5}$.

Solution in the parametric form:

$$x = aC_1^7\wp(f \pm 2\tau\wp^2)^{-1/2}, \quad y = bC_1^{-8}(\tau f + 2\wp), \quad \text{where} \quad A = -\frac{10}{3} a^{-1} b^{-2} \Big(\frac{3b}{a}\Big)^{3/5}.$$

92. $y''_{xx} = Ax^{-5/3} y^{-1/2} (y'_x)^3$.

Solution in the parametric form:

$$x = aC_1^9(f \pm 2\tau\wp^2)^{3/2}, \quad y = bC_1^{16}(\tau f + 2\wp)^2, \qquad \text{where} \quad A = \tfrac{1}{6} a^{8/3} b^{-3/2}.$$

93. $y''_{xx} = Ax^{-5/3} y^{-5/6} (y'_x)^3$.

Solution in the parametric form:

$$x = aC_1^7(f \pm 2\tau\wp^2)^{3/2}(\tau f + 2\wp)^{-2}, \quad y = bC_1^{16}(\tau f + 2\wp)^{-2},$$

where $A = \tfrac{1}{6} a^{8/3} b^{-7/6}$.

94. $y''_{xx} = Ax^5 y (y'_x)^{8/5}$.

Solution in the parametric form:

$$x = aC_1^8(\tau f + 2\wp)^{-1/3}, \quad y = bC_1^{-27}(\tau^2\wp \mp 1)(f \pm 2\tau\wp^2)^{-1/2},$$

where $A = 10a^{-5}b^{-2}\Big(\dfrac{b}{a}\Big)^{2/5}$.

95. $y''_{xx} = Ax^{-5/2} y^{-1/2} (y'_x)^{11/5}$.

Solution in the parametric form:

$$x = aC_1^7(f \pm 2\tau\wp^2)^{-1}(\tau f + 2\wp)^{4/3}, \quad y = bC_1^{27}(\tau^2\wp \mp 1)^2(f \pm 2\tau\wp^2)^{-1},$$

where $A = 5a^{7/2}b^{-3/2}\Big(\dfrac{b}{2a}\Big)^{4/5}$.

▶ *In the solutions of equations 96–97, the following notation is used:*

$$Z = \begin{cases} C_1 J_\nu(\tau) + C_2 Y_\nu(\tau) & \text{for the upper sign,} \\ C_1 I_\nu(\tau) + C_2 K_\nu(\tau) & \text{for the lower sign,} \end{cases}$$

where J_ν and Y_ν are Bessel functions, I_ν and K_ν are modified Bessel functions.

96. $y''_{xx} = Axy^m (y'_x)^3$.

Solution in the parametric form with $m \neq -2$:

$$x = \tau^\nu Z, \quad y = b\tau^{2\nu}, \qquad \text{where} \quad \nu = \frac{1}{m+2}, \quad A = \pm\Bigl(\frac{m+2}{2b}\Bigr)^2.$$

See 2.5.2.28 for the case $m = -2$.

97. $y''_{xx} = Ax^{-1/2} y^{-1/2} (y'_x)^l$.

Solution in the parametric form with $l \neq 3/2$:

$$x = a\tau^{2\nu} Z^2, \quad y = b\tau^{-2\nu}(\tau Z'_\tau + \nu Z)^2,$$

where $\nu = \dfrac{1-l}{3-2l}$, $A = \dfrac{1}{3-2l}\Bigl(\mp\dfrac{b}{a}\Bigr)^{\frac{3}{2}-l}$. See 2.5.2.29 for the case $l = 3/2$.

▶ *In the solutions of equations 98–106, the following notation is used:*

$$Z = \begin{cases} C_1 J_{1/3}(\tau) + C_2 Y_{1/3}(\tau) & \text{for the upper sign,} \\ C_1 I_{1/3}(\tau) + C_2 K_{1/3}(\tau) & \text{for the lower sign,} \end{cases}$$

where $J_{1/3}$ and $Y_{1/3}$ are Bessel functions, $I_{1/3}$ and $K_{1/3}$ are modified Bessel functions;

$$U_1 = \tau Z'_\tau + \tfrac{1}{3} Z, \quad U_2 = U_1^2 \pm \tau^2 Z^2, \quad U_3 = \pm\tfrac{2}{3}\tau^2 Z^3 - 2U_1 U_2.$$

98. $y''_{xx} = Axy^{-1/2} (y'_x)^{3/2}$.

Solution in the parametric form:

$$x = a\tau^{-2/3} Z^{-1} U_1, \quad y = \tau^{-4/3} U_2^2, \qquad \text{where} \quad A = -\frac{2}{a}\Bigl(\mp\frac{3}{a}\Bigr)^{1/2}.$$

99. $y''_{xx} = Axy^{-2} (y'_x)^{3/2}$.

Solution in the parametric form:

$$x = a\tau^{-4/3} Z^{-1} U_3, \quad y = b\tau^{-2/3} U_2, \qquad \text{where} \quad A = -\frac{b}{a^2}\Bigl(\pm 3ab\Bigr)^{1/2}.$$

100. $y''_{xx} = Ax^{-2} y^{-1/2} (y'_x)^{3/2}$.

Solution in the parametric form:

$$x = a\tau^{-4/3} Z^{-2} U_2, \quad y = \tau^{-8/3} Z^{-2} U_3^2, \qquad \text{where} \quad A = \pm\tfrac{2}{3} a^{3/2}.$$

101. $y''_{xx} = Ax^{-1/2}y^{-2}(y'_x)^3.$

Solution in the parametric form:

$$x = a\tau^{-4/3}Z^{-2}U_1^2, \quad y = b\tau^{-2/3}Z^{-2}, \qquad \text{where} \quad A = \pm\tfrac{1}{3}a^{3/2}.$$

102. $y''_{xx} = Ax^{-2}y(y'_x)^3.$

Solution in the parametric form:

$$x = a\tau^{2/3}Z^2, \quad y = b\tau^{-2/3}U_2, \qquad \text{where} \quad A = \frac{9}{2}\left(\frac{a}{b}\right)^3.$$

103. $y''_{xx} = Ax^{-1/2}y(y'_x)^{3/2}.$

Solution in the parametric form:

$$x = \tau^{-4/3}U_2^2, \quad y = b\tau^{-2/3}Z^{-1}U_1, \qquad \text{where} \quad A = \frac{2}{b}\left(\mp\frac{3}{b}\right)^{1/2}.$$

104. $y''_{xx} = Ax^{-2}y^{-2}(y'_x)^3.$

Solution in the parametric form:

$$x = a\tau^{4/3}Z^2U_2^{-1}, \quad y = \tau^{2/3}Z^{-1}U_2^{-1}, \qquad \text{where} \quad A = \tfrac{9}{2}a^3.$$

105. $y''_{xx} = Ax^{-2}y(y'_x)^{3/2}.$

Solution in the parametric form:

$$x = a\tau^{-2/3}U_2, \quad y = b\tau^{-4/3}Z^{-1}U_3, \qquad \text{where} \quad A = \frac{a}{b^2}\left(\pm 3ab\right)^{1/2}.$$

106. $y''_{xx} = Ax^{-1/2}y^{-2}(y'_x)^{3/2}.$

Solution in the parametric form:

$$x = \tau^{-8/3}Z^{-2}U_3^2, \quad y = b\tau^{-4/3}Z^{-2}U_2, \qquad \text{where} \quad A = \mp\tfrac{2}{3}b^{3/2}.$$

107. $y''_{xx} = Ax^n y^{-n-3}(y'_x)^{\frac{3n+4}{2n+3}}.$

In the books by Zaitsev & Polyanin (1993, 1994), it was shown that this equation is reducible to the Riccati equation whose solution is expressed in terms of associated Legendre functions.

2.5.3. Some Formulae and Transformations

For the sake of visualization, we use the symbolic notation

$$\{n,\ m,\ l\}$$

to denote the generalized Emden—Fowler equation

$$y''_{xx} = Ax^n y^m (y'_x)^l.$$

Hereinafter we omit the insignificant parameter A (which can be reduced to ± 1 by scaling the variables in accordance with the rule $x \to ax$, $y \to by$, selecting appropriate constants a and b).

1. With $m+l \neq 1$, the generalized Emden—Fowler equation has a particular solution:

$$y = Bx^{\frac{n+2-l}{1-m-l}}, \qquad \text{where} \quad B = \left(\frac{n+2-l}{1-m-l}\right)^{\frac{1-l}{m+l-1}} \left[\frac{n+m+1}{A(1-m-l)}\right]^{\frac{1}{m+l-1}}.$$

2. Assuming y as the independent variable and x as the dependent one, we obtain the generalized Emden—Fowler equation for function $x = x(y)$ with the parameters changed:

$$x''_{yy} = -Ay^m x^n (x'_y)^{3-l}.$$

Denote this transformation as $\mathcal{F}$ and represent it as follows:

$$\{n,\ m,\ l\} \quad \leftarrow - - - - \rightarrow \quad \{m,\ n,\ 3-l\} \qquad \textit{transformation } \mathcal{F}.$$

Twofold transformation $\mathcal{F}$ yields the original equation.

3. With $m \neq 0$, $n \neq -1$, and $l \neq 1$, the transformation

$$t = (y'_x)^{1-l}, \qquad w = x^{n+1}$$

leads to the generalized Emden—Fowler equation for function $w = w(t)$ with the parameters changed:

$$w''_{tt} = Bt^{\frac{1}{1-l}} w^{-\frac{n}{n+1}} (w'_t)^{\frac{2m+1}{m}},$$

where $B = -\dfrac{m}{n+1}\left[\dfrac{A(1-l)}{n+1}\right]^{\frac{1}{m}}$. Denote this transformation as $\mathcal{G}$ and represent it as follows:

$$\{n,\ m,\ l\} \quad \longmapsto \quad \left\{\frac{1}{1-l},\ -\frac{n}{n+1},\ \frac{2m+1}{m}\right\} \qquad \textit{transformation } \mathcal{G}.$$

Threefold transformation $\mathcal{G}$ yields the original equation.

When obtained the solution of the transformed equation in the form $w = w(t)$, the solution of the original equation can be written in the parametric form as

$$x = w^{\frac{1}{n+1}}, \quad y = k(w'_t)^{-\frac{1}{m}},$$

where $k = \left[\dfrac{n+1}{A(1-l)}\right]^{\frac{1}{m}}$.

Different compositions of transformations $\mathcal{F}$ and $\mathcal{G}$ generate six different generalized Emden—Fowler equations whose parameters are shown in Figure 1.

4. In the particular case $l = 0$, the transformation $y = w/t$, $x = 1/t$ leads to the Emden—Fowler equation with the independent variable to a different power:

$$w''_{tt} = At^{-n-m-3} w^m.$$

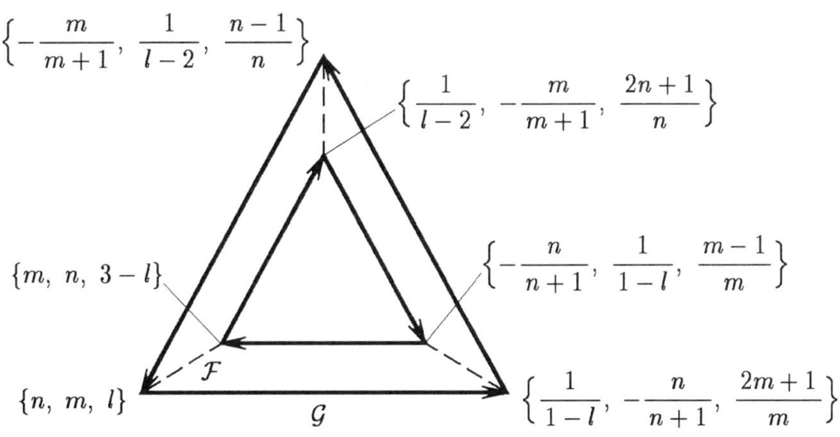

FIGURE 1

Denote this transformation as $\mathcal{G}$ and represent it as follows:

$$\{n,\ m,\ 0\} \longleftrightarrow \{-n-m-3,\ m,\ 0\} \qquad \textit{transformation } \mathcal{H}.$$

With $l = 0$, different compositions of transformations $\mathcal{F}$, $\mathcal{G}$, and $\mathcal{H}$ generate twelve different generalized Emden—Fowler equations whose parameters are shown in Figure 2.

With $l = 0$ and $n = 1$, different compositions of transformations $\mathcal{F}$, $\mathcal{G}$, and $\mathcal{H}$ generate twenty four different generalized Emden—Fowler equations whose parameters are presented in Figure 3.

5. The substitution

$$z = \frac{x}{y} y'_x, \qquad v = A x^{n-l+2} y^{m+l-1}$$

reduces the generalized Emden—Fowler equation to the equation

$$(z^l v - z^2 + z) v'_z = [(m+l-1)z + n - l + 2] v.$$

Furthermore, using the substitution

$$\xi = v - z^{2-l} + z^{1-l},$$

we obtain the Abel equation

$$\xi \xi'_z = [(m+2l-3)z + n - 2l + 3] z^{-l} \xi + [(m+l-1)z^2 + (n-m-2l+3)z - n + l - 2] z^{1-2l}.$$

2.6. Equations of the Form $y''_{xx} = A_1 x^{n_1} y^{m_1} (y'_x)^{l_1} + A_2 x^{n_2} y^{m_2} (y'_x)^{l_2}$

2.6.1. Modified Emden—Fowler Equation $y''_{xx} = A_1 x^{-1} y'_x + A_2 x^n y^m$

See Section 2.3 for the case $A_1 = 0$.

For the sake of clearness, below in this subsection we use the conventional notation $x y''_{xx} - k y'_x = A x^{n+1} y^m$ for the modified Emden—Fowler equation.

The classification Table 2.10 represents all solvable equations whose solutions are outlined in Subsection 2.6.1. Equations are arranged in accordance with the growth of parameter m. The number of the equation sought is indicated in the last column in this table.

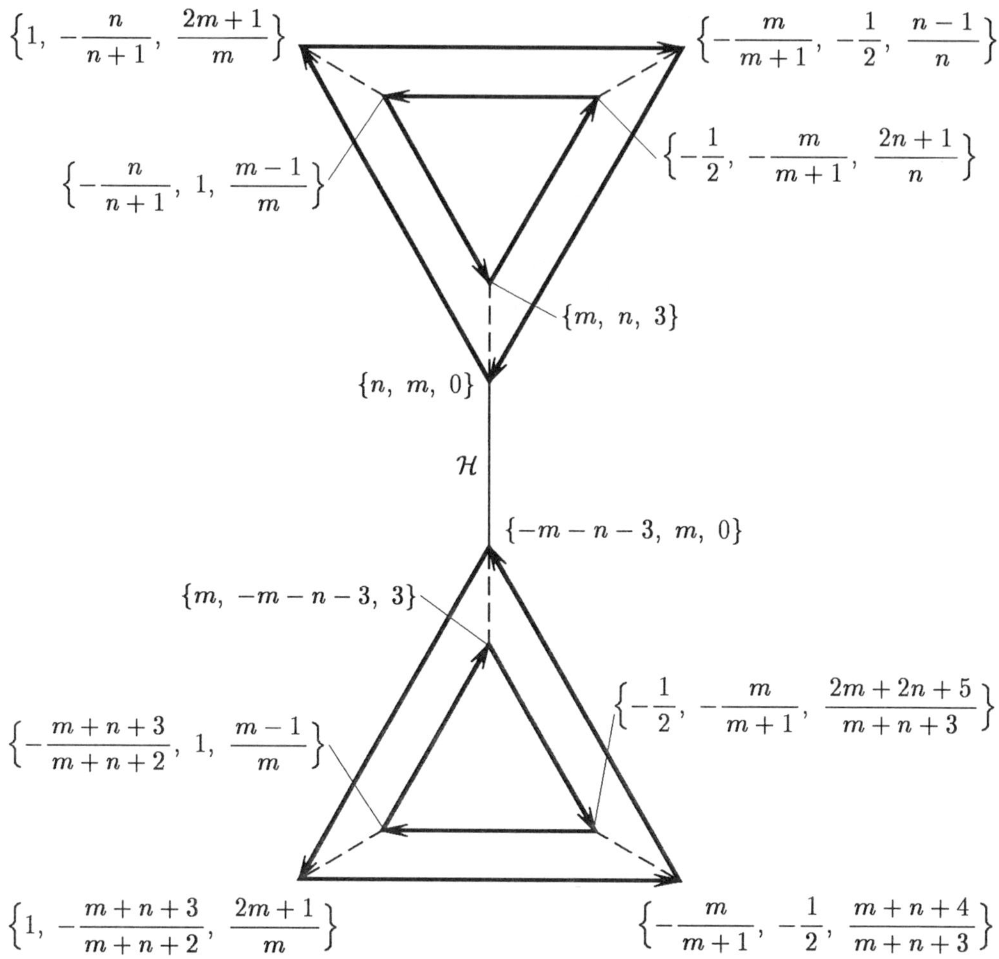

FIGURE 2

1. $\boldsymbol{xy''_{xx} - \frac{n}{2}y'_x = Ax^{n+1}y^m, \qquad m \neq -1, \quad n \neq -2.}$

Solution in the parametric form:

$$x = aC_1^{1-m}\left[\int (1 \pm \tau^{m+1})^{-1/2}\,d\tau + C_2\right]^{\frac{2}{n+2}}, \qquad y = bC_1^{n+2}\tau,$$

where $A = \pm\frac{1}{8}(m+1)(n+2)^2 a^{-n-2} b^{1-m}$.

2. $\boldsymbol{xy''_{xx} + \frac{n+m+3}{m+1}y'_x = Ax^{n+1}y^m, \qquad m \neq -1, \quad n \neq -2.}$

Solution in the parametric form:

$$x = aC_1^{1-m}\left[\int \left(1 \pm \tau^{m+1}\right)^{-1/2} d\tau + C_2\right]^{\frac{m+1}{n+2}},$$

$$y = bC_1^{n+2}\tau\left[\int \left(1 \pm \tau^{m+1}\right)^{-1/2} d\tau + C_2\right]^{-1},$$

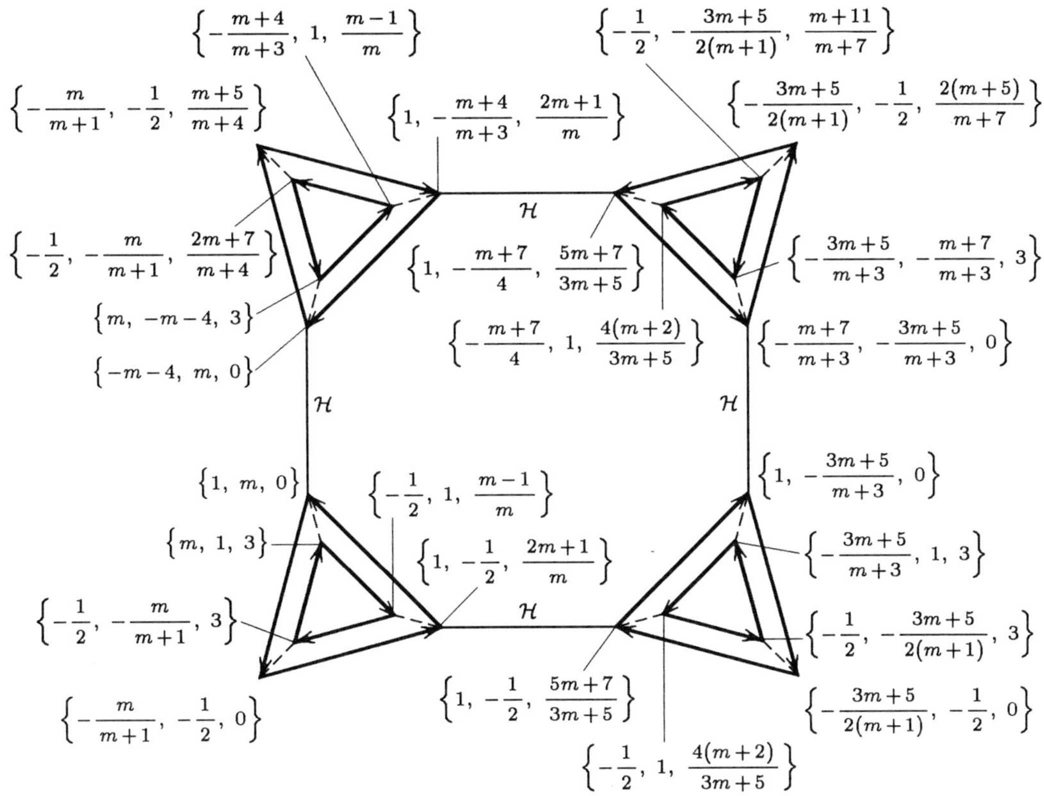

FIGURE 3

where $A = \pm\dfrac{(n+2)^2}{2(m+1)} a^{-n-2} b^{1-m}$.

3. $\displaystyle xy''_{xx} + \frac{2n+m+3}{m-1} y'_x = Ax^{n+1} y^m, \qquad m \neq -1, \quad n \neq -2.$

Solution in the parametric form:

$$x = \exp\left[\frac{1-m}{n+2} C_2 \int \left(C_1 + \frac{1}{4}\tau^2 + \frac{2B}{m+1}\tau^{m+1}\right)^{-1/2} d\tau\right],$$

$$y = \tau \exp\left[C_2 \int \left(C_1 + \frac{1}{4}\tau^2 + \frac{2B}{m+1}\tau^{m+1}\right)^{-1/2} d\tau\right],$$

where $A = \dfrac{4(n+2)^2}{(m-1)^2} B$.

4. $\displaystyle xy''_{xx} - \frac{n}{2} y'_x = Ax^{n+1} y^{-1}, \qquad n \neq -2.$

Solution in the parametric form:

$$x = aC_1^2 \left[\int \exp(\mp\tau^2)\, d\tau + C_2\right]^{\frac{2}{n+2}}, \qquad y = bC_1^{n+2} \exp(\mp\tau^2),$$

where $A = \mp\frac{1}{2}(n+2)^2 a^{-n-2} b^2$.

TABLE 2.10
Solvable cases of the modified Emden—Fowler equations
$$xy''_{xx} - ky'_x = Ax^{n+1}y^m$$

No	m	n	k	Equation
1	arbitrary ($m \neq -1$)	arbitrary ($n \neq -2$)	$\frac{1}{2}n$	2.6.1.1
2	arbitrary ($m \neq -1$)	arbitrary ($n \neq -2$)	$-\dfrac{n+m+3}{m+1}$	2.6.1.2
3	arbitrary ($m \neq -1$)	arbitrary ($n \neq -2$)	$\dfrac{2n+m+3}{1-m}$	2.6.1.3
4	arbitrary ($m \neq -1$)	-2	-1	2.6.1.6
5	-7	arbitrary ($n \neq -2$)	$\frac{1}{3}(n-1)$	2.6.1.45
6	-7	arbitrary ($n \neq -2$)	$\frac{1}{5}(n-3)$	2.6.1.46
7	-4	arbitrary ($n \neq -2$)	$\frac{1}{2}n$	2.6.1.40
8	-4	arbitrary ($n \neq -2$)	$\frac{1}{3}(n-1)$	2.6.1.42
9	-4	-2	-1	2.6.1.41
10	-2	arbitrary ($n \neq -2$)	$\frac{1}{3}(n-1)$	2.6.1.28
11	-2	-2	arbitrary ($k \neq -1$)	2.6.1.29
12	$-\frac{5}{2}$	arbitrary ($n \neq -2$)	$\frac{1}{2}n$	2.6.1.35
13	$-\frac{5}{2}$	arbitrary ($n \neq -2$)	$\frac{1}{3}(2n+1)$	2.6.1.37
14	$-\frac{5}{2}$	-2	-1	2.6.1.36
15	$-\frac{5}{3}$	arbitrary ($n \neq -2$)	$-3n-7$	2.6.1.14
16	$-\frac{5}{3}$	arbitrary ($n \neq -2$)	$\frac{1}{2}n$	2.6.1.8
17	$-\frac{5}{3}$	arbitrary ($n \neq -2$)	$\frac{1}{2}(3n+4)$	2.6.1.9
18	$-\frac{5}{3}$	arbitrary ($n \neq -2$)	$\frac{1}{3}(n-1)$	2.6.1.13
19	$-\frac{5}{3}$	arbitrary ($n \neq -2$)	$\frac{1}{3}(2n+1)$	2.6.1.38
20	$-\frac{5}{3}$	arbitrary ($n \neq -2$)	$\frac{1}{4}(n-2)$	2.6.1.18

TABLE 2.10 *Continued*
Solvable cases of the modified Emden—Fowler equation
$$xy''_{xx} - ky'_x = Ax^{n+1}y^m$$

No	m	n	k	Equation
21	$-\frac{5}{3}$	arbitrary ($n \neq -2$)	$-\frac{1}{4}(3n+10)$	2.6.1.19
22	$-\frac{5}{3}$	arbitrary ($n \neq -2$)	$\frac{1}{7}(6n+5)$	2.6.1.39
23	$-\frac{5}{3}$	-2	-1	2.6.1.22
24	$-\frac{7}{5}$	arbitrary ($n \neq -2$)	$\frac{1}{3}(n-1)$	2.6.1.24
25	$-\frac{7}{5}$	arbitrary ($n \neq -2$)	$-\frac{1}{3}(5n+13)$	2.6.1.25
26	-1	arbitrary ($n \neq -2$)	$n+1$	2.6.1.5
27	-1	arbitrary ($n \neq -2$)	$\frac{1}{2}n$	2.6.1.4
28	-1	-2	arbitrary ($k \neq -1$)	2.6.1.7
29	-1	-2	-1	2.6.1.20
30	$-\frac{1}{2}$	arbitrary ($n \neq -2$)	$-2n-5$	2.6.1.12
31	$-\frac{1}{2}$	arbitrary ($n \neq -2$)	$\frac{1}{2}n$	2.6.1.11
32	$-\frac{1}{2}$	arbitrary ($n \neq -2$)	$\frac{1}{2}(3n+4)$	2.6.1.43
33	$-\frac{1}{2}$	arbitrary ($n \neq -2$)	$\frac{1}{3}(n-1)$	2.6.1.16
34	$-\frac{1}{2}$	arbitrary ($n \neq -2$)	$\frac{1}{3}(2n+1)$	2.6.1.26
35	$-\frac{1}{2}$	arbitrary ($n \neq -2$)	$-\frac{1}{3}(2n+7)$	2.6.1.17
36	$-\frac{1}{2}$	arbitrary ($n \neq -2$)	$\frac{1}{5}(6n+7)$	2.6.1.44
37	$-\frac{1}{2}$	-2	arbitrary ($k \neq -1$)	2.6.1.27
38	$-\frac{1}{2}$	-2	-1	2.6.1.21
39	$\frac{1}{2}$	arbitrary ($n \neq -2$)	$\frac{1}{2}n$	2.6.1.15
40	$\frac{1}{2}$	arbitrary ($n \neq -2$)	$-\frac{1}{3}(2n+7)$	2.6.1.10
41	$\frac{1}{2}$	-2	-1	2.6.1.23

TABLE 2.10 *Continued*
Solvable cases of the modified Emden—Fowler equation
$xy''_{xx} - ky'_x = Ax^{n+1}y^m$

No	m	n	k	Equation
42	2	arbitrary $(n \neq -2)$	$-7n - 15$	2.6.1.33
43	2	arbitrary $(n \neq -2)$	$\frac{1}{2}n$	2.6.1.30
44	2	arbitrary $(n \neq -2)$	$-\frac{1}{3}(n+5)$	2.6.1.32
45	2	arbitrary $(n \neq -2)$	$-\frac{1}{6}(7n+20)$	2.6.1.34
46	2	-2	-1	2.6.1.31

5. $\quad xy''_{xx} - (n+1)y'_x = Ax^{n+1}y^{-1}, \qquad n \neq -2.$

Solution in the parametric form:

$$x = \exp\left\{\frac{2C_2}{n+2}\int\left[C_1 + \frac{1}{4}\tau^2 + \frac{2A}{(n+2)^2}\ln|\tau|\right]^{-1/2} d\tau\right\},$$
$$y = \tau\exp\left\{C_2\int\left[C_1 + \frac{1}{4}\tau^2 + \frac{2A}{(n+2)^2}\ln|\tau|\right]^{-1/2} d\tau\right\}.$$

6. $\quad xy''_{xx} + y'_x = Ax^{-1}y^m, \qquad m \neq -1.$

Solution in the parametric form:

$$x = C_2\exp\left[\int (C_1 \pm \tau^{m+1})^{-1/2}\,d\tau\right], \qquad y = b\tau,$$

where $A = \pm\frac{1}{2}b^{1-m}(m+1)$.

7. $\quad xy''_{xx} - ky'_x = Ax^{-1}y^{-1}, \qquad k \neq -1.$

Solution in the parametric form:

$$x = \left\{\int\left[\frac{2A}{(k+1)^2}\ln\tau + C_1\right]^{-1/2} d\tau + C_2\right\}^{-\frac{1}{k+1}},$$
$$y = \tau\left\{\int\left[\frac{2A}{(k+1)^2}\ln\tau + C_1\right]^{-1/2} d\tau + C_2\right\}^{-1}.$$

8. $\quad xy''_{xx} - \frac{n}{2}y'_x = Ax^{n+1}y^{-5/3}, \qquad n \neq -2.$

Solution in the parametric form:

$$x = aC_1^8(\tau^3 \pm 3\tau + C_2)^{\frac{2}{n+2}}, \qquad y = bC_1^{3n+6}(\tau^2 \pm 1)^{3/2},$$

where $A = \pm\frac{1}{12}a^{-n-2}b^{8/3}(n+2)^2$.

9. $xy''_{xx} - \dfrac{3n+4}{2} y'_x = Ax^{n+1} y^{-5/3}, \qquad n \neq -2.$

Solution in the parametric form:

$$x = aC_1^8 (\tau^3 \pm 3\tau + C_2)^{-\frac{2}{3n+6}}, \qquad y = bC_1^{3n+6} (\tau^2 \pm 1)^{3/2} (\tau^3 \pm 3\tau + C_2)^{-1},$$

where $A = \pm \frac{3}{4} a^{-n-2} b^{8/3} (n+2)^2$.

10. $xy''_{xx} + \dfrac{2n+7}{3} y'_x = Ax^{n+1} y^{1/2}, \qquad n \neq -2.$

Solution in the parametric form:

$$x = aC_1 \left[\int \frac{\tau\, d\tau}{\sqrt{\pm(4\tau^3 - 1)}} + C_2 \right]^{\frac{3}{2n+4}}, \qquad y = bC_1^{2n+4} \tau^2 \left[\int \frac{\tau\, d\tau}{\sqrt{\pm(4\tau^3 - 1)}} + C_2 \right]^{-1},$$

where $A = \pm \frac{16}{3} a^{-n-2} b^{1/2} (n+2)^2$.

11. $xy''_{xx} - \dfrac{n}{2} y'_x = Ax^{n+1} y^{-1/2}, \qquad n \neq -2.$

Solution in the parametric form:

$$x = aC_1^3 (\tau^3 - 3\tau + C_2)^{\frac{2}{n+2}}, \qquad y = bC_1^{2n+4} (\tau^2 - 1)^2,$$

where $A = \pm \frac{1}{9} (n+2)^2 a^{-n-2} b^{3/2}$.

12. $xy''_{xx} + (2n+5) y'_x = Ax^{n+1} y^{-1/2}, \qquad n \neq -2.$

Solution in the parametric form:

$$x = aC_1^3 (\tau^3 - 3\tau + C_2)^{\frac{1}{2n+4}}, \qquad y = bC_1^{2n+4} (\tau^2 - 1)^2 (\tau^3 - 3\tau + C_2)^{-1},$$

where $A = \pm \frac{16}{9} (n+2)^2 a^{-n-2} b^{3/2}$.

13. $xy''_{xx} - \dfrac{n-1}{3} y'_x = Ax^{n+1} y^{-5/3}, \qquad n \neq -2.$

Solution in the parametric form:

$$x = aC_1^8 \left[\pm(\tau^4 - 6\tau^2 + 4C_2\tau - 3) \right]^{\frac{3}{n+2}}, \qquad y = bC_1^{3n+6} (\tau^3 - 3\tau + C_2)^{3/2},$$

where $A = \pm \frac{1}{64} (n+2)^2 a^{-n-2} b^{8/3}$.

14. $xy''_{xx} + (3n+7) y'_x = Ax^{n+1} y^{-5/3}, \qquad n \neq -2.$

Solution in the parametric form:

$$x = aC_1^8 \left[\pm(\tau^4 - 6\tau^2 + 4C_2\tau - 3) \right]^{\frac{1}{3n+6}},$$
$$y = \pm bC_1^{3n+6} (\tau^3 - 3\tau + C_2)^{3/2} (\tau^4 - 6\tau^2 + 4C_2\tau - 3)^{-1},$$

where $A = \pm \frac{81}{64} (n+2)^2 a^{-n-2} b^{8/3}$.

15. $xy''_{xx} - \frac{n}{2}y'_x = Ax^{n+1}y^{1/2}, \qquad n \neq -2.$

Solution in the parametric form:

$$x = aC_1\left[\int \frac{\tau\,d\tau}{\sqrt{\pm(4\tau^3-1)}} + C_2\right]^{\frac{2}{n+2}}, \qquad y = bC_1^{2n+4}\tau^2,$$

where $A = \pm 3a^{-n-2}b^{1/2}(n+2)^2$.

16. $xy''_{xx} - \frac{n-1}{3}y'_x = Ax^{n+1}y^{-1/2}, \qquad n \neq -2.$

Solution in the parametric form:

$$x = \left[C_1 e^{2s\tau} + C_2 e^{-s\tau}\sin\left(\sqrt{3}\,s\tau\right)\right]^{\frac{3}{n+2}},$$
$$y = \left\{2C_1 s e^{2s\tau} + C_2 s e^{-s\tau}\left[\sqrt{3}\cos\left(\sqrt{3}\,s\tau\right) - \sin\left(\sqrt{3}\,s\tau\right)\right]\right\}^2,$$

where $A = \frac{16}{9}s^3(n+2)^2$.

17. $xy''_{xx} + \frac{2n+7}{3}y'_x = Ax^{n+1}y^{-1/2}, \qquad n \neq -2.$

Solution in the parametric form:

$$x = \left[C_1 e^{2s\tau} + C_2 e^{-s\tau}\sin\left(\sqrt{3}\,s\tau\right)\right]^{\frac{3}{2n+4}},$$
$$y = \frac{\left\{2C_1 s e^{2s\tau} + C_2 s e^{-s\tau}\left[\sqrt{3}\cos\left(\sqrt{3}\,s\tau\right) - \sin\left(\sqrt{3}\,s\tau\right)\right]\right\}^2}{C_1 e^{2s\tau} + C_2 e^{-s\tau}\sin\left(\sqrt{3}\,s\tau\right)},$$

where $A = \frac{64}{9}s^3(n+2)^2$.

18. $xy''_{xx} - \frac{n-2}{4}y'_x = Ax^{n+1}y^{-5/3}, \qquad n \neq -2.$

1°. Solution in the parametric form with $A < 0$:

$$x = aC_1^8\left[\cosh(\tau + C_2)\cos\tau\right]^{\frac{4}{n+2}}\left[\tanh(\tau + C_2) + \tan\tau\right]^{\frac{4}{n+2}},$$
$$y = bC_1^{3n+6}\left[\cosh(\tau + C_2)\cos\tau\right]^{3/2},$$

where $A = -\frac{3}{256}a^{-n-2}b^{8/3}(n+2)^2$.

2°. Solution in the parametric form with $A > 0$:

$$x = aC_1^8\left[\sinh\tau + \cos(\tau + C_2)\right]^{\frac{4}{n+2}},$$
$$y = bC_1^{3n+6}\left[\cosh\tau - \sin(\tau + C_2)\right]^{3/2},$$

where $A = \frac{3}{64}a^{-n-2}b^{8/3}(n+2)^2$.

19. $xy''_{xx} + \dfrac{3n+10}{4} y'_x = Ax^{n+1} y^{-5/3}, \qquad n \neq -2.$

1°. Solution in the parametric form with $A < 0$:

$$x = aC_1^8 \bigl[\cosh(\tau + C_2)\cos\tau\bigr]^{\frac{4}{3n+6}} \bigl[\tanh(\tau + C_2) + \tan\tau\bigr]^{\frac{4}{3n+6}},$$
$$y = bC_1^{3n+6} \bigl[\cosh(\tau + C_2)\cos\tau\bigr]^{1/2} \bigl[\tanh(\tau + C_2) + \tan\tau\bigr]^{-1},$$

where $A = -\frac{27}{256} a^{-n-2} b^{8/3} (n+2)^2$.

2°. Solution in the parametric form with $A > 0$:

$$x = aC_1^8 \bigl[\sinh\tau + \cos(\tau + C_2)\bigr]^{\frac{4}{3n+6}},$$
$$y = bC_1^{3n+6} \bigl[\cosh\tau - \sin(\tau + C_2)\bigr]^{3/2} \bigl[\sinh\tau + \cos(\tau + C_2)\bigr]^{-1},$$

where $A = \frac{27}{64} a^{-n-2} b^{8/3} (n+2)^2$.

20. $xy''_{xx} + y'_x = Ax^{-1} y^{-1}.$

Solution in the parametric form:

$$x = C_2 \exp\left[\int (2A\ln|\tau| + C_1)^{-1/2}\,d\tau\right], \qquad y = \tau.$$

21. $xy''_{xx} + y'_x = Ax^{-1} y^{-1/2}.$

Solution in the parametric form:

$$x = \exp(\pm\tau^3 - 3C_1\tau + C_2), \quad y = b(\pm\tau^2 - C_1)^2, \qquad \text{where} \quad A = \pm\tfrac{4}{9} b^{3/2}.$$

22. $xy''_{xx} + y'_x = Ax^{-1} y^{-5/3}.$

Solution in the parametric form:

$$x = \exp(C_1\tau^3 \pm 3\tau + C_2), \qquad y = (\pm 3A/C_1)^{3/8} (C_1\tau^2 \pm 1)^{3/2}.$$

23. $xy''_{xx} + y'_x = Ax^{-1} y^{1/2}.$

Solution in the parametric form:

$$x = C_1 \exp\left[\int \frac{\tau\,d\tau}{\sqrt{\pm(4\tau^3 - C_1)}}\right], \qquad y = b\tau^2,$$

where $A = \pm 12 b^{1/2}$.

► *In the solutions of equations 24–25, the following notation is used:*

$$S_1 = C_1 e^{2s\tau} + C_2 e^{-s\tau} \sin\bigl(\sqrt{3}\,s\tau\bigr),$$
$$S_2 = 2C_1 s e^{2s\tau} + C_2 s e^{-s\tau} \bigl[\sqrt{3}\cos\bigl(\sqrt{3}\,s\tau\bigr) - \sin\bigl(\sqrt{3}\,s\tau\bigr)\bigr],$$
$$S_3 = S_2^2 - 2S_1 (S_2)'_\tau.$$

24. $xy''_{xx} - \dfrac{n-1}{3} y'_x = Ax^{n+1} y^{-7/5}, \qquad n \neq -2.$

Solution in the parametric form:

$$x = aS_3^{\frac{3}{n+2}}, \quad y = bS_1^{5/2}, \qquad \text{where} \quad A = -\tfrac{5}{9216} a^{-n-2} b^{12/5} s^{-6} (n+2)^2.$$

25. $xy''_{xx} + \dfrac{5n+13}{3} y'_x = Ax^{n+1} y^{-7/5}, \qquad n \neq -2.$

Solution in the parametric form:

$$x = aS_3^{\frac{3}{5n+10}}, \quad y = bS_1^{5/2} S_3^{-1}, \qquad \text{where} \quad A = -\tfrac{125}{9216} a^{-n-2} b^{12/5} s^{-6} (n+2)^2.$$

▶ *In the solutions of equations 26–29, the following notation is used:*

$$Z = \begin{cases} C_1 J_{1/3}(\tau) + C_2 Y_{1/3}(\tau) & \textit{for the upper sign,} \\ C_1 I_{1/3}(\tau) + C_2 K_{1/3}(\tau) & \textit{for the lower sign,} \end{cases}$$

where $J_{1/3}$ and $Y_{1/3}$ are Bessel functions, $I_{1/3}$ and $K_{1/3}$ are modified Bessel functions.

26. $xy''_{xx} - \dfrac{2n+1}{3} y'_x = Ax^{n+1} y^{-1/2}, \qquad n \neq -2.$

Solution in the parametric form:

$$x = aC_1^3 \tau^{\frac{1}{n+2}} Z^{\frac{3}{n+2}}, \qquad y = bC_1^{2n+4} \tau^{-2/3} (\tau Z'_\tau + \tfrac{1}{3} Z)^2,$$

where $A = \mp \frac{4}{27} a^{-n-2} b^{3/2} (n+2)^2$.

27. $xy''_{xx} - ky'_x = Ax^{-1} y^{-1/2}, \qquad k \neq -1.$

Solution in the parametric form:

$$x = C_1 (\tau^{1/3} Z)^{-\frac{2}{k+1}}, \qquad y = b\tau^{-4/3} Z^{-2} (\tau Z'_\tau + \tfrac{1}{3} Z)^2,$$

where $A = \mp \frac{1}{3} b^{3/2} (k+1)^2$.

28. $xy''_{xx} - \dfrac{n-1}{3} y'_x = Ax^{n+1} y^{-2}, \qquad n \neq -2.$

Solution in the parametric form:

$$x = aC_1^3 \tau^{-\frac{2}{n+2}} \left[(\tau Z'_\tau + \tfrac{1}{3} Z)^2 \pm \tau^2 Z^2 \right]^{\frac{3}{n+2}}, \qquad y = bC_1^{n+2} \tau^{2/3} Z^2,$$

where $A = -\frac{1}{2} a^{-n-2} b^3 (n+2)^2$.

29. $xy''_{xx} - ky'_x = Ax^{-1} y^{-2}, \qquad k \neq -1.$

Solution in the parametric form:

$$x = C_1 \tau^{\frac{2}{3k+3}} \left[(\tau Z'_\tau + \tfrac{1}{3} Z)^2 \pm \tau^2 Z^2 \right]^{-\frac{1}{k+1}}, \qquad y = b\tau^{4/3} Z^2 \left[(\tau Z'_\tau + \tfrac{1}{3} Z)^2 \pm \tau^2 Z^2 \right]^{-1},$$

where $A = -\frac{9}{2} b^3 (k+1)^2$.

► *In the solutions of equations 30–39, function $\wp$ is defined implicitly:*

$$\tau = \int \frac{d\wp}{\sqrt{\pm(4\wp^3 - 1)}} - C_2, \qquad f = \sqrt{\pm(4\wp^3 - 1)}.$$

The upper sign in the formulae corresponds to the classical elliptic Weierstrass function $\wp = \wp(\tau + C_2,\ 0,\ 1)$. The solutions outlined below are written in the parametric form—one can assume both τ as the parameter, hence $\wp = \wp(\tau)$, and $\wp$, hence $\tau = \tau(\wp)$.

30. $xy''_{xx} - \frac{n}{2} y'_x = Ax^{n+1} y^2, \qquad n \neq -2.$

Solution in the parametric form:

$$x = aC_1^{-1} \tau^{\frac{2}{n+2}}, \quad y = bC_1^{n+2} \wp, \qquad \text{where} \quad A = \pm \tfrac{3}{2} a^{-n-2} b^{-1} (n+2)^2.$$

31. $xy''_{xx} + y'_x = Ax^{-1} y^2.$

Solution in the parametric form:

$$x = C_2 e^{\tau}, \qquad y = b\wp(\tau,\ 0,\ C_1),$$

where $A = \pm 6b^{-1}$, and the elliptic Weierstrass function $\wp = \wp(\tau,\ 0,\ C_1)$ is defined implicitly by the integral $\tau = \int_{\infty}^{\wp} (4z^3 - C_1)^{-1/2}\, dz$.

32. $xy''_{xx} + \frac{n+5}{3} y'_x = Ax^{n+1} y^2, \qquad n \neq -2.$

Solution in the parametric form:

$$x = aC_1^{-1} \tau^{\frac{3}{n+2}}, \quad y = bC_1^{n+2} \tau^{-1} \wp, \qquad \text{where} \quad A = \pm \tfrac{2}{3} a^{-n-2} b^{-1} (n+2)^2.$$

33. $xy''_{xx} + (7n + 15) y'_x = Ax^{n+1} y^2, \qquad n \neq -2.$

Solution in the parametric form:

$$x = aC_1^{-1} \tau^{-\frac{1}{n+2}}, \quad y = bC_1^{n+2} \tau (\tau^2 \wp \mp 1), \qquad \text{where} \quad A = \pm 6a^{-n-2} b^{-1} (n+2)^2.$$

34. $xy''_{xx} + \frac{7n+20}{6} y'_x = Ax^{n+1} y^2, \qquad n \neq -2.$

Solution in the parametric form:

$$x = aC_1^{-1} \tau^{\frac{6}{7(n+2)}}, \quad y = bC_1^{n+2} \tau^{-6} (\tau^2 \wp \mp 1),$$

where $A = \pm \frac{1}{6} a^{-n-2} b^{-1} (n+2)^2$.

35. $xy''_{xx} - \frac{n}{2} y'_x = Ax^{n+1} y^{-5/2}, \qquad n \neq -2.$

Solution in the parametric form:

$$x = aC_1^7 \wp^{-\frac{4}{n+2}} (f \pm 2\tau\wp^2)^{\frac{2}{n+2}}, \quad y = bC_1^{2n+4} \wp^{-2},$$

where $A = \mp \frac{3}{4} a^{-n-2} b^{7/2} (n+2)^2$.

36. $xy''_{xx} + y'_x = Ax^{-1}y^{-5/2}$.

Solution in the parametric form:

$$x = C_2 \exp[\wp^{-2}(f \pm 2\tau\wp^2)], \qquad y = b\wp^{-2},$$

where $A = \mp 3b^{7/2}$, and the elliptic Weierstrass function $\wp = \wp(\tau,\ 0,\ C_1)$ is defined implicitly by the integral $\tau = \int_{\infty}^{\wp} (4z^3 - C_1)^{-1/2}\,dz$.

37. $xy''_{xx} - \dfrac{2n+1}{3}y'_x = Ax^{n+1}y^{-5/2}, \qquad n \neq -2.$

Solution in the parametric form:

$$x = aC_1^7\wp^{\frac{3}{n+2}}(f \pm 2\tau\wp^2)^{-\frac{3}{2n+4}}, \quad y = bC_1^{2n+4}(f \pm 2\tau\wp^2)^{-1},$$

where $A = \mp\frac{4}{3}a^{-n-2}b^{7/2}(n+2)^2$.

38. $xy''_{xx} - \dfrac{2n+1}{3}y'_x = Ax^{n+1}y^{-5/3}, \qquad n \neq -2.$

Solution in the parametric form:

$$x = aC_1^8(\tau f + 2\wp)^{\frac{3}{n+2}}, \quad y = bC_1^{3n+6}(f \pm 2\tau\wp^2)^{3/2},$$

where $A = -\frac{2}{27}a^{-n-2}b^{8/3}(n+2)^2$.

39. $xy''_{xx} - \dfrac{6n+5}{7}y'_x = Ax^{n+1}y^{-5/3}, \qquad n \neq -2.$

Solution in the parametric form:

$$x = aC_1^8(\tau f + 2\wp)^{-\frac{7}{3n+6}}, \quad y = bC_1^{3n+6}(f \pm 2\tau\wp^2)^{3/2}(\tau f + 2\wp)^{-2},$$

where $A = -\frac{6}{49}a^{-n-2}b^{8/3}(n+2)^2$.

▶ *In the solutions of equations 40–46, the following notation is used:*

$$R = \sqrt{\pm(4\tau^3 - 1)}, \quad F_1 = 2\tau I(\tau) + C_2\tau \mp R, \quad F_2 = \tau^{-1}(RF_1 - 1),$$

where $I(\tau) = \int \tau R^{-1}\,d\tau$ *is the incomplete elliptic integral of the second kind in the Weierstrass form.*

40. $xy''_{xx} - \dfrac{n}{2}y'_x = Ax^{n+1}y^{-4}, \qquad n \neq -2.$

Solution in the parametric form:

$$x = aC_1^5(\tau^{-1}F_1)^{\frac{2}{n+2}}, \quad y = bC_1^{n+2}\tau^{-1}, \qquad \text{where} \quad A = \mp\tfrac{3}{2}a^{-n-2}b^5(n+2)^2.$$

41. $xy''_{xx} + y'_x = Ax^{-1}y^{-4}.$

Solution in the parametric form:

$$x = C_2 \exp\left[2\int \frac{\tau\,d\tau}{\sqrt{\pm(4\tau^3 - C_1)}} + C_2 \mp \frac{1}{\tau}\sqrt{\pm(4\tau^3 - C_1)}\right], \quad y = \mp(AC_1^2/6)^{1/5}\tau^{-1}.$$

42. $xy''_{xx} - \dfrac{n-1}{3}y'_x = Ax^{n+1}y^{-4}, \qquad n \neq -2.$

Solution in the parametric form:

$$x = aC_1^5(\tau F_1^{-1})^{\frac{3}{n+2}}, \quad y = bC_1^{n+2}F_1^{-1}, \qquad \text{where} \quad A = \mp\tfrac{2}{3}a^{-n-2}b^5(n+2)^2.$$

43. $xy''_{xx} - \dfrac{3n+4}{2}y'_x = Ax^{n+1}y^{-1/2}, \qquad n \neq -2.$

Solution in the parametric form:

$$x = aC_1^3 F_1^{\frac{2}{n+2}}, \quad y = bC_1^{2n+4}F_2^2, \qquad \text{where} \quad A = \pm 3a^{-n-2}b^{3/2}(n+2)^2.$$

44. $xy''_{xx} - \dfrac{6n+7}{5}y'_x = Ax^{n+1}y^{-1/2}, \qquad n \neq -2.$

Solution in the parametric form:

$$x = aC_1^3 F_1^{-\frac{5}{2n+4}}, \quad y = bC_1^{2n+4}F_1^{-3}F_2^2, \qquad \text{where} \quad A = \pm\tfrac{48}{25}a^{-n-2}b^{3/2}(n+2)^2.$$

45. $xy''_{xx} - \dfrac{n-1}{3}y'_x = Ax^{n+1}y^{-7}, \qquad n \neq -2.$

Solution in the parametric form:

$$x = aC_1^8(4\tau F_1^2 \mp F_2^2)^{\frac{3}{n+2}}, \quad y = bC_1^{n+2}F_1^{1/2}, \qquad \text{where} \quad A = \pm\tfrac{1}{192}a^{-n-2}b^8(n+2)^2.$$

46. $xy''_{xx} - \dfrac{n-3}{5}y'_x = Ax^{n+1}y^{-7}, \qquad n \neq -2.$

Solution in the parametric form:

$$x = aC_1^8(4\tau F_1^2 \mp F_2^2)^{-\frac{5}{n+2}}, \quad y = bC_1^{n+2}F_1^{1/2}(4\tau F_1^2 \mp F_2^2)^{-1},$$

where $A = \pm\frac{3}{1600}a^{-n-2}b^8(n+2)^2$.

2.6.2. Equations of the Form $y''_{xx} = (A_1 x^{n_1} y^{m_1} + A_2 x^{n_2} y^{m_2})(y'_x)^l$

See Section 2.4 for the case $l = 0$.

Table 2.11 represents all solvable equations whose solutions are outlined in Subsection 2.6.2. Equations are arranged in accordance with the growth of l, the growth of m_1 (for identical l), the growth of m_2 (for identical l and m_1, $m_1 \geq m_2$), the growth of n_1 (for identical l, m_1, and m_2), and the growth of n_2 (for identical l, m_1, m_2, and n_1). The number of the equation sought is indicated in the last column in this table.

TABLE 2.11
Solvable cases of the equation $y''_{xx} = (A_1 x^{n_1} y^{m_1} + A_2 x^{n_2} y^{m_2})(y'_x)^l$

l	m_1	m_2	n_1	n_2	A_1	A_2	Equation
Any ($l \neq 2$)	Any ($m_1 \neq -1$)	Any ($m_2 \neq -1$)	0	0	Any	Any	2.6.2.1
$\frac{m_1+2n_1+3}{m_1+n_1+2}$	Any	Any	Any	$\frac{m_2(n_1+1)-m_1+n_1}{m_1+1}$	Any	Any	2.6.2.98
Any ($l \neq 1$)	0	0	Any ($n_1 \neq -1$)	Any ($n_2 \neq -1$)	Any	Any	2.6.2.5
Any ($l \neq 2$)	Any ($m_1 \neq -1$)	-1	0	0	Any	Any	2.6.2.2
Any ($l \neq 2$)	0	0	Any ($n_1 \neq -1$)	-1	Any	Any	2.6.2.6
$\frac{3m_1+5}{2m_1+3}$	Any	$-m_1-2$	1	0	Any	Any	2.6.2.21
$\frac{m_1+5}{m_1+3}$	Any	$\frac{m_1-1}{2}$	1	0	Any	Any	2.6.2.94
$\frac{3n_1+4}{n_1+1}$	1	0	Any	$-n_1-2$	Any	Any	2.6.2.22
$\frac{2(n_1+2)}{n_1+3}$	1	0	Any	$\frac{n_1-1}{2}$	Any	Any	2.6.2.95
Any ($l \neq 1, 2$)	1	0	0	1	Any	Any	2.6.2.20
$\frac{1}{2}$	$-\frac{1}{4}$	$-\frac{7}{4}$	0	1	Any	Any	2.6.2.71
$\frac{1}{2}$	1	0	$-\frac{15}{8}$	$-\frac{7}{4}$	Any	Any	2.6.2.81
$\frac{1}{2}$	1	0	$-\frac{15}{8}$	$-\frac{13}{8}$	Any	Any	2.6.2.66
$\frac{1}{2}$	1	0	$-\frac{20}{13}$	$-\frac{15}{13}$	Any	Any	2.6.2.68
$\frac{1}{2}$	1	0	$-\frac{20}{13}$	$-\frac{14}{13}$	Any	Any	2.6.2.84
$\frac{1}{2}$	1	0	$-\frac{5}{4}$	$-\frac{3}{4}$	Any	Any	2.6.2.64
$\frac{1}{2}$	1	0	$-\frac{5}{4}$	$-\frac{1}{2}$	Any	Any	2.6.2.78
$\frac{1}{2}$	1	0	0	1	Any	Any	2.6.2.62
$\frac{1}{2}$	1	0	0	2	Any	Any	2.6.2.75
1	0	0	Any ($n_1 \neq -1$)	Any ($n_2 \neq -1$)	Any	Any	2.6.2.7
1	0	0	Any ($n_1 \neq -1$)	-1	Any	Any	2.6.2.8
1	0	-2	0	1	Any	Any	2.6.2.25
1	1	0	0	1	Any	Any	2.6.2.23
$\frac{3}{2}$	Any	Any	m_1	m_2	Any	Any	2.6.2.97

TABLE 2.11 *Continued*
Solvable cases of the equation $y''_{xx} = (A_1 x^{n_1} y^{m_1} + A_2 x^{n_2} y^{m_2})(y'_x)^l$

l	m_1	m_2	n_1	n_2	A_1	A_2	Equation
$\frac{3}{2}$	0	-2	0	1	Any	Any	2.6.2.107
$\frac{3}{2}$	0	$-\frac{1}{2}$	0	1	Any	Any	2.6.2.105
$\frac{3}{2}$	1	0	-2	0	Any	Any	2.6.2.108
$\frac{3}{2}$	1	0	$-\frac{1}{2}$	0	Any	Any	2.6.2.106
2	Any $(m_1 \neq -1)$	Any $(m_2 \neq -1)$	0	0	Any	Any	2.6.2.3
2	Any $(m_1 \neq -1)$	-1	0	0	Any	Any	2.6.2.4
2	1	0	-2	0	Any	Any	2.6.2.26
2	1	0	0	1	Any	Any	2.6.2.24
$\frac{5}{2}$	$-\frac{7}{4}$	$-\frac{15}{8}$	0	1	Any	Any	2.6.2.80
$\frac{5}{2}$	$-\frac{13}{8}$	$-\frac{15}{8}$	0	1	Any	Any	2.6.2.65
$\frac{5}{2}$	$-\frac{15}{13}$	$-\frac{20}{13}$	0	1	Any	Any	2.6.2.67
$\frac{5}{2}$	$-\frac{14}{13}$	$-\frac{20}{13}$	0	1	Any	Any	2.6.2.83
$\frac{5}{2}$	$-\frac{3}{4}$	$-\frac{5}{4}$	0	1	Any	Any	2.6.2.63
$\frac{5}{2}$	$-\frac{1}{2}$	$-\frac{5}{4}$	0	1	Any	Any	2.6.2.77
$\frac{5}{2}$	1	0	$-\frac{7}{4}$	$-\frac{1}{4}$	Any	Any	2.6.2.72
$\frac{5}{2}$	1	0	0	1	Any	Any	2.6.2.61
$\frac{5}{2}$	2	0	0	1	Any	Any	2.6.2.74
3	Any	Any	$-m_1-3$	$-m_2-3$	Any	Any	2.6.2.9
3	Any	Any	$-2m_1-3$	$-2m_2-3$	Any	Any	2.6.2.93
3	Any $(m_1 \neq -2)$	Any	1	0	Any	Any	2.6.2.49
3	Any	-3	$-m_1-3$	0	Any	Any	2.6.2.19
3	Any $(m_1 \neq -2)$	0	1	-3	Any	Any	2.6.2.51
3	-5	-6	1	3	Any	Any	2.6.2.100
3	-4	-5	0	2	Any	Any	2.6.2.76
3	-3	-5	0	1	Any	Any	2.6.2.44
3	-3	-5	0	2	Any	Any	2.6.2.58
3	-3	$-\frac{7}{2}$	0	$-\frac{1}{2}$	Any	Any	2.6.2.28
3	$-\frac{14}{5}$	$-\frac{18}{5}$	2	3	Any	Any	2.6.2.112

TABLE 2.11 *Continued*

Solvable cases of the equation $y''_{xx} = (A_1 x^{n_1} y^{m_1} + A_2 x^{n_2} y^{m_2})(y'_x)^l$

l	m_1	m_2	n_1	n_2	A_1	A_2	Equation
3	$-\frac{8}{3}$	$-\frac{10}{3}$	$-\frac{1}{3}$	$-\frac{5}{3}$	Any	Any	2.6.2.38
3	$-\frac{5}{2}$	-4	$-\frac{1}{2}$	0	Any	Any	2.6.2.86
3	$-\frac{5}{2}$	$-\frac{7}{2}$	$-\frac{1}{2}$	$-\frac{1}{2}$	Any	Any	2.6.2.13
3	$-\frac{5}{2}$	-3	$-\frac{1}{2}$	0	Any	Any	2.6.2.32
3	$-\frac{12}{5}$	$-\frac{13}{5}$	$-\frac{3}{5}$	$-\frac{7}{5}$	Any	Any	2.6.2.30
3	$-\frac{7}{3}$	$-\frac{10}{3}$	$-\frac{5}{3}$	$-\frac{5}{3}$	Any	Any	2.6.2.34
3	$-\frac{7}{3}$	$-\frac{10}{3}$	$-\frac{5}{3}$	$\frac{1}{3}$	Any	Any	2.6.2.88
3	$-\frac{7}{3}$	-3	$-\frac{2}{3}$	0	Any	Any	2.6.2.70
3	$-\frac{7}{3}$	$-\frac{8}{3}$	$-\frac{5}{3}$	$-\frac{1}{3}$	Any	Any	2.6.2.42
3	$-\frac{11}{5}$	$-\frac{12}{5}$	2	3	Any	Any	2.6.2.113
3	-2	$-n_2-1$	1	Any	$\frac{2(n_2+1)}{(n_2+3)^2}$	Any	2.6.2.132
3	-2	-3	-2	0	Any	Any	2.6.2.104
3	-2	-3	-1	0	Any	Any	2.6.2.12
3	-2	-3	1	2	$-\frac{6}{25}$	Any	2.6.2.145
3	-2	-3	1	2	$\frac{6}{25}$	Any	2.6.2.144
3	-2	-3	$-\frac{1}{2}$	0	Any	Any	2.6.2.102
3	-2	-2	1	Any	$\frac{2(n_2+1)}{(n_2+3)^2}$	Any	2.6.2.116
3	-2	-2	1	-7	$-\frac{15}{4}$	Any	2.6.2.117
3	-2	-2	1	-4	-6	Any	2.6.2.118
3	-2	-2	1	$-\frac{5}{2}$	-12	Any	2.6.2.119
3	-2	-2	1	-2	-2	Any	2.6.2.120
3	-2	-2	1	$-\frac{5}{3}$	$-\frac{63}{4}$	Any	2.6.2.124
3	-2	-2	1	$-\frac{5}{3}$	$-\frac{3}{4}$	Any	2.6.2.123
3	-2	-2	1	$-\frac{5}{3}$	$\frac{9}{100}$	Any	2.6.2.122
3	-2	-2	1	$-\frac{5}{3}$	$\frac{3}{16}$	Any	2.6.2.121
3	-2	-2	1	$-\frac{7}{5}$	$\frac{5}{36}$	Any	2.6.2.125
3	-2	-2	1	$-\frac{1}{2}$	-20	Any	2.6.2.128
3	-2	-2	1	$-\frac{1}{2}$	$\frac{4}{25}$	Any	2.6.2.127
3	-2	-2	1	$-\frac{1}{2}$	$\frac{2}{9}$	Any	2.6.2.126

TABLE 2.11 *Continued*

Solvable cases of the equation $y''_{xx} = (A_1 x^{n_1} y^{m_1} + A_2 x^{n_2} y^{m_2})(y'_x)^l$

l	m_1	m_2	n_1	n_2	A_1	A_2	Equation
3	-2	-2	1	$\frac{1}{2}$	$\frac{12}{49}$	Any	2.6.2.129
3	-2	-2	2	1	Any	$-\frac{6}{25}$	2.6.2.131
3	-2	-2	2	1	Any	$\frac{6}{25}$	2.6.2.130
3	$-\frac{13}{7}$	$-\frac{20}{7}$	0	2	Any	Any	2.6.2.82
3	$-\frac{12}{7}$	$-\frac{20}{7}$	0	2	Any	Any	2.6.2.60
3	$-\frac{5}{3}$	$-\frac{7}{3}$	$-\frac{4}{3}$	$-\frac{5}{3}$	Any	Any	2.6.2.92
3	$-\frac{8}{5}$	$-\frac{13}{5}$	$-\frac{7}{5}$	$-\frac{7}{5}$	Any	Any	2.6.2.109
3	$-\frac{3}{2}$	$-\frac{5}{2}$	0	$-\frac{1}{2}$	Any	Any	2.6.2.90
3	$-\frac{3}{2}$	-2	$-\frac{3}{2}$	-2	Any	Any	2.6.2.48
3	$-\frac{3}{2}$	-2	0	$-\frac{1}{2}$	Any	Any	2.6.2.46
3	$-\frac{3}{2}$	-2	$\frac{1}{2}$	1	Any	$\frac{12}{49}$	2.6.2.143
3	$-\frac{4}{3}$	$-\frac{10}{3}$	$-\frac{5}{3}$	$-\frac{5}{3}$	Any	Any	2.6.2.36
3	$-\frac{4}{3}$	$-\frac{8}{3}$	$-\frac{5}{3}$	$-\frac{1}{3}$	Any	Any	2.6.2.40
3	$-\frac{4}{3}$	$-\frac{7}{3}$	$-\frac{5}{3}$	$-\frac{5}{3}$	Any	Any	2.6.2.14
3	$-\frac{4}{3}$	$-\frac{4}{3}$	0	$-\frac{1}{2}$	Any	Any	2.6.2.15
3	$-\frac{9}{7}$	$-\frac{15}{7}$	0	2	Any	Any	2.6.2.59
3	$-\frac{7}{6}$	$-\frac{5}{3}$	$-\frac{1}{2}$	0	Any	Any	2.6.2.16
3	$-\frac{8}{7}$	$-\frac{15}{7}$	0	2	Any	Any	2.6.2.79
3	-1	-2	-2	-2	Any	Any	2.6.2.110
3	$-\frac{2}{3}$	$-\frac{4}{3}$	$-\frac{5}{3}$	$-\frac{5}{3}$	Any	Any	2.6.2.115
3	$-\frac{1}{2}$	-3	$-\frac{1}{2}$	0	Any	Any	2.6.2.53
3	$-\frac{1}{2}$	-2	$-\frac{1}{2}$	1	Any	-20	2.6.2.141
3	$-\frac{1}{2}$	-2	$-\frac{1}{2}$	1	Any	$\frac{4}{25}$	2.6.2.134
3	$-\frac{1}{2}$	-2	$-\frac{1}{2}$	1	Any	$\frac{2}{9}$	2.6.2.137
3	$-\frac{1}{2}$	$-\frac{3}{2}$	$-\frac{1}{2}$	0	Any	Any	2.6.2.45
3	0	-5	-3	1	Any	Any	2.6.2.52
3	0	-2	-3	-2	Any	Any	2.6.2.56
3	0	-2	0	$-\frac{1}{2}$	Any	Any	2.6.2.54
3	0	$-\frac{3}{2}$	$-\frac{1}{2}$	0	Any	Any	2.6.2.89
3	0	$-\frac{2}{3}$	$-\frac{5}{3}$	$-\frac{5}{3}$	Any	Any	2.6.2.114

TABLE 2.11 *Continued*
Solvable cases of the equation $y''_{xx} = (A_1 x^{n_1} y^{m_1} + A_2 x^{n_2} y^{m_2})(y'_x)^l$

l	m_1	m_2	n_1	n_2	A_1	A_2	Equation
3	0	$-\frac{1}{2}$	0	$-\frac{1}{2}$	Any	Any	2.6.2.101
3	0	0	$-\frac{1}{3}$	$-\frac{5}{3}$	Any	Any	2.6.2.39
3	0	0	0	-1	Any	Any	2.6.2.11
3	0	0	0	$-\frac{2}{3}$	Any	Any	2.6.2.69
3	0	0	0	$-\frac{1}{2}$	Any	Any	2.6.2.31
3	0	0	2	0	Any	Any	2.6.2.57
3	$\frac{2}{5}$	-2	$-\frac{7}{5}$	1	Any	$\frac{5}{36}$	2.6.2.139
3	$\frac{2}{3}$	-2	$-\frac{5}{3}$	1	Any	$-\frac{63}{4}$	2.6.2.147
3	$\frac{2}{3}$	-2	$-\frac{5}{3}$	1	Any	$-\frac{3}{4}$	2.6.2.136
3	$\frac{2}{3}$	-2	$-\frac{5}{3}$	1	Any	$\frac{9}{100}$	2.6.2.135
3	$\frac{2}{3}$	-2	$-\frac{5}{3}$	1	Any	$\frac{3}{16}$	2.6.2.138
3	1	-2	-2	1	Any	-2	2.6.2.133
3	1	0	-7	-3	Any	Any	2.6.2.17
3	1	0	-4	-3	Any	Any	2.6.2.10
3	1	0	-2	-3	Any	Any	2.6.2.55
3	1	0	-2	$-\frac{3}{2}$	Any	Any	2.6.2.47
3	1	0	-2	0	Any	Any	2.6.2.103
3	1	0	$-\frac{5}{3}$	$-\frac{4}{3}$	Any	Any	2.6.2.91
3	1	0	$-\frac{5}{3}$	$-\frac{1}{3}$	Any	Any	2.6.2.41
3	1	0	$-\frac{5}{3}$	$\frac{1}{3}$	Any	Any	2.6.2.87
3	1	0	$-\frac{7}{5}$	$-\frac{3}{5}$	Any	Any	2.6.2.29
3	1	0	$-\frac{1}{2}$	0	Any	Any	2.6.2.27
3	1	0	0	$-\frac{1}{2}$	Any	Any	2.6.2.85
3	1	0	0	2	Any	Any	2.6.2.73
3	1	0	1	-3	Any	Any	2.6.2.50
3	1	0	1	0	Any	Any	2.6.2.43
3	1	0	1	3	Any	Any	2.6.2.99
3	$\frac{3}{2}$	-2	$-\frac{5}{2}$	1	Any	-12	2.6.2.146
3	2	0	-5	-5	Any	Any	2.6.2.96
3	2	0	$-\frac{5}{3}$	$-\frac{5}{3}$	Any	Any	2.6.2.35

TABLE 2.11 *Continued*
Solvable cases of the equation $y''_{xx} = (A_1 x^{n_1} y^{m_1} + A_2 x^{n_2} y^{m_2})(y'_x)^l$

l	m_1	m_2	n_1	n_2	A_1	A_2	Equation
3	2	0	$-\frac{5}{3}$	$-\frac{1}{3}$	Any	Any	2.6.2.37
3	2	1	$-\frac{5}{3}$	$-\frac{5}{3}$	Any	Any	2.6.2.33
3	3	-2	-4	1	Any	-6	2.6.2.140
3	3	0	-7	-3	Any	Any	2.6.2.18
3	4	3	-7	-7	Any	Any	2.6.2.111
3	6	-2	-7	1	Any	$-\frac{15}{4}$	2.6.2.142

1. $y''_{xx} = (A_1 y^{m_1} + A_2 y^{m_2})(y'_x)^l, \qquad l \neq 2, \quad m_1 \neq -1, \quad m_2 \neq -1.$

1°. Solution in the parametric form:

$$x = a \int \left(C_1 + \tau^{m_1+1} \pm \tau^{m_2+1}\right)^{\frac{1}{l-2}} d\tau + C_2, \qquad y = b\tau,$$

where $A_1 = \dfrac{m_1+1}{2-l} a^{l-2} b^{1-m_1-l}$, $A_2 = \pm \dfrac{m_2+1}{2-l} a^{l-2} b^{1-m_2-l}$.

2°. Solution in the parametric form:

$$x = a \int \left(C_1 - \tau^{m_1+1} \pm \tau^{m_2+1}\right)^{\frac{1}{l-2}} d\tau + C_2, \qquad y = b\tau,$$

where $A_1 = \dfrac{m_1+1}{l-2} a^{l-2} b^{1-m_1-l}$, $A_2 = \pm \dfrac{m_2+1}{2-l} a^{l-2} b^{1-m_2-l}$.

2. $y''_{xx} = (A_1 y^m + A_2 y^{-1})(y'_x)^l, \qquad l \neq 2, \quad m \neq -1.$

Solution:

$$x = \int \left[C_1 + \frac{A_1(2-l)}{m+1} y^{m+1} + (2-l) A_2 \ln y\right]^{\frac{1}{l-2}} dy + C_2.$$

3. $y''_{xx} = (A_1 y^{m_1} + A_2 y^{m_2})(y'_x)^2, \qquad m_1 \neq -1, \quad m_2 \neq -1.$

Solution:

$$x = C_1 \int \exp\left(-\frac{A_1}{m_1+1} y^{m_1+1} - \frac{A_2}{m_2+1} y^{m_2+1}\right) dy + C_2.$$

4. $y''_{xx} = (A_1 y^m + A_2 y^{-1})(y'_x)^2, \qquad m \neq -1.$

Solution:

$$x = C_1 \int y^{-A_2} \exp\left(-\frac{A_1}{m+1} y^{m+1}\right) dy + C_2.$$

5. $y''_{xx} = (A_1x^{n_1} + A_2x^{n_2})(y'_x)^l, \qquad l \neq 1, \quad n_1 \neq -1, \quad n_2 \neq -1.$

1°. Solution in the parametric form:

$$x = a\tau, \qquad y = b\int \left(C_1 + \tau^{n_1+1} \pm \tau^{n_2+1}\right)^{\frac{1}{1-l}} d\tau + C_2,$$

where $A_1 = \dfrac{n_1+1}{1-l}a^{l-n_1-2}b^{1-l}$, $A_2 = \pm\dfrac{n_2+1}{1-l}a^{l-n_2-2}b^{1-l}$.

2°. Solution in the parametric form:

$$x = a\tau, \qquad y = b\int \left(C_1 - \tau^{n_1+1} \pm \tau^{n_2+1}\right)^{\frac{1}{1-l}} d\tau + C_2,$$

where $A_1 = \dfrac{n_1+1}{l-1}a^{l-n_1-2}b^{1-l}$, $A_2 = \pm\dfrac{n_2+1}{1-l}a^{l-n_2-2}b^{1-l}$.

6. $y''_{xx} = (A_1x^{n} + A_2x^{-1})(y'_x)^l, \qquad l \neq 1, \quad n \neq -1.$

Solution:

$$y = \int \left[C_1 + \frac{A_1(1-l)}{n+1}x^{n+1} + (1-l)A_2\ln x\right]^{\frac{1}{1-l}} dx + C_2.$$

7. $y''_{xx} = (A_1x^{n_1} + A_2x^{n_2})y'_x, \qquad n_1 \neq -1, \quad n_2 \neq -1.$

Solution:

$$y = C_1\int \exp\left(\frac{A_1}{n_1+1}x^{n_1+1} + \frac{A_2}{n_2+1}x^{n_2+1}\right) dx + C_2.$$

8. $y''_{xx} = (A_1x^{n} + A_2x^{-1})y'_x, \qquad n \neq -1.$

Solution:

$$y = C_1\int x^{A_2}\exp\left(\frac{A_1}{n+1}x^{n+1}\right) dx + C_2.$$

9. $y''_{xx} = (A_1x^{-m_1-3}y^{m_1} + A_2x^{-m_2-3}y^{m_2})(y'_x)^3, \qquad m_1 \neq -2, \quad m_2 \neq -2.$

1°. Solution in the parametric form:

$$x = a\tau\left[\int \left(C_1 + \tau^{-m_1-2} \pm \tau^{-m_2-2}\right)^{-1/2} d\tau + C_2\right]^{-1},$$

$$y = b\left[\int \left(C_1 + \tau^{-m_1-2} \pm \tau^{-m_2-2}\right)^{-1/2} d\tau + C_2\right]^{-1},$$

where $A_1 = \frac{1}{2}a^{m_1+4}b^{-m_1-2}(m_1+2)$, $A_2 = \pm\frac{1}{2}a^{m_2+4}b^{-m_2-2}(m_2+2)$.

2°. Solution in the parametric form:

$$x = a\tau\left[\int \left(C_1 - \tau^{-m_1-2} \pm \tau^{-m_2-2}\right)^{-1/2} d\tau + C_2\right]^{-1},$$

$$y = b\left[\int \left(C_1 - \tau^{-m_1-2} \pm \tau^{-m_2-2}\right)^{-1/2} d\tau + C_2\right]^{-1},$$

where $A_1 = -\frac{1}{2}a^{m_1+4}b^{-m_1-2}(m_1+2)$, $A_2 = \pm\frac{1}{2}a^{m_2+4}b^{-m_2-2}(m_2+2)$.

10. $y''_{xx} = (A_1 x^{-4} y + A_2 x^{-3})(y'_x)^3.$

1°. Solution in the parametric form:

$$x = a\tau \left[\int \left(C_1 + \tau^{-3} \pm \tau^{-2}\right)^{-1/2} d\tau + C_2 \right]^{-1},$$
$$y = b \left[\int \left(C_1 + \tau^{-3} \pm \tau^{-2}\right)^{-1/2} d\tau + C_2 \right]^{-1},$$

where $A_1 = \frac{3}{2} a^5 b^{-3}$, $A_2 = \pm a^4 b^{-2}$.

2°. Solution in the parametric form:

$$x = a\tau \left[\int \left(C_1 - \tau^{-3} \pm \tau^{-2}\right)^{-1/2} d\tau + C_2 \right]^{-1},$$
$$y = b \left[\int \left(C_1 - \tau^{-3} \pm \tau^{-2}\right)^{-1/2} d\tau + C_2 \right]^{-1},$$

where $A_1 = -\frac{3}{2} a^5 b^{-3}$, $A_2 = \pm a^4 b^{-2}$.

11. $y''_{xx} = (A_1 + A_2 x^{-1})(y'_x)^3.$

Solution:

$$y = \int (C_1 - 2A_1 x - 2A_2 \ln x)^{-1/2} \, dx + C_2.$$

12. $y''_{xx} = (A_1 x^{-1} y^{-2} + A_2 y^{-3})(y'_x)^3.$

Solution in the parametric form:

$$x = \tau \left[\int (C_1 - 2A_1 \ln \tau - 2A_2 \tau)^{-1/2} \, d\tau + C_2 \right]^{-1},$$
$$y = \left[\int (C_1 - 2A_1 \ln \tau - 2A_2 \tau)^{-1/2} \, d\tau + C_2 \right]^{-1}.$$

13. $y''_{xx} = (A_1 x^{-1/2} y^{-5/2} + A_2 x^{-1/2} y^{-7/2})(y'_x)^3.$

Solution in the parametric form:

$$x = \frac{k^2}{F} \left\{ 2C_1 e^{2k\tau} + C_2 e^{-k\tau} \left[\sqrt{3} \cos(\omega\tau) - \sin(\omega\tau) \right] \right\}^2, \qquad y = \frac{1}{F},$$

where $F = C_1 e^{2k\tau} + C_2 e^{-k\tau} \sin(\omega\tau) - \dfrac{A_1}{A_2}$, $A_2 = -16k^3$, $\omega = k\sqrt{3}$.

14. $y''_{xx} = (A_1 x^{-5/3} y^{-4/3} + A_2 x^{-5/3} y^{-7/3})(y'_x)^3.$

Solution in the parametric form:

$$x = \left(\tfrac{1}{36} A_2 \tau^4 + C_1 \tau^3 + C_2 \tau^2 + C_3 \tau\right)^{-1} \left(\tfrac{1}{9} A_2 \tau^3 + 3C_1 \tau^2 + 2C_2 \tau + C_3\right)^{3/2},$$
$$y = \left(\tfrac{1}{36} A_2 \tau^4 + C_1 \tau^3 + C_2 \tau^2 + C_3 \tau\right)^{-1},$$

where $A_1 = 9C_1 C_3 - 3C_2^2$.

► *In the solutions of equations 15–18, the following notation is used:*

$$R_1 = \left(C_1 + \tau^{-3} \pm \tau^{-2}\right)^{1/2}, \qquad R_2 = \left(C_1 - \tau^{-3} \pm \tau^{-2}\right)^{1/2},$$
$$E_1 = \int R_1^{-1}\,d\tau + C_2, \qquad E_2 = \int R_2^{-1}\,d\tau + C_2,$$
$$F_1 = \tau - R_1 E_1, \qquad F_2 = \tau - R_2 E_2,$$
$$H_1 = 3\tau^3 F_1^2 + 3(1 \pm \tau)E_1^2, \qquad H_2 = 3\tau^3 F_2^2 + 3(-1 \pm \tau)E_2^2.$$

15. $y''_{xx} = (A_1 y^{-4/3} + A_2 x^{-1/2} y^{-4/3})(y'_x)^3.$

Solution in the parametric form:

$$x = aF_k^2, \qquad y = b\tau^{-3}E_k^3,$$

where $A_1 = \pm\frac{2}{9}ab^{-2/3}$, $A_2 = \frac{1}{3}a^{3/2}b^{-2/3}(-1)^{k+1}$; $k = 1$ and $k = 2$.

16. $y''_{xx} = (A_1 x^{-1/2} y^{-7/6} + A_2 y^{-5/3})(y'_x)^3.$

Solution in the parametric form:

$$x = a\tau^3 E_k^{-3} F_k^2, \qquad y = b\tau^3 E_k^{-3},$$

where $A_1 = \frac{1}{3}a^{3/2}b^{-5/6}(-1)^k$, $A_2 = \mp\frac{2}{9}ab^{-1/3}$; $k = 1$ and $k = 2$.

17. $y''_{xx} = (A_1 x^{-7} y + A_2 x^{-3})(y'_x)^3.$

Solution in the parametric form:

$$x = a\tau^{-1/2}E_k^{1/2}, \qquad y = b\tau^{-3}H_k,$$

where $A_1 = \frac{1}{36}a^8 b^{-3}$, $A_2 = \pm\frac{1}{36}a^4 b^{-2}$; $k = 1$ and $k = 2$.

18. $y''_{xx} = (A_1 x^{-7} y^3 + A_2 x^{-3})(y'_x)^3.$

Solution in the parametric form:

$$x = a\tau^{5/2}E_k^{1/2}H_k^{-1}, \qquad y = b\tau^3 H_k^{-1},$$

where $A_1 = \frac{1}{36}a^8 b^{-5}$, $A_2 = \mp\frac{1}{36}a^4 b^{-2}$; $k = 1$ and $k = 2$.

► *In the solutions of equations 19–22, the following notation is used:*

$$R_1 = \left(C_1 \pm \tau^{\gamma+1} + \tau\right)^{1/2}, \qquad R_2 = \left(C_1 \pm \tau^{\gamma+1} - \tau\right)^{1/2},$$
$$E_1 = \int R_1^{-1}\,d\tau + C_2, \qquad E_2 = \int R_2^{-1}\,d\tau + C_2,$$
$$G_1 = 2R_1 - E_1, \qquad G_2 = 2R_2 + E_2,$$
$$H_1 = 4(\tau - R_1 F_1) + E_1^2, \qquad H_2 = 4(\tau - R_2 F_2) - E_2^2.$$

19. $y''_{xx} = (A_1 x^{-m-3} y^m + A_2 y^{-3})(y'_x)^3, \qquad m \neq -2.$

Solution in the parametric form:

$$x = a\tau E_k^{-1}, \quad y = bE_k^{-1}, \qquad \gamma = -m-3,$$

where $A_1 = \pm\frac{1}{2} a^{m+4} b^{-m-2}(m+2)$, $A_2 = \frac{1}{2}ab(-1)^k$; $k = 1$ and $k = 2$.

20. $y''_{xx} = (A_1 y + A_2 x)(y'_x)^l, \qquad l \neq 1, \quad l \neq 2.$

Solution in the parametric form:

$$x = aG_k, \quad y = bE_k, \qquad \gamma = \frac{1}{l-2},$$

where $A_1 = ab^{-1}A_2(-1)^{k+1}$, $A_2 = -\dfrac{\gamma}{2ab}\left[\pm\dfrac{(\gamma+1)a}{b}\right]^{1/\gamma}$; $k = 1$ and $k = 2$.

21. $y''_{xx} = (A_1 x y^m + A_2 y^{-m-2})(y'_x)^{\frac{3m+5}{2m+3}}.$

Solution in the parametric form:

$$x = aH_k, \quad y = bE_k^{\gamma+2}, \qquad \gamma = -\frac{2m+3}{m+1},$$

where $A_1 = \dfrac{\gamma}{4(\gamma+2)} a^{-1} b^{\frac{1}{\gamma+2}} \left[\mp\dfrac{2(\gamma+1)a}{(\gamma+2)b}\right]^{\frac{1}{\gamma}}$, $A_2 = ab^{-\frac{2}{\gamma+2}} A_1 (-1)^k$; $k = 1$ and $k = 2$.

22. $y''_{xx} = (A_1 x^n y + A_2 x^{-n-2})(y'_x)^{\frac{3n+4}{n+1}}.$

Solution in the parametric form:

$$x = aE_k^{\gamma+2}, \quad y = bH_k, \qquad \gamma = -\frac{2n+3}{n+1},$$

where $A_1 = -\dfrac{\gamma}{4(\gamma+2)} a^{\frac{1}{\gamma+2}} b^{-1} \left[\mp\dfrac{2(\gamma+1)b}{(\gamma+2)a}\right]^{\frac{1}{\gamma}}$, $A_2 = a^{-\frac{2}{\gamma+2}} bA_1(-1)^k$; $k = 1$ and $k = 2$.

▶ *In the solutions of equations 23–26, the following notation is used:*

$$R_1 = (C_1 + \tau \pm \ln\tau)^{1/2}, \qquad R_2 = (C_1 - \tau \pm \ln\tau)^{1/2},$$
$$E_1 = \int R_1^{-1}\, d\tau + C_2, \qquad E_2 = \int R_2^{-1}\, d\tau + C_2,$$
$$G_1 = 2R_1 - E_1, \qquad G_2 = 2R_2 + E_2,$$
$$H_1 = 4(\tau - R_1 F_1) + E_1^2, \qquad H_2 = 4(\tau - R_2 F_2) - E_2^2,$$

23. $y''_{xx} = (A_1 y + A_2 x) y'_x$.

Solution in the parametric form:

$$x = aG_k, \qquad y = bE_k,$$

where $A_1 = ab^{-1} A_2 (-1)^k$, $A_2 = \pm\frac{1}{2} a^{-2}$; $k = 1$ and $k = 2$.

24. $y''_{xx} = (A_1 y + A_2 x)(y'_x)^2$.

Solution in the parametric form:

$$x = aE_k, \qquad y = bG_k,$$

where $A_1 = \mp\frac{1}{2} b^{-2}$, $A_2 = a^{-1} b A_1 (-1)^k$; $k = 1$ and $k = 2$.

25. $y''_{xx} = (A_1 + A_2 x y^{-2}) y'_x$.

Solution in the parametric form:

$$x = aH_k, \qquad y = bE_k,$$

where $A_1 = ab^{-2} A_2 (-1)^k$, $A_2 = \pm\frac{1}{8} a^{-2} b^{-2}$; $k = 1$ and $k = 2$.

26. $y''_{xx} = (A_1 x^{-2} y + A_2)(y'_x)^2$.

Solution in the parametric form:

$$x = aE_k, \qquad y = bH_k,$$

where $A_1 = \mp\frac{1}{8} a^2 b^{-2}$, $A_2 = a^{-2} b A_1 (-1)^k$; $k = 1$ and $k = 2$.

► *In the solutions of equations 27–30, the following notation is used:*

$$\begin{aligned}
R_1 &= C_1 \tau^{k_1} + C_2 \tau^{k_2} + C_3 \tau^{k_3},\\
R_2 &= (C_1 + C_2 \tau) e^{k\tau} + C_3 e^{\omega\tau},\\
R_3 &= C_1 e^{k\tau} + e^{s\tau}(C_2 \sin \omega\tau + C_3 \cos \omega\tau),\\
Q_1 &= C_1 k_1 \tau^{k_1} + C_2 k_2 \tau^{k_2} + C_3 k_3 \tau^{k_3},\\
Q_2 &= (kC_1 + C_2 + kC_2 \tau) e^{k\tau} + \omega C_3 e^{\omega\tau},\\
Q_3 &= kC_1 e^{k\tau} + e^{s\tau}[(sC_2 - \omega C_3) \sin \omega\tau + (sC_3 + \omega C_2) \cos \omega\tau],\\
S_1 &= \tau (Q_1)'_\tau, \quad S_2 = (Q_2)'_\tau, \quad S_3 = (Q_3)'_\tau,
\end{aligned}$$

where k_1, k_2, and k_3 (real numbers) or k and $s \pm i\omega$ (one real and two complex numbers) are the roots of the cubic equation $\lambda^3 - \frac{1}{2} B_2 \lambda - \frac{1}{2} B_1 = 0$. Subscripts of functions R_m, Q_m, and S_m ($m = 1, 2, 3$) are selected depending on the sign of the following expression:

$$2B_2^3 - 27B_1^2 \quad \begin{cases} > 0 & \textit{subscript 1,} \\ = 0 & \textit{subscript 2,} \\ < 0 & \textit{subscript 3;} \end{cases}$$

if $2B_2^3 = 27B_1^2$ (subscript 2), then

$$\begin{aligned}
k &= (\tfrac{1}{6} B_2)^{1/2}, \quad \omega = -2(\tfrac{1}{6} B_2)^{1/2} && \textit{if} \ \ B_1 < 0,\\
k &= -(\tfrac{1}{6} B_2)^{1/2}, \quad \omega = 2(\tfrac{1}{6} B_2)^{1/2} && \textit{if} \ \ B_1 > 0.
\end{aligned}$$

Remark. *The expressions for R_m, and Q_m contain three constants C_1, C_2, and C_3. One of them may be arbitrarily fixed to let it be any nonzero number (for instance, we may set $C_3 = \pm 1$), while the other constants remain arbitrary.*

27. $y''_{xx} = (A_1 x^{-1/2} y + A_2)(y'_x)^3.$

Solution in the parametric form:

$$x = Q_m^2, \quad y = R_m, \qquad A_1 = -B_1, \quad A_2 = -B_2.$$

28. $y''_{xx} = (A_1 y^{-3} + A_2 x^{-1/2} y^{-7/2})(y'_x)^3.$

Solution in the parametric form:

$$x = R_m^{-1} Q_m^2, \quad y = R_m^{-1}, \qquad A_1 = -B_2, \quad A_2 = -B_1.$$

29. $y''_{xx} = (A_1 x^{-7/5} y + A_2 x^{-3/5})(y'_x)^3.$

Solution in the parametric form:

$$x = aR_m^{5/2}, \qquad y = b(2Q_m^2 - 4R_m S_m + B_2 R_m^2),$$

where $A_1 = \frac{5}{32} a^{12/5} b^{-3} B_1^{-2}$, $A_2 = -a^{-4/5} b A_1 B_2$.

30. $y''_{xx} = (A_1 x^{-3/5} y^{-12/5} + A_2 x^{-7/5} y^{-13/5})(y'_x)^3.$

Solution in the parametric form:

$$x = aR_m^{5/2}\bigl(2Q_m^2 - 4R_m S_m + B_2 R_m^2\bigr)^{-1}, \quad y = b\bigl(2Q_m^2 - 4R_m S_m + B_2 R_m^2\bigr)^{-1},$$

where $A_1 = -\frac{5}{32} a^{8/5} b^{2/5} B_1^{-2} B_2$, $A_2 = \frac{5}{32} a^{12/5} b^{3/5} B_1^{-2}$.

► *In the solutions of equations 31–32, the following notation is used:*

$$f_1 = \begin{cases} C_1 e^{k\tau} + C_2 e^{-k\tau} - \dfrac{B_1}{B_2}\tau & \text{if } B_2 > 0, \\ C_1 \sin(k\tau) + C_2 \cos(k\tau) - \dfrac{B_1}{B_2}\tau & \text{if } B_2 < 0, \end{cases}$$

$$f_2 = \begin{cases} k(C_1 e^{k\tau} - C_2 e^{-k\tau}) - \dfrac{B_1}{B_2} & \text{if } B_2 > 0, \\ k[C_1 \cos(k\tau) - C_2 \sin(k\tau)] - \dfrac{B_1}{B_2} & \text{if } B_2 < 0, \end{cases}$$

where $k = \sqrt{\frac{1}{2}|B_2|}$.

31. $y''_{xx} = (A_1 + A_2 x^{-1/2})(y'_x)^3.$

Solution in the parametric form:

$$x = f_2^2, \quad y = f_1, \qquad A_1 = -B_2, \quad A_2 = -B_1.$$

32. $y''_{xx} = (A_1 x^{-1/2} y^{-5/2} + A_2 y^{-3})(y'_x)^3.$

Solution in the parametric form:

$$x = f_1^{-1} f_2^2, \quad y = f_1^{-1}, \qquad A_1 = -B_1, \quad A_2 = -B_2.$$

► *In the solutions of equations 33–36, the following notation is used:*

For $B_1 > 0$:

$$T_1 = C_1 e^{k\tau} + C_2 e^{-k\tau} + C_3 \sin(k\tau), \quad k = \left(\tfrac{4}{3}B_1\right)^{1/4},$$
$$T_2 = k(C_1 e^{k\tau} - C_2 e^{-k\tau}) + kC_3 \cos(k\tau);$$

For $B_1 < 0$:

$$T_1 = e^{s\tau}[C_1 \sin(s\tau) + C_2 \cos(s\tau)] + C_3 e^{-s\tau} \sin(s\tau), \quad s = \left(-\tfrac{1}{3}B_1\right)^{1/4},$$
$$T_2 = se^{s\tau}[(C_1 - C_2)\sin(s\tau) + (C_1 + C_2)\cos(s\tau)] - sC_3 e^{-s\tau}[\sin(s\tau) - \cos(s\tau)].$$

33. $y''_{xx} = (A_1 x^{-5/3} y^2 + A_2 x^{-5/3} y)(y'_x)^3.$

Solution in the parametric form:

$$x = T_2^{3/2}, \qquad y = T_1 - \frac{A_2}{2A_1},$$

where $B_1 = -A_1$, $B_2 = -A_2$; arbitrary constants C_1, C_2, and C_3 are related by

$$C_1 C_3 = \tfrac{1}{16} A_1^{-2} A_2^2 \qquad \text{if } A_1 > 0,$$
$$4C_1 C_2 + C_3^2 = \tfrac{1}{4} A_1^{-2} A_2^2 \quad \text{if } A_1 < 0.$$

34. $y''_{xx} = (A_1 x^{-5/3} y^{-7/3} + A_2 x^{-5/3} y^{-10/3})(y'_x)^3.$

Solution in the parametric form:

$$x = \left(T_1 - \frac{A_1}{2A_2}\right)^{-1} T_2^{3/2}, \qquad y = \left(T_1 - \frac{A_1}{2A_2}\right)^{-1},$$

where $B_1 = -A_2$, $B_2 = -A_1$; arbitrary constants C_1, C_2, and C_3 are related by

$$C_1 C_3 = \tfrac{1}{16} A_1^2 A_2^{-2} \qquad \text{if } A_2 > 0,$$
$$4C_1 C_2 + C_3^2 = \tfrac{1}{4} A_1^2 A_2^{-2} \quad \text{if } A_2 < 0.$$

35. $y''_{xx} = (A_1 x^{-5/3} y^2 + A_2 x^{-5/3})(y'_x)^3.$

Solution in the parametric form:

$$x = T_2^{3/2}, \qquad y = T_1,$$

where $B_1 = -A_1$, $B_2 = -A_2$; arbitrary constants C_1, C_2, and C_3 are related by

$$C_1 C_3 = -\tfrac{1}{4} A_1^{-1} A_2 \qquad \text{if } A_1 > 0,$$
$$4C_1 C_2 + C_3^2 = -\tfrac{1}{2} A_1^{-1} A_2 \quad \text{if } A_1 < 0.$$

36. $y''_{xx} = (A_1 x^{-5/3} y^{-4/3} + A_2 x^{-5/3} y^{-10/3})(y'_x)^3.$

Solution in the parametric form:

$$x = T_1^{-1} T_2^{3/2}, \qquad y = T_1^{-1},$$

where $B_1 = -A_2$, $B_2 = -A_1$; arbitrary constants C_1, C_2, and C_3 are related by

$$C_1 C_3 = -\tfrac{1}{4} A_1 A_2^{-1} \qquad \text{if } A_2 > 0,$$
$$4C_1 C_2 + C_3^2 = -\tfrac{1}{2} A_1 A_2^{-1} \quad \text{if } A_2 < 0.$$

▶ *In the solutions of equations 37–38, the following notation is used:*

1. For $B_1 > 0$, $B_2 \neq 0$,

$$T_1 = C_1 e^{k\tau} + C_2 e^{-k\tau} + C_3 \sin \omega\tau,$$
$$T_2 = k(C_1 e^{k\tau} - C_2 e^{-k\tau}) + \omega C_3 \cos \omega\tau,$$

where

$$k = \left\{\tfrac{2}{3}[(B_2^2 + 3B_1)^{1/2} + B_2]\right\}^{1/2}, \quad \omega = \left\{\tfrac{2}{3}[(B_2^2 + 3B_1)^{1/2} - B_2]\right\}^{1/2},$$
$$4k^2 C_1 C_2 + \omega^2 C_3^2 = 0.$$

2. For $-B_2^2 < 3B_1 < 0$, $B_2 > 0$,

$$T_1 = C_1 \tau^{k_1} + C_2 \tau^{-k_1} + C_3 \tau^{k_2} + C_4 \tau^{-k_2},$$
$$T_2 = k_1(C_1 \tau^{k_1} - C_2 \tau^{-k_1}) + k_2(C_3 \tau^{k_2} - C_4 \tau^{-k_2}),$$

where

$$k_1 = \left\{\tfrac{2}{3}[B_2 + (B_2^2 + 3B_1)^{1/2}]\right\}^{1/2}, \quad k_2 = \left\{\tfrac{2}{3}[B_2 - (B_2^2 + 3B_1)^{1/2}]\right\}^{1/2},$$
$$(C_1 C_2 + C_3 C_4)(B_2^2 + 3B_1)^{1/2} + (C_1 C_2 - C_3 C_4) B_2 = 0.$$

3. For $-B_2^2 < 3B_1 < 0$, $B_2 < 0$,

$$T_1 = C_1 \sin \omega_1\tau + C_2 \cos \omega_1\tau + C_3 \sin \omega_2\tau,$$
$$T_2 = \omega_1(C_1 \cos \omega_1\tau - C_2 \sin \omega_1\tau) + \omega_2 C_3 \cos \omega_2\tau,$$

where

$$\omega_1 = \left\{-\tfrac{2}{3}[B_2 + (B_2^2 + 3B_1)^{1/2}]\right\}^{1/2}, \quad \omega_2 = \left\{-\tfrac{2}{3}[B_2 - (B_2^2 + 3B_1)^{1/2}]\right\}^{1/2},$$
$$\omega_1^2(C_1^2 + C_2^2) - \omega_2^2 C_3^2 = 0.$$

4. For $B_2^2 + 3B_1 = 0$, $B_2 > 0$,

$$T_1 = (C_1 + C_2\tau)e^{k\tau} + (C_3 + C_4\tau)e^{-k\tau},$$
$$T_2 = (kC_1 + C_2 + kC_2\tau)e^{k\tau} - (kC_3 - C_4 + kC_4\tau)e^{-k\tau},$$

where

$$k = \left(\tfrac{2}{3}B_2\right)^{1/2}, \quad (C_1 C_4 - C_2 C_3)\left(\tfrac{2}{3}B_2\right)^{1/2} + 2C_2 C_4 = 0.$$

5. For $B_2^2 + 3B_1 = 0$, $B_2 < 0$,

$$T_1 = (C_1 + C_2\tau) \sin \omega\tau + C_3\tau \cos \omega\tau,$$
$$T_2 = (\omega C_1 + C_3 + \omega C_2\tau) \cos \omega\tau + (C_2 - \omega C_3\tau) \sin \omega\tau,$$

where

$$\omega = \left(-\tfrac{2}{3}B_2\right)^{1/2}, \quad C_1 C_3 \left(-\tfrac{2}{3}B_2\right)^{1/2} + C_2^2 + C_3^2 = 0.$$

6. For $3B_1 < -B_2^2$,

$$T_1 = e^{k\tau}(C_1 \sin\omega\tau + C_2 \cos\omega\tau) + C_3 e^{-k\tau} \sin\omega\tau,$$
$$T_2 = e^{k\tau}[(kC_2 + \omega C_1)\cos\omega\tau + (kC_1 - \omega C_2)\sin\omega\tau)] + C_3 e^{-k\tau}(\omega\cos\omega\tau - k\sin\omega\tau),$$

where

$$k = \left\{\tfrac{1}{3}[B_2 + (-3B_1)^{1/2}]\right\}^{1/2}, \quad \omega = \left\{\tfrac{1}{3}[-B_2 + (-3B_1)^{1/2}]\right\}^{1/2},$$
$$C_2 B_2 + C_1(-B_2^2 - 3B_1)^{1/2} = 0.$$

37. $\boldsymbol{y''_{xx} = (A_1 x^{-5/3} y^2 + A_2 x^{-1/3})(y'_x)^3.}$

Solution in the parametric form:

$$x = T_2^{3/2}, \quad y = T_1, \qquad B_1 = -A_1, \quad B_2 = -A_2.$$

38. $\boldsymbol{y''_{xx} = (A_1 x^{-1/3} y^{-8/3} + A_2 x^{-5/3} y^{-10/3})(y'_x)^3.}$

Solution in the parametric form:

$$x = T_1^{-1} T_2^{3/2}, \quad y = T_1^{-1}, \qquad B_1 = -A_2, \quad B_2 = -A_1.$$

▶ *In the solutions of equations 39–42, the following notation is used:*

$$T_1 = \begin{cases} C_1 e^{\omega\tau} + C_2 e^{-\omega\tau} + C_3\tau & \text{if } B > 0, \\ C_1 \sin\omega\tau + C_2\cos\omega\tau + C_3\tau & \text{if } B < 0; \end{cases}$$

$$T_2 = \begin{cases} \omega(C_1 e^{\omega\tau} - C_2 e^{-\omega\tau}) + C_3 & \text{if } B > 0, \\ \omega(C_1\cos\omega\tau - C_2\sin\omega\tau) + C_3 & \text{if } B < 0; \end{cases}$$

where $\omega = |\tfrac{4}{3}B|^{1/2}$.

39. $\boldsymbol{y''_{xx} = (A_1 x^{-1/3} + A_2 x^{-5/3})(y'_x)^3.}$

Solution in the parametric form:

$$x = T_2^{3/2}, \quad y = T_1,$$

where $B = -A_1$, arbitrary constants C_1, C_2, and C_3 are related by

$$3(A_1 C_3^2 + A_2) - 4A_1^2(C_1^2 + C_2^2) = 0 \quad \text{if } A_1 > 0,$$
$$3(A_1 C_3^2 + A_2) - 16A_1^2 C_1 C_2 = 0 \quad \text{if } A_1 < 0.$$

40. $\boldsymbol{y''_{xx} = (A_1 x^{-5/3} y^{-4/3} + A_2 x^{-1/3} y^{-8/3})(y'_x)^3.}$

Solution in the parametric form:

$$x = T_1^{-1} T_2^{3/2}, \quad y = T_1^{-1},$$

where $B = -A_2$, arbitrary constants C_1, C_2, and C_3 are related by

$$3(A_1 + A_2 C_3^2) - 4A_2^2(C_1^2 + C_2^2) = 0 \quad \text{if } A_2 > 0,$$
$$3(A_1 + A_2 C_3^2) - 16A_2^2 C_1 C_2 = 0 \quad \text{if } A_2 < 0.$$

41. $y''_{xx} = (A_1 x^{-5/3} y + A_2 x^{-1/3})(y'_x)^3.$

Solution in the parametric form:

$$x = \left(T_2 - \frac{A_1}{2A_2}\tau\right)^{3/2}, \quad y = T_1 - \frac{A_1}{4A_2}\tau^2,$$

where $B = -A_2$, arbitrary constants C_1, C_2, and C_3 are related by

$$\begin{aligned} &3A_2C_3^2 - 4A_2^2(C_1^2 + C_2^2) - \tfrac{9}{16}A_1^2A_2^{-2} = 0 && \text{if } A_2 > 0,\\ &3A_2C_3^2 - 16A_2^2C_1C_2 - \tfrac{9}{16}A_1^2A_2^{-2} = 0 && \text{if } A_2 < 0.\end{aligned}$$

42. $y''_{xx} = (A_1 x^{-5/3} y^{-7/3} + A_2 x^{-1/3} y^{-8/3})(y'_x)^3.$

Solution in the parametric form:

$$x = \left(T_1 - \frac{A_1}{4A_2}\tau^2\right)^{-1}\left(T_2 - \frac{A_1}{2A_2}\tau\right)^{3/2}, \quad y = \left(T_1 - \frac{A_1}{4A_2}\tau^2\right)^{-1},$$

where $B = -A_2$, arbitrary constants C_1, C_2, and C_3 are related by

$$\begin{aligned} &3A_2C_3^2 - 4A_2^2(C_1^2 + C_2^2) - \tfrac{9}{16}A_1^2A_2^{-2} = 0 && \text{if } A_2 > 0,\\ &3A_2C_3^2 - 16A_2^2C_1C_2 - \tfrac{9}{16}A_1^2A_2^{-2} = 0 && \text{if } A_2 < 0.\end{aligned}$$

► *In the solutions of equations 43–48, the following notation is used:*

$$f = \begin{cases} J_{1/3}(\tau) & \textit{for the upper sign (Bessel function)},\\ I_{1/3}(\tau) & \textit{for the lower sign (modified Bessel function)},\end{cases}$$

$$g = \begin{cases} Y_{1/3}(\tau) & \textit{for the upper sign (Bessel function)},\\ K_{1/3}(\tau) & \textit{for the lower sign (modified Bessel function)},\end{cases}$$

$$H = C_1 f + C_2 g + \beta\omega\left(g\int f\,d\tau - f\int g\,d\tau\right), \qquad \omega = \begin{cases} \tfrac{1}{2}\pi & \textit{for the upper sign},\\ -1 & \textit{for the lower sign}.\end{cases}$$

43. $y''_{xx} = (A_1 xy + A_2)(y'_x)^3.$

Solution in the parametric form:

$$x = \tau^{1/3}H, \qquad y = b\tau^{2/3},$$

where $A_1 = \pm\frac{9}{4}b^{-3}$, $A_2 = -\frac{9}{4}b^{-2}\beta$.

44. $y''_{xx} = (A_1 y^{-3} + A_2 xy^{-5})(y'_x)^3.$

Solution in the parametric form:

$$x = \tau^{-1/3}H, \qquad y = b\tau^{-2/3},$$

where $A_1 = -\frac{9}{4}b\beta$, $A_2 = \pm\frac{9}{4}b^3$.

45. $y''_{xx} = (A_1 x^{-1/2} y^{-1/2} + A_2 y^{-3/2})(y'_x)^3.$

Solution in the parametric form:

$$x = a\tau^{-2/3}(\tau H'_\tau + \tfrac{1}{3}H)^2, \qquad y = b\tau^{2/3}H^2,$$

where $A_1 = \pm\frac{1}{3}a^{3/2}b^{-3/2}$, $A_2 = \frac{1}{2}ab^{-1/2}\beta$.

46. $y''_{xx} = (A_1 y^{-3/2} + A_2 x^{-1/2} y^{-2})(y'_x)^3.$

Solution in the parametric form:

$$x = a\tau^{-4/3}H^{-2}(\tau H'_\tau + \tfrac{1}{3}H)^2, \qquad y = b\tau^{-2/3}H^{-2},$$

where $A_1 = \frac{1}{2}ab^{-1/2}\beta$, $A_2 = \pm\frac{1}{3}a^{3/2}$.

47. $y''_{xx} = (A_1 x^{-2} y + A_2 x^{-3/2})(y'_x)^3.$

Solution in the parametric form:

$$x = a\tau^{2/3}H^2, \qquad y = b\tau^{-2/3}\bigl[\mp\tau^2H^2 + 2\beta\tau H - (\tau H'_\tau + \tfrac{1}{3}H)^2\bigr],$$

where $A_1 = -\frac{9}{2}a^3b^{-3}$, $A_2 = -a^{-1/2}b\beta A_1$.

48. $y''_{xx} = (A_1 x^{-3/2} y^{-3/2} + A_2 x^{-2} y^{-2})(y'_x)^3.$

Solution in the parametric form:

$$x = a\tau^{4/3}H^2\bigl[\mp\tau^2H^2 + 2\beta\tau H - (\tau H'_\tau + \tfrac{1}{3}H)^2\bigr]^{-1},$$
$$y = b\tau^{2/3}\bigl[\mp\tau^2H^2 + 2\beta\tau H - (\tau H'_\tau + \tfrac{1}{3}H)^2\bigr]^{-1},$$

where $A_1 = \frac{9}{2}a^{5/2}b^{-1/2}\beta$, $A_2 = -\frac{9}{2}a^3$.

49. $y''_{xx} = (A_1 x y^{m_1} + A_2 y^{m_2})(y'_x)^3, \qquad m_1 \neq -2.$

Solution in the parametric form:

$$x = \tau^\nu H, \quad y = b\tau^{2\nu}, \qquad \nu = \frac{1}{m_1+2},$$

where

$$H = C_1 f + C_2 g + \frac{4b^2\beta\omega}{(m_1+2)^2}\Bigl(g\int \tau^k f\,d\tau - f\int \tau^k g\,d\tau\Bigr), \qquad k = \frac{2m_2 - m_1 + 1}{m_1+2};$$

$$f = \begin{cases} J_\nu(\tau) & \text{for the upper sign (Bessel function),} \\ I_\nu(\tau) & \text{for the lower sign (modified Bessel function),} \end{cases}$$

$$g = \begin{cases} Y_\nu(\tau) & \text{for the upper sign (Bessel function),} \\ K_\nu(\tau) & \text{for the lower sign (modified Bessel function),} \end{cases}$$

$$A_1 = \pm\tfrac{1}{4}(m_1+2)^2b^{-m_1-2}, \quad A_2 = -b^{-m_2}\beta, \quad \omega = \begin{cases} \frac{1}{2}\pi, & \text{for the upper sign,} \\ -1 & \text{for the lower sign.} \end{cases}$$

► *In the solutions of equations 50–56, the following notation is used:*

$$U_\nu = \begin{cases} C_1 J_\nu(\tau) & \textit{for the upper sign (Bessel functions)}, \\ C_1 I_\nu(\tau) & \textit{for the lower sign (modified Bessel functions)}, \end{cases}$$

$$V_\nu = \begin{cases} C_2 Y_\nu(\tau) & \textit{for the upper sign (Bessel functions)}, \\ C_2 K_\nu(\tau) & \textit{for the lower sign (modified Bessel functions)}, \end{cases}$$

$$Z_\nu = \alpha_1 U_\nu + \alpha_2 V_\nu, \quad X_\nu = \beta_1 U_\nu + \beta_2 V_\nu, \quad F_\nu = \tau Z'_\nu + \nu Z_\nu, \quad G_\nu = \tau X'_\nu + \nu X_\nu,$$

$$N = \begin{cases} Z_\nu X_\nu & \textit{for } \Delta = -(\alpha_1\beta_2 - \alpha_2\beta_1)^2; \\ \alpha U_\nu^2 + \beta U_\nu V_\nu + \gamma V_\nu^2 & \textit{for } \Delta = 4\alpha\gamma - \beta^2, \end{cases}$$

$$N_1 = \begin{cases} Z_\nu G_\nu + X_\nu F_\nu & \textit{for } \Delta = -(\alpha_1\beta_2 - \alpha_2\beta_1)^2, \\ \tau N' + 2\nu N & \textit{for } \Delta = 4\alpha\gamma - \beta^2, \end{cases}$$

$$N_2 = N_1^2 \pm 4\tau^2 N^2 + \omega^2 \Delta, \qquad \omega = \begin{cases} 2/\pi & \textit{for the upper sign}, \\ -1 & \textit{for the lower sign}. \end{cases}$$

where prime stands for differentiation with respect to τ.

50. $\boldsymbol{y''_{xx} = (A_1 x y + A_2 x^{-3})(y'_x)^3}$.

Solution in the parametric form:

$$x = a\tau^{1/3} N^{1/2}, \qquad y = b\tau^{2/3},$$

where $\nu = \frac{1}{3}$, $A_1 = \pm\frac{9}{4} b^{-3}$, $A_2 = -\frac{9}{16} a^4 b^{-2} \omega^2 \Delta$.

51. $\boldsymbol{y''_{xx} = (A_1 x y^m + A_2 x^{-3})(y'_x)^3}$, $\quad \boldsymbol{m \neq -2}$.

Solution in the parametric form:

$$x = a\tau^\nu N^{1/2}, \qquad y = b\tau^{2\nu},$$

where $\nu = \dfrac{1}{m+2}$, $A_1 = \pm\frac{1}{4} b^{-m-2}(m+2)^2$, $A_2 = -\frac{1}{16} a^4 b^{-2} \omega^2 \Delta (m+2)^2$.

52. $\boldsymbol{y''_{xx} = (A_1 x^{-3} + A_2 x y^{-5})(y'_x)^3}$.

Solution in the parametric form:

$$x = a\tau^{-1/3} N^{1/2}, \qquad y = b\tau^{-2/3},$$

where $\nu = \frac{1}{3}$, $A_1 = -\frac{9}{16} a^4 b^{-2} \omega^2 \Delta$, $A_2 = \pm\frac{9}{4} b^3$.

53. $\boldsymbol{y''_{xx} = (A_1 x^{-1/2} y^{-1/2} + A_2 y^{-3})(y'_x)^3}$.

Solution in the parametric form:

$$x = a\tau^{-2/3} N^{-1} N_1^2, \qquad y = b\tau^{2/3} N,$$

where $\nu = \frac{1}{3}$, $A_1 = \pm\frac{8}{3} a^{3/2} b^{-3/2}$, $A_2 = 2ab\omega^2\Delta$.

54. $y''_{xx} = (A_1 + A_2 x^{-1/2} y^{-2})(y'_x)^3.$

Solution in the parametric form:

$$x = a\tau^{-4/3} N^{-2} N_1^2, \qquad y = b\tau^{-2/3} N^{-1},$$

where $\nu = \frac{1}{3}$, $A_1 = 2ab^{-2}\omega^2\Delta$, $A_2 = \pm\frac{8}{3}a^{3/2}$.

55. $y''_{xx} = (A_1 x^{-2} y + A_2 x^{-3})(y'_x)^3.$

Solution in the parametric form:

$$x = a\tau^{2/3} N, \qquad y = b\tau^{-2/3} N^{-1} N_2,$$

where $\nu = \frac{1}{3}$, $A_1 = \frac{9}{128}a^3 b^{-3}$, $A_2 = -\frac{9}{64}a^4 b^{-2}\omega^2\Delta$.

56. $y''_{xx} = (A_1 x^{-3} + A_2 x^{-2} y^{-2})(y'_x)^3.$

Solution in the parametric form:

$$x = a\tau^{4/3} N^2 N_2^{-1}, \qquad y = b\tau^{2/3} N N_2^{-1},$$

where $\nu = \frac{1}{3}$, $A_1 = -\frac{9}{64}a^4 b^{-2}\omega^2\Delta$, $A_2 = \frac{9}{128}a^3$.

► *In the solutions of equations 57–72, the following notation is used:*

$$f_1 = \sqrt{\pm 4\wp_1^3 - 2\wp_1 - C_2}, \qquad \tau = \int \frac{d\wp_1}{\sqrt{\pm 4\wp_1^3 - 2\wp_1 - C_2}} - C_1$$

and

$$f_2 = \sqrt{\pm 4\wp_2^3 + 2\wp_2 - C_2}, \qquad \tau = \int \frac{d\wp_2}{\sqrt{\pm 4\wp_2^3 + 2\wp_2 - C_2}} - C_1,$$

where functions $\wp_1 = \wp_1(\tau)$ and $\wp_2 = \wp_2(\tau)$ are defined implicitly by the above elliptic integrals. For the upper signs, they are the classical elliptic Weierstrass functions $\wp_1 = \wp(\tau + C_1,\ 2,\ C_2)$ and $\wp_2 = \wp(\tau + C_1,\ -2,\ C_2)$.

57. $y''_{xx} = (A_1 x^2 + A_2)(y'_x)^3.$

Solution in the parametric form:

$$x = a\wp_k, \qquad y = b\tau,$$

where $A_1 = \mp 6a^{-1}b^{-2}$, $A_2 = ab^{-2}(-1)^{k+1}$; $k = 1$ and $k = 2$.

58. $y''_{xx} = (A_1 y^{-3} + A_2 x^2 y^{-5})(y'_x)^3.$

Solution in the parametric form:

$$x = a\tau^{-1}\wp_k, \qquad y = b\tau^{-1},$$

where $A_1 = ab(-1)^{k+1}$, $A_2 = \mp 6a^{-1}b^3$; $k = 1$ and $k = 2$.

59. $y''_{xx} = (A_1 y^{-9/7} + A_2 x^2 y^{-15/7})(y'_x)^3.$

Solution in the parametric form:

$$x = a\tau(\tau^2 \wp_k \mp 1), \qquad y = b\tau^7,$$

where $A_1 = \frac{1}{49} ab^{-5/7}(-1)^{k+1}$, $A_2 = \mp\frac{6}{49} a^{-1} b^{1/7}$; $k = 1$ and $k = 2$.

60. $y''_{xx} = (A_1 y^{-12/7} + A_2 x^2 y^{-20/7})(y'_x)^3.$

Solution in the parametric form:

$$x = a\tau^{-6}(\tau^2 \wp_k \mp 1), \qquad y = b\tau^{-7},$$

where $A_1 = \frac{1}{49} ab^{-2/7}(-1)^{k+1}$, $A_2 = \mp\frac{6}{49} a^{-1} b^{6/7}$; $k = 1$ and $k = 2$.

61. $y''_{xx} = (A_1 y + A_2 x)(y'_x)^{5/2}.$

Solution in the parametric form:

$$x = a[f_k - (-1)^k \tau], \qquad y = b\tau,$$

where $A_1 = ab^{-1} A_2 (-1)^k$, $A_2 = -2a^{-1} b^{-1} \left(\pm\frac{6a}{b}\right)^{1/2}$; $k = 1$ and $k = 2$.

62. $y''_{xx} = (A_1 y + A_2 x)(y'_x)^{1/2}.$

Solution in the parametric form:

$$x = a\tau, \qquad y = b[f_k - (-1)^k \tau],$$

where $A_1 = 2a^{-1} b^{-1} \left(\pm\frac{6b}{a}\right)^{1/2}$, $A_2 = a^{-1} b A_1 (-1)^k$; $k = 1$ and $k = 2$.

63. $y''_{xx} = (A_1 y^{-3/4} + A_2 x y^{-5/4})(y'_x)^{5/2}.$

Solution in the parametric form:

$$x = a[2\tau f_k - 2\wp_k + (-1)^{k+1}\tau^2], \qquad y = b\tau^4,$$

where $A_1 = ab^{-1/2} A_2 (-1)^k$, $A_2 = -\frac{1}{4} a^{-1} b^{1/4} \left(\pm\frac{3a}{b}\right)^{1/2}$; $k = 1$ and $k = 2$.

64. $y''_{xx} = (A_1 x^{-5/4} y + A_2 x^{-3/4})(y'_x)^{1/2}.$

Solution in the parametric form:

$$x = a\tau^4, \qquad y = b[2\tau f_k - 2\wp_k + (-1)^{k+1}\tau^2],$$

where $A_1 = \frac{1}{4} a^{1/4} b^{-1} \left(\pm\frac{3b}{a}\right)^{1/2}$, $A_2 = a^{-1/2} b A_1 (-1)^k$; $k = 1$ and $k = 2$.

65. $y''_{xx} = (A_1 y^{-13/8} + A_2 x y^{-15/8})(y'_x)^{5/2}$.

Solution in the parametric form:

$$x = a\tau^{-6}[2\tau^3 f_k + 6\tau^2 \wp_k \mp 2 + (-1)^k \tau^4], \qquad y = b\tau^{-8},$$

where $A_1 = ab^{-1/4} A_2 (-1)^k$, $A_2 = \frac{1}{8} a^{-1} b^{7/8} \left(\mp \frac{3a}{2b}\right)^{1/2}$; $k = 1$ and $k = 2$.

66. $y''_{xx} = (A_1 x^{-15/8} y + A_2 x^{-13/8})(y'_x)^{1/2}$.

Solution in the parametric form:

$$x = a\tau^{-8}, \qquad y = b\tau^{-6}[2\tau^3 f_k + 6\tau^2 \wp_k \mp 2 + (-1)^k \tau^4],$$

where $A_1 = -\frac{1}{8} a^{7/8} b^{-1} \left(\mp \frac{3b}{2a}\right)^{1/2}$, $A_2 = a^{-1/4} b A_1 (-1)^k$; $k = 1$ and $k = 2$.

67. $y''_{xx} = (A_1 y^{-15/13} + A_2 x y^{-20/13})(y'_x)^{5/2}$.

Solution in the parametric form:

$$x = a\tau[5\tau^3 f_k - 20\tau^2 \wp_k \pm 30 - (-1)^k \tau^4], \qquad y = b\tau^{13},$$

where $A_1 = ab^{-5/13} A_2 (-1)^k$, $A_2 = -\frac{2}{65} a^{-1} b^{7/13} \left(\pm \frac{30a}{13b}\right)^{1/2}$; $k = 1$ and $k = 2$.

68. $y''_{xx} = (A_1 x^{-20/13} y + A_2 x^{-15/13})(y'_x)^{1/2}$.

Solution in the parametric form:

$$x = a\tau^{13}, \qquad y = b\tau[5\tau^3 f_k - 20\tau^2 \wp_k \pm 30 - (-1)^k \tau^4],$$

where $A_1 = \frac{2}{65} a^{7/13} b^{-1} \left(\pm \frac{30b}{13a}\right)^{1/2}$, $A_2 = a^{-5/13} b A_1 (-1)^k$; $k = 1$ and $k = 2$.

69. $y''_{xx} = (A_1 + A_2 x^{-2/3})(y'_x)^3$.

Solution in the parametric form:

$$x = a\wp_k^3, \qquad y = b[f_k - (-1)^k \tau],$$

where $A_1 = \mp\frac{1}{2} ab^{-2}$, $A_2 = \frac{1}{12} a^{5/3} b^{-2} (-1)^{k+1}$; $k = 1$ and $k = 2$.

70. $y''_{xx} = (A_1 x^{-2/3} y^{-7/3} + A_2 y^{-3})(y'_x)^3$.

Solution in the parametric form:

$$x = a\wp_k^3 \big[f_k - (-1)^k \tau\big]^{-1}, \qquad y = b\big[f_k - (-1)^k \tau\big]^{-1},$$

where $A_1 = \frac{1}{12} a^{5/3} b^{1/3} (-1)^{k+1}$, $A_2 = \mp\frac{1}{2} ab$; $k = 1$ and $k = 2$.

71. $y''_{xx} = (A_1 y^{-1/4} + A_2 x y^{-7/4})(y'_x)^{1/2}$.

Solution in the parametric form:

$$x = a\bigl(f_k^2 - \tau^2 \mp 4\wp_k^3\bigr), \qquad y = b\bigl[f_k - (-1)^k \tau\bigr]^{4/3},$$

$$\text{where } A_1 = ab^{-3/2} A_2, \quad A_2 = \begin{cases} \pm a^{-3} b^{11/4} \left(\mp \dfrac{a}{b}\right)^{1/2} & \text{if } k = 1, \\ \pm a^{-3} b^{11/4} \left(\pm \dfrac{a}{b}\right)^{1/2} & \text{if } k = 2. \end{cases}$$

72. $y''_{xx} = (A_1 x^{-7/4} y + A_2 x^{-1/4})(y'_x)^{5/2}$.

Solution in the parametric form:

$$x = a[f_k - (-1)^k \tau]^{4/3}, \qquad y = b(f_k^2 - \tau^2 \mp 4\wp_k^3),$$

$$\text{where } A_1 = \begin{cases} \mp a^{11/4} b^{-3} \left(\mp \dfrac{b}{a}\right)^{1/2} & \text{if } k = 1, \\ \mp a^{11/4} b^{-3} \left(\pm \dfrac{b}{a}\right)^{1/2} & \text{if } k = 2, \end{cases} \qquad A_2 = a^{-3/2} b A_1.$$

▶ *In the solutions of equations 73–92, the following notation is used: functions P_1 and P_2 are the general solutions of four modifications of the first Painlevé equation*

$$P_1'' = \pm 6P_1^2 + \tau, \qquad P_2'' = \pm 6P_2^2 - \tau$$

(in the case of the upper sign, the equation for P_1 is the canonical form of the first Painlevé equation; see Subsection 2.8.2). In addition,

$$\begin{aligned} Q_1 &= \pm 6P_1^2 + \tau, & Q_2 &= \pm 6P_2^2 - \tau, \\ R_1 &= 2P_1' - \tau^2, & R_2 &= 2P_2' + \tau^2, \\ S_1 &= 3\tau P_1' - 3P_1 - \tau^3, & S_2 &= 3\tau P_2' - 3P_2 + \tau^3, \\ T_1 &= \tau^2 P_1 \mp 1, & T_2 &= \tau^2 P_2 \mp 1, \\ U_1 &= (P_1')^2 - 2P_1 Q_1 \pm 8P_1^3, & U_2 &= (P_2')^2 - 2P_2 Q_2 \pm 8P_2^3, \\ V_1 &= P_1' Q_1' + P_1' - Q_1^2, & V_2 &= P_2' Q_2' - P_2' - Q_2^2, \\ W_1 &= \tau^3 P_1' + 3\tau^2 P_1 \mp 1 + \tau^5, & W_2 &= \tau^3 P_2' + 3\tau^2 P_2 \mp 1 - \tau^5, \\ Z_1 &= 6(\tau^3 P_1' - 4\tau^2 P_1 \pm 6) - \tau^5, & Z_2 &= 6(\tau^3 P_2' - 4\tau^2 P_2 \pm 6) + \tau^5, \end{aligned}$$

where primes stand for differentiation with respect to τ.

73. $y''_{xx} = (A_1 y + A_2 x^2)(y'_x)^3$.

Solution in the parametric form:

$$x = aP_k, \qquad y = b\tau,$$

where $A_1 = ab^{-3}(-1)^k$, $A_2 = \mp 6a^{-1}b^{-2}$; $k = 1$ and $k = 2$.

74. $y''_{xx} = (A_1 y^2 + A_2 x)(y'_x)^{5/2}$.

Solution in the parametric form:

$$x = aR_k, \qquad y = b\tau,$$

where $A_1 = ab^{-2}A_2(-1)^{k+1}$, $A_2 = -2a^{-1}b^{-1}\left(\pm\frac{3a}{b}\right)^{1/2}$; $k = 1$ and $k = 2$.

75. $y''_{xx} = (A_1 y + A_2 x^2)(y'_x)^{1/2}$.

Solution in the parametric form:

$$x = a\tau, \qquad y = bR_k,$$

where $A_1 = 2a^{-1}b^{-1}\left(\pm\frac{3b}{a}\right)^{1/2}$, $A_2 = a^{-2}bA_1(-1)^{k+1}$; $k = 1$ and $k = 2$.

76. $y''_{xx} = (A_1 y^{-4} + A_2 x^2 y^{-5})(y'_x)^3$.

Solution in the parametric form:

$$x = a\tau^{-1}P_k, \qquad y = b\tau^{-1},$$

where $A_1 = ab^2(-1)^k$, $A_2 = \mp 6a^{-1}b^3$; $k = 1$ and $k = 2$.

77. $y''_{xx} = (A_1 y^{-1/2} + A_2 x y^{-5/4})(y'_x)^{5/2}$.

Solution in the parametric form:

$$x = aS_k, \qquad y = b\tau^4,$$

where $A_1 = ab^{-3/4}A_2(-1)^{k+1}$, $A_2 = -\frac{1}{2}a^{-1}b^{-1/4}\left(\pm\frac{a}{2b}\right)^{1/2}$; $k = 1$ and $k = 2$.

78. $y''_{xx} = (A_1 x^{-5/4} y + A_2 x^{-1/2})(y'_x)^{1/2}$.

Solution in the parametric form:

$$x = a\tau^4, \qquad y = bS_k,$$

where $A_1 = \frac{1}{2}a^{-1/4}b^{-1}\left(\pm\frac{b}{2a}\right)^{1/2}$, $A_2 = a^{-3/4}bA_1(-1)^{k+1}$; $k = 1$ and $k = 2$.

79. $y''_{xx} = (A_1 y^{-8/7} + A_2 x^2 y^{-15/7})(y'_x)^3$.

Solution in the parametric form:

$$x = a\tau T_k, \qquad y = b\tau^7,$$

where $A_1 = \frac{1}{49}ab^{-6/7}(-1)^k$, $A_2 = \mp\frac{6}{49}a^{-1}b^{1/7}$; $k = 1$ and $k = 2$.

80. $y''_{xx} = (A_1 y^{-7/4} + A_2 x y^{-15/8})(y'_x)^{5/2}$.

Solution in the parametric form:

$$x = a\tau^{-6} W_k, \qquad y = b\tau^{-8},$$

where $A_1 = ab^{-1/8} A_2 (-1)^k$, $A_2 = \frac{1}{8} a^{-1} b^{7/8} \left(\mp \frac{3a}{b}\right)^{1/2}$; $k = 1$ and $k = 2$.

81. $y''_{xx} = (A_1 x^{-15/8} y + A_2 x^{-7/4})(y'_x)^{1/2}$.

Solution in the parametric form:

$$x = a\tau^{-8}, \qquad y = b\tau^{-6} W_k,$$

where $A_1 = -\frac{1}{8} a^{7/8} b^{-1} \left(\mp \frac{3b}{a}\right)^{1/2}$, $A_2 = a^{-1/8} b A_1 (-1)^k$; $k = 1$ and $k = 2$.

82. $y''_{xx} = (A_1 y^{-13/7} + A_2 x^2 y^{-20/7})(y'_x)^3$.

Solution in the parametric form:

$$x = a\tau^{-6} T_k, \qquad y = b\tau^{-7},$$

where $A_1 = \frac{1}{49} ab^{-1/7} (-1)^k$, $A_2 = \mp \frac{6}{49} a^{-1} b^{6/7}$; $k = 1$ and $k = 2$.

83. $y''_{xx} = (A_1 y^{-14/13} + A_2 x y^{-20/13})(y'_x)^{5/2}$.

Solution in the parametric form:

$$x = a\tau Z_k, \qquad y = b\tau^{13},$$

where $A_1 = ab^{-6/13} A_2 (-1)^{k+1}$, $A_2 = -\frac{2}{13} a^{-1} b^{7/13} \left(\pm \frac{a}{13b}\right)^{1/2}$; $k = 1$ and $k = 2$.

84. $y''_{xx} = (A_1 x^{-20/13} y + A_2 x^{-14/13})(y'_x)^{1/2}$.

Solution in the parametric form:

$$x = a\tau^{13}, \qquad y = b\tau Z_k,$$

where $A_1 = \frac{2}{13} a^{7/13} b^{-1} \left(\pm \frac{b}{13a}\right)^{1/2}$, $A_2 = a^{-6/13} b A_1 (-1)^{k+1}$; $k = 1$ and $k = 2$.

85. $y''_{xx} = (A_1 y + A_2 x^{-1/2})(y'_x)^3$.

Solution in the parametric form:

$$x = a(P'_k)^2, \qquad y = bP_k,$$

where $A_1 = \mp 24ab^{-3}$, $A_2 = 2a^{3/2} b^{-2} (-1)^k$; $k = 1$ and $k = 2$.

86. $y''_{xx} = (A_1 x^{-1/2} y^{-5/2} + A_2 y^{-4})(y'_x)^3$.

Solution in the parametric form:

$$x = aP_k^{-1}(P_k')^2, \qquad y = bP_k^{-1},$$

where $A_1 = 2a^{3/2}b^{1/2}(-1)^k$, $A_2 = \mp 24ab^2$; $k = 1$ and $k = 2$.

87. $y''_{xx} = (A_1 x^{-5/3} y + A_2 x^{1/3})(y'_x)^3$.

Solution in the parametric form:

$$x = aP_k^{3/2}, \qquad y = bU_k,$$

where $A_1 = \frac{3}{16}a^{8/3}b^{-3}$, $A_2 = \mp 8a^{-2}bA_1$; $k = 1$ and $k = 2$.

88. $y''_{xx} = (A_1 x^{-5/3} y^{-7/3} + A_2 x^{1/3} y^{-10/3})(y'_x)^3$.

Solution in the parametric form:

$$x = aP_k^{3/2}U_k^{-1}, \qquad y = bU_k^{-1},$$

where $A_1 = \frac{3}{16}a^{8/3}b^{1/3}$, $A_2 = \mp 8a^{-2}bA_1$; $k = 1$ and $k = 2$.

89. $y''_{xx} = (A_1 x^{-1/2} + A_2 y^{-3/2})(y'_x)^3$.

Solution in the parametric form:

$$x = aQ_k^2, \qquad y = b(P_k')^2,$$

where $A_1 = \mp 6a^{3/2}b^{-2}$, $A_2 = \frac{1}{2}ab^{-1/2}(-1)^{k+1}$; $k = 1$ and $k = 2$.

90. $y''_{xx} = (A_1 y^{-3/2} + A_2 x^{-1/2} y^{-5/2})(y'_x)^3$.

Solution in the parametric form:

$$x = a(P_k')^{-2}Q_k^2, \qquad y = b(P_k')^{-2},$$

where $A_1 = \frac{1}{2}ab^{-1/2}(-1)^{k+1}$, $A_2 = \mp 6a^{3/2}b^{1/2}$; $k = 1$ and $k = 2$.

91. $y''_{xx} = (A_1 x^{-5/3} y + A_2 x^{-4/3})(y'_x)^3$.

Solution in the parametric form:

$$x = a(P_k')^3, \qquad y = bV_k,$$

where $A_1 = -\frac{1}{36}a^{8/3}b^{-3}$, $A_2 = a^{-1/3}bA_1(-1)^k$; $k = 1$ and $k = 2$.

92. $y''_{xx} = (A_1 x^{-4/3} y^{-5/3} + A_2 x^{-5/3} y^{-7/3})(y'_x)^3$.

Solution in the parametric form:

$$x = a(P_k')^3V_k^{-1}, \qquad y = bV_k^{-1},$$

where $A_1 = \frac{1}{36}a^{7/3}b^{-1/3}(-1)^{k+1}$, $A_2 = -\frac{1}{36}a^{8/3}b^{1/3}$; $k = 1$ and $k = 2$.

▶ *In the solutions of equations 93–96, the following notation is used:*

$$r = \begin{cases} C_1 + \dfrac{1}{4}\tau^2 + \dfrac{2B_1}{k_1+1}\tau^{k_1+1} + \dfrac{2B_2}{k_2+1}\tau^{k_2+1} & \text{if } k_1 \neq -1,\ k_2 \neq -1; \\ C_1 + \dfrac{1}{4}\tau^2 + \dfrac{2B_1}{k+1}\tau^{k+1} + 2B_2 \ln|\tau| & \text{if } k = k_1 \neq -1,\ k_2 = -1; \end{cases}$$

$$F = C_2 \exp\biggl(\int \frac{d\tau}{\sqrt{r}}\biggr), \qquad G = \tau + 2\sqrt{r} + 4B_2,$$

$$H = \int \Bigl(C_1 - \frac{1}{2A_1A_2}\tau^{-4} - \tau^2\Bigr)^{-1/2} d\tau + C_2.$$

93. $y''_{xx} = (A_1 x^{-2m_1-3} y^{m_1} + A_2 x^{-2m_2-3} y^{m_2})(y'_x)^3.$

Solution in the parametric form:

$$x = \tau F^{1/2}, \qquad y = F,$$

where $k_1 = -2m_1 - 3,\ k_2 = -2m_2 - 3,\ A_1 = -B_1,\ A_2 = -B_2.$

94. $y''_{xx} = \Bigl(A_1 x y^m + A_2 y^{\frac{m-1}{2}}\Bigr)(y'_x)^{\frac{m+5}{m+3}}.$

Solution in the parametric form:

$$x = aF^{-1/2}G, \qquad y = bF^{\frac{1}{m+1}},$$

where $k_1 = k = -\dfrac{m+3}{m+1},\ k_2 = 0,\ A_1 = \dfrac{k}{k+1}a^{-1}b^{k+1}\Bigl[-\dfrac{4aB_1}{(k+1)b}\Bigr]^{1/k},\ A_2 = -4ab^{-\frac{1}{k+1}}A_1B_2.$

95. $y''_{xx} = \Bigl(A_1 x^n y + A_2 x^{\frac{n-1}{2}}\Bigr)(y'_x)^{\frac{2n+4}{n+3}}.$

Solution in the parametric form:

$$x = aF^{\frac{1}{n+1}}, \qquad y = F^{-1/2}G,$$

where $k_1 = k = -\dfrac{n+3}{n+1},\ k_2 = 0,\ A_1 = -\dfrac{k}{k+1}a^{\frac{2}{k+1}}b^{-1}\Bigl[-\dfrac{4bB_1}{(k+1)a}\Bigr]^{1/k},\ A_2 = -4a^{-\frac{1}{k+1}}bA_1B_2.$

96. $y''_{xx} = (A_1 x^{-5} y^2 + A_2 x^{-5})(y'_x)^3.$

Solution in the parametric form:

$$x = \frac{\tau\sqrt{-A_2}}{\cos H}, \qquad y = \sqrt{\frac{A_2}{A_1}}\tan H.$$

97. $y''_{xx} = (A_1 x^{m_1} y^{m_1} + A_2 x^{m_2} y^{m_2})(y'_x)^{3/2}$.

Solution in the parametric form:

$$x = C_1 \tau^{1/2} \exp\left(-\frac{1}{2}\int \frac{f\,d\tau}{\tau\sqrt{f^2+4}}\right), \qquad y = C_1^{-1}\tau^{1/2}\exp\left(\frac{1}{2}\int \frac{f\,d\tau}{\tau\sqrt{f^2+4}}\right),$$

where

$$f = \begin{cases} \tau^{-1/2}\left[C_2 + \dfrac{A_1}{2(m_1+1)}\tau^{m_1+1} + \dfrac{A_2}{2(m_2+1)}\tau^{m_2+1}\right] & \text{if } m_1 \neq -1,\ m_2 \neq -1, \\ \tau^{-1/2}\left[C_2 + \dfrac{A_1}{2(m+1)}\tau^{m+1} + \dfrac{1}{2}A_2 \ln\tau\right] & \text{if } m_1 = m \neq -1,\ m_2 = -1. \end{cases}$$

98. $y''_{xx} = \left(A_1 x^n y^{m_1} + A_2 x^{\frac{m_2(n+1)-m_1+n}{m_1+1}} y^{m_2}\right)(y'_x)^{\frac{m_1+2n+3}{m_1+n+2}}$.

Solution in the parametric form:

$$x = C_1 \exp\left(\int \frac{d\tau}{\tau z}\right), \qquad y = C_1^{-\frac{n+1}{m_1+1}}\,\tau \exp\left(-\frac{n+1}{m_1+1}\int \frac{d\tau}{\tau z}\right),$$

where $z = z(\tau)$ is the solution of the algebraic equation

$$\left(z - \frac{m_1+n+2}{m_1+1}\right)\left(z - \frac{n+1}{m_1+1}\right)^{\frac{n+1}{m_1+n+2}} = \tau^{-\frac{m_1+1}{m_1+n+2}}\left(C_2 + \frac{A_1}{m_1+n+2}\tau^{m_1+1} + F\right),$$

$$F = \begin{cases} \dfrac{A_2(m_1+1)}{(m_1+n+2)(m_2+1)}\tau^{m_2+1} & \text{if } m_2 \neq -1, \\ \dfrac{A_2}{m_1+n+2}\ln|\tau| & \text{if } m_2 = -1. \end{cases}$$

▶ *In the solutions of equations 99–108, the following notation is used: functions P_1 and P_2 are the general solutions of four modifications of the second Painlevé equation (with parameter $a = 0$)*

$$P_1'' = \tau P_1 \pm 2P_1^3, \qquad P_2'' = -\tau P_2 \pm 2P_2^3.$$

In the case of the upper sign, the equation for P_1 is the canonical form of the second Painlevé equation (with parameter $a = 0$; see Subsection 2.8.2);

$$\begin{aligned} Q_1 &= \tau P_1^2 \pm P_1^4 - (P_1')^2, & Q_2 &= \tau P_2^2 \pm P_2^4 - (P_2')^2, \\ R_1 &= P_1' \mp P_1 Q_1, & R_2 &= P_2' \pm P_2 Q_2, \\ S_1 &= 2P_1' Q_1 - P_1^3 \mp P_1 Q_1^2, & S_2 &= 2P_2' Q_2 + P_2^3 \pm P_2 Q_2^2, \end{aligned}$$

where primes stand for differentiation with respect to τ.

99. $y''_{xx} = (A_1 xy + A_2 x^3)(y'_x)^3$.

Solution in the parametric form:

$$x = aP_k, \qquad y = b\tau,$$

where $A_1 = b^3(-1)^k$, $A_2 = \mp 2a^{-2}b^{-2}$; $k = 1$ and $k = 2$.

100. $y''_{xx} = (A_1 x y^{-5} + A_2 x^3 y^{-6})(y'_x)^3$.

Solution in the parametric form:

$$x = a\tau^{-1} P_k, \qquad y = b\tau^{-1},$$

where $A_1 = b^3(-1)^k$, $A_2 = \mp 2a^{-2}b^4$; $k = 1$ and $k = 2$.

101. $y''_{xx} = (A_1 + A_2 x^{-1/2} y^{-1/2})(y'_x)^3$.

Solutions in the parametric form:

$$x = a(P'_k)^2, \quad y = bP_k^2, \qquad P'_k = (P_k)'_\tau,$$

where $A_1 = \mp 2ab^{-2}$, $A_2 = \frac{1}{2}a^{3/2}b^{-3/2}(-1)^k$; $k = 1$ and $k = 2$.

102. $y''_{xx} = (A_1 x^{-1/2} y^{-2} + A_2 y^{-3})(y'_x)^3$.

Solutions in the parametric form:

$$x = aP_k^{-2}(P'_k)^2, \quad y = bP_k^{-2}, \qquad P'_k = (P_k)'_\tau,$$

where $A_1 = \frac{1}{2}a^{3/2}(-1)^k$, $A_2 = \mp 2ab$; $k = 1$ and $k = 2$.

103. $y''_{xx} = (A_1 x^{-2} y + A_2)(y'_x)^3$.

Solutions in the parametric form:

$$x = aP_k^2, \quad y = b\big[\tau P_k^2 \pm P_k^4 - (P'_k)^2\big], \qquad P'_k = (P_k)'_\tau,$$

where $A_1 = 2a^3b^{-3}(-1)^k$, $A_2 = \pm 2ab^{-2}(-1)^k$; $k = 1$ and $k = 2$.

104. $y''_{xx} = (A_1 x^{-2} y^{-2} + A_2 y^{-3})(y'_x)^3$.

Solutions in the parametric form:

$$x = aP_k^2\big[\tau P_k^2 \pm P_k^4 - (P'_k)^2\big]^{-1}, \quad y = b\big[\tau P_k^2 \pm P_k^4 - (P'_k)^2\big]^{-1},$$

where $A_1 = -2a^3$, $A_2 = \mp 2ab$; $k = 1$ and $k = 2$.

105. $y''_{xx} = (A_1 + A_2 x y^{-1/2})(y'_x)^{3/2}$.

Solutions in the parametric form:

$$x = aP_k^{-1} R_k, \quad y = bQ_k^2,$$

$$\text{where} \quad A_1 = \mp ab^{-1/2} A_2 (-1)^k, \quad A_2 = \begin{cases} 4a^{-2}b^{1/2}\left(\dfrac{a}{2b}\right)^{1/2} & \text{if } k = 1, \\ -4a^{-2}b^{1/2}\left(-\dfrac{a}{2b}\right)^{1/2} & \text{if } k = 2, \end{cases}.$$

106. $y''_{xx} = (A_1 x^{-1/2} y + A_2)(y'_x)^{3/2}$.

Solutions in the parametric form:

$$x = aQ_k^2, \quad y = bP_k^{-1} R_k,$$

where $A_1 = \begin{cases} -4a^{1/2} b^{-2} \left(\dfrac{b}{2a}\right)^{1/2} & \text{if } k = 1, \\ 4a^{1/2} b^{-2} \left(-\dfrac{b}{2a}\right)^{1/2} & \text{if } k = 2, \end{cases} \qquad A_2 = \mp a^{-1/2} b A_1 (-1)^k.$

107. $y''_{xx} = (A_1 + A_2 x y^{-2})(y'_x)^{3/2}$.

Solutions in the parametric form:

$$x = aS_k, \quad y = bQ_k,$$

where $A_1 = \mp ab^{-2} A_2 (-1)^k, \quad A_2 = \begin{cases} a^{-2} b^2 \left(\dfrac{2a}{b}\right)^{1/2} & \text{if } k = 1, \\ a^{-2} b^2 \left(-\dfrac{2a}{b}\right)^{1/2} & \text{if } k = 2. \end{cases}$

108. $y''_{xx} = (A_1 x^{-2} y + A_2)(y'_x)^{3/2}$.

Solutions in the parametric form:

$$x = aQ_k, \quad y = bS_k,$$

where $A_1 = \begin{cases} -a^2 b^{-2} \left(\dfrac{2b}{a}\right)^{1/2} & \text{if } k = 1, \\ -a^2 b^{-2} \left(-\dfrac{2b}{a}\right)^{1/2} & \text{if } k = 2, \end{cases} \qquad A_2 = \mp a^{-2} b A_1 (-1)^k.$

109. $y''_{xx} = (A_1 x^{-7/5} y^{-8/5} + A_2 x^{-7/5} y^{-13/5})(y'_x)^3$.

Solution in the parametric form:

$$x = aC_1^5 S^{5/2} \left(bC_1^4 F - \frac{A_1}{A_2}\right)^{-1}, \qquad y = \left(bC_1^4 F - \frac{A_1}{A_2}\right)^{-1},$$

where $S = C_1 e^{2k\tau} + C_2 e^{-k\tau} \sin\left(\sqrt{3}\, k\tau\right)$, $F = (S'_\tau)^2 - 2SS''_{\tau\tau}$, $A_2 = \frac{5}{1024} a^{12/5} b^{-3} k^{-6}$.

110. $y''_{xx} = (A_1 x^{-2} y^{-1} + A_2 x^{-2} y^{-2})(y'_x)^3$.

Solution in the parametric form:

$$x = aC_1 \tau^{2/3} Z^2 \left\{ bC_1 \tau^{-2/3} \left[(\tau Z'_\tau + \tfrac{1}{3} Z)^2 \pm \tau^2 Z^2 \right] - \frac{A_1}{A_2} \right\}^{-1},$$

$$y = \left\{ bC_1 \tau^{-2/3} \left[(\tau Z'_\tau + \tfrac{1}{3} Z)^2 \pm \tau^2 Z^2 \right] - \frac{A_1}{A_2} \right\}^{-1},$$

where

$$Z = \begin{cases} C_1 J_{1/3}(\tau) + C_2 Y_{1/3}(\tau) & \text{for the upper sign,} \\ C_1 I_{1/3}(\tau) + C_2 K_{1/3}(\tau) & \text{for the lower sign,} \end{cases}$$

$J_{1/3}$ and $Y_{1/3}$ are Bessel functions, $I_{1/3}$ and $K_{1/3}$ are modified Bessel functions; $A_2 = \frac{9}{2} a^3 b^{-3}$.

111. $y''_{xx} = (A_1 x^{-7} y^4 + A_2 x^{-7} y^3)(y'_x)^3$.

Solution in the parametric form:

$$x = aC_1^3 F^{1/2}\Bigl(bC_1^8 G - \frac{A_1}{A_2}\Bigr)^{-1}, \qquad y = \Bigl(bC_1^8 G - \frac{A_1}{A_2}\Bigr)^{-1},$$

where $R = \sqrt{\pm(4\tau^3 - 1)}$, $F = 2\tau \int \tau R^{-1}\,d\tau + C_2\tau \mp R$, $G = 4\tau F^2 \mp \tau^{-2}(RF-1)^2$, $A_2 = \mp\frac{3}{64} a^8 b^{-3}$.

► *In the solutions of equations 112–113, the following notation is used:*

$$E = \int (1 \pm \tau^4)^{-1/2}\,d\tau + C_2, \qquad k^2 = \pm 1;$$

function E can be expressed in terms of the elliptic integrals or the lemniscate functions.

112. $y''_{xx} = (A_1 x^2 y^{-14/5} + A_2 x^3 y^{-18/5})(y'_x)^3$.

Solutions in the parametric form:

$$x = aC_1^4 E^{-4}(\tau E - k), \qquad y = bC_1^5 E^{-5},$$

where $A_1 = \mp\frac{6}{25} a^{-1} b^{4/5} k$, $A_2 = \mp\frac{2}{25} a^{-2} b^{8/5}$.

113. $y''_{xx} = (A_1 x^2 y^{-11/5} + A_2 x^3 y^{-12/5})(y'_x)^3$.

Solutions in the parametric form:

$$x = aC_1 E(\tau E - k), \qquad y = bC_1^5 E^5,$$

where $A_1 = \mp\frac{6}{25} a^{-1} b^{1/5} k$, $A_2 = \mp\frac{2}{25} a^{-2} b^{2/5}$.

► *In the solutions of equations 114–115, the following notation is used:*

$$\Delta = C_2^2 - 2C_1, \qquad R = (36\Delta + 54B\tau - 2\tau^3)^{1/2}, \qquad z = 3\int \tau^{-1} R^{-1}\,d\tau,$$

$$W(z) = \begin{cases} \dfrac{\sqrt{-\Delta}}{C_1}\tan\bigl(\pm\sqrt{-\Delta}\,z\bigr) + \dfrac{C_2}{C_1} & \textit{if } \Delta < 0; \\ \dfrac{\sqrt{\Delta}}{C_1}\tanh\bigl(\mp\sqrt{\Delta}\,z\bigr) + \dfrac{C_2}{C_1} & \textit{if } \Delta > 0; \\ \mp\dfrac{1}{C_1 z} - \sqrt{\dfrac{2}{|C_1|}} & \textit{if } \Delta = 0,\ C_2 < 0; \\ \mp\dfrac{1}{C_1 z} + \sqrt{\dfrac{2}{|C_1|}} & \textit{if } \Delta = 0,\ C_2 > 0. \end{cases}$$

114. $y''_{xx} = (A_1 x^{-5/3} + A_2 x^{-5/3} y^{-2/3})(y'_x)^3.$

Solutions in the parametric form:

$$x = a\tau^{-9/4}(C_1 W^2 - 2C_2 W + 2)^{3/4}(6C_1 W - 6C_2 \mp R)^{3/2},$$
$$y = b\tau^{-3/2}(C_1 W^2 - 2C_2 W + 2)^{3/2},$$

where $A_1 = -24a^{8/3}b^{-2}C_1$, $A_2 = 36a^{8/3}b^{-4/3}B$.

115. $y''_{xx} = (A_1 x^{-5/3} y^{-2/3} + A_2 x^{-5/3} y^{-4/3})(y'_x)^3.$

Solutions in the parametric form:

$$x = a\tau^{-3/4}(C_1 W^2 - 2C_2 W + 2)^{-3/4}(6C_1 W - 6C_2 \mp R)^{3/2},$$
$$y = b\tau^{3/2}(C_1 W^2 - 2C_2 W + 2)^{-3/2},$$

where $A_1 = 36a^{8/3}b^{-4/3}B$, $A_2 = -24a^{8/3}b^{-2/3}C_1$.

116. $y''_{xx} = \left[\frac{2(n+1)}{(n+3)^2}x + Ax^n\right]y^{-2}(y'_x)^3, \qquad n \neq -3, \quad n \neq -1.$

Assuming y as the independent variable, we obtain an equation of the form 2.4.2.4 for function $x = x(y)$:

$$x''_{yy} = y^{-2}\left[-\frac{2(n+1)}{(n+3)^2}x - Ax^n\right].$$

117. $y''_{xx} = (-\frac{15}{4}x + Ax^{-7})y^{-2}(y'_x)^3.$

Assuming y as the independent variable, we obtain an equation of the form 2.4.2.35 for function $x = x(y)$:

$$x''_{yy} = y^{-2}(\tfrac{15}{4}x - Ax^{-7}).$$

118. $y''_{xx} = (-6x + Ax^{-4})y^{-2}(y'_x)^3.$

Assuming y as the independent variable, we obtain an equation of the form 2.4.2.31 for function $x = x(y)$:

$$x''_{yy} = y^{-2}(6x - Ax^{-4}).$$

119. $y''_{xx} = (-12x + Ax^{-5/2})y^{-2}(y'_x)^3.$

Assuming y as the independent variable, we obtain an equation of the form 2.4.2.64 for function $x = x(y)$:

$$x''_{yy} = y^{-2}(12x - Ax^{-5/2}).$$

120. $y''_{xx} = (-2x + Ax^{-2})y^{-2}(y'_x)^3.$

Assuming y as the independent variable, we obtain an equation of the form 2.4.2.6 for function $x = x(y)$:

$$x''_{yy} = y^{-2}(2x - Ax^{-2}).$$

121. $y''_{xx} = (\frac{3}{16}x + Ax^{-5/3})y^{-2}(y'_x)^3$.

Assuming y as the independent variable, we obtain an equation of the form 2.4.2.26 for function $x = x(y)$:

$$x''_{yy} = y^{-2}(-\tfrac{3}{16}x - Ax^{-5/3}).$$

122. $y''_{xx} = (\frac{9}{100}x + Ax^{-5/3})y^{-2}(y'_x)^3$.

Assuming y as the independent variable, we obtain an equation of the form 2.4.2.10 for function $x = x(y)$:

$$x''_{yy} = y^{-2}(-\tfrac{9}{100}x - Ax^{-5/3}).$$

123. $y''_{xx} = (-\frac{3}{4}x + Ax^{-5/3})y^{-2}(y'_x)^3$.

Assuming y as the independent variable, we obtain an equation of the form 2.4.2.12 for function $x = x(y)$:

$$x''_{yy} = y^{-2}(\tfrac{3}{4}x - Ax^{-5/3}).$$

124. $y''_{xx} = (-\frac{63}{4}x + Ax^{-5/3})y^{-2}(y'_x)^3$.

Assuming y as the independent variable, we obtain an equation of the form 2.4.2.66 for function $x = x(y)$:

$$x''_{yy} = y^{-2}(\tfrac{63}{4}x - Ax^{-5/3}).$$

125. $y''_{xx} = (\frac{5}{36}x + Ax^{-7/5})y^{-2}(y'_x)^3$.

Assuming y as the independent variable, we obtain an equation of the form 2.4.2.29 for function $x = x(y)$:

$$x''_{yy} = y^{-2}(-\tfrac{5}{36}x - Ax^{-7/5}).$$

126. $y''_{xx} = (\frac{2}{9}x + Ax^{-1/2})y^{-2}(y'_x)^3$.

Assuming y as the independent variable, we obtain an equation of the form 2.4.2.14 for function $x = x(y)$:

$$x''_{yy} = y^{-2}(-\tfrac{2}{9}x - Ax^{-1/2}).$$

127. $y''_{xx} = (\frac{4}{25}x + Ax^{-1/2})y^{-2}(y'_x)^3$.

Assuming y as the independent variable, we obtain an equation of the form 2.4.2.8 for function $x = x(y)$:

$$x''_{yy} = y^{-2}(-\tfrac{4}{25}x - Ax^{-1/2}).$$

128. $y''_{xx} = (-20x + Ax^{-1/2})y^{-2}(y'_x)^3$.

Assuming y as the independent variable, we obtain an equation of the form 2.4.2.33 for function $x = x(y)$:

$$x''_{yy} = y^{-2}(20x - Ax^{-1/2}).$$

129. $y''_{xx} = (\frac{12}{49}x + Ax^{1/2})y^{-2}(y'_x)^3$.

Assuming y as the independent variable, we obtain an equation of the form 2.4.2.37 for function $x = x(y)$:

$$x''_{yy} = y^{-2}(-\tfrac{12}{49}x - Ax^{1/2}).$$

130. $y''_{xx} = (Ax^2 + \frac{6}{25}x)y^{-2}(y'_x)^3$.

Assuming y as the independent variable, we obtain an equation of the form 2.4.2.60 for function $x = x(y)$:

$$x''_{yy} = y^{-2}(-Ax^2 - \tfrac{6}{25}x).$$

131. $y''_{xx} = (Ax^2 - \frac{6}{25}x)y^{-2}(y'_x)^3$.

Assuming y as the independent variable, we obtain an equation of the form 2.4.2.62 for function $x = x(y)$:

$$x''_{yy} = y^{-2}(-Ax^2 + \tfrac{6}{25}x).$$

132. $y''_{xx} = \left[\dfrac{2(n+1)}{(n+3)^2}xy^{-2} + Ax^n y^{-n-1}\right](y'_x)^3, \qquad n \neq -3,\ n \neq -1.$

Assuming y as the independent variables, we obtain an equation of the form 2.4.2.5 for function $x = x(y)$:

$$x''_{yy} = \left[-\frac{2(n+1)}{(n+3)^2}y^{-2}x - Ay^{-n-1}x^n\right].$$

133. $y''_{xx} = (Ax^{-2}y - 2xy^{-2})(y'_x)^3$.

Assuming y as the independent variables, we obtain an equation of the form 2.4.2.7 for function $x = x(y)$:

$$x''_{yy} = (2y^{-2}x - Ayx^{-2}).$$

134. $y''_{xx} = (Ax^{-1/2}y^{-1/2} + \frac{4}{25}xy^{-2})(y'_x)^3$.

Assuming y as the independent variables, we obtain an equation of the form 2.4.2.9 for function $x = x(y)$:

$$x''_{yy} = (-\tfrac{4}{25}y^{-2}x - Ay^{-1/2}x^{-1/2}).$$

135. $y''_{xx} = (Ax^{-5/3}y^{2/3} + \frac{9}{100}xy^{-2})(y'_x)^3$.

Assuming y as the independent variables, we obtain an equation of the form 2.4.2.11 for function $x = x(y)$:

$$x''_{yy} = (-\tfrac{9}{100}y^{-2}x - Ay^{2/3}x^{-5/3}).$$

136. $y''_{xx} = (Ax^{-5/3}y^{2/3} - \frac{3}{4}xy^{-2})(y'_x)^3$.

Assuming y as the independent variables, we obtain an equation of the form 2.4.2.13 for function $x = x(y)$:

$$x''_{yy} = (\tfrac{3}{4}y^{-2}x - Ay^{2/3}x^{-5/3}).$$

137. $y''_{xx} = (Ax^{-1/2}y^{-1/2} + \frac{2}{9}xy^{-2})(y'_x)^3$.

Assuming y as the independent variables, we obtain an equation of the form 2.4.2.15 for function $x = x(y)$:

$$x''_{yy} = (-\tfrac{2}{9}y^{-2}x - Ay^{-1/2}x^{-1/2}).$$

138. $y''_{xx} = (Ax^{-5/3}y^{2/3} + \frac{3}{16}xy^{-2})(y'_x)^3$.

Assuming y as the independent variables, we obtain an equation of the form 2.4.2.27 for function $x = x(y)$:

$$x''_{yy} = (-\tfrac{3}{16}y^{-2}x - Ay^{2/3}x^{-5/3}).$$

139. $y''_{xx} = (Ax^{-7/5}y^{2/5} + \frac{5}{36}xy^{-2})(y'_x)^3$.

Assuming y as the independent variables, we obtain an equation of the form 2.4.2.30 for function $x = x(y)$:

$$x''_{yy} = (-\tfrac{5}{36}y^{-2}x - Ay^{2/5}x^{-7/5}).$$

140. $y''_{xx} = (Ax^{-4}y^3 - 6xy^{-2})(y'_x)^3$.

Assuming y as the independent variables, we obtain an equation of the form 2.4.2.32 for function $x = x(y)$:

$$x''_{yy} = (6y^{-2}x - Ay^3x^{-4}).$$

141. $y''_{xx} = (Ax^{-1/2}y^{-1/2} - 20xy^{-2})(y'_x)^3$.

Assuming y as the independent variables, we obtain an equation of the form 2.4.2.34 for function $x = x(y)$:

$$x''_{yy} = (20y^{-2}x - Ay^{-1/2}x^{-1/2}).$$

142. $y''_{xx} = (Ax^{-7}y^6 - \frac{15}{4}xy^{-2})(y'_x)^3$.

Assuming y as the independent variables, we obtain an equation of the form 2.4.2.36 for function $x = x(y)$:

$$x''_{yy} = (\tfrac{15}{4}y^{-2}x - Ay^6x^{-7}).$$

143. $y''_{xx} = (Ax^{1/2}y^{-3/2} + \frac{12}{49}xy^{-2})(y'_x)^3$.

Assuming y as the independent variables, we obtain an equation of the form 2.4.2.38 for function $x = x(y)$:

$$x''_{yy} = (-\tfrac{12}{49}y^{-2}x - Ay^{-3/2}x^{1/2}).$$

144. $y''_{xx} = (\frac{6}{25}xy^{-2} + Ax^2y^{-3})(y'_x)^3$.

Assuming y as the independent variables, we obtain an equation of the form 2.4.2.61 for function $x = x(y)$:

$$x''_{yy} = (-Ay^{-3}x^2 - \tfrac{6}{25}y^{-2}x).$$

145. $y''_{xx} = (-\frac{6}{25}xy^{-2} + Ax^2y^{-3})(y'_x)^3$.

Assuming y as the independent variables, we obtain an equation of the form 2.4.2.63 for function $x = x(y)$:

$$x''_{yy} = (-Ay^{-3}x^2 + \tfrac{6}{25}y^{-2}x).$$

146. $y''_{xx} = (Ax^{-5/2}y^{3/2} - 12xy^{-2})(y'_x)^3.$

Assuming y as the independent variables, we obtain an equation of the form 2.4.2.65 for function $x = x(y)$:

$$x''_{yy} = (12y^{-2}x - Ay^{3/2}x^{-5/2}).$$

147. $y''_{xx} = (Ax^{-5/3}y^{2/3} - \frac{63}{4}xy^{-2})(y'_x)^3.$

Assuming y as the independent variables, we obtain an equation of the form 2.4.2.67 for function $x = x(y)$:

$$x''_{yy} = (\tfrac{63}{4}y^{-2}x - Ay^{2/3}x^{-5/3}).$$

2.6.3. Equations of the Form $y''_{xx} = \sigma A x^n y^m (y'_x)^l + A x^{n-1} y^{m+1} (y'_x)^{l-1}$

Table 2.12 represents all solvable equations whose solutions are outlined in Subsection 2.6.3. The two-parameter families (in the space of parameters n, m, and l), one-parameter families, and isolated points are represented in a consecutive fashion. Equations are arranged in accordance with the growth of l. The number of the equation sought is indicated in the last column in this table.

1. $y''_{xx} = Ax^{-m-2}y^m y'_x - Ax^{-m-3}y^{m+1}, \qquad m \neq -1.$

Solution in the parametric form:

$$x = aC_1^m\left(\int \frac{d\tau}{1 \pm \tau^{m+1}} + C_2\right)^{-1}, \qquad y = bC_1^{m+1}\tau\left(\int \frac{d\tau}{1 \pm \tau^{m+1}} + C_2\right)^{-1},$$

where $A = \mp(m+1)a^{m+1}b^{-m}$.

2. $y''_{xx} = Ax^{-1}y^{-1}y'_x - Ax^{-2}.$

Solution in the parametric form:

$$x = C_1\left[\int \tau^{-1}\exp(\mp\tau^2)\,d\tau + C_2\right]^{-1},$$

$$y = -\frac{A}{2}\exp(\mp\tau^2)\left[\int \tau^{-1}\exp(\mp\tau^2)\,d\tau + C_2\right]^{-1}.$$

3. $y''_{xx} = Ax^{-m-2}y^m(y'_x)^3 - Ax^{-m-3}y^{m+1}(y'_x)^2, \qquad m \neq -2.$

Solution in the parametric form:

$$x = aC_1^{m+2}\tau\left(\int \frac{d\tau}{1 \pm \tau^{-m-2}} + C_2\right)^{-1}, \qquad y = bC_1^{m+3}\left(\int \frac{d\tau}{1 \pm \tau^{-m-2}} + C_2\right)^{-1},$$

where $A = \pm(m+2)a^{m+3}b^{-m-2}$.

TABLE 2.12
Solvable cases of the equation $y''_{xx} = \sigma A x^n y^m (y'_x)^l + A x^{n-1} y^{m+1} (y'_x)^{l-1}$

l	m	n	σ	Equation
arbitrary ($l \neq 2$)	arbitrary ($m \neq -1$)	$-m-1$	-1	2.6.3.75
arbitrary ($l \neq 2$)	$1-l$	$l-2$	-1	2.6.3.76
arbitrary ($l \neq 3$)	-2	1	-1	2.6.3.79
arbitrary ($l \neq 1$)	0	-1	-1	2.6.3.80
$\dfrac{m+3}{m+2}$	arbitrary ($m \neq -1, -2$)	1	$m+1$	2.6.3.74
$\dfrac{3n+2}{n+1}$	0	arbitrary ($n \neq 0, -1$)	$\dfrac{1}{n}$	2.6.3.73
1	arbitrary ($m \neq -1$)	$-m-2$	-1	2.6.3.1
1	0	arbitrary ($n \neq -1$)	$\dfrac{1}{n}$	2.6.3.23
1	1	arbitrary ($n \neq 0, -2$)	$\dfrac{2}{n}$	2.6.3.37
$\dfrac{3}{2}$	0	arbitrary ($n \neq 0, -1$)	$\dfrac{1}{n}$	2.6.3.41
2	arbitrary ($m \neq -1$)	arbitrary ($n \neq 0$)	-1	2.6.3.85
2	arbitrary ($m \neq -1$)	$m+1$	-1	2.6.3.82
2	arbitrary ($m \neq -1$)	$-2m-2$	-1	2.6.3.83
2	arbitrary ($m \neq -1$)	$-\dfrac{m+1}{2}$	-1	2.6.3.84
2	arbitrary ($m \neq -1$)	0	arbitrary	2.6.3.87
2	-1	arbitrary ($n \neq 0$)	arbitrary	2.6.3.86
$\dfrac{5}{2}$	arbitrary ($m \neq -1, -2$)	1	$m+1$	2.6.3.42
3	arbitrary ($m \neq -2$)	$-m-2$	-1	2.6.3.3
3	arbitrary ($m \neq -2$)	1	$m+1$	2.6.3.24
3	arbitrary ($m \neq -1, -3$)	2	$\dfrac{m+1}{2}$	2.6.3.38

TABLE 2.12 *Continued*

Solvable cases of the equation $y''_{xx} = \sigma A x^n y^m (y'_x)^l + A x^{n-1} y^{m+1} (y'_x)^{l-1}$

l	m	n	σ	Equation
0	-3	-1	2	2.6.3.65
0	-3	$\frac{1}{2}$	-4	2.6.3.61
0	-3	2	-1	2.6.3.35
0	0	-2	-1	2.6.3.48
0	0	-1	-2	2.6.3.50
0	0	-1	-1	2.6.3.33
0	0	$-\frac{2}{5}$	$-\frac{5}{2}$	2.6.3.59
0	0	2	$\frac{1}{2}$	2.6.3.63
1	-3	1	-1	2.6.3.15
1	-2	-2	$\frac{1}{2}$	2.6.3.51
1	-2	-1	arbitrary	2.6.3.71
1	-2	-1	-1	2.6.3.7
1	-2	-1	1	2.6.3.5
1	-2	$-\frac{1}{2}$	2	2.6.3.55
1	-2	$\frac{1}{2}$	-1	2.6.3.13
1	-2	1	arbitrary	2.6.3.69
1	-2	1	-2	2.6.3.11
1	-2	1	-1	2.6.3.29
1	-1	-1	-1	2.6.3.2
1	$-\frac{1}{2}$	-2	$-\frac{1}{4}$	2.6.3.53
1	$-\frac{1}{2}$	-1	-1	2.6.3.45
1	$-\frac{1}{2}$	$-\frac{1}{2}$	-1	2.6.3.31
1	$-\frac{1}{2}$	1	$\frac{1}{2}$	2.6.3.57
1	0	-1	-1	2.6.3.77
1	0	1	arbitrary	2.6.3.67
1	0	1	-1	2.6.3.9
1	1	-4	$-\frac{1}{2}$	2.6.3.21
1	1	-1	-2	2.6.3.39
1	1	-2	-1	2.6.3.25
1	1	1	1	2.6.3.17

TABLE 2.12 *Continued*

Solvable cases of the equation $y''_{xx} = \sigma A x^n y^m (y'_x)^l + A x^{n-1} y^{m+1} (y'_x)^{l-1}$

l	m	n	σ	Equation
$\frac{3}{2}$	0	−1	−1	2.6.3.27
$\frac{3}{2}$	0	$-\frac{1}{2}$	−2	2.6.3.43
$\frac{3}{2}$	0	1	1	2.6.3.19
2	−1	0	arbitrary	2.6.3.81
$\frac{5}{2}$	−2	1	−1	2.6.3.28
$\frac{5}{2}$	$-\frac{3}{2}$	1	$-\frac{1}{2}$	2.6.3.44
$\frac{5}{2}$	0	1	1	2.6.3.20
3	−5	2	−2	2.6.3.22
3	−3	−1	2	2.6.3.52
3	−3	$\frac{1}{2}$	−4	2.6.3.54
3	−3	2	−1	2.6.3.26
3	−2	−1	arbitrary	2.6.3.72
3	−2	−1	−1	2.6.3.8
3	−2	−1	1	2.6.3.6
3	−2	0	−1	2.6.3.4
3	−2	$\frac{1}{2}$	−1	2.6.3.46
3	−2	1	−1	2.6.3.78
3	−2	2	$-\frac{1}{2}$	2.6.3.40
3	$-\frac{3}{2}$	−1	$\frac{1}{2}$	2.6.3.56
3	$-\frac{3}{2}$	$\frac{1}{2}$	−1	2.6.3.32
3	$-\frac{1}{2}$	−1	−1	2.6.3.14
3	0	−2	−1	2.6.3.16
3	0	−1	arbitrary	2.6.3.70
3	0	−1	−1	2.6.3.30
3	0	−1	$-\frac{1}{2}$	2.6.3.12
3	0	$\frac{1}{2}$	2	2.6.3.58
3	0	1	arbitrary	2.6.3.68
3	0	1	−1	2.6.3.10
3	0	2	1	2.6.3.18
4	−3	1	−1	2.6.3.47

TABLE 2.12 *Continued*

Solvable cases of the equation $y''_{xx} = \sigma A x^n y^m (y'_x)^l + A x^{n-1} y^{m+1} (y'_x)^{l-1}$

l	m	n	σ	Equation
4	-2	-2	$\frac{1}{2}$	2.6.3.66
4	-2	1	-1	2.6.3.34
4	-2	1	$-\frac{1}{2}$	2.6.3.49
4	$-\frac{1}{2}$	-2	$-\frac{1}{4}$	2.6.3.62
4	$-\frac{2}{5}$	1	$-\frac{2}{5}$	2.6.3.60
4	1	-2	-1	2.6.3.36
4	1	1	2	2.6.3.64

4. $y''_{xx} = Ay^{-2}(y'_x)^3 - Ax^{-1}y^{-2}(y'_x)^2$.

Solution in the parametric form:

$$x = -\frac{A}{2}\exp(\mp\tau^2)\left[\int \tau^{-1}\exp(\mp\tau^2)\,d\tau + C_2\right]^{-1},$$

$$y = C_1\left[\int \tau^{-1}\exp(\mp\tau^2)\,d\tau + C_2\right]^{-1}.$$

▶ *In the solutions of equations 5–12, the following notation is used:*

$$f = \int \exp(\mp\tau^2)\,d\tau + C_2.$$

5. $y''_{xx} = Ax^{-1}y^{-2}y'_x + Ax^{-2}y^{-1}$.

Solution in the parametric form:

$$x = C_1\exp(\mp\tau^2)f^{-1}, \qquad y = b[2\tau \pm \exp(\mp\tau^2)f^{-1}],$$

where $A = \pm 2b^2$.

6. $y''_{xx} = Ax^{-1}y^{-2}(y'_x)^3 + Ax^{-2}y^{-1}(y'_x)^2$.

Solution in the parametric form:

$$x = a[2\tau \pm \exp(\mp\tau^2)f^{-1}], \qquad y = C_1\exp(\mp\tau^2)f^{-1},$$

where $A = \mp 2a^2$.

7. $y''_{xx} = Ax^{-1}y^{-2}y'_x - Ax^{-2}y^{-1}$.

Solution in the parametric form:

$$x = C_1\left[2\tau f \pm \exp(\mp\tau^2)\right]^{-1}, \qquad y = bf\left[2\tau f \pm \exp(\mp\tau^2)\right]^{-1},$$

where $A = \pm\frac{1}{2}b^2$.

8. $y''_{xx} = Ax^{-1}y^{-2}(y'_x)^3 - Ax^{-2}y^{-1}(y'_x)^2$.

Solution in the parametric form:

$$x = af\left[2\tau f \pm \exp(\mp\tau^2)\right]^{-1}, \qquad y = C_1\left[2\tau f \pm \exp(\mp\tau^2)\right]^{-1},$$

where $A = \pm\frac{1}{2}a^2$.

9. $y''_{xx} = Axy'_x - Ay$.

Solution in the parametric form:

$$x = a\tau, \qquad y = C_1[2\tau f \pm \exp(\mp\tau^2)],$$

where $A = \mp 2a^{-2}$.

10. $y''_{xx} = Ax(y'_x)^3 - Ay(y'_x)^2$.

Solution in the parametric form:

$$x = C_1[2\tau f \pm \exp(\mp\tau^2)], \qquad y = b\tau,$$

where $A = \mp 2b^{-2}$.

11. $y''_{xx} = 2Axy^{-2}y'_x - Ay^{-1}$.

Solution in the parametric form:

$$x = aC_1[2\tau^2 f \pm \tau\exp(\mp\tau^2) \pm f], \qquad y = bC_1[2\tau f \pm \exp(\mp\tau^2)],$$

where $A = \mp\frac{1}{2}a^{-2}b^2$.

12. $y''_{xx} = Ax^{-1}(y'_x)^3 - 2Ax^{-2}y(y'_x)^2$.

Solution in the parametric form:

$$x = aC_1[2\tau f \pm \exp(\mp\tau^2)], \qquad y = bC_1[2\tau^2 f \pm \tau\exp(\mp\tau^2) \pm f],$$

where $A = \mp\frac{1}{2}a^2b^{-2}$.

► *In the solutions of equations 13–22, the following notation is used:*

$$E = \sqrt{\tau(\tau+1)} - \ln\left(\sqrt{\tau} + \sqrt{\tau+1}\right) + C_2, \qquad F = E\sqrt{\frac{\tau+1}{\tau}} - \tau.$$

13. $y''_{xx} = Ax^{1/2}y^{-2}y'_x - Ax^{-1/2}y^{-1}$.

Solution in the parametric form:

$$x = aC_1^4 F^{-2}, \qquad y = bC_1^3\tau^{-1}EF^{-2},$$

where $A = -a^{-3/2}b^2$.

14. $y''_{xx} = Ax^{-1}y^{-1/2}(y'_x)^3 - Ax^{-2}y^{1/2}(y'_x)^2.$

Solution in the parametric form:

$$x = aC_1^3\tau^{-1}EF^{-2}, \qquad y = bC_1^4F^{-2},$$

where $A = -a^2b^{-3/2}$.

15. $y''_{xx} = Axy^{-3}y'_x - Ay^{-2}.$

Solution in the parametric form:

$$x = aC_1^3F^{-1}\sqrt{\frac{\tau+1}{\tau}}, \qquad y = bC_1^2F^{-1},$$

where $A = -2a^{-2}b^3$.

16. $y''_{xx} = Ax^{-2}(y'_x)^3 - Ax^{-3}y(y'_x)^2.$

Solution in the parametric form:

$$x = aC_1^2F^{-1}, \qquad y = bC_1^3F^{-1}\sqrt{\frac{\tau+1}{\tau}},$$

where $A = -2a^3b^{-2}$.

17. $y''_{xx} = Axyy'_x + Ay^2.$

Solution in the parametric form:

$$x = aC_1^{-1}\tau^{-1}E^{-1}(\tau F^2 + \tau^2F - E^2), \qquad y = bC_1^2F^{-1},$$

where $A = a^{-2}b^{-1}$.

18. $y''_{xx} = Ax^2(y'_x)^3 + Axy(y'_x)^2.$

Solution in the parametric form:

$$x = aC_1^2F^{-1}, \qquad y = bC_1^{-1}\tau^{-1}E^{-1}(\tau F^2 + \tau^2F - E^2),$$

where $A = -a^{-1}b^{-2}$.

19. $y''_{xx} = Ax(y'_x)^{3/2} + Ay(y'_x)^{1/2}.$

Solution in the parametric form:

$$x = aC_1^{-1}\left(F\sqrt{\frac{\tau+1}{\tau}} - E\tau^{-1}\right), \qquad y = bC_1^3F^{-1}\sqrt{\frac{\tau+1}{\tau}},$$

where $A = 2a^{-2}\left(-\frac{a}{b}\right)^{1/2}$.

20. $y''_{xx} = Ax(y'_x)^{5/2} + Ay(y'_x)^{3/2}$.

Solution in the parametric form:

$$x = aC_1^3 F^{-1}\sqrt{\frac{\tau+1}{\tau}}, \qquad y = bC_1^{-1}\left(F\sqrt{\frac{\tau+1}{\tau}} - E\tau^{-1}\right),$$

where $A = -2b^{-2}\left(-\frac{b}{a}\right)^{1/2}$.

21. $y''_{xx} = Ax^{-4}yy'_x - 2Ax^{-5}y^2$.

Solution in the parametric form:

$$x = aC_1\tau E\bigl(\tau F^2 + \tau^2 F - E^2\bigr)^{-1}, \qquad y = bC_1^3\tau EF^{-1}\bigl(\tau F^2 + \tau^2 F - E^2\bigr)^{-1},$$

where $A = -2a^3b^{-1}$.

22. $y''_{xx} = 2Ax^2y^{-5}(y'_x)^3 - Axy^{-4}(y'_x)^2$.

Solution in the parametric form:

$$x = aC_1^3\tau EF^{-1}\bigl(\tau F^2 + \tau^2 F - E^2\bigr)^{-1}, \qquad y = bC_1\tau E\bigl(\tau F^2 + \tau^2 F - E^2\bigr)^{-1},$$

where $A = -2a^{-1}b^3$.

23. $y''_{xx} = Ax^n y'_x + nAx^{n-1}y, \qquad n \neq -1$.

Solution in the parametric form:

$$x = a\tau^{\frac{1}{n+1}}, \qquad y = C_1 e^{\beta\tau}\left(\int \tau^{-\frac{n}{n+1}} e^{-\beta\tau}\,d\tau + C_2\right),$$

where $A = (n+1)a^{-n-1}\beta$.

24. $y''_{xx} = Axy^m(y'_x)^3 + \dfrac{A}{m+1}y^{m+1}(y'_x)^2, \qquad m \neq -2$.

Solution in the parametric form:

$$x = C_1 e^{\beta\tau}\left(\int \tau^{-\frac{m+1}{m+2}} e^{-\beta\tau}\,d\tau + C_2\right), \qquad y = b\tau^{\frac{1}{m+2}},$$

where $A = -\dfrac{m+2}{m+1}b^{-m-2}\beta$.

► *In the solutions of equations 25–36, the following notation is used:*

$$R = \begin{cases} \tau^\nu + C_2\tau^{-\nu} & \textit{for the upper sign,} \\ \sin(\nu\ln\tau) + C_2\cos(\nu\ln\tau) & \textit{for the lower sign,} \\ \ln\tau + C_2 & \textit{for } \nu = 0, \end{cases}$$

$$Q = \begin{cases} (1+\nu)\tau^\nu + (1-\nu)C_2\tau^{-\nu} & \textit{for the upper sign,} \\ (1-\nu C_2)\sin(\nu\ln\tau) + (C_2+\nu)\cos(\nu\ln\tau) & \textit{for the lower sign,} \\ \ln\tau + 1 + C_2 & \textit{for } \nu = 0. \end{cases}$$

25. $y''_{xx} = Ax^{-2}yy'_x - Ax^{-3}y^2.$

Solution in the parametric form:

$$x = a\tau^{-2}, \qquad y = b\tau^{-2}R^{-1}Q,$$

where $\nu = C_1$, $A = ab^{-1}$; the solution is valid for all three cases of functions R and Q given above.

26. $y''_{xx} = Ax^2y^{-3}(y'_x)^3 - Axy^{-2}(y'_x)^2.$

Solution in the parametric form:

$$x = a\tau^{-2}R^{-1}Q, \qquad y = b\tau^{-2},$$

where $\nu = C_1$, $A = a^{-1}b$; the solution is valid for all three cases of functions R and Q given above.

27. $y''_{xx} = Ax^{-1}(y'_x)^{3/2} - Ax^{-2}y(y'_x)^{1/2}.$

Solution in the parametric form:

$$x = a\tau^{-2}, \qquad y = b\tau^{-2}\left(QR^{-1} - \frac{1 \mp \nu^2}{2}\right),$$

where $\nu = C_1$, $A = \left(\frac{2a}{b}\right)^{1/2}$.

28. $y''_{xx} = Axy^{-2}(y'_x)^{5/2} - Ay^{-1}(y'_x)^{3/2}.$

Solution in the parametric form:

$$x = a\tau^{-2}\left(QR^{-1} - \frac{1 \mp \nu^2}{2}\right), \qquad y = b\tau^{-2},$$

where $\nu = C_1$, $A = \left(\frac{2b}{a}\right)^{1/2}$.

29. $y''_{xx} = Axy^{-2}y'_x - Ay^{-1}.$

Solution in the parametric form:

$$x = aC_1\tau R, \qquad y = bC_1\tau Q,$$

where $A = a^{-2}b^2(1 \mp \nu^2)$.

30. $y''_{xx} = Ax^{-1}(y'_x)^3 - Ax^{-2}y(y'_x)^2.$

Solution in the parametric form:

$$x = aC_1\tau Q, \qquad y = bC_1\tau R,$$

where $A = a^2b^{-2}(1 \mp \nu^2)$.

31. $y''_{xx} = Ax^{-1/2}y^{-1/2}y'_x - Ax^{-3/2}y^{1/2}$.

Solution in the parametric form:

$$x = a\tau^2 R^2, \qquad y = b\tau^2 Q^2,$$

where $\nu = C_1$, $A = a^{-1/2}b^{1/2}$.

32. $y''_{xx} = Ax^{1/2}y^{-3/2}(y'_x)^3 - Ax^{-1/2}y^{-1/2}(y'_x)^2$.

Solution in the parametric form:

$$x = a\tau^2 Q^2, \qquad y = b\tau^2 R^2,$$

where $\nu = C_1$, $A = a^{1/2}b^{-1/2}$.

33. $y''_{xx} = Ax^{-1} - Ax^{-2}y(y'_x)^{-1}$.

Solution in the parametric form:

$$x = a\tau^2 R^2, \qquad y = b\tau^2 [Q^2 + (1 \mp \nu^2)R^2],$$

where $\nu = C_1$, $A = 2a^{-1}b$.

34. $y''_{xx} = Axy^{-2}(y'_x)^4 - Ay^{-1}(y'_x)^3$.

Solution in the parametric form:

$$x = a\tau^2 [Q^2 + (1 \mp \nu^2)R^2], \qquad y = b\tau^2 R^2,$$

where $\nu = C_1$, $A = 2ab^{-1}$.

35. $y''_{xx} = Ax^2y^{-3} - Axy^{-2}(y'_x)^{-1}$.

Solution in the parametric form:

$$x = aC_1\tau R, \qquad y = bC_1\tau \big[Q^2 + (1 \mp \nu^2)R^2\big]^{1/2},$$

where $A = 4(1 \mp \nu^2)a^{-4}b^4$.

36. $y''_{xx} = Ax^{-2}y(y'_x)^4 - Ax^{-3}y^2(y'_x)^3$.

Solution in the parametric form:

$$x = aC_1\tau \big[Q^2 + (1 \mp \nu^2)R^2\big]^{1/2}, \qquad y = bC_1\tau R,$$

where $A = 4(1 \mp \nu^2)a^4b^{-4}$.

▶ *In the solutions of equations 37–50, the following notation is used:*

$$Z = \begin{cases} J_\nu(\tau) + C_2 Y_\nu(\tau) & \text{for the upper sign,} \\ I_\nu(\tau) + C_2 K_\nu(\tau) & \text{for the lower sign,} \end{cases}$$

where J_ν and Y_ν are Bessel functions, I_ν and K_ν are modified Bessel functions.

37. $y''_{xx} = 2Ax^n yy'_x + nAx^{n-1}y^2, \qquad n \neq 0, \quad n \neq -2.$

Solution in the parametric form:

$$x = aC_1^{-1}\tau^{2-2\nu}, \qquad y = bC_1^{n+1}\tau^{-2\nu}Z^{-1}(\tau Z'_\tau + \nu Z),$$

where $\nu = \dfrac{n+1}{n+2}$, $A = -\dfrac{n+2}{2}a^{-n-1}b^{-1}$.

38. $y''_{xx} = (m+1)Ax^2y^m(y'_x)^3 + 2Axy^{m+1}(y'_x)^2, \qquad m \neq -1, \quad m \neq -3.$

Solution in the parametric form:

$$x = aC_1^{-m-2}\tau^{-2\nu}Z^{-1}(\tau Z'_\tau + \nu Z), \qquad y = bC_1\tau^{2-2\nu},$$

where $\nu = \dfrac{m+2}{m+3}$, $A = \dfrac{m+3}{2}a^{-1}b^{-m-2}$.

39. $y''_{xx} = 2Ax^{-1}yy'_x - Ax^{-2}y^2.$

Solution in the parametric form:

$$x = C_1\tau^2, \qquad y = b\tau Z^{-1}Z'_\tau,$$

where $\nu = 0$, $A = -\frac{1}{2}b^{-1}$.

40. $y''_{xx} = Ax^2y^{-2}(y'_x)^3 - 2Axy^{-1}(y'_x)^2.$

Solution in the parametric form:

$$x = a\tau Z^{-1}Z'_\tau, \qquad y = C_1\tau^2,$$

where $\nu = 0$, $A = -\frac{1}{2}a^{-1}$.

41. $y''_{xx} = Ax^n(y'_x)^{3/2} + nAx^{n-1}y(y'_x)^{1/2}, \qquad n \neq 0, \ n \neq -1.$

Solution in the parametric form:

$$x = aC_1^{-1}\tau^{4-2\nu}, \qquad y = bC_1^{2n+1}\tau^{-2\nu}\left[Z^{-1}(\tau Z'_\tau + \nu Z) \pm \frac{1}{2(1-\nu)}\tau^2\right],$$

where $\nu = \dfrac{2n+1}{n+1}$, $A = -(n+1)a^{-n-1}\left[-\dfrac{2a}{(n+1)b}\right]^{1/2}$.

42. $y''_{xx} = (m+1)Axy^m (y'_x)^{5/2} + Ay^{m+1}(y'_x)^{3/2}, \qquad m \neq -1, \quad m \neq -2.$

Solution in the parametric form:

$$x = aC_1^{2m-3}\tau^{-2\nu}\left[Z^{-1}(\tau Z'_\tau + \nu Z) \pm \frac{1}{2(1-\nu)}\tau^2\right], \qquad y = bC_1\tau^{4-2\nu},$$

where $\nu = \dfrac{2m+3}{m+2}$, $A = (m+2)b^{-m-2}\left[-\dfrac{2b}{(m+2)a}\right]^{1/2}$.

43. $y''_{xx} = 2Ax^{-1/2}(y'_x)^{3/2} - Ax^{-3/2}y(y'_x)^{1/2}.$

Solution in the parametric form:

$$x = C_1\tau^4, \qquad y = b(\tau Z^{-1}Z'_\tau \pm \tfrac{1}{2}\tau^2),$$

where $\nu = 0$, $A = -\frac{1}{2}(-b)^{-1/2}$.

44. $y''_{xx} = Axy^{-3/2}(y'_x)^{5/2} - 2Ay^{-1/2}(y'_x)^{3/2}.$

Solution in the parametric form:

$$x = a(\tau Z^{-1}Z'_\tau \pm \tfrac{1}{2}\tau^2), \qquad y = C_1\tau^4,$$

where $\nu = 0$, $A = -\frac{1}{2}(-a)^{-1/2}$.

45. $y''_{xx} = Ax^{-1}y^{-1/2}y'_x - Ax^{-2}y^{1/2}.$

Solution in the parametric form:

$$x = C_1 Z^{-2}, \qquad y = b\tau^2 Z^{-2}(Z'_\tau)^2,$$

where $\nu = 0$, $A = -b^{1/2}$.

46. $y''_{xx} = Ax^{1/2}y^{-2}(y'_x)^3 - Ax^{-1/2}y^{-1}(y'_x)^2.$

Solution in the parametric form:

$$x = a\tau^2 Z^{-2}(Z'_\tau)^2, \qquad y = C_1 Z^{-2},$$

where $\nu = 0$, $A = -a^{1/2}$.

47. $y''_{xx} = Axy^{-3}(y'_x)^4 - Ay^{-2}(y'_x)^3.$

Solution in the parametric form:

$$x = aZ^{-1}(2\tau Z'_\tau \pm \tau^2 Z), \qquad y = C_1 Z^{-1},$$

where $\nu = 0$, $A = 4a$.

48. $y''_{xx} = Ax^{-2} - Ax^{-3}y(y'_x)^{-1}.$

Solution in the parametric form:

$$x = C_1 Z^{-1}, \qquad y = bZ^{-1}(2\tau Z'_\tau \pm \tau^2 Z),$$

where $\nu = 0$, $A = 4b$.

49. $y''_{xx} = Axy^{-2}(y'_x)^4 - 2Ay^{-1}(y'_x)^3.$

Solution in the parametric form:

$$x = aC_1[\tau^2(Z'_\tau)^2 + 2\tau ZZ'_\tau \pm \tau^2 Z^2], \qquad y = bC_1 Z^2,$$

where $\nu = 0$, $A = \frac{1}{2}ab^{-1}$.

50. $y''_{xx} = 2Ax^{-1} - Ax^{-2}y(y'_x)^{-1}.$

Solution in the parametric form:

$$x = aC_1 Z^2, \qquad y = bC_1[\tau^2(Z'_\tau)^2 + 2\tau ZZ'_\tau \pm \tau^2 Z^2],$$

where $\nu = 0$, $A = \frac{1}{2}a^{-1}b$.

► *In the solutions of equations 51–66, the following notation is used:*

$$Z = \begin{cases} J_{1/3}(\tau) + C_2 Y_{1/3}(\tau) & \text{for the upper sign,} \\ I_{1/3}(\tau) + C_2 K_{1/3}(\tau) & \text{for the lower sign,} \end{cases}$$

$$U_1 = \tau Z'_\tau + \tfrac{1}{3}Z, \quad U_2 = U_1^2 \pm \tau^2 Z^2, \quad U_3 = \pm\tfrac{2}{3}\tau^2 Z^3 - 2U_1U_2,$$

where $J_{1/3}$ and $Y_{1/3}$ are Bessel functions, $I_{1/3}$ and $K_{1/3}$ are modified Bessel functions.

51. $y''_{xx} = Ax^{-2}y^{-2}y'_x + 2Ax^{-3}y^{-1}.$

Solution in the parametric form:

$$x = aC_1^{-2}\tau^{4/3}Z^2U_2^{-1}, \qquad y = bC_1\tau^{-2/3}Z^{-1}U_2^{-1}U_3,$$

where $A = 2ab^2$.

52. $y''_{xx} = 2Ax^{-1}y^{-3}(y'_x)^3 + Ax^{-2}y^{-2}(y'_x)^2.$

Solution in the parametric form:

$$x = aC_1^{-1}\tau^{-2/3}Z^{-1}U_2^{-1}U_3, \qquad y = bC_1^2\tau^{4/3}Z^2U_2^{-1},$$

where $A = -2a^2b$.

53. $y''_{xx} = Ax^{-2}y^{-1/2}y'_x - 4Ax^{-3}y^{1/2}.$

Solution in the parametric form:

$$x = aC_1^{-1}\tau^{-4/3}Z^{-2}U_2, \qquad y = bC_1^2\tau^{-4/3}Z^{-2}U_2^{-2}U_3^2,$$

where $A = \mp\frac{2}{3}ab^{1/2}$.

54. $y''_{xx} = 4Ax^{1/2}y^{-3}(y'_x)^3 - Ax^{-1/2}y^{-2}(y'_x)^2.$

Solution in the parametric form:

$$x = aC_1^{-2}\tau^{-4/3}Z^{-2}U_2^{-2}U_3^2, \qquad y = bC_1\tau^{-4/3}Z^{-2}U_2,$$

where $A = \mp\frac{2}{3}a^{1/2}b$.

55. $y''_{xx} = 2Ax^{-1/2}y^{-2}y'_x + Ax^{-3/2}y^{-1}$.

Solution in the parametric form:

$$x = aC_1^4 \tau^{-4/3} Z^{-2} U_1^2, \qquad y = bC_1 \tau^{-4/3} Z^{-2} U_2,$$

where $A = \pm\frac{1}{6}a^{-1/2}b^2$.

56. $y''_{xx} = Ax^{-1}y^{-3/2}(y'_x)^3 + 2Ax^{-2}y^{-1/2}(y'_x)^2$.

Solution in the parametric form:

$$x = aC_1 \tau^{-4/3} Z^{-2} U_2, \qquad y = bC_1^4 \tau^{-4/3} Z^{-2} U_1^2,$$

where $A = \mp\frac{1}{6}a^2 b^{-1/2}$.

57. $y''_{xx} = Axy^{-1/2}y'_x + 2Ay^{1/2}$.

Solution in the parametric form:

$$x = aC_1 \tau^{-2/3} Z^{-1} U_1, \qquad y = bC_1^4 \tau^{-8/3} Z^{-4} U_2^2,$$

where $A = 2a^{-2}b^{1/2}$.

58. $y''_{xx} = 2Ax^{1/2}(y'_x)^3 + Ax^{-1/2}y(y'_x)^2$.

Solution in the parametric form:

$$x = aC_1^4 \tau^{-8/3} Z^{-4} U_2^2, \qquad y = bC_1 \tau^{-2/3} Z^{-1} U_1,$$

where $A = -2a^{1/2}b^{-2}$.

59. $y''_{xx} = 5Ax^{-2/5} - 2Ax^{-7/5}y(y'_x)^{-1}$.

Solution in the parametric form:

$$x = aC_1^5 \tau^{-5/3} Z^{-5/2} U_1^{5/2}, \qquad y = bC_1^8 \tau^{-8/3} Z^{-4} (U_2^2 \pm \tfrac{4}{3}\tau^2 Z^3 U_1),$$

where $A = \frac{32}{125}a^{-8/5}b$.

60. $y''_{xx} = 2Axy^{-2/5}(y'_x)^4 - 5Ay^{-7/5}(y'_x)^3$.

Solution in the parametric form:

$$x = aC_1^8 \tau^{-8/3} Z^{-4} (U_2^2 \pm \tfrac{4}{3}\tau^2 Z^3 U_1), \qquad y = bC_1^5 \tau^{-5/3} Z^{-5/2} U_1^{5/2},$$

where $A = \frac{32}{125}ab^{-8/5}$.

61. $y''_{xx} = 4Ax^{1/2}y^{-3} - Ax^{-1/2}y^{-2}(y'_x)^{-1}$.

Solution in the parametric form:

$$x = aC_1^8 \tau^{-4/3} Z^{-2} U_1^2, \qquad y = bC_1^5 \tau^{-4/3} Z^{-2} (U_2^2 \pm \tfrac{4}{3}\tau^2 Z^3 U_1)^{1/2},$$

where $A = \pm\frac{1}{3}a^{-5/2}b^4$.

62. $y''_{xx} = Ax^{-2}y^{-1/2}(y'_x)^4 - 4Ax^{-3}y^{1/2}(y'_x)^3.$

Solution in the parametric form:

$$x = aC_1^5\tau^{-4/3}Z^{-2}\big(U_2^2 \pm \tfrac{4}{3}\tau^2 Z^3 U_1\big)^{1/2}, \qquad y = bC_1^8\tau^{-4/3}Z^{-2}U_1^2,$$

where $A = \pm\frac{1}{3}a^4b^{-5/2}$.

63. $y''_{xx} = Ax^2 + 2Axy(y'_x)^{-1}.$

Solution in the parametric form:

$$x = aC_1\tau^{2/3}ZU_2^{-1/2}, \qquad y = bC_1^4\tau^{-4/3}Z^{-2}U_2^{-2}(U_3^2 - 4U_2^3),$$

where $A = \frac{32}{9}a^{-4}b$.

64. $y''_{xx} = 2Axy(y'_x)^4 + Ay^2(y'_x)^3.$

Solution in the parametric form:

$$x = aC_1^4\tau^{-4/3}Z^{-2}U_2^{-2}(U_3^2 - 4U_2^3), \qquad y = bC_1\tau^{2/3}ZU_2^{-1/2},$$

where $A = -\frac{32}{9}ab^{-4}$.

65. $y''_{xx} = 2Ax^{-1}y^{-3} + Ax^{-2}y^{-2}(y'_x)^{-1}.$

Solution in the parametric form:

$$x = aC_1^4\tau^{4/3}Z^2U_2^{-1}, \qquad y = bC_1\tau^{-2/3}Z^{-1}U_2^{-1}\big(U_3^2 - 4U_2^3\big)^{1/2},$$

where $A = -\frac{8}{9}a^{-1}b^4$.

66. $y''_{xx} = Ax^{-2}y^{-2}(y'_x)^4 + 2Ax^{-3}y^{-1}(y'_x)^3.$

Solution in the parametric form:

$$x = aC_1\tau^{-2/3}Z^{-1}U_2^{-1}\big(U_3^2 - 4U_2^3\big)^{1/2}, \qquad y = bC_1^4\tau^{4/3}Z^2U_2^{-1},$$

where $A = \frac{8}{9}a^4b^{-1}$.

▶ *In the solutions of equations 67–72, the following notation is used:*

$$M = C_1\Phi(\lambda, \tfrac{1}{2}, \pm\tau) + C_2\Psi(\lambda, \tfrac{1}{2}, \pm\tau),$$

where Φ and Ψ are linearly-independent solutions of the degenerate hypergeometric equation

$$\tau M''_{\tau\tau} + (\tfrac{1}{2} \pm \tau)M'_\tau - \lambda M = 0.$$

The function $\Phi = \Phi(\lambda, \frac{1}{2}, \pm\tau)$ can be expressed in terms of a degenerate hypergeometric series (see Equation 2.1.2.65).

67. $y''_{xx} = A_1 x y'_x + A_2 y.$

Solution in the parametric form:

$$x = a\tau^{1/2}, \qquad y = M,$$

where $A_1 = \pm 2a^{-2}$, $A_2 = \pm 4a^{-2}\lambda$.

68. $y''_{xx} = A_1 x (y'_x)^3 + A_2 y (y'_x)^2.$

Solution in the parametric form:

$$x = M, \qquad y = b\tau^{1/2},$$

where $A_1 = \mp 4b^{-2}\lambda$, $A_2 = \mp 2b^{-2}$.

69. $y''_{xx} = A_1 x y^{-2} y'_x + A_2 y^{-1}.$

Solution in the parametric form:

$$x = M, \qquad y = \pm b\tau^{1/2} M'_\tau,$$

where $A_1 = \mp b^2\lambda$, $A_2 = \pm b^2(\lambda + \frac{1}{2})$.

70. $y''_{xx} = A_1 x^{-1} (y'_x)^3 + A_2 x^{-2} y (y'_x)^2.$

Solution in the parametric form:

$$x = \pm a\tau^{1/2} M'_\tau, \qquad y = M,$$

where $A_1 = \mp a^2(\lambda + \frac{1}{2})$, $A_2 = \pm a^2\lambda$.

71. $y''_{xx} = A_1 x^{-1} y^{-2} y'_x + A_2 x^{-2} y^{-1}.$

Solution in the parametric form:

$$x = M^{-1}, \qquad y = \pm b\tau^{1/2} M^{-1} M'_\tau,$$

where $A_1 = \pm b^2\lambda$, $A_2 = \pm \frac{1}{2} b^2$.

72. $y''_{xx} = A_1 x^{-1} y^{-2} (y'_x)^3 + A_2 x^{-2} y^{-1} (y'_x)^2.$

Solution in the parametric form:

$$x = \pm a\tau^{1/2} M^{-1} M'_\tau, \qquad y = M^{-1},$$

where $A_1 = \mp \frac{1}{2} a^2$, $A_2 = \mp a^2\lambda$.

73. $y''_{xx} = A x^n (y'_x)^{\frac{3n+2}{n+1}} + nA x^{n-1} y (y'_x)^{\frac{2n+1}{n+1}}, \qquad n \neq 0, \quad n \neq -1.$

Solution in the parametric form:

$$x = aC_1^{-2n-1} \left(\int \frac{d\tau}{\beta\tau^k + 1} + C_2 \right)^{-1/n}, \qquad y = bC_1^{n^2} \left(\tau - \int \frac{d\tau}{\beta\tau^k + 1} - C_2 \right),$$

where $k = -\dfrac{n+1}{n}$, $A = \dfrac{n+1}{n^2\beta} a^{1-n} b^{-2} \left(-\dfrac{nb\beta}{a} \right)^{\frac{1}{n+1}}$.

74. $y''_{xx} = A(m+1)xy^m(y'_x)^{\frac{m+3}{m+2}} + Ay^{m+1}(y'_x)^{\frac{1}{m+2}}, \qquad m \neq -1, \quad m \neq -2.$

Solution in the parametric form:

$$x = aC_1^{(m+1)^2}\left(\tau - \int \frac{d\tau}{\beta\tau^k + 1} - C_2\right), \qquad y = bC_1^{-2m-3}\left(\int \frac{d\tau}{\beta\tau^k + 1} + C_2\right)^{-\frac{1}{m+1}},$$

where $k = -\dfrac{m+2}{m+1}$, $A = -\dfrac{m+2}{(m+1)^2\beta}a^{-2}b^{-m}\left[-\dfrac{a(m+1)\beta}{b}\right]^{\frac{1}{m+2}}$.

75. $y''_{xx} = Ax^{-m-1}y^m(y'_x)^l - Ax^{-m-2}y^{m+1}(y'_x)^{l-1}, \quad m \neq -1, \; l \neq 2, \; m+l-1 \neq 0.$

Solution in the parametric form:

$$x = aC_1 \exp\left(\frac{l-2}{m+l-1}\int \frac{d\tau}{F}\right), \qquad y = bC_1\tau^{\frac{l-2}{m+l-1}} \exp\left(\frac{l-2}{m+l-1}\int \frac{d\tau}{F}\right),$$

where

$$F = \frac{m+l-1}{l-2}\left(\beta + C_2\tau^k\right)^{\frac{1}{2-l}} - \tau, \qquad k = \frac{(m+1)(l-2)}{m+l-1},$$
$$A = -\frac{(m+1)(m+l-1)}{(l-2)^3}a^{m-1}b^{1-m}\beta\left[\frac{(l-2)a}{(m+l-1)b}\right]^l.$$

76. $y''_{xx} = Ax^{l-2}y^{1-l}(y'_x)^l - Ax^{l-3}y^{2-l}(y'_x)^{l-1}, \qquad l \neq 2.$

Solution in the parametric form:

$$x = C_1 \exp\left(\int \frac{d\tau}{F}\right), \qquad y = C_2 \exp\left(\tau + \int \frac{d\tau}{F}\right),$$

where

$$F = (2-l)\left[\beta + e^{(l-2)\tau}\right]^{\frac{1}{2-l}} - 1, \quad A = (2-l)^{2-l}\beta.$$

77. $y''_{xx} = Ax^{-1}y'_x - Ax^{-2}y.$

1°. Solution with $A \neq 1$:

$$y = C_1x^A + \frac{C_2}{1-A}x.$$

2°. Solution with $A = 1$:

$$y = x(C_1 + C_2 \ln|x|).$$

78. $y''_{xx} = Axy^{-2}(y'_x)^3 - Ay^{-1}(y'_x)^2.$

1°. Solution with $A \neq 1$:

$$x = C_1y^A + \frac{C_2}{1-A}y.$$

2°. Solution with $A = 1$:

$$x = y(C_1 + C_2 \ln|y|).$$

▶ *In the solutions of equations 79–80, the following notation is used:*

$$f = \begin{cases} \dfrac{1}{\beta+1}\tau^{\beta+1} + \dfrac{1}{\beta}\tau^{\beta} + C_2 & \text{if } \beta \neq 0, \\ \tau + \ln|\tau| + C_2 & \text{if } \beta = 0. \end{cases}$$

79. $y''_{xx} = Axy^{-2}(y'_x)^l - Ay^{-1}(y'_x)^{l-1}, \qquad l \neq 3.$

Solution in the parametric form:

$$x = aC_1[\tau^{\beta+1} - (\beta+1)f]\exp\left(-\int \tau^{\beta-1} f^{-1}\,d\tau\right), \qquad y = bC_1 \exp\left(-\int \tau^{\beta-1} f^{-1}\,d\tau\right),$$

where $\beta = \dfrac{2-l}{l-3}$, $A = -a^{l-3}b^{3-l}$.

80. $y''_{xx} = Ax^{-1}(y'_x)^l - Ax^{-2}y(y'_x)^{l-1}, \qquad l \neq 1.$

Solution in the parametric form:

$$x = aC_1 \exp\left(-\int \tau^{\beta-1} f^{-1}\,d\tau\right), \qquad y = bC_1[\tau^{\beta+1} - (\beta+1)f]\exp\left(-\int \tau^{\beta-1} f^{-1}\,d\tau\right),$$

where $\beta = \dfrac{l-2}{1-l}$, $A = -a^{l-1}b^{1-l}$.

81. $y''_{xx} = A_1 y^{-1}(y'_x)^2 + A_2 x^{-1} y'_x.$

1°. Solution with $A_1 \neq 1$, $A_2 \neq -1$:

$$y = \pm(C_1|x|^{A_2+1} + C_2)^{A_1-1}.$$

2°. Solution with $A_1 \neq 1$, $A_2 = -1$:

$$y = \pm(C_1 \ln|x| + C_2)^{A_1-1}.$$

3°. Solution with $A_1 = 1$, $A_2 \neq -1$:

$$y = C_2 \exp(C_1|x|^{A_2+1}).$$

4°. Solution with $A_1 = 1$, $A_2 = -1$:

$$y = C_2|x|^{C_1}.$$

▶ *In the solutions of equations 82–84, the following notation is used:*

$$U = \exp\left(\int \frac{W\,d\tau}{\tau\sqrt{W^2+4}}\right), \qquad W = C_2\tau^{-1/2}\exp\left[\frac{A}{2(k+1)}\tau^{k+1}\right].$$

82. $y''_{xx} = Ax^{m+1}y^m(y'_x)^2 - Ax^m y^{m+1}y'_x, \qquad m \neq -1.$

Solution in the parametric form:

$$x = C_1\tau^{1/2}U^{-1/2}, \quad y = C_1^{-1}\tau^{1/2}U^{1/2}, \qquad k = m.$$

83. $y''_{xx} = Ax^{-2m-2}y^m(y'_x)^2 - Ax^{-2m-3}y^{m+1}y'_x, \qquad m \neq -1.$

Solution in the parametric form:

$$x = C_1\tau^{-1/2}U^{1/2}, \quad y = C_1^2 U, \qquad k = m.$$

84. $y''_{xx} = Ax^{-\frac{m+1}{2}}y^m(y'_x)^2 - Ax^{-\frac{m+3}{2}}y^{m+1}y'_x, \qquad m \neq -1.$

Solution in the parametric form:

$$x = C_1^2 U, \quad y = C_1\tau^{-1/2}U^{1/2}, \qquad k = -\frac{m+3}{2}.$$

85. $y''_{xx} = Ax^n y^m(y'_x)^2 - Ax^{n-1}y^{m+1}y'_x, \qquad m \neq -1, \quad n \neq 0.$

Solution in the parametric form:

$$x = C_1 \exp\biggl(\int \frac{d\tau}{\tau F}\biggr), \quad y = C_1^k \tau \exp\biggl(k\int \frac{d\tau}{\tau F}\biggr), \qquad k = -\frac{n}{m+1},$$

where $F = F(\tau)$ is the solution of the transcendental equation

$$\frac{(F+k)^k}{(F+k-1)^{k-1}} = C_2\tau^{-1}\exp\Bigl(\frac{A}{m+1}\tau^{m+1}\Bigr).$$

86. $y''_{xx} = A_1x^n y^{-1}(y'_x)^2 + A_2x^{n-1}y'_x, \qquad n \neq 0.$

Solution:

$$y = C_1 \exp\biggl[\int \exp\Bigl(\frac{A_2}{n}x^n\Bigr)(F+C_2)^{-1}dx\biggr],$$

where $F = \displaystyle\int (1 - A_1x^n)\exp\Bigl(\frac{A_2}{n}x^n\Bigr)\,dx.$

87. $y''_{xx} = A_1y^m(y'_x)^2 + A_2x^{-1}y^{m+1}y'_x, \qquad m \neq -1.$

Solution:

$$x = C_1 \exp\biggl[\int \exp\Bigl(-\frac{A_1}{m+1}y^{m+1}\Bigr)(F+C_2)^{-1}\,dy\biggr],$$

where $F = \displaystyle\int (1 + A_2y^{m+1})\exp\Bigl(-\frac{A_1}{m+1}y^{m+1}\Bigr)\,dy.$

TABLE 2.13

Solvable cases of the equation $y''_{xx} = A_1 x^{n_1} y^{m_1} (y'_x)^{l_1} + A_2 x^{n_2} y^{m_2} (y'_x)^{l_2}$, $l_1 \neq l_2$

l_1	l_2	m_1	m_2	n_1	n_2	Equation
arbitrary	arbitrary	arbitrary	m_1	0	0	2.6.4.3
arbitrary	arbitrary	0	0	arbitrary	n_1	2.6.4.4
arbitrary	arbitrary	$1-l_1$	$1-l_2$	l_1-2	l_2-2	2.6.4.9
arbitrary	$3-l_1$	$1-l_1$	l_1-2	l_1-2	$1-l_1$	2.6.4.8
1	0	$-\frac{1}{2}$	$-\frac{1}{2}$	0	0	2.6.4.1
1	0	$-\frac{1}{2}$	$-\frac{1}{2}$	0	1	2.6.4.14
1	0	$-\frac{1}{2}$	0	0	0	2.6.4.16
2	0	-1	1	arbitrary	arbitrary	2.6.4.18
2	1	arbitrary	arbitrary	0	-1	2.6.4.13
2	1	arbitrary ($m_1 \neq -1$)	0	0	arbitrary ($n_2 \neq -1$)	2.6.4.5
2	1	arbitrary ($m_1 \neq -1$)	0	0	-1	2.6.4.7
2	1	-1	0	arbitrary	arbitrary	2.6.4.12
2	1	-1	0	0	arbitrary ($n_2 \neq -1$)	2.6.4.6
$\frac{5}{2}$	$\frac{1}{2}$	arbitrary	m_1+2	m_1+2	m_1	2.6.4.11
3	0	arbitrary* ($m_1 \neq -2$)	m_1+3	m_1+3	m_1	2.6.4.10
3	1	arbitrary	arbitrary	1	-1	2.6.4.19
3	2	0	0	$-\frac{1}{2}$	$-\frac{1}{2}$	2.6.4.2
3	2	0	0	0	$-\frac{1}{2}$	2.6.4.17
3	2	1	0	$-\frac{1}{2}$	$-\frac{1}{2}$	2.6.4.15

* For $m_1 = -2$, see Equation 2.6.4.8 with $l = 3$

2.6.4. Other Equations ($l_1 \neq l_2$)

Table 2.13 represents all solvable equations whose solutions are outlined in Subsection 2.6.4. Equations are arranged in accordance with the growth of l_1. The number of the equation sought is indicated in the last column in this table.

1. $y''_{xx} = A_1 y^{-1/2} y'_x + A_2 y^{-1/2}$.

Solution in the parametric form:

$$x = C_1 \exp(A_1 \tau) - \frac{A_2}{4A_1}\tau^2 + C_2, \qquad y = \left[A_1 C_1 \exp(A_1 \tau) - \frac{A_2}{2A_1}\tau\right]^2$$

2. $y''_{xx} = A_1 x^{-1/2}(y'_x)^3 + A_2 x^{-1/2}(y'_x)^2.$

Solution in the parametric form:

$$x = \left[A_2 C_1 \exp(-A_2\tau) + \frac{A_1}{2A_2}\tau\right]^2, \qquad y = C_1 \exp(-A_2\tau) - \frac{A_1}{4A_2}\tau^2 + C_2.$$

3. $y''_{xx} = A_1 y^m (y'_x)^{l_1} + A_2 y^m (y'_x)^{l_2}.$

1°. Solution in the parametric form with $m \neq -1$:

$$x = C_2 + \int \left(A_1\tau^{l_1} + A_2\tau^{l_2}\right)^{-1} f^{-\frac{m}{m+1}}\,d\tau, \qquad y = f^{\frac{1}{m+1}},$$

where $f = C_1 + (m+1)\int \tau\left(A_1\tau^{l_1} + A_2\tau^{l_2}\right)^{-1} d\tau$.

2°. Solution in the parametric form with $m = -1$:

$$x = C_2 + \int \left(A_1\tau^{l_1} + A_2\tau^{l_2}\right)^{-1} e^f\,d\tau, \qquad y = e^f,$$

where $f = C_1 \int \tau\left(A_1\tau^{l_1} + A_2\tau^{l_2}\right)^{-1} d\tau$.

4. $y''_{xx} = A_1 x^n (y'_x)^{l_1} + A_2 x^n (y'_x)^{l_2}.$

1°. Solution in the parametric form with $n \neq -1$:

$$x = f^{\frac{1}{n+1}}, \qquad y = C_2 + \int \tau\left(A_1\tau^{l_1} + A_2\tau^{l_2}\right)^{-1} f^{-\frac{n}{n+1}}\,d\tau,$$

where $f = C_1 + (n+1)\int \left(A_1\tau^{l_1} + A_2\tau^{l_2}\right)^{-1} d\tau$.

2°. Solution in the parametric form with $n = -1$:

$$x = e^f, \qquad y = C_2 + \int \tau\left(A_1\tau^{l_1} + A_2\tau^{l_2}\right)^{-1} e^f\,d\tau,$$

where $f = C_1 \int \left(A_1\tau^{l_1} + A_2\tau^{l_2}\right)^{-1} d\tau$.

5. $y''_{xx} = A_1 y^m (y'_x)^2 + A_2 x^n y'_x, \qquad n \neq -1, \quad m \neq -1.$

Solution:

$$\int \exp\left(-\frac{A_1}{m+1}y^{m+1}\right) dy = C_1 \int \exp\left(\frac{A_2}{n+1}x^{n+1}\right) dx + C_2.$$

6. $y''_{xx} = A_1 y^{-1}(y'_x)^2 + A_2 x^n y'_x, \qquad n \neq -1.$

1°. Solution with $A_1 \neq 1$:

$$y = \left[C_1 \int \exp\left(\frac{A_2}{n+1}x^{n+1}\right) dx + C_2\right]^{A_1 - 1}.$$

2°. Solution with $A_1 = 1$:

$$y = C_2 \exp\left[C_1 \int \exp\left(\frac{A_2}{n+1}x^{n+1}\right) dx\right].$$

7. $y''_{xx} = A_1 y^m (y'_x)^2 + A_2 x^{-1} y'_x, \qquad m \neq -1.$

1°. Solution with $A_2 \neq -1$:

$$x = \left[C_1 \int \exp\left(-\frac{A_1}{m+1} y^{m+1}\right) dy + C_2\right]^{-A_2-1}.$$

2°. Solution with $A_2 = -1$:

$$y = C_2 \exp\left[C_1 \int \exp\left(-\frac{A_1}{m+1} y^{m+1}\right) dy\right].$$

8. $y''_{xx} = A_1 x^{l-2} y^{1-l} (y'_x)^l + A_2 x^{1-l} y^{l-2} (y'_x)^{3-l}.$

Solution in the parametric form:

$$x = C_2 \exp\left[\int \left(A_1 \tau^l + A_2 \tau^{3-l} - \tau^2 + \tau\right)^{-1} d\tau\right],$$
$$y = C_1 \exp\left[\int \tau \left(A_1 \tau^l + A_2 \tau^{3-l} - \tau^2 + \tau\right)^{-1} d\tau\right].$$

9. $y''_{xx} = A_1 x^{l_1-2} y^{1-l_1} (y'_x)^{l_1} + A_2 x^{l_2-2} y^{1-l_2} (y'_x)^{l_2}.$

Solution in the parametric form:

$$x = C_2 \exp\left[\int \left(A_1 \tau^{l_1} + A_2 \tau^{l_2} - \tau^2 + \tau\right)^{-1} d\tau\right],$$
$$y = C_1 \exp\left[\int \tau \left(A_1 \tau^{l_1} + A_2 \tau^{l_2} - \tau^2 + \tau\right)^{-1} d\tau\right].$$

10. $y''_{xx} = A x^{m+3} y^m (y'_x)^3 - A x^m y^{m+3}, \qquad m \neq -2.$

Solution in the parametric form:

$$x = C_1 \tau^{1/2} \exp\left(-\frac{1}{2} \int \frac{V\, d\tau}{\tau\sqrt{V^2+4}}\right), \qquad y = C_1^{-1} \tau^{1/2} \exp\left(\frac{1}{2} \int \frac{V\, d\tau}{\tau\sqrt{V^2+4}}\right),$$

where $V = \tau^{-1/2} \exp\left(\frac{3A}{2m+4} \tau^{m+2}\right) \left[C_2 - A \int \tau^m \exp\left(\frac{3A}{m+2} \tau^{m+2}\right) d\tau\right]^{-1/2}.$

11. $y''_{xx} = A x^{m+2} y^m (y'_x)^{5/2} + A x^m y^{m+2} (y'_x)^{1/2}.$

Solution in the parametric form:

$$x = C_1 \tau^{1/2} \exp\left(-\frac{1}{2} \int \frac{V\, d\tau}{\tau\sqrt{V^2+4}}\right), \qquad y = C_1^{-1} \tau^{1/2} \exp\left(\frac{1}{2} \int \frac{V\, d\tau}{\tau\sqrt{V^2+4}}\right),$$

where function $V = V(\tau)$ is defined in the parametric form $\tau = \tau(u)$, $V = V(u)$ as follows:

1) For $m \neq -1, \; m \neq -3/2$,

$$\tau = a u^{\frac{2}{2m+3}}, \qquad V = \frac{b(2m+3)}{2(m+1)} u^{-1} Z^{-1} (\tau Z'_u + \nu Z),$$

where

$$Z = \begin{cases} C_2 J_\nu(u) + Y_\nu(u) & \text{for the upper sign,} \\ C_2 I_\nu(u) + K_\nu(u) & \text{for the lower sign,} \end{cases}$$

J_ν and Y_ν are Bessel functions, I_ν and K_ν are modified Bessel functions,

$$\nu = \frac{m+1}{2m+3}, \quad a = \Big(-\frac{2m+2}{Ab}\Big)^{-\frac{1}{m+1}}, \quad \pm\frac{(2m+3)^2}{8(m+1)^2} b^{\frac{2m+3}{m+1}} = \Big(-\frac{2m+2}{A}\Big)^{\frac{1}{m+1}};$$

2) For $m = -1$,

$$\tau = \frac{u^2}{2A^2}, \qquad V = \frac{A}{\sqrt{2}} Z^{-1} Z'_u,$$

where $Z = C_2 J_0(u) + Y_0(u)$, J_0 and Y_0 are Bessel functions;

3) For $m = -3/2$,

$$\tau = A^2 u^{-4}, \quad V = \begin{cases} \dfrac{1}{2A}\dfrac{(1+k)C_2 u^k + (1-k)u^{-k}}{C_2 u^k + u^{-k}} & \text{if } A^2 < \frac{1}{8}, \\ \dfrac{1}{2A}\dfrac{C_2 \ln u + C_2 + 1}{C_2 \ln u + 1} & \text{if } A^2 = \frac{1}{8}, \\ \dfrac{1}{2A}\dfrac{(C_2 - k)\sin(k\ln u) + (1 + kC_2)\cos(k\ln u)}{C_2 \sin(k\ln u) + \cos(k\ln u)} & \text{if } A^2 > \frac{1}{8}, \end{cases}$$

where $k = \sqrt{|1 - 8A^2|}$.

12. $\boldsymbol{y''_{xx} = A_1 x^{n_1} y^{-1} (y'_x)^2 + A_2 x^{n_2} y'_x}$.

1°. Solution with $n_2 \neq -1$:

$$y = C_1 \exp\Big(\int u\,dx\Big),$$

where

$$u = \exp\Big(\frac{A_2}{n_2+1} x^{n_2+1}\Big)\Big[C_2 + \int (1 - A_1 x^{n_1}) \exp\Big(\frac{A_2}{n_2+1} x^{n_2+1}\Big)\,dx\Big]^{-1}.$$

2°. Solution with $n_2 = -1$, $A_2 \neq -1$, $A_2 \neq -n_1 - 1$:

$$y = C_1 \exp\Big[\int x^{A_2}\Big(C_2 + \frac{1}{A_2+1} x^{A_2+1} - \frac{A_1}{n_1 + A_2 + 1} x^{n_1+A_2+1}\Big)^{-1} dx\Big].$$

3°. Solution with $n_2 = -1$, $A_2 = -1$:

$$y = C_1 \exp\Big[\int x^{-1}\Big(C_2 + \ln x - \frac{A_1}{n_1} x^{n_1}\Big)^{-1} dx\Big].$$

4°. Solution with $n_2 = -1$, $A_2 = -n_1 - 1$:

$$y = C_1 \exp\Big[\int x^{-n_1-1}\Big(C_2 - \frac{1}{n_1} x^{-n_1} - A_1 \ln x\Big)^{-1} dx\Big].$$

13. $y''_{xx} = A_1 y^{m_1} (y'_x)^2 + A_2 x^{-1} y^{m_2} y'_x$.

1°. Solution with $m_1 \neq -1$:

$$x = C_1 \exp\biggl(\int u\,dy\biggr),$$

where

$$u = \exp\Bigl(-\frac{A_1}{m_1+1} y^{m_1+1}\Bigr)\biggl[C_2 + \int (1 + A_2 y^{m_2}) \exp\Bigl(-\frac{A_1}{m_1+1} y^{m_1+1}\Bigr)\,dy\biggr]^{-1}.$$

2°. Solution with $m_1 = -1$, $A_1 \neq 1$, $A_1 \neq m_2 + 1$:

$$x = C_1 \exp\biggl[\int y^{-A_1}\Bigl(C_2 + \frac{1}{1-A_1} y^{1-A_1} + \frac{A_2}{m_2 - A_1 + 1} y^{m_2 - A_1 + 1}\Bigr)^{-1} dy\biggr].$$

3°. Solution with $m_1 = -1$, $A_1 = 1$:

$$x = C_1 \exp\biggl[\int y^{-1}\Bigl(C_2 + \ln y + \frac{A_2}{m_2} y^{m_2}\Bigr)^{-1} dy\biggr].$$

4°. Solution with $m_1 = -1$, $A_1 = m_2 + 1$:

$$x = C_1 \exp\biggl[\int y^{-m_2-1}\Bigl(C_2 - \frac{1}{m_2} y^{-m_2} + A_2 \ln y\Bigr)^{-1} dy\biggr].$$

► *In the solutions of equations 14–15, the following notation is used:*

$$R = \begin{cases} C_1 \tau^{k_1} + C_2 \tau^{k_2} + C_3 \tau^{k_3} & \textit{if } B_2(8B_1^3 + 27B_2) < 0, \\ C_1 \tau e^{k\tau} + C_2 e^{\sigma\tau} & \textit{if } 8B_1^3 + 27B_2 = 0, \\ C_1 e^{k\tau} + C_2 e^{\rho\tau} \cos \omega\tau & \textit{if } B_2(8B_1^3 + 27B_2) > 0, \end{cases}$$

$$Q = \begin{cases} C_1 k_1 \tau^{k_1} + C_2 k_2 \tau^{k_2} + C_3 k_3 \tau^{k_3} & \textit{if } B_2(8B_1^3 + 27B_2) < 0, \\ C_1 (1 + k\tau) e^{k\tau} + C_2 \sigma e^{\sigma\tau} & \textit{if } 8B_1^3 + 27B_2 = 0, \\ C_1 k e^{k\tau} + C_2 e^{\rho\tau} (\rho \cos \omega\tau - \omega \sin \omega\tau) & \textit{if } B_2(8B_1^3 + 27B_2) > 0, \end{cases}$$

where k_1, k_2, and k_3 (real numbers) or k and $\rho \pm i\omega$ (one real and two complex numbers) are the roots of the cubic equation

$$\lambda^3 - B_1 \lambda^2 - \tfrac{1}{2} B_2 = 0.$$

In the special case $8B_1^3 = -27B_2$, we have $k = \frac{2}{3} B_1$ (multiple root) and $\sigma = -\frac{1}{3} B_1$ (simple root).

Remark. *In the expressions for R and Q, constant C_3 may be set to any nonzero number (for example, one may set $C_3 = \pm 1$).*

14. $y''_{xx} = A_1 y^{-1/2} y'_x + A_2 x y^{-1/2}$.

Solution in the parametric form:

$$x = R, \quad y = Q^2; \qquad B_1 = A_1, \quad B_2 = A_2.$$

15. $y''_{xx} = A_1 x^{-1/2} y (y'_x)^3 + A_2 x^{-1/2} (y'_x)^2$.

Solution in the parametric form:

$$x = Q^2, \quad y = R; \qquad B_1 = -A_2, \quad B_2 = -A_1.$$

▶ *In the solutions of equations 16–17, the following notation is used:*

$$R = \begin{cases} \tau^{B_1/2}(C_1\tau^k + C_2\tau^{-k}) + C_3 & \text{if } B_1^2 + 2B_2 > 0, \\ C_1\tau\exp(\tfrac{1}{2}B_1\tau) + C_2 & \text{if } B_1^2 + 2B_2 = 0, \\ C_1\exp(\tfrac{1}{2}B_1\tau)\cos(\omega\tau) + C_2 & \text{if } B_1^2 + 2B_2 < 0, \end{cases}$$

$$Q = \begin{cases} \tau^{B_1/2}[(C_1(B_1 + 2k)\tau^k + C_2(B_1 - 2k)\tau^{-k}] & \text{if } B_1^2 + 2B_2 > 0, \\ C_1(B_1\tau + 2)\exp(\tfrac{1}{2}B_1\tau) & \text{if } B_1^2 + 2B_2 = 0, \\ C_1\exp(\tfrac{1}{2}B_1\tau)[B_1\cos(\omega\tau) - 2\omega\sin(\omega\tau)] & \text{if } B_1^2 + 2B_2 < 0, \end{cases}$$

where $k = \frac{1}{2}\sqrt{B_1^2 + 2B_2}$, $\omega = \frac{1}{2}\sqrt{-(B_1^2 + 2B_2)}$.

16. $y''_{xx} = A_1 y^{-1/2} y'_x + A_2$.

Solution in the parametric form:

$$x = R, \quad y = \tfrac{1}{4}Q^2; \qquad B_1 = A_1, \quad B_2 = A_2.$$

17. $y''_{xx} = A_1(y'_x)^3 + A_2 x^{-1/2}(y'_x)^2$.

Solution in the parametric form:

$$x = \tfrac{1}{4}Q^2, \quad y = R; \qquad B_1 = -A_2, \quad B_2 = -A_1.$$

18. $y''_{xx} = A_1 x^{n_1} y^{-1}(y'_x)^2 + A_2 x^{n_2} y$.

Solution:

$$y = C_1 \exp\left[-\int \frac{w'_x\,dx}{(A_1x^{n_1} - 1)w}\right],$$

where $w = w(x)$ is the general solution of the second order linear equation

$$(A_1x^{n_1} - 1)w''_{xx} - A_1n_1x^{n_1-1}w'_x + A_1x^{n_2}(A_1x^{n_1} - 1)^2 w = 0.$$

19. $y''_{xx} = A_1 x y^{m_1}(y'_x)^3 + A_2 x^{-1} y^{m_2} y'_x$.

Solution:

$$x = C_1 \exp\left[\int \frac{w'_y\,dy}{(A_2y^{m_2} + 1)w}\right],$$

where $w = w(y)$ is the general solution of the second order linear equation

$$(A_2y^{m_2} + 1)w''_{yy} - A_2m_2y^{m_2-1}w'_y - A_1y^{m_1}(A_2y^{m_2} + 1)^2 w = 0.$$

2.7. Equations of the Form $y''_{xx} = f(x)g(y)h(y'_x)$

See Section 2.3 for the case

$$f(x) = \text{const}\,x^n, \quad g(y) = \text{const}\,y^m, \quad h(w) = \text{const}.$$

See Section 2.5 for the case

$$f(x) = \text{const}\,x^n, \quad g(y) = \text{const}\,y^m, \quad h(w) = \text{const}\,w^l.$$

2.7.1. Equations of the Form $y''_{xx} = f(x)g(y)$

1. $y''_{xx} = x^{-2}\left[-\dfrac{2(m+1)}{(m+3)^2}y + Ay^m\right], \qquad m \neq -3, \quad m \neq -1.$

See equation 2.4.2.4.

2. $y''_{xx} = x^{-2}(\frac{15}{4}y + Ay^{-7})$.

See equation 2.4.2.35.

3. $y''_{xx} = x^{-2}(6y + Ay^{-4})$.

See equation 2.4.2.31.

4. $y''_{xx} = x^{-2}(12y + Ay^{-5/2})$.

See equation 2.4.2.64.

5. $y''_{xx} = x^{-2}(2y + Ay^{-2})$.

See equation 2.4.2.6.

6. $y''_{xx} = x^{-2}(-\frac{3}{16}y + Ay^{-5/3})$.

See equation 2.4.2.26.

7. $y''_{xx} = x^{-2}(-\frac{9}{100}y + Ay^{-5/3})$.

See equation 2.4.2.10.

8. $y''_{xx} = x^{-2}(\frac{3}{4}y + Ay^{-5/3})$.

See equation 2.4.2.12.

9. $y''_{xx} = x^{-2}(\frac{63}{4}y + Ay^{-5/3})$.

See equation 2.4.2.66.

10. $y''_{xx} = x^{-2}(-\frac{5}{36}y + Ay^{-7/5})$.

See equation 2.4.2.29.

11. $y''_{xx} = x^{-2}(-\frac{2}{9}y + Ay^{-1/2})$.

See equation 2.4.2.14.

12. $y''_{xx} = x^{-2}(-\frac{4}{25}y + Ay^{-1/2})$.

See equation 2.4.2.8.

13. $y''_{xx} = x^{-2}(20y + Ay^{-1/2})$.

See equation 2.4.2.33.

14. $y''_{xx} = x^{-2}(-\frac{12}{49}y + Ay^{1/2})$.

See equation 2.4.2.37.

15. $y''_{xx} = x^{-2}(Ay^2 - \frac{6}{25}y)$.

See equation 2.4.2.60.

16. $y''_{xx} = x^{-2}(Ay^2 + \frac{6}{25}y)$.

See equation 2.4.2.62.

17. $y''_{xx} = x^{-4/3}(A + By^{-1/2})$.

See equation 2.4.2.40.

18. $y''_{xx} = (Ax^4 + Bx^3)y^{-7}$.

See equation 2.4.2.39.

19. $y''_{xx} = (Ax^2 + B)y^{-5}$.

See equation 2.4.2.16.

20. $y''_{xx} = (Ax^{-1} + Bx^{-2})y^{-2}$.

See equation 2.4.2.28.

21. $y''_{xx} = (Ax^{-7/3} + Bx^{-10/3})y^{-5/3}$.

See equation 2.4.2.48.

22. $y''_{xx} = (Ax^{-4/3} + Bx^{-10/3})y^{-5/3}$.

See equation 2.4.2.49.

23. $y''_{xx} = (Ax^{-4/3} + Bx^{-7/3})y^{-5/3}$.

See equation 2.4.2.24.

24. $y''_{xx} = (Ax^{-2/3} + Bx^{-4/3})y^{-5/3}$.

See equation 2.4.2.90.

25. $y''_{xx} = (A + Bx^{-2/3})y^{-5/3}$.

See equation 2.4.2.89.

26. $y''_{xx} = (Ax^2 + B)y^{-5/3}$.

See equation 2.4.2.47.

27. $y''_{xx} = (Ax^2 + Bx)y^{-5/3}$.

See equation 2.4.2.46.

28. $y''_{xx} = A(ax^{-2/3} + bx^{-5/3})^2 y^{-5/3}$.

This is a special case of equation 2.7.1.37 with $c = 1$, $d = 0$.

29. $y''_{xx} = (Ax^{-8/5} + Bx^{-13/5}) y^{-7/5}$.

See equation 2.4.2.25.

30. $y''_{xx} = (Ax^{-5/2} + Bx^{-7/2}) y^{-1/2}$.

See equation 2.4.2.23.

31. $y''_{xx} = A(ax^5 + bx^4)^{-1/2} y^{-1/2}$.

This is a special case of equation 2.7.1.38 with $c = 1$, $d = 0$.

32. $y''_{xx} = A(ax^{15/8} + bx^{7/8})^{-4/3} y^{-1/2}$.

This is a special case of equation 2.7.1.39 with $c = 1$, $d = 0$.

33. $y''_{xx} = A(ax^{7/3} + bx^{4/3})^{-15/7} y^2$.

This is a special case of equation 2.7.1.40 with $c = 1$, $d = 0$.

34. $y''_{xx} = (ax^2 + bx + c) y^{-5/3}$.

The transformation $x = x(t)$, $y = (x'_t)^{3/2}$ leads to a third order equation:

$$2x'_t x'''_{ttt} - (x''_{tt})^2 = \tfrac{4}{3}(ax^2 + bx + c).$$

Differentiating the latter equation with respect to t and dividing it by x'_t, we obtain a constant-coefficient fourth-order linear equation: $3x''''_{tttt} = 4ax + 2b$.

35. $y''_{xx} = (ax^{-10/3} + bx^{-7/3} + cx^{-4/3}) y^{-5/3}$.

The transformation $x = 1/t$, $y = w/t$ leads to an equation of the form 2.7.1.34: $w''_{tt} = (at^2 + bt + c) w^{-5/3}$.

36. $y''_{xx} = (ax^2 + bx + c)^n y^{-2n-3}$.

This is a special case of equation 2.9.1.9 with $f(\xi) = \xi^{-2n}$.

37. $y''_{xx} = A(ax + b)^2 (cx + d)^{-10/3} y^{-5/3}$.

The transformation $\xi = \dfrac{ax+b}{cx+d}$, $w = \dfrac{y}{cx+d}$ leads to the Emden—Fowler equation of the form 2.3.1.9: $w''_{\xi\xi} = A\Delta^{-2}\xi^2 w^{-5/3}$, where $\Delta = ad - bc$.

38. $y''_{xx} = A(ax + b)^{-1/2} (cx + d)^{-2} y^{-1/2}$.

The transformation $\xi = \dfrac{ax+b}{cx+d}$, $w = \dfrac{y}{cx+d}$ leads to the Emden—Fowler equation of the form 2.3.1.25: $w''_{\xi\xi} = A\Delta^{-2}\xi^{-1/2} w^{-1/2}$, where $\Delta = ad - bc$.

39. $y''_{xx} = A(ax+b)^{-4/3}(cx+d)^{-7/6}y^{-1/2}$.

The transformation $\xi = \dfrac{ax+b}{cx+d}$, $w = \dfrac{y}{cx+d}$ leads to the Emden—Fowler equation of the form 2.3.1.17: $w''_{\xi\xi} = A\Delta^{-2}\xi^{-4/3}w^{-1/2}$, where $\Delta = ad - bc$.

40. $y''_{xx} = A(ax+b)^{-15/7}(cx+d)^{-20/7}y^2$.

The transformation $\xi = \dfrac{ax+b}{cx+d}$, $w = \dfrac{y}{cx+d}$ leads to the Emden—Fowler equation of the form 2.3.1.20: $w''_{\xi\xi} = A\Delta^{-2}\xi^{-15/7}w^2$, where $\Delta = ad - bc$.

41. $y''_{xx} = A\exp(ax^2+bx)\exp(ky)$.

The substitution $kw = ky + ax^2 + bx$ leads to an equation of the form 2.9.1.1: $w''_{xx} = Ae^{kw} + 2ak^{-1}$.

2.7.2. Equations Containing Power Functions ($h \not\equiv$ const)

1. $y''_{xx} = \left[\dfrac{2(n+1)}{(n+3)^2}x + Ax^n\right]y^{-2}(y'_x)^3, \qquad n \neq -3, \quad n \neq -1.$

See equation 2.6.2.116.

2. $y''_{xx} = (-\frac{15}{4}x + Ax^{-7})y^{-2}(y'_x)^3$.

See equation 2.6.2.117.

3. $y''_{xx} = (-6x + Ax^{-4})y^{-2}(y'_x)^3$.

See equation 2.6.2.118.

4. $y''_{xx} = (-12x + Ax^{-5/2})y^{-2}(y'_x)^3$.

See equation 2.6.2.119.

5. $y''_{xx} = (-2x + Ax^{-2})y^{-2}(y'_x)^3$.

See equation 2.6.2.120.

6. $y''_{xx} = (\frac{3}{16}x + Ax^{-5/3})y^{-2}(y'_x)^3$.

See equation 2.6.2.121.

7. $y''_{xx} = (\frac{9}{100}x + Ax^{-5/3})y^{-2}(y'_x)^3$.

See equation 2.6.2.122.

8. $y''_{xx} = (-\frac{3}{4}x + Ax^{-5/3})y^{-2}(y'_x)^3$.

See equation 2.6.2.123.

9. $y''_{xx} = (-\frac{63}{4}x + Ax^{-5/3})y^{-2}(y'_x)^3$.

See equation 2.6.2.124.

10. $y''_{xx} = (\frac{5}{36}x + Ax^{-7/5})y^{-2}(y'_x)^3.$

See equation 2.6.2.125.

11. $y''_{xx} = (\frac{2}{9}x + Ax^{-1/2})y^{-2}(y'_x)^3.$

See equation 2.6.2.126.

12. $y''_{xx} = (\frac{4}{25}x + Ax^{-1/2})y^{-2}(y'_x)^3.$

See equation 2.6.2.127.

13. $y''_{xx} = (-20x + Ax^{-1/2})y^{-2}(y'_x)^3.$

See equation 2.6.2.128.

14. $y''_{xx} = (\frac{12}{49}x + Ax^{1/2})y^{-2}(y'_x)^3.$

See equation 2.6.2.129.

15. $y''_{xx} = (Ax^2 + \frac{6}{25}x)y^{-2}(y'_x)^3.$

See equation 2.6.2.130.

16. $y''_{xx} = (Ax^2 - \frac{6}{25}x)y^{-2}(y'_x)^3.$

See equation 2.6.2.131.

17. $y''_{xx} = (A + Bx^{-1/2})y^{-4/3}(y'_x)^3.$

See equation 2.6.2.15.

18. $y''_{xx} = x^{-7}(Ay^4 + By^3)(y'_x)^3.$

See equation 2.6.2.111.

19. $y''_{xx} = x^{-5}(Ay^2 + B)(y'_x)^3.$

See equation 2.6.2.96.

20. $y''_{xx} = x^{-2}(Ay^{-1} + By^{-2})(y'_x)^3.$

See equation 2.6.2.110.

21. $y''_{xx} = x^{-5/3}(Ay^{-7/3} + By^{-10/3})(y'_x)^3.$

See equation 2.6.2.34.

22. $y''_{xx} = x^{-5/3}(Ay^{-4/3} + By^{-10/3})(y'_x)^3.$

See equation 2.6.2.36.

23. $y''_{xx} = x^{-5/3}(Ay^{-4/3} + By^{-7/3})(y'_x)^3.$

See equation 2.6.2.14.

24. $y''_{xx} = x^{-5/3}(Ay^{-2/3} + By^{-4/3})(y'_x)^3$.

See equation 2.6.2.115.

25. $y''_{xx} = x^{-5/3}(A + By^{-2/3})(y'_x)^3$.

See equation 2.6.2.114.

26. $y''_{xx} = x^{-5/3}(Ay^2 + B)(y'_x)^3$.

See equation 2.6.2.35.

27. $y''_{xx} = x^{-5/3}(Ay^2 + By)(y'_x)^3$.

See equation 2.6.2.33.

28. $y''_{xx} = Ax^{-5/3}(ay^{-2/3} + by^{-5/3})^2(y'_x)^3$.

This is a special case of equation 2.7.2.37 with $c = 1$, $d = 0$.

29. $y''_{xx} = x^{-7/5}(Ay^{-8/5} + By^{-13/5})(y'_x)^3$.

See equation 2.6.2.109.

30. $y''_{xx} = x^{-1/2}(Ay^{-5/2} + By^{-7/2})(y'_x)^3$.

See equation 2.6.2.13.

31. $y''_{xx} = Ax^{-1/2}(ay^5 + by^4)^{-1/2}(y'_x)^3$.

This is a special case of equation 2.7.2.38 with $c = 1$, $d = 0$.

32. $y''_{xx} = Ax^{-1/2}(ay^{15/8} + by^{7/8})^{-4/3}(y'_x)^3$.

This is a special case of equation 2.7.2.39 with $c = 1$, $d = 0$.

33. $y''_{xx} = Ax^2(ay^{7/3} + by^{4/3})^{-15/7}(y'_x)^3$.

This is a special case of equation 2.7.2.40 with $c = 1$, $d = 0$.

34. $y''_{xx} = x^{-5/3}(ay^2 + by + c)(y'_x)^3$.

Assuming y as the independent variable, we obtain an equation of the form 2.7.1.34 for function $x = x(y)$: $x''_{yy} = -(ay^2 + by + c)x^{-5/3}$.

35. $y''_{xx} = x^{-5/3}(ay^{-10/3} + by^{-7/3} + cy^{-4/3})(y'_x)^3$.

Assuming y as the independent variable, we obtain an equation of the form 2.7.1.35 for function $x = x(y)$: $x''_{yy} = -(ay^{-10/3} + by^{-4/3} + cy^{-4/3})x^{-5/3}$.

36. $y''_{xx} = x^{-2n-3}(ay^2 + by + c)^n(y'_x)^3$.

Assuming y as the independent variable, we obtain an equation of the form 2.9.1.9 with $f(\xi) = -\xi^{-2n}$ for function $x = x(y)$: $x''_{yy} = -(ay^2 + by + c)^n x^{-2n-3}$.

37. $y''_{xx} = Ax^{-5/3}(ay+b)^2(cy+d)^{-10/3}(y'_x)^3.$

Assuming y as the independent variable, we obtain an equation of the form 2.7.1.37 for function $x = x(y)$: $x''_{yy} = -A(ay+b)^2(cy+d)^{-10/3}x^{-5/3}$.

38. $y''_{xx} = Ax^{-1/2}(ay+b)^{-1/2}(cy+d)^{-2}(y'_x)^3.$

Assuming y as the independent variable, we obtain an equation of the form 2.7.1.38 for function $x = x(y)$: $x''_{yy} = -A(ay+b)^{-1/2}(cy+d)^{-2}x^{-1/2}$.

39. $y''_{xx} = Ax^{-1/2}(ay+b)^{-4/3}(cy+d)^{-7/6}(y'_x)^3.$

Assuming y as the independent variable, we obtain an equation of the form 2.7.1.39 for function $x = x(y)$: $x''_{yy} = -A(ay+b)^{-4/3}(cy+d)^{-7/6}x^{-1/2}$.

40. $y''_{xx} = Ax^2(ay+b)^{-15/7}(cy+d)^{-20/7}(y'_x)^3.$

Assuming y as the independent variable, we obtain an equation of the form 2.7.1.40 for function $x = x(y)$: $x''_{yy} = -A(ay+b)^{-15/7}(cy+d)^{-20/7}x^2$.

41. $y''_{xx} = Ax^{-1/2}y^{-2}\big[(y'_x)^2 + B^2\big]^{1/2}.$

Solution in the parametric form:

$$x = a(u^2-1)^{-1}(\tau u \pm R)^2, \qquad y = b\tau^{-1}(u^2-1)^{-1/2},$$

where $R = \sqrt{\tau^2 - 2\tau^{-1} + C_1}$, $u = \mp\tanh\Big(C_2 + \int R^{-1}\,d\tau\Big)$, $A = -\frac{1}{2}a^{-1/2}b^2$, $B = \frac{1}{2}a^{-1}b$.

42. $y''_{xx} = Ax^{-1/2}y^{-2}\big[(y'_x)^2 - B^2\big]^{1/2}.$

Solution in the parametric form:

$$x = a(u^2+1)^{-1}(\tau u \pm R)^2, \qquad y = b\tau^{-1}(u^2+1)^{-1/2},$$

where $R = \sqrt{-\tau^2 - 2\tau^{-1} + C_1}$, $u = \pm\tan\Big(C_2 + \int R^{-1}\,d\tau\Big)$, $A = -\frac{1}{2}a^{-1/2}b^2$, $B = \frac{1}{2}a^{-1}b$.

43. $y''_{xx} = Ax^{-1/2}y^{-2}\big[B^2 - (y'_x)^2\big]^{1/2}.$

Solution in the parametric form:

$$x = a(1-u^2)^{-1}(\tau u \mp R)^2, \qquad y = b\tau^{-1}(1-u^2)^{-1/2},$$

where $R = \sqrt{\tau^2 - 2\tau^{-1} + C_1}$, $u = \pm\tanh\Big(C_2 + \int R^{-1}\,d\tau\Big)$, $A = -\frac{1}{2}a^{-1/2}b^2$, $B = \frac{1}{2}a^{-1}b$.

44. $y''_{xx} = Ax^{-2}y^{-1/2}(y'_x)^2\left[(y'_x)^2 + B^2\right]^{1/2}$.

Solution in the parametric form:

$$x = a\tau^{-1}(u^2-1)^{-1/2}, \qquad y = b(u^2-1)^{-1}(\tau u \pm R)^2,$$

where $R = \sqrt{\tau^2 - 2\tau^{-1} + C_1}$, $u = \mp\tanh\Bigl(C_2 + \int R^{-1}\,d\tau\Bigr)$, $A = -\frac{1}{4}a^3b^{-3/2}$, $B = 2a^{-1}b$.

45. $y''_{xx} = Ax^{-2}y^{-1/2}(y'_x)^2\left[(y'_x)^2 - B^2\right]^{1/2}$.

Solution in the parametric form:

$$x = a\tau^{-1}(1-u^2)^{-1/2}, \qquad y = b(1-u^2)^{-1}(\tau u \mp R)^2,$$

where $R = \sqrt{\tau^2 - 2\tau^{-1} + C_1}$, $u = \pm\tanh\Bigl(C_2 + \int R^{-1}\,d\tau\Bigr)$, $A = -\frac{1}{4}a^3b^{-3/2}$, $B = 2a^{-1}b$.

46. $y''_{xx} = Ax^{-2}y^{-1/2}(y'_x)^2\left[B^2 - (y'_x)^2\right]^{1/2}$.

Solution in the parametric form:

$$x = a\tau^{-1}(1+u^2)^{-1/2}, \qquad y = b(1+u^2)^{-1}(\tau u \pm R)^2,$$

where $R = \sqrt{-\tau^2 - 2\tau^{-1} + C_1}$, $u = \pm\tan\Bigl(C_2 + \int R^{-1}\,d\tau\Bigr)$, $A = -\frac{1}{4}a^3b^{-3/2}$, $B = 2a^{-1}b$.

2.7.3. Equations Containing Exponential Functions ($h \not\equiv$ const)

Preliminary Comments.

1. With $l \neq 1 - m$, the equation

$$y''_{xx} = Ae^x y^m (y'_x)^l \tag{1}$$

has the particular solution

$$y = Be^{\lambda x}, \qquad \text{where} \quad \lambda = \frac{1}{1-m-l}, \qquad B = (A\lambda^{l-2})^{\lambda}.$$

2. With $m \neq 0$ and $l \neq 1$, equation (1) can be reduced, with the aid of the transformation

$$t = (y'_x)^{1-l}, \qquad w = e^x,$$

to the generalized Emden—Fowler equation with respect to $w = w(t)$:

$$w''_{tt} = Bt^{\frac{1}{1-l}} w^{-1} (w'_t)^{\frac{2m+1}{m}}, \tag{2}$$

where $B = -m\left[A(1-l)\right]^{\frac{1}{m}}$. Equations of the form (2) are outlined in Section 2.5.

When obtained the general solution $w = w(t)$ of the Emden—Fowler equation (2), the solution of the original equation (1) can be written in the parametric form with the formulae

$$x = \ln w, \quad y = k(w'_t)^{-\frac{1}{m}},$$

where $k = \big[A(1-l)\big]^{-\frac{1}{m}}$.

3. With $l \neq n+2$, the equation

$$y''_{xx} = Ax^n e^y (y'_x)^l \tag{3}$$

has the particular solution

$$y = \lambda \ln(Bx), \qquad \text{where} \quad \lambda = l - n - 2, \quad B = \left(-\frac{\lambda^{1-l}}{A}\right)^{\frac{1}{\lambda}}.$$

4. Taking y as the independent variable and x as the dependent one, we obtain from equation (3) an equation of the form (1) for $x = x(y)$:

$$x''_{yy} = -Ae^y x^n (x'_y)^{3-l}.$$

5. With $n \neq -1$ and $l \neq 1$, equation (3) can be reduced, with the aid of the transformation

$$t = (y'_x)^{1-l}, \qquad u = x^{n+1},$$

to the generalized Emden—Fowler equation for $u = u(t)$:

$$u''_{tt} = -\frac{1}{n+1} t^{\frac{1}{1-l}} u^{-\frac{n}{n+1}} (u'_t)^2. \tag{4}$$

Equations of this form are outlined in Section 2.5.

When obtained the general solution $u = u(t)$ of the Emden—Fowler equation (4), the solution of the original equation (3) can be written in the parametric form with the formulae

$$x = u^{\frac{1}{n+1}}, \quad y = -\ln(u'_t) + \ln\frac{n+1}{A(1-l)}.$$

1. $\boldsymbol{y''_{xx} = Ae^x (y'_x)^l}$.

1°. Solution in the parametric form with $l \neq 1$:

$$x = \ln\left[\pm\frac{1}{A(1-l)} C_1^{1-l}\tau\right], \qquad y = C_1 \int \frac{1}{\tau}(1\pm\tau)^{\frac{1}{1-l}}\,d\tau + C_2.$$

2°. Solution in the parametric form with $l = 1$:

$$x = \ln\left(\pm\frac{\tau}{A}\right), \qquad y = C_1 \int \frac{1}{\tau}\exp(\pm\tau)\,d\tau + C_2.$$

2. $y''_{xx} = Ae^{x}y^{m}(y'_{x})^{2}$.

1°. Solution in the parametric form with $m \neq -1$:

$$x = \int \frac{d\tau}{f} + C_2, \qquad y = \tau \exp\left[-\frac{1}{m+1}\left(\int \frac{d\tau}{f} + C_2\right)\right],$$

where function $f = f(\tau)$ is defined implicitly by the relation

$$\ln\left(\frac{f}{\tau} - \frac{1}{m+1}\right) - \frac{\tau}{(m+1)f - \tau} = \frac{A}{m+1}\tau^{m+1} - \ln\tau + C_1.$$

2°. Solution with $m = -1$:

$$y = C_2 \exp\left(\int \frac{dx}{x + Ae^{x} + C_1}\right).$$

3. $y''_{xx} = Ae^{x}y$.

1°. Solution with $A > 0$:

$$y = C_1 I_0\big(2\sqrt{A}\,e^{x/2}\big) + C_2 K_0\big(2\sqrt{A}\,e^{x/2}\big),$$

where I_0 and K_0 are modified Bessel functions.

2°. Solution with $A < 0$:

$$y = C_1 J_0\big(2\sqrt{-A}\,e^{x/2}\big) + C_2 Y_0\big(2\sqrt{-A}\,e^{x/2}\big),$$

where J_0 and Y_0 are Bessel functions.

4. $y''_{xx} = Ae^{x}y^{-1/2}(y'_{x})^{3/2}$.

Solution in the parametric form:

$$x = \tau^2 - \ln(Af), \qquad y = C_1\big[2\tau f - \exp(\tau^2)\big]^2,$$

where $f = \int \exp(\tau^2)\,d\tau + C_2$.

5. $y''_{xx} = Ae^{x}y(y'_{x})^{3/2}$.

Solution in the parametric form:

$$x = -\ln\big[AC_1^3\big(\sqrt{\tau^2 + \tau} - f\big)\big], \qquad y = 2C_1^2\left(1 - f\sqrt{\frac{\tau+1}{\tau}}\right),$$

where $f = \ln\big(\sqrt{\tau} + \sqrt{\tau+1}\big) + C_2$.

6. $y''_{xx} = Ae^{y}(y'_{x})^{l}$.

1°. Solution in the parametric form with $l \neq 2$:

$$x = C_1 \int \frac{1}{\tau}(1 \pm \tau)^{\frac{1}{l-2}}\,d\tau + C_2, \qquad y = \ln\left[\pm\frac{1}{A(2-l)}C_1^{l-2}\tau\right].$$

2°. Solution in the parametric form with $l = 2$:

$$x = C_1 \int \frac{1}{\tau}\exp(\mp\tau)\,d\tau + C_2, \qquad y = \ln\left(\pm\frac{\tau}{A}\right).$$

7. $y''_{xx} = Ax^n e^y y'_x$.

1°. Solution in the parametric form with $n \neq -1$:

$$x = \tau \exp\left[-\frac{1}{n+1}\left(\int \frac{d\tau}{f} + C_2\right)\right], \qquad y = \int \frac{d\tau}{f} + C_2,$$

where function $f = f(\tau)$ is defined implicitly by the relation

$$\ln\left(\frac{f}{\tau} - \frac{1}{n+1}\right) - \frac{\tau}{(n+1)f - \tau} = -\frac{A}{n+1}\tau^{n+1} - \ln \tau + C_1.$$

2°. Solution with $n = -1$:

$$x = C_2 \exp\left(\int \frac{dy}{y - Ae^y + C_1}\right).$$

8. $y''_{xx} = Ax^{-1/2} e^y (y'_x)^{3/2}$.

Solution in the parametric form:

$$x = C_1\left[2\tau f - \exp(\tau^2)\right]^2, \qquad y = \tau^2 - \ln(-Af),$$

where $f = \int \exp(\tau^2)\, d\tau + C_2$.

9. $y''_{xx} = Axe^y (y'_x)^{3/2}$.

Solution in the parametric form:

$$x = 2C_1^2\left(1 - f\sqrt{\frac{\tau+1}{\tau}}\right), \qquad y = -\ln\left[AC_1^3\left(f - \sqrt{\tau^2 + \tau}\right)\right],$$

where $f = \ln\left(\sqrt{\tau} + \sqrt{\tau + 1}\right) + C_2$.

10. $y''_{xx} = Axe^y (y'_x)^3$.

1°. Solution in the parametric form with $A > 0$:

$$x = C_1 J_0(2\tau) + C_2 Y_0(2\tau), \qquad y = \ln\left(\tau/\sqrt{A}\right),$$

where J_0 and Y_0 are Bessel functions.

2°. Solution in the parametric form with $A < 0$:

$$x = C_1 I_0(2\tau) + C_2 K_0(2\tau), \qquad y = \ln\left(\tau/\sqrt{-A}\right),$$

where I_0 and K_0 are modified Bessel functions.

11. $y''_{xx} = Ae^x e^y (y'_x)^l$.

Solution in the parametric form:

$$x = \int \frac{d\tau}{f\tau^l} + C_2, \qquad y = \ln\left(\frac{f}{A}\right) - \int \frac{d\tau}{f\tau^l} - C_2,$$

$$\text{where } f = \begin{cases} \dfrac{1}{2-l}\tau^{2-l} + \dfrac{1}{1-l}\tau^{1-l} + C_1 & \text{if } l \neq 1, 2; \\ \tau + \ln|\tau| + C_1 & \text{if } l = 1; \\ \ln|\tau| - \dfrac{1}{\tau} + C_1 & \text{if } l = 2. \end{cases}$$

12. $y''_{xx} = A\exp(kx)\exp(ay^2 + by)(y'_x)^3$.

Taking y as the independent variable, we obtain an equation of the form 2.7.1.41 for function $x = x(y)$: $x''_{yy} = -A\exp(ay^2 + by)\exp(kx)$.

13. $y''_{xx} = Ae^x y^{-1/2}(y'_x)^{3/2}\sqrt{y'_x - 2B}$.

Solution in the parametric form:

$$x = \ln[a\tau(\cosh u)^{-1}], \qquad y = B\cosh^2 u\,(\tau\tanh u \pm R)^2,$$

where $a = -A^{-1}B^{-1/2}$, $R = \sqrt{2\ln\tau + \tau^2 + C_1}$, $u = C_2 \mp \int R^{-1}\,d\tau$.

14. $y''_{xx} = Ae^x y^{-1/2}(y'_x)^{3/2}\sqrt{2B - y'_x}$.

Solution in the parametric form:

$$x = \ln[a\tau(\cos u)^{-1}], \qquad y = B\cos^2 u\,(\tau\tan u \pm R)^2,$$

where $a = -A^{-1}B^{-1/2}$, $R = \sqrt{2\ln\tau - \tau^2 + C_1}$, $u = C_2 \pm \int R^{-1}\,d\tau$.

15. $y''_{xx} = Ax^{-1/2}e^y y'_x\sqrt{y'_x - B}$.

Solution in the parametric form:

$$x = \frac{1}{2B}\cos^2 u\,(\tau\tan u \pm R)^2, \qquad y = \ln[b\tau(\cos u)^{-1}],$$

where $b = A^{-1}\sqrt{2}$, $R = \sqrt{2\ln\tau - \tau^2 + C_1}$, $u = C_2 \pm \int R^{-1}\,d\tau$.

16. $y''_{xx} = Ax^{-1/2}e^y y'_x\sqrt{B - y'_x}$.

Solution in the parametric form:

$$x = \frac{1}{2B}\cosh^2 u\,(\tau\tanh u \pm R)^2, \qquad y = \ln[b\tau(\cosh u)^{-1}],$$

where $b = A^{-1}\sqrt{2}$, $R = \sqrt{2\ln\tau + \tau^2 + C_1}$, $u = C_2 \mp \int R^{-1}\,d\tau$.

2.7.4. Equations Containing Hyperbolic Functions ($h \not\equiv \text{const}$)

1. $y''_{xx} = Ax[\cosh(\omega y)]^{-2}y'_x$.

Solution in the parametric form:

$$x = a\cosh u\,(\tau\tanh u \pm R), \qquad y = u/\omega,$$

where $A = a^{-2}$, $R = \sqrt{2\ln\tau + \tau^2 + C_1}$, $u = C_2 \mp \int R^{-1}\,d\tau$.

2. $y''_{xx} = Ax[\sinh(\omega y)]^{-2}y'_x$.

Solution in the parametric form:

$$x = a\sinh u\,(\tau\coth u \pm R), \qquad y = u/\omega,$$

where $A = a^{-2}$, $R = \sqrt{2\ln\tau + \tau^2 + C_1}$, $u = C_2 \mp \int R^{-1}\,d\tau$.

3. $y''_{xx} = Ax\cosh(\omega y)(y'_x)^{3/2}$.

Solution in the parametric form:

$$x = a(u^2+1)^{-1/2}(\tau u \pm R), \qquad y = \omega^{-1}\ln\bigl(u+\sqrt{u^2+1}\,\bigr),$$

where $A = 2a^{-2}\sqrt{a\omega}$, $R = \sqrt{C_1-\tau^2-2\tau^{-1}}$, $u = \pm\tan\bigl(C_2+\int R^{-1}\,d\tau\bigr)$.

4. $y''_{xx} = Ax\sinh(\omega y)(y'_x)^{3/2}$.

Solution in the parametric form:

$$x = a(u^2-1)^{-1/2}(\tau u \pm R), \qquad y = \pm\omega^{-1}\ln\bigl(u+\sqrt{u^2-1}\,\bigr),$$

where $A = \pm 2a^{-2}\sqrt{a\omega}$, $R = \sqrt{C_1+\tau^2-2\tau^{-1}}$, $u = \mp\tanh\bigl(C_2+\int R^{-1}\,d\tau\bigr)$.

5. $y''_{xx} = A\cosh(\omega x)y(y'_x)^{3/2}$.

Solution in the parametric form:

$$x = \omega^{-1}\ln\bigl(u+\sqrt{u^2+1}\bigr), \qquad y = b(u^2+1)^{-1/2}(\tau u \pm R),$$

where $A = -2b^{-2}\sqrt{b\omega}$, $R = \sqrt{C_1-\tau^2-2\tau^{-1}}$, $u = \pm\tan\bigl(C_2+\int R^{-1}\,d\tau\bigr)$.

6. $y''_{xx} = A\sinh(\omega x)y(y'_x)^{3/2}$.

Solution in the parametric form:

$$x = \pm\omega^{-1}\ln\bigl(u+\sqrt{u^2-1}\bigr), \qquad y = b(u^2-1)^{-1/2}(\tau u \pm R),$$

where $A = \mp 2b^{-2}\sqrt{b\omega}$, $R = \sqrt{C_1+\tau^2-2\tau^{-1}}$, $u = \mp\tanh\bigl(C_2+\int R^{-1}\,d\tau\bigr)$.

7. $y''_{xx} = A[\cosh(\omega x)]^{-2}y(y'_x)^2$.

Solution in the parametric form:

$$x = u/\omega, \qquad y = b\cosh u\,(\tau\tanh u \pm R),$$

where $A = -b^{-2}$, $R = \sqrt{2\ln\tau+\tau^2+C_1}$, $u = C_2 \mp \int R^{-1}\,d\tau$.

8. $y''_{xx} = A[\sinh(\omega x)]^{-2}y(y'_x)^2$.

Solution in the parametric form:

$$x = u/\omega, \qquad y = b\sinh u\,(\tau\coth u \pm R),$$

where $A = -b^{-2}$, $R = \sqrt{2\ln\tau+\tau^2+C_1}$, $u = C_2 \mp \int R^{-1}\,d\tau$.

2.7.5. Equations Containing Trigonometric Functions ($h \not\equiv$ const)

► *In the solutions of equations 1–4, the following notation is used:*

$$R = \sqrt{2\ln\tau - \tau^2 + C_1}, \qquad u = C_2 \pm \int R^{-1}\,d\tau.$$

1. $y''_{xx} = Ax[\cos(\omega y)]^{-2} y'_x$.

Solution in the parametric form:

$$x = a\cos u\,(\tau\tan u \pm R), \quad y = u/\omega, \qquad \text{where} \quad A = a^{-2}.$$

2. $y''_{xx} = Ax[\sin(\omega y)]^{-2} y'_x$.

Solution in the parametric form:

$$x = a\sin u\,(\tau\cot u \mp R), \quad y = u/\omega, \qquad \text{where} \quad A = a^{-2}.$$

3. $y''_{xx} = A[\cos(\omega x)]^{-2} y (y'_x)^2$.

Solution in the parametric form:

$$x = \omega^{-1}u, \quad y = b\cos u\,(\tau\tan u \pm R), \qquad \text{where} \quad A = -b^2.$$

4. $y''_{xx} = A[\sin(\omega x)]^{-2} y (y'_x)^2$.

Solution in the parametric form:

$$x = \omega^{-1}u, \quad y = b\sin u\,(\tau\cot u \mp R), \qquad \text{where} \quad A = -b^2.$$

► *In the solutions of equations 5–8, the following notation is used:*

$$R = \sqrt{\tau^2 - 2\tau^{-1} + C_1}, \qquad u = \pm\tanh\Bigl(C_2 + \int R^{-1}\,d\tau\Bigr).$$

5. $y''_{xx} = Ax\cos(\omega y)(y'_x)^{3/2}$.

Solution in the parametric form:

$$x = a(1-u^2)^{-1/2}(\tau u \mp R), \quad y = \omega^{-1}\arccos u, \qquad \text{where} \quad A = 2a^{-2}(-a\omega)^{1/2}.$$

6. $y''_{xx} = Ax\sin(\omega y)(y'_x)^{3/2}$.

Solution in the parametric form:

$$x = a(1-u^2)^{-1/2}(\tau u \mp R), \quad y = \omega^{-1}\arccos u, \qquad \text{where} \quad A = 2a^{-2}(a\omega)^{1/2}.$$

7. $y''_{xx} = A\cos(\omega x) y (y'_x)^{3/2}$.

Solution in the parametric form:

$$x = \omega^{-1}\arccos u, \quad y = b(1-u^2)^{-1/2}(\tau u \mp R), \qquad \text{where} \quad A = -2b^{-2}(-b\omega)^{1/2}.$$

8. $y''_{xx} = A\sin(\omega x) y (y'_x)^{3/2}$.

Solution in the parametric form:

$$x = \omega^{-1}\arccos u, \quad y = b(1-u^2)^{-1/2}(\tau u \mp R), \qquad \text{where} \quad A = -2b^{-2}(b\omega)^{1/2}.$$

2.7.6. Some Transformations

For the sake of visualization, we also use the symbolic notation $\{f,\ g,\ h\}$ to denote the equation

$$y''_{xx} = f_1(x)g_1(y)h_1(y'_x). \tag{1}$$

1. Taking y as the independent variable and x as the dependent one, we obtain an equation of the similar form for $x = x(y)$:

$$x''_{yy} = g_1(y)f_1(x)h_1^*(x'_y), \qquad \text{where} \quad h_1^*(w) = -w^3 h_1(1/w).$$

Denote this transformation by $\mathcal{F}$.

2. The Bäcklund transformation

$$\bar{x} = \int \frac{dw}{h_1(w)}, \qquad \bar{y} = \int f_1(x)\,dx, \qquad \text{where} \quad w = y'_x \tag{2}$$

leads to an equation of the similar form for function $\bar{y} = \bar{y}(\bar{x})$:

$$\bar{y}''_{\bar{x}\bar{x}} = f_2(\bar{x})g_2(\bar{y})h_2(\bar{y}'_{\bar{x}}),$$

where the functions f_2, g_2, and h_2 are defined in terms of the original functions f_1, g_1, and h_1 parametrically by the formulae

$$f_2(\bar{x}) = w, \qquad \bar{x} = \int \frac{dw}{h_1(w)},$$
$$g_2(\bar{y}) = \frac{1}{f_1(x)}, \qquad \bar{y} = \int f_1(x)\,dx,$$
$$h_2(\bar{w}) = -\frac{1}{[g_1(y)]^3}\frac{dg_1}{dy}, \qquad \bar{w} = \frac{1}{g_1(y)}.$$

Denote transformation (2) by $\mathcal{G}$. For equations of the form (1) wherein f_1, g_1, and h_1 are power functions of their arguments, transformation $\mathcal{G}$ (to a precision of constant factors) is considered in Subsection 2.5.3. For equations (1) with exponential functions f_1 and g_1, transformation $\mathcal{G}$ is discussed in Subsection 2.7.3.

When found the solution $\bar{y} = \bar{y}(\bar{x})$ of the transformed equation, the formulae

$$\bar{y} = \int f_1(x)\,dx, \qquad \bar{y}'_{\bar{x}} = \frac{1}{g_1(y)},$$

make it possible to obtain the solution of the original equation (1) in the parametric form $x = x(\bar{x})$, $y = y(\bar{x})$.

3. The twofold application of transformation $\mathcal{G}$ to the original equation yields an equation of the similar form:

$$\bar{\bar{y}}''_{\bar{\bar{x}}\bar{\bar{x}}} = f_3(\bar{\bar{x}})g_3(\bar{\bar{y}})h_3(\bar{\bar{y}}'_{\bar{\bar{x}}}),$$

where the functions f_3, g_3, and h_3 are defined in terms of the original functions f_1, g_1, and h_1 parametrically by the formulae

$$f_3(\bar{\bar{x}}) = \frac{1}{g_1(y)}, \qquad \bar{\bar{x}} = \int g_1(y)\,dy,$$
$$g_3(\bar{\bar{y}}) = \frac{1}{w}, \qquad \bar{\bar{y}} = \int \frac{w\,dw}{h_1(w)},$$
$$h_3(\bar{\bar{w}}) = \frac{df_1}{dx}, \qquad \bar{\bar{w}} = f_1(x).$$

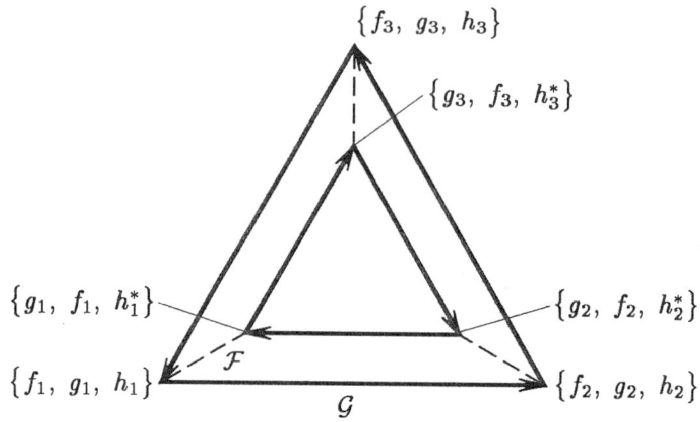

FIGURE 4

The threefold transformation $\mathcal{G}$ yields the original equation.

Different compositions of transformations $\mathcal{F}$ and $\mathcal{G}$ generate six different equations of the analogous form which are shown in Figure 4.

4. In the special case $g(y) = y^m$, $h = 1$, the transformation $x = \dfrac{1}{\tau}$, $y = \dfrac{u}{\tau}$ leads to an equation of the similar form:

$$u''_{\tau\tau} = \tau^{-m-3} f\Big(\frac{1}{\tau}\Big) u^m.$$

Denote this transformation by $\mathcal{H}$.

For $g(y) = y^m$ and $h = 1$, different compositions of transformations $\mathcal{F}$, $\mathcal{G}$, and $\mathcal{H}$ generate twelve different equations of the form (1).

2.8. Some Nonlinear Equations with Arbitrary Parameters

2.8.1. Equations Containing Power Functions

1. $\boldsymbol{y''_{xx} = (ay + bx + c)^n}$.

This is a special case of equation 2.9.1.2 with $f(\xi) = \xi^n$.

2. $\boldsymbol{y''_{xx} = (ay + bx^2)^n + c}$.

The substitution $aw = ay + bx^2$ leads to an equation of the form 2.9.1.1:

$$w''_{xx} = a^n w^n + c + \frac{2b}{a}.$$

3. $\boldsymbol{y''_{xx} = \lambda x^{-2n-3}(xy + a)^n}$.

This is a special case of equation 2.9.1.7 with $f(\xi) = \lambda\xi^n$, $b = c = 0$.

4. $\boldsymbol{y''_{xx} = (ax^2 + bx + c)y^{-5}}$.

This is a special case of equation 2.9.1.9 with $f(\xi) = \xi^{-2}$.

5. $\boldsymbol{y''_{xx} = A(ax+b)^n(cx+d)^{-n-m-3}y^m}$.

The transformation $\xi = \dfrac{ax+b}{cx+d}$, $w = \dfrac{y}{cx+d}$ leads to the Emden—Fowler equation (see Section 2.3): $w''_{\xi\xi} = A\Delta^{-2}\xi^n w^m$, where $\Delta = ad - bc$.

6. $\boldsymbol{y''_{xx} = cx^m y^{-nk-m-3}(ay^n + bx^n)^k}$.

This is a special case of equation 2.9.1.5 with $f(\xi) = c\xi^{-nk-m-3}(a\xi^n + b)^k$.

7. $\boldsymbol{y''_{xx} = cx^m y^{-2nk-2m-3}(ay^{2n} + bx^n)^k}$.

This is a special case of equation 2.9.1.6 with $f(\xi) = c\xi^{-2nk-2m-3}(a\xi^{2n} + b)^k$.

8. $\boldsymbol{y''_{xx} = (ay^2 + bxy + cx^2 + \alpha y + \beta x + \gamma)^{-3/2}, \qquad a \neq 0}$.

The substitution $2au = 2ay + bx + \alpha$ leads to an equation of the form 2.9.1.9:

$$u''_{xx} = u^{-3} f\left(\frac{u}{\sqrt{Ax^2 + Bx + C}}\right),$$

where $f(\xi) = \xi^3(a\xi^2+1)^{-3/2}$, $A = \dfrac{4ac - b^2}{4a}$, $B = \dfrac{2a\beta - b\alpha}{2a}$, $C = \dfrac{4a\gamma - \alpha^2}{4a}$.

9. $\boldsymbol{y''_{xx} = \lambda y^{-1/3} + (ax^2 + bx + c)y^{-5/3}}$.

The transformation $x = x(t)$, $y = (x'_t)^{3/2}$ leads to a third order equation:

$$2x'_t x'''_{ttt} - (x''_{tt})^2 = \tfrac{4}{3}\lambda(x'_t)^2 + \tfrac{4}{3}(ax^2 + bx + c).$$

Differentiating the latter equation with respect to t and dividing it by x'_t, we arrive at a constant-coefficient fourth-order linear equation: $3x''''_{tttt} = 2\lambda x''_{tt} + 4ax + 2b$.

10. $\boldsymbol{y''_{xx} = \lambda x^{-8/3} y^{-1/3} + (ax^{-10/3} + bx^{-7/3} + cx^{-4/3})y^{-5/3}}$.

The transformation $x = 1/t$, $y = w/t$ leads to an equation of the form 2.8.1.9:

$$w''_{tt} = \lambda w^{-1/3} + (at^2 + bt + c)w^{-5/3}.$$

11. $\boldsymbol{y''_{xx} + 3yy'_x + y^3 + ax^n y = 0}$.

This is a special case of equation 2.9.2.1 with $f(x) = ax^n$.

12. $\boldsymbol{y''_{xx} + (2ay + bx^n)y'_x + bnx^{n-1}y = 0}$.

This is a special case of equation 2.9.2.13 with $f(x) = bx^n$.

13. $\boldsymbol{y''_{xx} = ax^n(xy'_x - y)^2 + bx^m}$.

This is a special case of equation 2.9.3.4 with $f(x) = bx^m$, $g(x) = 0$, $h(x) = ax^n$.

14. $\boldsymbol{y''_{xx} = ax^n(xy'_x - y)^m}$.

This is a special case of equation 2.9.4.30 with $f(x) = ax^n$, $g(\xi) = \xi^m$.

15. $y''_{xx} = ax^{-n-3}y^n(xy'_x - y)^m$.

This is a special case of equation 2.9.4.31 with $f(\xi) = a\xi^m$.

16. $y''_{xx} = ax^{-1}y^n y'_x(xy'_x - y)^m$.

This is a special case of equation 2.9.4.33 with $f(y) = ay^n$, $g(\xi) = \xi^m$.

17. $y''_{xx} = ax^{nk-1}y^{mk-1}y'_x(xy'_x - y)^{\frac{2n+m}{n}}$.

This is a special case of equation 2.9.4.32 with $f(\xi) = a\xi^k$.

18. $y''_{xx} = kx^\alpha(y'_x)^\beta(xy'_x - y)^\gamma$.

The Legendre transformation $x = w'_t$, $y = tw'_t - w$, where $w = w(t)$, leads to the generalized Emden—Fowler equation

$$w''_{tt} = \frac{1}{k}t^{-\beta}w^{-\gamma}(w'_t)^{-\alpha}.$$

A good deal of solvable equations of this type are outlined in Section 2.3 and Section 2.5.

19. $y''_{xx} = ax^{n-1}y^{m-1}(y'_x)^{\frac{2n+m-nk}{n+m}}(xy'_x - y)^k$.

This is a special case of equation 2.9.4.36 with $f(\xi) = a\xi$.

20. $y''_{xx} = ax^n(xy'_x - y) + bx^m(xy'_x - y)^k$.

This is a special case of equation 2.9.4.39 with $f(x) = ax^n$, $g(x) = bx^m$.

21. $xy''_{xx} = ny'_x + ax^{2n+1} + bx^{2n+1}y^m$.

This is a special case of equation 2.9.2.4 with $f(y) = a + by^m$.

22. $xy''_{xx} = -(n+1)y'_x + ax^{n-1} + bx^{nm+n-1}y^m$.

This is a special case of equation 2.9.2.5 with $f(\xi) = a + b\xi^m$.

23. $yy''_{xx} - \frac{1}{4}(y'_x)^2 = ax^2 + bx + c$.

The substitution $y = w^{4/3}$ leads to a special case of the equation 2.8.1.9 with $\lambda = 0$:

$$4w''_{xx} = 3(ax^2 + bx + c)w^{-5/3}.$$

24. $3yy''_{xx} - 2(y'_x)^2 = ax^2 + bx + c$.

The substitution $y = w^3$ leads to the equation 2.8.1.4: $9w''_{xx} = (ax^2 + bx + c)w^{-5}$.

25. $2yy''_{xx} = (y'_x)^2 + bx^ny^2 - a$.

This is a special case of equation 2.9.3.7 with $f(x) = -bx^n$.

26. $yy''_{xx} = n(y'_x)^2 - ay^{4n-2} + bx^m y^2.$

This is a special case of equation 2.9.3.8 with $f(x) = -bx^m$.

27. $yy''_{xx} = n(y'_x)^2 + ax^k y^2 + bx^m y^{n+1}.$

This is a special case of equation 2.9.3.9 with $f(x) = -ax^k$, $g(x) = -bx^m$.

28. $(n+2)yy''_{xx} - (n+1)(y'_x)^2 = (ax^2 + bx + c)^n.$

The substitution $y = w^{n+2}$ leads to an equation of the form 2.7.1.36:

$$w''_{xx} = \frac{1}{(n+2)^2}(ax^2 + bx + c)^n w^{-2n-3}.$$

29. $yy''_{xx} = (y'_x)^2 + ax^n yy'_x + bx^m y^2.$

This is a special case of equation 2.9.3.6 with $f(x) = -ax^n$, $g(x) = -bx^m$.

30. $ayy''_{xx} + b(y'_x)^2 + (x^n + \lambda)^m yy'_x = 0.$

Solution: $y^{\frac{a+b}{a}} = C_1 \int \exp\left[-\frac{1}{a}\int (x^n + \lambda)^m \, dx\right] dx + C_2.$

31. $(y + ax)y''_{xx} = bx^n(xy'_x - y)^2.$

The substitution $y = -ax + xz$ leads to the equation

$$xzz''_{xx} + 2zz'_x - bx^{n+3}(z'_x)^2 = 0.$$

Having set $w = z'_x/z$, we obtain the Bernoulli equation

$$xw'_x + 2w + x(1 - bx^{n+2})w^2 = 0.$$

32. $x^2 y''_{xx} = n(n+1)y + ax^{3n+2} + bx^{nm+3n+2}y^m.$

This is a special case of equation 2.9.1.11 with $f(\xi) = a + b\xi^m$.

33. $x^2 y''_{xx} = k(k+1)y + ax^{km+3k+2}(bx^{2k+1} + c)^n y^m.$

The transformation $\xi = bx^{2k+1} + c$, $w = yx^k$ leads to the Emden—Fowler equation (see Section 2.3)

$$w''_{\xi\xi} = ab^{-2}(2k+1)^{-2}\xi^n w^m.$$

34. $x^2 y''_{xx} + xy'_x = ay^n + b.$

This is a special case of equation 2.9.2.8 with $f(y) = ay^n + b$.

35. $x^2 y''_{xx} = -(n+m+1)xy'_x - nmy + ax^{nk+n-2m}y^k.$

This is a special case of equation 2.9.2.9 with $f(\xi) = a\xi^k$.

36. $x^2y''_{xx} + axy'_x + by = cx^ny^m.$

The transformation $x = \xi^\alpha$, $y = \xi^\beta w$, where $\alpha = \pm\dfrac{1}{\sqrt{D}}$, $\beta = \pm\dfrac{1-a}{2\sqrt{D}} - \dfrac{1}{2}$, $D = (1-a)^2 - 4b$, leads to the Emden—Fowler equation

$$w''_{\xi\xi} = c\alpha^2\xi^{n\alpha+m\beta-\beta-2}w^m$$

whose solvable cases are outlined in Section 2.3.

37. $x^2y''_{xx} = n(n-1)y + ax^n(xy'_x - ny)^m.$

This is a special case of equation 2.9.4.37 with $f(x) = ax^{n-2}$.

38. $(ax^2 + b)y''_{xx} + axy'_x + cy^n = 0.$

This is a special case of equation 2.9.2.11 with $f(y) = cy^n$.

39. $(ay + bx^2)y''_{xx} = 1.$

This is a special case of equation 2.8.1.2 with $n = -1$, $c = 0$.

40. $xyy''_{xx} = x(y'_x)^2 - yy'_x + ax^ky^s.$

This is a special case of equation 2.9.4.14 with $f(\xi) = a\xi$, $g(\xi) = 1$, $k = n-1$, $s = m+2$.

41. $y^2y''_{xx} + y(y'_x)^2 = ax + b.$

Having set $\dfrac{1}{y} = u'_x(x)$, we obtain

$$-u'_xu'''_{xxx} + 3(u''_{xx})^2 = (ax+b)(u'_x)^5.$$

Taking u as the independent variable, we obtain a constant coefficient linear equation for $x = x(u)$: $x'''_{uuu} = ax + b$.

42. $(x + a)^2y^2y''_{xx} = bx.$

The transformation $\xi = \ln\left|\dfrac{x+a}{x}\right|$, $w = \dfrac{y}{x}$ leads to an equation of the form 2.2.1.7: $w''_{\xi\xi} - w'_\xi = a^{-2}w^{-2}$.

43. $(y^2 + ax^2 + 2bx + c)^2y''_{xx} + sy = 0.$

Dividing by the coefficient of y''_{xx} and multiplying by $ax(xy'_x - y) + b(2xy'_x - y) + cy'_x$, we arrive at a total differential equation. Integrating the latter, we obtain

$$(ax^2 + 2bx + c)(y'_x)^2 - 2(ax+b)yy'_x + ay^2 + \frac{sy^2}{y^2 + ax^2 + 2bx + c} = C.$$

44. $(ax+b)^2(cx+d)^2 y''_{xx} = sy + A(ax+b)^k(cx+d)^{1-m-k}y^m.$

The transformation

$$\xi = \ln\left(\frac{ax+b}{cx+d}\right), \qquad w = \left(\frac{ax+b}{cx+d}\right)^{\frac{k}{m-1}} \frac{y}{cx+d}$$

leads to an autonomous equation:

$$w''_{\xi\xi} - (2n+1)w'_\xi + (n^2+n-s\Delta^{-2})w = A\Delta^{-2}w^m,$$

where $n = \dfrac{k}{m-1}$, $\Delta = ad - bc$.

45. $(a^2-x^2)(b^2-y^2)y''_{xx} + (a^2-x^2)y(y'_x)^2 = x(b^2-y^2)y'_x.$

Solution: $\arcsin\dfrac{y}{b} = C_1 + C_2 \arcsin\dfrac{x}{b}.$

46. $(ay^n + bx^n)y''_{xx} + cx^{n-3} = 0.$

This is a special case of equation 2.9.1.5 with $f(\xi) = -c(a\xi^n+b)^{-1}$.

47. $(ay^n + bx^n)y''_{xx} + cy^{n-3} = 0.$

This is a special case of equation 2.9.1.5 with $f(\xi) = -c\xi^{n-3}(a\xi^n+b)^{-1}$.

48. $(ay^{2n} + bx^n)y''_{xx} + cy^{2n-3} = 0.$

This is a special case of equation 2.9.1.6 with $f(\xi) = -c\xi^{2n-3}(a\xi^{2n}+b)^{-1}$.

49. $(ay^n + bx^n)y''_{xx} + cx^m y^{n-m-3} = 0.$

This is a special case of equation 2.9.1.5 with $f(\xi) = -c\xi^{n-m-3}(a\xi^n+b)^{-1}$.

50. $(ay^{2n} + bx^n)y''_{xx} + cx^m y^{2n-2m-3} = 0.$

This is a special case of equation 2.9.1.6 with $f(\xi) = -c\xi^{2n-2m-3}(a\xi^{2n}+b)^{-1}$.

51. $(y''_{xx})^2 = \alpha(xy'_x - y) + \beta y'_x + \gamma.$

Differentiating the equation with respect to x, we obtain

$$y''_{xx}(2y'''_{xxx} - \alpha x - \beta) = 0. \tag{1}$$

Equating the second factor to zero and integrating, we find

$$y = \frac{\alpha x^4}{48} + \frac{\beta x^3}{12} + C_2x^2 + C_1x + C_0. \tag{2}$$

The integration constants C_i and the parameters α, β, and γ are related by the constraint $4C_2^2 = \beta C_1 - \alpha C_0 + \gamma$ which is obtained by substituting the above solution (2) into the original equation.

In addition, there is the solution that corresponds to the first factor in (1):

$$y = \widetilde{C}_1 x + \widetilde{C}_0, \qquad \text{where} \quad \beta\widetilde{C}_1 - \alpha\widetilde{C}_0 + \gamma = 0.$$

2.8.2. Painlevé Equations

Preliminary comments.

Painlevé equations are met with in different fields of contemporary physics. They resulted from solving the problem of extracting equation classes of the form

$$\frac{d^2y}{dx^2} = R\Big(x,\ y,\ \frac{dy}{dx}\Big),$$

where R is a rational function of y and dy/dx with analytic coefficients whose integrals have no movable critical points. This problem was solved in works by P. Painlevé and B. Gambier. A total of 50 equations were extracted, of which 44 equations have the general solutions that are expressed in terms of elementary functions, solutions of some linear equations, or solutions of the other six equations.

The six equations not integrated by Painlevé and Gambier are referred to as Painlevé transcendents (or irreducible Painlevé equations), and their solutions are called transcendental Painlevé functions. The absence of movable critical points makes it possible to use solutions of Painlevé transcendents as basic functions for representing solutions of other nonlinear differential equations, along with quadratures and solutions of variable-coefficient linear differential equations. The canonical forms of Painlevé transcendents are specified below.

1. First Painlevé transcendent:

$$y''_{xx} = 6y^2 + x. \tag{1}$$

The solution of the first Painlevé equation is a unique function of x. It can be presented, in the vicinity of movable pole x_0, in terms of the series

$$y = (x-x_0)^{-2} + \frac{x_0}{10}(x-x_0)^2 - \frac{1}{6}(x-x_0)^3 + C(x-x_0)^4 + \frac{x_0^2}{300}(x-x_0)^6 + \sum_{j=7}^{\infty} a_j(x-x_0)^j,$$

where x_0 and C are arbitrary constants; coefficients a_j $(j \geq 7)$ are uniquely defined in terms of x_0 and C.

For large values of $|z|$, the following asymptotic formulae holds:

$$y(x) \sim x^{1/2}\wp\big(\tfrac{4}{5}x^{5/4} - a;\ 12,\ b\big),$$

where the elliptic Weierstrass function $\wp(\zeta; 12, b)$ is defined implicitly by the integral

$$\zeta = \int \frac{d\wp}{\sqrt{4\wp^3 - 12\wp - b}};$$

a and b are some constants.

The first Painlevé transcendent (1) is invariant with respect to stretching variables $x = \lambda\bar{x}$, $y = \lambda^3\bar{y}$, where $\lambda^5 = 1$, i.e., it admits discrete symmetry of the fifth order.

2. Second Painlevé transcendent:

$$y''_{xx} = 2y^3 + xy + \alpha. \tag{2}$$

The solution of the second Painlevé equation is a unique function of x. Denote the solution by $y(x,\alpha)$ with fixed parameter α. Then, the following relation holds:

$$y(x,-\alpha) = -y(x,\alpha),$$

while solutions $y(x,\alpha)$ and $y(x,\alpha-1)$ are related by the Bäcklund transformations

$$y(x,\alpha-1) = -y(x,\alpha) + \frac{2\alpha-1}{2y'_x(x,\alpha) - 2y^2(x,\alpha) - x},$$

$$y(x,\alpha) = -y(x,\alpha-1) - \frac{2\alpha-1}{2y'_x(x,\alpha-1) + 2y^2(x,\alpha-1) + x}.$$

Therefore, in order to study the general solution of equation (2) with arbitrary α it is sufficient to construct the solution for all α out of the band $0 \le \operatorname{Re}\alpha < \frac{1}{2}$.

Three solution corresponding to α and $\alpha \pm 1$ are related by the rational formulae

$$y_{\alpha+1} = -\frac{(y_{\alpha-1}+y_\alpha)(4y_\alpha^3 + 2xy_\alpha + 2\alpha + 1) + (2\alpha-1)y_\alpha}{2(y_{\alpha-1}+y_\alpha)(2y_\alpha^2 + x) + 2\alpha - 1},$$

where y_α stands for $y(x,\alpha)$.

Solutions $y(x,\alpha)$ and $y(x,-\alpha-1)$ are related by the Bäcklunds transformations

$$y(x,-\alpha-1) = y(x,\alpha) + \frac{2\alpha+1}{2y'_x(x,\alpha) + 2y^2(x,\alpha) + x},$$

$$y(x,\alpha) = y(x,-\alpha-1) - \frac{2\alpha+1}{2y'_x(x,-\alpha-1) + 2y^2(x,-\alpha-1) + x}.$$

The second Painlevé transcendent (2) for all $\alpha = n + \frac{1}{2}$ (n is an integer) possesses the one-parameter family of solutions which are generated by the general solution of the Fuchsian equation

$$(y'_x)^n + \sum_{j=1}^{n} P_j(x,y)(y'_x)^{n-j} = 0$$

and are expressed in terms of Airy functions and their derivatives.

There exist rational solutions of the form $y = R_1/R_2$, where R_1 and R_2 are polynomials in x, only when $a = 0,\ \pm1,\ \pm2,\ \pm3,\ \dots$ A rational solution can be written as

$$y_a = \frac{1}{P_a}\frac{dP_a}{dx} - \frac{1}{Q_a}\frac{dQ_a}{dx},$$

where P_a and Q_a are polynomials, a is an integer.

In the special case $a = 1$, we have

$$P_1 = 1, \quad Q_1 = x \qquad \text{hence} \qquad y_1 = -\frac{1}{x}.$$

In the special case $a = 2$, we have

$$P_2 = x, \quad Q_2 = x^3 + 4 \qquad \text{hence} \qquad y_2 = \frac{1}{x} - \frac{3x^2}{x^3+4}.$$

The following recurrence formulae take place:

$$P_{a+1} = Q_a, \quad Q_{a+1} = [xQ_a^2 + (Q'_a)^2 - 4Q_aQ''_a]/P_a,$$

where prime denotes differentiation with respect to x.

If we stretch and shift the variables in accordance with the rule $x = \varepsilon^2\bar{x} - 6\varepsilon^{-10}$, $y = \varepsilon\bar{y} + \varepsilon^{-5}$ and let $\alpha = 4\varepsilon^{-15}$, equation (2), in the limit $\varepsilon \to 0$, transforms to equation (1).

3. Third Painlevé transcendent:

$$y''_{xx} = \frac{(y'_x)^2}{y} - \frac{y'_x}{y} + \frac{1}{x}(\alpha y^2 + \beta) + \gamma y^3 + \frac{\delta}{y}. \tag{3}$$

The solution of the third Painlevé equation is a unique function of x.

Any solution of the Riccati equation

$$y'_x = ky^2 + \frac{\alpha - k}{kx} y + c, \tag{4}$$

where $k^2 = \gamma$, $c^2 = -\delta$, $k\beta + c(\alpha - 2k) = 0$, is a solution of equation (3). Substituting $x = \lambda\tau$, $y = -\dfrac{u'_x}{ku}$, where $\lambda^2 = \dfrac{1}{kc}$, into (4), we obtain a linear equation:

$$u''_{\tau\tau} + \frac{k - \alpha}{k\tau} u'_\tau + u = 0,$$

whose general solution is expressed in terms of Bessel functions:

$$u = \tau^{\frac{\alpha}{2k}} \Big[C_1 J_{\frac{\alpha}{2k}}(\tau) + C_2 Y_{\frac{\alpha}{2k}}(\tau)\Big].$$

In some special cases, equation (3) can be integrated by quadrature. Rewrite equation (3) in the form of integro-differential relations by two ways

$$\left(\frac{y'_z}{y}\right)^2 + \left(\frac{\delta}{y^2} - \gamma y^2\right)e^{2z} + 2\left(\frac{\beta}{y} - \alpha y\right)e^z = 2\int\left[\left(\frac{\delta}{y^2} - \gamma y^2\right)e^{2z} + \left(\frac{\beta}{y} - \alpha y\right)e^z\right]dz \tag{5}$$

and

$$\frac{y'_z}{y} = \int\left[\left(\frac{\delta}{y^2} + \gamma y^2\right)e^{2z} + \left(\frac{\beta}{y} + \alpha y\right)e^z\right]dz, \qquad x = e^z. \tag{6}$$

It is obvious from (5) that for $\alpha = \beta = \gamma = \delta = 0$, the general solution has the form $y = C_1 x^{C_2}$. Adding (6) multiplied by 2 to (5), we obtain

$$\left(\frac{y'_z}{y}\right)^2 + 2\frac{y'_z}{y} + \left(\frac{\delta}{y^2} - \gamma y^2\right)e^{2z} + 2\left(\frac{\beta}{y} - \alpha y\right)e^z = 4\int \frac{\delta e^{2z} + \beta e^z y}{y^2}\,dz, \tag{7}$$

Subtracting (6) times 2 from (5), we find

$$\left(\frac{y'_z}{y}\right)^2 - 2\frac{y'_z}{y} + \left(\frac{\delta}{y^2} - \gamma y^2\right)e^{2z} + 2\left(\frac{\beta}{y} - \alpha y\right)e^z = -4\int(\gamma e^{2z}y^2 + \alpha e^z y)\,dz. \tag{8}$$

Substituting $\delta = \beta = 0$ into equation (7) and $\gamma = \alpha = 0$ into equation (8), we arrive at

$$\left(\frac{y'_z}{y}\right)^2 + 2\frac{y'_z}{y} - 2\alpha y e^z - \gamma y^2 e^{2z} = C_1, \tag{9}$$

$$\left(\frac{y'_z}{y}\right)^2 - 2\frac{y'_z}{y} + \frac{\delta}{y^2}e^{2z} + \frac{2\beta}{y}e^z = C_2. \tag{10}$$

Equations (9) and (10) are integrable by elementary functions. Substituting $y = e^{-z}/v$ into (9), we obtain

$$(v'_z)^2 = 2\alpha v + \gamma + (1 + C_1)v^2. \tag{11}$$

As a result, we find

$$y = \begin{cases} \dfrac{2\alpha}{x(\alpha^2 \ln^2 x + 2\alpha C \ln x + C^2 - \gamma)} & \text{if } C_1 = -1,\ \beta = \delta = 0; \\ \dfrac{1}{x\left(\sqrt{\gamma}\ln x + C\right)} & \text{if } C_1 = -1,\ \alpha = \beta = \delta = 0; \\ \dfrac{x^{m-1}}{C_2 x^{2m} + K_1 x^m + K_2} & \text{if } C_1 \neq -1,\ \beta = \delta = 0, \end{cases}$$

where $C_2 \neq 0$, $K_1 = -\dfrac{\alpha}{C_1 + 1}$, $K_2 = \dfrac{\alpha^2 - \gamma(1 + C_1)}{4C_2(1 + C_1)^2}$, $m^2 = 1 + C_1$. Accordingly, equation (10) is reduced to equation (11) with the substitution $y = ve^z$.

If $\beta = -\alpha$ and $\delta = -\gamma$, the transformation $y = e^{-iw}$ brings equation (3) to the equivalent form

$$w''_{xx} + \frac{1}{x} w'_x = \frac{2\alpha}{x} \sin w + 2\gamma \sin 2w.$$

If we perform the transformation $x = 1 + \varepsilon^2 \bar{x}$, $y = 1 + 2\varepsilon \bar{y}$, $\alpha = -\frac{1}{2}\varepsilon^{-6}$, $\beta = \frac{1}{2}\varepsilon^{-6} + 2\bar{\alpha}\varepsilon^{-3}$, $\gamma = \frac{1}{4}\varepsilon^{-6}$, $\delta = -\frac{1}{4}\varepsilon^{-6}$, then equation (3), in the limit $\varepsilon \to 0$, transforms to equation (2).

4. Fourth Painlevé transcendent:

$$y''_{xx} = \frac{(y'_x)^2}{2y} + \frac{3}{2} y^3 + 4xy^2 + 2(x^2 - \alpha)y + \frac{\beta}{y}. \tag{12}$$

If we pass on to the new independent variable $x = e^z$, the solutions are unique functions of z.

The Laurent-series expansion of the solution of equation (12) in the vicinity of any pole $x = x_0$ has the form

$$y(x) = \frac{m}{x - x_0} - x_0 - \frac{m}{3}(x_0^2 + 2\alpha - 4m)(x - x_0) + C(x - x_0)^2 + \sum_{j=3}^{\infty} a_j (x - x_0)^j,$$

where $m = \pm 1$, C is an arbitrary number, and a_j ($j \geq 3$) are uniquely defined in terms of α, β, x_0, and C.

If the condition $\beta + 2(1 + \alpha m)^2 = 0$, where $m = \pm 1$, is satisfied, then every solution of the Riccati equation

$$y'_x = my^2 + 2mxy - 2(1 + \alpha m)$$

is simultaneously a solution of the fourth Painlevé transcendent (12).

Equation (12) is invariant with respect to the transformation $y = \lambda \bar{y}$, $x = \lambda \bar{x}$, $\alpha = \bar{\alpha}\lambda^2$, $\beta = \bar{\beta}$, where $\lambda^4 = 1$. Two solutions of equation (12) corresponding to different values of parameters α and β are related to each other by the Bäcklund transformations

$$\begin{aligned} \tilde{y} &= \frac{1}{2sy}(y'_x - q - 2sxy - sy^2), & q^2 &= -2\beta, \\ y &= -\frac{1}{2s\tilde{y}}(\tilde{y}'_x - p + 2sx\tilde{y} + s\tilde{y}^2), & p^2 &= -2\tilde{\beta}, \\ 2\beta &= -\left(\tilde{\alpha}s - 1 - \tfrac{1}{2}p\right)^2, & 4\alpha &= -2s - 2\tilde{\alpha} - 3sp, \end{aligned}$$

where $y = y(x, \alpha, \beta)$, $\tilde{y} = \tilde{y}(x, \tilde{\alpha}, \tilde{\beta})$, s is an arbitrary parameter.

If we perform the transformation $x = 2^{-2/3}\varepsilon\bar{x} - \varepsilon^{-3}$, $y = 2^{2/3}\varepsilon\bar{y} + \varepsilon^{-3}$, $\alpha = -\bar{\alpha} - \frac{1}{2}\varepsilon^{-6}$, $\beta = -\frac{1}{2}\varepsilon^{-12}$, equation (12), in the limit $\varepsilon \to 0$, transforms to equation (3).

5. Fifth Painlevé transcendent:

$$y''_{xx} = \frac{3y-1}{2y(y-1)}(y'_x)^2 - \frac{y'_x}{x} + \frac{(y-1)^2}{x^2}\left(\alpha y + \frac{\beta}{y}\right) + \gamma\frac{y}{x} + \frac{\delta y(y+1)}{y-1}. \quad (13)$$

If we pass on to the new independent variable $x = e^z$, the solutions are unique functions of z.

Solutions of the fifth Painlevé transcendent (13) corresponding to different values of the parameters are related by two equalities

$$y(x, \alpha, \beta, \gamma, \delta) = y(-x, \alpha, \beta, -\gamma, \delta),$$
$$y(x, \alpha, \beta, \gamma, \delta) = \frac{1}{y(x, -\beta, -\alpha, -\gamma, \delta)}.$$

Having set $x = e^t$ in (13), we obtain

$$y''_{tt} = \frac{3y-1}{2y(y-1)}(y'_t)^2 + (y-1)^2\left(\alpha y + \frac{\beta}{y}\right) + \gamma y e^t + \frac{\delta y(y+1)}{y-1}e^{2t}. \quad (14)$$

If $\gamma = \delta = 0$, equation (14) is reduced, by means of integration, to a first order autonomous equation:

$$y'_t = (y-1)\sqrt{2\alpha y^2 + Cy - 2\beta},$$

which is readily integrable by quadrature.

If the condition

$$\gamma = \sqrt{-2\delta}\,\big(1 + \sqrt{-2\beta} - \sqrt{2\alpha}\,\big)$$

is satisfied, any solution of the Riccati equation

$$xy'_x = \sqrt{2\alpha}\,y^2 + \big(\sqrt{-2\delta}\,x - \sqrt{2\alpha} - \sqrt{-2\beta}\,\big)y + \sqrt{-2\beta} \quad (15)$$

is simultaneously a solution of the fifth Painlevé transcendent (13). Equation (15) can be reduced to the degenerate hypergeometric equation 2.1.2.65.

If we perform the transformation $y = 1 + \varepsilon\bar{y}$, $\beta = -\varepsilon^{-2}\bar{\beta}$, $\alpha = \varepsilon^{-2}\bar{\beta} + \varepsilon^{-1}\bar{\alpha}$, $\gamma = \varepsilon\bar{\gamma}$, $\delta = \varepsilon\bar{\delta}$, equation (13), in the limit $\varepsilon \to 0$, transforms to equation (3). In a similar manner, as a result of the transformation $y = \sqrt{2}\,\varepsilon\bar{y}$, $x = 1 + \sqrt{2}\,\varepsilon\bar{x}$, $\alpha = \frac{1}{2}\varepsilon^{-4}$, $\gamma = -\varepsilon^{-4}$, $\delta = -\frac{1}{2}\varepsilon^{-4} - \varepsilon^{-2}\bar{\delta}$, equation (13), in the limit $\varepsilon \to 0$, transforms to equation (12).

6. Sixth Painlevé transcendent:

$$y''_{xx} = \frac{1}{2}\left(\frac{1}{y} + \frac{1}{y-1} + \frac{1}{y-x}\right)(y'_x)^2 - \left(\frac{1}{x} + \frac{1}{x-1} + \frac{1}{y-x}\right)y'_x$$
$$+ \frac{y(y-1)(y-x)}{x^2(x-1)^2}\left[\alpha + \beta\frac{x}{y^2} + \gamma\frac{x-1}{(y-1)^2} + \delta\frac{x(x-1)}{(y-x)^2}\right]. \quad (16)$$

In equation (16), the points $x = 0$, $x = 1$, and $x = \infty$ are critical. Painlevé found two integrable case of the equation. First, if $\alpha = \beta = \gamma = \delta = 0$, the general solution of equation (16) has the form

$$y = E(C_1\omega_1 + C_2\omega_2,\ x),$$

where $E(u, x)$ is the elliptic function, defined by the integral

$$u = \int_0^E \frac{dy}{\sqrt{y(y-1)(y-x)}}, \tag{17}$$

with periods $2\omega_1$ and $2\omega_1$ which are functions of x. Second, if $\alpha = \beta = \gamma = 0$, $\delta = \frac{1}{2}$, the general solution of equation (16) has the form

$$y = E(w + C_1\omega_1 + C_2\omega_2,\ x),$$

where $w \not\equiv 0$ is any particular solution of the linear equation

$$w''_{xx} - \frac{2x-1}{x(x-1)}w'_x + \frac{1}{4x(x-1)}w = 0,$$

E is the elliptic function defined by formula (17).

Solutions of the sixth Painlevé transcendent (16) corresponding to different values of the parameters are related by three equalities

$$y(x,\ -\beta,\ -\alpha,\ \gamma,\ \delta) = \frac{1}{y\Big(\dfrac{1}{x},\ \alpha,\ \beta,\ \gamma,\ \delta\Big)},$$

$$y(x,\ -\beta,\ -\gamma,\ \alpha,\ \delta) = 1 - \frac{1}{y\Big(\dfrac{1}{1-x},\ \alpha,\ \beta,\ \gamma,\ \delta\Big)},$$

$$y\Big(x,\ -\beta,\ -\alpha,\ -\delta + \frac{1}{2},\ -\gamma + \frac{1}{2}\Big) = \frac{x}{y(x,\ \alpha,\ \beta,\ \gamma,\ \delta)}.$$

The consecutive use of these equalities yields 24 equations of the form (16) with different values of the parameters related by known transformations.

Every solution of the Riccati equation

$$y'_x = \frac{\sqrt{2\alpha}}{x(x-1)}y^2 + \frac{\lambda x + \mu}{x(x-1)}y + \frac{\sqrt{-2\beta}}{x-1} \tag{18}$$

is simultaneously a solutions of equation (16) if $\sqrt{2\alpha} - \sqrt{-2\beta} \neq 1$ and the condition

$$2\sqrt{2\alpha}\,(3\beta - \alpha + \gamma - \delta) + 2\sqrt{-2\beta}\,(3\alpha - \beta - \gamma + \delta) + 4\sqrt{-\alpha\beta}\,(\beta - \alpha + \gamma - \delta - 1)$$
$$+ (\alpha + \beta + \gamma + \delta)^2 + 2(\alpha - \beta - \gamma - 4\alpha\beta - 2\alpha\gamma - 2\beta\delta) = 0$$

is satisfied (one should take such a value of $\sqrt{-\alpha\beta}$ which coincides with $\sqrt{\alpha}\sqrt{-\beta}$). In equation (18),

$$\lambda = \frac{\sqrt{2\alpha} - (\alpha + \beta + \gamma + \delta)}{\sqrt{2\alpha} - \sqrt{-2\beta} - 1}, \qquad \mu = \frac{\sqrt{-2\beta} - (\alpha + \beta - \gamma - \delta)}{\sqrt{2\alpha} - \sqrt{-2\beta} - 1}.$$

If we perform the transformation $x = 1 + \varepsilon\bar{x}$, $\delta = \varepsilon^{-2}\bar{\delta}$, $\gamma = \varepsilon^{-1}\bar{\gamma} - \varepsilon^{-2}\bar{\delta}$, equation (16), in the limit $\varepsilon \to 0$, transforms to equation (13).

2.8.3. Equation Containing Exponential Functions

1. $y''_{xx} = \lambda^2 y + a\exp[\lambda(n+3)x]y^n.$

This is a special case of equation 2.9.1.15 with $f(\xi) = a\xi^n$.

2. $y''_{xx} = \lambda^2 y + ae^{\mu x}y^m, \qquad \lambda \neq 0.$

The transformation $\xi = e^{2\lambda x}$, $w = ye^{\lambda x}$ leads to the Emden—Fowler equation (see Section 2.3)

$$w''_{\xi\xi} = \frac{a}{4\lambda^2}\xi^n w^m, \qquad \text{where} \quad n = \frac{\mu - 3\lambda - m\lambda}{2\lambda}.$$

3. $y''_{xx} = \lambda^2 y + ae^{\lambda(m+3)x}(be^{2\lambda x} + c)^n y^m, \qquad \lambda \neq 0.$

The transformation $\xi = be^{2\lambda x} + c$, $w = ye^{\lambda x}$ leads to the Emden—Fowler equation (see Section 2.3)

$$w''_{\xi\xi} = \frac{a}{4b^2\lambda^2}\xi^n w^m.$$

4. $y''_{xx} = \lambda^2 y + Ae^{\lambda(m+3)x}(ae^{2\lambda x} + b)^n(ce^{2\lambda x} + d)^{-n-m-3}y^m.$

The transformation

$$\xi = \frac{ae^{2\lambda x} + b}{ce^{2\lambda x} + d}, \qquad w = \frac{ye^{\lambda x}}{ce^{2\lambda x} + d}$$

leads to the Emden—Fowler equation (see Section 2.3)

$$w''_{\xi\xi} = A(2\Delta\lambda)^{-2}\xi^n w^m, \qquad \text{where} \quad \Delta = ad - bc.$$

5. $y''_{xx} = ay'_x + be^{2ax}y^n.$

This is a special case of equation 2.9.2.17 with $f(y) = by^n$.

6. $y''_{xx} = -ay'_x + be^{anx}y^{n-1}.$

This is a special case of equation 2.9.2.18 with $f(\xi) = b\xi^{n-1}$.

7. $y''_{xx} + ay'_x + by = ce^{\lambda x}y^m.$

The substitution $\xi = e^x$ leads to an equation of the form 2.8.1.36:

$$\xi^2 y''_{\xi\xi} + (a+1)\xi y'_\xi + by = c\xi^\lambda y^m.$$

8. $y''_{xx} = -(\mu + \nu)y'_x - \nu\mu y + ae^{(n\mu - 2\nu)x}y^{n-1}.$

This is a special case of equation 2.9.2.19 with $f(\xi) = a\xi^{n-1}$.

9. $y''_{xx} = \lambda y'_x + bxy + ae^{2\lambda x}y^{-3}.$

This is a special case of equation 2.9.2.20 with $f(x) = -bx$.

10. $y''_{xx} = \lambda y'_x + be^{\mu x}y + ae^{2\lambda x}y^{-3}.$

This is a special case of equation 2.9.2.20 with $f(x) = -be^{\mu x}$.

11. $y''_{xx} = ay'_x + b\exp(2ax + cy^n)$.

This is a special case of equation 2.9.2.17 with $f(y) = b\exp(cy^n)$.

12. $y''_{xx} + 3yy'_x + y^3 + ae^{\lambda x}y = 0$.

This is a special case of equation 2.9.2.1 with $f(x) = ae^{\lambda x}$.

13. $y''_{xx} = axe^y y'_x + ae^y$.

Solution: $y = C_1 x - \ln(-a\int xe^{C_1 x}\,dx + C_2)$.

14. $y''_{xx} = 2ae^x yy'_x + ae^x y^2$.

Solution in the parametric form:

$$x = \ln\left(\frac{\tau^2}{2C_1}\right), \qquad y = -a^{-1}C_1\tau^{-2}Z^{-1}(\tau Z'_\tau + Z),$$

where $Z = C_1 J_1(\tau) + C_2 Y_1(\tau)$ or $Z = C_1 I_1(\tau) + C_2 K_1(\tau)$, J_1 and Y_1 are Bessel functions, I_1 and K_1 are modified Bessel functions.

15. $y''_{xx} = ax^n e^y y'_x + anx^{n-1}e^y$.

Solution: $y = C_1 x - \ln\left[C_2 - a\int x^n \exp(C_1 x)\,dx\right]$.

16. $y''_{xx} = ae^x y^{-1/2} y'_x + 2ae^x y^{1/2}$.

Solution in the parametric form:

$$x = \ln\left(\pm\frac{C_1}{a}f\right) \mp \tau^2, \qquad y = C_1^2\left[2\tau \pm \exp(\mp\tau^2)f\right]^2,$$

where $f = \left[\int \exp(\mp\tau^2)\,d\tau + C_2\right]^{-1}$.

17. $y''_{xx} + (2ay + be^{\lambda x})y'_x + \lambda be^{\lambda x}y = 0$.

Integrating the equation, we obtain the Riccati equation:

$$y'_x + ay^2 + be^{\lambda x}y = C.$$

18. $y''_{xx} = ae^{x+y}(y'_x + 1)$.

Solution:

$$y = -\ln\left(C_1 e^{-C_2 x} - \frac{a}{1 + C_2}e^x\right).$$

To the limiting case $C_2 \to -1$ corresponds $y = -x - \ln(C_1 - ax)$.

19. $y''_{xx} = a(y'_x)^2 - be^{4ay} + cx^n$.

This is a special case of equation 2.9.3.16 with $f(x) = -cx^n$.

20. $y''_{xx} = a(y'_x)^2 - be^{4ay} + ce^{\lambda x}$.

This is a special case of equation 2.9.3.16 with $f(x) = -ce^{\lambda x}$.

21. $y''_{xx} = a(y'_x)^2 + bx^n e^{ay} + cx^m$.

This is a special case of equation 2.9.3.15 with $f(x) = -bx^n$, $g(x) = -cx^m$.

22. $y''_{xx} = a(y'_x)^2 + be^{ay+cx} + kx^m$.

This is a special case of equation 2.9.3.15 with $f(x) = -be^{cx}$, $g(x) = -kx^m$.

23. $y''_{xx} = a(y'_x)^2 + be^{ay+\lambda x} + ce^{\mu x}$.

This is a special case of equation 2.9.3.15 with $f(x) = -be^{\lambda x}$, $g(x) = -ce^{\mu x}$.

24. $y''_{xx} + ay^n(y'_x)^2 + be^{\lambda y} + c = 0$.

This is a special case of equation 2.9.3.1 with $f(y) = ay^n$, $g(y) = be^{\lambda y} + c$.

25. $y''_{xx} = ay^n(y'_x)^2 + be^{\lambda x}y'_x$.

This is a special case of equation 2.9.3.2 with $f(y) = ay^n$, $g(x) = be^{\lambda x}$.

26. $y''_{xx} + ae^{\lambda y}(y'_x)^2 + by^n + c = 0$.

This is a special case of equation 2.9.3.1 with $f(y) = ae^{\lambda y}$, $g(y) = by^n + c$.

27. $y''_{xx} + ae^{\lambda y}(y'_x)^2 + be^{\mu y} + c = 0$.

This is a special case of equation 2.9.3.1 with $f(y) = ae^{\lambda y}$, $g(y) = be^{\mu y} + c$.

28. $y''_{xx} = ae^{\lambda y}(y'_x)^2 + bx^n y'_x$.

This is a special case of equation 2.9.3.2 with $f(y) = ae^{\lambda y}$, $g(x) = bx^n$.

29. $y''_{xx} = ae^{\lambda y}(y'_x)^2 + be^{\mu x}y'_x$.

This is a special case of equation 2.9.3.2 with $f(y) = ae^{\lambda y}$, $g(x) = be^{\mu x}$.

30. $y''_{xx} = ae^{\lambda x}(xy'_x - y)^2 + be^{\mu x}$.

This is a special case of equation 2.9.3.4 with $f(x) = be^{\mu x}$, $g(x) = 0$, $h(x) = ae^{\lambda x}$.

31. $y''_{xx} + be^{ax}y^m(y'_x)^3 + ay'_x = 0$.

This is a special case of equation 2.9.3.24 with $f(y) = by^m$.

32. $y''_{xx} + be^{ax+\lambda y}(y'_x)^3 + ay'_x = 0$.

This is a special case of equation 2.9.3.24 with $f(y) = be^{\lambda y}$.

33. $y''_{xx} = ae^x(y'_x)^3 + ae^x y(y'_x)^2$.

Solution: $\quad x = C_1 y - \ln\left(a \int ye^{C_1 y}\, dy + C_2\right)$.

34. $y''_{xx} = axe^y (y'_x)^3 + ae^y (y'_x)^2.$

Solution in the parametric form:

$$x = C_1 e^{-a\tau} \left(\int \tau^{-1} e^{a\tau} \, d\tau + C_2 \right), \qquad y = \ln \tau.$$

35. $y''_{xx} = ax^2 e^y (y'_x)^3 + 2axe^y (y'_x)^2.$

Solution in the parametric form:

$$x = a^{-1} C_1 \tau^{-2} Z^{-1} (\tau Z'_\tau + Z), \qquad y = \ln\left(\frac{\tau^2}{2C_1}\right),$$

where $Z = C_1 J_1(\tau) + C_2 Y_1(\tau)$ or $Z = C_1 I_1(\tau) + C_2 K_1(\tau)$, J_1 and Y_1 are Bessel functions, I_1 and K_1 are modified Bessel functions.

36. $y''_{xx} = 2ax^{1/2} e^y (y'_x)^3 + ax^{-1/2} e^y (y'_x)^2.$

Solution in the parametric form:

$$x = C_1^2 \left[2\tau \pm \exp(\mp\tau^2) f\right]^2, \qquad y = \ln\left(\mp \frac{C_1}{a} f\right) \mp \tau^2,$$

where $f = \left[\int \exp(\mp\tau^2) \, d\tau + C_2 \right]^{-1}$.

37. $y''_{xx} = ane^x y^{n-1} (y'_x)^3 + ae^x y^n (y'_x)^2.$

Solution: $x = C_1 y - \ln\left(a \int y^n e^{C_1 y} \, dy + C_2 \right).$

38. $y''_{xx} = ae^{x+y} [(y'_x)^3 + (y'_x)^2].$

Solution:

$$x = -\ln\left(C_1 e^{C_2 y} + \frac{a}{1 - C_2} e^y \right).$$

To the limiting case $C_2 \to 1$ corresponds $x = -y - \ln(C_1 + ay)$.

39. $y''_{xx} = ae^x (y'_x)^{3/2} + ae^x y (y'_x)^{1/2}.$

Solution in the parametric form:

$$x = \ln \tau^2, \qquad y = -2a^{-2} \tau^{-4} \left[Z^{-1} (\tau Z'_\tau + 2Z) \mp \tfrac{1}{2} \tau^2 \right],$$

where

$$Z = \begin{cases} C_1 J_2(\tau) + C_2 Y_2(\tau) & \text{for the upper sign,} \\ C_1 I_2(\tau) + C_2 K_2(\tau) & \text{for the lower sign,} \end{cases}$$

J_2 and Y_2 are Bessel functions, I_2 and K_2 are modified Bessel functions.

40. $y''_{xx} = axe^y (y'_x)^{5/2} + ae^y (y'_x)^{3/2}$.

Solution in the parametric form:

$$x = -2a^{-2}\tau^{-4}\bigl[Z^{-1}(\tau Z'_\tau + 2Z) \mp \tfrac{1}{2}\tau^2\bigr], \qquad y = \ln \tau^2,$$

where

$$Z = \begin{cases} C_1 J_2(\tau) + C_2 Y_2(\tau) & \text{for the upper sign,} \\ C_1 I_2(\tau) + C_2 K_2(\tau) & \text{for the lower sign,} \end{cases}$$

J_2 and Y_2 are Bessel functions, I_2 and K_2 are modified Bessel functions.

41. $y''_{xx} = -ay'_x + be^{amx} y^k (y'_x)^{m+2}$.

This is a special case of equation 2.9.4.49 with $f(y) = -by^k$, $n = m + 2$.

42. $y''_{xx} = -\dfrac{a}{m}\dfrac{2-k}{1-k} y'_x + be^{ax} y^{m-k+1} (y'_x)^k$.

This is a special case of equation 2.9.4.52 with $f(\xi) = b\xi$.

43. $y''_{xx} = y'_x + A \exp[(n + 2 - l)x] y^m (y'_x)^l$.

The substitution $\xi = e^x$ leads to the generalized Emden—Fowler equation

$$y''_{\xi\xi} = A\xi^n y^m (y'_\xi)^l$$

which is outlined in Section 2.5.

44. $y''_{xx} = -(y'_x)^2 + Ax^n \exp[(m + l - 1)y]\,(y'_x)^l$.

The substitution $w = e^y$ leads to the generalized Emden—Fowler equation

$$w''_{xx} = Ax^n w^m (w'_x)^l$$

which is outlined in Section 2.5.

45. $y''_{xx} = \dfrac{a}{n}\dfrac{1-k}{2-k}(y'_x)^2 + bx^{n+k-2} e^{ay} (y'_x)^k$.

This is a special case of equation 2.9.4.51 with $f(\xi) = b\xi$.

46. $y''_{xx} = ay^n (y'_x)^2 + be^{\lambda y} (y'_x)^k$.

This is a special case of equation 2.9.4.10 with $f(y) = ay^n$, $g(y) = be^{\lambda y}$.

47. $y''_{xx} = ae^{\lambda y} (y'_x)^2 + by^n (y'_x)^k$.

This is a special case of equation 2.9.4.10 with $f(y) = ae^{\lambda y}$, $g(y) = by^n$.

48. $y''_{xx} = ae^{\lambda y} (y'_x)^2 + be^{\mu y} (y'_x)^k$.

This is a special case of equation 2.9.4.10 with $f(y) = ae^{\lambda y}$, $g(y) = be^{\mu y}$.

49. $y''_{xx} = ax^n(xy'_x - y) + be^{\lambda x}(xy'_x - y)^k.$

This is a special case of equation 2.9.4.39 with $f(x) = ax^n$, $g(x) = be^{\lambda x}$.

50. $y''_{xx} = ae^{\lambda x}(xy'_x - y) + bx^n(xy'_x - y)^k.$

This is a special case of equation 2.9.4.39 with $f(x) = ae^{\lambda x}$, $g(x) = bx^n$.

51. $y''_{xx} = ae^{x+y}[(y'_x)^l + (y'_x)^{l-1}], \quad l \neq 2.$

Solution in the parametric form:

$$x = \tau - \int \frac{d\tau}{F} - C_2, \qquad y = \int \frac{d\tau}{F} + C_2,$$

where $F = \left[a(2-l)e^{\tau} + C_1\right]^{\frac{1}{l-2}} + 1$.

52. $y''_{xx} = ae^{\lambda x}(xy'_x - y) + be^{\mu x}(xy'_x - y)^k.$

This is a special case of equation 2.9.4.39 with $f(x) = ae^{\lambda x}$, $g(x) = be^{\mu y}$.

53. $xy''_{xx} + y'_x = ax^n e^{\lambda y}.$

This equation is encountered in the combustion theory and hydrodynamics.

The transformation $\xi = \ln x$, $w = \lambda y + (n+1)\ln x$ leads to an autonomous equation of the form 2.9.1.1: $w''_{\xi\xi} = a\lambda e^w$. Having integrated the latter equation, we obtain the solution of the original equation in the parametric form:

$$x = \exp[C_1 \pm f(t)], \qquad y = \frac{t}{\lambda} - \frac{n+1}{\lambda}[C_1 \pm f(t)],$$

where

$$f(t) = \begin{cases} \dfrac{1}{\sqrt{C_2}} \ln \dfrac{\sqrt{C_2 + 2a\lambda e^t} - \sqrt{C_2}}{\sqrt{C_2 + 2a\lambda e^t} + \sqrt{C_2}} & \text{if } C_2 > 0, \\ -\dfrac{2}{\sqrt{2a\lambda e^t}} & \text{if } C_2 = 0, \\ \dfrac{2}{\sqrt{-C_2}} \arctan \dfrac{\sqrt{C_2 + 2a\lambda e^t}}{\sqrt{-C_2}} & \text{if } C_2 < 0. \end{cases}$$

54. $xy''_{xx} = ny'_x + ax^{2n+1}e^{\lambda y}.$

This is a special case of equation 2.9.2.4 with $f(y) = ae^{\lambda y}$.

55. $xy''_{xx} = ny'_x + ax^{2n+1}\exp(\lambda y^m).$

This is a special case of equation 2.9.2.4 with $f(y) = a\exp(\lambda y^m)$.

56. $xy''_{xx} + y'_x = ax^n e^{\lambda y}(y'_x)^m.$

The transformation $\zeta = xy'_x$, $w = x^{n-m+1}e^{\lambda y}$ after dividing by w leads to a first order linear equation:

$$a\zeta^m w'_\zeta = \lambda\zeta + n - m + 1.$$

57. $xy''_{xx} + my'_x + ax^{nm-2m+1}e^{\lambda y}(y'_x)^n = 0.$

This is a special case of equation 2.9.4.11 with $f(y) = ae^{\lambda y}$.

58. $xy''_{xx} + y'_x = (ax^n e^{\lambda y} + bx^{m-1})(y'_x)^m.$

The transformation $\zeta = xy'_x$, $w = x^{n-m+1}e^{\lambda y}$ leads to a first order equation with separation of variables:

$$\zeta^m(aw + b)w'_\zeta = (\lambda\zeta + n - m + 1)w.$$

59. $2yy''_{xx} = (y'_x)^2 + be^{\lambda x}y^2 - a.$

This is a special case of equation 2.9.3.7 with $f(x) = -be^{\lambda x}$.

60. $yy''_{xx} = n(y'_x)^2 - ay^{4n-2} + be^{\lambda x}y^2.$

This is a special case of equation 2.9.3.8 with $f(x) = -be^{\lambda x}$.

61. $yy''_{xx} = n(y'_x)^2 + ax^m y^2 + be^{\lambda x}y^{n+1}.$

This is a special case of equation 2.9.3.9 with $f(x) = -ax^m$, $g(x) = -be^{\lambda x}$.

62. $yy''_{xx} = n(y'_x)^2 + ae^{\lambda x}y^2 + bx^m y^{n+1}.$

This is a special case of equation 2.9.3.9 with $f(x) = -ae^{\lambda x}$, $g(x) = -bx^m$.

63. $yy''_{xx} = n(y'_x)^2 + ae^{\lambda x}y^2 + be^{\mu x}y^{n+1}.$

This is a special case of equation 2.9.3.9 with $f(x) = -ae^{\lambda x}$, $g(x) = -be^{\mu x}$.

64. $yy''_{xx} = (y'_x)^2 + ax^n yy'_x + be^{\lambda x}y^2.$

This is a special case of equation 2.9.3.6 with $f(x) = -ax^n$, $g(x) = -be^{\lambda x}$.

65. $yy''_{xx} = (y'_x)^2 + ae^{\lambda x}yy'_x + bx^n y^2.$

This is a special case of equation 2.9.3.6 with $f(x) = -ae^{\lambda x}$, $g(x) = -bx^n$.

66. $yy''_{xx} = (y'_x)^2 + ae^{\lambda x}yy'_x + be^{\mu x}y^2.$

This is a special case of equation 2.9.3.6 with $f(x) = -ae^{\lambda x}$, $g(x) = -be^{\mu x}$.

67. $yy''_{xx} = (y'_x)^2 + be^{ax}y^n(y'_x)^k.$

This is a special case of equation 2.9.4.53 with $f(\xi) = b\xi$, $g(\zeta) = \zeta^k$, $n = m - k + 2$.

68. $yy''_{xx} = (y'_x)^2 + (ae^{\lambda x}y^n + by^{2-m})(y'_x)^m.$

The transformation $\xi = y'_x/y$, $w = e^{\lambda x}y^{n+m-2}$ leads to a first order equation with separation of variables:

$$\xi^m(aw + b)w'_\xi = [(n + m - 2)\xi + \lambda]w.$$

69. $x^2 y''_{xx} = a x^{n+2} e^y + n.$

This is a special case of equation 2.9.1.17 with $f(\xi) = a\xi$.

70. $x^2 y''_{xx} + x y'_x = a e^{\lambda y} + b.$

This is a special case of equation 2.9.2.8 with $f(y) = a e^{\lambda y} + b$.

71. $x^2 y''_{xx} + x y'_x = k x^n e^{ay} + b.$

This is a special case of equation 2.9.2.23 with $f(\xi) = k\xi + b$.

72. $(ax^2 + b) y''_{xx} + a x y'_x + c e^{\lambda y} = 0.$

This is a special case of equation 2.9.2.11 with $f(y) = c e^{\lambda y}$.

73. $(a e^{2x} + b) y''_{xx} + a e^{2x} y'_x + c y^n = 0.$

This is a special case of equation 2.9.2.14 with $g(x) = a e^{2x} + b$, $f(y) = -c y^n$.

74. $(a e^{2x} + b) y''_{xx} + a e^{2x} y'_x + c e^{\lambda y} = 0.$

This is a special case of equation 2.9.2.14 with $g(x) = a e^{2x} + b$, $f(y) = -c e^{\lambda y}$.

2.8.4. Equations Containing Hyperbolic Functions

1. $y''_{xx} = \lambda^2 y + a(\cosh \lambda x)^{-n-3} y^n.$

This is a special case of equation 2.9.1.21 with $f(\xi) = a\xi^n$.

2. $y''_{xx} = \lambda^2 y + a(\sinh \lambda x)^{-n-3} y^n.$

This is a special case of equation 2.9.1.20 with $f(\xi) = a\xi^n$.

3. $y''_{xx} = \lambda^2 y + a \sinh^n(\lambda x) \cosh^{-n-m-3}(\lambda x) y^m.$

The transformation $\xi = \tanh(\lambda x)$, $w = \dfrac{y}{\cosh(\lambda x)}$ leads to the Emden—Fowler equation

$$w''_{\xi\xi} = a\lambda^{-2} \xi^n w^m$$

which is outlined in Section 2.3.

4. $y''_{xx} = \lambda^2 y + a \cosh^n(\lambda x) \sinh^{-n-m-3}(\lambda x) y^m.$

The transformation $\xi = \coth(\lambda x)$, $w = \dfrac{y}{\sinh(\lambda x)}$ leads to the Emden—Fowler equation

$$w''_{\xi\xi} = a\lambda^{-2} \xi^n w^m$$

which is outlined in Section 2.3.

5. $y''_{xx} = b \cosh(\lambda x) y + a y^{-3}.$

This is a special case of equation 2.9.1.12 with $f(x) = -b\cosh(\lambda x)$.

6. $y''_{xx} = b\sinh(\lambda x)y + ay^{-3}$.

This is a special case of equation 2.9.1.12 with $f(x) = -b\sinh(\lambda x)$.

7. $y''_{xx} + 3yy'_x + y^3 + a\cosh(\lambda x)y = 0$.

This is a special case of equation 2.9.2.1 with $f(x) = a\cosh(\lambda x)$.

8. $y''_{xx} + 3yy'_x + y^3 + (a\sinh x + b)y = 0$.

This is a special case of equation 2.9.2.1 with $f(x) = a\sinh x + b$.

9. $y''_{xx} + ay^n(y'_x)^2 + b\cosh^m y + c = 0$.

This is a special case of equation 2.9.3.1 with $f(y) = ay^n$, $g(y) = b\cosh^m y + c$.

10. $y''_{xx} + ay^n(y'_x)^2 + b\tanh^m y + c = 0$.

This is a special case of equation 2.9.3.1 with $f(y) = ay^n$, $g(y) = b\tanh^m y + c$.

11. $y''_{xx} = ay^n(y'_x)^2 + b\sinh^m(\lambda x)y'_x$.

This is a special case of equation 2.9.3.2 with $f(y) = ay^n$, $g(x) = b\sinh^m(\lambda x)$.

12. $y''_{xx} = ay^n(y'_x)^2 + b\tanh^m(\lambda x)y'_x$.

This is a special case of equation 2.9.3.2 with $f(y) = ay^n$, $g(x) = b\tanh^m(\lambda x)$.

13. $y''_{xx} + a\cosh^n y\,(y'_x)^2 + by^m + c = 0$.

This is a special case of equation 2.9.3.1 with $f(y) = a\cosh^n y$, $g(y) = by^m + c$.

14. $y''_{xx} + a\cosh^n y\,(y'_x)^2 + b\cosh^m(\lambda y) + c = 0$.

This is a special case of equation 2.9.3.1 with $f(y) = a\cosh^n y$, $g(y) = b\cosh^m(\lambda y) + c$.

15. $y''_{xx} = a\sinh^n y\,(y'_x)^2 + bx^m y'_x$.

This is a special case of equation 2.9.3.2 with $f(y) = a\sinh^n y$, $g(x) = bx^m$.

16. $y''_{xx} = a\sinh^n y\,(y'_x)^2 + b\sinh^m(\lambda x)y'_x$.

This is a special case of equation 2.9.3.2 with $f(y) = a\sinh^n y$, $g(x) = b\sinh^m(\lambda x)$.

17. $y''_{xx} + a\tanh^n y\,(y'_x)^2 + by^m + c = 0$.

This is a special case of equation 2.9.3.1 with $f(y) = a\tanh^n y$, $g(y) = by^m + c$.

18. $y''_{xx} + a\tanh^n y\,(y'_x)^2 + b\tanh^m(\lambda y) + c = 0$.

This is a special case of equation 2.9.3.1 with $f(y) = a\tanh^n y$, $g(y) = b\tanh^m(\lambda y) + c$.

19. $y''_{xx} = a\tanh^n y\,(y'_x)^2 + bx^m y'_x$.

This is a special case of equation 2.9.3.2 with $f(y) = a\tanh^n y$, $g(x) = bx^m$.

20. $y''_{xx} = a\tanh^n y\,(y'_x)^2 + b\tanh^m(\lambda x) y'_x$.

This is a special case of equation 2.9.3.2 with $f(y) = a\tanh^n y$, $g(x) = b\tanh^m(\lambda x)$.

21. $y''_{xx} = ay^n(y'_x)^2 + b\cosh^m y\,(y'_x)^k$.

This is a special case of equation 2.9.4.10 with $f(y) = ay^n$, $g(y) = b\cosh^m y$.

22. $y''_{xx} = ay^n(y'_x)^2 + b\tanh^m y\,(y'_x)^k$.

This is a special case of equation 2.9.4.10 with $f(y) = ay^n$, $g(y) = b\tanh^m y$.

23. $y''_{xx} = a\cosh^n y\,(y'_x)^2 + by^m(y'_x)^k$.

This is a special case of equation 2.9.4.10 with $f(y) = a\cosh^n y$, $g(y) = by^m$.

24. $y''_{xx} = a\tanh^n y\,(y'_x)^2 + by^m(y'_x)^k$.

This is a special case of equation 2.9.4.10 with $f(y) = a\tanh^n y$, $g(y) = by^m$.

25. $xy''_{xx} = ny'_x + ax^{2n+1}\cosh^m(\lambda y)$.

This is a special case of equation 2.9.2.4 with $f(y) = a\cosh^m(\lambda y)$.

26. $xy''_{xx} = ny'_x + ax^{2n+1}\sinh^m(\lambda y)$.

This is a special case of equation 2.9.2.4 with $f(y) = a\sinh^m(\lambda y)$.

27. $xy''_{xx} = ny'_x + ax^{2n+1}\tanh^m(\lambda y)$.

This is a special case of equation 2.9.2.4 with $f(y) = a\tanh^m(\lambda y)$.

28. $xy''_{xx} = ny'_x + ax^{2n+1}\coth^m(\lambda y)$.

This is a special case of equation 2.9.2.4 with $f(y) = a\coth^m(\lambda y)$.

29. $2yy''_{xx} = (y'_x)^2 + b\cosh^m(\lambda x)y^2 - a$.

This is a special case of equation 2.9.3.7 with $f(x) = -b\cosh^m(\lambda x)$.

30. $yy''_{xx} = n(y'_x)^2 - ay^{4n-2} + b\cosh^m(\lambda x)y^2$.

This is a special case of equation 2.9.3.8 with $f(x) = -b\cosh^m(\lambda x)$.

31. $yy''_{xx} = n(y'_x)^2 + ax^m y^2 + b\cosh^k(\lambda x)y^{n+1}$.

This is a special case of equation 2.9.3.9 with $f(x) = -ax^m$, $g(x) = -b\cosh^k(\lambda x)$.

32. $yy''_{xx} = n(y'_x)^2 + ax^m y^2 + b\sinh^k(\lambda x)y^{n+1}$.

This is a special case of equation 2.9.3.9 with $f(x) = -ax^m$, $g(x) = -b\sinh^k(\lambda x)$.

33. $yy''_{xx} = n(y'_x)^2 + ax^m y^2 + b\tanh^k(\lambda x)y^{n+1}$.

This is a special case of equation 2.9.3.9 with $f(x) = -ax^m$, $g(x) = -b\tanh^k(\lambda x)$.

34. $yy''_{xx} = (y'_x)^2 + ax^n yy'_x + b\cosh^m(\lambda x)y^2.$

This is a special case of equation 2.9.3.6 with $f(x) = -ax^n$, $g(x) = -b\cosh^m(\lambda x)$.

35. $yy''_{xx} = (y'_x)^2 + a\cosh^n(\lambda x)yy'_x + bx^m y^2.$

This is a special case of equation 2.9.3.6 with $f(x) = -a\cosh^n(\lambda x)$, $g(x) = -bx^m$.

36. $x^2 y''_{xx} + xy'_x = a\cosh^n(\lambda y) + b.$

This is a special case of equation 2.9.2.8 with $f(y) = a\cosh^n(\lambda y) + b$.

37. $x^2 y''_{xx} + xy'_x = a\sinh^n(\lambda y) + b.$

This is a special case of equation 2.9.2.8 with $f(y) = a\sinh^n(\lambda y) + b$.

38. $x^2 y''_{xx} + xy'_x = a\tanh^n(\lambda y) + b.$

This is a special case of equation 2.9.2.8 with $f(y) = a\tanh^n(\lambda y) + b$.

39. $x^2 y''_{xx} + xy'_x = a\coth^n(\lambda y) + b.$

This is a special case of equation 2.9.2.8 with $f(y) = a\coth^n(\lambda y) + b$.

40. $(ax^2 + b)y''_{xx} + axy'_x + \cosh^n(\lambda y) + c = 0.$

This is a special case of equation 2.9.2.11 with $f(y) = \cosh^n(\lambda y) + c$.

41. $(ax^2 + b)y''_{xx} + axy'_x + \sinh^n(\lambda y) + c = 0.$

This is a special case of equation 2.9.2.11 with $f(y) = \sinh^n(\lambda y) + c$.

42. $(ax^2 + b)y''_{xx} + axy'_x + \tanh^n(\lambda y) + c = 0.$

This is a special case of equation 2.9.2.11 with $f(y) = \tanh^n(\lambda y) + c$.

43. $(ax^2 + b)y''_{xx} + axy'_x + \coth^n(\lambda y) + c = 0.$

This is a special case of equation 2.9.2.11 with $f(y) = \coth^n(\lambda y) + c$.

2.8.5. Equations Containing Logarithmic Functions

1. $y''_{xx} = ax^{-3}(\ln y - \ln x).$

This is a special case of equation 2.9.1.5 with $f(\xi) = a\ln\xi$.

2. $y''_{xx} = ax^{-3/2}(2\ln y - \ln x).$

This is a special case of equation 2.9.1.6 with $f(\xi) = 2a\ln\xi$.

3. $y''_{xx} + ay^n(y'_x)^2 + b\ln^m y + c = 0.$

This is a special case of equation 2.9.3.1 with $f(y) = ay^n$, $g(y) = b\ln^m y + c$.

4. $y''_{xx} = ay^n (y'_x)^2 + b\ln^m(\lambda x) y'_x$.

This is a special case of equation 2.9.3.2 with $f(y) = ay^n$, $g(x) = b\ln^m(\lambda x)$.

5. $y''_{xx} + a\ln^n y\,(y'_x)^2 + by^m + c = 0$.

This is a special case of equation 2.9.3.1 with $f(y) = a\ln^n y$, $g(y) = by^m + c$.

6. $y''_{xx} = a\ln^n y\,(y'_x)^2 + bx^m y'_x$.

This is a special case of equation 2.9.3.2 with $f(y) = a\ln^n y$, $g(x) = bx^m$.

7. $y''_{xx} = a\ln^n y\,(y'_x)^2 + b\ln^m(\lambda x) y'_x$.

This is a special case of equation 2.9.3.2 with $f(y) = a\ln^n y$, $g(x) = b\ln^m(\lambda x)$.

8. $y''_{xx} = ax^{-2}y^{-1}(y'_x)^4 - 2ax^{-3}\ln y\,(y'_x)^3$.

Solution in the parametric form:

$$x = \lambda\bigl[(F \pm 2\tau)^2 \pm 4\ln(C_1 F)\bigr]^{1/2}, \qquad y = C_1 F,$$

where $F = \exp(\mp\tau^2)\left[\int \exp(\mp\tau^2)\,d\tau + C_2\right]^{-1}$, $\lambda = \bigl(\pm\frac{1}{2}aC_1^2\bigr)^{1/4}$.

9. $y''_{xx} = 2a\ln x\,y^{-3} - ax^{-1}y^{-2}(y'_x)^{-1}$.

Solution in the parametric form:

$$x = C_1 F, \qquad y = \lambda\bigl[(F \pm 2\tau)^2 \pm 4\ln(C_1 F)\bigr]^{1/2},$$

where $F = \exp(\mp\tau^2)\left[\int \exp(\mp\tau^2)\,d\tau + C_2\right]^{-1}$, $\lambda = \bigl(\pm\frac{1}{2}aC_1^2\bigr)^{1/4}$.

10. $y''_{xx} = ay^n (y'_x)^2 + b\ln^m y\,(y'_x)^k$.

This is a special case of equation 2.9.4.10 with $f(y) = ay^n$, $g(y) = b\ln^m y$.

11. $y''_{xx} = a\ln^n y\,(y'_x)^2 + by^m (y'_x)^k$.

This is a special case of equation 2.9.4.10 with $f(y) = a\ln^n y$, $g(y) = by^m$.

12. $xy''_{xx} = ny'_x + ax^{2n+1}\ln^m(\lambda y)$.

This is a special case of equation 2.9.2.4 with $f(y) = a\ln^m(\lambda y)$.

13. $xy''_{xx} = -(n+1)y'_x + ax^{n-1}(\ln y + n\ln x)$.

This is a special case of equation 2.9.2.5 with $f(\xi) = a\ln\xi$.

14. $xy''_{xx} = (ay + n\ln x)y'_x$.

This is a special case of equation 2.9.2.22 with $f(\xi) = \ln\xi$.

15. $xy''_{xx} + x(2ay + \ln x + b)y'_x + y = 0.$

Integrating the equation, we obtain Riccati equation

$$y'_x + ay^2 + (\ln x + b)y = C.$$

16. $yy''_{xx} = n(y'_x)^2 + ax^m y^2 + b\ln^k(\lambda x) y^{n+1}.$

This is a special case of equation 2.9.3.9 with $f(x) = -ax^m$, $g(x) = -b\ln^k(\lambda x)$.

17. $yy''_{xx} = (y'_x)^2 + a\ln^n(\lambda x) yy'_x + bx^m y^2.$

This is a special case of equation 2.9.3.6 with $f(x) = -a\ln^n(\lambda x)$, $g(x) = -bx^m$.

18. $yy''_{xx} = (y'_x)^2 + ax^n yy'_x + b\ln^m(\lambda x) y^2.$

This is a special case of equation 2.9.3.6 with $f(x) = -ax^n$, $g(x) = -b\ln^m(\lambda x)$.

19. $yy''_{xx} = (ax + n\ln y)(y'_x)^2.$

This is a special case of equation 2.9.3.17 with $f(\xi) = \ln\xi$.

20. $x^2 y''_{xx} = x^2(y + a\ln x + b)^n + a.$

This is a special case of equation 2.9.1.22 with $f(\xi) = \xi^n$.

21. $x^2 y''_{xx} = n(n+1)y + ax^{3n+2}(\ln y + n\ln x).$

This is a special case of equation 2.9.1.11 with $f(\xi) = a\ln\xi$.

22. $x^2 y''_{xx} + \frac{1}{4}y + Ax^{\frac{1-m}{2}}(a\ln x + b)^n y^m = 0.$

The transformation $\xi = a\ln x + b$, $w = \dfrac{y}{\sqrt{x}}$ leads to the Emden—Fowler equation (see Section 2.3)

$$w''_{\xi\xi} + Aa^{-2}\xi^n w^m = 0.$$

23. $x^2 y''_{xx} + xy'_x = a\ln^n(\lambda y) + b.$

This is a special case of equation 2.9.2.8 with $f(y) = a\ln^n(\lambda y) + b$.

24. $(ax^2 + b)y''_{xx} + axy'_x + c\ln^n(\lambda y) = 0.$

This is a special case of equation 2.9.2.11 with $f(y) = c\ln^n(\lambda y)$.

2.8.6. Equations Containing Trigonometric Functions

1. $y''_{xx} = -\lambda^2 y + a(\cos\lambda x)^n y^{-n-3}.$

This is a special case of equation 2.9.1.27 with $f(\xi) = a\xi^{-n-3}$.

2. $y''_{xx} = -\lambda^2 y + a(\sin\lambda x)^n y^{-n-3}.$

This is a special case of equation 2.9.1.26 with $f(\xi) = a\xi^{-n-3}$.

3. $y''_{xx} = -\lambda^2 y + a\cos^n(\lambda x)\sin^m(\lambda x)y^{-n-m-3}$.

The transformation $\xi = \cot(\lambda x)$, $w = \dfrac{y}{\sin(\lambda x)}$ leads to the Emden—Fowler equation (see Section 2.3)

$$w''_{\xi\xi} = a\lambda^{-2}\xi^n w^{-n-m-3}.$$

4. $y''_{xx} = -\lambda^2 y + a\sin^n(\lambda x)[\sin(\lambda x) + b\cos(\lambda x)]^m y^{-n-m-3}$.

The transformation $\xi = 1 + b\cot(\lambda x)$, $w = \dfrac{y}{\sin(\lambda x)}$ leads to the Emden—Fowler equation (see Section 2.3)

$$w''_{\xi\xi} = a(b\lambda)^{-2}\xi^m w^{-n-m-3}.$$

5. $y''_{xx} = -\lambda^2 y + A\sin^n(\lambda x + a)\sin^m(\lambda x + b)y^{-n-m-3}$.

The transformation

$$\xi = \frac{\sin(\lambda x + a)}{\sin(\lambda x + b)}, \qquad w = \frac{y}{\sin(\lambda x + b)}$$

leads to the Emden—Fowler equation (see Section 2.3)

$$w''_{\xi\xi} = A[\lambda\sin(b-a)]^{-2}\xi^n w^{-n-m-3}.$$

6. $y''_{xx} = b\cos(\lambda x)y + ay^{-3}$.

This is a special case of equation 2.9.1.12 with $f(x) = -b\cos(\lambda x)$.

7. $y''_{xx} = b\sin(\lambda x)y + ay^{-3}$.

This is a special case of equation 2.9.1.12 with $f(x) = -b\sin(\lambda x)$.

8. $y''_{xx} = 2(\cos x)^{-2}y + a(\cot x)^{n+3}y^n$.

This is a special case of equation 2.9.1.29 with $f(\xi) = a\xi^n$.

9. $y''_{xx} = 2(\sin x)^{-2}y + a(\tan x)^{n+3}y^n$.

This is a special case of equation 2.9.1.28 with $f(\xi) = a\xi^n$.

10. $y''_{xx} = (n+1)(\tan x)y'_x + ny + a(\cos x)^{nm-2}y^{m-1}$.

This is a special case of equation 2.9.2.30 with $f(\xi) = a\xi^{m-1}$.

11. $y''_{xx} + 3yy'_x + y^3 + a\sin(\lambda x)y = 0$.

This is a special case of equation 2.9.2.1 with $f(x) = a\sin(\lambda x)$.

12. $y''_{xx} + 3yy'_x + y^3 + (a\cos x + b)y = 0$.

This is a special case of equation 2.9.2.1 with $f(x) = a\cos x + b$.

13. $y''_{xx} + (2ay + b\sin x)y'_x + b(\cos x)y = 0$.

Integrating, we obtain the Riccati equation $y'_x + ay^2 + b(\sin x)y = C$.

14. $y''_{xx} + ay^n(y'_x)^2 + b\sin^m y + c = 0.$

This is a special case of equation 2.9.3.1 with $f(y) = ay^n$, $g(y) = b\sin^m y + c$.

15. $y''_{xx} + ay^n(y'_x)^2 + b\tan^m y + c = 0.$

This is a special case of equation 2.9.3.1 with $f(y) = ay^n$, $g(y) = b\tan^m y + c$.

16. $y''_{xx} = ay^n(y'_x)^2 + b\sin^m(\lambda x)y'_x.$

This is a special case of equation 2.9.3.2 with $f(y) = ay^n$, $g(x) = b\sin^m(\lambda x)$.

17. $y''_{xx} = ay^n(y'_x)^2 + b\tan^m(\lambda x)y'_x.$

This is a special case of equation 2.9.3.2 with $f(y) = ay^n$, $g(x) = b\tan^m(\lambda x)$.

18. $y''_{xx} + a\sin^n y\,(y'_x)^2 + by^m + c = 0.$

This is a special case of equation 2.9.3.1 with $f(y) = a\sin^n y$, $g(y) = by^m + c$.

19. $y''_{xx} + a\sin^n y\,(y'_x)^2 + b\sin^m(\lambda y) + c = 0.$

This is a special case of equation 2.9.3.1 with $f(y) = a\sin^n y$, $g(y) = b\sin^m(\lambda y) + c$.

20. $y''_{xx} = a\sin^n y\,(y'_x)^2 + bx^m y'_x.$

This is a special case of equation 2.9.3.2 with $f(y) = a\sin^n y$, $g(x) = bx^m$.

21. $y''_{xx} = a\sin^n y\,(y'_x)^2 + b\sin^m(\lambda x)y'_x.$

This is a special case of equation 2.9.3.2 with $f(y) = a\sin^n y$, $g(x) = b\sin^m(\lambda x)$.

22. $y''_{xx} + a\tan^n y\,(y'_x)^2 + by^m + c = 0.$

This is a special case of equation 2.9.3.1 with $f(y) = a\tan^n y$, $g(y) = by^m + c$.

23. $y''_{xx} + a\tan^n y\,(y'_x)^2 + b\tan^m(\lambda y) + c = 0.$

This is a special case of equation 2.9.3.1 with $f(y) = a\tan^n y$, $g(y) = b\tan^m(\lambda y) + c$.

24. $y''_{xx} = a\tan^n y\,(y'_x)^2 + bx^m y'_x.$

This is a special case of equation 2.9.3.2 with $f(y) = a\tan^n y$, $g(x) = bx^m$.

25. $y''_{xx} = a\tan^n y\,(y'_x)^2 + b\tan^m(\lambda x)y'_x.$

This is a special case of equation 2.9.3.2 with $f(y) = a\tan^n y$, $g(x) = b\tan^m(\lambda x)$.

26. $y''_{xx} = ay^n(y'_x)^2 + b\sin^m y\,(y'_x)^k.$

This is a special case of equation 2.9.4.10 with $f(y) = ay^n$, $g(y) = b\sin^m y$.

27. $y''_{xx} = ay^n(y'_x)^2 + b\tan^m y\,(y'_x)^k.$

This is a special case of equation 2.9.4.10 with $f(y) = ay^n$, $g(y) = b\tan^m y$.

28. $y''_{xx} = a \sin^n y \, (y'_x)^2 + by^m (y'_x)^k.$

This is a special case of equation 2.9.4.10 with $f(y) = a\sin^n y$, $g(y) = by^m$.

29. $y''_{xx} = a \tan^n y \, (y'_x)^2 + by^m (y'_x)^k.$

This is a special case of equation 2.9.4.10 with $f(y) = a\tan^n y$, $g(y) = by^m$.

30. $xy''_{xx} = ny'_x + ax^{2n+1} \cos^m(\lambda y).$

This is a special case of equation 2.9.2.4 with $f(y) = a\cos^m(\lambda y)$.

31. $xy''_{xx} = ny'_x + ax^{2n+1} \sin^m(\lambda y).$

This is a special case of equation 2.9.2.4 with $f(y) = a\sin^m(\lambda y)$.

32. $xy''_{xx} = ny'_x + ax^{2n+1} \tan^m(\lambda y).$

This is a special case of equation 2.9.2.4 with $f(y) = a\tan^m(\lambda y)$.

33. $xy''_{xx} = ny'_x + ax^{2n+1} \cot^m(\lambda y).$

This is a special case of equation 2.9.2.4 with $f(y) = a\cot^m(\lambda y)$.

34. $2yy''_{xx} = (y'_x)^2 + b\sin(\lambda x) y^2 - a.$

This is a special case of equation 2.9.3.7 with $f(x) = -b\sin(\lambda x)$.

35. $yy''_{xx} = n(y'_x)^2 - ay^{4n-2} + b\sin(\lambda x) y^2.$

This is a special case of equation 2.9.3.8 with $f(x) = -b\sin(\lambda x)$.

36. $yy''_{xx} = n(y'_x)^2 + ax^m y^2 + b\cos^k(\lambda x) y^{n+1}.$

This is a special case of equation 2.9.3.9 with $f(x) = -ax^m$, $g(x) = -b\cos^k(\lambda x)$.

37. $yy''_{xx} = n(y'_x)^2 + ax^m y^2 + b\sin^k(\lambda x) y^{n+1}.$

This is a special case of equation 2.9.3.9 with $f(x) = -ax^m$, $g(x) = -b\sin^k(\lambda x)$.

38. $yy''_{xx} = n(y'_x)^2 + ax^m y^2 + b\tan^k(\lambda x) y^{n+1}.$

This is a special case of equation 2.9.3.9 with $f(x) = -ax^m$, $g(x) = -b\tan^k(\lambda x)$.

39. $yy''_{xx} = n(y'_x)^2 + a\sin(\lambda x) y^2 + b\sin(\mu x) y^{n+1}.$

This is a special case of equation 2.9.3.9 with $f(x) = -a\sin(\lambda x)$, $g(x) = -b\sin(\mu x)$.

40. $yy''_{xx} = (y'_x)^2 + ax^n yy'_x + b\sin^m(\lambda x) y^2.$

This is a special case of equation 2.9.3.6 with $f(x) = -ax^n$, $g(x) = -b\sin^m(\lambda x)$.

41. $yy''_{xx} = (y'_x)^2 + ax^n yy'_x + b\tan^m(\lambda x) y^2.$

This is a special case of equation 2.9.3.6 with $f(x) = -ax^n$, $g(x) = -b\tan^m(\lambda x)$.

42. $yy''_{xx} = (y'_x)^2 + a\sin^n(\lambda x)yy'_x + bx^m y^2.$

This is a special case of equation 2.9.3.6 with $f(x) = -a\sin^n(\lambda x)$, $g(x) = -bx^m$.

43. $x^2 y''_{xx} + xy'_x = a\sin^n(\lambda y) + b.$

This is a special case of equation 2.9.2.8 with $f(y) = a\sin^n(\lambda y) + b$.

44. $x^2 y''_{xx} + xy'_x = a\tan^n(\lambda y) + b.$

This is a special case of equation 2.9.2.8 with $f(y) = a\tan^n(\lambda y) + b$.

45. $x^2 y''_{xx} + ax^2\tan x\, y'_x + n(ax\tan x - n - 1)y = bx^{nm+2}(\cos x)^{2a} y^{m-3}.$

This is a special case of equation 2.9.2.33 with $f(\xi) = b\xi^{m-3}$.

46. $(ax^2 + b)y''_{xx} + axy'_x + \cos^n(\lambda y) + c = 0.$

This is a special case of equation 2.9.2.11 with $f(y) = \cos^n(\lambda y) + c$.

47. $(ax^2 + b)y''_{xx} + axy'_x + \sin^n(\lambda y) + c = 0.$

This is a special case of equation 2.9.2.11 with $f(y) = \sin^n(\lambda y) + c$.

48. $(ax^2 + b)y''_{xx} + axy'_x + \tan^n(\lambda y) + c = 0.$

This is a special case of equation 2.9.2.11 with $f(y) = \tan^n(\lambda y) + c$.

49. $(ax^2 + b)y''_{xx} + axy'_x + \cot^n(\lambda y) + c = 0.$

This is a special case of equation 2.9.2.11 with $f(y) = \cot^n(\lambda y) + c$.

50. $\sin x\, y''_{xx} + \frac{1}{2}\cos x\, y'_x = ay^n + b.$

This is a special case of equation 2.9.2.14 with $g(x) = \sin x$, $f(y) = ay^n + b$.

51. $\sin x\, y''_{xx} + \frac{1}{2}\cos x\, y'_x = a\sin^n(\lambda y) + b.$

This is a special case of equation 2.9.2.14 with $g(x) = \sin x$, $f(y) = a\sin^n(\lambda y) + b$.

52. $\sin x\, y''_{xx} + \frac{1}{2}\cos x\, y'_x = a\cos^n(\lambda y) + b.$

This is a special case of equation 2.9.2.14 with $g(x) = \sin x$, $f(y) = a\cos^n(\lambda y) + b$.

53. $\sin x\, y''_{xx} + \frac{1}{2}\cos x\, y'_x = a\tan^n(\lambda y) + b.$

This is a special case of equation 2.9.2.14 with $g(x) = \sin x$, $f(y) = a\tan^n(\lambda y) + b$.

54. $\sin^2 x\, y''_{xx} = n(n + 1 - n\sin^2 x)y + a(\sin x)^{nm+3n+2} y^m.$

This is a special case of equation 2.9.1.30 with $f(\xi) = a\xi^m$.

55. $\cos^2 x\, y''_{xx} = n(n + 1 - n\cos^2 x)y + a(\cos x)^{nm+3n+2} y^m.$

This is a special case of equation 2.9.1.31 with $f(\xi) = a\xi^m$.

2.8.7. Equations Containing the Combinations of Exponential, Hyperbolic, Logarithmic, and Trigonometric Functions

1. $y''_{xx} = \lambda^2 y + ae^{3\lambda x}(\ln y + \lambda x)$.

This is a special case of equation 2.9.1.15 with $f(\xi) = a \ln \xi$.

2. $y''_{xx} = -ay'_x + be^{ax}(\ln y + ax)$.

This is a special case of equation 2.9.2.18 with $f(\xi) = b \ln \xi$.

3. $y''_{xx} = ay'_x + be^{2ax} \ln^n(\lambda y)$.

This is a special case of equation 2.9.2.17 with $f(y) = b \ln^n(\lambda y)$.

4. $y''_{xx} = ay'_x + be^{2ax} \sin^n(\lambda y)$.

This is a special case of equation 2.9.2.17 with $f(y) = b \sin^n(\lambda y)$.

5. $y''_{xx} = ay'_x + be^{2ax} \tan^n(\lambda y)$.

This is a special case of equation 2.9.2.17 with $f(y) = b \tan^n(\lambda y)$.

6. $y''_{xx} + a \tan x \, y'_x + b(a \tan x - b)y = ce^{bmx}(\cos x)^{2a} y^{m-3}$.

This is a special case of equation 2.9.2.32 with $f(\xi) = c\xi^{m-3}$.

7. $y''_{xx} = a(y'_x)^2 - be^{4ay} + c \sinh(\lambda x)$.

This is a special case of equation 2.9.3.16 with $f(x) = -c \sinh(\lambda x)$.

8. $y''_{xx} = a(y'_x)^2 + b \cosh^n(\lambda x) e^{ay} + cx^m$.

This is a special case of equation 2.9.3.15 with $f(x) = -b \cosh^n(\lambda x)$, $g(x) = -cx^m$.

9. $y''_{xx} = a(y'_x)^2 + b \ln^n(\lambda x) e^{ay} + cx^m$.

This is a special case of equation 2.9.3.15 with $f(x) = -b \ln^n(\lambda x)$, $g(x) = -cx^m$.

10. $y''_{xx} = a(y'_x)^2 + b \ln^n(\lambda x) e^{ay} + ce^{\nu x}$.

This is a special case of equation 2.9.3.15 with $f(x) = -b \ln^n(\lambda x)$, $g(x) = -ce^{\nu x}$.

11. $y''_{xx} = a(y'_x)^2 - be^{4ay} + c \sin(\lambda x)$.

This is a special case of equation 2.9.3.16 with $f(x) = -c \sin(\lambda x)$.

12. $y''_{xx} = a(y'_x)^2 + b \sin^n(\lambda x) e^{ay} + cx^m$.

This is a special case of equation 2.9.3.15 with $f(x) = -b \sin^n(\lambda x)$, $g(x) = -cx^m$.

13. $y''_{xx} = a(y'_x)^2 + b \sin^n(\lambda x) e^{ay} + ce^{\nu x}$.

This is a special case of equation 2.9.3.15 with $f(x) = -b \sin^n(\lambda x)$, $g(x) = -ce^{\nu x}$.

14. $y''_{xx} + ae^{\lambda y}(y'_x)^2 + b\ln^n y + c = 0.$

This is a special case of equation 2.9.3.1 with $f(y) = ae^{\lambda y}$, $g(y) = b\ln^n y + c$.

15. $y''_{xx} + ae^{\lambda y}(y'_x)^2 + b\sin^n y + c = 0.$

This is a special case of equation 2.9.3.1 with $f(y) = ae^{\lambda y}$, $g(y) = b\sin^n y + c$.

16. $y''_{xx} = ae^{\lambda y}(y'_x)^2 + b\ln^n(\mu x)y'_x.$

This is a special case of equation 2.9.3.2 with $f(y) = ae^{\lambda y}$, $g(x) = b\ln^n(\mu x)$.

17. $y''_{xx} = ae^{\lambda y}(y'_x)^2 + b\sin^n(\mu x)y'_x.$

This is a special case of equation 2.9.3.2 with $f(y) = ae^{\lambda y}$, $g(x) = b\sin^n(\mu x)$.

18. $y''_{xx} = ae^{\lambda y}(y'_x)^2 + b\tan^n(\mu x)y'_x.$

This is a special case of equation 2.9.3.2 with $f(y) = ae^{\lambda y}$, $g(x) = b\tan^n(\mu x)$.

19. $y''_{xx} + a\ln^n y\,(y'_x)^2 + be^{\lambda y} + c = 0.$

This is a special case of equation 2.9.3.1 with $f(y) = a\ln^n y$, $g(y) = be^{\lambda y} + c$.

20. $y''_{xx} = a\ln^n y\,(y'_x)^2 + be^{\lambda x}y'_x.$

This is a special case of equation 2.9.3.2 with $f(y) = a\ln^n y$, $g(x) = be^{\lambda x}$.

21. $y''_{xx} = a\ln^n y\,(y'_x)^2 + b\sin^m(\lambda x)y'_x.$

This is a special case of equation 2.9.3.2 with $f(y) = a\ln^n y$, $g(x) = b\sin^m(\lambda x)$.

22. $y''_{xx} + a\sin^n y\,(y'_x)^2 + be^{\lambda y} + c = 0.$

This is a special case of equation 2.9.3.1 with $f(y) = a\sin^n y$, $g(y) = be^{\lambda y} + c$.

23. $y''_{xx} = a\sin^n y\,(y'_x)^2 + be^{\lambda x}y'_x.$

This is a special case of equation 2.9.3.2 with $f(y) = a\sin^n y$, $g(x) = be^{\lambda x}$.

24. $y''_{xx} = a\sin^n y\,(y'_x)^2 + b\ln^m(\lambda x)y'_x.$

This is a special case of equation 2.9.3.2 with $f(y) = a\sin^n y$, $g(x) = b\ln^m(\lambda x)$.

25. $y''_{xx} = a\tan^n y\,(y'_x)^2 + be^{\lambda x}y'_x.$

This is a special case of equation 2.9.3.2 with $f(y) = a\tan^n y$, $g(x) = be^{\lambda x}$.

26. $y''_{xx} + be^{ax}\cosh(\lambda y)(y'_x)^3 + ay'_x = 0.$

This is a special case of equation 2.9.3.24 with $f(y) = b\cosh(\lambda y)$.

27. $y''_{xx} + be^{ax}\sin(\lambda y)(y'_x)^3 + ay'_x = 0.$

This is a special case of equation 2.9.3.24 with $f(y) = b\sin(\lambda y)$.

28. $xy''_{xx} = ax \ln x \, e^y y'_x + ae^y.$

Solution: $y = -\ln\left[e^{C_1 x}\left(C_2 - \frac{a}{C_1}\int x^{-1} e^{-C_1 x}\,dx\right) + \frac{a}{C_1}\ln x\right].$

29. $yy''_{xx} = ae^x (y'_x)^3 + ae^x y \ln y \,(y'_x)^2.$

Solution: $x = -\ln\left[e^{C_1 y}\left(C_2 + \frac{a}{C_1}\int y^{-1} e^{-C_1 y}\,dy\right) - \frac{a}{C_1}\ln y\right].$

30. $yy''_{xx} = (y'_x)^2 + ae^{\lambda x} yy'_x + b\sin^n(\nu x) y^2.$

This is a special case of equation 2.9.3.6 with $f(x) = -ae^{\lambda x}$, $g(x) = -b\sin^n(\nu x)$.

31. $(ae^{2x} + b)y''_{xx} + ae^{2x} y'_x = \cosh^n(\lambda y) + c.$

This is a special case of equation 2.9.2.14 with $g(x) = ae^{2x} + b$, $f(y) = \cosh^n(\lambda y) + c$.

32. $(ae^{2x} + b)y''_{xx} + ae^{2x} y'_x = \tanh^n(\lambda y) + c.$

This is a special case of equation 2.9.2.14 with $g(x) = ae^{2x} + b$, $f(y) = \tanh^n(\lambda y) + c$.

33. $(ae^{2x} + b)y''_{xx} + ae^{2x} y'_x = \ln^n(\lambda y) + c.$

This is a special case of equation 2.9.2.14 with $g(x) = ae^{2x} + b$, $f(y) = \ln^n(\lambda y) + c$.

34. $(ae^{2x} + b)y''_{xx} + ae^{2x} y'_x = \sin^n(\lambda y) + c.$

This is a special case of equation 2.9.2.14 with $g(x) = ae^{2x} + b$, $f(y) = \sin^n(\lambda y) + c$.

35. $(ae^{2x} + b)y''_{xx} + ae^{2x} y'_x = \tan^n(\lambda y) + c.$

This is a special case of equation 2.9.2.14 with $g(x) = ae^{2x} + b$, $f(y) = \tan^n(\lambda y) + c$.

36. $\sin x\, y''_{xx} + \frac{1}{2}\cos x\, y'_x = ae^{\lambda y} + b.$

This is a special case of equation 2.9.2.14 with $g(x) = \sin x$, $f(y) = ae^{\lambda y} + b$.

37. $\sin x\, y''_{xx} + \frac{1}{2}\cos x\, y'_x = a\cosh^n(\lambda y) + b.$

This is a special case of equation 2.9.2.14 with $g(x) = \sin x$, $f(y) = a\cosh^n(\lambda y) + b$.

38. $\sin x\, y''_{xx} + \frac{1}{2}\cos x\, y'_x = a\sinh^n(\lambda y) + b.$

This is a special case of equation 2.9.2.14 with $g(x) = \sin x$, $f(y) = a\sinh^n(\lambda y) + b$.

39. $\sin x\, y''_{xx} + \frac{1}{2}\cos x\, y'_x = a\tanh^n(\lambda y) + b.$

This is a special case of equation 2.9.2.14 with $g(x) = \sin x$, $f(y) = a\tanh^n(\lambda y) + b$.

40. $\sin x\, y''_{xx} + \frac{1}{2}\cos x\, y'_x = a\ln^n y + b.$

This is a special case of equation 2.9.2.14 with $g(x) = \sin x$, $f(y) = a\ln^n y + b$.

2.9. Equations Containing Arbitrary Functions

Notation: f, g, h, φ, and ψ are arbitrary composite functions of their arguments indicated in parentheses just after the function name (the arguments may depend on x, y, y'_x).

2.9.1. Equations of the Form $F(x,y)y''_{xx} + G(x,y) = 0$

1. $y''_{xx} = f(y)$.

This autonomous equation is met with in mechanics, the combustion theory, and the theory of mass transfer with chemical reactions.

The substitution $u(y) = y'_x$ leads to a first order equation with separated variables: $uu'_y = f(y)$.

Solution: $\displaystyle\int\Big[C_1 + 2\int f(y)\,dy\Big]^{-1/2} dy = C_2 \pm x.$

Particular solutions: $y = A_k$, where A_k are roots of the equation $f(A_k) = 0$.

2. $y''_{xx} = f(ay + bx + c)$.

The substitution $w = ay + bx + c$ leads to an equation of the form 2.9.1.1: $w''_{xx} = af(w)$.

3. $y''_{xx} = f(y + ax^2 + bx + c)$.

The substitution $w = y + ax^2 + bx + c$ leads to an equation of the form 2.9.1.1: $w''_{xx} = f(w) + 2a$.

4. $y''_{xx} = f(y + ax^n + bx^2 + cx) - an(n-1)x^{n-2}$.

The substitution $w = y + ax^n + bx^2 + cx$ leads to an equation of the form 2.9.1.1: $w''_{xx} = f(w) + 2b$.

5. $y''_{xx} = \dfrac{1}{x^3} f\Big(\dfrac{y}{x}\Big)$.

The transformation $\xi = 1/x$, $w = y/x$ leads to an equation of the form 2.9.1.1: $w''_{\xi\xi} = f(w)$.

6. $y''_{xx} = x^{-3/2} f(yx^{-1/2})$.

Having set $w = yx^{-1/2}$, we obtain

$$\frac{d}{dx}(xw'_x)^2 = \tfrac{1}{2}ww'_x + 2f(w)w'_x.$$

Integrating the latter equation, we arrive at an equation with separation of variables.

Solution: $\displaystyle\int\Big[C_1 + \tfrac{1}{4}w^2 + 2\int f(w)\,dw\Big]^{-1/2} dw = C_2 \pm \ln|x|.$

7. $y''_{xx} = \dfrac{1}{x^3} f\Big(\dfrac{y}{x} + \dfrac{a}{x^2} + \dfrac{b}{x} + c\Big)$.

The transformation $\xi = 1/x$, $w = y/x$ leads to an equation of the form 2.9.1.3: $w''_{\xi\xi} = f(w + a\xi^2 + b\xi + c)$.

8. $y''_{xx} = x^{-3/2} f(ayx^{-1/2} + bx^{1/2})$.

The substitution $w = ay + bx$ leads to an equation of the form 2.9.1.6: $w''_{xx} = ax^{-3/2} f(wx^{-1/2})$.

9. $y''_{xx} = y^{-3} f\left(\dfrac{y}{\sqrt{ax^2 + bx + c}}\right)$.

Setting $u(x) = y(ax^2 + bx + c)^{-1/2}$ and integrating the equation, we obtain a first order equation with separation of variables:

$$(ax^2 + bx + c)^2 (u'_x)^2 = (\tfrac{1}{4}b^2 - ac)u^2 + 2\int u^{-3} f(u)\, du + C_1.$$

10. $(ax^2 + bx + c)^{3/2} y''_{xx} = f\left(\dfrac{\alpha y + \beta x + \gamma}{\sqrt{ax^2 + bx + c}}\right)$.

Setting $w = \alpha y + \beta x + \gamma$ and denoting $f(z) = \dfrac{1}{\alpha z^3}\varphi(z)$, we obtain an equation of the form 2.9.1.9:

$$w''_{xx} = w^{-3}\varphi\left(\frac{w}{\sqrt{ax^2 + bx + c}}\right).$$

11. $y''_{xx} = n(n+1)x^{-2} y + x^{3n} f(yx^n)$.

This is a special case of equation 2.9.1.14 with $\psi = x^{-n}$.

12. $y''_{xx} + f(x)y = ay^{-3}$.

Yermakov's equation.

Let $w = w(x)$ be a nontrivial solution of the linear equation $w''_{xx} + f(x)w = 0$. The transformation $\xi = \displaystyle\int \frac{dx}{w^2}$, $z = \dfrac{y}{w}$ leads to an autonomous equation of the form 2.9.1.1: $z''_{\xi\xi} = az^{-3}$.

Solution: $\quad C_1 y^2 = aw^2 + w^2\left(C_2 + C_1 \displaystyle\int \frac{dx}{w^2}\right)^2$.

13. $(ax^n + b)y''_{xx} = an(n-1)x^{n-2} y + y^{-2} f\left(\dfrac{y}{ax^n + b}\right)$.

The transformation $\xi = \displaystyle\int \frac{dx}{(ax^n + b)^2}$, $w = \dfrac{y}{ax^n + b}$ leads to an equation of the form 2.9.1.1: $w''_{\xi\xi} = w^{-2} f(w)$.

14. $y''_{xx} = \dfrac{\psi''_{xx}}{\psi} y + \psi^{-3} f\left(\dfrac{y}{\psi}\right), \qquad \psi = \psi(x)$.

The transformation $\xi = \displaystyle\int \frac{dx}{\psi^2}$, $w = \dfrac{y}{\psi}$ leads to an equation of the form 2.9.1.1: $w''_{\xi\xi} = f(w)$.

Solution: $\quad \displaystyle\int \left[C_1 + 2\int f(w)\, dw\right]^{-1/2} dw = C_2 \pm \int \frac{dx}{\psi^2(x)}$.

15. $y''_{xx} = \lambda^2 y + e^{3\lambda x} f(ye^{\lambda x})$.

This is a special case of equation 2.9.1.14 with $\psi = e^{-\lambda x}$.

16. $y''_{xx} = f(y + ae^{\lambda x} + b) - a\lambda^2 e^{\lambda x}$.

The substitution $w = y + ae^{\lambda x} + b$ leads to an equation of the form 2.9.1.1: $w''_{xx} = f(w)$.

17. $x^2 y''_{xx} = x^2 f(x^n e^y) + n$.

The substitution $y = w - n \ln x$ leads to an equation of the form 2.9.1.1: $w''_{xx} = f(e^w)$.

18. $y''_{xx} = f(y + a \sinh x + b) - a \sinh x$.

The substitution $w = y + a \sinh x + b$ yields an equation of the form 2.9.1.1: $w''_{xx} = f(w)$.

19. $y''_{xx} = f(y + a \cosh x + b) - a \cosh x$.

The substitution $w = y + a \cosh x + b$ yields an equation of the form 2.9.1.1: $w''_{xx} = f(w)$.

20. $y''_{xx} = \lambda^2 y + (\sinh \lambda x)^{-3} f\left(\dfrac{y}{\sinh \lambda x}\right)$.

This is a special case of equation 2.9.1.14 with $\psi = \sinh \lambda x$.

21. $y''_{xx} = \lambda^2 y + (\cosh \lambda x)^{-3} f\left(\dfrac{y}{\cosh \lambda x}\right)$.

This is a special case of equation 2.9.1.14 with $\psi = \cosh \lambda x$.

22. $x^2 y''_{xx} = x^2 f(y + a \ln x + b) + a$.

The substitution $w = y + a \ln x + b$ leads to an equation of the form 2.9.1.1: $w''_{xx} = f(w)$.

23. $y''_{xx} = -\dfrac{y}{x^2 \ln x} + \dfrac{1}{(\ln x)^3} f\left(\dfrac{y}{\ln x}\right)$.

This is a special case of equation 2.9.1.14 with $\psi = \ln x$.

24. $y''_{xx} = f(y + a \sin x + b) + a \sin x$.

The substitution $w = y + a \sin x + b$ leads to an equation of the form 2.9.1.1: $w''_{xx} = f(w)$.

25. $y''_{xx} = f(y + a \cos x + b) + a \cos x$.

The substitution $w = y + a \cos x + b$ leads to an equation of the form 2.9.1.1: $w''_{xx} = f(w)$.

26. $y''_{xx} = -\lambda^2 y + (\sin \lambda x)^{-3} f\left(\dfrac{y}{\sin \lambda x}\right)$.

This is a special case of equation 2.9.1.14 with $\psi = \sin \lambda x$.

27. $y''_{xx} = -\lambda^2 y + (\cos \lambda x)^{-3} f\left(\dfrac{y}{\cos \lambda x}\right)$.

This is a special case of equation 2.9.1.14 with $\psi = \cos \lambda x$.

28. $y''_{xx} = 2(\sin x)^{-2} y + (\tan x)^3 f(y \tan x)$.

This is a special case of equation 2.9.1.14 with $\psi = \cot x$.

29. $y''_{xx} = 2(\cos x)^{-2} y + (\cot x)^3 f(y \cot x)$.

This is a special case of equation 2.9.1.14 with $\psi = \tan x$.

30. $\sin^2 x\, y''_{xx} = n(n+1-n\sin^2 x)y + \sin^{3n+2} x\, f(y \sin^n x)$.

The substitution $x = \xi + \frac{\pi}{2}$ leads to an equation of the form 2.9.1.31:

$$\cos^2 \xi\, y''_{\xi\xi} = n(n+1-n\cos^2 \xi)y + \cos^{3n+2} \xi\, f(y \cos^n \xi).$$

31. $\cos^2 x\, y''_{xx} = n(n+1-n\cos^2 x)y + \cos^{3n+2} x\, f(y \cos^n x)$.

The transformation $\xi = \int \cos^{2n} x\, dx$, $w = y \cos^n x$ leads to an autonomous equation of the form 2.9.1.1: $w''_{\xi\xi} = f(w)$.

32. $y''_{xx} = \varphi^{-3} f\Big(\dfrac{y}{\varphi} + \psi\Big) + \dfrac{\varphi''_{xx}}{\varphi} y - \varphi\psi''_{xx} - 2\varphi'_x\psi'_x, \qquad \varphi = \varphi(x),\ \psi = \psi(x)$.

The transformation

$$t = \int \frac{dx}{\varphi^2}, \qquad w = \frac{y}{\varphi} + \psi$$

leads to an autonomous equation of the form 2.9.1.1: $w''_{tt} = f(w)$.

Solution:

$$\int \frac{dw}{\sqrt{2F(w) + C_1}} = \pm \int \frac{dx}{\varphi^2} + C_2, \qquad \text{where} \quad F(w) = \int f(w)\, dw.$$

2.9.2. Equations of the Form $F(x,y)y''_{xx} + G(x,y)y'_x + H(x,y) = 0$

1. $y''_{xx} + 3yy'_x + y^3 + f(x)y = 0$.

The substitution $y = w\left(\int w\, dx\right)^{-1}$ leads to the linear equation: $w''_{xx} + f(x)w = 0$.

2. $y''_{xx} + [2ay + f(x)]y'_x + af(x)y^2 = g(x)$.

Having set $u = y'_x + ay^2$, we obtain $u'_x + f(x)u = g(x)$. Thus, the original equation is reduced to the first order linear equation and the Riccati equation.

3. $y''_{xx} + [3y + f(x)]y'_x + y^3 + f(x)y^2 + g(x)y + h(x) = 0$.

The substitution $y = u'_x/u$ leads to a third order linear equation:

$$u'''_{xxx} + f(x)u''_{xx} + g(x)u'_x + h(x)u = 0.$$

4. $xy''_{xx} = ny'_x + x^{2n+1}f(y).$

Multiplying both sides by x^{-2n-1}, we obtain an equation of the form 2.9.2.14.

1°. Solution with $n \neq -1$:

$$\int \left[C_1 + 2\int f(y)\,dy\right]^{-1/2} dy = \pm\frac{x^{n+1}}{n+1} + C_2.$$

2°. Solution with $n = -1$:

$$\int \left[C_1 + 2\int f(y)\,dy\right]^{-1/2} dy = \pm\ln|x| + C_2.$$

5. $xy''_{xx} + (n+1)y'_x = x^{n-1}f(yx^n).$

The transformation $\xi = x^n$, $w = yx^n$ leads to an autonomous equation of the form 2.9.1.1: $n^2 w''_{\xi\xi} = f(w)$.

6. $xy''_{xx} - ny'_x + f(x)y = ax^{2n+1}y^{-3}.$

The substitution $w = yx^{-n/2}$ leads to Yermakov's equation 2.9.1.12:

$$w''_{xx} + x^{-2}[xf(x) - \tfrac{1}{4}n(n+2)]w = aw^{-3}.$$

7. $xy''_{xx} = f(y)y'_x.$

The substitution $w(y) = xy'_x/y$ leads to a first order linear equation: $yw'_y = -w + 1 + f(y)$.

8. $x^2y''_{xx} + xy'_x = f(y).$

Substitutions $x = \pm e^t$ leads to an equation of the form 2.9.1.1: $y''_{tt} = f(y)$.

9. $x^2y''_{xx} + (n+m+1)xy'_x + nmy = x^{n-2m}f(yx^n).$

1°. For $n \neq m$, the transformation $\xi = x^{n-m}$, $w = yx^n$ leads to an autonomous equation of the form 2.9.1.1: $(n-m)^2 w''_{\xi\xi} = f(w)$.

2°. For $n = m$, the transformation $\xi = \ln x$, $w = yx^n$ leads to an autonomous equation of the form 2.9.1.1: $w''_{\xi\xi} = f(w)$.

10. $x^2y''_{xx} = f(y/x)(xy'_x - y).$

This is a special case of equation 2.9.3.23 with $g(x) = h(x) = 0$.

11. $(ax^2 + b)y''_{xx} + axy'_x + f(y) = 0.$

The substitution $\xi = \displaystyle\int \frac{dx}{\sqrt{ax^2+b}}$ leads to an autonomous equation of the form 2.9.1.1: $y''_{\xi\xi} + f(y) = 0$.

12. $x^3y''_{xx} = f(y/x)(xy'_x - y).$

This is a special case of equation 2.9.4.34 with $g(z) = z$.

13. $y''_{xx} + [2ay + f(x)]y'_x + f'_x(x)y = 0.$

Integrating, we obtain the Riccati equation: $y'_x + f(x)y + ay^2 = C$.

14. $gy''_{xx} + \frac{1}{2}g'_x y'_x = f(y), \qquad g = g(x).$

Integrating, we obtain an equation with separation of variables:

$$g(x)(y'_x)^2 = 2\int f(y)\,dy + C_1.$$

With $g(x) \geq 0$, the solution is

$$\int \left[C_1 + 2\int f(y)\,dy\right]^{-1/2} dy = C_2 \pm \int \frac{dx}{\sqrt{g(x)}}.$$

15. $y''_{xx} + (2fy + g)y'_x + f'_x y^2 + g'_x y = 0, \qquad f = f(x), \ \ g = g(x).$

Integrating, we obtain the Riccati equation: $y'_x + fy^2 + gy = C$.

16. $y''_{xx} - \dfrac{\varphi'_x}{\varphi}y'_x + \left(\dfrac{\varphi'_x}{\varphi}\dfrac{\psi'_x}{\psi} - \dfrac{\psi''_{xx}}{\psi}\right)y = \dfrac{\varphi^2}{\psi^3} f\left(\dfrac{y}{\psi}\right), \qquad \varphi = \varphi(x), \ \ \psi = \psi(x).$

The transformation $\xi = \displaystyle\int \frac{\varphi}{\psi^2}\,dx$, $w = \dfrac{y}{\psi}$ leads to an autonomous equation of the form 2.9.1.1: $w''_{\xi\xi} = f(w)$.

17. $y''_{xx} = ay'_x + e^{2ax} f(y).$

Multiplying both sides by e^{-2ax}, we obtain an equation of the form 2.9.2.14.

Solution:

$$\int \left[C_1 + 2\int f(y)\,dy\right]^{-1/2} dy = C_2 \pm \frac{1}{a}e^{ax}.$$

18. $y''_{xx} = -ay'_x + e^{ax} f(ye^{ax}).$

The transformation $\xi = e^{ax}$, $w = ye^{ax}$ leads to an equation of the form 2.9.1.1: $w''_{\xi\xi} = a^{-2} f(w)$.

19. $y''_{xx} + (\mu + \nu)y'_x + \nu\mu y = e^{(\mu - 2\nu)x} f(ye^{\mu x}).$

1°. For $\mu \neq \nu$, the transformation $\xi = e^{(\mu-\nu)x}$, $w = ye^{\mu x}$ leads to an autonomous equation of the form 2.9.1.1: $(\mu - \nu)^2 w''_{\xi\xi} = f(w)$.

2°. For $\mu = \nu$, the substitution $w = ye^{\mu x}$ leads to an autonomous equation of the form 2.9.1.1: $w''_{\xi\xi} = f(w)$.

20. $y''_{xx} - \lambda y'_x + f(x)y = ae^{2\lambda x} y^{-3}.$

The substitution $w = ye^{-\lambda x/2}$ leads to Yermakov's equation 2.9.1.12:

$$w''_{xx} + \left[f(x) - \tfrac{1}{4}\lambda^2\right]w = aw^{-3}.$$

21. $xy''_{xx} - ny'_x - a(ax+n)y = x^{2n+1}e^{3ax}f(ye^{ax})$.

This is a special case of equation 2.9.2.16 with $\varphi = x^n$, $\psi = e^{-ax}$.

22. $xy''_{xx} = f(x^n e^{ay})y'_x$.

The transformation $z = x^n e^{ay}$, $w = xy'_x$ leads to a first order equation with separation of variables: $z(aw+n)w'_z = [f(z)+1]w$.

23. $x^2y''_{xx} + xy'_x = f(x^n e^{ay})$.

The transformation $z = x^n e^{ay}$, $w = xy'_x$ leads to a first order equation with separation of variables: $z(aw+n)w'_z = f(z)$.

24. $x^2y''_{xx} - ax^2y'_x - n(ax+n+1)y = x^{3n+2}e^{2ax}f(yx^n)$.

This is a special case of equation 2.9.2.16 with $\varphi = e^{ax}$, $\psi = x^{-n}$.

25. $(ae^{2\lambda x}+b)y''_{xx} + a\lambda e^{2\lambda x}y'_x + f(y) = 0$.

This is a special case of equation 2.9.2.14 with $g(x) = ae^{2\lambda x} + b$.

26. $y''_{xx} + g'_x y'_x + fy = ae^{-2g}y^{-3}, \qquad f = f(x),\ \ g = g(x)$.

The substitution $w = ye^{g/2}$ leads to Yermakov's equation 2.9.1.12:

$$w''_{xx} + [f - \tfrac{1}{4}(g'_x)^2 - \tfrac{1}{2}g''_{xx}]w = aw^{-3}.$$

27. $y''_{xx} - \dfrac{\varphi'_x}{\varphi}y'_x - a\left(\dfrac{\varphi'_x}{\varphi} + a\right)y = e^{3ax}\varphi^2 f(ye^{ax}), \qquad \varphi = \varphi(x)$.

The transformation $\xi = \int \varphi e^{2ax}\,dx$, $w = ye^{ax}$ leads to an equation of the form 2.9.1.1: $w''_{\xi\xi} = f(w)$.

28. $x^2y''_{xx} + xy'_x = f(y + a\ln x + b\ln^2 x)$.

The substitution $x = e^t$ leads to an equation of the form 2.9.1.3: $y''_{tt} = f(y + at + bt^2)$.

29. $y''_{xx} + \lambda m\tan(\lambda x)y'_x + f(x)y = a[\cos(\lambda x)]^{2m}y^{-3}$.

This is a special case of equation 2.9.2.26 with $g = -m\ln\cos(\lambda x)$.

30. $y''_{xx} - (n+1)\tan x\,y'_x - ny = \cos^{n-2}x\,f(y\cos^n x)$.

This is a special case of equation 2.9.2.16 with $\varphi = \cos^{-n-1}x$, $\psi = \cos^{-n}x$.

31. $y''_{xx} + (m-n)\tan x\,y'_x - n[(m+1)\tan^2 x + 1]y = \cos^{2m+n}x\,f(y\cos^n x)$.

This is a special case of equation 2.9.2.16 with $\varphi = \cos^{m-n}x$, $\psi = \cos^{-n}x$.

32. $y''_{xx} + a\tan x\,y'_x + b(a\tan x - b)y = \cos^{2a}x\,e^{3bx}f(ye^{bx})$.

This is a special case of equation 2.9.2.16 with $\varphi = \cos^a x$, $\psi = e^{-bx}$.

33. $x^2 y''_{xx} + a x^2 \tan x \, y'_x + n(ax \tan x - n - 1) y = x^{3n+2} \cos^{2a} x \, f(y x^n)$.

This is a special case of equation 2.9.2.16 with $\varphi = \cos^a x$, $\psi = x^{-n}$.

34. $x^2 y''_{xx} - a x^2 \cot x \, y'_x - n(ax \cot x + n + 1) y = x^{3n+2} \sin^{2a} x \, f(y x^n)$.

This is a special case of equation 2.9.2.16 with $\varphi = \sin^a x$, $\psi = x^{-n}$.

35. $\sin x \, y''_{xx} + \frac{1}{2} \cos x \, y'_x = f(y)$.

This is a special case of equation 2.9.2.14 with $g = \sin x$.

36. $\cos x \, y''_{xx} - \frac{1}{2} \sin x \, y'_x = f(y)$.

This is a special case of equation 2.9.2.14 with $g = \cos x$.

2.9.3. Equations of the Form $F(x,y) y''_{xx} + \sum\limits_{m=0}^{M} G_m(x,y)(y'_x)^m = 0$ $(M = 2,\ 3,\ 4)$

1. $y''_{xx} + f(y)(y'_x)^2 + g(y) = 0$.

The substitution $w(y) = (y'_x)^2$ leads to a first order linear equation:

$$w'_y + 2f(y)w + 2g(y) = 0.$$

2. $y''_{xx} = f(y)(y'_x)^2 + g(x) y'_x$.

Dividing by y'_x, we obtain a total differential equation. Its solution is found from the equation

$$\ln |y'_x| = \int f(y)\, dy + \int g(x)\, dx + C.$$

Solving the latter for y'_x, we arrive at an equation with separation of variables. In addition, $y = C_1$ is the solution with arbitrary constant C_1.

3. $y''_{xx} = \dfrac{f(x^n y^m)}{xy} (x y'_x - y) y'_x$.

The transformation $z = x^n y^m$, $w = \dfrac{x y'_x}{y}$ leads to an equation with separation of variables: $z(mw + n) w'_z = [f(z) - 1](w^2 - w)$.

4. $y''_{xx} = f(x) + g(x)(x y'_x - y) + h(x)(x y'_x - y)^2$.

The substitution $w(x) = x y'_x - y$ leads to the Riccati equation:

$$w'_x = x f(x) + x g(x) w + x h(x) w^2.$$

5. $y y''_{xx} + (y'_x)^2 + f(x) y y'_x + g(x) = 0$.

The substitution $u = y^2$ leads to a linear equation: $u''_{xx} + f(x) u'_x + 2g(x) = 0$.

6. $\boldsymbol{yy''_{xx} - (y'_x)^2 + f(x)yy'_x + g(x)y^2 = 0.}$

The substitution $u = y'_x/y$ leads to a first order linear equation: $u'_x + f(x)u + g(x) = 0$.

7. $\boldsymbol{2yy''_{xx} - (y'_x)^2 + f(x)y^2 + a = 0, \qquad a > 0.}$

If u and v are two solutions of the linear equation $4y''_{xx} + f(x)y = 0$, which satisfy the condition $(uv'_x - u'_x v)^2 = a$, then $y = uv$ is a solution of the original equation.

8. $\boldsymbol{yy''_{xx} - n(y'_x)^2 + f(x)y^2 + ay^{4n-2} = 0.}$

1°. With $n = 1$, this is an equation of the form 2.9.3.6.

2°. With $n \neq 1$, the substitution $w = y^{1-n}$ leads to Yermakov's equation 2.9.1.12:

$$w''_{xx} + (1-n)f(x)w + a(1-n)w^{-3} = 0.$$

9. $\boldsymbol{yy''_{xx} - n(y'_x)^2 + f(x)y^2 + g(x)y^{n+1} = 0.}$

The substitution $w = y^{1-n}$ leads to a nonhomogeneous linear equation:

$$w''_{xx} + (1-n)f(x)w + (1-n)g(x) = 0.$$

10. $\boldsymbol{yy''_{xx} + a(y'_x)^2 + f(x)yy'_x + g(x)y^2 = 0.}$

The substitution $w = y^{a+1}$ leads to a linear equation: $w''_{xx} + f(x)w'_x + (a+1)g(x)w = 0$.

11. $\boldsymbol{yy''_{xx} - 2(y'_x)^2 - (fy + 2g)y'_x + f'_x y^2 + g'_x y = 0, \qquad f = f(x), \quad g = g(x).}$

Integrating, we obtain the Riccati equation: $y'_x + Cy^2 + fy + g = 0$.

12. $\boldsymbol{yy''_{xx} - (y'_x)^2 + (fy^2 + g)y'_x + f'_x y^3 - g'_x y = 0, \qquad f = f(x), \quad g = g(x).}$

Integrating, we obtain the Riccati equation: $y'_x + fy^2 + Cy - g = 0$, where C is an arbitrary constant.

13. $\boldsymbol{yy''_{xx} = f(x)(y'_x)^2.}$

The substitution $w(x) = xy'_x/y$ leads to the Bernoulli equation 1.1.5:

$$xw'_x = w + [f(x) - 1]w^2.$$

14. $\boldsymbol{yy''_{xx} + f(x)(y'_x)^2 + g(x)yy'_x + h(x)y^2 = 0.}$

The substitution $u = y'_x/y$ leads to the Riccati equation:

$$u'_x + (1+f)u^2 + gu + h = 0.$$

15. $\boldsymbol{y''_{xx} - a(y'_x)^2 + f(x)e^{ay} + g(x) = 0.}$

The substitution $w = e^{-ay}$ leads to a nonhomogeneous linear equation:

$$w''_{xx} - ag(x)w = af(x).$$

16. $y''_{xx} - a(y'_x)^2 + be^{4ay} + f(x) = 0.$

The substitution $w = e^{-ay}$ leads to Yermakov's equation 2.9.1.12:

$$w''_{xx} - af(x)w = abw^{-3}.$$

17. $yy''_{xx} = f(e^{ax}y^n)(y'_x)^2.$

The transformation $z = e^{ax}y^n$, $w = y'_x/y$ leads to a first order separable equation: $z(nw + a)w'_z = [f(z) - 1]w^2$.

18. $y''_{xx} = xf(y)(y'_x)^3.$

Taking y as the independent variable, we obtain a linear equation for $x = x(y)$: $x''_{yy} = -f(y)x$.

19. $y''_{xx} + [xf(y) + g(y)](y'_x)^3 + h(y)(y'_x)^2 = 0.$

Taking y as the independent variable, we obtain a linear equation for $x = x(y)$: $x''_{yy} - h(y)x'_y - f(y)x - g(y) = 0$.

20. $x^2y''_{xx} = f\left(\frac{y}{x}\right)(xy'_x - y)(y'_x)^2.$

This is a special case of equation 2.9.4.35 with $k = 2$.

21. $x^3y''_{xx} + [x^4f(y) + a](y'_x)^3 = 0.$

Taking y as the independent variable, we obtain an equation of the form 2.9.1.12 for $x = x(y)$: $x''_{yy} - f(y)x - ax^{-3} = 0$.

22. $y''_{xx} = x^{-1}[f(y) + g(y)(xy'_x - y) + h(y)(xy'_x - y)^2]y'_x.$

The substitution $w(y) = xy'_x - y$ leads to the Riccati equation:

$$w'_y = f(y) + g(y)w + h(y)w^2.$$

23. $y''_{xx} = x^{-2}(xy'_x - y)\left[f\left(\frac{y}{x}\right) + g\left(\frac{y}{x}\right)y'_x + h\left(\frac{y}{x}\right)(y'_x)^2\right].$

The transformation $z = y/x$, $w = xy'_x/y$ leads to the Riccati equation:

$$zw'_z = f(z) + [zg(z) - 1]w + z^2h(z)w^2.$$

24. $y''_{xx} + e^{ax}f(y)(y'_x)^3 + ay'_x = 0.$

The substitution $\xi = e^{-ax}$ leads to an equation of the form 2.9.4.2 with $g(z) = a^3z^3$: $y''_{\xi\xi} - af(y)(y'_\xi)^3 = 0$.

25. $y''_{xx} = x^{-5/2}\varphi(yx^{-3/2})(y'_x)^4.$

This is a special case of equation 2.9.4.8 with $n = -3/2$, $m = 1$, $f(z) = z\varphi(z)$.

26. $y''_{xx} + f(y)(y'_x)^4 + g(y)(y'_x)^2 + h(y) = 0.$

The substitution $w(y) = (y'_x)^2$ leads to the Riccati equation:

$$w'_y + 2f(y)w^2 + 2g(y)w + 2h(y) = 0.$$

27. $xy''_{xx} + x^{2m+1}f(y)(y'_x)^4 + my'_x = 0.$

This is a special case of equation 2.9.4.46 with $n = 4$, $\varphi = x^{-m}$.

28. $(y + ax)y''_{xx} = f(x)(xy'_x - y)^2.$

The substitution $y = -ax + xz$ leads to the equation

$$xzz''_{xx} + 2zz'_x - x^3 f(x)(z'_x)^2 = 0.$$

Setting $w = z'_x/z$, we obtain the Bernoulli equation:

$$xw'_x + 2w + [x - x^3 f(x)]w^2 = 0.$$

2.9.4. Equations of the Form $F(x, y, y'_x)y''_{xx} + G(x, y, y'_x) = 0$

1. $y''_{xx} = f(x)g(y'_x).$

The substitution $u(x) = y'_x$ leads to a first order equation with separation of variables: $u'_x = f(x)g(u)$.

2. $y''_{xx} = f(y)g(y'_x).$

The substitution $u(y) = y'_x$ leads to a first order equation with separation of variables: $uu'_y = f(y)g(u)$.

In addition, there may exist solutions of the form $y = Ax + C$, where A are roots of the equation $g(A) = 0$, C is an arbitrary number, or $y = B$, where B are roots of the equation $f(B) = 0$.

3. $y''_{xx} = f(ax + by + c)g(y'_x).$

With $b = 0$, we have an equation of the form 2.9.4.1. With $b \neq 0$, the substitution $u(x) = y + (ax + c)/b$ leads to an equation of the form 2.9.4.2: $u''_{xx} = f(bu)g\left(u'_x - \frac{a}{b}\right)$.

4. $y''_{xx} = x^{-n-1}f(x^n y'_x).$

The substitution $w(x) = x^n y'_x$ leads to a first order equation with separation of variables: $xw'_x = f(w) + nw$.

5. $y''_{xx} = y^{-2n-1}f(y^n y'_x).$

The substitution $w(y) = y^n y'_x$ leads to a first order equation with separation of variables: $yww'_y = f(w) + nw^2$.

6. $y''_{xx} = \dfrac{y}{x^2} f\left(\dfrac{xy'_x}{y}\right).$

The substitution $w(x) = xy'_x/y$ leads to a first order equation with separation of variables: $xw'_x = f(w) + w - w^2$.

7. $y''_{xx} = \dfrac{f(y'_x)}{\sqrt{xy}}$.

Setting $u = y'_x$ and passing on to new variables

$$t = \int \frac{du}{f(u)}, \qquad w = 2\sqrt{x},$$

we have $y = (w'_t)^2$. Differentiating the latter with respect to x, we obtain a second order linear equation $w''_{tt} = g(t)w$, where function $g(t)$ is defined parametrically:

$$g = \tfrac{1}{4}u, \quad t = \int \frac{du}{f(u)}.$$

8. $y''_{xx} = \dfrac{f(x^n y^m)}{xy}(y'_x)^{\frac{2n+m}{n+m}}$.

The transformation $z = x^n y^m$, $w = xy'_x/y$ yields

$$z(mw+n)w'_z = z^{-\frac{1}{m+1}} f(z) w^{\frac{2n+m}{n+m}} + w - w^2.$$

Divide both sides of this equation by $w^{\frac{2n+m}{n+m}}$ and introduce a new variable

$$\zeta = w^{\frac{m}{n+m}} - w^{-\frac{n}{n+m}}.$$

As a result we obtain a first order linear equation:

$$(n+m)z\zeta'_z = -\zeta + z^{-\frac{1}{n+m}} f(z).$$

9. $y''_{xx} + f(x)y'_x + g(x)(y'_x)^k = 0$.

The substitution $u(x) = y'_x$ leads to the Bernoulli equation: $u'_x + f(x)u + g(x)u^k = 0$.

10. $y''_{xx} = f(y)(y'_x)^2 + g(y)(y'_x)^k$.

The substitution $w(y) = y'_x$ leads to the Bernoulli equation: $w'_y = f(y)w + g(y)w^{k-1}$.

11. $xy''_{xx} + x^{nm-2m+1} f(y)(y'_x)^n + my'_x = 0$.

This is a special case of equation 2.9.4.46 with $\varphi = x^{-m}$.

12. $y''_{xx} = \dfrac{n - kn - km}{km} x^{-1} y'_x + x^{k-1} y^{-k} f(x^n y^m)(y'_x)^{k+1}$.

Passing on to new variables $z = x^n y^m$, $w = xy'_x/y$, we arrive at

$$z(mw+n)w'_z = \frac{n(1-k)}{km} w - w^2 + f(z)w^{k+1}.$$

Multiplying the latter by w^{-k-1}, we obtain, with the aid of the substitution

$$\zeta = \frac{m}{1-k} w^{1-k} - \frac{n}{k} w^{-k},$$

a first order linear equation:

$$z\zeta'_z = \frac{k-1}{m}\zeta + f(z).$$

13. $y''_{xx} = \dfrac{m+km+kn}{kn} y^{-1}(y'_x)^2 + x^k y^{-k-1} f(x^n y^m)(y'_x)^{k+2}$.

Passing on to new variables $z = x^n y^m$, $w = xy'_x/y$, we arrive at

$$z(mw+n)w'_z = w + \frac{m(1+k)}{kn} w^2 + f(z)w^{k+2}.$$

Multiplying the latter by w^{-k-2}, we obtain, with the aid of the substitution

$$\zeta = \frac{m}{k} w^{-k} + \frac{n}{k+1} w^{-k-1},$$

a first order linear equation:

$$z\zeta'_z = -\frac{k+1}{n}\zeta - f(z).$$

14. $y''_{xx} = y^{-1}(y'_x)^2 - x^{-1}y'_x + x^{-2} y f(x^n y^m) g\left(\dfrac{xy'_x}{y}\right)$.

The transformation $z = x^n y^m$, $w = xy'_x/y$ leads to a first order separable equation:

$$z(mw+n)w'_z = f(z)g(w).$$

15. $y''_{xx} = \dfrac{f(y'_x)}{\sqrt{axy+b}}$.

Setting $u = y'_x$, we rewrite the equation as follows:

$$\left[\sqrt{x}\, u'_x / f(u)\right]^{-2} = ay + bx^{-1}.$$

Differentiating both sides with respect to x and passing on to new variables

$$t = \frac{1}{2}\int \frac{du}{f(u)}, \qquad z = \sqrt{x},$$

we obtain an equation of the form 2.9.1.12:

$$z''_{tt} = au(t)z - bz^{-3}.$$

16. $y''_{xx} = \dfrac{f(y'_x)}{\sqrt{ay+bx^2}}$.

Setting $u = y'_x$, we rewrite the equation as follows:

$$\left[u'_x / f(u)\right]^{-2} = ay + bx^2.$$

Differentiating both sides with respect to x and introducing a new independent variable $t = \displaystyle\int \frac{du}{f(u)}$, we obtain a second order linear equation for $x = x(t)$, integrable by quadrature:

$$2x''_{tt} = 2bx + au(t),$$

where function $u = u(t)$ is defined implicitly: $t = \displaystyle\int \frac{du}{f(u)}$.

17. $y''_{xx} = \dfrac{f(y'_x)}{\sqrt{ax + by^2}}$.

Taking y as the independent variable, we obtain an equation of the form 2.9.4.16 for $x = x(y)$:

$$x''_{yy} = -(ax + by^2)^{-1/2} f(1/x'_y)(x'_y)^3.$$

18. $y''_{xx} = (ax^2 + bxy + cy^2 + \alpha x + \beta y + \gamma)^{-1/2} f(y'_x)$.

The transformation $x = At + Bu + C$, $y = Dt + Pu + Q$, where $u = u(t)$, reduces this equation, by selecting appropriate constants A, B, C, D, P, and Q, to an equation of the form 2.9.4.15, 2.9.4.16, or 2.9.4.17.

19. $y''_{xx} = \dfrac{f(y'_x)}{\sqrt{axy + bx^{3/2} + cx}}$.

Setting $u = y'_x$, we rewrite the equation as follows:

$$\left[\sqrt{x}\, u'_x / f(u)\right]^{-2} = ay + b\sqrt{x} + c.$$

Differentiating both sides with respect to x and changing to new variables

$$t = \frac{1}{2}\int \frac{du}{f(u)}, \qquad z = \sqrt{x},$$

we obtain a second order linear equation:

$$2z''_{tt} = 2au(t)z + b,$$

where function $u = u(t)$ is defined in the implicit form: $t = \dfrac{1}{2}\displaystyle\int \frac{du}{f(u)}$.

20. $y''_{xx} = \dfrac{f(y'_x)}{\sqrt{axy + by^{3/2} + cy}}$.

Taking y as the independent variable, we obtain an equation of the form 2.9.4.19 for $x = x(y)$:

$$x''_{yy} = -(axy + by^{3/2} + cy)^{-1/2} f(1/x'_y)(x'_y)^3.$$

21. $y''_{xx} = \dfrac{f(y'_x)}{\sqrt{axy + bx^2 + cx^{3/2} + dx}}$.

The substitution $aw = ay + bx$ leads to an equation of the form 2.9.4.19:

$$w''_{xx} = \frac{f(w'_x - b/a)}{\sqrt{axw + cx^{3/2} + dx}}.$$

22. $y''_{xx} = \dfrac{f(y'_x)}{\sqrt{axy + by^2 + cy^{3/2} + dy}}$.

Taking y as the independent variable, we obtain an equation of the form 2.9.4.21 for $x = x(y)$:

$$x''_{yy} = -(axy + by^2 + cy^{3/2} + dy)^{-1/2} f(1/x'_y)(x'_y)^3.$$

23. $y''_{xx} = \left[xf(y'_x) + yg(y'_x) + h(y'_x)\right]^{-1}.$

The Legendre transformation $x = w'_t$, $y = tw'_t - w$ ($y'_x = t$, $y''_{xx} = 1/w''_{tt}$) leads to a second order linear equation:

$$w''_{tt} = [f(t) + tg(t)]w'_t - g(t)w + h(t).$$

24. $y''_{xx} = a + f(x)\sqrt{(y'_x)^2 - 2ay}.$

Setting $u = y'_x$, we rewrite the equation as follows:

$$(u'_x - a)^2[f(x)]^{-2} = u^2 - 2ay.$$

Differentiating both sides with respect to x and dividing by $(u'_x - a)$, we obtain a linear equation:

$$fu''_{xx} = f'_x u'_x + f^3 u - af'_x.$$

25. $y''_{xx} = f(x)\sqrt{y'_x(xy'_x - y)}.$

The Legendre transformation $x = w'_t$, $y = tw'_t - w$ leads to an equation of the form 2.9.4.7: $w''_{tt} = \dfrac{1}{f(w'_t)\sqrt{tw}}$.

26. $y''_{xx} = f(y)(xy'_x - y)^{1/2}(y'_x)^2.$

The transformation $x = tw'_t - w$, $y = -w'_t$, where $w = w(t)$, leads to an equation of the form 2.9.4.7: $w''_{tt} = \dfrac{1}{f(-w'_t)\sqrt{tw}}$.

27. $y''_{xx} = x^{-2}(xy'_x - y)f(y'_x).$

The Legendre transformation $x = w'_t$, $y = tw'_t - w$ ($y'_x = t$, $y''_{xx} = 1/w''_{tt}$) leads to an equation of the form 2.9.3.13: $w''_{tt} = \dfrac{(w'_t)^2}{f(t)w}$.

28. $y''_{xx} = x^{-3}f(xy'_x - y)(y'_x)^{-1}.$

The Legendre transformation $x = w'_t$, $y = tw'_t - w$ leads to an equation of the form 2.9.3.18: $w''_{tt} = t[f(w)]^{-1}(w'_t)^3$.

29. $y''_{xx} = x^{-1}f\left(y'_x - \dfrac{y}{x}\right).$

The substitution $w = y'_x - \dfrac{y}{x}$ leads to an equation with separation of variables: $xw'_x = -w + f(w)$.

30. $y''_{xx} = f(x)g(xy'_x - y).$

The substitution $w = xy'_x - y$ leads to an equation with separation of variables: $w'_x = xf(x)g(w)$.

31. $y''_{xx} = x^{-n-3}y^n f(xy'_x - y).$

The transformation $x = 1/t$, $y = w/t$ leads to an autonomous equation of the form 2.9.4.2: $w''_{tt} = w^n f(-w'_t)$.

32. $y''_{xx} = \dfrac{f(x^n y^m)}{xy}(xy'_x - y)^{\frac{2n+m}{n}}.$

This is a special case of equation 2.9.4.36 with $k = \dfrac{2n+m}{n}$.

33. $y''_{xx} = x^{-1} f(y) g(xy'_x - y) y'_x.$

The substitution $w(y) = xy'_x - y$ leads to an equation with separation of variables: $w'_y = f(y)g(w)$.

34. $y''_{xx} = x^{-3} f(y/x) g(xy'_x - y).$

The transformation $x = 1/t$, $y = w/t$ leads to an equation of the form 2.9.4.2: $w''_{tt} = f(w)g(-w'_t)$.

35. $y''_{xx} = x^{-2} f(y/x)(xy'_x - y)(y'_x)^k.$

The transformation $z = y/x$, $w = xy'_x/y$ leads to the Bernoulli equation

$$zw'_z = -w + z^k f(z) w^k.$$

There are particular solutions $y = x$ and $y = Ax$, where A are roots of the equation $f(A) = 0$; with $k > 0$ we also have $y = C$, where C is an arbitrary number.

36. $y''_{xx} = \dfrac{f(x^n y^m)}{xy}(y'_x)^{\frac{2n+m-nk}{n+m}}(xy'_x - y)^k.$

The transformation $z = x^n y^m$, $w = xy'_x/y$ leads to the equation

$$z(mw+n)w'_z = z^{\frac{k-1}{n+m}} f(z) w^{\frac{2n+m-nk}{n+m}}(w-1)^k + w - w^2.$$

Multiplying both sides by $w^{-\frac{2n+m}{n+m}}$ and passing on to a new variable

$$\zeta = w^{\frac{m}{n+m}} - w^{-\frac{n}{n+m}},$$

we arrive at the Bernoulli equation

$$(n+m)z\zeta'_z = -\zeta + z^{\frac{k-1}{n+m}} f(z)\zeta^k.$$

37. $y''_{xx} = n(n-1)x^{-2}y + f(x)(xy'_x - ny)^m.$

The substitution $w = x^n y'_x - nx^{n-1}y$ leads to a first order equation with separation of variables: $w'_x = x^{n+m-nm} f(x) w^m$.

38. $y''_{xx} = n(n-1)x^{-2}y + f(x)g(x^n y'_x - nx^{n-1}y).$

The substitution $w = x^n y'_x - nx^{n-1}y$ leads to a first order equation with separation of variables: $w'_x = x^n f(x) g(w)$.

39. $y''_{xx} = f(x)(xy'_x - y) + g(x)(xy'_x - y)^k.$

The substitution $w(x) = xy'_x - y$ leads to the Bernoulli equation

$$w'_x = xf(x)w + xg(x)w^k.$$

40. $y''_{xx} = x^{-1}\big[f(y)(xy'_x - y) + g(y)(xy'_x - y)^k\big]y'_x$.

The substitution $w(y) = xy'_x - y$ leads to the Bernoulli equation $w'_y = f(y)w + g(y)w^k$.

41. $y''_{xx} = x^{-2}(xy'_x - y)\left[f\left(\frac{y}{x}\right)y'_x + g\left(\frac{y}{x}\right)(y'_x)^k\right]$.

The transformation $z = y/x$, $w = xy'_x/y$ leads to the Bernoulli equation: $zw'_z = [zf(z) - 1]w + z^k g(z)w^k$.

42. $y''_{xx} = x^{-3}(xy'_x - y)^2 f\left(\frac{y}{x}\right) + x^{-3}(xy'_x - y)^k g\left(\frac{y}{x}\right)$.

The transformation $x = -1/t$, $y = -w/t$ leads to an equation of the form 2.9.4.10:

$$w''_{tt} = f(w)(w'_t)^2 + g(w)(w'_t)^k.$$

43. $y''_{xx} = \dfrac{f(x^n y^m)}{xy}(y'_x)^{\frac{2n+m-nk}{n+m}}(xy'_x - y)^k + \dfrac{g(x^n y^m)}{xy}y'_x(xy'_x - y)$.

The transformation $z = x^n y^m$, $w = xy'_x/y$ followed by the substitution

$$\zeta = w^{\frac{m}{n+m}} - w^{-\frac{n}{n+m}}$$

leads to the Bernoulli equation

$$(n+m)z\zeta'_z = [g(z) - 1]\zeta + z^{\frac{k-1}{n+m}} f(z)\zeta^k.$$

44. $y''_{xx} = x^n y^m (y'_x)^{\frac{2n+m+3}{n+m+2}} F(\zeta), \qquad \zeta = (xy'_x - y)(y'_x)^{-\frac{n+1}{n+m+2}}$.

The Legendre transformation $x = w'_t$, $y = tw'_t - w$ leads to an equation of the form 2.9.4.36:

$$w''_{tt} = \frac{1}{tw}\left[\frac{t^a w}{F(t^a w)}\right](w'_t)^{\frac{2a+1-ab}{a+1}}(tw'_t - w)^b,$$

where $a = -\frac{n+1}{n+m+2}$, $b = -m$.

45. $y''_{xx} = \dfrac{f(x^n y^m)}{xy}(y'_x)^{\frac{2n+m}{n+m}} + \dfrac{g(x^n y^m)}{xy}y'_x(xy'_x - y)$
$$+ \frac{h(x^n y^m)}{xy}(y'_x)^{\frac{m}{n+m}}(xy'_x - y)^2.$$

The transformation $z = x^n y^m$, $w = xy'_x/y$ followed by the substitution

$$\zeta(z) = w^{\frac{m}{n+m}} - w^{-\frac{n}{n+m}}$$

leads to the Riccati equation

$$(n+m)z\zeta'_z = z^{\frac{1}{n+m}} h(z)\zeta^2 + [g(z) - 1]\zeta + z^{-\frac{1}{n+m}} f(z).$$

46. $y''_{xx} + \varphi^{2-n} f(y)(y'_x)^n - \dfrac{\varphi'_x}{\varphi} y'_x = 0, \qquad \varphi = \varphi(x).$

The substitution $\xi = \int \varphi(x)\,dx$ leads to an equation of the form 2.9.4.2:

$$y''_{\xi\xi} + f(y)(y'_\xi)^n = 0.$$

47. $gy''_{xx} + \frac{1}{2} g'_x y'_x = f(y) h\big(y'_x \sqrt{g}\,\big), \qquad g = g(x).$

The substitution $w(y) = y'_x \sqrt{g}$ leads to an equation with separation of variables: $ww'_y = f(y)h(w)$.

48. $fy''_{xx} + \frac{1}{2} f'_x y'_x = fg(y)(y'_x)^2 + f^n h(y)(y'_x)^{2n}, \qquad f = f(x).$

The substitution $\xi = \displaystyle\int \frac{dx}{\sqrt{f(x)}}$ leads to an autonomous equation of the form 2.9.4.10: $y''_{\xi\xi} = g(y)(y'_\xi)^2 + h(y)(y'_\xi)^{2n}$.

49. $y''_{xx} + e^{a(n-2)x} f(y)(y'_x)^n + ay'_x = 0.$

This is a special case of equation 2.9.4.46 with $\varphi = e^{-ax}$.

50. $y''_{xx} = -x^{-1} y'_x + x^{-2} f(x^n e^{ay}) g(xy'_x).$

The transformation $z = x^n e^{ay}$, $w = xy'_x$ leads to an equation with separation of variables: $z(aw+n)w'_z = f(z)g(w)$.

51. $y''_{xx} = \dfrac{a}{n} \dfrac{1-k}{2-k} (y'_x)^2 + x^{k-2} f(x^n e^{ay})(y'_x)^k.$

Passing on to new variables $z = x^n e^{ay}$, $w = xy'_x$, we have

$$z(aw+n)w'_z = \frac{a}{n}\frac{1-k}{2-k} w^2 + w + f(z)w^k.$$

Multiplying both sides by w^{-k} and introducing a new variable

$$v = \frac{a}{2-k} w^{2-k} + \frac{n}{1-k} w^{1-k},$$

we obtain a first order linear equation:

$$zv'_z = \frac{1-k}{n} v + f(z).$$

52. $y''_{xx} = -\dfrac{a}{m} \dfrac{2-k}{1-k} y'_x + y^{1-k} f(e^{ax} y^m)(y'_x)^k.$

Passing on to new variables $z = e^{ax} y^m$, $w = y'_x / y$, we have

$$z(mw+a)w'_z = -w^2 - \frac{a}{m}\frac{2-k}{1-k} w + f(z)w^k.$$

Multiplying both sides by w^{-k} and introducing a new variable

$$v = \frac{m}{2-k} w^{2-k} + \frac{a}{1-k} w^{1-k},$$

we obtain a first order linear equation:

$$mzv'_z = (k-2)v + mf(z).$$

53. $y''_{xx} = y^{-1}(y'_x)^2 + yf(e^{ax}y^m)g(y'_x/y)$.

The transformation $z = e^{ax}y^m$, $w = y'_x/y$ leads to an equation with separation of variables: $z(mw + a)w'_z = f(z)g(w)$.

54. $y''_{xx} = x^{-2}f(x^n e^{ay})\exp\Bigl(-\dfrac{a}{n}xy'_x\Bigr)$.

The transformation $z = x^n e^{ay}$, $w = xy'_x$ leads to the equation

$$z(aw + n)w'_z = w + f(z)\exp\Bigl(-\frac{a}{n}w\Bigr)$$

which can be reduced, with the aid of the substitution $\zeta = w\exp\Bigl(\dfrac{a}{n}w\Bigr)$, to a first order linear equation: $nz\zeta'_z = \zeta + f(z)$.

55. $y''_{xx} = -\dfrac{a}{m}y'_x \ln\Bigl(\dfrac{y'_x}{y}\Bigr) + f(e^{ax}y^m)y'_x$.

The transformation $z = e^{ax}y^m$, $w = y'_x/y$ leads to an equation

$$z(mw + a)w'_z = -\frac{a}{m}w\ln w - w^2 + f(z)w.$$

Dividing both sides by w and passing on to a new variable $v = mw + a\ln w$, we obtain a first order linear equation: $mzv'_z = -v + mf(z)$.

56. $y''_{xx} = -\dfrac{a}{n}(y'_x)^2\ln(xy'_x) + f(x^n e^{ay})(y'_x)^2$.

The transformation $z = x^n e^{ay}$, $w = xy'_x$ leads to the equation

$$z(aw + n)w'_z = w - \frac{a}{n}w^2\ln w + w^2 f(z).$$

Dividing both sides by w^2 and passing on to a new variable $v = a\ln w - nw^{-1}$, we obtain a first order linear equation: $nzv'_z = -v + nf(z)$.

2.9.5. Equations of the Form $F(x,\ y,\ y'_x,\ y''_{xx}) = 0$

1. $y = f(y''_{xx})$.

The substitution $w(y) = \frac{1}{2}(y'_x)^2$ leads to an equation of the form 1.8.1.2: $y = f(w'_y)$.

2. $y'_x = yf\Bigl(\dfrac{y''_{xx}}{y}\Bigr)$.

The transformation $t = y^2$, $w = (y'_x)^2$ leads to an equation of the form 1.8.1.7: $w = tf^2(w'_t)$.

3. $y = xf(x^3 y''_{xx})$.

The transformation $x = 1/t$, $y = w/t$ leads to an equation of the form 2.9.5.1: $w = f(w''_{tt})$.

4. $\boldsymbol{y = ax^2 + bx + c + f(y''_{xx})}$.

The substitution $w = y - ax^2 - bx - c$ leads to an equation of the form 2.9.5.1: $w = f(w''_{xx} + 2a)$.

5. $\boldsymbol{xy'_x - y = f(x^n y''_{xx})}$.

This is a special case of equation 2.9.5.10 with $\varphi = x^n$.

6. $\boldsymbol{xy'_x = y + a(y'_x)^2 + by'_x + c + f(y''_{xx})}$.

The Legendre transformation $x = w'_t$, $y = tw'_t - w$ ($y'_x = t$, $y''_{xx} = 1/w''_{tt}$) leads to an equation of the form 2.9.5.4: $w = at^2 + bt + c + f(1/w''_{tt})$.

7. $\boldsymbol{f(y''_{xx}) + xy''_{xx} = y'_x}$.

Solution: $2y = C_1x^2 + 2xf(C_1) + C_2$.

8. $\boldsymbol{F(y''_{xx} + y) = (y''_{xx})^2 + (y'_x)^2}$.

Differentiating with respect to x, we obtain

$$[f'(y''_{xx} + y) - 2y''_{xx}](y'''_{xxx} + y'_x) = 0.$$

From the equation $y'''_{xxx} + y'_x = 0$, we find

$$y = A\sin(x + C_1) + C_2, \qquad \text{where} \quad A^2 = f(C_2).$$

Equating the expression in the square brackets to zero, we arrive at another solution in the parametric form:

$$x = \int \frac{[2 - f''_{uu}(u)]\,du}{\sqrt{4f(u) - [f'_u(u)]^2}}, \qquad y = u - \tfrac{1}{2}f'_u(u).$$

9. $\boldsymbol{f(x)(y''_{xx} - a)^2 = (y'_x)^2 - 2ay}$.

Differentiating with respect to x, we obtain

$$(y''_{xx} - a)(2fy'''_{xxx} + f'_x y''_{xx} - 2y'_x - af'_x) = 0. \tag{1}$$

Equating the second factor to zero and making the transformation $\xi = \int \frac{dx}{\sqrt{f}}$, $w = y'_x$, we arrive at a second order linear equation of the form 2.1.9.1:

$$w''_{\xi\xi} - w = \tfrac{1}{2}af'_x,$$

whose right-hand side is to be expressed in terms of ξ. Substituting the solution of the latter equation into the original one, we obtain a relation connecting integration constants.

Equating the first factor in (1) to zero, we find the second solution $y = \frac{1}{2}a(x+C)^2$.

10. $\boldsymbol{xy'_x - y = f(\varphi y''_{xx})}, \qquad \varphi = \varphi(x)$.

The transformation $\xi = \int \frac{x}{\varphi}\,dx$, $w = xy'_x - y$ leads to an equation of the form 1.8.1.2: $w = f(w'_\xi)$.

11. $F\left(y+\dfrac{(y'_x)^2}{y''_{xx}},\ x+\dfrac{y'_x-(y'_x)^3}{2y''_{xx}}\right)=0.$

Solutions can be found from the equality

$$(y-C_1)^2=2C_2(x-A)+C_2^2, \qquad \text{where} \quad F(C_1,\,A)=0.$$

The question of whether there are other solutions calls for further investigation.

12. $F(y''_{xx},\ xy''_{xx}-y'_x,\ x^2y''_{xx}-2xy'_x+2y)=0.$

It is known that all the functions of the form

$$y=\tfrac{1}{2}Ax^2-C_1x+\tfrac{1}{2}C_2,$$

where $A=A(C_1,C_2)$ is determined from the equation $F(A,C_1,C_2)=0$, are the solutions of the original equation.

13. $F\left(x-\dfrac{(y'_x)^2+1}{y''_{xx}}y'_x,\ y+\dfrac{(y'_x)^2+1}{y''_{xx}},\ \dfrac{\left((y'_x)^2+1\right)^{3/2}}{y''_{xx}}\right)=0.$

It is known that all the functions of the form

$$(x-C_1)^2+(y-C_2)^2=A^2,$$

where $A=A(C_1,C_2)$ is determined from the equation $F(C_1,C_2,A)=0$, are the solutions of the original equation.

14. $xy'_x-y=f(e^{\lambda x}y''_{xx}).$

This is a special case of equation 2.9.5.10 with $\varphi=e^{\lambda x}$.

2.9.6. General Equations Admitting the Order Reduction

1. $y''_{xx}=F(x,\ y'_x).$

The substitution $w(x)=y'_x$ leads to a first order equation: $w'_x=F(x,w)$.

2. $y''_{xx}=F(y,\ y'_x).$

The substitution $w(y)=y'_x$ leads to a first order equation: $ww'_y=F(y,w)$.

3. $y''_{xx}=F(ax+by,\ y'_x).$

The substitution $bw=ax+by$ leads to the equation of the form 2.9.6.2:

$$w''_{xx}=F\left(bw,\ w'_x-\frac{a}{b}\right).$$

4. $y''_{xx}=x^{k-2}F(x^{-k}y,\ x^{1-k}y'_x).$

Homogeneous equation in the extended sense.

The transformation $t=\ln x$, $w=x^{-k}y$ leads to an equation of the form 2.9.6.2:

$$w''_{tt}+(2k-1)w'_t+k(k-1)w=F(w,\ w'_t+kw).$$

5. $y''_{xx} = \dfrac{y}{x^2} F\Big(x^n y^m,\ \dfrac{x}{y} y'_x\Big).$

Homogeneous equation in the extended sense.

The transformation $z = x^n y^m$, $w = xy'_x/y$ leads to a first order equation:

$$z(mw+n)w'_z = F(z,w) + w - w^2.$$

6. $y''_{xx} = F(x,\ xy'_x - y).$

The substitution $w(x) = xy'_x - y$ leads to a first order equation: $w'_x = xF(x,w)$.

7. $y''_{xx} = F\Big(x,\ y'_x - \dfrac{y}{x}\Big).$

The substitution $w(x) = y'_x - \dfrac{y}{x}$ leads to a first order equation: $xw'_x = -w + xF(x,w)$.

8. $y''_{xx} = x^{-2} F(y,\ xy'_x - y).$

The substitution $w(y) = xy'_x - y$ leads to a first order equation: $(y+w)w'_y = F(y,w)$.

9. $y''_{xx} + [f(x) + g(y)]y'_x + f'_x(x)y = 0.$

Integrating yields a first order equation: $y'_x + f(x)y + \int g(y)\,dy = C$.

10. $f_1 y'_x y''_{xx} + f_2 y y''_{xx} + f_3 (y'_x)^2 + f_4 y y'_x + f_5 y^2 = 0, \qquad f_k = f_k(x).$

The substitution $w(x) = y'_x/y$ leads to the Abel equation

$$(f_1 w + f_2)w'_x + f_1 w^3 + (f_2 + f_3)w^2 + f_4 w + f_5 = 0.$$

11. $f(y'_x)y''_{xx} + g(y)y'_x + h(x) = 0.$

Integrating, we obtain

$$\int f(u)\,du + \int g(y)\,dy + \int h(x)\,dx = C, \qquad \text{where} \quad u = y'_x.$$

12. $f^2 y''_{xx} + f f'_x y'_x = \Phi(y,\ f y'_x), \qquad f = f(x).$

The substitution $w(y) = fy'_x$ leads to a first order equation: $ww'_y = \Phi(y,w)$.

13. $x(x-a)^2 y''_{xx} = f(y/x).$

The transformation $\xi = \ln\Big(\dfrac{x-a}{x}\Big)$, $w = \dfrac{y}{x}$ leads to an equation of the form 2.9.6.2: $w''_{\xi\xi} - w'_\xi = a^{-2} f(w)$. For some specific functions f, the solutions of this equation are given in Subsection 2.2.1.

14. $y''_{xx} = \dfrac{(cx+d)^{n-1}}{(ax+b)^{n+2}} f\Big(\dfrac{(ax+b)^n y}{(cx+d)^{n+1}}\Big).$

The transformation

$$\xi = \ln\Big(\frac{ax+b}{cx+d}\Big), \qquad w = \frac{(ax+b)^n y}{(cx+d)^{n+1}}$$

leads to an equation of the form 2.9.6.2:

$$w''_{\xi\xi} - (2n+1)w'_\xi + n(n+1)w = \Delta^{-2} f(w), \qquad \text{where} \quad \Delta = ad - bc.$$

15. $y''_{xx} = 2ayy'_x + F(x,\ y'_x - ay^2)$.

This equation can be derived by eliminating subsidiary function z from the simultaneous first-order equations

$$z'_x = F(x, z), \qquad (1)$$
$$y'_x - ay^2 = z. \qquad (2)$$

If one succeed in finding the general solution $z = z(x, C_1)$ of equation (1), the original equation is reducible to the Riccati equation (2) with a known right-hand side.

16. $y''_{xx} = e^{-ax} F(e^{ax} y,\ e^{ax} y'_x)$.

The substitution $w = e^{ax} y$ leads to an equation of the form 2.9.6.2:

$$w''_{xx} - 2aw'_x + a^2 w = F(w,\ w'_x - aw).$$

17. $y''_{xx} = yF(e^{ax} y^m,\ y'_x/y)$.

Exponential homogeneous equation.
The transformation $z = e^{ax} y^m$, $w = y'_x/y$ leads to a first order equation:

$$z(mw + a)w'_z = F(z, w) - w^2.$$

18. $y''_{xx} = x^{-2} F(x^n e^{ay},\ xy'_x)$.

Exponential homogeneous equation.
The transformation $z = x^n e^{ay}$, $w = xy'_x$ leads to a first order equation:

$$z(aw + n)w'_z = F(z, w) + w.$$

19. $y''_{xx} = e^{2ay} F(xe^{ay},\ e^{-ay} y'_x)$.

The transformation $z = xe^{ay}$, $w = e^{-ay} y'_x$ leads to a first order equation:

$$(azw + 1)w'_z = F(z, w) - aw^2.$$

20. $y''_{xx} = ae^y y'_x + F(x,\ y'_x - ae^y)$.

This equation can be derived by eliminating subsidiary function z from the simultaneous first-order equations

$$z'_x = F(x, z), \qquad (1)$$
$$y'_x - ae^y = z. \qquad (2)$$

If one succeed in finding the general solution $z = z(x, C_1)$ of equation (1), the original equation is equation (2) of the form 1.7.2.5 which is readily integrable.

21. $y''_{xx} = x^{-2} F(ay + b\ln x,\ xy'_x)$.

The transformation $z = ay + b\ln x$, $w = xy'_x$ leads to a first order equation:

$$(aw + b)w'_z = F(z, w) + w.$$

22. $y''_{xx} = yF(ax + b\ln y,\ y'_x/y)$.

The transformation $z = ax + b\ln y$, $w = y'_x/y$ leads to a first order equation:

$$(bw + a)w'_z = F(z, w) - w^2.$$

23. $y''_{xx} = F(x, y)$.

Let $F \neq \varphi(x)y + \psi(x)$, i.e., the equation is a nonlinear one. Then, its order can be lowered by one if the right-hand side of the equation has the following form:

$$F(x,y) = f^{-3/2}E\Big\{\Phi(u) + \int\Big[\frac{1}{2}ff'''_{xxx}(u+V) + f^{1/2}g''_{xx}E^{-1}\Big]\,dx\Big\}, \qquad (1)$$

where

$$E = \exp\Big(k\int f^{-1}\,dx\Big), \quad V = \int f^{-3/2}gE^{-1}\,dx, \quad u = f^{-1/2}E^{-1}y - V;$$

$\Phi = \Phi(u)$, $f = f(x)$, $g = g(x)$ are arbitrary functions, k is an arbitrary constant.

The integral in (1) may always be expressed in terms of E and V. The following cases are possible:

Case 1. For $f'''_{xxx} \neq 0$,

$$F(x,y) = f^{-3/2}E\Phi(u) + \frac{1}{4f^2}\big[2ff''_{xx} - (f'_x)^2\big]y + \frac{1}{2f^2}\big[2fg'_x - f'_xg + 2kg\big] + k^2f^{-3/2}EV.$$

Case 2. For $f = ax^2 + bx + c$, $f'_x \neq -2k$, $f'_x \neq \frac{2}{3}k$,

$$F(x,y) = f^{-3/2}E\Phi(u) + \frac{1}{2f^2}\big[2fg'_x - f'_xg + 2kg\big] + \Big(k^2 + \frac{1}{4}\Delta\Big)f^{-3/2}EV,$$

where $\Delta = 4ac - b^2$.

Case 3. For $f = \beta - 2kx$,

$$F(x,y) = f^{-2}[\Phi(y+W) + fg'_x + 2kg], \qquad W = -\int f^{-1}g\,dx.$$

Case 4. For $f = \frac{2}{3}kx + \beta$,

$$F(x,y) = \Phi(f^{-2}y - U) + f^{-2}\Big(fg'_x + \frac{2}{3}kg\Big) + \frac{8}{9}k^2U, \qquad U = \int f^{-3}g\,dx.$$

In all these cases, the transformation

$$t = \int f^{-1}dx, \qquad u = f^{-1/2}E^{-1}y - V$$

leads to an autonomous equation:

$$u''_{tt} + 2ku'_t + k^2u = \Phi(u),$$

which is reducible, with the aid of the substitution $z(u) = u'_t$, to the Abel equation:

$$zz'_u + 2kz + k^2u = \Phi(u)$$

(see Subsection 1.3.1).

With $k = 0$, the solution of the original equation for the first and second cases is as follows:

for case 1,

$$\int \frac{du}{\sqrt{2\Psi(u) + C_1}} = \pm \int \frac{dx}{f} + C_2, \qquad \text{where} \quad \Psi(u) = \int \Phi(u)\, du.$$

for case 2,

$$\int \frac{du}{\sqrt{2\Psi(u) - \frac{1}{4}\Delta u^2 + C_1}} = \pm \int \frac{dx}{ax^2 + bx + c} + C_2, \qquad \text{where} \quad \Psi(u) = \int \Phi(u)\, du.$$

Remark. The original equation can be reduced, with the aid of point transformations, to an autonomous form *only* for function F of the form (1).

2.9.7. Some Transformations

1. $\boldsymbol{y''_{xx} + x^{-3} F\Big(\dfrac{1}{x},\ \dfrac{y}{x}\Big) = 0.}$

The transformation $\xi = 1/x$, $w = y/x$ leads to the equation $w''_{\xi\xi} + F(\xi, w) = 0$.

2. $\boldsymbol{y''_{xx} = n(n+1)x^{-2}y + x^{3n} F(x^{2n+1},\ x^n y) = 0.}$

The transformation $\xi = x^{2n+1}$, $w = x^n y$ leads to the equation $(2n+1)^2 w''_{\xi\xi} = F(\xi, w)$.

3. $\boldsymbol{y''_{xx} + (ax+b)^{-3} F\Big(\dfrac{cx+d}{ax+b},\ \dfrac{y}{ax+b}\Big) = 0.}$

The transformation $\xi = \dfrac{cx+d}{ax+b}$, $w = \dfrac{y}{ax+b}$ leads to the equation

$$w''_{\xi\xi} + \Delta^{-2} F(\xi, w) = 0, \qquad \text{where} \quad \Delta = ad - bc.$$

4. $\boldsymbol{x^2 y''_{xx} + axy'_x + by + F(x,\ y) = 0.}$

The transformation $x = \xi^\nu$, $y = \xi^\mu w$, where parameters ν and μ are found from the simultaneous algebraic equations

$$2\mu + 1 + (a-1)\nu = 0, \qquad \mu^2 + (a-1)\mu\nu + b\nu^2 = 0,$$

leads to an equation of the form

$$w''_{\xi\xi} + \nu^2 \xi^{-\mu-2} F(\xi^\nu,\ \xi^\mu w) = 0.$$

5. $y''_{xx} = n(n+1)x^{-2}y + x^{3n}F(ax^{2n+1}+b,\ x^n y)$.

The transformation $\xi = ax^{2n+1}+b$, $w = x^n y$ leads to the equation

$$a^2(2n+1)^2 w''_{\xi\xi} = F(\xi, w).$$

6. $y''_{xx} = \lambda^2 y + e^{3\lambda x}F(ae^{2\lambda x}+b,\ e^{\lambda x}y)$.

The transformation $\xi = ae^{2\lambda x}+b$, $w = e^{\lambda x}y$ leads to the equation

$$w''_{\xi\xi} = (2a\lambda)^{-2}F(\xi, w).$$

7. $y''_{xx} = \lambda^2 y + \dfrac{e^{3\lambda x}}{(ce^{2\lambda x}+d)^3}F\left(\dfrac{ae^{2\lambda x}+b}{ce^{2\lambda x}+d},\ \dfrac{e^{\lambda x}y}{ce^{2\lambda x}+d}\right)$.

The transformation

$$\xi = \frac{ae^{2\lambda x}+b}{ce^{2\lambda x}+d}, \qquad w = \frac{e^{\lambda x}y}{ce^{2\lambda x}+d}$$

leads to the equation

$$w''_{\xi\xi} = (2\Delta\lambda)^{-2}F(\xi, w), \qquad \text{where} \quad \Delta = ad - bc.$$

8. $y''_{xx} = \lambda^2 y + \sinh^{-3}(\lambda x)F\left(\coth(\lambda x),\ \dfrac{y}{\sinh(\lambda x)}\right)$.

The transformation $\xi = \coth(\lambda x)$, $w = \dfrac{y}{\sinh(\lambda x)}$ leads to the equation

$$w''_{\xi\xi} = \lambda^{-2}F(\xi, w).$$

9. $y''_{xx} = \lambda^2 y + \cosh^{-3}(\lambda x)F\left(\tanh(\lambda x),\ \dfrac{y}{\cosh(\lambda x)}\right)$.

The transformation $\xi = \tanh(\lambda x)$, $w = \dfrac{y}{\cosh(\lambda x)}$ leads to the equation

$$w''_{\xi\xi} = \lambda^{-2}F(\xi, w).$$

10. $x^2 y''_{xx} + \frac{1}{4}y + \sqrt{x}\,F\left(a\ln x + b,\ \dfrac{y}{\sqrt{x}}\right) = 0$.

The transformation $\xi = a\ln x + b$, $w = \dfrac{y}{\sqrt{x}}$ leads to the equation

$$w''_{\xi\xi} + a^{-2}F(\xi, w) = 0.$$

11. $|x^2-1|^{3/2}y''_{xx} = F\left(\ln\dfrac{ax-a}{x+1},\ \dfrac{y}{\sqrt{|x^2-1|}}\right)$.

The transformation $\xi = \ln\dfrac{ax-a}{x+1}$, $w = \dfrac{y}{\sqrt{|x^2-1|}}$ leads to the equation

$$4w''_{\xi\xi} = F(\xi, w) + w.$$

12. $y''_{xx} + \lambda^2 y + \sin^{-3}(\lambda x) F\left(\cot(\lambda x), \dfrac{y}{\sin(\lambda x)}\right) = 0.$

The transformation $\xi = \cot(\lambda x)$, $w = \dfrac{y}{\sin(\lambda x)}$ leads to the equation

$$w''_{\xi\xi} + \lambda^{-2} F(\xi, w) = 0.$$

13. $y''_{xx} + \lambda^2 y + \cos^{-3}(\lambda x) F\left(\tan(\lambda x), \dfrac{y}{\cos(\lambda x)}\right) = 0.$

The transformation $\xi = \tan(\lambda x)$, $w = \dfrac{y}{\cos(\lambda x)}$ leads to the equation

$$w''_{\xi\xi} + \lambda^{-2} F(\xi, w) = 0.$$

14. $y''_{xx} + \lambda^2 y + \sin^{-3}(\lambda x + b) F\left(\dfrac{\sin(\lambda x + a)}{\sin(\lambda x + b)}, \dfrac{y}{\sin(\lambda x + b)}\right) = 0.$

The transformation $\xi = \dfrac{\sin(\lambda x + a)}{\sin(\lambda x + b)}$, $w = \dfrac{y}{\sin(\lambda x + b)}$ leads to the equation

$$w''_{\xi\xi} + [\lambda \sin(b - a)]^{-2} F(\xi, w) = 0.$$

15. $(x^2 + 1)^{3/2} y''_{xx} + F\left(\arctan x + b, \dfrac{y}{\sqrt{x^2 + 1}}\right) = 0.$

The transformation $\xi = \arctan x + b$, $w = \dfrac{y}{\sqrt{x^2 + 1}}$ leads to the equation

$$w''_{\xi\xi} + w + F(\xi, w) = 0.$$

16. $(x^2 + 1)^{3/2} y''_{xx} + F\left(\operatorname{arccot} x + b, \dfrac{y}{\sqrt{x^2 + 1}}\right) = 0.$

The transformation $\xi = \operatorname{arccot} x + b$, $w = \dfrac{y}{\sqrt{x^2 + 1}}$ leads to the equation

$$w''_{\xi\xi} + w + F(\xi, w) = 0.$$

17. $y''_{xx} + F(x, y) = 0.$

The transformation $x = \varphi(\xi)$, $y = w\sqrt{a\varphi'_\xi}$ leads to the equation

$$w''_{\xi\xi} + \left[\frac{1}{2}\frac{\varphi'''_{\xi\xi\xi}}{\varphi'_\xi} - \frac{3}{4}\left(\frac{\varphi''_{\xi\xi}}{\varphi'_\xi}\right)^2\right] w + a^{-2}(a\varphi'_\xi)^{3/2} F\left(\varphi, w\sqrt{a\varphi'_\xi}\right) = 0.$$

The sign of parameter a must coincide with that of derivative φ'_ξ.

18. $y''_{xx} + f(x, y)(y'_x)^3 + g(x, y)(y'_x)^2 = 0.$

Taking y as the independent variable, we obtain the following equation for $x = x(y)$: $x''_{yy} - g(x, y)x'_y - f(x, y) = 0.$

19. $\boldsymbol{F(x,\ y,\ y'_x,\ y''_{xx}) = 0}$.

The Legendre transformation $x = w'_t$, $y = tw'_t - w$, where $w = w(t)$, in view of the relations $y'_x = t$ and $y''_{xx} = 1/w''_{tt}$, leads to the equation

$$F\left(w'_t,\ tw'_t - w,\ t,\ \frac{1}{w''_{tt}}\right) = 0.$$

Given the solution of the original equation, the solution of the transformed equation is written in the parametric form:

$$t = y'_x, \quad w = xy'_x - y, \qquad \text{where} \quad y = y(x).$$

Chapter 3
Third Order Differential Equations

3.1. Linear Equations

3.1.1. Preliminary Comments

1. A homogeneous linear equation of the third order has the form

$$f_3(x)y'''_{xxx} + f_2(x)y''_{xx} + f_1(x)y'_x + f_0(x)y = 0. \tag{1}$$

Let $y_0 = y_0(x)$ be a nontrivial particular solution of this equation. The substitution

$$y = y_0(x) \int z(x)\,dx$$

yields a linear equation of the second order:

$$f_3 y_0 z'' + (3f_3 y_0' + f_2 y_0)z' + (3f_3 y_0'' + 2f_2 y_0' + f_1 y_0)z = 0, \tag{2}$$

where prime denotes differentiation with respect to x.

2. Let $y_1 = y_1(x)$ and $y_2 = y_2(x)$ be two nontrivial linearly-independent particular solutions of equation (1). Then, the general solution of this equation can be written in the form

$$y = C_1 y_1 + C_2 y_2 + C_3\Big(y_2 \int y_1 \psi\,dx - y_1 \int y_2 \psi\,dx\Big), \tag{3}$$

where

$$\psi = \exp\Big(-\int \frac{f_2}{f_3}\,dx\Big)\,(y_1 y_2' - y_1' y_2)^{-2}.$$

For specific equations described below in 3.1.2–3.1.8, often only particular solutions will be given, while the generalized solution can be obtained with formula (3).

3. A nonhomogeneous linear equation of the third order has the form:

$$f_3(x)y'''_{xxx} + f_2(x)y''_{xx} + f_1(x)y'_x + f_0(x)y = g(x). \tag{4}$$

Let $y_1 = y_1(x)$ and $y_2 = y_2(x)$ be two linearly-independent particular solutions of the corresponding homogeneous equation (1). Then, the general solution of equation (4) is defined by formula (3) with

$$\psi = \Delta^{-2} e^{-F}\Big(1 + \frac{1}{C_3}\int \frac{g}{f_3}\Delta e^F\,dx\Big), \qquad \text{where} \quad F = \int \frac{f_2}{f_3}\,dx, \quad \Delta = (y_1' y_2 - y_1 y_2').$$

4. The substitution

$$y = z\exp\Big(-\frac{1}{3}\int\frac{f_2}{f_3}\,dx\Big)$$

reduces equation (1) to the form wherein the second derivative is absent:

$$z''' + \Big(-\varphi_2' - \frac{1}{3}\varphi_2^2 + \varphi_1\Big)z' + \Big(-\frac{1}{3}\varphi_2'' - \frac{1}{3}\varphi_1\varphi_2 + \frac{2}{27}\varphi_2^3 + \varphi_0\Big)z = 0,$$

where $\varphi_k = f_k/f_3$ ($k = 0,\ 1,\ 2$).

3.1.2. Equations Containing Power Functions

1. $\boldsymbol{y'''_{xxx} + \lambda y = 0.}$

Solution:

$$y = \begin{cases} C_1 + C_2x + C_3x^2 & \text{if } \lambda = 0,\\ C_1e^{-kx} + e^{kx/2}\Big(C_2\cos\dfrac{kx\sqrt{3}}{2} + C_3\sin\dfrac{kx\sqrt{3}}{2}\Big) & \text{if } \lambda \neq 0,\end{cases}$$

where k is the real root of the equation $\lambda = k^3$.

2. $\boldsymbol{y'''_{xxx} + \lambda y = ax^2 + bx + c.}$

Solution: $y = w + \dfrac{1}{\lambda}(ax^2 + bx + c)$, where w is the general solution of the equation 3.1.2.1: $w'''_{xxx} + \lambda w = 0$.

3. $\boldsymbol{y'''_{xxx} = axy + b.}$

This is a special case of equation 5.1.2.4.

4. $\boldsymbol{y'''_{xxx} + (ax + b)y = 0.}$

For $a = 0$, this is an equation of the form 3.1.2.1.

For $a \neq 0$, the substitution $a\xi = ax + b$ leads to an equation of the form 3.1.2.3: $y'''_{\xi\xi\xi} + a\xi y = 0$.

5. $\boldsymbol{y'''_{xxx} + ax^3y = bx.}$

The substitution $\xi = x^2$ leads to an equation of the form 3.1.2.90: $2\xi y'''_{\xi\xi\xi} + 3y''_{\xi\xi} + \frac{1}{4}a\xi y = \frac{1}{4}b$.

6. $\boldsymbol{y'''_{xxx} + (3a^2x - a^3x^3)y = 0.}$

By integrating, we obtain a nonhomogeneous second-order linear equation:

$$y''_{xx} + axy'_x + (a^2x^2 - a)y = C\exp\Big(\frac{ax^2}{2}\Big)$$

(see 2.1.2.28 for the solution of the corresponding homogeneous equation).

7. $y'''_{xxx} = ax^n y.$

1°. For $n = -9,\ -7,\ -6,\ -9/2,\ -3,\ -3/2, 1$, and 3, see equations 3.1.2.183,3.1.2.180, 3.1.2.175, 3.1.2.185, 3.1.2.150, 3.1.2.184, 3.1.2.3, and 3.1.2.5, respectively.

2°. The transformation $x = t^{-1}$, $y = wt^{-2}$ leads to an equation of the similar form: $w'''_{xxx} = -at^{-n-6}w$.

3°. For $n \neq -3$, the transformation $\xi = x^{(n+3)/3}$, $u = x^{n/3}y$ leads to an equation of the form 3.1.2.151:

$$\xi^3 u'''_{\xi\xi\xi} + (1-\nu^2)\xi u'_\xi + (\nu^2 - 1 - a\nu^3\xi^3)u = 0, \qquad \text{where} \quad \nu = \frac{3}{n+3}.$$

8. $y'''_{xxx} + [a^3x^{3n} - 3a^2nx^{2n-1} + an(n-1)x^{n-2}]y = 0.$

Particular solution: $y_0 = \exp\Bigl(-\dfrac{ax^{n+1}}{n+1}\Bigr)$.

The substitution $y = \exp\Bigl(-\dfrac{ax^{n+1}}{n+1}\Bigr)\int z(x)\,dx$ leads to a second order equation of the form 2.1.2.44: $z''_{xx} - 3ax^n z'_x + (3a^2x^{2n} - 3anx^{n-1})z = 0$.

9. $y'''_{xxx} + aby'_x + a^2x(3 - b - ax^2)y = 0.$

By integrating, we obtain a nonhomogeneous second-order linear equation:

$$y''_{xx} + axy'_x + (a^2x^2 + ab - a)y = C\exp\Bigl(\frac{ax^2}{2}\Bigr)$$

(see 2.1.2.28 for the solution of the corresponding homogeneous equation).

10. $y'''_{xxx} + axy'_x + any = 0, \qquad n = 1,\ 2,\ 3,\ \ldots$

Solution: $y = w_x^{(n-1)}$, where w is the solution of the second order equation $w''_{xx} + axw = C$.

11. $y'''_{xxx} + axy'_x - 2ay = 0.$

The substitution $w = xy'_x - 2y$ leads to a second order equation of the form 2.1.2.2: $w''_{xx} + axw = 0$.

12. $y'''_{xxx} + axy'_x + b(ax + b^2)y = 0.$

Particular solution: $y_0 = e^{-bx}$.

The substitution $w = y'_x + by$ leads to an equation of the form 2.1.2.12: $w''_{xx} - bw'_x + (ax + b^2)w = 0$.

13. $y'''_{xxx} + axy'_x + (abx + a + b^3)y = 0.$

By integrating, we obtain a nonhomogeneous second-order linear equation:

$$y''_{xx} - by'_x + (ax + b^2)y = Ce^{-bx}$$

(see 2.1.2.103 for the solution of the corresponding homogeneous equation).

14. $y'''_{xxx} + (ax + b)y'_x + ay = 0.$

By integrating, we obtain a nonhomogeneous second-order linear equation:

$$y''_{xx} + (ax + b)y = C$$

(see 2.1.2.2 for the solution of the corresponding homogeneous equation).

15. $y'''_{xxx} + (ax + b)y'_x - ay = 0.$

Particular solution: $y_0 = ax + b$.

The transformation

$$\xi = ax + b, \quad z = \frac{y'_x}{ax + b} - \frac{ay}{(ax + b)^2}$$

leads to a second order equation of the form 2.1.2.62: $\xi z''_{\xi\xi} + 3z'_\xi + a^{-2}\xi^2 z = 0$.

16. $y'''_{xxx} + (ax + b)y'_x + 3ay = 0.$

The substitution $a\xi = ax + b$ leads to an equation of the form 3.1.2.35 with $m = 0$: $y'''_{\xi\xi\xi} + a\xi y'_\xi + 3ay = 0$.

17. $y'''_{xxx} + (2ax + b)y'_x + ay = 0.$

The substitution $a\xi = ax + \frac{1}{2}b$ leads to an equation of the form 3.1.2.37 with $n = 1$: $y'''_{\xi\xi\xi} + 2a\xi y'_\xi + ay = 0$.

18. $y'''_{xxx} + (ax - b^2)y'_x + abxy = 0.$

The substitution $w = y'_x + by$ leads to an equation of the form 2.1.2.103: $w''_{xx} - bw'_x + axw = 0$.

19. $y'''_{xxx} + (ax - b^2)y'_x + a(bx + 1)y = 0.$

By integrating, we obtain a nonhomogeneous second-order linear equation:

$$y''_{xx} - by'_x + axy = Ce^{-bx}$$

(see 2.1.2.103 for the solution of the corresponding homogeneous equation).

20. $y'''_{xxx} + (ax + b)y'_x + c(ax + b + c^2)y = 0.$

The substitution $w = y'_x + cy$ leads to a second order linear equation of the form 2.1.2.12: $w''_{xx} - cw'_x + (ax + b + c^2)w = 0$.

21. $y'''_{xxx} + (ax + b)y'_x + cx(c^2x^2 + ax + b - 3c)y = 0.$

Particular solution: $y_0 = \exp\left(-\frac{cx^2}{2}\right)$.

The substitution $y = \exp\left(-\frac{cx^2}{2}\right) \int z(x)\, dx$ leads to an equation of the form 2.1.2.28: $z''_{xx} - 3cxz'_x + (3c^2x^2 + ax + b - 3c)z = 0$.

22. $y'''_{xxx} + ax^2y'_x + axy = 0.$

This is a special case of equation 3.1.2.36 with $n = 1$.

23. $y'''_{xxx} + ax^2y'_x - 2axy = 0.$

The substitution $w = xy'_x - 2y$ leads to a second order linear equation of the form 2.1.2.7: $w''_{xx} + ax^2w = 0$.

24. $y'''_{xxx} - a^2x^2y'_x + a^2xy = 0.$

By integrating, we obtain a nonhomogeneous second-order linear equation:

$$y''_{xx} + axy'_x - ay = C\exp\left(\frac{ax^2}{2}\right)$$

(see 2.1.2.103 for the solution of the corresponding homogeneous equation).

25. $y'''_{xxx} + ax^2y'_x + b(ax^2 + b^2)y = 0.$

The substitution $w = y'_x + by$ leads to an equation of the form 2.1.2.28: $w''_{xx} - bw'_x + (ax^2 + b^2)w = 0$.

26. $y'''_{xxx} + (a-1)b^2x^2y'_x + b^2x(abx^2 + 2a + 1)y = 0.$

By integrating, we obtain a nonhomogeneous second-order linear equation:

$$y''_{xx} - bxy'_x + (ab^2x^2 + b)y = C\exp\left(-\frac{bx^2}{2}\right)$$

(see 2.1.2.28 for the solution of the corresponding homogeneous equation).

27. $y'''_{xxx} + (ax^2 + b)y'_x + 2axy = 0.$

By integrating, we obtain a nonhomogeneous second-order linear equation:

$$y''_{xx} + (ax^2 + b)y = C$$

(see 2.1.2.4 for the solution of the corresponding homogeneous equation).

28. $y'''_{xxx} + (ax^2 - b^2)y'_x + ax(2 - bx)y = 0.$

By integrating, we obtain a nonhomogeneous second-order linear equation:

$$y''_{xx} + by'_x + ax^2y = Ce^{bx}$$

(see 2.1.2.13 for the solution of the corresponding homogeneous equation).

29. $y'''_{xxx} + (ax^2 + b)y'_x + c(ax^2 + b + c^2)y = 0.$

The substitution $w = y'_x + cy$ leads to a second order equation of the form 2.1.2.13: $w''_{xx} - cw'_x + (ax^2 + b + c^2)w = 0$.

30. $y'''_{xxx} - (3b^2x^2 + a + 3b)y'_x + 2bx(b^2x^2 - a)y = 0.$

1°. Particular solutions with $a > 0$:

$$y_1 = \exp\left(\frac{bx^2}{2} + x\sqrt{a}\right), \qquad y_2 = \exp\left(\frac{bx^2}{2} - x\sqrt{a}\right).$$

2°. Particular solutions with $a < 0$:

$$y_1 = \cos\left(x\sqrt{-a}\right)\exp\left(\frac{bx^2}{2}\right), \qquad y_2 = \sin\left(x\sqrt{-a}\right)\exp\left(\frac{bx^2}{2}\right).$$

3°. Particular solutions with $a = 0$:

$$y_1 = \exp\left(\frac{bx^2}{2}\right), \qquad y_2 = x\exp\left(\frac{bx^2}{2}\right).$$

31. $y'''_{xxx} + (ax^2 + bx + c)y'_x + kx[(a + k^2)x^2 + bx + c - 3k]y = 0.$

Particular solution: $y_0 = \exp\left(-\frac{kx^2}{2}\right)$.

The substitution $y = \exp\left(-\frac{kx^2}{2}\right)\int z(x)\,dx$ leads to a second order linear equation of the form 2.1.2.28:

$$z''_{xx} - 3kxz'_x + [(a + 3k^2)x^2 + bx + c - 3k]z = 0.$$

32. $y'''_{xxx} + (ax^4 + bx)y'_x - 2(ax^3 + b)y = 0.$

This is a special case of equation 3.1.2.38 with $n = 2$.

33. $y'''_{xxx} + ax^n y'_x - 2ax^{n-1}y = 0.$

The substitution $w = xy'_x - 2y$ leads to a second order equation of the form 2.1.2.7: $w''_{xx} + ax^n w = 0$.

34. $y'''_{xxx} + ax^n y'_x + anx^{n-1}y = 0.$

By integrating, we obtain a nonhomogeneous second-order linear equation:

$$y''_{xx} + ax^n y = C$$

(see 2.1.2.7 for the solution of the corresponding homogeneous equation).

35. $y'''_{xxx} + ax^{m+1}y'_x + a(m+3)x^m y = 0.$

The substitution $x = t^{-1}$, $y = wt^{-2}$ leads to an equation of the form 3.1.2.34 with $n = -m - 5$:

$$w'''_{ttt} + at^{-m-5}w'_t - a(m+5)t^{-m-6}w = 0.$$

36. $y'''_{xxx} + ax^{2n}y'_x + anx^{2n-1}y = 0.$

Solution:

$$y = C_1 x J_\nu^2(u) + C_2 x J_\nu(u) Y_\nu(u) + C_3 x Y_\nu^2(u),$$

where $\nu = \dfrac{1}{2(n+1)}$, $u = \dfrac{\sqrt{a}}{2(n+1)}x^{n+1}$; J_ν and Y_ν are Bessel functions.

37. $y'''_{xxx} + 2ax^n y'_x + anx^{n-1}y = 0.$

Solution:

$$y = C_1 w_1^2 + C_2 w_1 w_2 + C_3 w_2^2,$$

where w_1 and w_2 form a fundamental set of solutions of a second order equation of the form 2.1.2.7: $2w''_{xx} + ax^n w = 0$.

38. $y'''_{xxx} + (ax^{2n} + bx^{n-1})y'_x - 2(ax^{2n-1} + bx^{n-2})y = 0.$

The substitution $w = xy'_x - 2y$ leads to a second order equation of the form 2.1.2.10: $w''_{xx} + (ax^{2n} + bx^{n-1})w = 0$.

39. $y'''_{xxx} + (ax^{2n} + bx^{n-1})y'_x - c[(a+c^2)x^{3n} + (b+3cn)x^{2n-1} + n(n-1)x^{n-2}]y = 0.$

Particular solution: $y_0 = \exp\left(\dfrac{cx^{n+1}}{n+1}\right)$.

The substitution $y = \exp\left(\dfrac{cx^{n+1}}{n+1}\right)\displaystyle\int z(x)\,dx$ leads to a second order linear equation of the form 2.1.2.44:

$$z''_{xx} + 3cx^n z'_x + [(a + 3c^2)x^{2n} + (b + 3cn)x^{n-1}]z = 0.$$

40. $y'''_{xxx} + 3ay''_{xx} + 3a^2 y'_x + a^3 y = 0.$

Solution: $y = e^{-ax}(C_1 + C_2 x + C_3 x^2)$.

41. $y'''_{xxx} + a_2 y''_{xx} + a_1 y'_x + a_0 y = 0.$

Denote $P(\lambda) = \lambda^3 + a_2\lambda^2 + a_1\lambda + a_0$.

1°. Let the characteristic polynomial $P(\lambda)$ be factorizable:

$$P(\lambda) = (\lambda - \lambda_1)(\lambda - \lambda_2)(\lambda - \lambda_3),$$

where λ_1, λ_2, and λ_3 are real numbers. The following cases may take place:

a) all the roots λ_k are different:

$$y = C_1 e^{\lambda_1 x} + C_2 e^{\lambda_2 x} + C_3 e^{\lambda_3 x}$$

b) $\lambda_1 = \lambda_2 \neq \lambda_3$:

$$y = (C_1 + C_2 x)e^{\lambda_1 x} + C_3 e^{\lambda_3 x}.$$

c) $\lambda_1 = \lambda_2 = \lambda_3$: see 3.1.2.40 for this case.

2°. Let $P(\lambda) = (\lambda - \lambda_1)(\lambda^2 + 2b_1\lambda + b_0)$, where $b_1^2 < b_0$. Then

$$y = C_1 e^{\lambda_1 x} + e^{-b_1 x}(C_2 \cos \mu x + C_3 \sin \mu x), \quad \mu = \sqrt{b_0 - b_1^2}.$$

42. $y'''_{xxx} + ay''_{xx} + (bx + c)y'_x + (abx + ac + b)y = 0.$

By integrating, we obtain a nonhomogeneous second-order linear equation:

$$y''_{xx} + (bx + c)y = Ce^{-ax}$$

(see 2.1.2.2 for the solution of the corresponding homogeneous equation).

43. $y'''_{xxx} + 3ay''_{xx} + 2(bx + a^2)y'_x + b(2ax + 1)y = 0.$

This is a special case of equation 3.1.2.71 with $n = 0$, $m = 1$.

44. $y'''_{xxx} + ay''_{xx} + (bx^2 + cx + d)y'_x + a(bx^2 + cx + d)y = 0.$

The substitution $w = y'_x + ay$ leads to a second order equation of the form 2.1.2.6: $w''_{xx} + (bx^2 + cx + d)w = 0$.

45. $y'''_{xxx} + ay''_{xx} + bx^n y'_x + abx^n y = 0.$

The substitution $w = y'_x + ay$ leads to a second order equation of the form 2.1.2.7: $w''_{xx} + bx^n w = 0$.

46. $y'''_{xxx} + 3axy''_{xx} + 3a^2x^2y'_x + (a^3x^3 + b)y = 0.$

The substitution $y = w\exp\Bigl(-\dfrac{ax^2}{2}\Bigr)$ leads to a constant coefficient equation: $w'''_{xxx} - 3aw'_x + bw = 0$.

47. $y'''_{xxx} + axy''_{xx} + (abx + a - b^2)y'_x + aby = 0.$

Particular solutions: $y_1 = e^{-bx}$, $y_2 = e^{-bx}\displaystyle\int \exp\Bigl(2bx - \frac{ax^2}{2}\Bigr)\,dx$.

48. $y'''_{xxx} + 3axy''_{xx} + (2a^2x^2 + a + b)y'_x + abxy = 0.$

Solution:

$$y = C_1w_1^2 + C_2w_1w_2 + C_3w_2^2,$$

where w_1 and w_2 are linearly-independent solutions of a second order equation of the form 2.1.2.25: $w''_{xx} + axw'_x + \frac{1}{4}bw = 0$.

49. $y'''_{xxx} + 3axy''_{xx} + (2a^2x^2 + 2bx + a)y'_x + b(2ax^2 + 1)y = 0.$

This is a special case of equation 3.1.2.71 with $n = 1$, $m = 1$.

50. $y'''_{xxx} + 3axy''_{xx} + 3(a^2x^2 + a)y'_x + (a^3x^3 + bx + c)y = 0.$

The substitution $y = \exp\Bigl(-\dfrac{ax^2}{2}\Bigr)w$ leads to an equation of the form 3.1.2.4: $w'''_{xxx} + [(b - 3a^2)x + c]w = 0$.

51. $y'''_{xxx} + 3axy''_{xx} + [2(a^2 + b)x^2 + a]y'_x + 2bx(ax^2 + 1)y = 0.$

This is a special case of equation 3.1.2.71 with $n = 1$, $m = 2$.

52. $y'''_{xxx} + (ax + b)y''_{xx} + (abx + a + c)y'_x + bcy = 0.$

By integrating, we obtain a nonhomogeneous second-order linear equation:

$$y''_{xx} + axy'_x + cy = Ce^{-bx}$$

(see 2.1.2.25 for the solution of the corresponding homogeneous equation).

53. $y'''_{xxx} + (abx + a + b)y''_{xx} + ab^2xy'_x - ab^2y = 0.$

Particular solutions: $y_1 = x, \quad y_2 = e^{-bx}.$

54. $y'''_{xxx} + (ax + b)y''_{xx} + [(ab + c)x + a]y'_x + c(bx + 1)y = 0.$

By integrating, we obtain a nonhomogeneous second-order linear equation:

$$y''_{xx} + axy'_x + cxy = Ce^{-bx}$$

(see 2.1.2.25 for the solution of the corresponding homogeneous equation).

55. $y'''_{xxx} + (ax + b + c)y''_{xx} + (acx + bc + s)y'_x + s(ax + b)y = 0.$

Particular solutions: $y_1 = e^{\lambda_1 x}, \quad y_2 = e^{\lambda_2 x},$ where λ_1 and λ_2 are the roots of the quadratic equation $\lambda^2 + c\lambda + s = 0$.

56. $y'''_{xxx} + (ax + b)y''_{xx} + (cx + 2a)y'_x + a[(c - ab)x^2 + b]y = 0.$

Particular solution: $y_0 = \exp\Big(-\dfrac{ax^2}{2}\Big).$

The substitution $y = \exp\Big(-\dfrac{ax^2}{2}\Big)\displaystyle\int z(x)\,dx$ leads to a second order linear equation of the form 2.1.2.28:

$$z''_{xx} + (b - 2ax)z'_x + [a^2x^2 + (c - 2ab)x - a]z = 0.$$

57. $y'''_{xxx} + (ax + b)y''_{xx} + (cx + d)y'_x + [acx^2 + (ad + bc)x + c + bd]y = 0.$

By integrating, we obtain a nonhomogeneous second-order linear equation:

$$y''_{xx} + (cx + d)y = C\exp\Big(-\frac{ax^2}{2} - bx\Big)$$

(see 2.1.2.2 for the solution of the corresponding homogeneous equation).

58. $y'''_{xxx} + (ax + b)y''_{xx} + (\alpha x^2 + \beta x + \gamma)y'_x - k[\alpha x^2 + (ak + \beta)x + k^2 + bk + \gamma]y = 0.$

The substitution $w = y'_x - ky$ leads to a second order equation of the form 2.1.2.28:

$$w''_{xx} + (ax + b + k)w'_x + [\alpha x^2 + (ak + \beta)x + k^2 + bk + \gamma]w = 0.$$

59. $y'''_{xxx} - x^2y''_{xx} + (a + b - 1)xy'_x - aby = 0.$

The following three series, converging for any x, make up a fundamental set of solutions:

$$y_1 = 1 + \sum_{n=1}^{\infty} \frac{ab(a-3)(b-3)\dots(a-3n+3)(b-3n+3)}{(3n)!}x^{3n},$$

$$y_2 = x + \sum_{n=1}^{\infty} \frac{(a-1)(b-1)(a-4)(b-4)\dots(a-3n+2)(b-3n+2)}{(3n+1)!}x^{3n+1},$$

$$y_3 = \frac{x^2}{2} + \sum_{n=1}^{\infty} \frac{(a-2)(b-2)(a-5)(b-5)\dots(a-3n+1)(b-3n+1)}{(3n+2)!}x^{3n+2}.$$

60. $y'''_{xxx} + ax^n y''_{xx} - 2ax^{n-2}y = 0.$

The substitution $w = xy'_x - 2y$ leads to a second order equation of the form 2.1.2.42: $w''_{xx} + ax^n w'_x + ax^{n-1}w = 0$.

61. $y'''_{xxx} + ax^n y''_{xx} - by'_x - abx^n y = 0.$

1°. Particular solutions with $b > 0$: $y_1 = \exp(-x\sqrt{b})$, $y_2 = \exp(x\sqrt{b})$.

2°. Particular solutions with $b < 0$: $y_1 = \cos(x\sqrt{-b})$, $y_2 = \sin(x\sqrt{-b})$.

62. $y'''_{xxx} + ax^n y''_{xx} - 2ax^{n-1}y'_x + 2ax^{n-2}y = 0.$

Particular solutions: $y_1 = x$, $y_2 = x^2$.

63. $y'''_{xxx} + ax^n y''_{xx} + bx^{n-1}y'_x - 2(a+b)x^{n-2}y = 0.$

The substitution $w = xy'_x - 2y$ leads to a second order equation of the form 2.1.2.42: $w''_{xx} + ax^n w'_x + (a+b)x^{n-1}w = 0$.

64. $y'''_{xxx} + ax^n y''_{xx} - (ax^{n-1} - bx^2)y'_x + bx(ax^{n+1} + 3)y = 0.$

Particular solutions: $y_1 = \cos\left(\dfrac{x^2\sqrt{b}}{2}\right)$, $y_2 = \sin\left(\dfrac{x^2\sqrt{b}}{2}\right)$.

65. $y'''_{xxx} + ax^n y''_{xx} + (abx^n + anx^{n-1} - b^2)y'_x + abnx^{n-1}y = 0.$

Particular solutions: $y_1 = e^{-bx}$, $y_2 = e^{-bx}\displaystyle\int \exp\left(2bx - \frac{a}{n+1}x^{n+1}\right) dx$.

66. $y'''_{xxx} + ax^n y''_{xx} + bx^m y'_x - bx^{m-1}y = 0.$

Particular solution: $y_0 = x$.

67. $y'''_{xxx} + ax^n y''_{xx} + bx^m y'_x + bx^{m-1}(ax^{n+1} + m)y = 0.$

The substitution $w = y''_{xx} + bx^m y$ leads to a first order linear equation: $w'_x + ax^n w = 0$.

68. $y'''_{xxx} + ax^n y''_{xx} - b(2ax^n + 3b)y'_x + b^2(ax^n + 2b)y = 0.$

Particular solutions: $y_1 = e^{bx}$, $y_2 = xe^{bx}$.

69. $y'''_{xxx} + ax^n y''_{xx} + (abx^n - b^2 + c)y'_x + c(ax^n - b)y = 0.$

Particular solutions: $y_1 = e^{\lambda_1 x}$, $y_2 = e^{\lambda_2 x}$, where λ_1 and λ_2 are the roots of the quadratic equation $\lambda^2 + b\lambda + c = 0$.

70. $y'''_{xxx} + ax^n y''_{xx} + (bx^m - c^2)y'_x - c(acx^n + bx^m)y = 0.$

Particular solution: $y_0 = e^{cx}$.

71. $y'''_{xxx} + 3ax^n y''_{xx} + (2a^2x^{2n} + 2bx^m + anx^{n-1})y'_x + b(2ax^{n+m} + mx^{m-1})y = 0.$

Solution:

$$y = C_1 w_1^2 + C_2 w_1 w_2 + C_3 w_2^2,$$

where w_1 and w_2 form a fundamental set of solutions of the second order linear equation $w''_{xx} + ax^n w'_x + \frac{1}{2}bx^m w = 0$.

72. $y'''_{xxx} = (x^n - a)y''_{xx} + (ax^n - b)y'_x + bx^n y.$

Particular solutions: $y_1 = e^{\lambda_1 x}$, $y_2 = e^{\lambda_2 x}$, where λ_1 and λ_2 are the roots of the quadratic equation $\lambda^2 + a\lambda + b = 0$.

73. $y'''_{xxx} + (ax^n + b)y''_{xx} + (acx^n + bc + m)y'_x + (m + c^2)(ax^n + b - c)y = 0.$

Particular solutions: $y_1 = e^{\lambda_1 x}$, $y_2 = e^{\lambda_2 x}$, where λ_1 and λ_2 are the roots of the quadratic equation $\lambda^2 + c\lambda + m + c^2 = 0$.

74. $y'''_{xxx} + (ax^n - b)y''_{xx} + cx^m y'_x - b(abx^n + cx^m)y = 0.$

Particular solution: $y_0 = e^{bx}$.

75. $y'''_{xxx} + (x^n + a)y''_{xx} + (ax^n + bx^m)y'_x + abx^m y = 0.$

Particular solution: $y_0 = e^{-ax}$.

76. $y'''_{xxx} + (ax^n + c)y''_{xx} + [acx^n + (an + b)x^{n-1}]y'_x + b[cx^{n-1} + (n-1)x^{n-2}]y = 0.$

By integrating, we obtain a nonhomogeneous second-order linear equation:

$$y''_{xx} + ax^n y'_x + bx^{n-1} y = Ce^{-cx}$$

(see 2.1.2.42 for the solution of the corresponding homogeneous equation).

77. $y'''_{xxx} + (ax^n + bx)y''_{xx} + b(ax^{n+1} + 2)y'_x + abx^n y = 0.$

Particular solutions: $y_1 = \exp\Bigl(-\dfrac{bx^2}{2}\Bigr)$, $y_2 = \exp\Bigl(-\dfrac{bx^2}{2}\Bigr)\displaystyle\int \exp\Bigl(\dfrac{bx^2}{2}\Bigr)\,dx$.

78. $y'''_{xxx} + (ax^n + bx)y''_{xx} + (abx^{n+1} + bcx + b - c^2)y'_x + c(abx^{n+1} - acx^n + b)y = 0.$

Particular solutions: $y_1 = e^{-cx}$, $y_2 = e^{-cx}\displaystyle\int \exp\Bigl(2cx - \dfrac{bx^2}{2}\Bigr)\,dx$.

79. $y'''_{xxx} + (abx^n + ax^{n-1} + b)y''_{xx} + ab^2x^n y'_x - ab^2x^{n-1}y = 0.$

Particular solutions: $y_1 = x$, $y_2 = e^{-bx}$.

80. $y'''_{xxx} + (ax^n + bx^m)y''_{xx} + cy'_x + c(ax^n + bx^m)y = 0.$

1°. Particular solutions with $c > 0$: $y_1 = \cos(x\sqrt{c})$, $y_2 = \sin(x\sqrt{c})$.

2°. Particular solutions with $c < 0$: $y_1 = \exp(-x\sqrt{-c})$, $y_2 = \exp(x\sqrt{-c})$.

81. $y'''_{xxx} + (ax^n + bx^m)y''_{xx} + (abx^{n+m} + bcx^m + bmx^{m-1} - c^2)y'_x + c(abx^{n+m} - acx^n + bmx^{m-1})y = 0.$

Particular solutions: $y_1 = e^{-cx}$, $y_2 = e^{-cx}\displaystyle\int \exp\Bigl(2cx - \dfrac{bx^{m+1}}{m+1}\Bigr)\,dx$.

82. $xy'''_{xxx} - a^2(ax + 3)y = 0.$

Particular solution: $y_0 = xe^{ax}$.

The substitution $y = xe^{ax}\int z(x)\,dx$ leads to a second order equation of the form 2.1.2.103: $xz''_{xx} + 3(ax + 1)z'_x + 3a(ax + 2)z = 0$.

83. $xy'''_{xxx} + ay'_x + b(b^2x + a)y = 0.$

The substitution $w = y'_x + by$ leads to a second order equation of the form 2.1.2.103: $xw''_{xx} - bxw'_x + (b^2x + a)w = 0$.

84. $xy'''_{xxx} + axy'_x - [b(a + b^2)x + a + 3b^2]y = 0.$

Particular solution: $y_0 = xe^{bx}$.

The substitution $y = xe^{bx} \int z(x)\,dx$ leads to a second order equation of the form 2.1.2.103: $xz''_{xx} + 3(bx + 1)z'_x + [(a + 3b^2)x + 6b]z = 0$.

85. $xy'''_{xxx} + (b - a^2x)y'_x + aby = 0.$

The substitution $w = y'_x + ay$ leads to an equation of the form 2.1.2.103: $xw''_{xx} - axw'_x + bw = 0$.

86. $xy'''_{xxx} + (ax^2 + bx)y'_x - 2(ax + b)y = 0.$

The substitution $w = xy'_x - 2y$ leads to a second order equation of the form 2.1.2.2: $w''_{xx} + (ax + b)w = 0$.

87. $xy'''_{xxx} + (ax^3 + bx)y'_x - 2(ax^2 + b)y = 0.$

The substitution $w = xy'_x - 2y$ leads to a second order equation of the form 2.1.2.4 : $w''_{xx} + (ax^2 + b)w = 0$.

88. $xy'''_{xxx} + 3y''_{xx} + axy = 0.$

The substitution $w = xy$ leads to a constant coefficient equation of the form 3.1.2.1: $w'''_{xxx} + aw = 0$.

89. $xy'''_{xxx} - 3ny''_{xx} + axy = 0, \qquad n = 0,\ 1,\ 2,\ \dots$

Solution:

$$y = x^{3n+2}\left(\frac{1}{x^2}\frac{d}{dx}\right)^n\left(\frac{w}{x^2}\right),$$

where w is the solution of the equation 3.1.2.1: $w'''_{xxx} + aw = 0$.

90. $2xy'''_{xxx} + 3y''_{xx} + axy = b, \qquad a \neq 0.$

Solution:

$$y = \sum_{\nu=1}^{4} C_\nu \int_0^{\lambda_\nu} \frac{e^{xz}\,dz}{\sqrt{2z^3 + a}},$$

where λ_1, λ_2, and λ_3 are the roots of the cubic equation $2\lambda^3 + a = 0$; $\lambda_4 = -\infty$ with $x > 0$ and $\lambda_4 = +\infty$ with $x < 0$. In addition, constants C_ν are related by the constraint $\sqrt{a}\,(C_1 + C_2 + C_3 + C_4) + b = 0$, and the integrals are taken along straight lines.

91. $xy'''_{xxx} + 3y''_{xx} + ax^2y = b.$

The substitution $w = xy$ leads to an equation of the form 3.1.2.3: $w'''_{xxx} + axw = b$.

92. $xy'''_{xxx} + 3y''_{xx} + ax^4y = bx.$

The substitution $w = xy$ leads to an equation of the form 3.1.2.5: $w'''_{xxx} + ax^3w = bx$.

93. $xy'''_{xxx} + ay''_{xx} + aby'_x + b^3xy = 0.$

The substitution $w = y'_x + by$ leads to a second order linear equation of the form 2.1.2.103: $xw''_{xx} + (a - bx)w'_x + b^2xw = 0$.

94. $xy'''_{xxx} + (a+b)y''_{xx} - xy'_x - ay = 0, \qquad a > 0, \quad b > 0.$

Solution:

$$y = \sum_{\nu=1}^{3} C_\nu \int_{\gamma_\nu}^{\beta_\nu} |t|^{a-1}|t^2 - 1|^{(b-2)/2} e^{-tx}\,dt,$$

where $\gamma_1 = -1$, $\beta_1 = \gamma_2 = 0$, $\beta_2 = 1$; for $x > 0$, $\gamma_3 = 1$ and $\beta_3 = +\infty$; for $x < 0$, $\gamma_3 = -\infty$ and $\beta_3 = -1$.

95. $xy'''_{xxx} + ay''_{xx} + (b - c^2)xy'_x - c(ac + bx)y = 0.$

The substitution $w = y'_x - cy$ leads to a second order equation of the form 2.1.2.103: $xw''_{xx} + (cx + a)w'_x + (bx + ac)w = 0$.

96. $xy'''_{xxx} + ay''_{xx} + [(c - b^2)x + ab]y'_x + c(a - bx)y = 0.$

Particular solutions: $y_1 = e^{\lambda_1 x}$, $y_2 = e^{\lambda_2 x}$, where λ_1 and λ_2 are the roots of the quadratic equation $\lambda^2 + b\lambda + c = 0$.

97. $xy'''_{xxx} + ay''_{xx} + bx^n y'_x + b(a + n - 1)x^{n-1}y = 0.$

The substitution $w = y''_{xx} + bx^{n-1}y$ leads to a first order linear equation: $xw'_x + aw = 0$.

98. $xy'''_{xxx} + (ax + b)y''_{xx} - a^2by = 0.$

The substitution $w = y'_x + ay$ leads to a second order equation of the form 2.1.2.103: $xw''_{xx} + bw'_x - abw = 0$.

99. $xy'''_{xxx} + (ax + b)y''_{xx} + cxy'_x - cy = 0.$

The substitution $w = xy'_x - y$ leads to a second order equation of the form 2.1.2.103: $xw''_{xx} + (ax + b - 1)w'_x + cxw = 0$.

100. $xy'''_{xxx} + (ax + 3)y''_{xx} + (bx + 2a)y'_x + (cx + b)y = 0.$

The substitution $w = xy$ leads to a constant coefficient equation: $w'''_{xxx} + aw''_{xx} + bw'_x + cw = 0$.

101. $xy'''_{xxx} + (ax + 3)y''_{xx} + a(bx + 2)y'_x + b[b(a - b)x + a]y = 0.$

Solution:

$$y = \frac{1}{x}\left[C_1 e^{(b-a)x} + C_2 e^{-bx/2}\cos\left(\frac{bx\sqrt{3}}{2}\right) + C_3 e^{-bx/2}\sin\left(\frac{bx\sqrt{3}}{2}\right)\right].$$

102. $xy'''_{xxx} + [a(b + 1)x + b]y''_{xx} + a^2bxy'_x - a^2by = 0.$

Particular solutions: $y_1 = x$, $y_2 = e^{-ax}$.

103. $xy'''_{xxx} - (x+2a)y''_{xx} - (x-2a-1)y'_x + (x-1)y = 0.$

Solution:

$$y = C_1 e^x + x^{a+1}[C_2 I_{a+1}(x) + C_3 K_{a+1}(x)],$$

where I_a and K_a are modified Bessel functions.

104. $2xy'''_{xxx} - 4(x+a-1)y''_{xx} + (2x+6a-5)y'_x + (1-2a)y = 0.$

Solution:

$$y = C_1 e^x + x^a e^{x/2}[C_2 I_a(x/2) + C_3 K_a(x/2)],$$

where I_a and K_a are modified Bessel functions.

105. $2xy'''_{xxx} + 3(2ax+k)y''_{xx} + 6(bx+ak)y'_x + (2cx+3bk)y = 0, \qquad k > 0.$

Solution:

$$y = \sum_{\nu=1}^{4} C_\nu \int_0^{\lambda_\nu} e^{xz}[P(z)]^{(k-2)/2}\,dz,$$

where $P(z) = z^3 + 3az^2 + 3bz + c$; λ_1, λ_2, and λ_3 are the roots of this polynomial that are assumed to be different; $\lambda_4 = -\infty$ with $x > 0$ and $\lambda_4 = +\infty$ with $x < 0$; $C_4 = -C_1 - C_2 - C_3$.

106. $xy'''_{xxx} + (ax+b)y''_{xx} + [(ac+s-c^2)x + bc]y'_x + s[(a-c)x + b]y = 0.$

Particular solutions: $y_1 = e^{\lambda_1 x}$, $y_2 = e^{\lambda_2 x}$, where λ_1 and λ_2 are the roots of the quadratic equation $\lambda^2 + c\lambda + s = 0$.

107. $xy'''_{xxx} + (ax^2+b+2)y''_{xx} - ab(b+1)y = 0.$

This is a special case of equation 3.1.2.109 with $n = 2$.

108. $xy'''_{xxx} + (ax^2+b)y''_{xx} + 4axy'_x + 2ay = 0.$

Integrating the equation twice, we arrive at a first order linear equation: $xy'_x + (ax^2+b-2)y = C_1 + C_2 x$.

109. $xy'''_{xxx} + (ax^n+b+2)y''_{xx} - ab(b+1)x^{n-2}y = 0.$

The substitution $w = xy'_x + by$ leads to a second order equation of the form 2.1.2.42: $w''_{xx} + ax^{n-1}w'_x - a(b+1)x^{n-2}w = 0$.

110. $xy'''_{xxx} + (ax^n+3)y''_{xx} + (2ax^{n-1}+bx)y'_x + b(ax^n+1)y = 0.$

Particular solutions: $y_1 = \dfrac{1}{x}\cos\bigl(x\sqrt{b}\bigr)$, $y_2 = \dfrac{1}{x}\sin\bigl(x\sqrt{b}\bigr)$.

111. $xy'''_{xxx} + (ax^{n+1}+3)y''_{xx} + a(bx+2)x^n y'_x + b(abx^{n+1} + ax^n - b^2x)y = 0.$

Particular solutions:

$$y_1 = \frac{1}{x}\exp\Bigl(-\frac{bx}{2}\Bigr)\cos\biggl(\frac{bx\sqrt{3}}{2}\biggr), \qquad y_2 = \frac{1}{x}\exp\Bigl(-\frac{bx}{2}\Bigr)\sin\biggl(\frac{bx\sqrt{3}}{2}\biggr).$$

112. $xy'''_{xxx} + (ax^n + 3)y''_{xx} + (abx^n + 2ax^{n-1} - b^2x)y'_x + b(ax^{n-1} - b)y = 0.$

Particular solutions: $y_1 = \dfrac{1}{x}, \quad y_2 = \dfrac{1}{x}e^{-bx}.$

113. $(ax + b)y'''_{xxx} + cy'_x + k(ak^2x + bk^2 + c)y = 0.$

The substitution $w = y'_x + ky$ leads to a second order equation of the form 2.1.2.103:

$$(ax + b)w''_{xx} - k(ax + b)w'_x + (ak^2x + bk^2 + c)w = 0.$$

114. $(ax + 2)y'''_{xxx} - a^3xy'_x + 2a^3y = 0.$

Particular solutions: $y_1 = x^2, \quad y_2 = e^{-ax}.$

115. $(acx + bc - a)y'''_{xxx} - c^3(ax + b)y'_x + ac^3y = 0.$

Particular solutions: $y_1 = ax + b, \quad y_2 = e^{cx}.$

116. $(ax + b)y'''_{xxx} + (cx + d)y'_x + s[(as^2 + c)x + bs^2 + d]y = 0.$

The substitution $w = y'_x + sy$ leads to a second order equation of the form 2.1.2.103:

$$(ax + b)w''_{xx} - s(ax + b)w'_x + [(as^2 + c)x + bs^2 + d]w = 0.$$

117. $(ax + b)y'''_{xxx} + [(c - ak^2)x + d - bk^2]y'_x + k(cx + d)y = 0.$

The substitution $w = y'_x + ky$ leads to a second order equation of the form 2.1.2.103:

$$(ax + b)w''_{xx} - k(ax + b)w'_x + (cx + d)w = 0.$$

118. $(ax + b)y'''_{xxx} + [b(a + 1)x + b^2 + 1]y''_{xx} + b^2xy'_x - b^2y = 0.$

Particular solutions: $y_1 = x, \quad y_2 = e^{-bx}.$

119. $(ax + b)y'''_{xxx} + k(ax + b)y''_{xx} + (cx + d)y'_x + k(cx + d)y = 0.$

The substitution $w = y'_x + ky$ leads to a second order equation of the form 2.1.2.103: $(ax + b)w''_{xx} + (cx + d)w = 0.$

120. $(ax + b)y'''_{xxx} + (cx + d)y''_{xx} + [(a\lambda + c\mu)x + b\lambda + d\mu]y'_x + (\lambda + \mu^2)[(c - a\mu)x + d - b\mu]y = 0.$

Particular solutions: $y_1 = \exp(s_1x), \quad y_2 = \exp(s_2x),$ where s_1 and s_2 are the roots of the quadratic equation $s^2 + \mu s + \lambda + \mu^2 = 0.$

121. $(ax + b)y'''_{xxx} + (cx + d)y''_{xx} - k[(3ak + 2c)x + 3bk + 2d]y'_x + k^2[(2ak + c)x + 2bk + d]y = 0.$

Particular solutions: $y_1 = e^{kx}, \quad y_2 = xe^{kx}.$

122. $(ax + b)y'''_{xxx} + (cx + d)y''_{xx} + sx(ax + b)y'_x + s[cx^2 + (a + d)x + b]y = 0.$

The substitution $w = y''_{xx} + sxy$ leads to a first order linear equation: $(ax + b)w'_x + (cx + d)w = 0.$

123. $(1-x)y'''_{xxx} + x(ax-2a+1)y''_{xx} + (-ax^2+2a-1)y'_x + 2a(x-1)y = 0.$

Particular solutions: $y_1 = x^2$, $y_2 = e^x$.

124. $(ax+b)y'''_{xxx} - (a^3x^3 - 3a^2x + b^3)y'_x + abx(a^2x^2 - 3a - b^2)y = 0.$

Particular solutions: $y_1 = e^{bx}$, $y_2 = \exp\Bigl(-\dfrac{ax^2}{2}\Bigr)$.

125. $(ax+b)y'''_{xxx} + (cx+d)y''_{xx} + sx^n(ax+b)y'_x + sx^{n-1}[cx^2 + (an+d)x + bn]y = 0.$

The substitution $w = y''_{xx} + sx^n y$ leads to a first order linear equation: $(ax+b)w'_x + (cx+d)w = 0$.

126. $(ax-1)y'''_{xxx} + x(abx^{n+1} - 2bx^n - a^2)y''_{xx}$
$+ (2bx^n - a^2bx^{n+2} + a^2)y'_x + 2ab(ax-1)x^n y = 0.$

Particular solutions: $y_1 = x^2$, $y_2 = e^{ax}$.

127. $x^2y'''_{xxx} - 6y'_x + ax^2y + 2bx = 0.$

The substitution $y = x^2 w$ leads to the equation 3.1.2.159 wherein w should be substituted for y.

128. $x^2y'''_{xxx} + (ax^2 + bx - m^2 - m)y'_x + (m-1)(ax+b)y = 0.$

The substitution $w = xy'_x + (m-1)y$ leads to a second order equation of the form 2.1.2.103: $xw''_{xx} - (m+1)w'_x + (ax+b)w = 0$.

129. $x^2y'''_{xxx} + (ax^2 + bx + c)y'_x - k[(a+k^2)x^2 + bx + c]y = 0.$

The substitution $w = y'_x - ky$ leads to a second order equation of the form 2.1.2.130:

$$x^2w''_{xx} + kx^2w'_x + [(a+k^2)x^2 + bx + c]w = 0.$$

130. $x^2y'''_{xxx} + (ax^n - b^2 - b)y'_x + a(b-1)x^{n-1}y = 0.$

The substitution $w = xy'_x + (b-1)y$ leads to a second order equation of the form 2.1.2.62: $xw''_{xx} - (b+1)w'_x + ax^{n-1}w = 0$.

131. $x^2y'''_{xxx} + 3xy''_{xx} - 3y'_x + ax^2y + b = 0.$

Solution: $y = \dfrac{d}{dx}\Bigl(\dfrac{w}{x}\Bigr)$, where function $w = w(x)$ satisfies a constant coefficient equation of the form 3.1.2.2: $w'''_{xxx} + aw = b$.

132. $x^2y'''_{xxx} + 6xy''_{xx} + 6y'_x + ax^2y = b.$

The substitution $w = x^2y$ leads to a constant coefficient equation of the form 3.1.2.2: $w'''_{xxx} + aw = b$.

133. $x^2y'''_{xxx} - 3(n+m)xy''_{xx} + 3n(3m+1)y'_x - x^2y = 0, \qquad m, n = 1,\ 2,\ 3,\ \ldots$

Solution:

$$y = \prod_{\mu=0}^{n-1}(\delta - 3\mu - 1)\prod_{\nu=0}^{m-1}(\delta - 3\nu - 2)\sum_{k=1}^{3} C_k e^{\omega_k x},$$

where $\delta = x\dfrac{d}{dx}$, ω_k are three roots of the cubic equation $\omega^3 = 1$.

134. $x^2 y'''_{xxx} + 6xy''_{xx} + 6y'_x + ax^3 y = b.$

The substitution $w = x^2 y$ leads to an equation of the form 3.1.2.3: $w'''_{xxx} + axw = b$.

135. $x^2 y'''_{xxx} - 2(n+1)xy''_{xx} - (ax^2 - 6n)y'_x + 2axy = 0, \qquad n = 1,\ 2,\ 3,\ \dots$

Solution:

$$y = \begin{cases} C_1 + C_2 x^4 + C_3 x^{2n+1} & \text{if } a = 0, \\ C_1(ax^2 - 4n + 2) + C_2 e^{x\sqrt{a}} P(x) + C_3 e^{-x\sqrt{a}} Q(x) & \text{if } a \neq 0, \end{cases}$$

where P and Q are some polynomials of the degree $\leq 2n+2$.

136. $x^2 y'''_{xxx} + 3xy''_{xx} + (4a^2 x^{2a} + 1 - 4a^2 b^2)y'_x + 4a^3 x^{2a-1} y = 0.$

Solution:

$$y = C_1 J_b^2(x^a) + C_2 J_b(x^a) Y_b(x^a) + C_3 Y_b^2(x^a),$$

where J_b and Y_b are Bessel functions.

137. $x^2 y'''_{xxx} + ax^2 y''_{xx} + (bx + c)y'_x + a(bx + c)y = 0.$

The substitution $w = y'_x + ay$ leads to a second order equation of the form 2.1.2.106: $x^2 w''_{xx} + (bx + c)w = 0$.

138. $x^2 y'''_{xxx} + ax^2 y''_{xx} + (bx^n + c)y'_x + a(bx^n + c)y = 0.$

The substitution $w = y'_x + ay$ leads to a second order equation of the form 2.1.2.113: $x^2 w''_{xx} + (bx^n + c)w = 0$.

139. $x^2 y'''_{xxx} - (x + a)xy''_{xx} + a(2x + 1)y'_x - a(x + 1)y = 0.$

Solution:

$$y = C_1 e^x + x^{(a+1)/2} \bigl[C_2 J_{a+1}(2\sqrt{ax}\,) + C_3 Y_{a+1}(2\sqrt{ax}\,)\bigr],$$

where J_a and Y_a are Bessel functions.

140. $x^2 y'''_{xxx} - (x^2 - 2x)y''_{xx} - (x^2 + a^2 - \frac{1}{4})y'_x + (x^2 - 2x + a^2 - \frac{1}{4})y = 0.$

Solution:

$$y = C_1 e^x + \sqrt{x}\,\bigl[C_2 I_a(x) + C_3 K_a(x)\bigr],$$

where I_a and K_a are modified Bessel functions.

141. $x^2 y'''_{xxx} - 2x(x - 1)y''_{xx} + (x^2 - 2x + \frac{1}{4} - a^2)y'_x + (a^2 - \frac{1}{4})y = 0.$

Solution:

$$y = C_1 e^x + \sqrt{x}\, e^{x/2} \bigl[C_2 I_a(x/2) + C_3 K_a(x/2)\bigr],$$

where I_a and K_a are modified Bessel functions.

142. $x^2 y'''_{xxx} - 3(x - a)xy''_{xx} + [2x^2 + 4(b - a)x + a(2a - 1)]y'_x - 2b(2x - 2a + 1)y = 0.$

Solution:

$$y = C_1 w_1^2 + C_2 w_1 w_2 + C_3 w_2^2,$$

where w_1 and w_2 form a fundamental set of solutions of a second order equation of the form 2.1.2.103: $xw''_{xx} + (a - x)w'_x + bw = 0$.

143. $x^2y'''_{xxx}+x[(a+c)x+b]y''_{xx}+[(ac+\alpha)x^2+(bc+\beta)x+\gamma]y'_x + c(\alpha x^2+\beta x+\gamma)y=0.$

The substitution $w=y'_x+cy$ leads to an equation of the form 2.1.2.141 with $n=1$: $x^2w''_{xx}+x(ax+b)w'_x+(\alpha x^2+\beta x+\gamma)w=0$.

144. $x^2y'''_{xxx}+(ax^{n+1}-b^2-b)y'_x+a(b-1)x^ny=0.$

The substitution $w=xy'_x+(b-1)y$ leads to a second order equation of the form 2.1.2.62: $xw''_{xx}-(b+1)w'_x+ax^nw=0$.

145. $x^2y'''_{xxx}-3ax^{n+1}(n+ax^{n+1})y'_x+ax^n(n-n^2+2a^2x^{2n+2})y=0.$

1°. Particular solutions with $n\neq-1$: $y_1=\exp\Bigl(\dfrac{ax^{n+1}}{n+1}\Bigr)$, $y_2=x\exp\Bigl(\dfrac{ax^{n+1}}{n+1}\Bigr)$.

2°. Particular solutions with $n=-1$: $y_1=x^a$, $y_2=x^{a+1}$.

146. $x^2y'''_{xxx}+(ax^{n+1}+bx)y''_{xx}+[a(b-2)x^n+c]y'_x+a(c-b+2)x^{n-1}y=0.$

Particular solutions: $y_1=x^{m_1}$, $y_2=x^{m_2}$, where m_1 and m_2 are the roots of the quadratic equation: $m^2+(b-3)m+c-b+2=0$.

147. $x(ax+b)y'''_{xxx}+x(cx+d)y'_x-2(cx+d)y=0.$

The substitution $w=xy'_x-2y$ leads to a second order equation of the form 2.1.2.103: $(ax+b)w''_{xx}+(cx+d)w=0$.

148. $2x(x-1)y'''_{xxx}+3(2x-1)y''_{xx}+(2ax+b)y'_x+ay=0.$

Solution:

$$y=C_1w_1^2+C_2w_1w_2+C_3w_2^2,$$

where w_1 and w_2 form a fundamental set of solutions of the equation

$$2x(x-1)w''_{xx}+(2x-1)w'_x+\Bigl(\frac{a}{2}x+\frac{b}{4}-\frac{1}{2}\Bigr)w=0,$$

which is reduced, by means of the substitution $x=\cos^2\xi$, to the Mathieu equation 2.1.6.4: $2w''_{\xi\xi}=(a+b-2+a\cos 2\xi)w$.

149. $(a_2x^2+a_1x+a_0)y'''_{xxx}+(b_1x+b_0)y''_{xx}+(c_1x+c_0)y'_x-mc_1y=0,\qquad c_1\neq0,$ where m is a positive integer.

A solution of this equation is a polynomial of the degree m, which can presented as follows:

$$y=\sum_{k=0}^{m}\Bigl(-\frac{1}{c_1}\Bigr)^k\bigl\{x^mIx^{-m-1}[(ax^2+a_1x+a_0)D^3+(b_1x+b_0)D^2+c_0D]\bigr\}^kx^m,$$

where $D=\dfrac{d}{dx}$, $Ix^\nu=\dfrac{x^{\nu+1}}{\nu+1}$ with $\nu\neq-1$.

150. $x^3 y'''_{xxx} = a(a^2 - 1)y.$

This is a special case of equation 3.1.2.161.

Solution:

$$y = x(C_1 x^{n_1} + C_2 x^{n_2} + C_3 x^a),$$

where n_1 and n_2 are the roots of the quadratic equation $n^2 + an + a^2 - 1 = 0$.

151. $x^3 y'''_{xxx} + (1 - a^2)xy'_x + (bx^3 + a^2 - 1)y = 0.$

For $a = \pm 1$, we have a constant coefficient equation of the form 3.1.2.1. For $b = 0$, we obtain the Euler equation 3.1.2.161.

If $b \neq 0$ and a is a positive integer greater than 1, then

$$y = x^{1-a} \sum_{k=1}^{3} C_k \exp(-\lambda_k x) P_k(x),$$

where λ_1, λ_2, λ_3 are the roots of the cubic equation $\lambda^3 = b$ and $P_k(x)$ are polynomials of the degree $\leq 3(a - 1)$.

Denote the solution of the original equation for arbitrary (including complex) a by y_a. Then, the recurrence relation holds

$$y_{a+3} = by_a + (2a + 3)x^{-1} y''_a - (a + 1)(2a + 3)(x^{-2} y'_a - x^{-3} y_a). \qquad (1)$$

Since $y_{\pm 1} = e^{-\lambda x}$, corresponding to three values of λ which satisfy the condition $\lambda^3 = b$, make up a fundamental set of solutions, then formula (1) makes it possible to find all y_n for any integer values of n not divisible by 3. In particular, $y_2 = (x^{-1} + \lambda)e^{-\lambda x}$ $(\lambda^3 = b)$.

152. $x^3 y'''_{xxx} + (4x^3 + ax)y'_x - ay = 0.$

Solution:

$$y = C_1 x J_\nu^2(x) + C_2 x J_\nu(x) Y_\nu(x) + C_3 x Y_\nu^2(x),$$

where J_ν and Y_ν are Bessel functions; $4\nu^2 = 1 - a$.

153. $x^3 y'''_{xxx} + x[ax^2 + 3b(1 - b)]y'_x + 2b(ax^2 + b^2 - 1)y = 0.$

1°. Particular solutions with $a > 0$: $y_1 = x^b \sin(x\sqrt{a})$, $y_2 = x^b \cos(x\sqrt{a})$.

2°. Particular solutions with $a < 0$: $y_1 = x^b \exp(-x\sqrt{-a})$, $y_2 = x^b \exp(x\sqrt{-a})$.

3°. Particular solutions with $a = 0$: $y_1 = x^b$, $y_2 = x^{b+1}$.

154. $x^3 y'''_{xxx} + x(ax^2 + bx + c)y'_x + (k - 1)(ax^2 + bx + c + k^2 + k)y = 0.$

The substitution $w = xy'_x + (k - 1)y$ leads to a second order linear equation of the form 2.1.2.126:

$$x^2 w''_{xx} - (k + 1)xw'_x + (ax^2 + bx + c + k^2 + k)w = 0.$$

155. $x^3 y'''_{xxx} + ax^n y'_x + (b - 1)(ax^{n-1} + b^2 + b)y = 0.$

The substitution $w = xy'_x + (b - 1)y$ leads to a second order linear equation of the form 2.1.2.127:

$$x^2 w''_{xx} - (b + 1)xw'_x + (ax^{n-1} + b^2 + b)w = 0.$$

156. $x^3y'''_{xxx} + x(ax^n + b)y'_x - 2(ax^n + b)y = 0.$

The substitution $w = xy'_x - 2y$ leads to a second order linear equation of the form 2.1.2.113: $x^2w''_{xx} + (ax^n + b)w = 0$.

157. $x^3y'''_{xxx} + x(ax^n + b - c)y'_x + (c - 1)(ax^n + b + c^2)y = 0.$

The substitution $w = xy'_x + (c - 1)y$ leads to a second order linear equation of the form 2.1.2.127:

$$x^2w''_{xx} - (c+1)xw'_x + (ax^n + b + c^2)w = 0.$$

158. $x^3y'''_{xxx} + (ax^{2n} + 1 - n^2)xy'_x + [bx^{3n} + a(n - 1)x^{2n} + n^2 - 1]y = 0.$

The transformation $\xi = \dfrac{1}{n}x^n$, $z = x^{n-1}y$ leads to a constant coefficient equation: $z'''_{\xi\xi\xi} + az'_\xi + bz = 0$.

159. $x^3y'''_{xxx} + 6x^2y''_{xx} + (ax^3 - 12)y + 2b = 0.$

Solution: $y = \dfrac{d}{dx}\Bigl(\dfrac{w}{x^2}\Bigr)$, where $w = w(x)$ satisfies a constant coefficient equation of the form 3.1.2.2: $w'''_{xxx} + aw = b$.

160. $x^3y'''_{xxx} + ax^2y''_{xx} + bxy'_x + (a - 2)by = 0.$

This is a special case of equation 3.1.2.161.

Solution:

$$y = C_1x^{2-a} + C_2x^{n_1} + C_3x^{n_2},$$

where n_1 and n_2 are the roots of the quadratic equation $n^2 - n + b = 0$.

161. $x^3y'''_{xxx} + ax^2y''_{xx} + bxy'_x + cy = 0.$

The Euler equation.

The substitution $t = \ln|x|$ leads to a constant coefficient equation of the form 3.1.2.41: $y'''_{ttt} + (a - 3)y''_{tt} + (2 - a + b)y'_t + cy = 0$.

162. $x^3y'''_{xxx} + 3ax^2y''_{xx} + 3a(a - 1)xy'_x + [bx^3 + a(a - 1)(a - 2)]y = 0.$

The substitution $w = x^a y$ leads to a constant coefficient equation of the form 3.1.2.1: $w'''_{xxx} + bw = 0$.

163. $x^3y'''_{xxx} + 3ax^2y''_{xx} + 3a(a - 1)xy'_x + [bx^n + a(a - 1)(a - 2)]y = 0.$

The substitution $w = x^a y$ leads to an equation of the form 3.1.2.7: $w'''_{xxx} + bx^{n-3}w = 0$.

164. $x^3y'''_{xxx} + 3(1 - a)x^2y''_{xx} + x[4b^2c^2x^{2c} + 1 - 4\nu^2c^2 + 3a(a - 1)]y'_x$
$+ [4b^2c^2(c - a)x^{2c} + a(4\nu^2c^2 - a^2)]y = 0.$

Solution:

$$y = C_1x^aJ^2_\nu(u) + C_2x^aJ_\nu(u)Y_\nu(u) + C_3x^aY^2_\nu(u),$$

where $u = bx^c$; J_ν and Y_ν are Bessel functions.

165. $x^3y'''_{xxx} + (ax^2+b)y''_{xx} + 2(2a-9)xy'_x + 2(a-6)y = 0.$

Integrating the equation twice, we arrive at a first order linear equation: $x^3y'_x + [(a-6)x^2+b]y = C_1 + C_2x$.

166. $x^3y'''_{xxx} + x^2(ax+b)y''_{xx} + cxy'_x + c(ax+b-2)y = 0.$

Particular solutions: $y_1 = x^{n_1}$, $y_2 = x^{n_2}$, where n_1 and n_2 are the roots of the quadratic equation $n^2 - n + c = 0$.

167. $x^3y'''_{xxx} + x^2(2ax+b)y''_{xx} + x(a^2x^2+2abx+c)y'_x + (a^2bx^2+bc-2c)y = 0.$

Particular solutions: $y_1 = e^{-ax}x^{n_1}$, $y_2 = e^{-ax}x^{n_2}$, where n_1 and n_2 are the roots of the quadratic equation $n^2 - n + c = 0$.

168. $x^3y'''_{xxx} + ax^ny''_{xx} + bxy'_x + b(ax^{n-2}-2)y = 0.$

Particular solutions: $y_1 = x^{m_1}$, $y_2 = x^{m_2}$, where m_1 and m_2 are the roots of the quadratic equation $m^2 - m + b = 0$.

169. $x^3y'''_{xxx} + x^2(ax^n+b)y''_{xx} + x(ax^n+b-1)y'_x + (ax^n+b-3)y = 0.$

Particular solutions: $y_1 = \cos(\ln x)$, $y_2 = \sin(\ln x)$.

170. $x^3y'''_{xxx} + x^2(ax^n+b+c+1)y''_{xx} + x[\alpha x^{2n} + (ac+\beta)x^n + \gamma + bc]y'_x$
$+ (c-1)(\alpha x^{2n} + \beta x^n + \gamma)y = 0.$

The substitution $w = xy'_x + (c-1)y$ leads to a second order equation of the form 2.1.2.141: $x^2w''_{xx} + x(ax^n+b)w'_x + (\alpha x^{2n} + \beta x^n + \gamma)w = 0$.

171. $x^2(ax+b)y'''_{xxx} + (cx-bm^2-bm)y'_x + (m-1)(c+am^2+am)y = 0.$

The substitution $w = xy'_x + (m-1)y$ leads to a hypergeometric equation of the form 2.1.2.159: $x(ax+b)w''_{xx} - (m+1)(ax+b)w'_x + (c+am^2+am)w = 0$.

172. $x(ax^2+bx+c)y'''_{xxx} + xy'_x - 2y = 0.$

The substitution $w = xy'_x - 2y$ leads to a second order equation of the form 2.1.2.166: $(ax^2+bx+c)w''_{xx} + w = 0$.

173. $(ax+b)x^3y'''_{xxx} + (cx+d)x^2y''_{xx} + s(ax+b)xy'_x + s[(c-2a)x+d-2b]y = 0.$

Particular solutions: $y_1 = x^{n_1}$, $y_2 = x^{n_2}$, where n_1 and n_2 are the roots of the quadratic equation $n^2 - n + s = 0$.

174. $x^5y'''_{xxx} = a(xy'_x - 2y).$

Solution:

$$y = \begin{cases} x^2\left[C_1 + C_2\exp\left(\dfrac{\sqrt{a}}{x}\right) + C_3\exp\left(-\dfrac{\sqrt{a}}{x}\right)\right] & \text{if } a > 0, \\ x^2\left[C_1 + C_2\cos\left(\dfrac{\sqrt{-a}}{x}\right) + C_3\sin\left(\dfrac{\sqrt{-a}}{x}\right)\right] & \text{if } a < 0. \end{cases}$$

175. $x^6 y'''_{xxx} = ay + bx^2.$

The transformation $x = t^{-1}$, $y = wt^{-2}$ leads to a constant coefficient equation of the form 3.1.2.2: $w'''_{ttt} + aw + b = 0$.

176. $x^6 y'''_{xxx} + ax^2 y'_x + (b - 2ax)y = 0.$

The transformation $x = t^{-1}$, $y = wt^{-2}$ leads to a constant coefficient equation: $w'''_{ttt} + aw'_t - bw = 0$.

177. $x^6 y'''_{xxx} + 6x^5 y''_{xx} - ay + 2bx = 0.$

The substitution $x = t^{-1}$ leads to an equation of the form 3.1.2.127: $t^2 y'''_{ttt} - 6y'_t + at^2 y - 2bt = 0$.

178. $(x - a)^3 (x - b)^3 y'''_{xxx} - cy = 0, \qquad a \neq b.$

The transformation

$$t = \ln\left|\frac{x - a}{x - b}\right|, \qquad w = \frac{y}{(x - b)^2}$$

leads to a constant coefficient equation: $(a - b)^3 (w'''_{ttt} - 3w''_{tt} + 2w'_t) - cw = 0$.

179. $(ax^2 + bx + c)^3 y'''_{xxx} = ky.$

The transformation

$$\xi = \int \frac{dx}{ax^2 + bx + c}, \qquad w = \frac{y}{ax^2 + bx + c}$$

leads to a constant coefficient equation: $w'''_{\xi\xi\xi} + (4ac - b^2)w'_\xi = kw$.

180. $x^7 y'''_{xxx} = ay + bx^3.$

The transformation $x = t^{-1}$, $y = wt^{-2}$ leads to an equation of the form 3.1.2.3: $w'''_{ttt} + atw + b = 0$.

181. $x^7 y'''_{xxx} + (ax + b)y = 0.$

The transformation $x = t^{-1}$, $y = wt^{-2}$ leads to an equation of the form 3.1.2.4: $w'''_{ttt} - (bt + a)w = 0$.

182. $x^9 y'''_{xxx} + (a^3 - 3a^2 x^2)y = 0.$

The transformation $x = t^{-1}$, $y = wt^{-2}$ leads to an equation of the form 3.1.2.6: $w'''_{ttt} + (3a^2 t - a^3 t^3)w = 0$.

183. $x^9 y'''_{xxx} = ay.$

The transformation $x = t^{-1}$, $y = wt^{-2}$ leads to an equation of the form 3.1.2.5: $w'''_{ttt} + at^3 w = 0$.

184. $x^{3/2} y'''_{xxx} = ay.$

This is a special case of equation 5.1.2.25 with $n = 1$.

185. $x^{9/2}y'''_{xxx} = ay.$

This is a special case of equation 5.1.2.26 with $n = 1$.

186. $x^2(x^n + a)y'''_{xxx} + x(bx^{m+1} + 2nx^n + cx)y''_{xx}$
$+ [2bmx^{m+1} + n(n-1)x^n]y'_x + bm(m-1)x^m y = 0.$

The twofold integration yields a first order linear equation:

$$(x^n + a)y'_x + (bx^m + c)y = C_1 + C_2 x.$$

3.1.3. Equations Containing Exponential Functions

1. $y'''_{xxx} - ae^{\lambda x}(a^2e^{2\lambda x} + 3a\lambda e^{\lambda x} + \lambda^2)y = 0.$

Particular solution: $y_0 = \exp\Bigl(\dfrac{a}{\lambda}e^{\lambda x}\Bigr)$.

The substitution $y = \exp\Bigl(\dfrac{a}{\lambda}e^{\lambda x}\Bigr)\displaystyle\int z(x)\,dx$ leads to a second order equation of the form 2.1.3.29: $z''_{xx} + 3ae^{\lambda x}z'_x + (3a^2e^{2\lambda x} + 3a\lambda e^{\lambda x})z = 0$.

2. $y'''_{xxx} + ae^{\lambda x}y'_x + a\lambda e^{\lambda x}y = be^{\mu x}.$

By integrating, we obtain a nonhomogeneous second-order linear equation:

$$y''_{xx} + ae^{\lambda x}y = \frac{b}{\mu}e^{\mu x} + C$$

(see 2.1.3.1 for the solution of the corresponding homogeneous equation).

3. $y'''_{xxx} + ae^{\lambda x}y'_x + b(ae^{\lambda x} + b^2)y = 0.$

The substitution $w = y'_x + by$ leads to a second order equation of the form 2.1.3.10: $w''_{xx} - bw'_x + (ae^{\lambda x} + b^2)w = 0$.

4. $y'''_{xxx} + ae^{\lambda x}y'_x + bx(ae^{\lambda x} + b^2x^2 - 3b)y = 0.$

Particular solution: $y_0 = \exp\Bigl(-\dfrac{bx^2}{2}\Bigr)$.

5. $y'''_{xxx} + ae^{\lambda x}y'_x + [a(\lambda - b)e^{\lambda x} - b^3]y = 0.$

By integrating, we obtain a nonhomogeneous second-order linear equation:

$$y''_{xx} + by'_x + (ae^{\lambda x} + b^2)y = Ce^{bx}$$

(see 2.1.3.10 for the solution of the corresponding homogeneous equation).

6. $y'''_{xxx} + (ae^{\lambda x} - b^2)y'_x + abe^{\lambda x}y = 0.$

The substitution $w = y'_x + by$ leads to a second order equation of the form 2.1.3.10: $w''_{xx} - bw'_x + ae^{\lambda x}w = 0$.

7. $y'''_{xxx} + (ae^{\lambda x} - b^2)y'_x + a(\lambda - b)e^{\lambda x}y = 0.$

By integrating, we obtain a nonhomogeneous second-order linear equation:

$$y''_{xx} + by'_x + ae^{\lambda x}y = Ce^{bx}$$

(see 2.1.3.10 for the solution of the corresponding homogeneous equation).

8. $y'''_{xxx} + (ae^{\lambda x} + b)y'_x + c(ae^{\lambda x} + b + c^2)y = 0.$

The substitution $w = y'_x + cy$ leads to a second order equation of the form 2.1.3.10: $w''_{xx} - cw'_x + (ae^{\lambda x} + b + c^2)w = 0$.

9. $y'''_{xxx} + (ax + b)e^{\lambda x}y'_x - ae^{\lambda x}y = 0.$

Particular solution: $y_0 = ax + b$.

10. $y'''_{xxx} + (ae^{2\lambda x} + be^{\lambda x})y'_x - c(ae^{2\lambda x} + be^{\lambda x} + c^2)y = 0.$

The substitution $w = y'_x - cy$ leads to a second order equation of the form 2.1.3.29: $w''_{xx} + cw'_x + (ae^{2\lambda x} + be^{\lambda x} + c^2)w = 0$.

11. $y'''_{xxx} - 3ae^{\lambda x}(ae^{\lambda x} + \lambda)y'_x + ae^{\lambda x}(2a^2e^{2\lambda x} - \lambda^2)y = 0.$

Particular solutions: $y_1 = \exp\left(\frac{a}{\lambda}e^{\lambda x}\right)$, $y_2 = x\exp\left(\frac{a}{\lambda}e^{\lambda x}\right)$.

12. $y'''_{xxx} - (3a^2e^{2\lambda x} + 3a\lambda e^{\lambda x} + b)y'_x + ae^{\lambda x}(2a^2e^{2\lambda x} - 2b - \lambda^2)y = 0.$

1°. Particular solutions with $b > 0$:

$$y_1 = \exp\left(\frac{a}{\lambda}e^{\lambda x} - x\sqrt{b}\right), \quad y_2 = \exp\left(\frac{a}{\lambda}e^{\lambda x} + x\sqrt{b}\right).$$

2°. Particular solutions with $b < 0$:

$$y_1 = \cos\left(x\sqrt{-b}\right)\exp\left(\frac{a}{\lambda}e^{\lambda x}\right), \quad y_2 = \sin\left(x\sqrt{-b}\right)\exp\left(\frac{a}{\lambda}e^{\lambda x}\right).$$

13. $y'''_{xxx} + ay''_{xx} + be^{\lambda x}y'_x + abe^{\lambda x}y = 0.$

The substitution $w = y'_x + ay$ leads to a second order equation of the form 2.1.3.1: $w''_{xx} + be^{\lambda x}w = 0$.

14. $y'''_{xxx} + ay''_{xx} + (be^{\lambda x} + c)y'_x + [b(a + \lambda)e^{\lambda x} + ac]y = 0.$

By integrating, we obtain a nonhomogeneous second-order linear equation:

$$y''_{xx} + (be^{\lambda x} + c)y = Ce^{-ax}$$

(see 2.1.3.2 for the solution of the corresponding homogeneous equation).

15. $y'''_{xxx} + ay''_{xx} + (be^{2\lambda x} + ce^{\lambda x})y'_x + a(be^{2\lambda x} + ce^{\lambda x})y = 0.$

The substitution $w = y'_x + ay$ leads to a second order equation of the form 2.1.3.29: $w''_{xx} + (be^{2\lambda x} + ce^{\lambda x})w = 0$.

16. $\boldsymbol{y'''_{xxx} + ae^{\lambda x}y''_{xx} - b^2(ae^{\lambda x} + b)y = 0.}$

The substitution $w = y'_x - by$ leads to a second order equation of the form 2.1.3.29: $w''_{xx} + (ae^{\lambda x} + b)w'_x + b(ae^{\lambda x} + b)w = 0$.

17. $\boldsymbol{y'''_{xxx} + ae^{\lambda x}y''_{xx} - by'_x - abe^{\lambda x}y = 0.}$

1°. Particular solutions with $b > 0$: $y_1 = \exp(-x\sqrt{b})$, $y_2 = \exp(x\sqrt{b})$.

2°. Particular solutions with $b < 0$: $y_1 = \cos(x\sqrt{-b})$, $y_2 = \sin(x\sqrt{-b})$.

18. $\boldsymbol{y'''_{xxx} + ae^{\lambda x}y''_{xx} + abe^{\lambda x}y'_x + b^3y = 0.}$

The substitution $w = y'_x + by$ leads to a second order equation of the form 2.1.3.29: $w''_{xx} + (ae^{\lambda x} - b)w'_x + b^2w = 0$.

19. $\boldsymbol{y'''_{xxx} + ae^{\lambda x}y''_{xx} - b(2ae^{\lambda x} + 3b)y'_x + b^2(ae^{\lambda x} + 2b)y = 0.}$

Particular solutions: $y_1 = e^{bx}$, $y_2 = xe^{bx}$.

20. $\boldsymbol{y'''_{xxx} + ae^{\lambda x}y''_{xx} + (be^{\mu x} - c^2)y'_x - c(ace^{\lambda x} + be^{\mu x})y = 0.}$

Particular solution: $y_0 = e^{cx}$.

21. $\boldsymbol{y'''_{xxx} + ae^{\lambda x}y''_{xx} + (abe^{\lambda x} - b^2 + c)y'_x + c(ae^{\lambda x} - b)y = 0.}$

Particular solutions: $y_1 = e^{\beta_1 x}$, $y_2 = e^{\beta_2 x}$, where β_1 and β_2 are the roots of the quadratic equation $\beta^2 + b\beta + c = 0$.

22. $\boldsymbol{y'''_{xxx} + ae^{\lambda x}y''_{xx} + [a(b + \lambda)e^{\lambda x} - b^2]y'_x + ab\lambda e^{\lambda x}y = 0.}$

Particular solutions: $y_1 = e^{-bx}$, $y_2 = e^{-bx}\displaystyle\int \exp\Bigl(2bx - \frac{a}{\lambda}e^{\lambda x}\Bigr)\,dx$.

23. $\boldsymbol{y'''_{xxx} + ae^{\lambda x}y''_{xx} + bx^ny'_x + bx^{n-1}(axe^{\lambda x} + n)y = 0.}$

The substitution $w = y''_{xx} + bx^ny$ leads to a first order linear equation: $w'_x + ae^{\lambda x}w = 0$.

24. $\boldsymbol{y'''_{xxx} + axe^{\lambda x}y''_{xx} + (bx^2 - ae^{\lambda x})y'_x + bx(ax^2e^{\lambda x} + 3)y = 0.}$

1°. Particular solutions with $b > 0$:

$$y_1 = \cos\biggl(\frac{x^2\sqrt{b}}{2}\biggr), \quad y_2 = \sin\biggl(\frac{x^2\sqrt{b}}{2}\biggr).$$

2°. Particular solutions with $b < 0$:

$$y_1 = \exp\biggl(\frac{-x^2\sqrt{-b}}{2}\biggr), \quad y_2 = \exp\biggl(\frac{x^2\sqrt{-b}}{2}\biggr).$$

25. $\boldsymbol{y'''_{xxx} + ax^2e^{\lambda x}y''_{xx} - 2axe^{\lambda x}y'_x + 2ae^{\lambda x}y = 0.}$

Particular solutions: $y_1 = x$, $y_2 = x^2$.

26. $y'''_{xxx} + (ae^{\lambda x} + b)y''_{xx} - ab^2 e^{\lambda x} y = 0.$

The substitution $w = y'_x + by$ leads to a second order equation of the form 2.1.3.29: $w''_{xx} + ae^{\lambda x} w'_x - abe^{\lambda x} w = 0$.

27. $y'''_{xxx} + (ae^{\lambda x} + b)y''_{xx} - c^2(ae^{\lambda x} + b + c)y = 0.$

The substitution $w = y'_x - cy$ leads to a second order equation of the form 2.1.3.29: $w''_{xx} + (ae^{\lambda x} + b + c)w'_x + c(ae^{\lambda x} + b + c)w = 0$.

28. $y'''_{xxx} + (ae^{\lambda x} + b)y''_{xx} + c(ae^{\lambda x} + b)y'_x + c^3 y = 0.$

The substitution $w = y'_x + cy$ leads to a second order equation of the form 2.1.3.29: $w''_{xx} + (ae^{\lambda x} + b - c)w'_x + c^2 w = 0$.

29. $y'''_{xxx} + (be^{ax} + 2a)y''_{xx} - a(be^{ax} + a)y'_x - 2a^3 y = 0.$

Particular solutions: $y_1 = e^{ax}$, $y_2 = e^{-ax} + \dfrac{b}{a}$.

30. $y'''_{xxx} = (e^{\lambda x} - a)y''_{xx} + (ae^{\lambda x} - b)y'_x + be^{\lambda x} y.$

Particular solutions: $y_1 = e^{\beta_1 x}$, $y_2 = e^{\beta_2 x}$, where β_1 and β_2 are the roots of the quadratic equation $\beta^2 + a\beta + b = 0$.

31. $y'''_{xxx} + (ae^{\lambda x} + b)y''_{xx} + (ce^{\lambda x} + d)y'_x - s[(as + c)e^{\lambda x} + bs + d + s^2]y = 0.$

The substitution $w = y'_x - sy$ leads to a second order equation of the form 2.1.3.29:

$$w''_{xx} + (ae^{\lambda x} + b + s)w'_x + [(as + c)e^{\lambda x} + bs + d + s^2]w = 0.$$

32. $y'''_{xxx} + (ae^{\lambda x} + b)y''_{xx} + (ce^{2\lambda x} + d)y'_x - s(ce^{2\lambda x} + ase^{\lambda x} + bs + d + s^2)y = 0.$

The substitution $w = y'_x - sy$ leads to a second order equation of the form 2.1.3.29:

$$w''_{xx} + (ae^{\lambda x} + b + s)w'_x + (ce^{2\lambda x} + ase^{\lambda x} + bs + d + s^2)w = 0.$$

33. $y'''_{xxx} + (ae^{\lambda x} + b)y''_{xx} + (ce^{\lambda x} + d)y'_x - ke^{\lambda x}[k(a + k)e^{2\lambda x}$
$+ (a\lambda + 3k\lambda + bk + c)e^{\lambda x} + \lambda^2 + b\lambda + d]y = 0.$

Particular solution: $y_0 = \exp\Bigl(\dfrac{k}{\lambda} e^{\lambda x}\Bigr)$.

The substitution $y = \exp\Bigl(\dfrac{k}{\lambda} e^{\lambda x}\Bigr) \displaystyle\int z(x)\,dx$ lead to a second order linear equation of the form 2.1.3.29.

34. $y'''_{xxx} + (2ae^{\lambda x} + b)y''_{xx} + ae^{\lambda x}(ae^{\lambda x} + 2b + 3\lambda)y'_x$
$+ ae^{\lambda x}[a(b + 2\lambda)e^{\lambda x} + b\lambda + \lambda^2]y = 0.$

Particular solutions: $y_1 = \exp\Bigl(-\dfrac{a}{\lambda} e^{\lambda x}\Bigr)$, $y_2 = x\exp\Bigl(-\dfrac{a}{\lambda} e^{\lambda x}\Bigr)$.

35. $y'''_{xxx} + (ae^{\lambda x} - b)y''_{xx} + ce^{\mu x} y'_x - b(abe^{\lambda x} + ce^{\mu x})y = 0.$

Particular solution: $y_0 = e^{bx}$.

36. $y'''_{xxx} + (e^{\lambda x} + a)y''_{xx} + (ae^{\lambda x} + be^{\mu x})y'_x + abe^{\mu x}y = 0.$

Particular solution: $y_0 = e^{-ax}$.

37. $y'''_{xxx} + (ax + be^{\lambda x})y''_{xx} + a(bxe^{\lambda x} + 2)y'_x + abe^{\lambda x}y = 0.$

Particular solutions: $y_1 = \exp\Bigl(-\dfrac{ax^2}{2}\Bigr)$, $y_2 = \exp\Bigl(-\dfrac{ax^2}{2}\Bigr)\displaystyle\int \exp\Bigl(\dfrac{ax^2}{2}\Bigr)\,dx$.

38. $y'''_{xxx} + (abxe^{\lambda x} + be^{\lambda x} + a)y''_{xx} + a^2bxe^{\lambda x}y'_x - a^2be^{\lambda x}y = 0.$

Particular solutions: $y_1 = x$, $y_2 = e^{-ax}$.

39. $y'''_{xxx} + ax^n(be^{\lambda x} + 2\lambda)y''_{xx} - \lambda(abx^ne^{\lambda x} + \lambda)y'_x - 2a\lambda^3x^ny = 0.$

Particular solutions: $y_1 = e^{\lambda x}$, $y_2 = e^{-\lambda x} + \dfrac{b}{\lambda}$.

40. $y'''_{xxx} + (ax^n - 2be^{\lambda x})y''_{xx} - be^{\lambda x}(2ax^n - be^{\lambda x} + 3\lambda)y'_x$
$+ be^{\lambda x}[ax^n(be^{\lambda x} - \lambda) + 2b\lambda e^{\lambda x} - \lambda^2]y = 0.$

Particular solutions: $y_1 = \exp\Bigl(\dfrac{b}{\lambda}e^{\lambda x}\Bigr)$, $y_2 = x\exp\Bigl(\dfrac{b}{\lambda}e^{\lambda x}\Bigr)$.

41. $y'''_{xxx} + (ae^{\lambda x} + be^{\mu x})y''_{xx} + cy'_x + c(ae^{\lambda x} + be^{\mu x})y = 0.$

The substitution $w = y''_{xx} + cy$ leads to a first order linear equation: $w'_x + (ae^{\lambda x} + be^{\mu x})w = 0$.

42. $y'''_{xxx} + (ae^{\lambda x} + be^{\mu x})y''_{xx} + [abe^{(\lambda+\mu)x} + b(c+\mu)e^{\mu x} - c^2]y'_x$
$+ c[abe^{(\lambda+\mu)x} - ace^{\lambda x} + b\mu e^{\mu x}]y = 0.$

Particular solutions: $y_1 = e^{-cx}$, $y_2 = e^{-cx}\displaystyle\int \exp\Bigl(2cx - \dfrac{b}{\mu}e^{\mu x}\Bigr)\,dx$.

43. $y'''_{xxx} + ae^{\lambda x}(be^{\mu x} + 2\mu)y''_{xx} - \mu(abe^{(\lambda+\mu)x} + \mu)y'_x - 2a\mu^3e^{\lambda x}y = 0.$

Particular solutions: $y_1 = e^{\mu x}$, $y_2 = e^{-\mu x} + \dfrac{b}{\mu}$.

44. $xy'''_{xxx} + ay''_{xx} + x(be^{\lambda x} + c)y'_x + [b(\lambda x + a)e^{\lambda x} + ac]y = 0.$

The substitution $w = y''_{xx} + (be^{\lambda x} + c)y$ yields a first order linear equation: $xw'_x + aw = 0$.

45. $xy'''_{xxx} + axe^{\lambda x}y'_x - 2ae^{\lambda x}y = 0.$

The substitution $w = xy'_x - 2y$ leads to a second order equation of the form 2.1.3.1: $w''_{xx} + ae^{\lambda x}w = 0$.

46. $xy'''_{xxx} = (e^{\lambda x} - ax)y''_{xx} + (ae^{\lambda x} - bx)y'_x + be^{\lambda x}y.$

Particular solutions: $y_1 = e^{\beta_1 x}$, $y_2 = e^{\beta_2 x}$, where β_1 and β_2 are the roots of the quadratic equation $\beta^2 + a\beta + b = 0$.

47. $xy'''_{xxx} + (axe^{\lambda x} + 3)y''_{xx} + a(bx + 2)e^{\lambda x}y'_x + b(abxe^{\lambda x} + ae^{\lambda x} - b^2x)y = 0.$

Particular solutions:

$$y_1 = \frac{1}{x}\exp\left(-\frac{bx}{2}\right)\cos\left(\frac{bx\sqrt{3}}{2}\right), \quad y_2 = \frac{1}{x}\exp\left(-\frac{bx}{2}\right)\sin\left(\frac{bx\sqrt{3}}{2}\right).$$

48. $x^2y'''_{xxx} + x(axe^{\lambda x} + b)y''_{xx} + [a(b - 2)xe^{\lambda x} + c]y'_x + a(c - b + 2)e^{\lambda x}y = 0.$

Particular solutions: $y_1 = x^{n_1}$, $y_2 = x^{n_2}$, where n_1 and n_2 are the roots of the quadratic equation $n^2 + (b - 3)n + c - b + 2 = 0$.

49. $x^3y'''_{xxx} + bx^2e^{\lambda x}y''_{xx} + axy'_x + a(be^{\lambda x} - 2)y = 0.$

Particular solutions: $y_1 = x^{n_1}$, $y_2 = x^{n_2}$, where n_1 and n_2 are the roots of the quadratic equation $n^2 - n + a = 0$.

50. $x^3y'''_{xxx} + x^2(ae^{\lambda x} + b)y''_{xx} + x(abe^{\lambda x} + c - b)y'_x + c(ae^{\lambda x} - 2)y = 0.$

Particular solutions: $y_1 = x^{n_1}$, $y_2 = x^{n_2}$, where n_1 and n_2 are the roots of the quadratic equation $n^2 + (b - 1)n + c = 0$.

51. $(ae^x + b)y'''_{xxx} - ae^xy = 0.$

Particular solution: $y_0 = ae^x + b$.

52. $(bce^{ax} + a + c)y'''_{xxx} - (bc^3e^{ax} + a^3 + c^3)y'_x + ac(a^2 - c^2)y = 0.$

Particular solutions: $y_1 = e^{cx}$, $y_2 = e^{-ax} + b$.

53. $(ae^{\lambda x} + b)y'''_{xxx} + (ce^{\lambda x} + d)y''_{xx} + k(ae^{\lambda x} + b)y'_x + k(ce^{\lambda x} + d)y = 0.$

1°. Particular solutions with $k > 0$: $y_1 = \cos(x\sqrt{k})$, $y_2 = \sin(x\sqrt{k})$.

2°. Particular solutions with $k < 0$: $y_1 = \exp(-x\sqrt{-k})$, $y_2 = \exp(x\sqrt{-k})$.

54. $(ae^x + bx)y'''_{xxx} - ae^xy = 0.$

Particular solution: $y_0 = ae^x + bx$.

55. $(ae^x + bx^2)y'''_{xxx} - ae^xy = 0.$

Particular solution: $y_0 = ae^x + bx^2$.

56. $(axe^x + b)y'''_{xxx} + by = 0.$

Particular solution: $y_0 = ax + be^{-x}$.

57. $(ax^2e^x + b)y'''_{xxx} + by = 0.$

Particular solution: $y_0 = ax^2 + be^{-x}$.

3.1.4. Equations Containing Hyperbolic Functions

1. $y'''_{xxx} - a^3 \tanh(ax) y = 0.$

Particular solution: $y_0 = \cosh(ax)$.

The substitution $y = \cosh(ax) \int z(x)\,dx$ leads to a second order equation of the form 2.1.4.44: $z''_{xx} + 3a\tanh(ax) z'_x + 3a^2 z = 0$.

2. $y'''_{xxx} - a^3 \coth(ax) y = 0.$

Particular solution: $y_0 = \sinh(ax)$.

The substitution $y = \sinh(ax) \int z(x)\,dx$ leads to a second order equation of the form 2.1.4.45: $z''_{xx} + 3a\coth(ax) z'_x + 3a^2 z = 0$.

3. $y'''_{xxx} - 3a^2 y'_x + 2a^3 \tanh(ax) y = 0.$

Particular solutions: $y_1 = \cosh(ax)$, $y_2 = x\cosh(ax)$.

4. $y'''_{xxx} - 3a^2 y'_x + 2a^3 \coth(ax) y = 0.$

Particular solutions: $y_1 = \sinh(ax)$, $y_2 = x\sinh(ax)$.

5. $y'''_{xxx} + [a\cosh(2x) + b] y'_x + a\sinh(2x) y = 0.$

Solution:

$$y = C_1 w_1^2 + C_2 w_1 w_2 + C_3 w_2^2,$$

where w_1 and w_2 form a fundamental set of solutions of the modified Mathieu equation 2.1.4.1: $4w''_{xx} + [a\cosh(2x) + b]w = 0$.

6. $y'''_{xxx} + a y''_{xx} + b\cosh(2x) y'_x + ab\cosh(2x) y = 0.$

The substitution $w = y'_x + ay$ leads to a second order equation of the form 2.1.4.1: $w''_{xx} + b\cosh(2x) w = 0$.

7. $y'''_{xxx} + a y''_{xx} + b\cosh^2 x\, y'_x + ab\cosh^2 x\, y = 0.$

The substitution $w = y'_x + ay$ leads to a second order equation of the form 2.1.4.2: $w''_{xx} + b\cosh^2 x\, w = 0$.

8. $y'''_{xxx} + a y''_{xx} + b\sinh^2 x\, y'_x + ab\sinh^2 x\, y = 0.$

The substitution $w = y'_x + ay$ leads to a second order equation of the form 2.1.4.5: $w''_{xx} + b\sinh^2 x\, w = 0$.

9. $y'''_{xxx} + a y''_{xx} + [b\tanh(\lambda x) + c] y'_x + a[b\tanh(\lambda x) + c] y = 0.$

The substitution $w = y'_x + ay$ leads to a second order equation of the form 2.1.4.6: $w''_{xx} + [b\tanh(\lambda x) + c]w = 0$.

10. $y'''_{xxx} + a y''_{xx} - \lambda[2a\tanh(\lambda x) + 3\lambda] y'_x + \lambda^2[2a\tanh^2(\lambda x) + 2\lambda\tanh(\lambda x) - a] y = 0.$

Particular solutions: $y_1 = \cosh(\lambda x)$, $y_2 = x\cosh(\lambda x)$.

11. $y'''_{xxx} + ay''_{xx} + [b\coth(\lambda x) + c]y'_x + a[b\coth(\lambda x) + c]y = 0.$

The substitution $w = y'_x + ay$ leads to a second order equation of the form 2.1.4.10: $w''_{xx} + [b\coth(\lambda x) + c]w = 0$.

12. $y'''_{xxx} + ay''_{xx} - \lambda[2a\coth(\lambda x) + 3\lambda]y'_x + \lambda^2[2a\coth^2(\lambda x) + 2\lambda\coth(\lambda x) - a]y = 0.$

Particular solutions: $y_1 = \sinh(\lambda x)$, $y_2 = x\sinh(\lambda x)$.

13. $y'''_{xxx} + a\cosh^n(\lambda x)y''_{xx} + by'_x + ab\cosh^n(\lambda x)y = 0.$

1°. Particular solutions with $b > 0$: $y_1 = \cos(x\sqrt{b})$, $y_2 = \sin(x\sqrt{b})$.

2°. Particular solutions with $b < 0$: $y_1 = \exp(-x\sqrt{-b})$, $y_2 = \exp(x\sqrt{-b})$.

14. $y'''_{xxx} + a\cosh^n(\lambda x)y''_{xx} + bx^m y'_x + bx^{m-1}[ax\cosh^n(\lambda x) + m]y = 0.$

The substitution $w = y''_{xx} + bx^m y$ leads to a first order linear equation: $w'_x + a\cosh^n(\lambda x)w = 0$.

15. $y'''_{xxx} + a\cosh^n(\lambda x)y''_{xx} + ab\cosh^n(\lambda x)y'_x + b^2[a\cosh^n(\lambda x) - b]y = 0.$

Particular solutions: $y_1 = e^{-bx/2}\cos\left(\dfrac{bx\sqrt{3}}{2}\right)$, $y_2 = e^{-bx/2}\sin\left(\dfrac{bx\sqrt{3}}{2}\right)$.

16. $y'''_{xxx} + a\cosh^n(\lambda x)y''_{xx} - b[2a\cosh^n(\lambda x) + 3b]y'_x + b^2[a\cosh^n(\lambda x) + 2b]y = 0.$

Particular solutions: $y_1 = e^{bx}$, $y_2 = xe^{bx}$.

17. $y'''_{xxx} + a\cosh^n x\, y''_{xx} + (ab\cosh^n x + c - b^2)y'_x + c(a\cosh^n x - b)y = 0.$

Particular solutions: $y_1 = \exp(\lambda_1 x)$, $y_2 = \exp(\lambda_2 x)$, where λ_1 and λ_2 are the roots of the quadratic equation $\lambda^2 + b\lambda + c = 0$.

18. $y'''_{xxx} + ax\cosh^n x\, y''_{xx} + (bx^2 - a\cosh^n x)y'_x + bx(ax^2\cosh^n x + 3)y = 0.$

Particular solutions: $y_1 = \cos\left(\tfrac{1}{2}x^2\sqrt{b}\right)$, $y_2 = \sin\left(\tfrac{1}{2}x^2\sqrt{b}\right)$.

19. $y'''_{xxx} + ax^2\cosh^n(\lambda x)y''_{xx} - 2ax\cosh^n(\lambda x)y'_x + 2a\cosh^n(\lambda x)y = 0.$

Particular solutions: $y_1 = x$, $y_2 = x^2$.

20. $y'''_{xxx} = (\cosh^n x - a)y''_{xx} + (a\cosh^n x - b)y'_x + b\cosh^n x\, y.$

Particular solutions: $y_1 = \exp(\lambda_1 x)$, $y_2 = \exp(\lambda_2 x)$, where λ_1 and λ_2 are the roots of the quadratic equation $\lambda^2 + a\lambda + b = 0$.

21. $y'''_{xxx} + (a\cosh^n x + bx)y''_{xx} + b(ax\cosh^n x + 2)y'_x + ab\cosh^n x\, y = 0.$

Particular solutions: $y_1 = \exp\left(-\tfrac{1}{2}bx^2\right)$, $y_2 = \exp\left(-\tfrac{1}{2}bx^2\right)\displaystyle\int \exp\left(\tfrac{1}{2}bx^2\right)dx$.

22. $y'''_{xxx} + a\sinh^n(\lambda x)y''_{xx} + by'_x + ab\sinh^n(\lambda x)y = 0.$

1°. Particular solutions with $b > 0$: $y_1 = \cos(x\sqrt{b})$, $y_2 = \sin(x\sqrt{b})$.

2°. Particular solutions with $b < 0$: $y_1 = \exp(-x\sqrt{-b})$, $y_2 = \exp(x\sqrt{-b})$.

23. $y'''_{xxx} + a\sinh^n(\lambda x)y''_{xx} + bx^m y'_x + bx^{m-1}[ax\sinh^n(\lambda x) + m]y = 0.$

The substitution $w = y''_{xx} + bx^m y$ leads to a first order linear equation: $w'_x + a\sinh^n(\lambda x)w = 0$.

24. $y'''_{xxx} + a\sinh^n(\lambda x)y''_{xx} + ab\sinh^n(\lambda x)y'_x + b^2[a\sinh^n(\lambda x) - b]y = 0.$

Particular solutions: $y_1 = e^{-bx/2}\cos\left(\dfrac{bx\sqrt{3}}{2}\right)$, $y_2 = e^{-bx/2}\sin\left(\dfrac{bx\sqrt{3}}{2}\right)$.

25. $y'''_{xxx} + a\sinh^n(\lambda x)y''_{xx} - b[2a\sinh^n(\lambda x) + 3b]y'_x + b^2[a\sinh^n(\lambda x) + 2b]y = 0.$

Particular solutions: $y_1 = e^{bx}$, $y_2 = xe^{bx}$.

26. $y'''_{xxx} + a\sinh^n x\, y''_{xx} + (ab\sinh^n x + c - b^2)y'_x + c(a\sinh^n x - b)y = 0.$

Particular solutions: $y_1 = \exp(\lambda_1 x)$, $y_2 = \exp(\lambda_2 x)$, where λ_1 and λ_2 are the roots of the quadratic equation $\lambda^2 + b\lambda + c = 0$.

27. $y'''_{xxx} + ax\sinh^n x\, y''_{xx} + (bx^2 - a\sinh^n x)y'_x + bx(ax^2\sinh^n x + 3)y = 0.$

Particular solutions: $y_1 = \cos\left(\tfrac{1}{2}x^2\sqrt{b}\right)$, $y_2 = \sin\left(\tfrac{1}{2}x^2\sqrt{b}\right)$.

28. $y'''_{xxx} + ax^2\sinh^n(\lambda x)y''_{xx} - 2ax\sinh^n(\lambda x)y'_x + 2a\sinh^n(\lambda x)y = 0.$

Particular solutions: $y_1 = x$, $y_2 = x^2$.

29. $y'''_{xxx} = (\sinh^n x - a)y''_{xx} + (a\sinh^n x - b)y'_x + b\sinh^n x\, y.$

Particular solutions: $y_1 = \exp(\lambda_1 x)$, $y_2 = \exp(\lambda_2 x)$, where λ_1 and λ_2 are the roots of the quadratic equation $\lambda^2 + a\lambda + b = 0$.

30. $y'''_{xxx} + (a\sinh^n x + bx)y''_{xx} + b(ax\sinh^n x + 2)y'_x + ab\sinh^n x\, y = 0.$

Particular solutions: $y_1 = \exp\left(-\tfrac{1}{2}bx^2\right)$, $y_2 = \exp\left(-\tfrac{1}{2}bx^2\right)\displaystyle\int \exp\left(\tfrac{1}{2}bx^2\right)dx$.

31. $y'''_{xxx} - \tanh x\, y''_{xx} - ay'_x + a\tanh x\, y = 0.$

1°. Solution with $a > 0$: $y = C_1\exp(-x\sqrt{a}) + C_2\exp(x\sqrt{a}) + C_3\cosh x$.

2°. Solution with $a < 0$: $y = C_1\cos(x\sqrt{-a}) + C_2\sin(x\sqrt{-a}) + C_3\cosh x$.

32. $y'''_{xxx} + a\tanh^n(\lambda x)y''_{xx} + bx^m y'_x + bx^{m-1}[ax\tanh^n(\lambda x) + m]y = 0.$

The substitution $w = y''_{xx} + bx^m y$ leads to a first order linear equation: $w'_x + a\tanh^n(\lambda x)w = 0$.

33. $y'''_{xxx} + a\tanh^n(\lambda x)y''_{xx} + ab\tanh^n(\lambda x)y'_x + b^2[a\tanh^n(\lambda x) - b]y = 0.$

Particular solutions: $y_1 = e^{-bx/2}\cos\left(\dfrac{bx\sqrt{3}}{2}\right)$, $y_2 = e^{-bx/2}\sin\left(\dfrac{bx\sqrt{3}}{2}\right)$.

34. $y'''_{xxx} + a\tanh^n(\lambda x)y''_{xx} - b[2a\tanh^n(\lambda x) + 3b]y'_x + b^2[a\tanh^n(\lambda x) + 2b]y = 0.$

Particular solutions: $y_1 = e^{bx}$, $y_2 = xe^{bx}$.

35. $y'''_{xxx} + a\tanh^n(\lambda x)y''_{xx} + [ab\tanh^n(\lambda x) + c - b^2]y'_x + c[a\tanh^n(\lambda x) - b]y = 0.$

Particular solutions: $y_1 = \exp(\beta_1 x)$, $y_2 = \exp(\beta_2 x)$, where β_1 and β_2 are the roots of the quadratic equation $\beta^2 + b\beta + c = 0$.

36. $y'''_{xxx} + ax^n y''_{xx} - (2ax^n\tanh x + 3)y'_x + [ax^n(2\tanh^2 x - 1) + 2\tanh x]y = 0.$

Particular solutions: $y_1 = \cosh x$, $y_2 = x\cosh x$.

37. $y'''_{xxx} + a\tanh^n x\, y''_{xx} - (2a\tanh^{n+1} x + 3)y'_x$
$+ (2a\tanh^{n+2} x - a\tanh^n x + 2\tanh x)y = 0.$

Particular solutions: $y_1 = \cosh x$, $y_2 = x\cosh x$.

38. $y'''_{xxx} + ax\tanh^n x\, y''_{xx} + (bx^2 - a\tanh^n x)y'_x + bx(ax^2\tanh^n x + 3)y = 0.$

Particular solutions: $y_1 = \cos\left(\frac{1}{2}x^2\sqrt{b}\right)$, $y_2 = \sin\left(\frac{1}{2}x^2\sqrt{b}\right)$.

39. $y'''_{xxx} + ax^2\tanh^n(\lambda x)y''_{xx} - 2ax\tanh^n(\lambda x)y'_x + 2a\tanh^n(\lambda x)y = 0.$

Particular solutions: $y_1 = x$, $y_2 = x^2$.

40. $y'''_{xxx} = (\tanh^n x - a)y''_{xx} + (a\tanh^n x - b)y'_x + b\tanh^n x\, y.$

Particular solutions: $y_1 = \exp(\lambda_1 x)$, $y_2 = \exp(\lambda_2 x)$, where λ_1 and λ_2 are the roots of the quadratic equation $\lambda^2 + a\lambda + b = 0$.

41. $y'''_{xxx} + (a\tanh^n x + bx)y''_{xx} + b(ax\tanh^n x + 2)y'_x + ab\tanh^n x\, y = 0.$

Particular solutions: $y_1 = \exp\left(-\frac{1}{2}bx^2\right)$, $y_2 = \exp\left(-\frac{1}{2}bx^2\right)\int\exp\left(\frac{1}{2}bx^2\right)dx$.

42. $y'''_{xxx} + [ax^n(\tanh x - b) - b]y''_{xx} + [a(b^2 - 1)x^n - 1]y'_x$
$+ b[ax^n(1 - b\tanh x) + 1]y = 0.$

Particular solutions: $y_1 = e^{bx}$, $y_2 = \cosh x$.

43. $y'''_{xxx} + [\lambda\tanh(\lambda x)(ax^n - 1) - ax^{n-1}]y''_{xx} - a\lambda^2 x^n y'_x + a\lambda^2 x^{n-1}y = 0.$

Particular solutions: $y_1 = x$, $y_2 = \cosh(\lambda x)$.

44. $y'''_{xxx} + (a\tanh^{n+1} x - ab\tanh^n x - b)y''_{xx}$
$+ [a(b^2 - 1)\tanh^n x - 1]y'_x + b(-ab\tanh^{n+1} x + a\tanh^n x + 1)y = 0.$

Particular solutions: $y_1 = e^{bx}$, $y_2 = \cosh x$.

45. $y'''_{xxx} - \coth x\, y''_{xx} - ay'_x + a\coth x\, y = 0.$

1°. Solution with $a > 0$: $y = C_1\exp(-x\sqrt{a}) + C_2\exp(x\sqrt{a}) + C_3\sinh x$.

2°. Solution with $a < 0$: $y = C_1\cos(x\sqrt{-a}) + C_2\sin(x\sqrt{-a}) + C_3\sinh x$.

46. $y'''_{xxx} + (a\coth x - ab - b)y''_{xx} + (ab^2 - a - 1)y'_x + b(-ab\coth x + a + 1)y = 0.$

Particular solutions: $y_1 = e^{bx}$, $y_2 = \sinh x$.

47. $y'''_{xxx} + a\coth^n(\lambda x)y''_{xx} + bx^m y'_x + bx^{m-1}[ax\coth^n(\lambda x) + m]y = 0.$

The substitution $w = y''_{xx} + bx^m y$ leads to a first order linear equation: $w'_x + a\coth^n(\lambda x)w = 0$.

48. $y'''_{xxx} + a\coth^n(\lambda x)y''_{xx} + ab\coth^n(\lambda x)y'_x + b^2[a\coth^n(\lambda x) - b]y = 0.$

Particular solutions: $y_1 = e^{-bx/2}\cos\left(\dfrac{bx\sqrt{3}}{2}\right)$, $y_2 = e^{-bx/2}\sin\left(\dfrac{bx\sqrt{3}}{2}\right)$.

49. $y'''_{xxx} + a\coth^n(\lambda x)y''_{xx} - b[2a\coth^n(\lambda x) + 3b]y'_x + b^2[a\coth^n(\lambda x) + 2b]y = 0.$

Particular solutions: $y_1 = e^{bx}$, $y_2 = xe^{bx}$.

50. $y'''_{xxx} + ax\coth^n x\, y''_{xx} + (bx^2 - a\coth^n x)y'_x + bx(ax^2\coth^n x + 3)y = 0.$

Particular solutions: $y_1 = \cos\left(\frac{1}{2}x^2\sqrt{b}\right)$, $y_2 = \sin\left(\frac{1}{2}x^2\sqrt{b}\right)$.

51. $y'''_{xxx} + ax^2\coth^n(\lambda x)y''_{xx} - 2ax\coth^n(\lambda x)y'_x + 2a\coth^n(\lambda x)y = 0.$

Particular solutions: $y_1 = x$, $y_2 = x^2$.

52. $y'''_{xxx} + ax^n y''_{xx} - (2ax^n\coth x + 3)y'_x + [ax^n(2\coth^2 x - 1) + 2\coth x]y = 0.$

Particular solutions: $y_1 = \sinh x$, $y_2 = x\sinh x$.

53. $y'''_{xxx} + a\coth^n x\, y''_{xx} - (2a\coth^{n+1} x + 3)y'_x$
$\qquad + (2a\coth^{n+2} x - a\coth^n x + 2\coth x)y = 0.$

Particular solutions: $y_1 = \sinh x$, $y_2 = x\sinh x$.

54. $y'''_{xxx} + (a\coth^n x + bx)y''_{xx} + b(ax\coth^n x + 2)y'_x + ab\coth^n x\, y = 0.$

Particular solutions: $y_1 = \exp\left(-\frac{1}{2}bx^2\right)$, $y_2 = \exp\left(-\frac{1}{2}bx^2\right)\displaystyle\int \exp\left(\frac{1}{2}bx^2\right)dx$.

55. $y'''_{xxx} + [\lambda\coth(\lambda x)(ax^n - 1) - ax^{n-1}]y''_{xx} - a\lambda^2 x^n y'_x + a\lambda^2 x^{n-1}y = 0.$

Particular solutions: $y_1 = x$, $y_2 = \sinh(\lambda x)$.

56. $xy'''_{xxx} + x[a\cosh(2x) + b]y'_x - 2[a\cosh(2x) + b]y = 0.$

The substitution $w = xy'_x - 2y$ leads to a second order equation of the form 2.1.4.1: $w''_{xx} + [a\cosh(2x) + b]w = 0$.

57. $xy'''_{xxx} + x(a\cosh^2 x + b)y'_x - 2(a\cosh^2 x + b)y = 0.$

The substitution $w = xy'_x - 2y$ leads to a second order equation of the form 2.1.4.2: $w''_{xx} + (a\cosh^2 x + b)w = 0$.

58. $xy'''_{xxx} + x(a\sinh^2 x + b)y'_x - 2(a\sinh^2 x + b)y = 0.$

The substitution $w = xy'_x - 2y$ leads to a second order equation of the form 2.1.4.5: $w''_{xx} + (a\sinh^2 x + b)w = 0$.

59. $xy'''_{xxx} = (\cosh^n x - ax)y''_{xx} + (a\cosh^n x - bx)y'_x + b\cosh^n x\, y.$

Particular solutions: $y_1 = \exp(\lambda_1 x)$, $y_2 = \exp(\lambda_2 x)$, where λ_1 and λ_2 are the roots of the quadratic equation $\lambda^2 + a\lambda + b = 0$.

60. $$x^2 y'''_{xxx} + (ax^2\cosh^n x + bx)y''_{xx} + [a(b-2)x\cosh^n x + c]y'_x + a(c-b+2)\cosh^n x\, y = 0.$$

Particular solutions: $y_1 = x^{m_1}$, $y_2 = x^{m_2}$, where m_1 and m_2 are the roots of the quadratic equation $m^2 + (b-3)m + c - b + 2 = 0$.

61. $$x^2 y'''_{xxx} + (ax^2\sinh^n x + bx)y''_{xx} + [a(b-2)x\sinh^n x + c]y'_x + a(c-b+2)\sinh^n x\, y = 0.$$

Particular solutions: $y_1 = x^{m_1}$, $y_2 = x^{m_2}$, where m_1 and m_2 are the roots of the quadratic equation $m^2 + (b-3)m + c - b + 2 = 0$.

62. $$x^2 y'''_{xxx} + (ax^2\tanh^n x + bx)y''_{xx} + [a(b-2)x\tanh^n x + c]y'_x + a(c-b+2)\tanh^n x\, y = 0.$$

Particular solutions: $y_1 = x^{m_1}$, $y_2 = x^{m_2}$, where m_1 and m_2 are the roots of the quadratic equation $m^2 + (b-3)m + c - b + 2 = 0$.

63. $$x^2 y'''_{xxx} + (ax^2\coth^n x + bx)y''_{xx} + [a(b-2)x\coth^n x + c]y'_x + a(c-b+2)\coth^n x\, y = 0.$$

Particular solutions: $y_1 = x^{m_1}$, $y_2 = x^{m_2}$, where m_1 and m_2 are the roots of the quadratic equation $m^2 + (b-3)m + c - b + 2 = 0$.

64. $x^3 y'''_{xxx} + x^2(a\cosh^n x + b)y''_{xx} + x(ab\cosh^n x + c - b)y'_x + c(a\cosh^n x - 2)y = 0.$

Particular solutions: $y_1 = x^{m_1}$, $y_2 = x^{m_2}$, where m_1 and m_2 are the roots of the quadratic equation $m^2 + (b-1)m + c = 0$.

65. $x^3 y'''_{xxx} + x^2(a\sinh^n x + b)y''_{xx} + x(ab\sinh^n x + c - b)y'_x + c(a\sinh^n x - 2)y = 0.$

Particular solutions: $y_1 = x^{m_1}$, $y_2 = x^{m_2}$, where m_1 and m_2 are the roots of the quadratic equation $m^2 + (b-1)m + c = 0$.

66. $x^3 y'''_{xxx} + x^2(a\tanh^n x + b)y''_{xx} + x(ab\tanh^n x + c - b)y'_x + c(a\tanh^n x - 2)y = 0.$

Particular solutions: $y_1 = x^{m_1}$, $y_2 = x^{m_2}$, where m_1 and m_2 are the roots of the quadratic equation $m^2 + (b-1)m + c = 0$.

67. $x^3 y'''_{xxx} + x^2(a\coth^n x + b)y''_{xx} + x(ab\coth^n x + c - b)y'_x + c(a\coth^n x - 2)y = 0.$

Particular solutions: $y_1 = x^{m_1}$, $y_2 = x^{m_2}$, where m_1 and m_2 are the roots of the quadratic equation $m^2 + (b-1)m + c = 0$.

68. $\cosh^n x\, y'''_{xxx} + ay''_{xx} + aby'_x + b^2(a - b\cosh^n x)y = 0.$

Particular solutions: $y_1 = e^{-bx/2}\cos\left(\dfrac{bx\sqrt{3}}{2}\right)$, $y_2 = e^{-bx/2}\sin\left(\dfrac{bx\sqrt{3}}{2}\right)$.

69. $\cosh^n x\, y'''_{xxx} + ay''_{xx} + b\cosh^n x\, y'_x + aby = 0.$

1°. Particular solutions with $b > 0$: $y_1 = \cos\left(x\sqrt{b}\right)$, $y_2 = \sin\left(x\sqrt{b}\right)$.

2°. Particular solutions with $b < 0$: $y_1 = \exp\left(-x\sqrt{-b}\right)$, $y_2 = \exp\left(x\sqrt{-b}\right)$.

70. $\cosh^n x\, y'''_{xxx} + ay''_{xx} - b(2a + 3b\cosh^n x)y'_x + b^2(a + 2b\cosh^n x)y = 0.$

Particular solutions: $y_1 = e^{bx}$, $y_2 = xe^{bx}$.

71. $\cosh^n x\, y'''_{xxx} + y''_{xx} + [(b - a^2)\cosh^n x + a]y'_x + b(1 - a\cosh^n x)y = 0.$

Particular solutions: $y_1 = e^{\lambda_1 x}$, $y_2 = e^{\lambda_2 x}$, where λ_1 and λ_2 are the roots of the quadratic equation $\lambda^2 + a\lambda + b = 0$.

72. $\cosh^n(\lambda x)y'''_{xxx} + ax^2y''_{xx} - 2axy'_x + 2ay = 0.$

Particular solutions: $y_1 = x$, $y_2 = x^2$.

73. $\cosh^n x\, y'''_{xxx} + (a\cosh^n x + ax + 1)y''_{xx} + a^2xy'_x - a^2y = 0.$

Particular solutions: $y_1 = x$, $y_2 = e^{-ax}$.

74. $\cosh^n x\, y'''_{xxx} + (ax\cosh^n x + 1)y''_{xx} + a(x + 2\cosh^n x)y'_x + ay = 0.$

Particular solutions: $y_1 = \exp\left(-\dfrac{ax^2}{2}\right)$, $y_2 = \exp\left(-\dfrac{ax^2}{2}\right)\displaystyle\int \exp\left(\dfrac{ax^2}{2}\right) dx$.

75. $x\cosh^n x\, y'''_{xxx} + (3\cosh^n x + x)y''_{xx} + (ax\cosh^n x + 2)y'_x + a(\cosh^n x + x)y = 0.$

Particular solutions: $y_1 = \dfrac{1}{x}\cos\left(x\sqrt{a}\right)$, $y_2 = \dfrac{1}{x}\sin\left(x\sqrt{a}\right)$.

76. $x^3\cosh^n x\, y'''_{xxx} + ax^2y''_{xx} - 2x\cosh^n x\, y'_x + 2(2\cosh^n x - a)y = 0.$

Particular solutions: $y_1 = x^{-1}$, $y_2 = x^2$.

77. $x^3\cosh^n x\, y'''_{xxx} + ax^2y''_{xx} - 6x\cosh^n x\, y'_x + 6(2\cosh^n x - a)y = 0.$

Particular solutions: $y_1 = x^{-2}$, $y_2 = x^3$.

78. $x^3\cosh^n x\, y'''_{xxx} + ax^2y''_{xx} + x(a - \cosh^n x)y'_x + (a - 3\cosh^n x)y = 0.$

Particular solutions: $y_1 = \cos(\ln|x|)$, $y_2 = \sin(\ln|x|)$.

79. $x^3\cosh^n x\, y'''_{xxx} + x^2(\cosh^n x + a)y''_{xx} + x[a - (b + 1)\cosh^n x]y'_x + b(2\cosh^n x - a)y = 0.$

Particular solutions: $y_1 = x^{-\sqrt{b}}$, $y_2 = x^{\sqrt{b}}$.

80. $\sinh^n x\, y'''_{xxx} + ay''_{xx} + aby'_x + b^2(a - b\sinh^n x)y = 0.$

Particular solutions: $y_1 = e^{-bx/2}\cos\left(\dfrac{bx\sqrt{3}}{2}\right)$, $y_2 = e^{-bx/2}\sin\left(\dfrac{bx\sqrt{3}}{2}\right)$.

81. $\sinh^n x\, y'''_{xxx} + a y''_{xx} + b \sinh^n x\, y'_x + aby = 0.$

1°. Particular solutions with $b > 0$: $y_1 = \cos(x\sqrt{b})$, $y_2 = \sin(x\sqrt{b})$.

2°. Particular solutions with $b < 0$: $y_1 = \exp(-x\sqrt{-b})$, $y_2 = \exp(x\sqrt{-b})$.

82. $\sinh^n x\, y'''_{xxx} + a y''_{xx} - b(2a + 3b \sinh^n x) y'_x + b^2(a + 2b \sinh^n x) y = 0.$

Particular solutions: $y_1 = e^{bx}$, $y_2 = xe^{bx}$.

83. $\sinh^n x\, y'''_{xxx} + y''_{xx} + [(b - a^2) \sinh^n x + a] y'_x + b(1 - a \sinh^n x) y = 0.$

Particular solutions: $y_1 = e^{\lambda_1 x}$, $y_2 = e^{\lambda_2 x}$, where λ_1 and λ_2 are the roots of the quadratic equation $\lambda^2 + a\lambda + b = 0$.

84. $\sinh^n(\lambda x) y'''_{xxx} + a x^2 y''_{xx} - 2axy'_x + 2ay = 0.$

Particular solutions: $y_1 = x$, $y_2 = x^2$.

85. $\sinh^n x\, y'''_{xxx} + (a \sinh^n x + ax + 1) y''_{xx} + a^2 x y'_x - a^2 y = 0.$

Particular solutions: $y_1 = x$, $y_2 = e^{-ax}$.

86. $\sinh^n x\, y'''_{xxx} + (ax \sinh^n x + 1) y''_{xx} + a(x + 2\sinh^n x) y'_x + ay = 0.$

Particular solutions: $y_1 = \exp\left(-\dfrac{ax^2}{2}\right)$, $y_2 = \exp\left(-\dfrac{ax^2}{2}\right) \displaystyle\int \exp\left(\dfrac{ax^2}{2}\right) dx$.

87. $x \sinh^n x\, y'''_{xxx} + (3\sinh^n x + x) y''_{xx} + (ax \sinh^n x + 2) y'_x + a(\sinh^n x + x) y = 0.$

Particular solutions: $y_1 = \dfrac{1}{x} \cos(x\sqrt{a})$, $y_2 = \dfrac{1}{x} \sin(x\sqrt{a})$.

88. $x^3 \sinh^n x\, y'''_{xxx} + a x^2 y''_{xx} - 2x \sinh^n x\, y'_x + 2(2\sinh^n x - a) y = 0.$

Particular solutions: $y_1 = x^{-1}$, $y_2 = x^2$.

89. $x^3 \sinh^n x\, y'''_{xxx} + a x^2 y''_{xx} - 6x \sinh^n x\, y'_x + 6(2\sinh^n x - a) y = 0.$

Particular solutions: $y_1 = x^{-2}$, $y_2 = x^3$.

90. $x^3 \sinh^n x\, y'''_{xxx} + a x^2 y''_{xx} + x(a - \sinh^n x) y'_x + (a - 3\sinh^n x) y = 0.$

Particular solutions: $y_1 = \cos(\ln x)$, $y_2 = \sin(\ln x)$.

91. $x^3 \sinh^n x\, y'''_{xxx} + x^2(\sinh^n x + a) y''_{xx} +$
$$x[a - (b+1)\sinh^n x] y'_x + b(2\sinh^n x - a) y = 0.$$

Particular solutions: $y_1 = x^{-\sqrt{b}}$, $y_2 = x^{\sqrt{b}}$.

92. $\tanh^n x\, y'''_{xxx} + a y''_{xx} + aby'_x + b^2(a - b \tanh^n x) y = 0.$

Particular solutions: $y_1 = e^{-bx/2} \cos\left(\dfrac{bx\sqrt{3}}{2}\right)$, $y_2 = e^{-bx/2} \sin\left(\dfrac{bx\sqrt{3}}{2}\right)$.

93. $\tanh^n x\, y'''_{xxx} + a y''_{xx} + b \tanh^n x\, y'_x + aby = 0.$

1°. Particular solutions with $b > 0$: $y_1 = \cos(x\sqrt{b})$, $y_2 = \sin(x\sqrt{b})$.

2°. Particular solutions with $b < 0$: $y_1 = \exp(-x\sqrt{-b})$, $y_2 = \exp(x\sqrt{-b})$.

94. $\tanh^n x\, y'''_{xxx} + a y''_{xx} - b(2a + 3b \tanh^n x) y'_x + b^2(a + 2b \tanh^n x) y = 0.$

Particular solutions: $y_1 = e^{bx}$, $y_2 = xe^{bx}$.

95. $\tanh^n x\, y'''_{xxx} + y''_{xx} + [(b - a^2)\tanh^n x + a] y'_x + b(1 - a \tanh^n x) y = 0.$

Particular solutions: $y_1 = e^{\lambda_1 x}$, $y_2 = e^{\lambda_2 x}$, where λ_1 and λ_2 are the roots of the quadratic equation $\lambda^2 + a\lambda + b = 0$.

96. $\tanh^n(\lambda x) y'''_{xxx} + ax^2 y''_{xx} - 2axy'_x + 2ay = 0.$

Particular solutions: $y_1 = x$, $y_2 = x^2$.

97. $\tanh^n x\, y'''_{xxx} + (a \tanh^n x + ax + 1) y''_{xx} + a^2 x y'_x - a^2 y = 0.$

Particular solutions: $y_1 = x$, $y_2 = e^{-ax}$.

98. $\tanh^n x\, y'''_{xxx} + (ax \tanh^n x + 1) y''_{xx} + a(x + 2\tanh^n x) y'_x + ay = 0.$

Particular solutions: $y_1 = \exp\Bigl(-\dfrac{ax^2}{2}\Bigr)$, $y_2 = \exp\Bigl(-\dfrac{ax^2}{2}\Bigr)\displaystyle\int \exp\Bigl(\dfrac{ax^2}{2}\Bigr)\,dx$.

99. $x \tanh^n x\, y'''_{xxx} + (3\tanh^n x + x) y''_{xx} + (ax \tanh^n x + 2) y'_x + a(\tanh^n x + x) y = 0.$

Particular solutions: $y_1 = \dfrac{1}{x}\cos(x\sqrt{a})$, $y_2 = \dfrac{1}{x}\sin(x\sqrt{a})$.

100. $x^3 \tanh^n x\, y'''_{xxx} + ax^2 y''_{xx} - 2x \tanh^n x\, y'_x + 2(2\tanh^n x - a) y = 0.$

Particular solutions: $y_1 = x^{-1}$, $y_2 = x^2$.

101. $x^3 \tanh^n x\, y'''_{xxx} + ax^2 y''_{xx} - 6x \tanh^n x\, y'_x + 6(2\tanh^n x - a) y = 0.$

Particular solutions: $y_1 = x^{-2}$, $y_2 = x^3$.

102. $x^3 \tanh^n x\, y'''_{xxx} + ax^2 y''_{xx} + x(a - \tanh^n x) y'_x + (a - 3\tanh^n x) y = 0.$

Particular solutions: $y_1 = \cos(\ln|x|)$, $y_2 = \sin(\ln|x|)$.

103. $x^3 \tanh^n x\, y'''_{xxx} + x^2(\tanh^n x + a) y''_{xx} +$
$x[a - (b+1)\tanh^n x] y'_x + b(2\tanh^n x - a) y = 0.$

Particular solutions: $y_1 = x^{-\sqrt{b}}$, $y_2 = x^{\sqrt{b}}$.

104. $\coth^n x\, y'''_{xxx} + a y''_{xx} + aby'_x + b^2(a - b\coth^n x) y = 0.$

Particular solutions: $y_1 = e^{-bx/2}\cos\Bigl(\dfrac{bx\sqrt{3}}{2}\Bigr)$, $y_2 = e^{-bx/2}\sin\Bigl(\dfrac{bx\sqrt{3}}{2}\Bigr)$.

105. $\coth^n x\, y'''_{xxx} + a y''_{xx} + b\coth^n x\, y'_x + aby = 0.$

1°. Particular solutions with $b > 0$: $y_1 = \cos(x\sqrt{b})$, $y_2 = \sin(x\sqrt{b})$.

2°. Particular solutions with $b < 0$: $y_1 = \exp(-x\sqrt{-b})$, $y_2 = \exp(x\sqrt{-b})$.

106. $\coth^n x\, y'''_{xxx} + a y''_{xx} - b(2a + 3b\coth^n x)y'_x + b^2(a + 2b\coth^n x)y = 0.$

Particular solutions: $y_1 = e^{bx}$, $y_2 = xe^{bx}$.

107. $\coth^n x\, y'''_{xxx} + y''_{xx} + [(b - a^2)\coth^n x + a]y'_x + b(1 - a\coth^n x)y = 0.$

Particular solutions: $y_1 = e^{\lambda_1 x}$, $y_2 = e^{\lambda_2 x}$, where λ_1 and λ_2 are the roots of the quadratic equation $\lambda^2 + a\lambda + b = 0$.

108. $\coth^n(\lambda x) y'''_{xxx} + ax^2 y''_{xx} - 2axy'_x + 2ay = 0.$

Particular solutions: $y_1 = x$, $y_2 = x^2$.

109. $\coth^n x\, y'''_{xxx} + (a\coth^n x + ax + 1)y''_{xx} + a^2 x y'_x - a^2 y = 0.$

Particular solutions: $y_1 = x$, $y_2 = e^{-ax}$.

110. $\coth^n x\, y'''_{xxx} + (ax\coth^n x + 1)y''_{xx} + a(x + 2\coth^n x)y'_x + ay = 0.$

Particular solutions: $y_1 = \exp\left(-\dfrac{ax^2}{2}\right)$, $y_2 = \exp\left(-\dfrac{ax^2}{2}\right)\displaystyle\int \exp\left(\dfrac{ax^2}{2}\right) dx$.

111. $x\coth^n x\, y'''_{xxx} + (3\coth^n x + x)y''_{xx} + (ax\coth^n x + 2)y'_x + a(\coth^n x + x)y = 0.$

Particular solutions: $y_1 = \dfrac{1}{x}\cos(x\sqrt{a})$, $y_2 = \dfrac{1}{x}\sin(x\sqrt{a})$.

112. $x^3\coth^n x\, y'''_{xxx} + ax^2 y''_{xx} - 2x\coth^n x\, y'_x + 2(2\coth^n x - a)y = 0.$

Particular solutions: $y_1 = x^{-1}$, $y_2 = x^2$.

113. $x^3\coth^n x\, y'''_{xxx} + ax^2 y''_{xx} - 6x\coth^n x\, y'_x + 6(2\coth^n x - a)y = 0.$

Particular solutions: $y_1 = x^{-2}$, $y_2 = x^3$.

114. $x^3\coth^n x\, y'''_{xxx} + ax^2 y''_{xx} + x(a - \coth^n x)y'_x + (a - 3\coth^n x)y = 0.$

Particular solutions: $y_1 = \cos(\ln x)$, $y_2 = \sin(\ln x)$.

115. $x^3\coth^n x\, y'''_{xxx} + x^2(\coth^n x + a)y''_{xx} +$
$x[a - (b + 1)\coth^n x]y'_x + b(2\coth^n x - a)y = 0.$

Particular solutions: $y_1 = x^{-\sqrt{b}}$, $y_2 = x^{\sqrt{b}}$.

3.1.5. Equations Containing Logarithmic Functions

1. $y'''_{xxx} + a\ln^n(\lambda x)y''_{xx} + by'_x + ab\ln^n(\lambda x)y = 0.$

1°. Particular solutions with $b > 0$: $y_1 = \cos(x\sqrt{b})$, $y_2 = \sin(x\sqrt{b})$.

2°. Particular solutions with $b < 0$: $y_1 = \exp(-x\sqrt{-b})$, $y_2 = \exp(x\sqrt{-b})$.

2. $y'''_{xxx} + a\ln^n x\, y''_{xx} + ab\ln^n x\, y'_x + b^2(a\ln^n x - b)y = 0.$

Particular solutions: $y_1 = e^{-bx/2}\cos\left(\dfrac{bx\sqrt{3}}{2}\right)$, $y_2 = e^{-bx/2}\sin\left(\dfrac{bx\sqrt{3}}{2}\right)$.

3. $y'''_{xxx} + a\ln^n x\, y''_{xx} + (bx + c)\, y'_x + (abx\ln^n x + ac\ln^n x + b)y = 0.$

By integrating, we obtain a nonhomogeneous second-order linear equation:

$$y''_{xx} + (bx + c)y = C\exp\Big(-a\int \ln^n x\, dx\Big)$$

(see 2.1.2.2 for the solution of the corresponding homogeneous equation).

4. $y'''_{xxx} + a\ln^n(\lambda x)y''_{xx} - b[2a\ln^n(\lambda x) + 3b]y'_x + b^2[a\ln^n(\lambda x) + 2b]y = 0.$

Particular solutions: $y_1 = e^{bx}$, $y_2 = xe^{bx}$.

5. $y'''_{xxx} + a\ln^n x\, y''_{xx} + (ab\ln^n x + c - b^2)y'_x + c(a\ln^n x - b)y = 0.$

Particular solutions: $y_1 = \exp(\lambda_1 x)$, $y_2 = \exp(\lambda_2 x)$, where λ_1 and λ_2 are the roots of the quadratic equation $\lambda^2 + b\lambda + c = 0$.

6. $y'''_{xxx} + a\ln^n(\lambda x)y''_{xx} + bx^m y'_x + bx^{m-1}[ax\ln^n(\lambda x) + m]y = 0.$

Assuming $w = y''_{xx} + bx^m y$, we obtain a first order linear equation: $w'_x + a\ln^n(\lambda x)w = 0$.

7. $y'''_{xxx} + (a\ln^n x + b)y''_{xx} + c\, y'_x + c(a\ln^n x + b)y = 0.$

1°. Particular solutions with $c > 0$: $y_1 = \cos(x\sqrt{c})$, $y_2 = \sin(x\sqrt{c})$.

2°. Particular solutions with $c < 0$: $y_1 = \exp(-x\sqrt{-c}), y_2 = \exp(x\sqrt{-c})$.

8. $y'''_{xxx} = (\ln^n x - a)y''_{xx} + (a\ln^n x - b)y'_x + b\ln^n x\, y.$

Particular solutions: $y_1 = \exp(\lambda_1 x)$, $y_2 = \exp(\lambda_2 x)$, where λ_1 and λ_2 are the roots of the quadratic equation $\lambda^2 + a\lambda + b = 0$.

9. $y'''_{xxx} + ax\ln^n x\, y''_{xx} + (bx^2 - a\ln^n x)y'_x + bx(ax^2\ln^n x + 3)y = 0.$

Particular solutions: $y_1 = \cos\big(\tfrac{1}{2}x^2\sqrt{b}\big)$, $y_2 = \sin\big(\tfrac{1}{2}x^2\sqrt{b}\big)$.

10. $y'''_{xxx} + (a\ln^n x + bx)y''_{xx} + b(ax\ln^n x + 2)y'_x + ab\ln^n x\, y = 0.$

Particular solutions: $y_1 = \exp\big(-\tfrac{1}{2}bx^2\big)$, $y_2 = \exp\big(-\tfrac{1}{2}bx^2\big)\int \exp\big(\tfrac{1}{2}bx^2\big)\, dx$.

11. $y'''_{xxx} + (ax\ln^n x + b)y''_{xx} + a(bx-1)\ln^n x\, y'_x - ab\ln^n x\, y = 0.$

The substitution $w = y'_x + by$ leads to a second order linear equation of the form 2.1.5.5: $w''_{xx} + ax\ln^n x\, w'_x - a\ln^n x\, w = 0$.

12. $y'''_{xxx} + (abx\ln^n x + a\ln^n x + b)y''_{xx} + ab^2 x\ln^n x\, y'_x - ab^2\ln^n x\, y = 0.$

Particular solutions: $y_1 = x, \quad y_2 = e^{-bx}$.

13. $y'''_{xxx} + ax^2\ln^n(\lambda x)y''_{xx} - 2ax\ln^n(\lambda x)y'_x + 2a\ln^n(\lambda x)y = 0.$

Particular solutions: $y_1 = x, \quad y_2 = x^2$.

14. $xy'''_{xxx} + axy''_{xx} - b(bx\ln^2 x + 1)y'_x - ab(bx\ln^2 x + 1)y = 0.$

The substitution $w = y'_x + ay$ leads to an equation of the form 2.1.5.7: $xw''_{xx} - (b^2 x\ln^2 x + b)w = 0$.

15. $xy'''_{xxx} + a\ln^n(\lambda x)y''_{xx} + bxy'_x + ab\ln^n(\lambda x)y = 0.$

1°. Particular solutions with $b > 0$: $y_1 = \cos(x\sqrt{b}), \quad y_2 = \sin(x\sqrt{b})$.

2°. Particular solutions with $b < 0$: $y_1 = \exp(-x\sqrt{-b}), \quad y_2 = \exp(x\sqrt{-b})$.

16. $xy'''_{xxx} + ax\ln x\, y''_{xx} + (abx\ln x - b^2 x + a)y'_x + aby = 0.$

Particular solutions: $y_1 = e^{-bx}, \quad y_2 = e^{-bx}\int x^{-ax}e^{(a+2b)x}\,dx$.

17. $xy'''_{xxx} = (\ln^n x - ax)y''_{xx} + (a\ln^n x - bx)y'_x + b\ln^n x\, y.$

Particular solutions: $y_1 = \exp(\lambda_1 x), \quad y_2 = \exp(\lambda_2 x)$, where λ_1 and λ_2 are the roots of the quadratic equation $\lambda^2 + a\lambda + b = 0$.

18. $xy'''_{xxx} + (ax\ln^n x + 3)y''_{xx} + (2a\ln^n x + bx)y'_x + b(ax\ln^n x + 1)y = 0.$

Particular solutions: $y_1 = \dfrac{1}{x}\cos(x\sqrt{b}), \quad y_2 = \dfrac{1}{x}\sin(x\sqrt{b})$.

19. $xy'''_{xxx} + (ax\ln^n x + 3)y''_{xx} + (abx\ln^n x + 2a\ln^n x - b^2 x)y'_x + b(a\ln^n x - b)y = 0.$

Particular solutions: $y_1 = \dfrac{1}{x}, \quad y_2 = \dfrac{1}{x}e^{-bx}$.

20. $xy'''_{xxx} + [a(b - \ln x)x^n + 2]y''_{xx} + ax^{n-1}y'_x - ax^{n-2}y = 0.$

Particular solutions: $y_1 = x, \quad y_2 = \ln x - b + 1$.

21. $x^2 y'''_{xxx} + a\ln^n(\lambda x)y''_{xx} + bx^2 y'_x + ab\ln^n(\lambda x)y = 0.$

1°. Particular solutions with $b > 0$: $y_1 = \cos(x\sqrt{b}), \quad y_2 = \sin(x\sqrt{b})$.

2°. Particular solutions with $b < 0$: $y_1 = \exp(-x\sqrt{-b}), \quad y_2 = \exp(x\sqrt{-b})$.

22. $x^2 y'''_{xxx} + x^2(a \ln x + b) y''_{xx} + 2axy'_x - ay = 0.$

Integrating the equation twice, we obtain a first order linear equation:

$$y'_x + (a \ln x + b) y = C_1 + C_2 x.$$

23. $x^2 y'''_{xxx} - 3ax[ax \ln^2(\lambda x) + 1] y'_x + a[2a^2 x^2 \ln^3(\lambda x) + 1] y = 0.$

Particular solutions: $y_1 = \exp\left[a \int \ln(\lambda x)\, dx\right], \quad y_2 = x \exp\left[a \int \ln(\lambda x)\, dx\right].$

24. $x^2 y'''_{xxx} + x^2(a \ln x + bx) y''_{xx} + 2x(bx + a) y'_x - ay = 0.$

Integrating the equation twice, we obtain a first order linear equation:

$$y'_x + (a \ln x + bx) y = C_1 + C_2 x.$$

25. $x^2 y'''_{xxx} + (ax^2 \ln^n x + bx) y''_{xx} + [a(b-2)x \ln^n x + c] y'_x + a(c-b+2) \ln^n x\, y = 0.$

Particular solutions: $y_1 = x^{m_1}, \quad y_2 = x^{m_2}$, where m_1 and m_2 are the roots of the quadratic equation $m^2 + (b-3)m + c - b + 2 = 0$.

26. $x^3 y'''_{xxx} + x^2(a \ln x + b) y''_{xx} + 2axy'_x - ay = 0.$

Integrating the equation twice, we obtain a first order linear equation:

$$xy'_x + (a \ln x + b - 2) y = C_1 + C_2 x.$$

27. $x^3 y'''_{xxx} + a \ln^n(\lambda x) y''_{xx} + bx^3 y'_x + ab \ln^n(\lambda x) y = 0.$

1°. Particular solutions with $b > 0$: $y_1 = \cos\bigl(x\sqrt{b}\bigr), \quad y_2 = \sin\bigl(x\sqrt{b}\bigr).$

2°. Particular solutions with $b < 0$: $y_1 = \exp\bigl(-x\sqrt{-b}\bigr), \quad y_2 = \exp\bigl(x\sqrt{-b}\bigr).$

28. $x^3 y'''_{xxx} + ax^2 \ln^n x\, y''_{xx} - 2xy'_x + 2(2 - a \ln^n x) y = 0.$

Particular solutions: $y_1 = x^{-1}, \quad y_2 = x^2.$

29. $x^3 y'''_{xxx} + ax^2 \ln^n x\, y''_{xx} - 6xy'_x + 6(2 - a \ln^n x) y = 0.$

Particular solutions: $y_1 = x^{-2}, \quad y_2 = x^3.$

30. $x^3 y'''_{xxx} + x^2(a \ln x + bx) y''_{xx} + 2x(bx + a) y'_x - ay = 0.$

Integrating the equation twice, we obtain a first order linear equation:

$$xy'_x + (a \ln x + bx - 2) y = C_1 + C_2 x.$$

31. $x^3 y'''_{xxx} + x^2(a \ln^n x + b) y''_{xx} + x(ab \ln^n x + c - b) y'_x + c(a \ln^n x - 2) y = 0.$

Particular solutions: $y_1 = x^{m_1}, \quad y_2 = x^{m_2}$, where m_1 and m_2 are the roots of the quadratic equation $m^2 + (b-1)m + c = 0$.

32. $\ln^n x\, y'''_{xxx} + ay''_{xx} + aby'_x + b^2(a - b \ln^n x) y = 0.$

Particular solutions: $y_1 = e^{-bx/2} \cos\left(\dfrac{bx\sqrt{3}}{2}\right), \quad y_2 = e^{-bx/2} \sin\left(\dfrac{bx\sqrt{3}}{2}\right).$

33. $\ln^n x\, y'''_{xxx} + (ax\ln^n x + 1)y''_{xx} + a(x + 2\ln^n x)y'_x + ay = 0.$

Particular solutions: $y_1 = \exp\Bigl(-\dfrac{ax^2}{2}\Bigr)$, $y_2 = \exp\Bigl(-\dfrac{ax^2}{2}\Bigr)\int \exp\Bigl(\dfrac{ax^2}{2}\Bigr)\,dx.$

34. $\ln^n x\, y'''_{xxx} + (a\ln^n x + ax + 1)y''_{xx} + a^2x\,y'_x - a^2y = 0.$

Particular solutions: $y_1 = x$, $y_2 = e^{-ax}$.

35. $\ln^n(\lambda x)y'''_{xxx} + ax^2y''_{xx} - 2axy'_x + 2ay = 0.$

Particular solutions: $y_1 = x$, $y_2 = x^2$.

36. $\ln^n x\, y'''_{xxx} + y''_{xx} + [(a - b^2)\ln^n x + b]y'_x + a(1 - b\ln^n x)y = 0.$

Particular solutions: $y_1 = e^{\lambda_1 x}$, $y_2 = e^{\lambda_2 x}$, where λ_1 and λ_2 are the roots of the quadratic equation $\lambda^2 + b\lambda + a = 0$.

37. $\ln^n x\, y'''_{xxx} + ay''_{xx} - b(2a + 3b\ln^n x)y'_x + b^2(a + 2b\ln^n x)y = 0.$

Particular solutions: $y_1 = e^{bx}$, $y_2 = xe^{bx}$.

38. $\ln^n(\lambda x)y'''_{xxx} + ay''_{xx} + b\ln^n(\lambda x)y'_x + aby = 0.$

1°. Particular solutions with $b > 0$: $y_1 = \cos\bigl(x\sqrt{b}\bigr)$, $y_2 = \sin\bigl(x\sqrt{b}\bigr)$.

2°. Particular solutions with $b < 0$: $y_1 = \exp\bigl(-x\sqrt{-b}\bigr)$, $y_2 = \exp\bigl(x\sqrt{-b}\bigr)$.

3.1.6. Equations Containing Trigonometric Functions

1. $y'''_{xxx} + a^3\tan(ax)\,y = 0.$

By integrating, we obtain a nonhomogeneous second-order linear equation:

$$y''_{\xi\xi} + \tan\xi\, y'_\xi - y = \frac{C}{\cos\xi}, \qquad \xi = ax$$

(see 2.1.6.29 for the solution of the corresponding homogeneous equation).

2. $y'''_{xxx} - a^3\tan(ax)\,y = 0.$

Particular solution: $y_0 = \cos(ax)$.

The substitution $y = \cos(ax)\int z(x)\,dx$ leads to a second order linear equation of the form 2.1.6.29: $z''_{\xi\xi} - 3\tan\xi\, z'_\xi - 3z = 0$, where $\xi = ax$.

3. $y'''_{xxx} + a^3\cot(ax)\,y = 0.$

The substitution $x = t + \dfrac{\pi}{2a}$ leads to a linear equation of the form 3.1.6.2: $y'''_{ttt} - a^3\tan(at)\,y = 0$.

4. $y'''_{xxx} - a^3\cot(ax)\,y = 0.$

The substitution $x = t + \dfrac{\pi}{2a}$ leads to a linear equation of the form 3.1.6.1: $y'''_{ttt} + a^3\tan(at)\,y = 0$.

5. $\boldsymbol{y'''_{xxx} + 3a^2 y'_x + 2a^3 \tan(ax)\, y = 0.}$

Particular solutions: $y_1 = \cos(ax)$, $y_2 = x\cos(ax)$.

6. $\boldsymbol{y'''_{xxx} + a y'_x + \lambda(a - \lambda^2)\tan(\lambda x)\, y = 0.}$

Particular solution: $y_0 = \cos(\lambda x)$.

The substitution $y = \cos(\lambda x) \int z(x)\,dx$ leads to a second order equation of the form 2.2.6.29: $z''_{\xi\xi} - 3\tan\xi\, z'_\xi + (a\lambda^{-2} - 3)z = 0$, where $\xi = \lambda x$.

7. $\boldsymbol{y'''_{xxx} + 3a^2 y'_x - 2a^3 \cot(ax)\, y = 0.}$

Particular solutions: $y_1 = \sin(ax)$, $y_2 = x\sin(ax)$.

8. $\boldsymbol{y'''_{xxx} + a\cos(2x) y'_x - b[a\cos(2x) + b^2]y = 0.}$

The substitution $w = e^{bx/2}(y'_x - by)$ leads to the Mathieu equation 2.1.6.4: $w''_{xx} + [a\cos(2x) + \frac{3}{4}b^2]w = 0$.

9. $\boldsymbol{y'''_{xxx} + [a\cos(2x) + b]y'_x - a\sin(2x)\, y = 0.}$

Solution:

$$y = C_1 w_1^2 + C_2 w_1 w_2 + C_3 w_2^2,$$

where w_1 and w_2 is the fundamental set of solutions of the Mathieu equation 2.1.6.4: $4w''_{xx} + [a\cos(2x) + b]w = 0$.

10. $\boldsymbol{y'''_{xxx} + [a\cos(2x) + b]y'_x - 2a\sin(2x)\, y = 0.}$

By integrating, we obtain a nonhomogeneous Mathieu equation:

$$y''_{xx} + [a\cos(2x) + b]y = C.$$

11. $\boldsymbol{y'''_{xxx} + [a\cos(\lambda x) + b]y'_x - c[a\cos(\lambda x) + b + c^2]y = 0.}$

The transformation $\xi = \frac{1}{2}\lambda x$, $w = e^{cx/2}(y'_x - cy)$ leads to the Mathieu equation 2.1.6.4: $w''_{\xi\xi} + 4\lambda^{-2}[a\cos(2\xi) + b + \frac{3}{4}c^2]w = 0$.

12. $\boldsymbol{y'''_{xxx} + [a\cos(2x) - b^2]y'_x - a[b\cos(2x) + 2\sin(2x)]y = 0.}$

By integrating and substituting $w = ye^{bx/2}$, we obtain a nonhomogeneous Mathieu equation: $w''_{xx} + [a\cos(2x) - \frac{1}{4}b^2]w = Ce^{3bx/2}$.

13. $\boldsymbol{y'''_{xxx} + a\sin x\, y'_x - b(a\sin x + b^2)y = 0.}$

The substitution $w = e^{bx/2}(y'_x - by)$ leads to an equation of the form 2.1.6.3: $w''_{xx} + (a\sin x + \frac{3}{4}b^2)w = 0$.

14. $\boldsymbol{y'''_{xxx} + a\sin^2 x\, y'_x - b(a\sin^2 x + b^2)y = 0.}$

The substitution $w = e^{bx/2}(y'_x - by)$ leads to an equation of the form 2.1.6.5: $w''_{xx} + (a\sin^2 x + \frac{3}{4}b^2)w = 0$.

15. $\boldsymbol{y'''_{xxx} + [a\sin(\lambda x) + b]y'_x + a\lambda\cos(\lambda x)\,y = 0.}$

By integrating, we obtain a nonhomogeneous second-order linear equation:

$$y''_{xx} + [a\sin(\lambda x) + b]y = C$$

(see 2.1.6.3 for the solution of the corresponding homogeneous equation).

16. $\boldsymbol{y'''_{xxx} + [a\sin(\lambda x) - b^2]y'_x + a[\lambda\cos(\lambda x) - b\sin(\lambda x)]y = 0.}$

By integrating and substituting $w = ye^{bx/2}$, we obtain a nonhomogeneous second-order linear equation:

$$w''_{xx} + [a\sin(\lambda x) - \tfrac{1}{4}b^2]w = Ce^{3bx/2}$$

(see 2.1.6.3 for the solution of the corresponding homogeneous equation).

17. $\boldsymbol{y'''_{xxx} + [a\sin(\lambda x) + b]y'_x - c[a\sin(\lambda x) + b + c^2]y = 0.}$

The substitution $w = e^{cx/2}(y'_x - cy)$ leads to an equation of the form 2.1.6.3: $w''_{xx} + [a\sin(\lambda x) + b + \frac{3}{4}c^2]w = 0$.

18. $\boldsymbol{y'''_{xxx} - 3a[a\sin^2(bx) + b\cos(bx)]y'_x + a\sin(bx)[b^2 + 2a^2\sin^2(bx)]y = 0.}$

Particular solutions: $y_1 = \exp\Big[-\dfrac{a}{b}\cos(bx)\Big]$, $y_2 = x\exp\Big[-\dfrac{a}{b}\cos(bx)\Big]$.

19. $\boldsymbol{y'''_{xxx} + a\tan^2 x\, y'_x - b(a\tan^2 x + b^2)y = 0.}$

The substitution $w = e^{bx/2}(y'_x - by)$ leads to an equation of the form 2.1.6.10: $w''_{xx} + (a\tan^2 x + \frac{3}{4}b^2)w = 0$.

20. $\boldsymbol{y'''_{xxx} + [a\tan^2(\lambda x) + b]y'_x - c[a\tan^2(\lambda x) + b + c^2]y = 0.}$

The transformation $\xi = \lambda x$, $w = e^{cx/2}(y'_x - cy)$ leads to an equation of the form 2.1.6.10: $w''_{\xi\xi} + \lambda^{-2}(a\tan^2\xi + b + \frac{3}{4}c^2)w = 0$.

21. $\boldsymbol{y'''_{xxx} + a\cot^2 x\, y'_x - b(a\cot^2 x + b^2)y = 0.}$

The substitution $w = e^{bx/2}(y'_x - by)$ leads to an equation of the form 2.1.6.12: $w''_{xx} + (a\cot^2 x + \frac{3}{4}b^2)w = 0$.

22. $\boldsymbol{y'''_{xxx} + ay''_{xx} + (b\cos 2x + c)y'_x + a(b\cos 2x + c)y = 0.}$

The substitution $w = y'_x + ay$ reduces the original equation to the Mathieu equation 2.1.6.4: $w''_{xx} + (b\cos 2x + c)w = 0$.

23. $\boldsymbol{y'''_{xxx} + ay''_{xx} + [b\sin(\lambda x) + c]y'_x + a[b\sin(\lambda x) + c]y = 0.}$

The substitution $w = y'_x + ay$ leads to a second order linear equation of the form 2.1.6.3: $w''_{xx} + [b\sin(\lambda x) + c]w = 0$.

24. $\boldsymbol{y'''_{xxx} + ay''_{xx} + b\sin^2(\lambda x)y'_x + ab\sin^2(\lambda x)\,y = 0.}$

The substitution $w = y'_x + ay$ leads to a second order linear equation of the form 2.1.6.5: $w''_{xx} + b\sin^2(\lambda x)\,w = 0$.

25. $y'''_{xxx} + ay''_{xx} + (b\tan^2 x + c)y'_x + a(b\tan^2 x + c)y = 0.$

The substitution $w = y'_x + ay$ leads to a second order linear equation of the form 2.1.6.10: $w''_{xx} + (b\tan^2 x + c)w = 0$.

26. $y'''_{xxx} + ay''_{xx} + \lambda[3\lambda + 2a\tan(\lambda x)]y'_x + \lambda^2\{a[1 + 2\tan^2(\lambda x)] + 2\lambda\tan(\lambda x)\}y = 0.$

Particular solutions: $y_1 = \cos(\lambda x)$, $y_2 = x\cos(\lambda x)$.

27. $y'''_{xxx} + ay''_{xx} + (b\cot^2 x + c)y'_x + a(b\cot^2 x + c)y = 0.$

The substitution $w = y'_x + ay$ leads to a second order linear equation of the form 2.1.6.12: $w''_{xx} + (b\cot^2 x + c)w = 0$.

28. $y'''_{xxx} + ay''_{xx} + \lambda[3\lambda - 2a\cot(\lambda x)]y'_x + \lambda^2\{a[1 + 2\cot^2(\lambda x)] - 2\lambda\cot(\lambda x)\}y = 0.$

Particular solutions: $y_1 = \sin(\lambda x)$, $y_2 = x\sin(\lambda x)$.

29. $y'''_{xxx} + a\cos^n(\lambda x)y''_{xx} + by'_x + ab\cos^n(\lambda x)\, y = 0.$

1°. Particular solutions with $b > 0$: $y_1 = \cos\bigl(x\sqrt{b}\,\bigr)$, $y_2 = \sin\bigl(x\sqrt{b}\,\bigr)$.

2°. Particular solutions with $b < 0$: $y_1 = \exp\bigl(-x\sqrt{-b}\,\bigr)$, $y_2 = \exp\bigl(x\sqrt{-b}\,\bigr)$.

30. $y'''_{xxx} + a\cos^n(\lambda x)y''_{xx} + ab\cos^n(\lambda x)y'_x + b^2[a\cos^n(\lambda x) - b]y = 0.$

Particular solutions: $y_1 = e^{-bx/2}\cos\left(\dfrac{bx\sqrt{3}}{2}\right)$, $y_2 = e^{-bx/2}\sin\left(\dfrac{bx\sqrt{3}}{2}\right)$.

31. $y'''_{xxx} + a\cos^n(\lambda x)y''_{xx} - b[2a\cos^n(\lambda x) + 3b]y'_x + b^2[a\cos^n(\lambda x) + 2b]y = 0.$

Particular solutions: $y_1 = e^{bx}$, $y_2 = xe^{bx}$.

32. $y'''_{xxx} + a\cos^n x\, y''_{xx} + (ab\cos^n x + c - b^2)y'_x + c(a\cos^n x - b)y = 0.$

Particular solutions: $y_1 = \exp(\lambda_1 x)$, $y_2 = \exp(\lambda_2 x)$, where λ_1 and λ_2 are the roots of the quadratic equation $\lambda^2 + b\lambda + c = 0$.

33. $y'''_{xxx} + a\cos^n(\lambda x)y''_{xx} + bx^m y'_x + bx^{m-1}[ax\cos^n(\lambda x) + m]y = 0.$

Assuming $w = y''_{xx} + bx^m y$ yields a first order linear equation: $w'_x + a\cos^n(\lambda x)\, w = 0$.

34. $y'''_{xxx} + (a\cos^n x + b)y''_{xx} + cy'_x + c(a\cos^n x + b)y = 0.$

1°. Particular solutions with $c > 0$: $y_1 = \cos\bigl(x\sqrt{c}\,\bigr)$, $y_2 = \sin\bigl(x\sqrt{c}\,\bigr)$.

2°. Particular solutions with $c < 0$: $y_1 = \exp\bigl(-x\sqrt{-c}\,\bigr)$, $y_2 = \exp\bigl(x\sqrt{-c}\,\bigr)$.

35. $y'''_{xxx} = (\cos^n x - a)y''_{xx} + (a\cos^n x - b)y'_x + b\cos^n x\, y.$

Particular solutions: $y_1 = \exp(\lambda_1 x)$, $y_2 = \exp(\lambda_2 x)$, where λ_1 and λ_2 are the roots of the quadratic equation $\lambda^2 + a\lambda + b = 0$.

36. $y'''_{xxx} + (a\cos^n x + bx)y''_{xx} + b(ax\cos^n x + 2)y'_x + ab\cos^n x\, y = 0.$

Particular solutions: $y_1 = \exp(-\tfrac{1}{2}bx^2)$, $y_2 = \exp(-\tfrac{1}{2}bx^2)\displaystyle\int \exp(\tfrac{1}{2}bx^2)\, dx$.

37. $y'''_{xxx} + ax\cos^n x\, y''_{xx} + (bx^2 - a\cos^n x)y'_x + bx(ax^2\cos^n x + 3)y = 0.$

Particular solutions: $y_1 = \cos\bigl(\tfrac{1}{2}x^2\sqrt{b}\bigr)$, $y_2 = \sin\bigl(\tfrac{1}{2}x^2\sqrt{b}\bigr)$.

38. $y'''_{xxx} + (abx\cos^n x + a\cos^n x + b)y''_{xx} + ab^2x\cos^n x\, y'_x - ab^2\cos^n x\, y = 0.$

Particular solutions: $y_1 = x$, $y_2 = e^{-bx}$.

39. $y'''_{xxx} + ax^2\cos^n(\lambda x)y''_{xx} - 2ax\cos^n(\lambda x)y'_x + 2a\cos^n(\lambda x)\, y = 0.$

Particular solutions: $y_1 = x$, $y_2 = x^2$.

40. $y'''_{xxx} + a\sin^n(\lambda x)y''_{xx} - by'_x - ab\sin^n(\lambda x)\, y = 0.$

1°. Particular solutions with $b > 0$: $y_1 = \exp\bigl(-x\sqrt{b}\bigr)$, $y_2 = \exp\bigl(x\sqrt{b}\bigr)$.

2°. Particular solutions with $b < 0$: $y_1 = \cos\bigl(x\sqrt{-b}\bigr)$, $y_2 = \sin\bigl(x\sqrt{-b}\bigr)$.

41. $y'''_{xxx} + a\sin^n(\lambda x)y''_{xx} + ab\sin^n(\lambda x)y'_x + b^2[a\sin^n(\lambda x) - b]y = 0.$

Particular solutions: $y_1 = e^{-bx/2}\cos\left(\dfrac{bx\sqrt{3}}{2}\right)$, $y_2 = e^{-bx/2}\sin\left(\dfrac{bx\sqrt{3}}{2}\right)$.

42. $y'''_{xxx} + a\sin^n(\lambda x)y''_{xx} - b[2a\sin^n(\lambda x) + 3b]y'_x + b^2[a\sin^n(\lambda x) + 2b]y = 0.$

Particular solutions: $y_1 = e^{bx}$, $y_2 = xe^{bx}$.

43. $y'''_{xxx} + a\sin^n x\, y''_{xx} + (ab\sin^n x + c - b^2)y'_x + c(a\sin^n x - b)y = 0.$

Particular solutions: $y_1 = \exp(\lambda_1 x)$, $y_2 = \exp(\lambda_2 x)$, where λ_1 and λ_2 are the roots of the quadratic equation $\lambda^2 + b\lambda + c = 0$.

44. $y'''_{xxx} + a\sin^n(\lambda x)y''_{xx} + bx^m y'_x + bx^{m-1}[ax\sin^n(\lambda x) + m]y = 0.$

Assuming $w = y''_{xx} + bx^m y$ yields a first order linear equation: $w'_x + a\sin^n(\lambda x)\, w = 0$.

45. $y'''_{xxx} + (a\sin^n x + b)y''_{xx} + cy'_x + c(a\sin^n x + b)y = 0.$

1°. Particular solutions with $c > 0$: $y_1 = \cos\bigl(x\sqrt{c}\bigr)$, $y_2 = \sin\bigl(x\sqrt{c}\bigr)$.

2°. Particular solutions with $c < 0$: $y_1 = \exp\bigl(-x\sqrt{-c}\bigr)$, $y_2 = \exp\bigl(x\sqrt{-c}\bigr)$.

46. $y'''_{xxx} = (\sin^n x - a)y''_{xx} + (a\sin^n x - b)y'_x + b\sin^n x\, y.$

Particular solutions: $y_1 = \exp(\lambda_1 x)$, $y_2 = \exp(\lambda_2 x)$, where λ_1 and λ_2 are the roots of the quadratic equation $\lambda^2 + a\lambda + b = 0$.

47. $y'''_{xxx} + (a\sin^n x + bx)y''_{xx} + b(ax\sin^n x + 2)y'_x + ab\sin^n x\, y = 0.$

Particular solutions: $y_1 = \exp\bigl(-\tfrac{1}{2}bx^2\bigr)$, $y_2 = \exp\bigl(-\tfrac{1}{2}bx^2\bigr)\displaystyle\int \exp\bigl(\tfrac{1}{2}bx^2\bigr)\,dx$.

48. $y'''_{xxx} + ax\sin^n x\, y''_{xx} + (bx^2 - a\sin^n x)y'_x + bx(ax^2\sin^n x + 3)y = 0.$

Particular solutions: $y_1 = \cos\bigl(\tfrac{1}{2}x^2\sqrt{b}\bigr)$, $y_2 = \sin\bigl(\tfrac{1}{2}x^2\sqrt{b}\bigr)$.

49. $y'''_{xxx} + (abx\sin^n x + a\sin^n x + b)y''_{xx} + ab^2 x\sin^n x\, y'_x - ab^2\sin^n x\, y = 0.$

Particular solutions: $y_1 = x, \quad y_2 = e^{-bx}.$

50. $y'''_{xxx} + ax^2\sin^n(\lambda x)y''_{xx} - 2ax\sin^n(\lambda x)y'_x + 2a\sin^n(\lambda x)\, y = 0.$

Particular solutions: $y_1 = x, \quad y_2 = x^2.$

51. $y'''_{xxx} + \lambda\tan(\lambda x)y''_{xx} - ay'_x - a\lambda\tan(\lambda x)\, y = 0.$

1°. Solution with $a > 0$: $y = C_1\exp(-x\sqrt{a}) + C_2\exp(x\sqrt{a}) + C_3\cos(\lambda x).$

2°. Solution with $a < 0$: $y = C_1\cos(x\sqrt{-a}) + C_2\sin(x\sqrt{-a}) + C_3\cos(\lambda x).$

52. $y'''_{xxx} + a\tan(\lambda x)y''_{xx} + by'_x + \lambda(a\lambda + b - \lambda^2)\tan(\lambda x)\, y = 0.$

Particular solution: $y_0 = \cos(\lambda x).$

The transformation $x = \dfrac{\xi}{\lambda}$, $y = \cos(\lambda x)\displaystyle\int z\,dx$ leads to a second order equation of the form 2.1.6.55:

$$z''_{\xi\xi} + \Big(\frac{a}{\lambda} - 3\Big)\tan\xi\, z'_\xi + \Big(\frac{b}{\lambda^2} - 3 - \frac{2a}{\lambda}\tan^2\xi\Big)z = 0.$$

53. $y'''_{xxx} + a\tan^n(\lambda x)y''_{xx} + by'_x + ab\tan^n(\lambda x)\, y = 0.$

1°. Particular solutions with $b > 0$: $y_1 = \cos(x\sqrt{b}), \quad y_2 = \sin(x\sqrt{b}).$

2°. Particular solutions with $b < 0$: $y_1 = \exp(-x\sqrt{-b}), \quad y_2 = \exp(x\sqrt{-b}).$

54. $y'''_{xxx} + a\tan^n(\lambda x)y''_{xx} + ab\tan^n(\lambda x)y'_x + b^2[a\tan^n(\lambda x) - b]y = 0.$

Particular solutions: $y_1 = e^{-bx/2}\cos\Big(\dfrac{bx\sqrt{3}}{2}\Big), \quad y_2 = e^{-bx/2}\sin\Big(\dfrac{bx\sqrt{3}}{2}\Big).$

55. $y'''_{xxx} + a\tan^n(\lambda x)y''_{xx} - b[2a\tan^n(\lambda x) + 3b]y'_x + b^2[a\tan^n(\lambda x) + 2b]y = 0.$

Particular solutions: $y_1 = e^{bx}, \quad y_2 = xe^{bx}.$

56. $y'''_{xxx} + a\tan^n x\, y''_{xx} + (ab\tan^n x + c - b^2)y'_x + c(a\tan^n x - b)y = 0.$

Particular solutions: $y_1 = \exp(\lambda_1 x)$, $y_2 = \exp(\lambda_2 x)$, where λ_1 and λ_2 are the roots of the quadratic equation $\lambda^2 + b\lambda + c = 0.$

57. $y'''_{xxx} + a\tan^n(\lambda x)y''_{xx} + bx^m y'_x + bx^{m-1}[ax\tan^n(\lambda x) + m]y = 0.$

Assuming $w = y''_{xx} + bx^m y$ yields a first order linear equation: $w'_x + a\tan^n(\lambda x)\, w = 0.$

58. $y'''_{xxx} + (a\tan^n x + b)y''_{xx} + cy'_x + c(a\tan^n x + b)y = 0.$

1°. Particular solutions with $c > 0$: $y_1 = \cos(x\sqrt{c}), \quad y_2 = \sin(x\sqrt{c}).$

2°. Particular solutions with $c < 0$: $y_1 = \exp(-x\sqrt{-c}), \quad y_2 = \exp(x\sqrt{-c}).$

59. $y'''_{xxx} = (\tan^n x - a)y''_{xx} + (a\tan^n x - b)y'_x + b\tan^n x\, y.$

Particular solutions: $y_1 = \exp(\lambda_1 x)$, $y_2 = \exp(\lambda_2 x)$, where λ_1 and λ_2 are the roots of the quadratic equation $\lambda^2 + a\lambda + b = 0.$

60. $y'''_{xxx} + (a\tan^n x + bx)y''_{xx} + b(ax\tan^n x + 2)y'_x + ab\tan^n x\, y = 0.$

Particular solutions: $y_1 = \exp\bigl(-\tfrac{1}{2}bx^2\bigr)$, $y_2 = \exp\bigl(-\tfrac{1}{2}bx^2\bigr)\int \exp\bigl(\tfrac{1}{2}bx^2\bigr)\,dx$.

61. $y'''_{xxx} + ax\tan^n x\, y''_{xx} + (bx^2 - a\tan^n x)y'_x + bx(ax^2\tan^n x + 3)y = 0.$

Particular solutions: $y_1 = \cos\bigl(\tfrac{1}{2}x^2\sqrt{b}\bigr)$, $y_2 = \sin\bigl(\tfrac{1}{2}x^2\sqrt{b}\bigr)$.

62. $y'''_{xxx} + (abx\tan^n x + a\tan^n x + b)y''_{xx} + ab^2x\tan^n x\, y'_x - ab^2\tan^n x\, y = 0.$

Particular solutions: $y_1 = x$, $y_2 = e^{-bx}$.

63. $y'''_{xxx} + ax^2\tan^n(\lambda x)y''_{xx} - 2ax\tan^n(\lambda x)y'_x + 2a\tan^n(\lambda x)\, y = 0.$

Particular solutions: $y_1 = x$, $y_2 = x^2$.

64. $y'''_{xxx} - [b(a+\tan x)x^n + a]y''_{xx} + [b(a^2+1)x^n + 1]y'_x + a[b(a\tan x - 1)x^n - 1]y = 0.$

Particular solutions: $y_1 = e^{ax}$, $y_2 = \cos x$.

65. $y'''_{xxx} - (ab\tan^n x + b\tan^{n+1} x + a)y''_{xx} + [b(a^2+1)\tan^n x + 1]y'_x + a(ab\tan^{n+1} x - b\tan^n x - 1)y = 0.$

Particular solutions: $y_1 = e^{ax}$, $y_2 = \cos x$.

66. $y'''_{xxx} + [\lambda\tan(\lambda x)(ax^n + 1) + ax^{n-1}]y''_{xx} - a\lambda^2x^ny'_x + a\lambda^2x^{n-1}y = 0.$

Particular solutions: $y_1 = x$, $y_2 = \cos(\lambda x)$.

67. $y'''_{xxx} - \lambda\cot(\lambda x)y''_{xx} - ay'_x + a\lambda\cot(\lambda x)\, y = 0.$

1°. Solution with $a > 0$: $y = C_1\exp\bigl(-x\sqrt{a}\bigr) + C_2\exp\bigl(x\sqrt{a}\bigr) + C_3\sin(\lambda x)$.

2°. Solution with $a < 0$: $y = C_1\cos\bigl(x\sqrt{-a}\bigr) + C_2\sin\bigl(x\sqrt{-a}\bigr) + C_3\sin(\lambda x)$.

68. $y'''_{xxx} + a\cot^n(\lambda x)y''_{xx} + ab\cot^n(\lambda x)y'_x + b^2[a\cot^n(\lambda x) - b]y = 0.$

Particular solutions: $y_1 = e^{-bx/2}\cos\left(\dfrac{bx\sqrt{3}}{2}\right)$, $y_2 = e^{-bx/2}\sin\left(\dfrac{bx\sqrt{3}}{2}\right)$.

69. $y'''_{xxx} + a\cot^n(\lambda x)y''_{xx} - b[2a\cot^n(\lambda x) + 3b]y'_x + b^2[a\cot^n(\lambda x) + 2b]y = 0.$

Particular solutions: $y_1 = e^{bx}$, $y_2 = xe^{bx}$.

70. $y'''_{xxx} + a\cot^n x\, y''_{xx} + (ab\cot^n x + c - b^2)y'_x + c(a\cot^n x - b)y = 0.$

Particular solutions: $y_1 = \exp(\lambda_1 x)$, $y_2 = \exp(\lambda_2 x)$, where λ_1 and λ_2 are the roots of the quadratic equation $\lambda^2 + b\lambda + c = 0$.

71. $y'''_{xxx} + a\cot^n(\lambda x)y''_{xx} + bx^my'_x + bx^{m-1}[ax\cot^n(\lambda x) + m]y = 0.$

Assuming $w = y''_{xx} + bx^my$ yields a first order linear equation: $w'_x + a\cot^n(\lambda x)\, w = 0$.

72. $y'''_{xxx} + (a\cot^n x + b)y''_{xx} + cy'_x + c(a\cot^n x + b)y = 0.$

1°. Particular solutions with $c > 0$: $y_1 = \cos(x\sqrt{c})$, $y_2 = \sin(x\sqrt{c})$.

2°. Particular solutions with $c < 0$: $y_1 = \exp(-x\sqrt{-c})$, $y_2 = \exp(x\sqrt{-c})$.

73. $y'''_{xxx} = (\cot^n x - a)y''_{xx} + (a\cot^n x - b)y'_x + b\cot^n x\, y.$

Particular solutions: $y_1 = \exp(\lambda_1 x)$, $y_2 = \exp(\lambda_2 x)$, where λ_1 and λ_2 are the roots of the quadratic equation $\lambda^2 + a\lambda + b = 0$.

74. $y'''_{xxx} + ax\cot^n x\, y''_{xx} + (bx^2 - a\cot^n x)y'_x + bx(ax^2\cot^n x + 3)y = 0.$

Particular solutions: $y_1 = \cos\left(\frac{1}{2}x^2\sqrt{b}\right)$, $y_2 = \sin\left(\frac{1}{2}x^2\sqrt{b}\right)$.

75. $y'''_{xxx} + (abx\cot^n x + a\cot^n x + b)y''_{xx} + ab^2 x\cot^n x\, y'_x - ab^2\cot^n x\, y = 0.$

Particular solutions: $y_1 = x$, $y_2 = e^{-bx}$.

76. $y'''_{xxx} + ax^2\cot^n(\lambda x)y''_{xx} - 2ax\cot^n(\lambda x)y'_x + 2a\cot^n(\lambda x)\, y = 0.$

Particular solutions: $y_1 = x$, $y_2 = x^2$.

77. $xy'''_{xxx} + x(a\cos 2x + b)y'_x - 2(a\cos 2x + b)y = 0.$

The substitution $w = xy'_x - 2y$ leads to the Mathieu equation 2.1.6.4:

$$w''_{xx} + (a\cos 2x + b)w = 0.$$

78. $xy'''_{xxx} + x(a\cos^2 x + b)y'_x - 2(a\cos^2 x + b)y = 0.$

The substitution $w = xy'_x - 2y$ leads to a second order linear equation of the form 2.1.6.6: $w''_{xx} + (a\cos^2 x + b)w = 0$.

79. $xy'''_{xxx} = (\cos^n x - ax)y''_{xx} + (a\cos^n x - bx)y'_x + b\cos^n x\, y.$

Particular solutions: $y_1 = \exp(\lambda_1 x)$, $y_2 = \exp(\lambda_2 x)$, where λ_1 and λ_2 are the roots of the quadratic equation $\lambda^2 + a\lambda + b = 0$.

80. $xy'''_{xxx} + (ax\cos^n x + 3)y''_{xx} + (2a\cos^n x + bx)y'_x + b(ax\cos^n x + 1)y = 0.$

Particular solutions: $y_1 = \dfrac{1}{x}\cos\left(x\sqrt{b}\right)$, $y_2 = \dfrac{1}{x}\sin\left(x\sqrt{b}\right)$.

81. $xy'''_{xxx} + x[a\sin(\lambda x) + b]y'_x - 2[a\sin(\lambda x) + b]y = 0.$

The substitution $w = xy'_x - 2y$ leads to an equation of the form 2.1.6.3: $w''_{xx} + [a\sin(\lambda x) + b]w = 0$.

82. $xy'''_{xxx} = (\sin^n x - ax)y''_{xx} + (a\sin^n x - bx)y'_x + b\sin^n x\, y.$

Particular solutions: $y_1 = \exp(\lambda_1 x)$, $y_2 = \exp(\lambda_2 x)$, where λ_1 and λ_2 are the roots of the quadratic equation $\lambda^2 + a\lambda + b = 0$.

83. $xy'''_{xxx} + (ax\sin^n x + 3)y''_{xx} + (2a\sin^n x + bx)y'_x + b(ax\sin^n x + 1)y = 0.$

Particular solutions: $y_1 = \dfrac{1}{x}\cos\bigl(x\sqrt{b}\bigr)$, $y_2 = \dfrac{1}{x}\sin\bigl(x\sqrt{b}\bigr)$.

84. $xy'''_{xxx} + x(a\tan^2 x + b)y'_x - 2(a\tan^2 x + b)y = 0.$

The substitution $w = xy'_x - 2y$ leads to an equation of the form 2.1.6.10: $w''_{xx} + (a\tan^2 x + b)w = 0$.

85. $xy'''_{xxx} = (\tan^n x - ax)y''_{xx} + (a\tan^n x - bx)y'_x + b\tan^n x\, y.$

Particular solutions: $y_1 = \exp(\lambda_1 x)$, $y_2 = \exp(\lambda_2 x)$, where λ_1 and λ_2 are the roots of the quadratic equation $\lambda^2 + a\lambda + b = 0$.

86. $xy'''_{xxx} + (ax\tan^n x + 3)y''_{xx} + (2a\tan^n x + bx)y'_x + b(ax\tan^n x + 1)y = 0.$

Particular solutions: $y_1 = \dfrac{1}{x}\cos\bigl(x\sqrt{b}\bigr)$, $y_2 = \dfrac{1}{x}\sin\bigl(x\sqrt{b}\bigr)$.

87. $axy'''_{xxx} + [1 - \lambda(a+1)x\cot(\lambda x)]y''_{xx} - \lambda^2 xy'_x + \lambda^2 y = 0.$

Particular solutions: $y_1 = x$, $y_2 = \sin(\lambda x)$.

88. $xy'''_{xxx} + x(a\cot^2 x + b)y'_x - 2(a\cot^2 x + b)y = 0.$

The substitution $w = xy'_x - 2y$ leads to an equation of the form 2.1.6.12: $w''_{xx} + (a\cot^2 x + b)w = 0$.

89. $xy'''_{xxx} + (ax\cot^n x + 3)y''_{xx} + (2a\cot^n x + bx)y'_x + b(ax\cot^n x + 1)y = 0.$

Particular solutions: $y_1 = \dfrac{1}{x}\cos\bigl(x\sqrt{b}\bigr)$, $y_2 = \dfrac{1}{x}\sin\bigl(x\sqrt{b}\bigr)$.

90. $x^2y'''_{xxx} = (\cos^n x - ax^2)y''_{xx} + (a\cos^n x - bx^2)y'_x + b\cos^n x\, y.$

Particular solutions: $y_1 = \exp(\lambda_1 x)$, $y_2 = \exp(\lambda_2 x)$, where λ_1 and λ_2 are the roots of the quadratic equation $\lambda^2 + a\lambda + b = 0$.

91. $x^2y'''_{xxx} = (\sin^n x - ax^2)y''_{xx} + (a\sin^n x - bx^2)y'_x + b\sin^n x\, y.$

Particular solutions: $y_1 = \exp(\lambda_1 x)$, $y_2 = \exp(\lambda_2 x)$, where λ_1 and λ_2 are the roots of the quadratic equation $\lambda^2 + a\lambda + b = 0$.

92. $x^2y'''_{xxx} = (\tan^n x - ax^2)y''_{xx} + (a\tan^n x - bx^2)y'_x + b\tan^n x\, y.$

Particular solutions: $y_1 = \exp(\lambda_1 x)$, $y_2 = \exp(\lambda_2 x)$, where λ_1 and λ_2 are the roots of the quadratic equation $\lambda^2 + a\lambda + b = 0$.

93. $x^2y'''_{xxx} = (\cot^n x - ax^2)y''_{xx} + (a\cot^n x - bx^2)y'_x + b\cot^n x\, y.$

Particular solutions: $y_1 = \exp(\lambda_1 x)$, $y_2 = \exp(\lambda_2 x)$, where λ_1 and λ_2 are the roots of the quadratic equation $\lambda^2 + a\lambda + b = 0$.

94. $x^3 y'''_{xxx} + ax^2 \cos^n(\lambda x) y''_{xx} + bxy'_x + b[a\cos^n(\lambda x) - 2]y = 0.$

Particular solutions: $y_1 = x^{m_1}$, $y_2 = x^{m_2}$, where m_1 and m_2 are the roots of the quadratic equation $m^2 - m + b = 0$.

95. $x^3 y'''_{xxx} + ax^2 \sin^n(\lambda x) y''_{xx} + bxy'_x + b[a\sin^n(\lambda x) - 2]y = 0.$

Particular solutions: $y_1 = x^{m_1}$, $y_2 = x^{m_2}$, where m_1 and m_2 are the roots of the quadratic equation $m^2 - m + b = 0$.

96. $x^3 y'''_{xxx} + ax^2 \tan^n(\lambda x) y''_{xx} + bxy'_x + b[a\tan^n(\lambda x) - 2]y = 0.$

Particular solutions: $y_1 = x^{m_1}$, $y_2 = x^{m_2}$, where m_1 and m_2 are the roots of the quadratic equation $m^2 - m + b = 0$.

97. $x^3 y'''_{xxx} + ax^2 \cot^n(\lambda x) y''_{xx} + bxy'_x + b[a\cot^n(\lambda x) - 2]y = 0.$

Particular solutions: $y_1 = x^{m_1}$, $y_2 = x^{m_2}$, where m_1 and m_2 are the roots of the quadratic equation $m^2 - m + b = 0$.

98. $\cos^2 x\, y'''_{xxx} + a\cos^2 x\, y''_{xx} + by'_x + aby = 0.$

The substitution $x = \xi + \frac{\pi}{2}$ leads to an equation of the form 3.1.6.112: $\sin^2\xi\, y'''_{\xi\xi\xi} + a\sin^2\xi\, y''_{\xi\xi} + by'_\xi + aby = 0$.

99. $\cos^n x\, y'''_{xxx} + ay''_{xx} + aby'_x + b^2(a - b\cos^n x)y = 0.$

Particular solutions: $y_1 = e^{-bx/2}\cos\left(\dfrac{bx\sqrt{3}}{2}\right)$, $y_2 = e^{-bx/2}\sin\left(\dfrac{bx\sqrt{3}}{2}\right)$.

100. $\cos^n x\, y'''_{xxx} + ay''_{xx} + b\cos^n x\, y'_x + aby = 0.$

1°. Particular solutions with $b > 0$: $y_1 = \cos(x\sqrt{b})$, $y_2 = \sin(x\sqrt{b})$.

2°. Particular solutions with $b < 0$: $y_1 = \exp(-x\sqrt{-b})$, $y_2 = \exp(x\sqrt{-b})$.

101. $\cos^n x\, y'''_{xxx} + ay''_{xx} - b(2a + 3b\cos^n x)y'_x + b^2(a + 2b\cos^n x)y = 0.$

Particular solutions: $y_1 = e^{bx}$, $y_2 = xe^{bx}$.

102. $\cos^n x\, y'''_{xxx} + y''_{xx} + [(b - a^2)\cos^n x + a]y'_x + b(1 - a\cos^n x)y = 0.$

Particular solutions: $y_1 = e^{\lambda_1 x}$, $y_2 = e^{\lambda_2 x}$, where λ_1 and λ_2 are the roots of the quadratic equation $\lambda^2 + a\lambda + b = 0$.

103. $\cos^n(\lambda x) y'''_{xxx} + ax^2 y''_{xx} - 2axy'_x + 2ay = 0.$

Particular solutions: $y_1 = x$, $y_2 = x^2$.

104. $\cos^n x\, y'''_{xxx} + (a\cos^n x + ax + 1)y''_{xx} + a^2 xy'_x - a^2 y = 0.$

Particular solutions: $y_1 = x$, $y_2 = e^{-ax}$.

105. $\cos^n x\, y'''_{xxx} + (ax\cos^n x + 1)y''_{xx} + a(x + 2\cos^n x)y'_x + ay = 0.$

Particular solutions: $y_1 = \exp\Bigl(-\dfrac{ax^2}{2}\Bigr), \quad y_2 = \exp\Bigl(-\dfrac{ax^2}{2}\Bigr)\displaystyle\int \exp\Bigl(\dfrac{ax^2}{2}\Bigr)\,dx.$

106. $x\cos^n x\, y'''_{xxx} + (3\cos^n x + x)y''_{xx} + (ax\cos^n x + 2)y'_x + a(\cos^n x + x)y = 0.$

Particular solutions: $y_1 = \dfrac{1}{x}\cos\bigl(x\sqrt{a}\bigr), \quad y_2 = \dfrac{1}{x}\sin\bigl(x\sqrt{a}\bigr).$

107. $x^3\cos^n x\, y'''_{xxx} + ax^2 y''_{xx} - 2x\cos^n x\, y'_x + 2(2\cos^n x - a)y = 0.$

Particular solutions: $y_1 = x^{-1}, \quad y_2 = x^2.$

108. $x^3\cos^n x\, y'''_{xxx} + ax^2 y''_{xx} - 6x\cos^n x\, y'_x + 6(2\cos^n x - a)y = 0.$

Particular solutions: $y_1 = x^{-2}, \quad y_2 = x^3.$

109. $x^3\cos^n x\, y'''_{xxx} + ax^2 y''_{xx} + x(a - \cos^n x)y'_x + (a - 3\cos^n x)y = 0.$

Particular solutions: $y_1 = \cos(\ln|x|), \quad y_2 = \sin(\ln|x|).$

110. $x^3\cos^n x\, y'''_{xxx} + x^2(\cos^n x + a)y''_{xx} + x[a - (b+1)\cos^n x]y'_x + b(2\cos^n x - a)y = 0.$

Particular solutions: $y_1 = x^{-\sqrt{b}}, \quad y_2 = x^{\sqrt{b}}.$

111. $\sin^2 x\, y'''_{xxx} + 3\sin x\cos x\, y''_{xx} + [\cos 2x + 4\nu(\nu+1)\sin^2 x]y'_x + 2\nu(\nu+1)\sin 2x\, y = 0.$

Solution:

$$y = C_1 y_1^2 + C_2 y_1 y_2 + C_3 y_2^2,$$

where y_1, y_2 form a fundamental set of solutions of the Legendre equation 2.1.2.148, with argument x of functions y_1 and y_2 substituted by $\cos x$.

112. $\sin^2 x\, y'''_{xxx} + a\sin^2 x\, y''_{xx} + by'_x + aby = 0.$

The substitution $w = y'_x + ay$ leads to a second order equation of the form 2.1.6.108: $\sin^2 x\, w''_{xx} + bw = 0$.

113. $\sin^n x\, y'''_{xxx} + ay''_{xx} + aby'_x + b^2(a - b\sin^n x)y = 0.$

Particular solutions: $y_1 = e^{-bx/2}\cos\Bigl(\dfrac{bx\sqrt{3}}{2}\Bigr), \quad y_2 = e^{-bx/2}\sin\Bigl(\dfrac{bx\sqrt{3}}{2}\Bigr).$

114. $\sin^n x\, y'''_{xxx} + ay''_{xx} + b\sin^n x\, y'_x + aby = 0.$

1°. Particular solutions with $b > 0$: $y_1 = \cos\bigl(x\sqrt{b}\bigr), \quad y_2 = \sin\bigl(x\sqrt{b}\bigr).$

2°. Particular solutions with $b < 0$: $y_1 = \exp\bigl(-x\sqrt{-b}\bigr), \quad y_2 = \exp\bigl(x\sqrt{-b}\bigr).$

115. $\sin^n x\, y'''_{xxx} + ay''_{xx} - b(2a + 3b\sin^n x)y'_x + b^2(a + 2b\sin^n x)y = 0.$

Particular solutions: $y_1 = e^{bx}, \quad y_2 = xe^{bx}.$

116. $\sin^n x\, y'''_{xxx} + y''_{xx} + [(b-a^2)\sin^n x + a]y'_x + b(1 - a\sin^n x)y = 0.$

Particular solutions: $y_1 = e^{\lambda_1 x}$, $y_2 = e^{\lambda_2 x}$, where λ_1 and λ_2 are the roots of the quadratic equation $\lambda^2 + a\lambda + b = 0$.

117. $\sin^n(\lambda x) y'''_{xxx} + ax^2 y''_{xx} - 2axy'_x + 2ay = 0.$

Particular solutions: $y_1 = x$, $y_2 = x^2$.

118. $\sin^n x\, y'''_{xxx} + (a\sin^n x + ax + 1)y''_{xx} + a^2 xy'_x - a^2 y = 0.$

Particular solutions: $y_1 = x$, $y_2 = e^{-ax}$.

119. $\sin^n x\, y'''_{xxx} + (ax\sin^n x + 1)y''_{xx} + a(x + 2\sin^n x)y'_x + ay = 0.$

Particular solutions: $y_1 = \exp\left(-\dfrac{ax^2}{2}\right)$, $y_2 = \exp\left(-\dfrac{ax^2}{2}\right)\displaystyle\int \exp\left(\dfrac{ax^2}{2}\right) dx$.

120. $x\sin^n x\, y'''_{xxx} + (3\sin^n x + x)y''_{xx} + (ax\sin^n x + 2)y'_x + a(\sin^n x + x)y = 0.$

Particular solutions: $y_1 = \dfrac{1}{x}\cos\left(x\sqrt{a}\right)$, $y_2 = \dfrac{1}{x}\sin\left(x\sqrt{a}\right)$.

121. $x^3\sin^n x\, y'''_{xxx} + ax^2 y''_{xx} - 2x\sin^n x\, y'_x + 2(2\sin^n x - a)y = 0.$

Particular solutions: $y_1 = x^{-1}$, $y_2 = x^2$.

122. $x^3\sin^n x\, y'''_{xxx} + ax^2 y''_{xx} - 6x\sin^n x\, y'_x + 6(2\sin^n x - a)y = 0.$

Particular solutions: $y_1 = x^{-2}$, $y_2 = x^3$.

123. $x^3\sin^n x\, y'''_{xxx} + ax^2 y''_{xx} + x(a - \sin^n x)y'_x + (a - 3\sin^n x)y = 0.$

Particular solutions: $y_1 = \cos(\ln x)$, $y_2 = \sin(\ln x)$.

124. $x^3\sin^n x\, y'''_{xxx} + x^2(\sin^n x + a)y''_{xx} + x[a - (b+1)\sin^n x]y'_x + b(2\sin^n x - a)y = 0.$

Particular solutions: $y_1 = x^{-\sqrt{b}}$, $y_2 = x^{\sqrt{b}}$.

125. $\tan^n x\, y'''_{xxx} + ay''_{xx} + aby'_x + b^2(a - b\tan^n x)y = 0.$

Particular solutions: $y_1 = e^{-bx/2}\cos\left(\dfrac{bx\sqrt{3}}{2}\right)$, $y_2 = e^{-bx/2}\sin\left(\dfrac{bx\sqrt{3}}{2}\right)$.

126. $\tan^n x\, y'''_{xxx} + ay''_{xx} + b\tan^n x\, y'_x + aby = 0.$

1°. Particular solutions with $b > 0$: $y_1 = \cos\left(x\sqrt{b}\right)$, $y_2 = \sin\left(x\sqrt{b}\right)$.

2°. Particular solutions with $b < 0$: $y_1 = \exp\left(-x\sqrt{-b}\right)$, $y_2 = \exp\left(x\sqrt{-b}\right)$.

127. $\tan^n x\, y'''_{xxx} + ay''_{xx} - b(2a + 3b\tan^n x)y'_x + b^2(a + 2b\tan^n x)y = 0.$

Particular solutions: $y_1 = e^{bx}$, $y_2 = xe^{bx}$.

128. $\tan^n x\, y'''_{xxx} + y''_{xx} + [(b-a^2)\tan^n x + a]y'_x + b(1 - a\tan^n x)y = 0.$

Particular solutions: $y_1 = e^{\lambda_1 x}, \quad y_2 = e^{\lambda_2 x},$ where λ_1 and λ_2 are the roots of the quadratic equation $\lambda^2 + a\lambda + b = 0$.

129. $\tan^n(\lambda x)y'''_{xxx} + ax^2 y''_{xx} - 2axy'_x + 2ay = 0.$

Particular solutions: $y_1 = x, \quad y_2 = x^2.$

130. $\tan^n x\, y'''_{xxx} + (a\tan^n x + ax + 1)y''_{xx} + a^2xy'_x - a^2y = 0.$

Particular solutions: $y_1 = x, \quad y_2 = e^{-ax}.$

131. $\tan^n x\, y'''_{xxx} + (ax\tan^n x + 1)y''_{xx} + a(x + 2\tan^n x)y'_x + ay = 0.$

Particular solutions: $y_1 = \exp\Bigl(-\dfrac{ax^2}{2}\Bigr), \quad y_2 = \exp\Bigl(-\dfrac{ax^2}{2}\Bigr)\displaystyle\int \exp\Bigl(\dfrac{ax^2}{2}\Bigr)\,dx.$

132. $x\tan^n x\, y'''_{xxx} + (3\tan^n x + x)y''_{xx} + (ax\tan^n x + 2)y'_x + a(\tan^n x + x)y = 0.$

Particular solutions: $y_1 = \dfrac{1}{x}\cos\bigl(x\sqrt{a}\,\bigr), \quad y_2 = \dfrac{1}{x}\sin\bigl(x\sqrt{a}\,\bigr).$

133. $x^3\tan^n x\, y'''_{xxx} + ax^2y''_{xx} - 2x\tan^n x\, y'_x + 2(2\tan^n x - a)y = 0.$

Particular solutions: $y_1 = x^{-1}, \quad y_2 = x^2.$

134. $x^3\tan^n x\, y'''_{xxx} + ax^2y''_{xx} - 6x\tan^n x\, y'_x + 6(2\tan^n x - a)y = 0.$

Particular solutions: $y_1 = x^{-2}, \quad y_2 = x^3.$

135. $x^3\tan^n x\, y'''_{xxx} + ax^2y''_{xx} + x(a - \tan^n x)y'_x + (a - 3\tan^n x)y = 0.$

Particular solutions: $y_1 = \cos(\ln|x|), \quad y_2 = \sin(\ln|x|).$

136. $x^3\tan^n x\, y'''_{xxx} + x^2(\tan^n x + a)y''_{xx} + x[a - (b+1)\tan^n x]y'_x + b(2\tan^n x - a)y = 0.$

Particular solutions: $y_1 = x^{-\sqrt{b}}, \quad y_2 = x^{\sqrt{b}}.$

137. $\cot^n x\, y'''_{xxx} + ay''_{xx} + aby'_x + b^2(a - b\cot^n x)y = 0.$

Particular solutions: $y_1 = e^{-bx/2}\cos\biggl(\dfrac{bx\sqrt{3}}{2}\biggr), \quad y_2 = e^{-bx/2}\sin\biggl(\dfrac{bx\sqrt{3}}{2}\biggr).$

138. $\cot^n x\, y'''_{xxx} + ay''_{xx} + b\cot^n x\, y'_x + aby = 0.$

1°. Particular solutions with $b > 0$: $y_1 = \cos\bigl(x\sqrt{b}\,\bigr), \quad y_2 = \sin\bigl(x\sqrt{b}\,\bigr).$

2°. Particular solutions with $b < 0$: $y_1 = \exp\bigl(-x\sqrt{-b}\,\bigr), \quad y_2 = \exp\bigl(x\sqrt{-b}\,\bigr).$

139. $\cot^n x\, y'''_{xxx} + ay''_{xx} - b(2a + 3b\cot^n x)y'_x + b^2(a + 2b\cot^n x)y = 0.$

Particular solutions: $y_1 = e^{bx}, \quad y_2 = xe^{bx}.$

140. $\cot^n x\, y'''_{xxx} + y''_{xx} + [(b-a^2)\cot^n x + a]y'_x + b(1 - a\cot^n x)y = 0.$

Particular solutions: $y_1 = e^{\lambda_1 x}$, $y_2 = e^{\lambda_2 x}$, where λ_1 and λ_2 are the roots of the quadratic equation $\lambda^2 + a\lambda + b = 0$.

141. $\cot^n(\lambda x) y'''_{xxx} + ax^2 y''_{xx} - 2axy'_x + 2ay = 0.$

Particular solutions: $y_1 = x$, $y_2 = x^2$.

142. $\cot^n x\, y'''_{xxx} + (a\cot^n x + ax + 1)y''_{xx} + a^2 xy'_x - a^2 y = 0.$

Particular solutions: $y_1 = x$, $y_2 = e^{-ax}$.

143. $\cot^n x\, y'''_{xxx} + (ax\cot^n x + 1)y''_{xx} + a(x + 2\cot^n x)y'_x + ay = 0.$

Particular solutions: $y_1 = \exp\left(-\dfrac{ax^2}{2}\right)$, $y_2 = \exp\left(-\dfrac{ax^2}{2}\right)\displaystyle\int \exp\left(\dfrac{ax^2}{2}\right)dx$.

144. $x\cot^n x\, y'''_{xxx} + (3\cot^n x + x)y''_{xx} + (ax\cot^n x + 2)y'_x + a(\cot^n x + x)y = 0.$

Particular solutions: $y_1 = \dfrac{1}{x}\cos\left(x\sqrt{a}\right)$, $y_2 = \dfrac{1}{x}\sin\left(x\sqrt{a}\right)$.

145. $x^3\cot^n x\, y'''_{xxx} + ax^2 y''_{xx} - 2x\cot^n x\, y'_x + 2(2\cot^n x - a)y = 0.$

Particular solutions: $y_1 = x^{-1}$, $y_2 = x^2$.

146. $x^3\cot^n x\, y'''_{xxx} + ax^2 y''_{xx} - 6x\cot^n x\, y'_x + 6(2\cot^n x - a)y = 0.$

Particular solutions: $y_1 = x^{-2}$, $y_2 = x^3$.

147. $x^3\cot^n x\, y'''_{xxx} + ax^2 y''_{xx} + x(a - \cot^n x)y'_x + (a - 3\cot^n x)y = 0.$

Particular solutions: $y_1 = \cos(\ln x)$, $y_2 = \sin(\ln x)$.

148. $x^3\cot^n x\, y'''_{xxx} + x^2(\cot^n x + a)y''_{xx} + x[a - (b+1)\cot^n x]y'_x + b(2\cot^n x - a)y = 0.$

Particular solutions: $y_1 = x^{-\sqrt{b}}$, $y_2 = x^{\sqrt{b}}$.

3.1.7. Equations Containing Inverse Trigonometric Functions

1. $y'''_{xxx} + ay''_{xx} + by'_x + cy = \arcsin^k x.$

This is a special case of equation 5.1.5.9.

2. $y'''_{xxx} + \arcsin^k x\, y''_{xx} + ay'_x + a\arcsin^k x\, y = 0.$

1°. Particular solutions with $a > 0$: $y_1 = \cos\left(x\sqrt{a}\right)$, $y_1 = \sin\left(x\sqrt{a}\right)$.

2°. Particular solutions with $a < 0$: $y_1 = \exp\left(-x\sqrt{-a}\right)$, $y_1 = \exp\left(x\sqrt{-a}\right)$.

The substitution $w = y''_{xx} + ay$ leads to a first order linear equation: $w'_x + \arcsin^k x\, w = 0$.

3. $y'''_{xxx} + \arcsin^k x\, y''_{xx} + ax^n y'_x + ax^{n-1}(x \arcsin^k x + n)y = 0.$

The substitution $w = y''_{xx} + ax^n y$ leads to a first order linear equation: $w'_x + \arcsin^k x\, w = 0$.

4. $y'''_{xxx} + \arcsin^k x\, y''_{xx} + a \arcsin^k x\, y'_x + a^2(\arcsin^k x - a)y = 0.$

Particular solutions: $y_1 = e^{-ax/2} \cos\left(\frac{a\sqrt{3}}{2}x\right)$, $y_2 = e^{-ax/2} \sin\left(\frac{a\sqrt{3}}{2}x\right)$.

5. $y'''_{xxx} + \arcsin^k x\, y''_{xx} - a(2\arcsin^k x + 3a)y'_x + a^2(\arcsin^k x + 2a)y = 0.$

Particular solutions: $y_1 = e^{ax}$, $y_2 = xe^{ax}$.

6. $y'''_{xxx} + \arcsin^k x\, y''_{xx} + (a \arcsin^k x + b - a^2)y'_x + b(\arcsin^k x - a)y = 0.$

Particular solutions: $y_1 = e^{\lambda_1 x}$, $y_2 = e^{\lambda_2 x}$, where λ_1 and λ_2 are the roots of the quadratic equation $\lambda^2 + a\lambda + b = 0$.

7. $y'''_{xxx} = (\arcsin^k x - a)y''_{xx} + (a \arcsin^k x - b)y'_x + b \arcsin^k x\, y.$

The substitution $w = y''_{xx} + ay'_x + by$ leads to a first order linear equation: $w'_x = \arcsin^k x\, w$.

8. $y'''_{xxx} + x \arcsin^k x\, y''_{xx} + (ax^2 - \arcsin^k x)y'_x + ax(x^2 \arcsin^k x + 3)y = 0.$

Particular solutions: $y_1 = \cos\left(\frac{x^2\sqrt{a}}{2}\right)$, $y_2 = \sin\left(\frac{x^2\sqrt{a}}{2}\right)$.

9. $y'''_{xxx} + (\arcsin^k x + ax)y''_{xx} + a(x \arcsin^k x + 2)y'_x + a \arcsin^k x\, y = 0.$

Particular solutions: $y_1 = \exp\left(-\frac{ax^2}{2}\right)$, $y_2 = \exp\left(-\frac{ax^2}{2}\right) \int \exp\left(\frac{ax^2}{2}\right) dx$.

10. $y'''_{xxx} + x^2 \arcsin^k x\, y''_{xx} - 2x \arcsin^k x\, y'_x + 2 \arcsin^k x\, y = 0.$

Solution:

$$y = C_1 x + C_2 x^2 + C_3\left(x^2 \int x^{-3}\psi\, dx - x \int x^{-2}\psi\, dx\right),$$

where $\psi = \exp(-\int x^2 \arcsin^k x\, dx)$.

11. $y'''_{xxx} + (ax \arcsin^k x + \arcsin^k x + a)y''_{xx} + a^2 x \arcsin^k x\, y'_x - a^2 \arcsin^k x\, y = 0.$

Particular solutions: $y_1 = x$, $y_2 = e^{-ax}$.

12. $y'''_{xxx} + \arccos^k x\, y''_{xx} + ay'_x + a \arccos^k x\, y = 0.$

1°. Particular solutions with $a > 0$: $y_1 = \cos(x\sqrt{a})$, $y_1 = \sin(x\sqrt{a})$.

2°. Particular solutions with $a < 0$: $y_1 = \exp(-x\sqrt{-a})$, $y_1 = \exp(x\sqrt{-a})$.

The substitution $w = y''_{xx} + ay$ leads to a first order linear equation: $w'_x + \arccos^k x\, w = 0$.

13. $y'''_{xxx} + \arccos^k x\, y''_{xx} + ax^n y'_x + ax^{n-1}(x \arccos^k x + n)y = 0.$

The substitution $w = y''_{xx} + ax^n y$ leads to a first order linear equation: $w'_x + \arccos^k x\, w = 0$.

14. $y'''_{xxx} + \arccos^k x\, y''_{xx} + a \arccos^k x\, y'_x + a^2(\arccos^k x - a)y = 0.$

Particular solutions: $y_1 = e^{-ax/2} \cos\left(\dfrac{a\sqrt{3}}{2}x\right)$, $y_2 = e^{-ax/2} \sin\left(\dfrac{a\sqrt{3}}{2}x\right)$.

15. $y'''_{xxx} + \arccos^k x\, y''_{xx} - a(2\arccos^k x + 3a)y'_x + a^2(\arccos^k x + 2a)y = 0.$

Particular solutions: $y_1 = e^{ax}$, $y_2 = xe^{ax}$.

16. $y'''_{xxx} + \arccos^k x\, y''_{xx} + (a \arccos^k x + b - a^2)y'_x + b(\arccos^k x - a)y = 0.$

Particular solutions: $y_1 = e^{\lambda_1 x}$, $y_2 = e^{\lambda_2 x}$, where λ_1 and λ_2 are the roots of the quadratic equation $\lambda^2 + a\lambda + b = 0$.

17. $y'''_{xxx} = (\arccos^k x - a)y''_{xx} + (a \arccos^k x - b)y'_x + b \arccos^k x\, y.$

The substitution $w = y''_{xx} + ay'_x + by$ leads to a first order linear equation: $w'_x = \arccos^k x\, w$.

18. $y'''_{xxx} + x \arccos^k x\, y''_{xx} + (ax^2 - \arccos^k x)y'_x + ax(x^2 \arccos^k x + 3)y = 0.$

Particular solutions: $y_1 = \cos\left(\dfrac{x^2\sqrt{a}}{2}\right)$, $y_2 = \sin\left(\dfrac{x^2\sqrt{a}}{2}\right)$.

19. $y'''_{xxx} + (\arccos^k x + ax)y''_{xx} + a(x \arccos^k x + 2)y'_x + a \arccos^k x\, y = 0.$

Particular solutions: $y_1 = \exp\left(-\dfrac{ax^2}{2}\right)$, $y_2 = \exp\left(-\dfrac{ax^2}{2}\right) \displaystyle\int \exp\left(\dfrac{ax^2}{2}\right) dx$.

20. $y'''_{xxx} + x^2 \arccos^k x\, y''_{xx} - 2x \arccos^k x\, y'_x + 2 \arccos^k x\, y = 0.$

Solution:

$$y = C_1 x + C_2 x^2 + C_3\left(x^2 \int x^{-3}\psi\, dx - x \int x^{-2}\psi\, dx\right),$$

where $\psi = \exp(-\int x^2 \arccos^k x\, dx)$.

21. $y'''_{xxx} + (ax \arccos^k x + \arccos^k x + a)y''_{xx} + a^2 x \arccos^k x\, y'_x - a^2 \arccos^k x\, y = 0.$

Particular solutions: $y_1 = x$, $y_2 = e^{-ax}$.

22. $y'''_{xxx} + \arctan^k x\, y''_{xx} + ay'_x + a \arctan^k x\, y = 0.$

1°. Particular solutions with $a > 0$: $y_1 = \cos(x\sqrt{a})$, $y_1 = \sin(x\sqrt{a})$.

2°. Particular solutions with $a < 0$: $y_1 = \exp(-x\sqrt{-a})$, $y_1 = \exp(x\sqrt{-a})$.

The substitution $w = y''_{xx} + ay$ leads to a first order linear equation: $w'_x + \arctan^k x\, w = 0$.

23. $y'''_{xxx} + \arctan^k x\, y''_{xx} + ax^n y'_x + ax^{n-1}(x\arctan^k x + n)y = 0.$

The substitution $w = y''_{xx} + ax^n y$ leads to a first order linear equation: $w'_x + \arctan^k x\, w = 0$.

24. $y'''_{xxx} + \arctan^k x\, y''_{xx} + a\arctan^k x\, y'_x + a^2(\arctan^k x - a)y = 0.$

Particular solutions: $y_1 = e^{-ax/2}\cos\Bigl(\dfrac{a\sqrt{3}}{2}x\Bigr)$, $y_2 = e^{-ax/2}\sin\Bigl(\dfrac{a\sqrt{3}}{2}x\Bigr)$.

25. $y'''_{xxx} + \arctan^k x\, y''_{xx} - a(2\arctan^k x + 3a)y'_x + a^2(\arctan^k x + 2a)y = 0.$

Particular solutions: $y_1 = e^{ax}$, $y_2 = xe^{ax}$.

26. $y'''_{xxx} + \arctan^k x\, y''_{xx} + (a\arctan^k x + b - a^2)y'_x + b(\arctan^k x - a)y = 0.$

Particular solutions: $y_1 = e^{\lambda_1 x}$, $y_2 = e^{\lambda_2 x}$, where λ_1 and λ_2 are the roots of the quadratic equation $\lambda^2 + a\lambda + b = 0$.

27. $y'''_{xxx} = (\arctan^k x - a)y''_{xx} + (a\arctan^k x - b)y'_x + b\arctan^k x\, y.$

The substitution $w = y''_{xx} + ay'_x + by$ leads to a first order linear equation: $w'_x = \arctan^k x\, w$.

28. $y'''_{xxx} + x\arctan^k x\, y''_{xx} + (ax^2 - \arctan^k x)y'_x + ax(x^2\arctan^k x + 3)y = 0.$

Particular solutions: $y_1 = \cos\biggl(\dfrac{x^2\sqrt{a}}{2}\biggr)$, $y_2 = \sin\biggl(\dfrac{x^2\sqrt{a}}{2}\biggr)$.

29. $y'''_{xxx} + (\arctan^k x + ax)y''_{xx} + a(x\arctan^k x + 2)y'_x + a\arctan^k x\, y = 0.$

Particular solutions: $y_1 = \exp\Bigl(-\dfrac{ax^2}{2}\Bigr)$, $y_2 = \exp\Bigl(-\dfrac{ax^2}{2}\Bigr)\displaystyle\int \exp\Bigl(\dfrac{ax^2}{2}\Bigr)\,dx$.

30. $y'''_{xxx} + x^2\arctan^k x\, y''_{xx} - 2x\arctan^k x\, y'_x + 2\arctan^k x\, y = 0.$

Solution:

$$y = C_1 x + C_2 x^2 + C_3\Bigl(x^2\int x^{-3}\psi\,dx - x\int x^{-2}\psi\,dx\Bigr),$$

where $\psi = \exp(-\int x^2\arctan^k x\,dx)$.

31. $y'''_{xxx} + (ax\arctan^k x + \arctan^k x + a)y''_{xx} + a^2 x\arctan^k x\, y'_x - a^2\arctan^k x\, y = 0.$

Particular solutions: $y_1 = x$, $y_2 = e^{-ax}$.

32. $y'''_{xxx} + \operatorname{arccot}^k x\, y''_{xx} + ay'_x + a\operatorname{arccot}^k x\, y = 0.$

1°. Particular solutions with $a > 0$: $y_1 = \cos(x\sqrt{a})$, $y_1 = \sin(x\sqrt{a})$.

2°. Particular solutions with $a < 0$: $y_1 = \exp(-x\sqrt{-a})$, $y_1 = \exp(x\sqrt{-a})$.

The substitution $w = y''_{xx} + ay$ leads to a first order linear equation: $w'_x + \operatorname{arccot}^k x\, w = 0$.

33. $y'''_{xxx} + \operatorname{arccot}^k x\, y''_{xx} + ax^n y'_x + ax^{n-1}(x \operatorname{arccot}^k x + n)y = 0.$

The substitution $w = y''_{xx} + ax^n y$ leads to a first order linear equation: $w'_x + \operatorname{arccot}^k x\, w = 0$.

34. $y'''_{xxx} + \operatorname{arccot}^k x\, y''_{xx} + a \operatorname{arccot}^k x\, y'_x + a^2(\operatorname{arccot}^k x - a)y = 0.$

Particular solutions: $y_1 = e^{-ax/2} \cos\left(\frac{a\sqrt{3}}{2}x\right)$, $y_2 = e^{-ax/2} \sin\left(\frac{a\sqrt{3}}{2}x\right)$.

35. $y'''_{xxx} + \operatorname{arccot}^k x\, y''_{xx} - a(2 \operatorname{arccot}^k x + 3a)y'_x + a^2(\operatorname{arccot}^k x + 2a)y = 0.$

Particular solutions: $y_1 = e^{ax}$, $y_2 = xe^{ax}$.

36. $y'''_{xxx} + \operatorname{arccot}^k x\, y''_{xx} + (a \operatorname{arccot}^k x + b - a^2)y'_x + b(\operatorname{arccot}^k x - a)y = 0.$

Particular solutions: $y_1 = e^{\lambda_1 x}$, $y_2 = e^{\lambda_2 x}$, where λ_1 and λ_2 are the roots of the quadratic equation $\lambda^2 + a\lambda + b = 0$.

37. $y'''_{xxx} = (\operatorname{arccot}^k x - a)y''_{xx} + (a \operatorname{arccot}^k x - b)y'_x + b \operatorname{arccot}^k x\, y.$

The substitution $w = y''_{xx} + ay'_x + by$ leads to a first order linear equation: $w'_x = \operatorname{arccot}^k x\, w$.

38. $y'''_{xxx} + x \operatorname{arccot}^k x\, y''_{xx} + (ax^2 - \operatorname{arccot}^k x)y'_x + ax(x^2 \operatorname{arccot}^k x + 3)y = 0.$

Particular solutions: $y_1 = \cos\left(\frac{x^2\sqrt{a}}{2}\right)$, $y_2 = \sin\left(\frac{x^2\sqrt{a}}{2}\right)$.

39. $y'''_{xxx} + (\operatorname{arccot}^k x + ax)y''_{xx} + a(x \operatorname{arccot}^k x + 2)y'_x + a \operatorname{arccot}^k x\, y = 0.$

Particular solutions: $y_1 = \exp\left(-\frac{ax^2}{2}\right)$, $y_2 = \exp\left(-\frac{ax^2}{2}\right) \int \exp\left(\frac{ax^2}{2}\right) dx$.

40. $y'''_{xxx} + x^2 \operatorname{arccot}^k x\, y''_{xx} - 2x \operatorname{arccot}^k x\, y'_x + 2 \operatorname{arccot}^k x\, y = 0.$

Solution:

$$y = C_1 x + C_2 x^2 + C_3\left(x^2 \int x^{-3}\psi\, dx - x \int x^{-2}\psi\, dx\right),$$

where $\psi = \exp(-\int x^2 \operatorname{arccot}^k x\, dx)$.

41. $y'''_{xxx} + (ax \operatorname{arccot}^k x + \operatorname{arccot}^k x + a)y''_{xx} + a^2 x \operatorname{arccot}^k x\, y'_x - a^2 \operatorname{arccot}^k x\, y = 0.$

Particular solutions: $y_1 = x$, $y_2 = e^{-ax}$.

42. $xy'''_{xxx} + (ax^2 + b)y''_{xx} + 4axy'_x + 2ay = \arcsin^k x.$

Twice integrating yields a first order linear equation:

$$xy'_x + (ax^2 + b - 2)y = C_1 + C_2 x + \int\left(\int \arcsin^k x\, dx\right) dx.$$

43. $xy'''_{xxx} + (x \arcsin^k x + 3)y''_{xx} + (2 \arcsin^k x + ax)y'_x + a(x \arcsin^k x + 1)y = 0.$

Particular solutions: $y_1 = \frac{1}{x}\cos(x\sqrt{a})$, $y_2 = \frac{1}{x}\sin(x\sqrt{a})$.

44. $xy'''_{xxx} + (x \arccos^k x + 3)y''_{xx} + (2\arccos^k x + ax)y'_x + a(x\arccos^k x + 1)y = 0.$

Particular solutions: $y_1 = \dfrac{1}{x}\cos(x\sqrt{a}), \quad y_2 = \dfrac{1}{x}\sin(x\sqrt{a}).$

45. $xy'''_{xxx} + (x \arctan^k x + 3)y''_{xx} + (2\arctan^k x + ax)y'_x + a(x\arctan^k x + 1)y = 0.$

Particular solutions: $y_1 = \dfrac{1}{x}\cos(x\sqrt{a}), \quad y_2 = \dfrac{1}{x}\sin(x\sqrt{a}).$

46. $xy'''_{xxx} + (x \operatorname{arccot}^k x + 3)y''_{xx} + (2\operatorname{arccot}^k x + ax)y'_x + a(x\operatorname{arccot}^k x + 1)y = 0.$

Particular solutions: $y_1 = \dfrac{1}{x}\cos(x\sqrt{a}), \quad y_2 = \dfrac{1}{x}\sin(x\sqrt{a}).$

47. $x^3y'''_{xxx} + [(a+6)x^2 + b]y''_{xx} + 2(2a+3)xy'_x + 2ay = \arcsin^k x.$

Twice integrating yields a first order linear equation:

$$x^3y'_x + (ax^2 + b)y = C_1 + C_2x + \int\left(\int \arcsin^k x\,dx\right)dx.$$

48. $x^3y'''_{xxx} + x^2\arcsin^k x y''_{xx} - 2xy'_x + 2(2 - \arcsin^k x)y = 0.$

Particular solutions: $y_1 = x^{-1}, \quad y_2 = x^2.$

49. $x^3y'''_{xxx} + x^2\arcsin^k x\, y''_{xx} - 6xy'_x + 6(2 - \arcsin^k x)y = 0.$

Particular solutions: $y_1 = x^{-2}, \quad y_2 = x^3.$

50. $x^3y'''_{xxx} + x^2\arcsin^k x\, y''_{xx} + x(\arcsin^k x - 1)y'_x + (\arcsin^k x - 3)y = 0.$

Particular solutions: $y_1 = \cos(\ln x), \quad y_2 = \sin(\ln x).$

51. $x^3y'''_{xxx} + x^2(\arcsin^k x + 1)y''_{xx} + x(\arcsin^k x - a - 1)y'_x - a(\arcsin^k x - 2)y = 0.$

Particular solutions: $y_1 = x^{-\sqrt{a}}, \quad y_2 = x^{\sqrt{a}}.$

52. $x^3y'''_{xxx} + x^2(\arcsin^k x + a)y''_{xx} + x(a\arcsin^k x + b - a)y'_x + b(\arcsin^k x - 2)y = 0.$

Particular solutions: $y_1 = x^{n_1}, \quad y_2 = x^{n_2},$ where n_1 and n_2 are the roots of the quadratic equation $n^2 + (a-1)n + b = 0.$

53. $x^3y'''_{xxx} + x^2\arccos^k x\, y''_{xx} - 2xy'_x + 2(2 - \arccos^k x)y = 0.$

Particular solutions: $y_1 = x^{-1}, \quad y_2 = x^2.$

54. $x^3y'''_{xxx} + x^2\arccos^k x\, y''_{xx} - 6xy'_x + 6(2 - \arccos^k x)y = 0.$

Particular solutions: $y_1 = x^{-2}, \quad y_2 = x^3.$

55. $x^3y'''_{xxx} + x^2\arccos^k x\, y''_{xx} + x(\arccos^k x - 1)y'_x + (\arccos^k x - 3)y = 0.$

Particular solutions: $y_1 = \cos(\ln x), \quad y_2 = \sin(\ln x).$

56. $x^3y'''_{xxx}+x^2(\arccos^k x+1)y''_{xx}+x(\arccos^k x-a-1)y'_x-a(\arccos^k x-2)y=0.$

Particular solutions: $y_1=x^{-\sqrt{a}}$, $y_2=x^{\sqrt{a}}$.

57. $x^3y'''_{xxx}+x^2(\arccos^k x+a)y''_{xx}+x(a\arccos^k x+b-a)y'_x+b(\arccos^k x-2)y=0.$

Particular solutions: $y_1=x^{n_1}$, $y_2=x^{n_2}$, where n_1 and n_2 are the roots of the quadratic equation $n^2+(a-1)n+b=0$.

58. $x^3y'''_{xxx}+x^2\arctan^k x\,y''_{xx}-2xy'_x+2(2-\arctan^k x)y=0.$

Particular solutions: $y_1=x^{-1}$, $y_2=x^2$.

59. $x^3y'''_{xxx}+x^2\arctan^k x\,y''_{xx}-6xy'_x+6(2-\arctan^k x)y=0.$

Particular solutions: $y_1=x^{-2}$, $y_2=x^3$.

60. $x^3y'''_{xxx}+x^2\arctan^k x\,y''_{xx}+x(\arctan^k x-1)y'_x+(\arctan^k x-3)y=0.$

Particular solutions: $y_1=\cos(\ln x)$, $y_2=\sin(\ln x)$.

61. $x^3y'''_{xxx}+x^2(\arctan^k x+1)y''_{xx}+x(\arctan^k x-a-1)y'_x-a(\arctan^k x-2)y=0.$

Particular solutions: $y_1=x^{-\sqrt{a}}$, $y_2=x^{\sqrt{a}}$.

62. $x^3y'''_{xxx}+x^2(\arctan^k x+a)y''_{xx}+x(a\arctan^k x+b-a)y'_x$
$+\,b(\arctan^k x-2)y=0.$

Particular solutions: $y_1=x^{n_1}$, $y_2=x^{n_2}$, where n_1 and n_2 are the roots of the quadratic equation $n^2+(a-1)n+b=0$.

63. $x^3y'''_{xxx}+x^2\,\mathrm{arccot}^k x\,y''_{xx}-2xy'_x+2(2-\mathrm{arccot}^k x)y=0.$

Particular solutions: $y_1=x^{-1}$, $y_2=x^2$.

64. $x^3y'''_{xxx}+x^2\,\mathrm{arccot}^k x\,y''_{xx}-6xy'_x+6(2-\mathrm{arccot}^k x)y=0.$

Particular solutions: $y_1=x^{-2}$, $y_2=x^3$.

65. $x^3y'''_{xxx}+x^2\,\mathrm{arccot}^k x\,y''_{xx}+x(\mathrm{arccot}^k x-1)y'_x+(\mathrm{arccot}^k x-3)y=0.$

Particular solutions: $y_1=\cos(\ln x)$, $y_2=\sin(\ln x)$.

66. $x^3y'''_{xxx}+x^2(\mathrm{arccot}^k x+1)y''_{xx}+x(\mathrm{arccot}^k x-a-1)y'_x-a(\mathrm{arccot}^k x-2)y=0.$

Particular solutions: $y_1=x^{-\sqrt{a}}$, $y_2=x^{\sqrt{a}}$.

67. $x^3y'''_{xxx}+x^2(\mathrm{arccot}^k x+a)y''_{xx}+x(a\,\mathrm{arccot}^k x+b-a)y'_x+b(\mathrm{arccot}^k x-2)y=0.$

Particular solutions: $y_1=x^{n_1}$, $y_2=x^{n_2}$, where n_1 and n_2 are the roots of the quadratic equation $n^2+(a-1)n+b=0$.

3.1.8. Equations Containing Combinations of Exponential, Logarithmic, Trigonometric, and Other Functions

1. $y'''_{xxx} + ae^{\lambda x}y''_{xx} + (2ae^{\lambda x}\tan x + 3)y'_x + [ae^{\lambda x}(2\tan^2 x + 1) + 2\tan x]y = 0.$

   Particular solutions: $y_1 = \cos x, \quad y_2 = x\cos x.$

2. $y'''_{xxx} + ae^{\lambda x}y''_{xx} + (3 - 2ae^{\lambda x}\cot x)y'_x + [ae^{\lambda x}(2\cot^2 x + 1) - 2\cot x]y = 0.$

   Particular solutions: $y_1 = \sin x, \quad y_2 = x\sin x.$

3. $y'''_{xxx} + a\cosh^n x\, y''_{xx} + (2a\cosh^n x\tan x + 3)y'_x + [a\cosh^n x\,(2\tan^2 x + 1) + 2\tan x]y = 0.$

   Particular solutions: $y_1 = \cos x, \quad y_2 = x\cos x.$

4. $y'''_{xxx} + a\cosh^n x\, y''_{xx} + (3 - 2a\cosh^n x\cot x)y'_x + [a\cosh^n x\,(2\cot^2 x + 1) - 2\cot x]y = 0.$

   Particular solutions: $y_1 = \sin x, \quad y_2 = x\sin x.$

5. $y'''_{xxx} + a\sinh^n x\, y''_{xx} + (2a\sinh^n x\tan x + 3)y'_x + [a\sinh^n x\,(2\tan^2 x + 1) + 2\tan x]y = 0.$

   Particular solutions: $y_1 = \cos x, \quad y_2 = x\cos x.$

6. $y'''_{xxx} + a\sinh^n x\, y''_{xx} + (3 - 2a\sinh^n x\cot x)y'_x + [a\sinh^n x\,(2\cot^2 x + 1) - 2\cot x]y = 0.$

   Particular solutions: $y_1 = \sin x, \quad y_2 = x\sin x.$

7. $y'''_{xxx} + a\tanh^n x\, y''_{xx} + (2a\tanh^n x\tan x + 3)y'_x + [a\tanh^n x\,(2\tan^2 x + 1) + 2\tan x]y = 0.$

   Particular solutions: $y_1 = \cos x, \quad y_2 = x\cos x.$

8. $y'''_{xxx} + a\tanh^n x\, y''_{xx} + (3 - 2a\tanh^n x\cot x)y'_x + [a\tanh^n x\,(2\cot^2 x + 1) - 2\cot x]y = 0.$

   Particular solutions: $y_1 = \sin x, \quad y_2 = x\sin x.$

9. $y'''_{xxx} + a\coth^n x\, y''_{xx} + (2a\coth^n x\tan x + 3)y'_x + [a\coth^n x\,(2\tan^2 x + 1) + 2\tan x]y = 0.$

   Particular solutions: $y_1 = \cos x, \quad y_2 = x\cos x.$

10. $y'''_{xxx} + a\coth^n x\, y''_{xx} + (3 - 2a\coth^n x\cot x)y'_x + [a\coth^n x\,(2\cot^2 x + 1) - 2\cot x]y = 0.$

    Particular solutions: $y_1 = \sin x, \quad y_2 = x\sin x.$

11. $y'''_{xxx} + a\ln^n x\, y''_{xx} - (2a\ln^n x\tanh x + 3)y'_x + [a\ln^n x\,(2\tanh^2 x - 1) + 2\tanh x]y = 0.$

    Particular solutions: $y_1 = \cosh x, \quad y_2 = x\cosh x.$

12. $y'''_{xxx} + a\ln^n x\, y''_{xx} - (2a\ln^n x \coth x + 3)y'_x$
$$+ [a\ln^n x\,(2\coth^2 x - 1) + 2\coth x]y = 0.$$

Particular solutions: $y_1 = \sinh x$, $y_2 = x\sinh x$.

13. $y'''_{xxx} + a\ln^n x\, y''_{xx} + (2a\ln^n x \tan x + 3)y'_x$
$$+ [a\ln^n x\,(2\tan^2 x + 1) + 2\tan x]y = 0.$$

Particular solutions: $y_1 = \cos x$, $y_2 = x\cos x$.

14. $y'''_{xxx} + a\ln^n x\, y''_{xx} + (3 - 2a\ln^n x \cot x)y'_x + [a\ln^n x\,(2\cot^2 x + 1) - 2\cot x]y = 0.$

Particular solutions: $y_1 = \sin x$, $y_2 = x\sin x$.

15. $y'''_{xxx} + a\cos^n x\, y''_{xx} - (2a\cos^n x \tanh x + 3)y'_x$
$$+ [a\cos^n x\,(2\tanh^2 x - 1) + 2\tanh x]y = 0.$$

Particular solutions: $y_1 = \cosh x$, $y_2 = x\cosh x$.

16. $y'''_{xxx} + a\cos^n x\, y''_{xx} - (2a\cos^n x \coth x + 3)y'_x$
$$+ [a\cos^n x\,(2\coth^2 x - 1) + 2\coth x]y = 0.$$

Particular solutions: $y_1 = \sinh x$, $y_2 = x\sinh x$.

17. $y'''_{xxx} + a\sin^n x\, y''_{xx} - (2a\sin^n x \tanh x + 3)y'_x$
$$+ [a\sin^n x\,(2\tanh^2 x - 1) + 2\tanh x]y = 0.$$

Particular solutions: $y_1 = \cosh x$, $y_2 = x\cosh x$.

18. $y'''_{xxx} + a\sin^n x\, y''_{xx} - (2a\sin^n x \coth x + 3)y'_x$
$$+ [a\sin^n x\,(2\coth^2 x - 1) + 2\coth x]y = 0.$$

Particular solutions: $y_1 = \sinh x$, $y_2 = x\sinh x$.

19. $y'''_{xxx} + a\tan^n x\, y''_{xx} - (2a\tan^n x \tanh x + 3)y'_x$
$$+ [a\tan^n x\,(2\tanh^2 x - 1) + 2\tanh x]y = 0.$$

Particular solutions: $y_1 = \cosh x$, $y_2 = x\cosh x$.

20. $y'''_{xxx} + a\tan^n x\, y''_{xx} - (2a\tan^n x \coth x + 3)y'_x$
$$+ [a\tan^n x\,(2\coth^2 x - 1) + 2\coth x]y = 0.$$

Particular solutions: $y_1 = \sinh x$, $y_2 = x\sinh x$.

21. $y'''_{xxx} + a\cot^n x\, y''_{xx} - (2a\cot^n x \tanh x + 3)y'_x$
$$+ [a\cot^n x\,(2\tanh^2 x - 1) + 2\tanh x]y = 0.$$

Particular solutions: $y_1 = \cosh x$, $y_2 = x\cosh x$.

22. $y'''_{xxx} + a\cot^n x\, y''_{xx} - (2a\cot^n x \coth x + 3)y'_x$
$$+ [a\cot^n x\,(2\coth^2 x - 1) + 2\coth x]y = 0.$$

Particular solutions: $y_1 = \sinh x$, $y_2 = x\sinh x$.

23. $y'''_{xxx} + (be^{ax} + 2a)\cosh^n x\, y''_{xx} - a(be^{ax}\cosh^n x + a)y'_x - 2a^3\cosh^n x\, y = 0.$

Particular solutions: $y_1 = e^{ax}, \quad y_2 = e^{-ax} + \dfrac{b}{a}.$

24. $y'''_{xxx} + (be^{ax} + 2a)\sinh^n x\, y''_{xx} - a(be^{ax}\sinh^n x + a)y'_x - 2a^3\sinh^n x\, y = 0.$

Particular solutions: $y_1 = e^{ax}, \quad y_2 = e^{-ax} + \dfrac{b}{a}.$

25. $y'''_{xxx} + (be^{ax} + 2a)\tanh^n x\, y''_{xx} - a(be^{ax}\tanh^n x + a)y'_x - 2a^3\tanh^n x\, y = 0.$

Particular solutions: $y_1 = e^{ax}, \quad y_2 = e^{-ax} + \dfrac{b}{a}.$

26. $y'''_{xxx} + (be^{ax} + 2a)\coth^n x\, y''_{xx} - a(be^{ax}\coth^n x + a)y'_x - 2a^3\coth^n x\, y = 0.$

Particular solutions: $y_1 = e^{ax}, \quad y_2 = e^{-ax} + \dfrac{b}{a}.$

27. $y'''_{xxx} + (be^{ax} + 2a)\ln^n x\, y''_{xx} - a(be^{ax}\ln^n x + a)y'_x - 2a^3\ln^n x\, y = 0.$

Particular solutions: $y_1 = e^{ax}, \quad y_2 = e^{-ax} + \dfrac{b}{a}.$

28. $$y'''_{xxx} + (a\ln^n x - 2be^x)y''_{xx} - be^x(2a\ln^n x - be^x + 3)y'_x + be^x[a\ln^n x\,(be^x - 1) + 2be^x - 1]y = 0.$$

Particular solutions: $y_1 = \exp(be^x), \quad y_2 = x\exp(be^x).$

29. $$y'''_{xxx} + (a\cos^n x - 2be^x)y''_{xx} - be^x(2a\cos^n x - be^x + 3)y'_x + be^x[a\cos^n x\,(be^x - 1) + 2be^x - 1]y = 0.$$

Particular solutions: $y_1 = \exp(be^x), \quad y_2 = x\exp(be^x).$

30. $y'''_{xxx} + (be^{ax} + 2a)\cos^n x\, y''_{xx} - a(be^{ax}\cos^n x + a)y'_x - 2a^3\cos^n x\, y = 0.$

Particular solutions: $y_1 = e^{ax}, \quad y_2 = e^{-ax} + \dfrac{b}{a}.$

31. $$y'''_{xxx} + (a\sin^n x - 2be^x)y''_{xx} - be^x(2a\sin^n x - be^x + 3)y'_x + be^x[a\sin^n x\,(be^x - 1) + 2be^x - 1]y = 0.$$

Particular solutions: $y_1 = \exp(be^x), \quad y_2 = x\exp(be^x).$

32. $y'''_{xxx} + (be^{ax} + 2a)\sin^n x\, y''_{xx} - a(be^{ax}\sin^n x + a)y'_x - 2a^3\sin^n x\, y = 0.$

Particular solutions: $y_1 = e^{ax}, \quad y_2 = e^{-ax} + \dfrac{b}{a}.$

33. $y'''_{xxx} - [e^{\lambda x}(\tan x + a) + a]y''_{xx} + [(a^2+1)e^{\lambda x} + 1]y'_x + a[e^{\lambda x}(a\tan x - 1) - 1]y = 0.$

Particular solutions: $y_1 = e^{ax}, \quad y_2 = \cos x.$

34. $y'''_{xxx} + [\tan x\,(axe^{\lambda x} + 1) + ae^{\lambda x}]y''_{xx} - axe^{\lambda x}y'_x + ae^{\lambda x}y = 0.$

Particular solutions: $y_1 = x, \quad y_2 = \cos x.$

35. $y'''_{xxx} + (a\tan^n x - 2be^x)y''_{xx} - be^x(2a\tan^n x - be^x + 3)y'_x$
$$+ be^x[a\tan^n x\,(be^x - 1) + 2be^x - 1]y = 0.$$

Particular solutions: $y_1 = \exp(be^x)$, $y_2 = x\exp(be^x)$.

36. $y'''_{xxx} + (be^{ax} + 2a)\tan^n x\, y''_{xx} - a(be^{ax}\tan^n x + a)y'_x - 2a^3\tan^n x\, y = 0.$

Particular solutions: $y_1 = e^{ax}$, $y_2 = e^{-ax} + \dfrac{b}{a}$.

37. $y'''_{xxx} + [e^{\lambda x}(\cot x + a) + a]y''_{xx} + [(a^2+1)e^{\lambda x} + 1]y'_x + a[e^{\lambda x}(1 - a\cot x) + 1]y = 0.$

Particular solutions: $y_1 = e^{-ax}$, $y_2 = \sin x$.

38. $y'''_{xxx} + [ae^{\lambda x} - \cot x\,(axe^{\lambda x} + 1)]y''_{xx} - axe^{\lambda x}y'_x + ae^{\lambda x}y = 0.$

Particular solutions: $y_1 = x$, $y_2 = \sin x$.

39. $y'''_{xxx} + (a\cot^n x - 2be^x)y''_{xx} - be^x(2a\cot^n x - be^x + 3)y'_x$
$$+ be^x[a\cot^n x\,(be^x - 1) + 2be^x - 1]y = 0.$$

Particular solutions: $y_1 = \exp(be^x)$, $y_2 = x\exp(be^x)$.

40. $y'''_{xxx} + (be^{ax} + 2a)\cot^n x\, y''_{xx} - a(be^{ax}\cot^n x + a)y'_x - 2a^3\cot^n x\, y = 0.$

Particular solutions: $y_1 = e^{ax}$, $y_2 = e^{-ax} + \dfrac{b}{a}$.

41. $y'''_{xxx} - [\cosh^n x\,(\tan x + a) + a]y''_{xx} + [(a^2 + 1)\cosh^n x + 1]y'_x$
$$+ a[\cosh^n x\,(a\tan x - 1) - 1]y = 0.$$

Particular solutions: $y_1 = e^{ax}$, $y_2 = \cos x$.

42. $y'''_{xxx} + [\cosh^n x\,(\cot x + a) + a]y''_{xx} + [(a^2 + 1)\cosh^n x + 1]y'_x$
$$+ a[\cosh^n x\,(1 - a\cot x) + 1]y = 0.$$

Particular solutions: $y_1 = e^{-ax}$, $y_2 = \sin x$.

43. $y'''_{xxx} - [\sinh^n x\,(\tan x + a) + a]y''_{xx} + [(a^2 + 1)\sinh^n x + 1]y'_x$
$$+ a[\sinh^n x\,(a\tan x - 1) - 1]y = 0.$$

Particular solutions: $y_1 = e^{ax}$, $y_2 = \cos x$.

44. $y'''_{xxx} + [\sinh^n x\,(\cot x + a) + a]y''_{xx} + [(a^2 + 1)\sinh^n x + 1]y'_x$
$$+ a[\sinh^n x\,(1 - a\cot x) + 1]y = 0.$$

Particular solutions: $y_1 = e^{-ax}$, $y_2 = \sin x$.

45. $y'''_{xxx} - [\tanh^n x\,(\tan x + a) + a]y''_{xx} + [(a^2 + 1)\tanh^n x + 1]y'_x$
$$+ a[\tanh^n x\,(a\tan x - 1) - 1]y = 0.$$

Particular solutions: $y_1 = e^{ax}$, $y_2 = \cos x$.

46. $y'''_{xxx} + [\tanh^n x\,(\cot x + a) + a]y''_{xx} + [(a^2+1)\tanh^n x + 1]y'_x + a[\tanh^n x\,(1 - a\cot x) + 1]y = 0.$

Particular solutions: $y_1 = e^{-ax}$, $y_2 = \sin x$.

47. $y'''_{xxx} - [\coth^n x\,(\tan x + a) + a]y''_{xx} + [(a^2+1)\coth^n x + 1]y'_x + a[\coth^n x\,(\tan x - 1) - 1]y = 0.$

Particular solutions: $y_1 = e^{ax}$, $y_2 = \cos x$.

48. $y'''_{xxx} + [\coth^n x\,(\cot x + a) + a]y''_{xx} + [(a^2+1)\coth^n x + 1]y'_x + a[\coth^n x\,(1 - a\cot x) + 1]y = 0.$

Particular solutions: $y_1 = e^{-ax}$, $y_2 = \sin x$.

49. $y'''_{xxx} + [a\tan^n x\,(\tanh x - b) - b]y''_{xx} + [a(b^2-1)\tan^n x - 1]y'_x + b[a\tan^n x\,(1 - b\tanh x) + 1]y = 0.$

Particular solutions: $y_1 = e^{bx}$, $y_2 = \cosh x$.

50. $y'''_{xxx} + (a\tan^n x + b\tanh^m x)y''_{xx} + cy'_x + c(a\tan^n x + b\tanh^m x)y = 0.$

1°. Particular solutions with $c > 0$: $y_1 = \cos\bigl(x\sqrt{c}\bigr)$, $y_2 = \sin\bigl(x\sqrt{c}\bigr)$.

2°. Particular solutions with $c < 0$: $y_1 = \exp\bigl(-x\sqrt{-c}\bigr)$, $y_2 = \exp\bigl(x\sqrt{-c}\bigr)$.

51. $y'''_{xxx} + [a\tan^n x\,(\coth x - b) - b]y''_{xx} + [a(b^2-1)\tan^n x - 1]y'_x + b[a\tan^n x\,(1 - b\coth x) + 1]y = 0.$

Particular solutions: $y_1 = e^{bx}$, $y_2 = \sinh x$.

52. $y'''_{xxx} + (a\tan^n x + b\coth^m x)y''_{xx} + cy'_x + c(a\tan^n x + b\coth^m x)y = 0.$

1°. Particular solutions with $c > 0$: $y_1 = \cos\bigl(x\sqrt{c}\bigr)$, $y_2 = \sin\bigl(x\sqrt{c}\bigr)$.

2°. Particular solutions with $c < 0$: $y_1 = \exp\bigl(-x\sqrt{-c}\bigr)$, $y_2 = \exp\bigl(x\sqrt{-c}\bigr)$.

53. $y'''_{xxx} + [a\cot^n x\,(\tanh x - b) - b]y''_{xx} + [a(b^2-1)\cot^n x - 1]y'_x + b[a\cot^n x\,(1 - b\tanh x) + 1]y = 0.$

Particular solutions: $y_1 = e^{bx}$, $y_2 = \cosh x$.

54. $y'''_{xxx} + a\cot^n x\tanh^m x\,y''_{xx} - by'_x - ab\cot^n x\tanh^m x\,y = 0.$

1°. Particular solutions with $b > 0$: $y_1 = \exp\bigl(-x\sqrt{b}\bigr)$, $y_2 = \exp\bigl(x\sqrt{b}\bigr)$.

2°. Particular solutions with $b < 0$: $y_1 = \cos\bigl(x\sqrt{-b}\bigr)$, $y_2 = \sin\bigl(x\sqrt{-b}\bigr)$.

55. $y'''_{xxx} + [a\cot^n x\,(\coth x - b) - b]y''_{xx} + [a(b^2-1)\cot^n x - 1]y'_x + b[a\cot^n x\,(1 - b\coth x) + 1]y = 0.$

Particular solutions: $y_1 = e^{bx}$, $y_2 = \sinh x$.

56. $y'''_{xxx} + (a\cot^n x + b\coth^m x)y''_{xx} + cy'_x + c(a\cot^n x + b\coth^m x)y = 0.$

1°. Particular solutions with $c > 0$: $y_1 = \cos(x\sqrt{c})$, $y_2 = \sin(x\sqrt{c})$.

2°. Particular solutions with $c < 0$: $y_1 = \exp(-x\sqrt{-c})$, $y_2 = \exp(x\sqrt{-c})$.

57. $$y'''_{xxx} + [a\ln^n x\,(\tanh x - b) - b]y''_{xx} + [a(b^2-1)\ln^n x - 1]y'_x + b[a\ln^n x\,(1 - b\tanh x) + 1]y = 0.$$

Particular solutions: $y_1 = e^{bx}$, $y_2 = \cosh x$.

58. $y'''_{xxx} + a\ln^n x\tanh^m x\,y''_{xx} - by'_x - ab\ln^n x\tanh^m x\,y = 0.$

1°. Particular solutions with $b > 0$: $y_1 = \exp(-x\sqrt{b})$, $y_2 = \exp(x\sqrt{b})$.

2°. Particular solutions with $b < 0$: $y_1 = \cos(x\sqrt{-b})$, $y_2 = \sin(x\sqrt{-b})$.

59. $$y'''_{xxx} + [a\ln^n x\,(\coth x - b) - b]y''_{xx} + [a(b^2-1)\ln^n x - 1]y'_x + b[a\ln^n x\,(1 - b\coth x) + 1]y = 0.$$

Particular solutions: $y_1 = e^{bx}$, $y_2 = \sinh x$.

60. $y'''_{xxx} + (a\ln^n x + b\coth^m x)y''_{xx} + cy'_x + c(a\ln^n x + b\coth^m x)y = 0.$

1°. Particular solutions with $c > 0$: $y_1 = \cos(x\sqrt{c})$, $y_2 = \sin(x\sqrt{c})$.

2°. Particular solutions with $c < 0$: $y_1 = \exp(-x\sqrt{-c})$, $y_2 = \exp(x\sqrt{-c})$.

61. $$y'''_{xxx} - [\ln^n x\,(\tan x + a) + a]y''_{xx} + [(a^2+1)\ln^n x + 1]y'_x + a[\ln^n x\,(a\tan x - 1) - 1]y = 0.$$

Particular solutions: $y_1 = e^{ax}$, $y_2 = \cos x$.

62. $$y'''_{xxx} + [\ln^n x\,(\cot x + a) + a]y''_{xx} + [(a^2+1)\ln^n x + 1]y'_x + a[\ln^n x\,(1 - a\cot x) + 1]y = 0.$$

Particular solutions: $y_1 = e^{-ax}$, $y_2 = \sin x$.

63. $$y'''_{xxx} + [a\cos^n x\,(\tanh x - b) - b]y''_{xx} + [a(b^2-1)\cos^n x - 1]y'_x + b[a\cos^n x\,(1 - b\tanh x) + 1]y = 0.$$

Particular solutions: $y_1 = e^{bx}$, $y_2 = \cosh x$.

64. $y'''_{xxx} + a\cos^n x\tanh^m x\,y''_{xx} + by'_x + ab\cos^n x\tanh^m x\,y = 0.$

1°. Particular solutions with $b > 0$: $y_1 = \cos(x\sqrt{b})$, $y_2 = \sin(x\sqrt{b})$.

2°. Particular solutions with $b < 0$: $y_1 = \exp(-x\sqrt{-b})$, $y_2 = \exp(x\sqrt{-b})$.

65. $$y'''_{xxx} + [a\cos^n x\,(\coth x - b) - b]y''_{xx} + [a(b^2-1)\cos^n x - 1]y'_x + b[a\cos^n x\,(1 - b\coth x) + 1]y = 0.$$

Particular solutions: $y_1 = e^{bx}$, $y_2 = \sinh x$.

66. $y'''_{xxx} + (a\cos^n x + b\coth^m x)y''_{xx} + cy'_x + c(a\cos^n x + b\coth^m x)y = 0.$

1°. Particular solutions with $c > 0$: $y_1 = \cos(x\sqrt{c})$, $y_2 = \sin(x\sqrt{c})$.

2°. Particular solutions with $c < 0$: $y_1 = \exp(-x\sqrt{-c})$, $y_2 = \exp(x\sqrt{-c})$.

67. $y'''_{xxx} + [a\sin^n x\,(\tanh x - b) - b]y''_{xx} + [a(b^2-1)\sin^n x - 1]y'_x + b[a\sin^n x\,(1 - b\tanh x) + 1]y = 0.$

Particular solutions: $y_1 = e^{bx}$, $y_2 = \cosh x$.

68. $y'''_{xxx} + a\sin^n x\tanh^m x\,y''_{xx} + by'_x + ab\sin^n x\tanh^m x\,y = 0.$

1°. Particular solutions with $b > 0$: $y_1 = \cos(x\sqrt{b})$, $y_2 = \sin(x\sqrt{b})$.

2°. Particular solutions with $b < 0$: $y_1 = \exp(-x\sqrt{-b})$, $y_2 = \exp(x\sqrt{-b})$.

69. $y'''_{xxx} + [a\sin^n x\,(\coth x - b) - b]y''_{xx} + [a(b^2-1)\sin^n x - 1]y'_x + b[a\sin^n x\,(1 - b\coth x) + 1]y = 0.$

Particular solutions: $y_1 = e^{bx}$, $y_2 = \sinh x$.

70. $y'''_{xxx} + (a\sin^n x + b\coth^m x)y''_{xx} + cy'_x + c(a\sin^n x + b\coth^m x)y = 0.$

1°. Particular solutions with $c > 0$: $y_1 = \cos(x\sqrt{c})$, $y_2 = \sin(x\sqrt{c})$.

2°. Particular solutions with $c < 0$: $y_1 = \exp(-x\sqrt{-c})$, $y_2 = \exp(x\sqrt{-c})$.

71. $xy'''_{xxx} + [ax^2e^{\lambda x}(b - \ln x) + 2]y''_{xx} + axe^{\lambda x}y'_x - ae^{\lambda x}y = 0.$

Particular solutions: $y_1 = x$, $y_2 = \ln x - b + 1$.

72. $(e^{\lambda x} - 1)y'''_{xxx} - (ae^{\lambda x} + \tan x)y''_{xx} + (e^{\lambda x} + a^2)y'_x + a(a\tan x - e^{\lambda x})y = 0.$

Particular solutions: $y_1 = e^{ax}$, $y_2 = \cos x$.

73. $a\cosh^n x\,y'''_{xxx} + [\tan x\,(a\cosh^n x + x) + 1]y''_{xx} - xy'_x + y = 0.$

Particular solutions: $y_1 = x$, $y_2 = \cos x$.

74. $a\cosh^n x\,y'''_{xxx} + [1 - \cot x\,(a\cosh^n x + x)]y''_{xx} - xy'_x + y = 0.$

Particular solutions: $y_1 = x$, $y_2 = \sin x$.

75. $a\sinh^n x\,y'''_{xxx} + [\tan x\,(a\sinh^n x + x) + 1]y''_{xx} - xy'_x + y = 0.$

Particular solutions: $y_1 = x$, $y_2 = \cos x$.

76. $a\sinh^n x\,y'''_{xxx} + [1 - \cot x\,(a\sinh^n x + x)]y''_{xx} - xy'_x + y = 0.$

Particular solutions: $y_1 = x$, $y_2 = \sin x$.

77. $a\tanh^n x\,y'''_{xxx} + [\tan x\,(a\tanh^n x + x) + 1]y''_{xx} - xy'_x + y = 0.$

Particular solutions: $y_1 = x$, $y_2 = \cos x$.

78. $a\tanh^n x\, y'''_{xxx} + [1 - \cot x\,(a\tanh^n x + x)]y''_{xx} - xy'_x + y = 0.$

Particular solutions: $y_1 = x, \quad y_2 = \sin x.$

79. $a\coth^n x\, y'''_{xxx} + [\tan x\,(a\coth^n x + x) + 1]y''_{xx} - xy'_x + y = 0.$

Particular solutions: $y_1 = x, \quad y_2 = \cos x.$

80. $a\coth^n x\, y'''_{xxx} + [1 - \cot x\,(a\coth^n x + x)]y''_{xx} - xy'_x + y = 0.$

Particular solutions: $y_1 = x, \quad y_2 = \sin x.$

81. $a\ln^n x\, y'''_{xxx} + [\tanh x\,(x - a\ln^n x) - 1]y''_{xx} - xy'_x + y = 0.$

Particular solutions: $y_1 = x, \quad y_2 = \cosh x.$

82. $a\ln^n x\, y'''_{xxx} + [\coth x\,(x - a\ln^n x) - 1]y''_{xx} - xy'_x + y = 0.$

Particular solutions: $y_1 = x, \quad y_2 = \sinh x.$

83. $a\ln^n x\, y'''_{xxx} + [\tan x\,(a\ln^n x + x) + 1]y''_{xx} - xy'_x + y = 0.$

Particular solutions: $y_1 = x, \quad y_2 = \cos x.$

84. $a\ln^n x\, y'''_{xxx} + [1 - \cot x\,(a\ln^n x + x)]y''_{xx} - xy'_x + y = 0.$

Particular solutions: $y_1 = x, \quad y_2 = \sin x.$

85. $a\cos^n x\, y'''_{xxx} + [\tanh x\,(x - a\cos^n x) - 1]y''_{xx} - xy'_x + y = 0.$

Particular solutions: $y_1 = x, \quad y_2 = \cosh x.$

86. $a\cos^n x\, y'''_{xxx} + [\coth x\,(x - a\cos^n x) - 1]y''_{xx} - xy'_x + y = 0.$

Particular solutions: $y_1 = x, \quad y_2 = \sinh x.$

87. $ax\cos^n x\, y'''_{xxx} + (2a\cos^n x - x^2\ln x + bx^2)y''_{xx} + xy'_x - y = 0.$

Particular solutions: $y_1 = x, \quad y_2 = \ln x - b + 1.$

88. $a\sin^n x\, y'''_{xxx} + [\tanh x\,(x - a\sin^n x) - 1]y''_{xx} - xy'_x + y = 0.$

Particular solutions: $y_1 = x, \quad y_2 = \cosh x.$

89. $a\sin^n x\, y'''_{xxx} + [\coth x\,(x - a\sin^n x) - 1]y''_{xx} - xy'_x + y = 0.$

Particular solutions: $y_1 = x, \quad y_2 = \sinh x.$

90. $ax\sin^n x\, y'''_{xxx} + (2a\sin^n x - x^2\ln x + bx^2)y''_{xx} + xy'_x - y = 0.$

Particular solutions: $y_1 = x, \quad y_2 = \ln x - b + 1.$

91. $a\tan^n x\, y'''_{xxx} + [\tanh x\,(x - a\tan^n x) - 1]y''_{xx} - xy'_x + y = 0.$

Particular solutions: $y_1 = x, \quad y_2 = \cosh x.$

92. $a\tan^n x\, y'''_{xxx} + [\coth x\,(x - a\tan^n x) - 1]y''_{xx} - xy'_x + y = 0.$

Particular solutions: $y_1 = x, \quad y_2 = \sinh x.$

93. $ax\tan^n x\, y'''_{xxx} + (2a\tan^n x - x^2\ln x + bx^2)y''_{xx} + xy'_x - y = 0.$

Particular solutions: $y_1 = x, \quad y_2 = \ln x - b + 1.$

94. $a\cot^n x\, y'''_{xxx} + [\tanh x\,(x - a\cot^n x) - 1]y''_{xx} - xy'_x + y = 0.$

Particular solutions: $y_1 = x, \quad y_2 = \cosh x.$

95. $a\cot^n x\, y'''_{xxx} + [\coth x\,(x - a\cot^n x) - 1]y''_{xx} - xy'_x + y = 0.$

Particular solutions: $y_1 = x, \quad y_2 = \sinh x.$

96. $ax\cot^n x\, y'''_{xxx} + (2a\cot^n x - x^2\ln x + bx^2)y''_{xx} + xy'_x - y = 0.$

Particular solutions: $y_1 = x, \quad y_2 = \ln x - b + 1.$

3.1.9. Equations Containing Arbitrary Functions

Notation: $f = f(x)$, $g = g(x)$, *and* $h = h(x)$ *are arbitrary function of argument* x; a, b, c, n, *and* λ *are parameters.*

1. $y'''_{xxx} + fy'_x - (af + a^3)y = 0.$

Particular solution: $y_0 = e^{ax}.$

The substitution $w = y'_x - ay$ leads to a second order linear equation: $w''_{xx} + aw'_x + (f + a^2)w = 0.$

2. $y'''_{xxx} + fy'_x + ax(f + a^2x^2 - 3a)y = 0.$

Particular solution: $y_0 = \exp\Bigl(-\dfrac{ax^2}{2}\Bigr).$

The substitution $y = \exp\Bigl(-\dfrac{ax^2}{2}\Bigr)\displaystyle\int z(x)\,dx$ leads to a second order linear equation: $z''_{xx} - 3axz'_x + (f + 3a^2x^2 - 3a)z = 0.$

3. $y'''_{xxx} + (f - a^2)y'_x + afy = 0.$

Particular solution: $y_0 = e^{-ax}.$

The substitution $w = y'_x + ay$ leads to a second order linear equation: $w''_{xx} - aw'_x + fw = 0.$

4. $y'''_{xxx} + xfy'_x - 2fy = 0.$

Particular solution: $y_0 = x^2.$

The substitution $w = xy'_x - 2y$ leads to a second order linear equation: $w''_{xx} + xfw = 0.$

5. $y'''_{xxx} + (ax + b)fy'_x - afy = 0.$

Particular solution: $y_0 = ax + b.$

6. $y'''_{xxx} + (f - a^2x^2)y'_x + ax(f - 3a)y = 0.$

Particular solution: $y_0 = \exp\left(-\dfrac{ax^2}{2}\right).$

The substitution $y = \exp\left(\dfrac{-ax^2}{2}\right)\int z(x)\,dx$ leads to a second order linear equation: $z''_{xx} - 3axz'_x + (2a^2x^2 - 3a + f)z = 0.$

7. $y'''_{xxx} + (f - a^2x^{2n})y'_x - a[x^n f + 3anx^{2n-1} + n(n-1)x^{n-2}]y = 0.$

Particular solution: $y_0 = \exp\left(\dfrac{a}{n+1}x^{n+1}\right).$

The substitution $y = \exp\left(\dfrac{a}{n+1}x^{n+1}\right)\int z(x)\,dx$ leads to a second order linear equation: $z''_{xx} + 3ax^n z'_x + (2a^2x^{2n} + 3anx^{n-1} + f)z = 0.$

8. $y'''_{xxx} + ay''_{xx} + by'_x + cy = f(x).$

This is a special case of equation 5.1.5.9.

9. $y'''_{xxx} + ay''_{xx} + fy'_x + afy = 0.$

The substitution $w = y'_x + ay$ leads to a second order linear equation: $w''_{xx} + fw = 0.$

10. $y'''_{xxx} + fy''_{xx} - a^2(f + a)y = 0.$

Particular solution: $y_0 = e^{ax}.$

The substitution $w = y'_x - ay$ leads to a second order linear equation: $w''_{xx} + (f + a)w'_x + a(f + a)w = 0.$

11. $y'''_{xxx} + fy''_{xx} + ay'_x + afy = 0.$

1°. Particular solutions with $a > 0$: $y_1 = \cos\left(x\sqrt{a}\right)$, $y_2 = \sin\left(x\sqrt{a}\right).$

2°. Particular solutions with $a < 0$: $y_1 = \exp\left(-x\sqrt{-a}\right)$, $y_2 = \exp\left(x\sqrt{-a}\right).$

The substitution $w = y''_{xx} + ay$ leads to a first order linear equation: $w'_x + fw = 0.$

12. $y'''_{xxx} + fy''_{xx} + ax^n y'_x + ax^{n-1}(xf + n)y = 0.$

The substitution $w = y''_{xx} + ax^n y$ leads to a first order linear equation: $w'_x + fw = 0.$

13. $y'''_{xxx} + fy''_{xx} + afy'_x + a^3y = 0.$

The substitution $w = y'_x + ay$ leads to a second order linear equation: $w''_{xx} + (f - a)w'_x + a^2w = 0.$

14. $y'''_{xxx} + fy''_{xx} + afy'_x + a^2(f - a)y = 0.$

Particular solutions: $y_1 = e^{-ax/2}\cos\left(\dfrac{a\sqrt{3}}{2}x\right)$, $y_2 = e^{-ax/2}\sin\left(\dfrac{a\sqrt{3}}{2}x\right).$

15. $y'''_{xxx} + fy''_{xx} + gy'_x + h = 0.$

The substitution $w = y'_x$ leads to a second order linear equation: $w''_{xx} + fw'_x + gw + h = 0.$

16. $y'''_{xxx} + fy''_{xx} - a(2f + 3a)y'_x + a^2(f + 2a)y = 0.$

Particular solutions: $y_1 = e^{ax}$, $y_2 = xe^{ax}$.

17. $y'''_{xxx} + fy''_{xx} + xgy'_x - gy = 0.$

The substitution $w = xy'_x - y$ leads to a second order linear equation: $xw''_{xx} + (xf - 1)w'_x + x^2 gw = 0$.

18. $y'''_{xxx} + fy''_{xx} + (g - a^2)y'_x - a(af + g)y = 0.$

Particular solution: $y_0 = e^{ax}$.

The substitution $w = y'_x - ay$ leads to a second order linear equation: $w''_{xx} + (f + a)w'_x + (af + g)w = 0$.

19. $y'''_{xxx} + fy''_{xx} + (af + b - a^2)y'_x + b(f - a)y = 0.$

Particular solutions: $y_1 = e^{\lambda_1 x}$, $y_2 = e^{\lambda_2 x}$, where λ_1 and λ_2 are the roots of the quadratic equation $\lambda^2 + a\lambda + b = 0$.

20. $y'''_{xxx} + (f - a)y''_{xx} - a^2 fy = 0.$

Particular solution: $y_0 = e^{ax}$.

The substitution $w = y'_x - ay$ leads to a second order equation: $w''_{xx} + fw'_x + afw = 0$.

21. $y'''_{xxx} = (f - a)y''_{xx} + (af - b)y'_x + bfy.$

Particular solutions: $y_1 = \exp(\lambda_1 x)$, $y_2 = \exp(\lambda_2 x)$, where λ_1 and λ_2 are roots of the quadratic equation $\lambda^2 + a\lambda + b = 0$.

The substitution $w = y''_{xx} + ay'_x + by$ leads to a first order linear equation: $w'_x = fw$.

22. $y'''_{xxx} + (f - a)y''_{xx} + gy'_x - a(af + g)y = 0.$

Particular solution: $y_0 = e^{ax}$.

23. $y'''_{xxx} + (f + a)y''_{xx} + (af + g)y'_x + agy = 0.$

Particular solution: $y_0 = e^{-ax}$.

24. $y'''_{xxx} + xfy''_{xx} + (ax^2 - f)y'_x + ax(x^2 f + 3)y = 0.$

Particular solutions: $y_1 = \cos\left(\frac{x^2\sqrt{a}}{2}\right)$, $y_2 = \sin\left(\frac{x^2\sqrt{a}}{2}\right)$.

25. $y'''_{xxx} + (ax + b)fy''_{xx} + xfy'_x - 2fy = 0.$

Particular solution: $y_0 = x^2 + 2ax + b$.

26. $y'''_{xxx} + (f + ax)y''_{xx} + a(xf + 2)y'_x + afy = 0.$

Particular solutions: $y_1 = \exp\left(-\frac{ax^2}{2}\right)$, $y_2 = \exp\left(-\frac{ax^2}{2}\right) \int \exp\left(\frac{ax^2}{2}\right) dx$.

27. $y'''_{xxx} + x^2 f y''_{xx} - 2xfy'_x + 2fy = 0.$

Particular solutions: $y_1 = x, \quad y_2 = x^2.$

Solution:

$$y = C_1 x + C_2 x^2 + C_3\Big(x^2 \int x^{-3}\psi\,dx - x \int x^{-2}\psi\,dx\Big),$$

where $\psi = \exp(-\int x^2 f\,dx)$.

28. $y'''_{xxx} + (f + ax)y''_{xx} + (g + 2a)y'_x + a[xg + (1 - ax^2)f]y = 0.$

Particular solution: $y_0 = \exp\Big(-\dfrac{ax^2}{2}\Big).$

The substitution $w = y'_x + axy$ leads to a second order linear equation: $w''_{xx} + fw'_x + (g - axf)w = 0.$

29. $y'''_{xxx} + (axf + f + a)y''_{xx} + a^2 x f y'_x - a^2 f y = 0.$

Particular solutions: $y_1 = x, \quad y_2 = e^{-ax}.$

30. $y'''_{xxx} + (ax^2 + bx + c)fy''_{xx} - 2afy = 0.$

Particular solution: $y_0 = ax^2 + bx + c.$

31. $y'''_{xxx} + x(xf + g)y''_{xx} - gy'_x - 2fy = 0.$

Particular solution: $y_0 = x^2$. The substitution $w = xy'_x - 2y$ leads to a second order linear equation: $w''_{xx} + x(xf + g)w'_x + xfw = 0.$

32. $y'''_{xxx} - x(ax + b)fy''_{xx} + (b - a^2)fy'_x + 2afy = 0.$

Particular solution: $y_0 = x^2 + ax + \dfrac{1}{2}(a^2 - b).$

33. $y'''_{xxx} - [(2x + a)f + (x^2 + ax + b)g]y''_{xx} + 2fy'_x + 2gy = 0.$

Particular solution: $y_0 = x^2 + ax + b.$

34. $xy'''_{xxx} + 3y''_{xx} + x(ax^2 + 1)fy'_x - (ax^2 - 1)fy = 0.$

Particular solution: $y_0 = ax + \dfrac{1}{x}.$

35. $xy'''_{xxx} + (ax^2 + b)y''_{xx} + 4axy'_x + 2ay = f.$

Integrating the equation twice, we obtain a first order linear equation:

$$xy'_x + (ax^2 + b - 2)y = C_1 + C_2 x + \int\Big(\int f\,dx\Big)\,dx.$$

36. $xy'''_{xxx} + xfy'_x - [(ax + 1)f + a^3 x + 3a^2]y = 0.$

Particular solution: $y_0 = xe^{ax}.$

37. $xy'''_{xxx} + x(f - 2a)y''_{xx} + x(g + a^2)y'_x - [a(ax + 2)f + (ax + 1)g]y = 0.$

Particular solution: $y_0 = xe^{ax}$.

38. $xy'''_{xxx} + (xf + 3)y''_{xx} + (2f + ax)y'_x + a(xf + 1)y = 0.$

Particular solutions: $y_1 = \dfrac{1}{x}\cos(x\sqrt{a}), \quad y_2 = \dfrac{1}{x}\sin(x\sqrt{a})$.

39. $xy'''_{xxx} + (xf + 3)y''_{xx} + (ax + 2)fy'_x + a(axf + f - a^2x)y = 0.$

Particular solutions: $y_1 = \dfrac{1}{x}e^{-ax/2}\cos\left(\dfrac{a\sqrt{3}}{2}x\right), \quad y_2 = \dfrac{1}{x}e^{-ax/2}\sin\left(\dfrac{a\sqrt{3}}{2}x\right)$.

40. $xy'''_{xxx} + (xf + 3)y''_{xx} + (axf + 2f - a^2x)y'_x + a(f - a)y = 0.$

Particular solutions: $y_1 = \dfrac{1}{x}, \quad y_2 = \dfrac{1}{x}e^{-ax}$.

41. $xy'''_{xxx} + (xf + 3)y''_{xx} + (2f + ax^{n+1})y'_x + ax^n(xf + n + 1)y = 0.$

The substitution $w = xy$ leads to an equation of the form 3.1.9.12: $w'''_{xxx} + fw''_{xx} + ax^n w'_x + ax^{n-1}(xf + n)w = 0$.

42. $xy'''_{xxx} + (x^2f + a + 2)y''_{xx} - a(a + 1)fy = 0.$

Particular solution: $y_0 = x^{-a}$.

The substitution $w = xy'_x + ay$ leads to a second order linear equation: $w''_{xx} + xfw'_x - (a + 1)fw = 0$.

43. $xy'''_{xxx} + [x^2(ax^2 + 1)f + 3]y''_{xx} - 2fy = 0.$

Particular solution: $y_0 = ax + \dfrac{1}{x}$.

44. $xy'''_{xxx} + [x(ax^2 - 1)f + x^2(ax^2 + 1)g + 3]y''_{xx} - 2fy'_x - 2gy = 0.$

Particular solution: $y_0 = ax + \dfrac{1}{x}$.

45. $(ax - 1)y'''_{xxx} + x[(ax - 2)f - a^2]y''_{xx} + [(2 - a^2x^2)f + a^2]y'_x + 2a(ax - 1)fy = 0.$

Particular solutions: $y_1 = x^2, \quad y_2 = e^{ax}$.

46. $x^2y'''_{xxx} + (xf - a^2 - a)y'_x + (a - 1)fy = 0.$

Particular solution: $y_0 = x^{1-a}$.

The substitution $w = xy'_x + (a - 1)y$ leads to a second order linear equation: $xw''_{xx} - (a + 1)w'_x + fw = 0$.

47. $x^2y'''_{xxx} + [x(ax + 1)f - 6]y'_x + fy = 0.$

Particular solution: $y_0 = a + \dfrac{1}{x}$.

48. $x^2 y'''_{xxx} + xfy''_{xx} + [x(ax+1)g + 2f - 6]y'_x + gy = 0.$

Particular solution: $y_0 = a + \dfrac{1}{x}$.

49. $x^2 y'''_{xxx} + x[x(ax+1)f + 3]y''_{xx} - 2fy = 0.$

Particular solution: $y_0 = a + \dfrac{1}{x}$.

50. $x^2 y'''_{xxx} + x(xf + a)y''_{xx} + [(a-2)xf + b]y'_x + (b - a + 2)fy = 0.$

By integrating, we obtain the nonhomogeneous Euler equation 2.1.9.15:

$$x^2 y''_{xx} + (a-2)xy'_x + (b-a+2)y = C\exp\Bigl(-\int f\,dx\Bigr).$$

51. $(ax+b)xy'''_{xxx} + (\alpha x + \beta)y''_{xx} + xy'_x + y = f.$

By integrating, we obtain a second order linear equation:

$$(ax+b)xy''_{xx} + [(\alpha - 2a)x + \beta - b]y'_x + (x + 2a - \alpha)y = \int f\,dx + C.$$

52. $x(x+1)y'''_{xxx} + x(f - x - 3)y'_x - (x+1)fy = 0.$

Particular solution: $y_0 = xe^x$.

53. $x^3 y'''_{xxx} + xfy'_x + (a-1)(f + a^2 + a)y = 0.$

Particular solution: $y_0 = x^{1-a}$.

The substitution $w = xy'_x + (a-1)y$ leads to a second order linear equation: $x^2 w''_{xx} - (a+1)xw'_x + (f + a^2 + a)w = 0$.

54. $x^3 y'''_{xxx} + ax^2 y''_{xx} + bxy'_x + cy = f(x).$

The nonhomogeneous Euler equation.

The substitution $t = \ln|x|$ leads to an equation of the form 3.1.9.8:

$$y'''_{ttt} + (a-3)y''_{tt} + (b-a+2)y'_t + cy = f(\pm e^t).$$

55. $x^3 y'''_{xxx} + (a+2)x^2 y''_{xx} + xfy'_x + afy = 0.$

Particular solution: $y_0 = x^{-a}$.

56. $x^3 y'''_{xxx} + [(a+6)x^2 + b]y''_{xx} + 2(2a+3)xy'_x + 2ay = f(x).$

Integrating the equation twice, we obtain a first order linear equation:

$$x^3 y'_x + (ax^2 + b)y = C_1 + C_2 x + \int\Bigl(\int f\,dx\Bigr)dx.$$

57. $x^3 y'''_{xxx} + x^2(bx^{2a+1} - 3a)y''_{xx} + 2a(a+1)(2a+1)y = f(x).$

Integrating the equation twice, we obtain a first order linear equation:

$$x^{-2a} y'_x + (ax^{-2a-1} + b)y = C_1 + C_2 x + \int\Bigl(\int fx^{-2a-3}\,dx\Bigr)dx.$$

58. $x^3 y'''_{xxx} + x^2 f y''_{xx} - 2x y'_x + 2(2 - f)y = 0.$

Particular solutions: $y_1 = x^{-1}, \quad y_2 = x^2.$

59. $x^3 y'''_{xxx} + x^2 f y''_{xx} - 6x y'_x + 6(2 - f)y = 0.$

Particular solutions: $y_1 = x^{-2}, \quad y_2 = x^3.$

60. $x^3 y'''_{xxx} + x^2 f y''_{xx} + x(f - 1)y'_x + (f - 3)y = 0.$

Particular solutions: $y_1 = \cos(\ln x), \quad y_2 = \sin(\ln x).$

61. $x^3 y'''_{xxx} + x^2 (f + 1) y''_{xx} + x(f - a - 1)y'_x - a(f - 2)y = 0.$

Particular solutions: $y_1 = x^{-\sqrt{a}}, \quad y_2 = x^{\sqrt{a}}.$

62. $x^3 y'''_{xxx} + x^2 (f + a) y''_{xx} + x(af + b - a)y'_x + b(f - 2)y = 0.$

Particular solutions: $y_1 = x^{n_1}, \quad y_2 = x^{n_2},$ where n_1 and n_2 are the roots of the quadratic equation $n^2 + (a - 1)n + b = 0.$

63. $x^3 y'''_{xxx} + x^2 (f + a) y''_{xx} + x[g + (a - 1)f]y'_x + (a - 2)gy = 0.$

Particular solution: $y_0 = x^{2-a}$. The substitution $w = xy'_x + (a-2)y$ leads to a second order linear equation: $x^2 w''_{xx} + xfw'_x + gw = 0.$

64. $x^3 y'''_{xxx} + x^2 (f + 2ax) y''_{xx} + x(2axf + a^2x^2 + b)y'_x + (a^2x^2 f + bf - 2b)y = 0.$

Particular solutions: $y_1 = e^{-ax} x^{n_1}, \quad y_2 = e^{-ax} x^{n_2},$ where n_1 and n_2 are the roots of the quadratic equation $n^2 - n + b = 0.$

65. $x^6 y'''_{xxx} + x^2 f y'_x + (a^3 + af - 2xf)y = 0.$

Particular solution: $y_0 = x^2 e^{a/x}.$

66. $y'''_{xxx} + ae^{\lambda x} y''_{xx} - 3\lambda^2 y'_x + 2\lambda^3 y = f(x).$

Integrating the equation twice, we obtain a first order linear equation:

$$e^{-\lambda x} y'_x + (a + 2\lambda e^{-\lambda x})y = C_1 + C_2 x + \int \left(\int f e^{-\lambda x}\, dx \right) dx.$$

67. $y'''_{xxx} + (f - a^2 e^{2\lambda x})y'_x - ae^{\lambda x}(f + 3a\lambda e^{\lambda x} + \lambda^2)y = 0.$

Particular solution: $y_0 = \exp\left(\dfrac{a}{\lambda} e^{\lambda x}\right).$

The substitution $y = \exp\left(\dfrac{a}{\lambda} e^{\lambda x}\right) \int z(x)\, dx$ leads to a second order linear equation: $z''_{xx} + 3ae^{\lambda x} z'_x + (f + 2a^2 e^{2\lambda x} + 3a\lambda e^{\lambda x})z = 0.$

68. $y'''_{xxx} + [(1 + be^{ax})f - a^2]y'_x + afy = 0.$

Particular solution: $y_0 = e^{-ax} + b.$

69. $y'''_{xxx} + (f + a)y''_{xx} + [af + (1 + be^{ax})g]y'_x + agy = 0.$

Particular solution: $y_0 = e^{-ax} + b.$

70. $y'''_{xxx} + (be^{ax} + 2a)fy''_{xx} - a(be^{ax}f + a)y'_x - 2a^3 fy = 0.$

Particular solutions: $y_1 = e^{ax}, \quad y_2 = e^{-ax} + \dfrac{b}{a}.$

71. $y'''_{xxx} + (f - 2ae^{\lambda x})y''_{xx} - ae^{\lambda x}(2f - ae^{\lambda x} + 3\lambda)y'_x$
$+ ae^{\lambda x}[(ae^{\lambda x} - \lambda)f + 2a\lambda e^{\lambda x} - \lambda^2]y = 0.$

Particular solutions: $y_1 = \exp\Bigl(\dfrac{a}{\lambda}e^{\lambda x}\Bigr), \quad y_2 = x\exp\Bigl(\dfrac{a}{\lambda}e^{\lambda x}\Bigr).$

72. $y'''_{xxx} + (f - ae^{\lambda x})y''_{xx} + (g - 2a\lambda e^{\lambda x})y'_x - ae^{\lambda x}[(ae^{\lambda x} + \lambda)f + g + \lambda^2]y = 0.$

Particular solution: $y_0 = \exp\Bigl(\dfrac{a}{\lambda}e^{\lambda x}\Bigr).$

The substitution $y = \exp\Bigl(\dfrac{a}{\lambda}e^{\lambda x}\Bigr)\displaystyle\int z(x)\,dx$ leads to a second order equation:

$$z''_{xx} + (f + 2ae^{\lambda x})z'_x + (2ae^{\lambda x}f + g + a^2e^{2\lambda x} + a\lambda e^{\lambda x})z = 0.$$

73. $y'''_{xxx} - [(a+c+be^{ax})f - a + c]y''_{xx} + [(c^2 - a^2 + bce^{ax})f - ac]y'_x + ac(a+c)fy = 0.$

Particular solutions: $y_1 = e^{cx}, \quad y_2 = e^{-ax} + \dfrac{b}{c}.$

74. $e^{\lambda x}y'''_{xxx} + (2\lambda e^{\lambda x} + \beta e^{\mu x} + \gamma)y''_{xx} + (\lambda^2 e^{\lambda x} + 2\beta\mu e^{\mu x})y'_x + \beta\mu^2 e^{\mu x}y = f(x).$

Integrating the equation twice, we obtain a first order linear equation:

$$e^{\lambda x}y'_x + (\beta e^{\mu x} + \gamma)y = C_1 + C_2 x + \int\Bigl(\int f\,dx\Bigr)dx.$$

75. $y'''_{xxx} + fy''_{xx} + gy'_x - \lambda[\lambda f + \tanh(\lambda x)(g + \lambda^2)]y = 0.$

Particular solution: $y_0 = \cosh(\lambda x).$

The substitution $y = \cosh(\lambda x)\int z(x)\,dx$ leads to a second order equation:

$$z''_{xx} + [f + 3\lambda\tanh(\lambda x)]z'_x + [g + 3\lambda^2 + 2\lambda f\tanh(\lambda x)]z = 0.$$

76. $y'''_{xxx} + fy''_{xx} - \lambda[2f\tanh(\lambda x) + 3\lambda]y'_x$
$+ \lambda^2\{[2\tanh^2(\lambda x) - 1]f + 2\lambda\tanh(\lambda x)\}y = 0.$

Particular solutions: $y_1 = \cosh(\lambda x), \quad y_2 = x\cosh(\lambda x).$

77. $y'''_{xxx} + fy''_{xx} - \lambda[2f\coth(\lambda x) + 3\lambda]y'_x$
$+ \lambda^2\{[2\coth^2(\lambda x) - 1]f + 2\lambda\coth(\lambda x)\}y = 0.$

Particular solutions: $y_1 = \sinh(\lambda x), \quad y_2 = x\sinh(\lambda x).$

78. $y'''_{xxx} + [(\tanh x - a)f - a]y''_{xx} + [(a^2 - 1)f - 1]y'_x + a[(1 - a\tanh x)f + 1]y = 0.$

Particular solutions: $y_1 = e^{ax}, \quad y_2 = \cosh x.$

79. $y'''_{xxx} + [(\coth x - a)f - a]y''_{xx} + [(a^2 - 1)f - 1]y'_x + a[(1 - a\coth x)f + 1]y = 0.$

Particular solutions: $y_1 = e^{ax}$, $y_2 = \sinh x$.

80. $y'''_{xxx} + [\lambda \tanh(\lambda x)(xf - 1) - f]y''_{xx} - \lambda^2 x f y'_x + \lambda^2 f y = 0.$

Particular solutions: $y_1 = x$, $y_2 = \cosh(\lambda x)$.

81. $y'''_{xxx} + [\lambda \coth(\lambda x)(xf - 1) - f]y''_{xx} - \lambda^2 x f y'_x + \lambda^2 f y = 0.$

Particular solutions: $y_1 = x$, $y_2 = \sinh(\lambda x)$.

82. $xy'''_{xxx} + [x^2(a - \ln x)f + 2]y''_{xx} + xfy'_x - fy = 0.$

Particular solutions: $y_1 = x$, $y_2 = \ln x - a + 1$.

83. $y'''_{xxx} + fy'_x + \tan x(f - 1)y = 0.$

Particular solution: $y_0 = \cos x$.

The substitution $y = \cos x \int z(x)\,dx$ leads to a second order linear equation: $z''_{xx} - 3\tan x\, z'_x + (f - 3)z = 0$.

84. $y'''_{xxx} + fy'_x + \cot x(1 - f)y = 0.$

Particular solution: $y_0 = \sin x$.

85. $y'''_{xxx} + fy''_{xx} + gy'_x + \lambda[\lambda f + \tan(\lambda x)(g - \lambda^2)]y = 0.$

Particular solution: $y_0 = \cos(\lambda x)$.

The substitution $y = \cos(\lambda x) \int z(x)\,dx$ leads to a second order linear equation:

$$z''_{xx} + [f - 3\lambda \tan(\lambda x)]z'_x + [g - 3\lambda^2 - 2\lambda f \tan(\lambda x)]z = 0.$$

86. $y'''_{xxx} + fy''_{xx} + \lambda[2f\tan(\lambda x) + 3\lambda]y'_x + \lambda^2\{[1 + 2\tan^2(\lambda x)]f + 2\lambda\tan(\lambda x)\}y = 0.$

Particular solutions: $y_1 = \cos(\lambda x)$, $y_2 = x\cos(\lambda x)$.

87. $y'''_{xxx} + fy''_{xx} + \lambda[3\lambda - 2f\cot(\lambda x)]y'_x + \lambda^2\{[1 + 2\cot^2(\lambda x)]f - 2\lambda\cot(\lambda x)\}y = 0.$

Particular solutions: $y_1 = \sin(\lambda x)$, $y_2 = x\sin(\lambda x)$.

88. $y'''_{xxx} - [(a + \tan x)f + a]y''_{xx} + [(a^2 + 1)f + 1]y'_x + a[(a\tan x - 1)f - 1]y = 0.$

Particular solutions: $y_1 = e^{ax}$, $y_2 = \cos x$.

89. $y'''_{xxx} + [(\cot x + a)f + a]y''_{xx} + [(a^2 + 1)f + 1]y'_x + a[(1 - a\cot x)f + 1]y = 0.$

Particular solutions: $y_1 = e^{-ax}$, $y_2 = \sin x$.

90. $y'''_{xxx} + [f + \lambda\tan(\lambda x)(xf + 1)]y''_{xx} - \lambda^2 x f y'_x + \lambda^2 f y = 0.$

Particular solutions: $y_1 = x$, $y_2 = \cos(\lambda x)$.

91. $y'''_{xxx} + [f - \lambda\cot(\lambda x)(xf + 1)]y''_{xx} - \lambda^2 x f y'_x + \lambda^2 f y = 0.$

Particular solutions: $y_1 = x$, $y_2 = \sin(\lambda x)$.

92. $a\sin(\lambda x)y'''_{xxx} + by''_{xx} + 3a\lambda^2\sin(\lambda x)y'_x + 2a\lambda^3\cos(\lambda x)y = f(x).$

Integrating the equation twice, we obtain a first order linear equation:

$$a\sin(\lambda x)y'_x + [b - 2a\lambda\cos(\lambda x)]y = C_1 + C_2x + \int\left(\int f\,dx\right)dx.$$

93. $\sin(\lambda x)y'''_{xxx} + [a + (2\lambda+1)\cos(\lambda x)]y''_{xx} - (\lambda^2+2\lambda)\sin(\lambda x)y'_x$
$- \lambda^2\cos(\lambda x)y = f(x).$

Integrating the equation twice, we obtain a first order linear equation:

$$\sin(\lambda x)y'_x + [a + \cos(\lambda x)]y = C_1 + C_2x + \int\left(\int f\,dx\right)dx.$$

94. $(f-1)y'''_{xxx} - [af + \lambda\tan(\lambda x)]y''_{xx} + (\lambda^2 f + a^2)y'_x + a\lambda[a\tan(\lambda x) - \lambda f]y = 0.$

Particular solutions: $y_1 = e^{ax}$, $y_2 = \cos(\lambda x)$.

95. $y'''_{xxx} + fy'_x + f'_x y = g.$

Integration yeilds a second order linear equation: $y''_{xx} + fy = \int g\,dx + C$.

96. $y'''_{xxx} + 2fy'_x + f'_x y = 0.$

Solution: $y = C_1w_1^2 + C_2w_1w_2 + C_3w_2^2$, where w_1 and w_2 are linearly-independent solutions of the second order linear equation $2w''_{xx} + fw = 0$.

97. $y'''_{xxx} + f'_x y'_x + f(2f'_x - f^2)y = 0.$

Integration yeilds a second order linear equation: $y''_{xx} + fy'_x + f^2y = C\exp(\int f\,dx)$.

98. $y'''_{xxx} + (a-1)f^2y'_x - [f''_{xx} - (2a+1)ff'_x + af^3]y = 0.$

Integration yeilds a second order linear equation:

$$y''_{xx} + fy'_x + (af^2 - f'_x)y = C\exp(\int f\,dx).$$

99. $y'''_{xxx} + (f - a^2)y'_x + (f'_x - af)y = 0.$

The substitution $w = y''_{xx} + ay'_x + fy$ leads to a first order linear equation: $w'_x - aw = 0$.

100. $y'''_{xxx} + fy''_{xx} + gy'_x + (fg + g'_x)y = 0.$

Integration yeilds a second order linear equation: $y''_{xx} + gy = C\exp(-\int f\,dx)$.

101. $y'''_{xxx} + 3fy''_{xx} + (f'_x + 2f^2 + 2g)y'_x + (2fg + g'_x)y = 0.$

Solution:

$$y = C_1w_1^2 + C_2w_1w_2 + C_3w_2^2,$$

where w_1 and w_2 is the fundamental set of solutions of the second order linear equation $w''_{xx} + fw'_x + \frac{1}{2}gw = 0$.

102. $y'''_{xxx} + (f+g)y''_{xx} + (f'_x + fg + h)y'_x + (h'_x + gh)y = 0.$

Integration yeilds a second order equation: $y''_{xx} + fy'_x + hy = C\exp(-\int g\,dx)$.

103. $y'''_{xxx} + (f+g)y''_{xx} + (2g'_x + fg + h)y'_x + (g''_{xx} + fg'_x + gh)y = 0.$

The substitution $w = y'_x + gy$ leads to a second order equation: $w''_{xx} + fw'_x + hw = 0$.

104. $fy'''_{xxx} - f'''_{xxx}y = 0.$

Particular solution: $y_0 = f$.

The substitution $y = f\int z\,dx$ leads to a second order equation: $fz''_{xx} + 3f'_x z'_x + 3f''_{xx}z = 0$.

105. $fy'''_{xxx} + f'''_{xxx}y = g.$

Integration yeilds a second order equation: $fy''_{xx} - f'_x y'_x + f''_{xx}y = \int g\,dx + C$.

106. $y'''_{xxx} = f(x)y.$

The transformation $x = t^{-1}$, $y = wt^{-2}$ leads to an equation of the similar form: $w'''_{ttt} = -t^{-6}f\Bigl(\dfrac{1}{t}\Bigr)w$.

3.2. Equations of the Form $y'''_{xxx} = Ax^\alpha y^\beta (y'_x)^\gamma (y''_{xx})^\delta$

3.2.1. Preliminary Comments. Classification Table

The value of the insignificant parameter A is in many cases defined in the form of a function of two (one) auxiliary coefficients a and b:

$$A = \varphi(a,\ b) \tag{1}$$

and the corresponding solutions are represented in the parametric form

$$x = f_1(\tau, C_1, C_2, C_3, a), \qquad y = f_2(\tau, C_1, C_2, C_3, b), \tag{2}$$

where τ is a parameter, C_1, C_2, and C_3 are arbitrary constants, f_1 and f_2 are some functions.

Having fixed the auxiliary coefficient sign $a > 0$ (or $b > 0$), the coefficient b should be expressed in terms of both A and a with the help of

$$b = \psi(A,\ a).$$

Substituting this formula into (2), we obtain a solution of the equation under consideration (where the concrete numerical value of the coefficient a may be chosen arbitrarily). The case $a < 0$ (or $b < 0$), which may lead to the branch of the solution or to a different domain of determining the variables x and y in (2), should be considered in a similar manner.

The following Table 3.1 represents all solvable equations whose solutions are outlined in Subsections 3.2.2–3.2.4. The two-parameter families (in the space of parameters α, β, γ, and δ), one-parameter families and isolated points are represented in a consecutive fashion. Equations are arranged in accordance with the growth of δ, the growth of γ (for identical δ), the growth of β (for identical δ and γ), and the growth of α (for identical δ, γ, and β). The number of the equation sought is indicated in the last column in this table.

TABLE 3.1
Solvable equations of the form $y'''_{xxx} = Ax^\alpha y^\beta (y'_x)^\gamma (y''_{xx})^\delta$

δ	γ	β	α	Equation
arbitrary	arbitrary	0	arbitrary	3.2.4.15*
arbitrary ($\delta \neq 2$)	arbitrary ($\gamma \neq -1$)	0	0	3.2.4.1
$\dfrac{\gamma+4\beta+5}{\gamma+2\beta+3}$	arbitrary ($\gamma \neq -1$)	arbitrary ($\beta \neq -1$)	0	3.2.4.174
$\dfrac{3\gamma+7}{2(\gamma+2)}$	arbitrary ($\gamma \neq -2$)	$-\frac{1}{2}$	0	3.2.4.10
$\dfrac{3\gamma+7}{2(\gamma+2)}$	arbitrary ($\gamma \neq -2$)	1	0	3.2.4.7
arbitrary ($\delta \neq 1,\ 2$)	-1	-1	0	3.2.4.175
arbitrary ($\delta \neq 2$)	-1	0	0	3.2.4.11
$\dfrac{3\beta+4}{2\beta+3}$	0	arbitrary ($\beta \neq -\frac{3}{2}$)	0	3.2.4.8
arbitrary ($\delta \neq \frac{3}{2}$)	0	$-\frac{1}{2}$	0	3.2.4.87
arbitrary ($\delta \neq 1$)	1	arbitrary ($\beta \neq -1$)	0	3.2.4.2
arbitrary ($\delta \neq 1$)	1	-1	0	3.2.4.13
arbitrary ($\delta \neq 2$)	1	1	0	3.2.4.4
$\dfrac{3\beta+4}{2\beta+3}$	3	arbitrary ($\beta \neq -\frac{3}{2}$)	0	3.2.4.9
-1	3	$-\frac{7}{5}$	0	3.2.4.168
-1	3	0	0	3.2.4.164
0	arbitrary ($\gamma \neq -1$)	0	0	3.2.4.3
0	arbitrary	$-\frac{1}{4}(\gamma+5)$	0	3.2.4.171
0	$-2\beta-5$	arbitrary ($\beta \neq -2$)	0	3.2.4.5
0	-13	1	0	3.2.4.153
0	-13	3	0	3.2.4.155
0	-7	0	0	3.2.4.141
0	-7	1	0	3.2.4.145

* given are formulae of reducing to the generalized Emden—Fowler equation

TABLE 3.1 *Continued*
Solvable equations of the form $y'''_{xxx} = Ax^{\alpha}y^{\beta}(y'_x)^{\gamma}(y''_{xx})^{\delta}$

δ	γ	β	α	Equation
0	-4	$-\frac{1}{2}$	0	3.2.4.127
0	-4	0	0	3.2.4.123
0	-3	-2	0	3.2.4.95
0	-3	-1	0	3.2.4.30
0	-3	0	0	3.2.4.26
0	-3	1	0	3.2.4.91
0	$-\frac{7}{3}$	$-\frac{10}{3}$	0	3.2.4.76
0	$-\frac{7}{3}$	$-\frac{7}{3}$	0	3.2.4.42
0	$-\frac{7}{3}$	$-\frac{4}{3}$	0	3.2.4.52
0	$-\frac{7}{3}$	$-\frac{5}{6}$	0	3.2.4.133
0	$-\frac{7}{3}$	$-\frac{1}{2}$	0	3.2.4.131
0	$-\frac{7}{3}$	0	0	3.2.4.48
0	$-\frac{7}{3}$	1	0	3.2.4.38
0	$-\frac{7}{3}$	2	0	3.2.4.70
0	$-\frac{9}{5}$	$-\frac{13}{5}$	0	3.2.4.64
0	$-\frac{9}{5}$	1	0	3.2.4.60
0	-1	-2	0	3.2.4.22
0	-1	0	0	3.2.4.18
0	0	$-\frac{7}{2}$	0	3.2.2.2
0	0	$-\frac{7}{2}$	3	3.2.3.3
0	0	$-\frac{5}{2}$	0	3.2.2.3
0	0	$-\frac{5}{2}$	1	3.2.3.4
0	0	-2	0	3.2.2.6
0	0	$-\frac{4}{3}$	$-\frac{4}{3}$	3.2.3.5
0	0	$-\frac{4}{3}$	0	3.2.2.4
0	0	$-\frac{5}{4}$	$-\frac{3}{2}$	3.2.3.7
0	0	$-\frac{5}{4}$	0	3.2.2.8
0	0	$-\frac{7}{6}$	$-\frac{5}{3}$	3.2.3.6
0	0	$-\frac{7}{6}$	0	3.2.2.5
0	0	$-\frac{1}{2}$	-3	3.2.3.8

TABLE 3.1 *Continued*
Solvable equations of the form $y'''_{xxx} = Ax^\alpha y^\beta (y'_x)^\gamma (y''_{xx})^\delta$

δ	γ	β	α	Equation
0	0	$-\frac{1}{2}$	$-\frac{3}{2}$	3.2.3.9
0	0	$-\frac{1}{2}$	0	3.2.2.7
0	0	0	arbitrary	3.2.3.1
0	0	0	0	3.2.2.1
0	0	1	arbitrary	3.2.3.2
0	1	1	0	3.2.4.35
0	2	$-\frac{7}{2}$	0	3.2.4.165
0	2	0	0	3.2.4.161
0	3	arbitrary ($\beta \neq -2$)	0	3.2.4.85
0	3	-2	0	3.2.4.82
0	5	-5	0	3.2.4.105
0	5	$-\frac{20}{7}$	0	3.2.4.117
0	5	$-\frac{15}{7}$	0	3.2.4.111
0	5	0	0	3.2.4.101
$\frac{1}{2}$	0	$-\frac{5}{2}$	0	3.2.4.74
$\frac{1}{2}$	3	$-\frac{15}{8}$	0	3.2.4.114
$\frac{1}{2}$	3	$-\frac{20}{13}$	0	3.2.4.120
$\frac{1}{2}$	3	$-\frac{5}{4}$	0	3.2.4.108
$\frac{1}{2}$	3	0	0	3.2.4.104
$\frac{2}{3}$	0	$-\frac{7}{6}$	0	3.2.4.157
$\frac{4}{5}$	-4	$-\frac{1}{2}$	0	3.2.4.137
1	arbitrary ($\gamma \neq 1$)	-1	0	3.2.4.140
1	-3	$-\frac{1}{2}$	0	3.2.4.32
1	-3	1	0	3.2.4.24
1	-1	-1	0	3.2.4.177
1	1	arbitrary ($\beta \neq -1$)	0	3.2.4.14
1	1	-1	0	3.2.4.17
1	1	1	0	3.2.4.21
$\frac{8}{7}$	3	$-\frac{3}{4}$	0	3.2.4.159

TABLE 3.1 *Continued*
Solvable equations of the form $y'''_{xxx} = Ax^\alpha y^\beta (y'_x)^\gamma (y''_{xx})^\delta$

δ	γ	β	α	Equation
$\frac{8}{7}$	3	$-\frac{1}{2}$	0	3.2.4.151
$\frac{6}{5}$	0	$-\frac{2}{3}$	0	3.2.4.109
$\frac{5}{4}$	-4	$-\frac{1}{2}$	0	3.2.4.57
$\frac{5}{4}$	3	$-\frac{1}{2}$	0	3.2.4.148
$\frac{5}{4}$	3	0	0	3.2.4.144
$\frac{9}{7}$	$-\frac{9}{4}$	1	0	3.2.4.66
$\frac{9}{7}$	0	$-\frac{1}{3}$	0	3.2.4.169
$\frac{9}{7}$	0	1	0	3.2.4.62
$\frac{13}{10}$	0	$-\frac{5}{2}$	0	3.2.4.80
$\frac{27}{20}$	0	$-\frac{2}{3}$	0	3.2.4.121
$\frac{18}{13}$	0	$-\frac{7}{2}$	0	3.2.4.68
$\frac{7}{5}$	-7	1	0	3.2.4.54
$\frac{7}{5}$	$-\frac{5}{2}$	1	0	3.2.4.45
$\frac{7}{5}$	$-\frac{13}{7}$	1	0	3.2.4.78
$\frac{7}{5}$	$-\frac{1}{3}$	1	0	3.2.4.72
$\frac{7}{5}$	0	1	0	3.2.4.40
$\frac{7}{5}$	1	1	0	3.2.4.50
$\frac{7}{5}$	3	0	0	3.2.4.126
$\frac{7}{5}$	3	1	0	3.2.4.130
$\frac{7}{5}$	11	1	0	3.2.4.135
$\frac{10}{7}$	0	$-\frac{5}{2}$	0	3.2.4.43
$\frac{22}{15}$	0	$-\frac{2}{3}$	0	3.2.4.115
$\frac{3}{2}$	arbitrary	$\frac{1}{2}(\gamma-1)$	0	3.2.4.173
$\frac{3}{2}$	-3	$-\frac{1}{2}$	0	3.2.4.100
$\frac{3}{2}$	-3	1	0	3.2.4.97
$\frac{3}{2}$	0	-2	0	3.2.4.99
$\frac{3}{2}$	0	$-\frac{1}{2}$	0	3.2.4.84
$\frac{3}{2}$	0	1	0	3.2.4.93
$\frac{3}{2}$	1	1	0	3.2.4.28
$\frac{3}{2}$	3	-2	0	3.2.4.98

TABLE 3.1 *Continued*
Solvable equations of the form $y'''_{xxx} = Ax^\alpha y^\beta (y'_x)^\gamma (y''_{xx})^\delta$

δ	γ	β	α	Equation
$\frac{3}{2}$	3	$-\frac{1}{2}$	0	3.2.4.94
$\frac{3}{2}$	3	0	0	3.2.4.29
$\frac{23}{15}$	$-\frac{1}{3}$	$-\frac{1}{2}$	0	3.2.4.116
$\frac{11}{7}$	-4	$-\frac{1}{2}$	0	3.2.4.47
$\frac{8}{5}$	1	1	0	3.2.4.125
$\frac{8}{5}$	3	-4	0	3.2.4.55
$\frac{8}{5}$	3	$-\frac{7}{4}$	0	3.2.4.46
$\frac{8}{5}$	3	$-\frac{10}{7}$	0	3.2.4.79
$\frac{8}{5}$	3	$-\frac{2}{3}$	0	3.2.4.73
$\frac{8}{5}$	3	$-\frac{1}{2}$	0	3.2.4.41
$\frac{8}{5}$	3	0	0	3.2.4.51
$\frac{8}{5}$	3	1	0	3.2.4.129
$\frac{8}{5}$	3	5	0	3.2.4.136
$\frac{21}{13}$	-6	$-\frac{1}{2}$	0	3.2.4.69
$\frac{33}{20}$	$-\frac{1}{3}$	$-\frac{1}{2}$	0	3.2.4.122
$\frac{17}{10}$	-4	$-\frac{1}{2}$	0	3.2.4.81
$\frac{12}{7}$	$\frac{1}{3}$	$-\frac{1}{2}$	0	3.2.4.170
$\frac{12}{7}$	3	$-\frac{13}{8}$	0	3.2.4.67
$\frac{12}{7}$	3	$-\frac{1}{2}$	0	3.2.4.63
$\frac{7}{4}$	0	$-\frac{5}{2}$	0	3.2.4.56
$\frac{7}{4}$	0	1	0	3.2.4.147
$\frac{7}{4}$	1	1	0	3.2.4.143
$\frac{9}{5}$	$-\frac{1}{3}$	$-\frac{1}{2}$	0	3.2.4.110
$\frac{13}{7}$	$-\frac{1}{2}$	1	0	3.2.4.160
$\frac{13}{7}$	0	1	0	3.2.4.152
2	arbitrary ($\gamma \neq -1$)	0	0	3.2.4.12
2	-1	arbitrary ($\beta \neq 0$)	0	3.2.4.139
2	-1	-1	0	3.2.4.176
2	-1	0	0	3.2.4.16

TABLE 3.1 *Continued*
Solvable equations of the form $y'''_{xxx} = Ax^\alpha y^\beta (y'_x)^\gamma (y''_{xx})^\delta$

δ	γ	β	α	Equation
2	0	-2	0	3.2.4.33
2	3	-2	0	3.2.4.25
2	3	0	0	3.2.4.19
$\frac{11}{5}$	0	$-\frac{5}{2}$	0	3.2.4.138
$\frac{7}{3}$	$-\frac{4}{3}$	$-\frac{1}{2}$	0	3.2.4.158
$\frac{5}{2}$	-4	$-\frac{1}{2}$	0	3.2.4.75
$\frac{5}{2}$	$-\frac{11}{4}$	1	0	3.2.4.113
$\frac{5}{2}$	$-\frac{27}{13}$	1	0	3.2.4.119
$\frac{5}{2}$	$-\frac{3}{2}$	1	0	3.2.4.107
$\frac{5}{2}$	1	1	0	3.2.4.103
3	arbitrary ($\gamma \neq -3$)	1	0	3.2.4.86
3	$-2\beta - 5$	arbitrary ($\beta \neq -2$)	0	3.2.4.6
3	arbitrary	$-\gamma - 2$	0	3.2.4.172
3	-9	2	0	3.2.4.106
3	-6	$-\frac{1}{2}$	0	3.2.4.59
3	-6	$\frac{1}{2}$	0	3.2.4.166
3	$-\frac{17}{3}$	$-\frac{5}{3}$	0	3.2.4.77
3	$-\frac{33}{7}$	2	0	3.2.4.118
3	$-\frac{21}{5}$	$-\frac{7}{5}$	0	3.2.4.65
3	-4	$-\frac{1}{2}$	0	3.2.4.37
3	$-\frac{11}{3}$	$-\frac{5}{3}$	0	3.2.4.44
3	$-\frac{23}{7}$	2	0	3.2.4.112
3	-3	-2	0	3.2.4.96
3	-3	-1	0	3.2.4.23
3	-3	$-\frac{1}{2}$	0	3.2.4.90
3	-3	1	0	3.2.4.83
3	$-\frac{5}{3}$	$-\frac{5}{3}$	0	3.2.4.53
3	$-\frac{5}{3}$	$-\frac{1}{2}$	0	3.2.4.149
3	$-\frac{4}{3}$	$-\frac{1}{2}$	0	3.2.4.150

TABLE 3.1 *Continued*
Solvable equations of the form $y'''_{xxx} = Ax^\alpha y^\beta (y'_x)^\gamma (y''_{xx})^\delta$

δ	γ	β	α	Equation
3	-1	-2	0	3.2.4.31
3	$-\frac{2}{3}$	$-\frac{5}{3}$	0	3.2.4.134
3	0	$-\frac{5}{2}$	0	3.2.4.128
3	0	$-\frac{5}{3}$	0	3.2.4.132
3	0	$-\frac{1}{2}$	0	3.2.4.89
3	0	1	-3	3.2.4.88
3	1	-4	0	3.2.4.142
3	1	$-\frac{5}{2}$	0	3.2.4.124
3	1	-2	0	3.2.4.27
3	1	$-\frac{5}{3}$	0	3.2.4.49
3	1	-1	0	3.2.4.20
3	1	$-\frac{1}{2}$	0	3.2.4.34
3	1	$\frac{1}{2}$	0	3.2.4.162
3	1	2	0	3.2.4.102
3	3	-7	0	3.2.4.154
3	3	-4	0	3.2.4.146
3	3	-2	0	3.2.4.92
3	3	$-\frac{5}{3}$	0	3.2.4.39
3	3	$-\frac{7}{5}$	0	3.2.4.61
3	3	$-\frac{1}{2}$	0	3.2.4.58
3	3	0	0	3.2.4.36
3	5	$-\frac{5}{3}$	0	3.2.4.71
3	7	-7	0	3.2.4.156
4	$-\frac{9}{5}$	1	0	3.2.4.167
4	1	1	0	3.2.4.163

3.2.2. Equations of the Form $y'''_{xxx} = Ay^\beta$

1. $y'''_{xxx} = A$.

Solution: $y = \frac{1}{6}Ax^3 + C_2x^2 + C_1x + C_0$.

2. $y'''_{xxx} = Ay^{-7/2}$.

Solution in the parametric form:

$$x = aC_1^3 \int \left[C_1 e^{2\sigma\tau} + C_2 e^{-\sigma\tau} \sin\left(\sqrt{3}\,\sigma\tau\right)\right]^{-3/2} d\tau + C_3,$$
$$y = bC_1^2 \left[C_1 e^{2\sigma\tau} + C_2 e^{-\sigma\tau} \sin\left(\sqrt{3}\,\sigma\tau\right)\right]^{-1},$$

where $A = -8a^{-3}b^{9/2}\sigma^3$.

3. $y'''_{xxx} = Ay^{-5/2}$.

Solution in the parametric form:

$$x = aC_1^7 \int (\tau^3 - 3\tau + C_2)^{-3/2} d\tau + C_3, \qquad y = bC_1^6 (\tau^3 - 3\tau + C_2)^{-1},$$

where $A = -6a^{-3}b^{7/2}$.

4. $y'''_{xxx} = Ay^{-4/3}$.

Solution in the parametric form:

$$x = aC_1^7 \int R^{-1}(2\tau I \mp R)^2 d\tau + C_3, \qquad y = bC_1^9 (2\tau I \mp R)^3,$$

where $R = \sqrt{\pm(4\tau^3 - 1)}$, $I = \int \tau R^{-1} d\tau + C_2$, $A = \pm 18a^{-3}b^{7/3}$.

5. $y'''_{xxx} = Ay^{-7/6}$.

Solution in the parametric form:

$$x = aC_1^{13} \int R^{-1}(2\tau I \mp R)^{-5/2} d\tau + C_3, \qquad y = bC_1^{18} (2\tau I \mp R)^{-3},$$

where $R = \sqrt{\pm(4\tau^3 - 1)}$, $I = \int \tau R^{-1} d\tau + C_2$, $A = \mp 18a^{-3}b^{13/6}$.

▶ *In the solutions of equations 6–7, the following notation is used:*

$$Z = \begin{cases} C_1 J_{1/3}(\tau) + C_2 Y_{1/3}(\tau) & \text{for the upper sign,} \\ C_1 I_{1/3}(\tau) + C_2 K_{1/3}(\tau) & \text{for the lower sign,} \end{cases}$$

where $J_{1/3}$ and $Y_{1/3}$ are Bessel functions, $I_{1/3}$ and $K_{1/3}$ are modified Bessel functions.

6. $y'''_{xxx} = Ay^{-2}$.

Solution in the parametric form:

$$x = aC_1 \int \tau^{-1} Z^{-2} d\tau + C_3, \qquad y = bC_1 \tau^{-2/3} Z^{-2},$$

where $A = \pm\frac{4}{3}a^{-3}b^3$.

7. $y'''_{xxx} = Ay^{-1/2}$.

Solution in the parametric form:

$$x = aC_1 \int Z\, d\tau + C_3, \qquad y = bC_1^2 \tau^{2/3} Z^2,$$

where $A = \mp\frac{4}{3}a^{-3}b^{3/2}$.

8. $y'''_{xxx} = Ay^{-5/4}$.

This is a special case of equation 3.2.4.171 with $\gamma = 0$.

3.2.3. Equations of the Form $y'''_{xxx} = Ax^\alpha y^\beta$

See Subsection 3.2.2 for the case $\alpha = 0$.

1. $y'''_{xxx} = Ax^\alpha.$

Solution: $y = Af(x) + C_2x^2 + C_1x + C_0,$ where

$$f(x) = \begin{cases} \dfrac{x^{\alpha+3}}{(\alpha+1)(\alpha+2)(\alpha+3)} & \text{if } \alpha \neq -1,\ -2,\ -3; \\ \frac{1}{2}x^2 \ln|x| - \frac{3}{4}x^2 & \text{if } \alpha = -1; \\ -x\ln|x| + x & \text{if } \alpha = -2; \\ \frac{1}{2}\ln|x| & \text{if } \alpha = -3. \end{cases}$$

2. $y'''_{xxx} = Ax^\alpha y.$

See equation 3.1.2.7.

3. $y'''_{xxx} = Ax^3 y^{-7/2}.$

Solution in the parametric form:

$$x = aC_1^3\left(\int f^{-3/2}\,d\tau + C_3\right)^{-1}, \qquad y = bC_1^4 f^{-1}\left(\int f^{-3/2}\,d\tau + C_3\right)^{-2},$$

where $f = C_1e^{2\sigma\tau} + C_2e^{-\sigma\tau}\sin\bigl(\sqrt{3}\,\sigma\tau\bigr)$, $A = 8a^{-6}b^{9/2}\sigma^3$.

4. $y'''_{xxx} = Axy^{-5/2}.$

Solution in the parametric form:

$$x = aC_1^7\left[\int (\tau^3 - 3\tau + C_2)^{-3/2}\,d\tau + C_3\right]^{-1},$$

$$y = bC_1^8(\tau^3 - 3\tau + C_2)^{-1}\left[\int (\tau^3 - 3\tau + C_2)^{-3/2}\,d\tau + C_3\right]^{-2},$$

where $A = 6a^{-4}b^{7/2}$.

► *In the solutions of equations 5–6, the following notation is used:*

$$R = \sqrt{\pm(4\tau^3 - 1)}, \qquad I = \int \tau R^{-1}\,d\tau + C_2.$$

5. $y'''_{xxx} = Ax^{-4/3}y^{-4/3}.$

Solution in the parametric form:

$$x = aC_1^7\left[\int R^{-1}(2\tau I \mp R)^2\,d\tau + C_3\right]^{-1},$$

$$y = bC_1^5(2\tau I \mp R)^3\left[\int R^{-1}(2\tau I \mp R)^2\,d\tau + C_3\right]^{-2},$$

where $A = \mp 18a^{-5/3}b^{7/3}$.

6. $y'''_{xxx} = Ax^{-5/3}y^{-7/6}$.

Solution in the parametric form:

$$x = aC_1^{13}\left[\int R^{-1}(2\tau I \mp R)^{-5/2}\,d\tau + C_3\right]^{-1},$$

$$y = bC_1^{8}(2\tau I \mp R)^{-3}\left[\int R^{-1}(2\tau I \mp R)^{-5/2}\,d\tau + C_3\right]^{-2},$$

where $A = \mp 18a^{-4/3}b^{13/6}$.

7. $y'''_{xxx} = Ax^{-3/2}y^{-5/4}$.

Solution in the parametric form:

$$x = aC_1^3\left(\int \tau^{-1/2}z^{-1/2}f^{3/4}\,d\tau + C_3\right)^{-1}, \qquad y = bC_1^2 f\left(\int \tau^{-1/2}z^{-1/2}f^{3/4}\,d\tau + C_3\right)^{-2},$$

where $z = C_2 + \frac{1}{4}\tau^2 + 4B\tau^{1/2}$, $f = \exp\left(\int z^{-1/2}\,d\tau\right)$, $A = \frac{1}{2}Ba^{-3/2}b^{9/4}$.

8. $y'''_{xxx} = Ax^{-3}y^{-1/2}$.

Solution in the parametric form:

$$x = C_1\left(\int Z\,d\tau + C_3\right)^{-1}, \qquad y = b\tau^{2/3}Z^2\left(\int Z\,d\tau + C_3\right)^{-2},$$

where

$$A = \pm\tfrac{4}{3}b^{3/2}, \qquad Z = \begin{cases} C_1 J_{1/3}(\tau) + C_2 Y_{1/3}(\tau) & \text{for the upper sign,} \\ C_1 I_{1/3}(\tau) + C_2 K_{1/3}(\tau) & \text{for the lower sign,} \end{cases}$$

($J_{1/3}$ and $Y_{1/3}$ are Bessel functions, $I_{1/3}$ and $K_{1/3}$ are modified Bessel functions).

9. $y'''_{xxx} = Ax^{-3/2}y^{-1/2}$.

Solution in the parametric form:

$$x = aC_3 \exp\left(2\int P\,d\tau\right), \qquad x = bC_3 P^2 \exp\left(2\int P\,d\tau\right),$$

where $P = P(\tau, C_1, C_2)$ is the general solution of the second Painlevé transcendent:

$$P''_{\tau\tau} = \pm\tau P + 2P^3, \qquad A = \pm\tfrac{1}{4}a^{-3/2}b^{3/2}.$$

3.2.4. Equations with $|\gamma| + |\delta| \neq 0$

1. $y'''_{xxx} = A(y'_x)^{\gamma}(y''_{xx})^{\delta}$, $\gamma \neq -1$, $\delta \neq 2$.

Solution in the parametric form:

$$x = aC_1^{\gamma+\delta-1}\int \tau^{-1/2}\left(1 \pm \tau^{\frac{\gamma+1}{2}}\right)^{\frac{1}{\delta-2}}d\tau + C_3,$$

$$y = bC_1^{\gamma+2\delta-3}\int \left(1 \pm \tau^{\frac{\gamma+1}{2}}\right)^{\frac{1}{\delta-2}}d\tau + C_2,$$

where $A = \pm\dfrac{\gamma+1}{2-\delta}2^{\delta-2}a^{\gamma+2\delta-3}b^{1-\gamma-\delta}$.

2. $y'''_{xxx} = Ay^\beta y'_x (y''_{xx})^\delta, \qquad \beta \neq -1, \quad \delta \neq 1.$

Solution in the parametric form:

$$x = aC_1^{\beta+\delta} \int \left[\int \left(1 \pm \tau^{\beta+1}\right)^{\frac{1}{1-\delta}} d\tau + C_2 \right]^{-1/2} d\tau + C_3, \qquad y = bC_1^{2\delta-2}\tau,$$

where $A = \pm \dfrac{\beta+1}{1-\delta} 2^{\delta-1} a^{2\delta-2} b^{-\beta-\delta}$.

▶ *In the solutions of equations 3–10, the following notation is used:*

$$R = \sqrt{1 \pm \tau^{m+1}}, \quad E = \int \left(1 \pm \tau^{m+1}\right)^{-1/2} d\tau + C_2, \quad F = RE - \tau.$$

3. $y'''_{xxx} = A(y'_x)^\gamma, \qquad \gamma \neq -1.$

Solution in the parametric form:

$$x = aC_1^m \int \tau^{-1/2} R^{-1}\, d\tau + C_3, \qquad y = bC_1^{m-1} E,$$

where $m = \dfrac{\gamma - 1}{2}$, $A = \pm \dfrac{m+1}{4} a^{2m-2} b^{-2m}$.

4. $y'''_{xxx} = Ayy'_x (y''_{xx})^\delta, \qquad \delta \neq 2.$

Solution in the parametric form:

$$x = aC_1^{3m+1} \int E^{-1/2} R^{-1}\, d\tau + C_3, \qquad y = bC_1^{2m+2} R,$$

where $m = \dfrac{1}{\delta - 2}$, $A = -8ma^2 b^{-3} \left[\pm \dfrac{2a^2}{(m+1)b} \right]^{1/m}$.

5. $y'''_{xxx} = Ay^\beta (y'_x)^{-2\beta-5}, \qquad \beta \neq -2.$

Solution in the parametric form:

$$x = aC_1^{m-3} \int \tau^{-1/2} E^{-3/2} R^{-1}\, d\tau + C_3, \qquad y = bC_1^{2m-2} E^{-1},$$

where $m = -\beta - 3$, $A = \pm \frac{1}{4} (-1)^{-2m} (m+1) a^{2m-2} b^{3-m}$.

6. $y'''_{xxx} = Ay^\beta (y'_x)^{-2\beta-5} (y''_{xx})^3, \qquad \beta \neq -2.$

Solution in the parametric form:

$$x = aC_1^{m+3} \int E^{-3/2} R^{-1} F\, d\tau + C_3, \qquad y = bC_1^{2m+2} \tau E^{-1},$$

where $m = \beta$, $A = \mp 2(m+1) a^{-2m-2} b^{m+3}$.

7. $y'''_{xxx} = Ay(y'_x)^\gamma (y''_{xx})^{\frac{3\gamma+7}{2\gamma+4}}, \qquad \gamma \neq -2.$

Solution in the parametric form:

$$x = aC_1^{m^2+m+2} \int E^{-m/2} R^{-1}\, d\tau + C_3, \qquad y = bC_1^4 F,$$

where $m = -\dfrac{2(\gamma+2)}{\gamma+1}$, $A = -\dfrac{2mb^{-1}}{m+2}\left[\pm\dfrac{(m+1)b}{2a}\right]^{\frac{2}{m+2}}\left[\pm\dfrac{4a^2}{(m+1)(m+2)b}\right]^{\frac{1}{m}}$.

8. $y'''_{xxx} = Ay^\beta (y''_{xx})^{\frac{3\beta+4}{2\beta+3}}, \qquad \beta \neq -3/2.$

Solution in the parametric form:

$$x = aC_1^{m^2+2m-1} \int \tau^m E^{-m-2} R^{-1}\, d\tau + C_3, \qquad y = bC_1^{(m+1)^2} \tau^{m+1} E^{-m-1},$$

where $m = -\dfrac{\beta}{\beta+1}$, $A = (m+3)a^{-1} b^{\frac{m}{m+1}}\left[\pm\dfrac{2a^2}{(m+1)^2 b}\right]^{\frac{1}{m+3}}$.

9. $y'''_{xxx} = Ay^\beta (y'_x)^3 (y''_{xx})^{\frac{3\beta+4}{2\beta+3}}, \qquad \beta \neq -3/2.$

Solution in the parametric form:

$$x = aC_1^{m^2+m-1} \int E^{m+1} F^{-1/2} R^{-1}\, d\tau + C_3, \qquad y = bC_1^{(m-1)(m+2)} E^{m+2},$$

where $m = -\dfrac{2\beta+3}{\beta+1}$, $A = \dfrac{m}{(m+2)^3} a^2 b^{-\frac{2m+1}{m+1}}\left[\pm\dfrac{(m+1)(m+2)b}{4a^2}\right]^{\frac{1}{m}}$.

10. $y'''_{xxx} = Ay^{-1/2}(y'_x)^\gamma (y''_{xx})^{\frac{3\gamma+7}{2\gamma+4}}, \qquad \gamma \neq -2.$

Solution in the parametric form:

$$x = aC_1^{m^2+2m-7} \int \tau^{\frac{m-1}{2}} R^{-1} E^{\frac{m+3}{2}} F\, d\tau + C_3, \qquad y = bC_1^{-8} F^2,$$

where $m = \dfrac{1-\gamma}{1+\gamma}$, $A = -\dfrac{2(m+3)}{m+1} b^{1/2}\left[\pm\dfrac{a}{(m+1)b}\right]^{\frac{2}{m+1}}\left[\pm\dfrac{(m+1)^2 b}{2a^2}\right]^{\frac{1}{m+3}}$.

11. $y'''_{xxx} = A(y'_x)^{-1}(y''_{xx})^\delta, \qquad \delta \neq 2.$

Solution in the parametric form:

$$x = aC_1 \int \tau^{\frac{\delta}{\delta-2}} \exp(\mp\tau^2)\, d\tau + C_2, \qquad y = bC_1^2 \int \tau^{\frac{\delta}{\delta-2}} \exp(\mp 2\tau^2)\, d\tau + C_3,$$

where $A = \mp\dfrac{4b^2}{(2-\delta)a^4}\left(\mp\dfrac{a^2}{2b}\right)^\delta$.

12. $y'''_{xxx} = A(y'_x)^\gamma (y''_{xx})^2, \qquad \gamma \neq -1.$

Solution in the parametric form:

$$x = aC_1 \int \tau^{\frac{1-\gamma}{1+\gamma}} \exp(\mp\tau^2)\,d\tau + C_2, \qquad y = bC_1 \int \tau^{\frac{3-\gamma}{1+\gamma}} \exp(\mp\tau^2)\,d\tau + C_3,$$

where $A = \pm(\gamma+1)a^{\gamma+1}b^{-\gamma-1}$.

13. $y'''_{xxx} = Ay^{-1}y'_x (y''_{xx})^\delta, \qquad \delta \neq 1.$

Solution in the parametric form:

$$x = aC_1 \int \tau \exp(\mp\tau^2) \left[\int \tau^{\frac{3-\delta}{1-\delta}} \exp(\mp\tau^2)\,d\tau + C_2\right]^{-1/2} d\tau + C_3, \qquad y = bC_1^2 \exp(\mp\tau^2),$$

where $A = \dfrac{1}{1-\delta}(\mp 1)^{-\delta} a^{2\delta-2} b^{1-\delta}$.

14. $y'''_{xxx} = Ay^\beta y'_x y''_{xx}, \qquad \beta \neq -1.$

Solution in the parametric form:

$$x = C_1 \int \tau^{\frac{1-\beta}{1+\beta}} \left[\int \tau^{\frac{1-\beta}{1+\beta}} \exp(\mp\tau^2)\,d\tau + C_2\right]^{-1/2} d\tau + C_3, \qquad y = b\tau^{\frac{2}{1+\beta}},$$

where $A = \mp(\beta+1)b^{-1-\beta}$.

15. $y'''_{xxx} = Ax^\alpha (y'_x)^\gamma (y''_{xx})^\delta.$

Solution in the parametric form:

$$x = aC_1^{\gamma+\delta-1} X(\tau), \qquad y = bC_1^{\gamma+2\delta-\alpha-3} \int Y(\tau) \frac{dX(\tau)}{d\tau}\,d\tau + C_3,$$

where $X = X(\tau)$, $Y = Y(\tau)$ is the general solution of the generalized Emden—Fowler equation

$$Y''_{XX} = BX^\alpha Y^\gamma (Y'_X)^\delta, \qquad A = Ba^{\gamma+2\delta-\alpha-3}b^{1-\gamma-\delta}.$$

16. $y'''_{xxx} = A(y'_x)^{-1}(y''_{xx})^2.$

Solution:

$$y = \begin{cases} \dfrac{1-A}{(2-A)C_1}(C_1x + C_2)^{\frac{2-A}{1-A}} + C_3 & \text{if } A \neq 1,\ A \neq 2, \\ \dfrac{C_2}{C_1}\exp(C_1x) + C_3 & \text{if } A = 1, \\ \dfrac{1}{C_1}\ln(C_1x + C_2) + C_3 & \text{if } A = 2. \end{cases}$$

17. $y'''_{xxx} = Ay^{-1}y'_x y''_{xx}.$

Solution:

$$x = \begin{cases} \displaystyle\int (C_1y^{A+1} + C_2)^{-1/2}\,dy + C_3 & \text{if } A \neq -1, \\ \displaystyle\int (C_1\ln y + C_2)^{-1/2}\,dy + C_3 & \text{if } A = -1. \end{cases}$$

18. $y'''_{xxx} = A(y'_x)^{-1}$.

Solution in the parametric form:

$$x = aC_1 \int \exp(\mp\tfrac{1}{2}\tau^2)\,d\tau + C_2, \qquad y = bC_1^2 \int \exp(\mp\tau^2)\,d\tau + C_3,$$

where $A = \mp a^{-4}b^2$.

19. $y'''_{xxx} = A(y'_x)^3(y''_{xx})^2$.

Solution in the parametric form:

$$x = aC_1 \int \tau^{-1/2} \exp(\mp\tau^2)\,d\tau + C_2, \qquad y = bC_1 \int \exp(\mp\tau^2)\,d\tau + C_3,$$

where $A = \pm 4a^4b^{-4}$.

▶ *In the solutions of equations 20–25, the following notation is used:*

$$E = \int \exp(\mp\tau^2)\,d\tau + C_2, \qquad F = 2\tau E \pm \exp(\mp\tau^2).$$

20. $y'''_{xxx} = Ay^{-1}y'_x(y''_{xx})^3$.

Solution in the parametric form:

$$x = aC_1 \int \tau \exp(\mp\tau^2) E^{-1/2}\,d\tau + C_3, \qquad y = bC_1^2 \exp(\mp\tau^2),$$

where $A = \pm\tfrac{1}{2}a^4b^{-2}$.

21. $y'''_{xxx} = Ayy'_xy''_{xx}$.

Solution in the parametric form:

$$x = C_1 \int E^{-1/2}\,d\tau + C_3, \qquad y = b\tau,$$

where $A = \mp 2b^{-2}$.

22. $y'''_{xxx} = Ay^{-2}(y'_x)^{-1}$.

Solution in the parametric form:

$$x = aC_1 \int E^{-3/2} \exp(\mp\tfrac{1}{2}\tau^2)\,d\tau + C_3, \qquad y = bC_1E^{-1},$$

where $A = \mp a^{-4}b^4$.

23. $y'''_{xxx} = Ay^{-1}(y'_x)^{-3}(y''_{xx})^3$.

Solution in the parametric form:

$$x = C_1 \int E^{-3/2} F \exp(\mp\tau^2)\,d\tau + C_3, \qquad y = bE^{-1}\exp(\mp\tau^2),$$

where $A = \mp 8b^2$.

24. $y'''_{xxx} = Ay(y'_x)^{-3}y''_{xx}$.

Solution in the parametric form:

$$x = aC_1 \int E^{1/2}\,d\tau + C_3, \qquad y = bC_1^2 F,$$

where $A = \mp 8a^{-4}b^2$.

25. $y'''_{xxx} = Ay^{-2}(y'_x)^3(y''_{xx})^2$.

Solution in the parametric form:

$$x = aC_1 \int F^{-1/2}\exp(\mp\tau^2)\,d\tau + C_3, \qquad y = bC_1^2 E,$$

where $A = \pm a^4 b^{-2}$.

▶ *In the solutions of equations 26–33, the following notation is used:*

$$E = \sqrt{\tau(\tau+1)} - \ln\bigl(\sqrt{\tau} + \sqrt{\tau+1}\bigr) + C_2, \qquad R = \sqrt{\frac{\tau+1}{\tau}}, \qquad F = RE - \tau.$$

26. $y'''_{xxx} = A(y'_x)^{-3}$.

Solution in the parametric form:

$$x = 2aC_1^2\sqrt{\tau+1} + C_3, \qquad y = bC_1^3 E,$$

where $A = -\frac{1}{4}a^{-6}b^4$.

27. $y'''_{xxx} = Ay^{-2}y'_x(y''_{xx})^3$.

Solution in the parametric form:

$$x = aC_1 \int E^{-1/2}\,d\tau + C_3, \qquad y = bC_1^4\tau,$$

where $A = 2a^4b^{-1}$.

28. $y'''_{xxx} = Ayy'_x(y''_{xx})^{3/2}$.

Solution in the parametric form:

$$x = aC_1^5 \int \tau^{-2}R^{-1}E^{-1/2}\,d\tau + C_3, \qquad y = bC_1^2 R,$$

where $A = -8a(-b)^{-5/2}$.

29. $y'''_{xxx} = A(y'_x)^3(y''_{xx})^{3/2}$.

Solution in the parametric form:

$$x = aC_1^7 \int R^{-3/2}\,d\tau + C_3, \qquad y = bC_1^6 E,$$

where $A = 4a^3(-b)^{-7/2}$.

30. $y'''_{xxx} = Ay^{-1}(y'_x)^{-3}$.

Solution in the parametric form:

$$x = aC_1^5 \int \tau^{-1/2} R^{-1} E^{-3/2}\, d\tau + C_3, \qquad y = bC_1^6 E^{-1},$$

where $A = -\frac{1}{4} a^{-6} b^5$.

31. $y'''_{xxx} = Ay^{-2}(y'_x)^{-1}(y''_{xx})^3$.

Solution in the parametric form:

$$x = aC_1^{-1} \int R^{-1} E^{-3/2} F\, d\tau + C_3, \qquad y = bC_1^2 \tau E^{-1},$$

where $A = 2a^2 b$.

32. $y'''_{xxx} = Ay^{-1/2}(y'_x)^{-3} y''_{xx}$.

Solution in the parametric form:

$$x = aC_1^7 \int \tau^{-3/2} R^{-1} E^{1/2} F\, d\tau + C_3, \qquad y = bC_1^8 F^2,$$

where $A = a^{-4} b^{7/2}$.

33. $y'''_{xxx} = Ay^{-2}(y''_{xx})^2$.

Solution in the parametric form:

$$x = aC_1^{-1} \int \tau^{-2} R^{-1}\, d\tau + C_3, \qquad y = bC_1 \tau^{-1} E,$$

where $A = 2ab$.

34. $y'''_{xxx} = Ay^{-1/2} y'_x (y''_{xx})^3$.

Solution in the parametric form:

$$x = \pm aC_1^5 \int \tau(\tau^2 - 1)(\tau^3 - 3\tau + C_2)^{-1/2}\, d\tau + C_3, \qquad y = bC_1^8 (\tau^2 - 1)^2,$$

where $A = \mp\frac{1}{144} a^4 b^{-5/2}$.

35. $y'''_{xxx} = Ayy'_x$.

Solution in the parametric form:

$$x = aC_1^{-1} \int (\tau^3 - 3\tau + C_2)^{-1/2}\, d\tau + C_3, \qquad y = bC_1^2 \tau,$$

where $A = 3a^{-2} b^{-1}$.

36. $y'''_{xxx} = A(y'_x)^3 (y''_{xx})^3$.

Solution in the parametric form:

$$x = \pm\tfrac{2}{5} a C_1^5 \tau^{1/2}(\tau^2 - 5) + C_3, \qquad y = bC_1^6(\tau^3 - 3\tau + C_2),$$

where $A = -\frac{8}{243} a^6 b^{-5}$.

37. $y'''_{xxx} = Ay^{-1/2}(y'_x)^{-4}(y''_{xx})^3$.

Solution in the parametric form:

$$x = aC_1^5 \int (\tau^2 - 1)(\tau^3 - 3\tau + C_2)^{-3/2}(\tau^4 - 6\tau^2 + 4C_2\tau - 3)\,d\tau + C_3,$$

$$y = bC_1^2(\tau^2 - 1)^2(\tau^3 - 3\tau + C_2)^{-1},$$

where $A = \mp\frac{16}{9} a^{-1} b^{5/2}$.

38. $y'''_{xxx} = Ay(y'_x)^{-7/3}$.

Solution in the parametric form:

$$x = aC_1^7 \int (\tau^3 - 3\tau + C_2)^{1/4}\,d\tau + C_3, \qquad y = \pm bC_1^{16}(\tau^4 - 6\tau^2 + 4C_2\tau - 3),$$

where $A = \pm 72 a^{-5} b^2 \left(\frac{4b}{a}\right)^{1/3}$.

39. $y'''_{xxx} = Ay^{-5/3}(y'_x)^3(y''_{xx})^3$.

Solution in the parametric form:

$$x = \pm aC_1^5 \int (\tau^2 - 1)(\tau^3 - 3\tau + C_2)^{1/2}[\pm(\tau^4 - 6\tau^2 + 4C_2\tau - 3)]^{-1/2}\,d\tau + C_3,$$

$$y = bC_1^9(\tau^3 - 3\tau + C_2)^{3/2},$$

where $A = \mp 8 \cdot 9^{-5} a^6 b^{-10/3}$.

40. $y'''_{xxx} = Ay(y''_{xx})^{7/5}$.

Solution in the parametric form:

$$x = aC_1^{-7} \int (\tau^3 - 3\tau + C_2)^{-3/2}\,d\tau + C_3, \qquad y = \pm bC_1(\tau^2 - 1)(\tau^3 - 3\tau + C_2)^{-1/2},$$

where $A = \pm\frac{15}{2} a^{-1} b^{-1} \left(\frac{a^2}{2b}\right)^{2/5}$.

41. $y'''_{xxx} = Ay^{-1/2}(y'_x)^3(y''_{xx})^{8/5}$.

Solution in the parametric form:

$$x = \pm aC_1^{31} \int [\pm(\tau^2 - 1)]^{-1/2}(\tau^3 - 3\tau + C_2)^{5/4}(\tau^4 - 6\tau^2 + 4C_2\tau - 3)\,d\tau + C_3,$$

$$y = bC_1^{32}(\tau^4 - 6\tau^2 + 4C_2\tau - 3)^2,$$

where $A = -15 \cdot 2^{-10} a^2 b^{-5/2} \left(\frac{a^2}{2b}\right)^{3/5}$.

42. $y'''_{xxx} = Ay^{-7/3}(y'_x)^{-7/3}$.

Solution in the parametric form:

$$x = aC_1^{17}\int (\tau^3 - 3\tau + C_2)^{1/4}[\pm(\tau^4 - 6\tau^2 + 4C_2\tau - 3)]^{-3/2}\,d\tau + C_3,$$
$$y = \pm bC_1^{16}(\tau^4 - 6\tau^2 + 4C_2\tau - 3)^{-1},$$

where $A = \pm 72a^{-5}b^{17/3}\left(\frac{a}{4}\right)^{-1/3}$.

43. $y'''_{xxx} = Ay^{-5/2}(y''_{xx})^{10/7}$.

Solution in the parametric form:

$$x = \pm aC_1^{29}\int (\tau^3 - 3\tau + C_2)^{-3/2}(\tau^4 - 6\tau^2 + 4C_2\tau - 3)^{-1/3}\,d\tau + C_3,$$
$$y = bC_1^2(\tau^3 - 3\tau + C_2)^{-1}(\tau^4 - 6\tau^2 + 4C_2\tau - 3)^{2/3},$$

where $A = -\frac{28}{3}a^{-1}b^{5/2}\left(\frac{2a^2}{3b}\right)^{3/7}$.

► *In the solutions of equations 44–47, the following notation is used:*

$$P_6(\tau) = \pm(\tau^6 - 15\tau^4 + 20C_2\tau^3 - 45\tau^2 + 12C_2\tau + 27 - 8C_2^2).$$

44. $y'''_{xxx} = Ay^{-5/3}(y'_x)^{-11/3}(y''_{xx})^3$.

Solution in the parametric form:

$$x = aC_1^5\int (\tau^3 - 3\tau + C_2)^{1/2}[\pm(\tau^4 - 6\tau^2 + 4C_2\tau - 3)]^{-3/2}P_6(\tau)\,d\tau + C_3,$$
$$y = \pm bC_1(\tau^3 - 3\tau + C_2)^{3/2}(\tau^4 - 6\tau^2 + 4C_2\tau - 3)^{-1},$$

where $A = \mp\frac{9}{16}b^{10/3}(2a)^{-2/3}$.

45. $y'''_{xxx} = Ay(y'_x)^{-5/2}(y''_{xx})^{7/5}$.

Solution in the parametric form:

$$x = aC_1^{11}\int (\tau^3 - 3\tau + C_2)^{-3/2}(\tau^4 - 6\tau^2 + 4C_2\tau - 3)^{4/3}\,d\tau + C_3,$$
$$y = bC_1^{27}(\tau^3 - 3\tau + C_2)^{-1/2}P_6(\tau),$$

where $A = \frac{405}{8}a^{-3}b\left(\pm\frac{b}{2a}\right)^{1/2}\left(\frac{a^2}{12b}\right)^{2/5}$.

46. $y'''_{xxx} = Ay^{-7/4}(y'_x)^3(y''_{xx})^{8/5}$.

Solution in the parametric form:

$$x = \pm aC_1^{37}\int (\tau^3 - 3\tau + C_2)^{5/4}(\tau^4 - 6\tau^2 + 4C_2\tau - 3)^{1/3}[P_6(\tau)]^{-1/2}\,d\tau + C_3,$$
$$y = bC_1^{64}(\tau^4 - 6\tau^2 + 4C_2\tau - 3)^{4/3},$$

where $A = -45\cdot 2^{-13}a^2b^{-5/4}\left(\frac{a^2}{12b}\right)^{3/5}$.

47. $y'''_{xxx} = Ay^{-1/2}(y'_x)^{-4}(y''_{xx})^{11/7}$.

Solution in the parametric form:

$$x = \pm aC_1^{55} \int (\tau^3 - 3\tau + C_2)^{-3/2} (\tau^4 - 6\tau^2 + 4C_2\tau - 3)^{5/3} P_6(\tau)\, d\tau + C_3,$$
$$y = bC_1^{54}(\tau^3 - 3\tau + C_2)^{-1}[P_6(\tau)]^2,$$

where $A = \mp 28 \cdot 3^7 a^{-5} b^{9/2} \left(\frac{2a^2}{3b}\right)^{4/7}$.

48. $y'''_{xxx} = A(y'_x)^{-7/3}$.

Solution in the parametric form:

$$x = aC_1^5 \int (\tau^2 \pm 1)^{1/4}\, d\tau + C_3, \qquad y = bC_1^8(\tau^3 \pm 3\tau + C_2),$$

where $A = \pm \frac{81}{2} a^{-5} b^3 \left(\frac{3b}{a}\right)^{1/3}$.

49. $y'''_{xxx} = Ay^{-5/3}y'_x(y''_{xx})^3$.

Solution in the parametric form:

$$x = aC_1 \int \tau(\tau^2 \pm 1)^{1/2}(\tau^3 \pm 3\tau + C_2)^{-1/2}\, d\tau + C_3, \qquad y = bC_1^3(\tau^2 \pm 1)^{3/2},$$

where $A = \mp \frac{4}{243} a^4 b^{-4/3}$.

50. $y'''_{xxx} = Ayy'_x(y''_{xx})^{7/5}$.

Solution in the parametric form:

$$x = aC_1^3 \int (\tau^2 \pm 1)^{-3/2}(\tau^3 \pm 3\tau + C_2)^{-1/2}\, d\tau + C_3, \qquad y = bC_1\tau(\tau^2 \pm 1)^{-1/2},$$

where $A = \pm 5b^{-2} \left(\frac{2a^2}{3b}\right)^{2/5}$.

51. $y'''_{xxx} = A(y'_x)^3(y''_{xx})^{8/5}$.

Solution in the parametric form:

$$x = aC_1^9 \int \tau^{-1/2}(\tau^2 \pm 1)^{5/4}\, d\tau + C_3, \qquad y = bC_1^8(\tau^3 \pm 3\tau + C_2),$$

where $A = \mp \frac{4}{27} a^2 b^{-3} \left(\frac{2a^2}{3b}\right)^{3/5}$.

52. $y'''_{xxx} = Ay^{-4/3}(y'_x)^{-7/3}$.

Solution in the parametric form:

$$x = aC_1^7 \int (\tau^2 \pm 1)^{1/4}(\tau^3 \pm 3\tau + C_2)^{-3/2}\, d\tau + C_3, \qquad y = bC_1^8(\tau^3 \pm 3\tau + C_2)^{-1},$$

where $A = \pm \frac{81}{2} a^{-5} b^{13/3} \left(\frac{3b}{a}\right)^{1/3}$.

53. $y'''_{xxx} = Ay^{-5/3}(y'_x)^{-5/3}(y''_{xx})^3.$

Solution in the parametric form:

$$x = aC_1^{-1}\int(\pm\tau^2 + C_2\tau - 1)(\tau^2 \pm 1)^{1/2}(\tau^3 \pm 3\tau + C_2)^{-3/2}\,d\tau + C_3,$$
$$y = bC_1(\tau^2 \pm 1)^{3/2}(\tau^3 \pm 3\tau + C_2)^{-1},$$

where $A = \mp\frac{4}{27}a^2b^{2/3}\left(\frac{3b}{a}\right)^{2/3}$.

54. $y'''_{xxx} = Ay(y'_x)^{-7}(y''_{xx})^{7/5}.$

Solution in the parametric form:

$$x = aC_1^7\int(\tau^2 \pm 1)^{-3/2}(\tau^3 \pm 3\tau + C_2)^{5/6}\,d\tau + C_3, \quad y = bC_1^9(\pm\tau^2 + C_2\tau - 1)(\tau^2 \pm 1)^{-1/2},$$

where $A = \pm 5a^{-8}b^6\left(\frac{2a^2}{b}\right)^{2/5}$.

55. $y'''_{xxx} = Ay^{-4}(y'_x)^3(y''_{xx})^{8/5}.$

Solution in the parametric form:

$$x = aC_1^{-1}\int(\tau^2 \pm 1)^{5/4}(\pm\tau^2 + C_2\tau - 1)^{-1/2}(\tau^3 \pm 3\tau + C_2)^{-2/3}\,d\tau + C_3,$$
$$y = bC_1^8(\tau^3 \pm 3\tau + C_2)^{1/3},$$

where $A = \mp 5a^2b\left(\frac{2a^2}{b}\right)^{3/5}$.

56. $y'''_{xxx} = Ay^{-5/2}(y''_{xx})^{7/4}.$

Solution in the parametric form:

$$x = aC_1^{-7}\int(\tau^2 \pm 1)^{-3/2}(\tau^3 \pm 3\tau + C_2)^{-1/3}\,d\tau + C_3,$$
$$y = bC_1^2(\tau^2 \pm 1)^{-1}(\tau^3 \pm 3\tau + C_2)^{2/3},$$

where $A = 4a^{-1}b^{5/2}\left(\mp\frac{a^2}{2b}\right)^{3/4}$.

57. $y'''_{xxx} = Ay^{-1/2}(y'_x)^{-4}(y''_{xx})^{5/4}.$

Solution in the parametric form:

$$x = aC_1^{17}\int(\pm\tau^2 + C_2\tau - 1)(\tau^2 \pm 1)^{-3/2}(\tau^3 \pm 3\tau + C_2)^{2/3}\,d\tau + C_3,$$
$$y = bC_1^{18}(\pm\tau^2 + C_2\tau - 1)^2(\tau^2 \pm 1)^{-1},$$

where $A = -64a^{-5}b^{9/2}\left(\mp\frac{a^2}{2b}\right)^{1/4}$.

▶ *In the solutions of equations 58–69, the following notation is used:*

$$S_1 = C_1e^{2k\tau} + C_2e^{-k\tau}\sin(\omega\tau), \qquad \omega = k\sqrt{3},$$
$$S_2 = 2kC_1e^{2k\tau} + kC_2e^{-k\tau}\big[\sqrt{3}\cos(\omega\tau) - \sin(\omega\tau)\big],$$
$$S_3 = 4k^2C_1e^{2k\tau} - 2k^2C_2e^{-k\tau}\big[\sqrt{3}\cos(\omega\tau) + \sin(\omega\tau)\big],$$
$$S_4 = S_2^2 - 2S_1S_3, \qquad S_5 = 5S_2S_4 + 32k^3S_1^3.$$

58. $y'''_{xxx} = Ay^{-1/2}(y'_x)^3(y''_{xx})^3.$

Solution in the parametric form:

$$x = aC_1^3\int S_1^{-1/2}S_2S_3\,d\tau + C_3, \qquad y = bC_1^4S_2^2,$$

where $A = -a^6b^{-9/2}k^3$.

59. $y'''_{xxx} = Ay^{-1/2}(y'_x)^{-6}(y''_{xx})^3.$

Solution in the parametric form:

$$x = aC_1^3\int S_1^{-3/2}S_2S_4\,d\tau + C_3, \qquad y = bC_1^2S_1^{-1}S_2^2,$$

where $A = 16a^{-3}b^{9/2}k^3$.

60. $y'''_{xxx} = Ay(y'_x)^{-9/5}.$

Solution in the parametric form:

$$x = aC_1^3\int S_1^{3/4}\,d\tau + C_3, \qquad y = bC_1^8S_4,$$

where $A = -160a^{-4}bk^6\left(\dfrac{16bk^3}{a}\right)^{4/5}$.

61. $y'''_{xxx} = Ay^{-7/5}(y'_x)^3(y''_{xx})^3.$

Solution in the parametric form:

$$x = aC_1^3\int S_1^{3/2}S_2S_4^{-1/2}\,d\tau + C_3, \qquad y = bC_1^5S_1^{5/2},$$

where $A = \frac{1}{4}\cdot 5^{-5}a^6b^{-18/5}k^{-6}$.

62. $y'''_{xxx} = Ay(y''_{xx})^{9/7}.$

Solution in the parametric form:

$$x = aC_1^{-3}\int S_1^{-3/2}\,d\tau + C_3, \qquad y = bC_1S_1^{-1/2}S_2,$$

where $A = \dfrac{7}{2}a^{-1}b^{-1}\left(\dfrac{a^2}{8bk^3}\right)^{2/7}$.

63. $y'''_{xxx} = Ay^{-1/2}(y'_x)^3(y''_{xx})^{12/7}$.

Solution in the parametric form:

$$x = aC_1^{15}\int S_1^{9/4}S_2^{-1/2}S_4\,d\tau + C_3, \qquad y = bC_1^{16}S_4^2,$$

where $A = 7\cdot 16^{-4}a^2b^{-5/2}k^{-9}\left(\dfrac{a^2}{8bk^3}\right)^{5/7}$.

64. $y'''_{xxx} = Ay^{-13/5}(y'_x)^{-9/5}$.

Solution in the parametric form:

$$x = aC_1^{9}\int S_1^{3/4}S_4^{-3/2}\,d\tau + C_3, \qquad y = bC_1^{8}S_4^{-1},$$

where $A = -160a^{-4}b^{23/5}k^6\left(\dfrac{16bk^3}{a}\right)^{4/5}$.

65. $y'''_{xxx} = Ay^{-7/5}(y'_x)^{-21/5}(y''_{xx})^3$.

Solution in the parametric form:

$$x = aC_1^{3}\int S_1^{3/2}S_4^{-3/2}S_5\,d\tau + C_3, \qquad y = bC_1S_1^{5/2}S_4^{-1},$$

where $A = \dfrac{5}{512}a^{-1}b^{17/5}k^{-6}\left(\dfrac{b}{2a}\right)^{1/5}$.

66. $y'''_{xxx} = Ay(y'_x)^{-9/4}(y''_{xx})^{9/7}$.

Solution in the parametric form:

$$x = aC_1^{9}\int S_1^{-3/2}S_4^{6/5}\,d\tau + C_3, \qquad y = bC_1^{25}S_1^{-1/2}S_5,$$

where $A = \dfrac{35}{8}a^{-3}b\left(-\dfrac{5b}{2a}\right)^{1/4}\left(\dfrac{a^2}{32bk^3}\right)^{2/7}$.

67. $y'''_{xxx} = Ay^{-13/8}(y'_x)^3(y''_{xx})^{12/7}$.

Solution in the parametric form:

$$x = aC_1^{39}\int S_1^{9/4}S_4^{3/5}S_5^{-1/2}\,d\tau + C_3, \qquad y = bC_1^{64}S_4^{8/5},$$

where $A = 175\cdot 2^{-22}a^2b^{-11/8}k^{-9}\left(\dfrac{a^2}{32bk^3}\right)^{5/7}$.

68. $y'''_{xxx} = Ay^{-7/2}(y''_{xx})^{18/13}$.

Solution in the parametric form:

$$x = aC_1^{27}\int S_1^{-3/2}S_4^{-3/5}\,d\tau + C_3, \qquad y = bC_1^{2}S_1^{-1}S_4^{2/5},$$

where $A = -\dfrac{208}{5}a^{-1}b^{7/2}k^3\left(\dfrac{2a^2}{b}\right)^{5/13}$.

69. $y'''_{xxx} = Ay^{-1/2}(y'_x)^{-6}(y''_{xx})^{21/13}.$

Solution in the parametric form:

$$x = aC_1^{51} \int S_1^{-3/2} S_4^{9/5} S_5 \, d\tau + C_3, \qquad y = bC_1^{50} S_1^{-1} S_5^2,$$

where $A = 208 \cdot 5^5 a^{-7} b^{13/2} k^3 \left(\frac{2a^2}{b}\right)^{8/13}$.

▶ *In the solutions of equations 70–81, the following notation is used:*

$$\begin{aligned} T_1 &= \cosh(\tau + C_2)\cos\tau, & \theta_1 &= \cosh\tau - \sin(\tau + C_2), \\ T_2 &= \tanh(\tau + C_2) + \tan\tau, & \theta_2 &= \sinh\tau + \cos(\tau + C_2), \\ T_3 &= \tanh(\tau + C_2) - \tan\tau, & \theta_3 &= \sinh\tau - \cos(\tau + C_2), \\ T_4 &= 3T_2T_3 - 4, & \theta_4 &= 3\theta_2\theta_3 - 2\theta_1^2. \end{aligned}$$

70. $y'''_{xxx} = Ay^2(y'_x)^{-7/3}.$

1°. Solution in the parametric form:

$$x = aC_1 \int T_1^{1/4} \, d\tau + C_3, \qquad y = bC_1^4 T_1 T_2,$$

where $A = -3a^{-5} b \left(\frac{2b}{a}\right)^{1/3}$.

2°. Solution in the parametric form:

$$x = aC_1 \int \theta_1^{1/4} \, d\tau + C_3, \qquad y = bC_1^4 \theta_2,$$

where $A = \frac{3}{8} a^{-5} b \left(\frac{b}{a}\right)^{1/3}$.

71. $y'''_{xxx} = Ay^{-5/3}(y'_x)^5(y''_{xx})^3.$

1°. Solution in the parametric form:

$$x = aC_1^2 \int T_1 T_2^{-1/2} T_3 \, d\tau + C_3, \qquad y = bC_1^3 T_1^{3/2},$$

where $A = 64 \cdot 3^{-7} a^8 b^{-16/3}$.

2°. Solution in the parametric form:

$$x = aC_1^2 \int \theta_1^{1/2} \theta_2^{-1/2} \theta_3 \, d\tau + C_3, \qquad y = bC_1^3 \theta_1^{3/2},$$

where $A = -256 \cdot 3^{-7} a^8 b^{-16/3}$.

72. $y'''_{xxx} = Ay(y'_x)^{-1/3}(y''_{xx})^{7/5}$.

1°. Solution in the parametric form:

$$x = aC_1^{-2}\int T_1^{-1}T_2^{1/2}\,d\tau + C_3, \qquad y = bC_1T_1^{1/2}T_3,$$

where $A = -\dfrac{5}{2}a^{-1}b^{-1}\Bigl(\dfrac{b}{2a}\Bigr)^{1/3}\Bigl(\dfrac{2a^2}{3b}\Bigr)^{2/5}$.

2°. Solution in the parametric form:

$$x = aC_1^{-2}\int \theta_1^{-3/2}\theta_2^{1/2}\,d\tau + C_3, \qquad y = bC_1\theta_1^{-1/2}\theta_3,$$

where $A = \dfrac{5}{2}a^{-1}b^{-1}\Bigl(\dfrac{b}{2a}\Bigr)^{1/3}\Bigl(\dfrac{4a^2}{3b}\Bigr)^{2/5}$.

73. $y'''_{xxx} = Ay^{-2/3}(y'_x)^3(y''_{xx})^{8/5}$.

1°. Solution in the parametric form:

$$x = aC_1^{11}\int T_1^{11/4}T_2^2T_3^{-1/2}\,d\tau + C_3, \qquad y = bC_1^{12}T_1^3T_2^3,$$

where $A = \dfrac{5}{432}a^2b^{-7/3}\Bigl(\dfrac{2a^2}{3b}\Bigr)^{3/5}$.

2°. Solution in the parametric form:

$$x = aC_1^{11}\int \theta_1^{5/4}\theta_2^2\theta_3^{-1/2}\,d\tau + C_3, \qquad y = bC_1^{12}\theta_2^3,$$

where $A = -\dfrac{5}{54}a^2b^{-7/3}\Bigl(\dfrac{4a^2}{3b}\Bigr)^{3/5}$.

74. $y'''_{xxx} = Ay^{-5/2}(y''_{xx})^{1/2}$.

1°. Solution in the parametric form:

$$x = aC_1^3\int T_1^{-3/2}\,d\tau + C_3, \qquad y = bC_1^2T_1^{-1},$$

where $A = 2a^{-2}b^{7/2}\Bigl(\dfrac{2}{b}\Bigr)^{1/2}$.

2°. Solution in the parametric form:

$$x = aC_1^3\int \theta_1^{-3/2}\,d\tau + C_3, \qquad y = bC_1^2\theta_1^{-1},$$

where $A = -a^{-2}b^{7/2}\Bigl(-\dfrac{2}{b}\Bigr)^{1/2}$.

75. $y'''_{xxx} = Ay^{-1/2}(y'_x)^{-4}(y''_{xx})^{5/2}$.

1°. Solution in the parametric form:

$$x = aC_1^3 \int T_1^{3/2} T_2^2 T_3 \, d\tau + C_3, \qquad y = bC_1^2 T_1 T_3^2,$$

where $A = -32a^{-2}b^{7/2}\left(\frac{b}{2}\right)^{-1/2}$.

2°. Solution in the parametric form:

$$x = aC_1^3 \int \theta_1^{-3/2} \theta_2^2 \theta_3 \, d\tau + C_3, \qquad y = bC_1^2 \theta_1^{-1} \theta_3^2,$$

where $A = 16a^{-2}b^{7/2}\left(-\frac{b}{2}\right)^{-1/2}$.

76. $y'''_{xxx} = Ay^{-10/3}(y'_x)^{-7/3}$.

1°. Solution in the parametric form:

$$x = aC_1^5 \int T_1^{-5/4} T_2^{-3/2} \, d\tau + C_3, \qquad y = bC_1^4 T_1^{-1} T_2^{-1},$$

where $A = -3a^{-5}b^{20/3}\left(\frac{2}{a}\right)^{1/3}$.

2°. Solution in the parametric form:

$$x = aC_1^5 \int \theta_1^{1/4} \theta_2^{-3/2} \, d\tau + C_3, \qquad y = bC_1^4 \theta_2^{-1},$$

where $A = \frac{3}{8}a^{-16/3}b^{20/3}$.

77. $y'''_{xxx} = Ay^{-5/3}(y'_x)^{-17/3}(y''_{xx})^3$.

1°. Solution in the parametric form:

$$x = aC_1^2 \int T_1 T_2^{-3/2} T_4 \, d\tau + C_3, \qquad y = bC_1 T_1^{1/2} T_2^{-1},$$

where $A = \frac{3}{16}a^{-2}b^{14/3}\left(\frac{b}{2a}\right)^{2/3}$.

2°. Solution in the parametric form:

$$x = aC_1^2 \int \theta_1^{1/2} \theta_2^{-3/2} \theta_4 \, d\tau + C_3, \qquad y = bC_1 \theta_1^{3/2} \theta_2^{-1},$$

where $A = -\frac{3}{4}a^{-2}b^{14/3}\left(\frac{b}{2a}\right)^{2/3}$.

78. $y'''_{xxx} = Ay(y'_x)^{-13/7}(y''_{xx})^{7/5}$.

1°. Solution in the parametric form:

$$x = aC_1^2 \int T_1^{1/3} T_2^{11/6}\, d\tau + C_3, \qquad y = bC_1^9 T_1^{3/2} T_4,$$

where $A = -\frac{5}{4}a^{-2}\Big(\frac{3b}{2a}\Big)^{6/7}\Big(\frac{2a^2}{7b}\Big)^{2/5}$.

2°. Solution in the parametric form:

$$x = aC_1^2 \int \theta_1^{-3/2} \theta_2^{11/6}\, d\tau + C_3, \qquad y = bC_1^9 \theta_1^{-1/2} \theta_4,$$

where $A = \frac{5}{4}a^{-2}\Big(\frac{3b}{2a}\Big)^{6/7}\Big(\frac{4a^2}{7b}\Big)^{2/5}$.

79. $y'''_{xxx} = Ay^{-10/7}(y'_x)^3(y''_{xx})^{8/5}$.

1°. Solution in the parametric form:

$$x = aC_1^{19} \int T_1^{19/12} T_2^{4/3} T_4^{-1/2}\, d\tau + C_3, \qquad y = bC_1^{28} T_1^{7/3} T_2^{7/3},$$

where $A = \frac{45}{16}\cdot 7^{-3}a^2 b^{-11/7}\Big(\frac{2a^2}{7b}\Big)^{3/5}$.

2°. Solution in the parametric form:

$$x = aC_1^{19} \int \theta_1^{5/4} \theta_2^{4/3} \theta_4^{-1/2}\, d\tau + C_3, \qquad y = bC_1^{28} \theta_2^{7/3},$$

where $A = -\frac{45}{2}\cdot 7^{-3}a^2 b^{-11/7}\Big(\frac{4a^2}{7b}\Big)^{3/5}$.

80. $y'''_{xxx} = Ay^{-5/2}(y''_{xx})^{13/10}$.

1°. Solution in the parametric form:

$$x = aC_1^{-11} \int T_1^{-11/6} T_2^{-1/3}\, d\tau + C_3, \qquad y = bC_1^4 T_1^{-1/3} T_2^{2/3},$$

where $A = \frac{20}{3}a^{-1}b^{5/2}\Big(\frac{2a^2}{b}\Big)^{3/10}$.

2°. Solution in the parametric form:

$$x = aC_1^{-11} \int \theta_1^{-3/2} \theta_2^{-1/3}\, d\tau + C_3, \qquad y = bC_1^4 \theta_1^{-1} \theta_2^{2/3},$$

where $A = \frac{10}{3}a^{-1}b^{5/2}\Big(-\frac{2a^2}{b}\Big)^{3/10}$.

81. $y'''_{xxx} = Ay^{-1/2}(y'_x)^{-4}(y''_{xx})^{17/10}$.

1°. Solution in the parametric form:

$$x = aC_1^{19} \int T_1^{19/6} T_2^{8/3} T_4\, d\tau + C_3, \qquad y = bC_1^{18} T_1^3 T_4^2,$$

where $A = -540a^{-5}b^{9/2}\Big(\frac{2a^2}{b}\Big)^{7/10}$.

2°. Solution in the parametric form:

$$x = aC_1^{19} \int \theta_1^{-3/2} \theta_2^{8/3} \theta_4\, d\tau + C_3, \qquad y = bC_1^{18} \theta_1^{-1} \theta_5^2,$$

where $A = -270a^{-5}b^{9/2}\Big(-\frac{2a^2}{b}\Big)^{7/10}$.

▶ *In the solutions of equations 82–84, the following notation is used:*

$$L_1 = C_1\tau^k + C_2\tau^{-k}, \qquad N_1 = (1+k)C_1\tau^k + (1-k)C_2\tau^{-k},$$
$$L_2 = C_1\ln\tau + C_2, \qquad N_2 = C_1\ln\tau + C_1 + C_2,$$
$$L_3 = C_1\sin(k\ln\tau) + C_2\cos(k\ln\tau), \qquad N_3 = (C_1 - kC_2)\sin(k\ln\tau) + (C_2 + kC_1)\cos(k\ln\tau).$$

82. $y'''_{xxx} = Ay^{-2}(y'_x)^3.$

Solution in the parametric form:

$$x = \int \tau^{1/2} L_m^{-1/2}\,d\tau + C_3, \qquad y = \tau^2,$$

where $k = \sqrt{|1+8A|}$, $m = \begin{cases} 1 & \text{if } A > -1/8, \\ 2 & \text{if } A = -1/8, \\ 3 & \text{if } A < -1/8. \end{cases}$

83. $y'''_{xxx} = Ay(y'_x)^{-3}(y''_{xx})^3.$

Solution in the parametric form:

$$x = \int \tau^{-1} N_m\,d\tau + C_3, \qquad y = \tau L_m,$$

where $k = \sqrt{|A-1|}$, $m = \begin{cases} 1 & \text{if } A < 1, \\ 2 & \text{if } A = 1, \\ 3 & \text{if } A > 1. \end{cases}$

84. $y'''_{xxx} = Ay^{-1/2}(y''_{xx})^{3/2}.$

Solution in the parametric form:

$$x = \mp 4\int \tau^2 L_1\,d\tau + C_3, \qquad y = \tau^2 L_1^2,$$

where $k = \sqrt{1+8A^{-2}}$.

▶ *In the solutions of equations 85–100, the following notation is used:*

$$Z = \begin{cases} C_1 J_\nu(\tau) + C_2 Y_\nu(\tau) & \text{for the upper sign,} \\ C_1 I_\nu(\tau) + C_2 K_\nu(\tau) & \text{for the lower sign,} \end{cases}$$

$$U_1 = \tau Z'_\tau + \nu Z, \qquad U_2 = U_1^2 \pm \tau^2 Z^2, \qquad U_3 = \pm\tfrac{2}{3}\tau^2 Z^3 - 2U_1U_2,$$

where J_ν and Y_ν are Bessel functions, I_ν and K_ν are modified Bessel functions.

85. $y'''_{xxx} = Ay^\beta(y'_x)^3, \qquad \beta \neq -2.$

Solution in the parametric form:

$$x = C_1\int \tau^{\frac{3\nu-2}{2}} Z^{-1/2}\,d\tau + C_3, \qquad y = b\tau^{2\nu},$$

where $\nu = \dfrac{1}{\beta+2}$, $A = \mp\dfrac{1}{8\nu^2}b^{-\beta-2}$.

86. $y'''_{xxx} = Ay(y'_x)^\gamma (y''_{xx})^3, \qquad \gamma \neq -3.$

Solution in the parametric form:

$$x = aC_1 \int \tau^{-1} U_1 \, d\tau + C_3, \qquad y = bC_1 \tau^\nu Z,$$

where $\nu = \dfrac{2}{\gamma+3}, \quad A = \pm \dfrac{1}{\nu^2} a^{\gamma+3} b^{-\gamma-3}.$

87. $y'''_{xxx} = Ay^{-1/2}(y''_{xx})^\delta, \qquad \delta \neq 3/2.$

Solution in the parametric form:

$$x = aC_1 \int \tau^{3\nu-1} Z \, d\tau + C_3, \qquad y = bC_1^2 \tau^{2\nu} Z^2,$$

where $\nu = \dfrac{1-\delta}{3-2\delta}, \quad A = \mp \dfrac{4}{3-2\delta} a^{-3} b^{3/2} \left(\mp \dfrac{a^2}{2b}\right)^\delta.$

88. $y'''_{xxx} = Ax^{-3}y(y''_{xx})^3.$

Solution in the parametric form:

$$x = aC_1 \int Z \, d\tau + C_3, \qquad y = bC_1^2 \tau^{-1/3} \left(\tau Z^2 - U_1 \int Z \, d\tau - C_3 U_1\right),$$

where $\nu = \frac{1}{3}, \quad A = \frac{9}{4} a^6 b^{-3}.$

89. $y'''_{xxx} = Ay^{-1/2}(y''_{xx})^3.$

Solution in the parametric form:

$$x = aC_1 \int \tau Z \, d\tau + C_3, \qquad y = bC_1^2 \tau^{4/3} Z^2,$$

where $\nu = \frac{2}{3}, \quad A = -\frac{1}{6} a^3 b^{-3/2}.$

90. $y'''_{xxx} = Ay^{-1/2}(y'_x)^{-3}(y''_{xx})^3.$

Solution in the parametric form:

$$x = C_1 \int \tau^{-2} Z^{-2} U_1 U_2 \, d\tau + C_3, \qquad y = b\tau^{-4/3} Z^{-2} U_1^2,$$

where $\nu = \frac{1}{3}, \quad A = \pm \frac{4}{3} b^{3/2}.$

91. $y'''_{xxx} = Ay(y'_x)^{-3}.$

Solution in the parametric form:

$$x = aC_1 \int Z \, d\tau + C_3, \qquad y = bC_1^2 \tau^{-2/3} U_2,$$

where $\nu = \frac{1}{3}, \quad A = -\frac{16}{81} a^{-6} b^3.$

92. $y'''_{xxx} = Ay^{-2}(y'_x)^3(y''_{xx})^3$.

Solution in the parametric form:

$$x = aC_1 \int ZU_1U_2^{-1/2}\,d\tau + C_3, \qquad y = bC_1^2\tau^{2/3}Z^2,$$

where $\nu = \frac{1}{3}$, $A = \frac{9}{32}a^6b^{-3}$.

93. $y'''_{xxx} = Ay(y''_{xx})^{3/2}$.

Solution in the parametric form:

$$x = C_1 \int \tau^{-1}Z^{-2}\,d\tau + C_3, \qquad y = b\tau^{-2/3}Z^{-1}U_1,$$

where $\nu = \frac{1}{3}$, $A = 4b^{-1}\left(\mp\frac{3}{2b}\right)^{1/2}$.

94. $y'''_{xxx} = Ay^{-1/2}(y'_x)^3(y''_{xx})^{3/2}$.

Solution in the parametric form:

$$x = aC_1 \int Z^{5/2}U_1^{-1/2}U_2\,d\tau + C_3, \qquad y = bC_1\tau^{-4/3}U_2^2,$$

where $\nu = \frac{1}{3}$, $A = \mp\dfrac{27}{16}a^3b^{-5/2}\left(\mp\dfrac{3}{2b}\right)^{1/2}$.

95. $y'''_{xxx} = Ay^{-2}(y'_x)^{-3}$.

Solution in the parametric form:

$$x = aC_1 \int \tau ZU_2^{-3/2}\,d\tau + C_3, \qquad y = bC_1\tau^{2/3}U_2^{-1},$$

where $\nu = \frac{1}{3}$, $A = -\frac{16}{81}a^{-6}b^6$.

96. $y'''_{xxx} = Ay^{-2}(y'_x)^{-3}(y''_{xx})^3$.

Solution in the parametric form:

$$x = C_1 \int ZU_2^{-3/2}U_3\,d\tau + C_3, \qquad y = b\tau^{4/3}Z^2U_2^{-1},$$

where $\nu = \frac{1}{3}$, $A = 18b^{-3}$.

97. $y'''_{xxx} = Ay(y'_x)^{-3}(y''_{xx})^{3/2}$.

Solution in the parametric form:

$$x = aC_1 \int \tau^{-2}Z^{-2}U_2^{3/2}\,d\tau + C_3, \qquad y = bC_1^2\tau^{-4/3}Z^{-1}U_3,$$

where $\nu = \frac{1}{3}$, $A = -16a^{-3}b^2\left(\pm\dfrac{3}{2b}\right)^{1/2}$.

98. $y'''_{xxx} = Ay^{-2}(y'_x)^3(y''_{xx})^{3/2}.$

Solution in the parametric form:

$$x = aC_1 \int \tau Z^{5/2} U_3^{-1/2}\, d\tau + C_3, \qquad y = bC_1^2 \tau^{-2/3} U_2,$$

where $\nu = \frac{1}{3}$, $A = \pm\frac{27}{4} a^3 b^{-1}\left(\pm\frac{3}{2b}\right)^{1/2}$.

99. $y'''_{xxx} = Ay^{-2}(y''_{xx})^{3/2}.$

Solution in the parametric form:

$$x = C_1 \int \tau^{-1} Z^{-2}\, d\tau + C_3, \qquad y = b\tau^{-4/3} Z^{-2} U_2,$$

where $\nu = \frac{1}{3}$, $A = \pm\frac{4}{3} b^2 (2b)^{-1/2}$.

100. $y'''_{xxx} = Ay^{-1/2}(y'_x)^{-3}(y''_{xx})^{3/2}.$

Solution in the parametric form:

$$x = aC_1 \int \tau^{-3} Z^{-2} U_2^{3/2} U_3\, d\tau + C_3, \qquad y = bC_1 \tau^{-8/3} Z^{-2} U_3^2,$$

where $\nu = \frac{1}{3}$, $A = \mp\frac{256}{3} a^{-3} b^{7/2} (2b)^{-1/2}$.

► *In the solutions of equations 101–138, the following notation is used:*

$$f = \sqrt{\pm(4\wp^3 - 1)}, \qquad \tau = \int \frac{d\wp}{\sqrt{\pm(4\wp^3 - 1)}} - C_2.$$

Function $\wp = \wp(\tau)$ is defined implicitly by the above elliptic integral of the first kind. For the upper sign, the function $\wp$ coincides with the classical elliptic Weierstrass function $\wp = \wp(\tau + C_2,\ 0,\ 1)$. In the solution given below, we can assume $\wp$ as the parameter instead of τ and use the explicit dependence $\tau = \tau(\wp)$.

101. $y'''_{xxx} = A(y'_x)^5.$

Solution in the parametric form:

$$x = aC_1^2 \int \wp^{-1/2}\, d\tau + C_3, \qquad y = bC_1 \tau,$$

where $A = \pm 3a^2 b^{-4}$.

102. $y'''_{xxx} = Ay^2 y'_x (y''_{xx})^3.$

Solution in the parametric form:

$$x = aC_1^5 \int \tau^{-1/2} f\, d\tau + C_3, \qquad y = bC_1^4 \wp,$$

where $A = \mp 24a^4 b^{-5}$.

103. $y'''_{xxx} = Ayy'_x(y''_{xx})^{5/2}$.

Solution in the parametric form:

$$x = aC_1^7 \int \tau^{-1/2} \wp^2 \, d\tau + C_3, \qquad y = bC_1^6 f,$$

where $A = -\frac{1}{9}a^3 b^{-3}(\pm 3b)^{-1/2}$.

104. $y'''_{xxx} = A(y'_x)^3(y''_{xx})^{1/2}$.

Solution in the parametric form:

$$x = aC_1^5 \int f^{-1/2} \, d\tau + C_3, \qquad y = bC_1^2 \tau,$$

where $A = \pm 6ab^{-2}(\pm 3b)^{-1/2}$.

105. $y'''_{xxx} = Ay^{-5}(y'_x)^5$.

Solution in the parametric form:

$$x = aC_1^{-1} \int \tau^{-3/2} \wp^{-1/2} \, d\tau + C_3, \qquad y = bC_1^2 \tau^{-1},$$

where $A = \pm 3a^2 b$.

106. $y'''_{xxx} = Ay^2(y'_x)^{-9}(y''_{xx})^3$.

Solution in the parametric form:

$$x = aC_1^5 \int \tau^{-3/2}(\tau f - \wp) \, d\tau + C_3, \qquad y = bC_1^6 \tau^{-1} \wp,$$

where $A = \mp 24a^{-6}b^5$.

107. $y'''_{xxx} = Ay(y'_x)^{-3/2}(y''_{xx})^{5/2}$.

Solution in the parametric form:

$$x = aC_1^2 \int \tau^{-1} \wp^2 \, d\tau + C_3, \qquad y = bC_1(\tau f - \wp),$$

where $A = \mp \dfrac{1}{2}ab\left(\pm \dfrac{2}{a}\right)^{1/2}$.

108. $y'''_{xxx} = Ay^{-5/4}(y'_x)^3(y''_{xx})^{1/2}$.

Solution in the parametric form:

$$x = aC_1^5 \int \tau^3(\tau f - \wp)^{-1/2} \, d\tau + C_3, \qquad y = bC_1^4 \tau^4,$$

where $A = \pm \frac{3}{16}ab^{-3/4}(\pm 3b)^{-1/2}$.

109. $y'''_{xxx} = Ay^{-2/3}(y''_{xx})^{6/5}$.

Solution in the parametric form:

$$x = aC_1^7 \int \tau^{-4}\wp^2\,d\tau + C_3, \qquad y = bC_1^9\tau^{-3}\wp^3,$$

where $A = \pm 5a^{-1}b^{-2/3}\left(\dfrac{a^2}{18b}\right)^{1/5}$.

110. $y'''_{xxx} = Ay^{-1/2}(y'_x)^{-1/3}(y''_{xx})^{9/5}$.

Solution in the parametric form:

$$x = aC_1^{-1} \int \tau^{5/2}\wp^{1/2}(\tau f - \wp)\,d\tau + C_3, \qquad y = bC_1^8(\tau f - \wp)^2,$$

where $A = \mp 5a^{-1}b^{1/2}\left(\dfrac{12b}{a}\right)^{1/3}\left(\dfrac{a^2}{18b}\right)^{4/5}$.

111. $y'''_{xxx} = Ay^{-15/7}(y'_x)^5$.

Solution in the parametric form:

$$x = aC_1^{13} \int \tau^{11/2}(\tau^2\wp \mp 1)^{-1/2}\,d\tau + C_3, \qquad y = bC_1^{14}\tau^7,$$

where $A = \pm 3\cdot 7^{-4}a^2b^{-13/7}$.

112. $y'''_{xxx} = Ay^2(y'_x)^{-23/7}(y''_{xx})^3$.

Solution in the parametric form:

$$x = aC_1^{-5} \int \tau^{-7/2}(\tau^3 f + 3\tau^2\wp \mp 1)\,d\tau + C_3, \qquad y = bC_1^2\tau(\tau^2\wp \mp 1),$$

where $A = \mp\frac{24}{49}a^{-2/7}b^{-5/7}$.

113. $y'''_{xxx} = Ay(y'_x)^{-11/4}(y''_{xx})^{5/2}$.

Solution in the parametric form:

$$x = aC_1 \int \tau^{-3}(\tau^2\wp \mp 1)^2\,d\tau + C_3, \qquad y = bC_1^3\tau^{-6}(\tau^3 f + 3\tau^2\wp \mp 1),$$

where $A = \pm\dfrac{3}{2}\left(\pm\dfrac{6b}{a}\right)^{3/4}\left(\mp 6b\right)^{-1/2}$.

114. $y'''_{xxx} = Ay^{-15/8}(y'_x)^3(y''_{xx})^{1/2}$.

Solution in the parametric form:

$$x = aC_1^5 \int \tau^{-6}(\tau^3 f + 3\tau^2\wp \mp 1)^{-1/2}\,d\tau + C_3, \qquad y = bC_1^8\tau^{-8},$$

where $A = \pm\frac{3}{64}ab^{-1/8}(\mp 6b)^{-1/2}$.

115. $y'''_{xxx} = Ay^{-2/3}(y''_{xx})^{22/15}$.

Solution in the parametric form:

$$x = aC_1^3 \int \tau^8 (\tau^2 \wp \mp 1)^2 \, d\tau + C_3, \qquad y = bC_1 \tau^3 (\tau^2 \wp \mp 1)^3,$$

where $A = \mp 15 a^{-1/15} b^{2/3} (18b)^{-7/15}$.

116. $y'''_{xxx} = Ay^{-1/2}(y'_x)^{-1/3}(y''_{xx})^{23/15}$.

Solution in the parametric form:

$$x = aC_1^9 \int \tau^{-29/2} (\tau^3 f + 3\tau^2 \wp \mp 1)(\tau^2 \wp \mp 1)^{1/2} \, d\tau + C_3,$$
$$y = bC_1^8 \tau^{-12} (\tau^3 f + 3\tau^2 \wp \mp 1)^2,$$

where $A = \pm 15 a^{-1} b^{1/2} \left(\dfrac{12b}{a}\right)^{1/3} \left(\dfrac{a^2}{18b}\right)^{8/15}$.

117. $y'''_{xxx} = Ay^{-20/7}(y'_x)^5$.

Solution in the parametric form:

$$x = aC_1^4 \int \tau^{-5} (\tau^2 \wp \mp 1)^{-1/2} \, d\tau + C_3, \quad y = bC_1^7 \tau^{-7},$$

where $A = \pm 3 \cdot 7^{-4} a^2 b^{-8/7}$.

118. $y'''_{xxx} = Ay^2(y'_x)^{-33/7}(y''_{xx})^3$.

Solution in the parametric form:

$$x = aC_1^5 \int \tau^{-7/2} (\tau^3 f - 4\tau^2 \wp \pm 6) \, d\tau + C_3, \quad y = bC_1^{12} \tau^{-6} (\tau^2 \wp \mp 1),$$

where $A = \mp \frac{24}{49} a^{-12/7} b^{5/7}$.

119. $y'''_{xxx} = Ay(y'_x)^{-27/13}(y''_{xx})^{5/2}$.

Solution in the parametric form:

$$x = aC_1^{-11} \int \tau^{-13/2} (\tau^2 \wp \mp 1)^2 \, d\tau + C_3, \quad y = bC_1^2 \tau (\tau^3 f - 4\tau^2 \wp \pm 6),$$

where $A = -\dfrac{24}{13} \left(\dfrac{6b}{a}\right)^{1/13} (\pm 39b)^{-1/2}$.

120. $y'''_{xxx} = Ay^{-20/13}(y'_x)^3(y''_{xx})^{1/2}$.

Solution in the parametric form:

$$x = aC_1^{25} \int \tau^{23/2} (\tau^3 f - 4\tau^2 \wp \pm 6)^{-1/2} \, d\tau + C_3, \quad y = bC_1^{26} \tau^{13},$$

where $A = \pm \frac{6}{169} a b^{-6/13} (\pm 39b)^{-1/2}$.

121. $y'''_{xxx} = Ay^{-2/3}(y''_{xx})^{27/20}$.

Solution in the parametric form:

$$x = aC_1^{19}\int \tau^{-20}(\tau^2\wp \mp 1)^2\,d\tau + C_3, \quad y = bC_1^{18}\tau^{-18}(\tau^2\wp \mp 1)^3,$$

where $A = 20a^{-1}b^{2/3}\left(\pm\dfrac{a^2}{18b}\right)^{7/20}$.

122. $y'''_{xxx} = Ay^{-1/2}(y'_x)^{-1/3}(y''_{xx})^{33/20}$.

Solution in the parametric form:

$$x = aC_1^{11}\int \tau^{10}(\tau^3 f - 4\tau^2\wp \pm 6)(\tau^2\wp \mp 1)^{1/2}\,d\tau + C_3, \quad y = bC_1^2\tau^2(\tau^3 f - 4\tau^2\wp \pm 6)^2,$$

where $A = \mp 20a^{-1}b^{1/2}\left(\dfrac{12b}{a}\right)^{1/3}\left(\pm\dfrac{a^2}{18b}\right)^{13/20}$.

123. $y'''_{xxx} = A(y'_x)^{-4}$.

Solution in the parametric form:

$$x = aC_1^5\int \wp^{-2}\,d\tau + C_3, \quad y = bC_1^7\wp^{-2}(f \pm 2\tau\wp^2),$$

where $A = -192a^{-7}b^5$.

124. $y'''_{xxx} = Ay^{-5/2}y'_x(y''_{xx})^3$.

Solution in the parametric form:

$$x = aC_1\int f\wp^{-2}(f \pm 2\tau\wp^2)^{-1/2}\,d\tau + C_3, \quad y = bC_1^8\wp^{-2},$$

where $A = \pm\frac{3}{4}a^4b^{-1/2}$.

125. $y'''_{xxx} = Ayy'_x(y''_{xx})^{8/5}$.

Solution in the parametric form:

$$x = aC_1^{13}\int \wp^3(f \pm 2\tau\wp^2)^{-1/2}\,d\tau + C_3, \quad y = bC_1^6 f,$$

where $A = \mp\dfrac{5}{6}b^{-2}\left(\dfrac{a^2}{6b}\right)^{3/5}$.

126. $y'''_{xxx} = A(y'_x)^3(y''_{xx})^{7/5}$.

Solution in the parametric form:

$$x = aC_1^{17}\int \wp^{-3}f^{-1/2}\,d\tau + C_3, \quad y = bC_1^{14}\wp^{-2}(f \pm 2\tau\wp^2),$$

where $A = \mp\dfrac{5}{8}a^2b^{-3}\left(\dfrac{a^2}{6b}\right)^{2/5}$.

127. $y'''_{xxx} = Ay^{-1/2}(y'_x)^{-4}$.

Solution in the parametric form:

$$x = aC_1^{11}\int \wp(f \pm 2\tau\wp^2)^{-3/2}\,d\tau + C_3, \quad y = bC_1^{14}\wp^2(f \pm 2\tau\wp^2)^{-1},$$

where $A = 192a^{-7}b^{11/2}$.

128. $y'''_{xxx} = Ay^{-5/2}(y''_{xx})^3$.

Solution in the parametric form:

$$x = aC_1^{-1}\int (f \pm 2\tau\wp^2)^{-3/2}(\tau f + 2\wp)\,d\tau + C_3, \quad y = bC_1^{6}(f \pm 2\tau\wp^2)^{-1},$$

where $A = -\frac{3}{16}a^3b^{1/2}$.

129. $y'''_{xxx} = Ay(y'_x)^3(y''_{xx})^{8/5}$.

Solution in the parametric form:

$$x = aC_1^{23}\int \wp^{-1/2}(f \pm 2\tau\wp^2)^{5/4}\,d\tau + C_3, \quad y = bC_1^{16}(\tau f + 2\wp),$$

where $A = \dfrac{10}{27}a^2b^{-4}\left(\dfrac{2a^2}{3b}\right)^{3/5}$.

130. $y'''_{xxx} = Ay(y'_x)^3(y''_{xx})^{7/5}$.

Solution in the parametric form:

$$x = aC_1^{11}\int (f \pm 2\tau\wp^2)^{-3/2}(\tau f + 2\wp)^{-1/2}\,d\tau + C_3, \quad y = bC_1^{7}\wp(f \pm 2\tau\wp^2)^{-1/2},$$

where $A = -10a^2b^{-4}\left(\dfrac{2a^2}{3b}\right)^{2/5}$.

131. $y'''_{xxx} = Ay^{-1/2}(y'_x)^{-7/3}$.

Solution in the parametric form:

$$x = aC_1^{23}\int (f \pm 2\tau\wp^2)^{1/4}(\tau f + 2\wp)\,d\tau + C_3, \quad y = bC_1^{32}(\tau f + 2\wp)^2,$$

where $A = -648a^{-5}b^{7/2}\left(\dfrac{6b}{a}\right)^{1/3}$.

132. $y'''_{xxx} = Ay^{-5/3}(y''_{xx})^3$.

Solution in the parametric form:

$$x = aC_1\int \wp(f \pm 2\tau\wp^2)^{1/2}\,d\tau + C_3, \quad y = bC_1^{9}(f \pm 2\tau\wp^2)^{3/2},$$

where $A = \pm\frac{1}{324}a^3b^{-1/3}$.

133. $y'''_{xxx} = Ay^{-5/6}(y'_x)^{-7/3}$.

Solution in the parametric form:

$$x = aC_1^{25}\int (f \pm 2\tau\wp^2)^{1/4}(\tau f + 2\wp)^{-2}\,d\tau + C_3, \quad y = bC_1^{32}(\tau f + 2\wp)^{-2},$$

where $A = -648a^{-5}b^{23/6}\left(\dfrac{6b}{a}\right)^{1/3}$.

134. $y'''_{xxx} = Ay^{-5/3}(y'_x)^{-2/3}(y''_{xx})^3$.

Solution in the parametric form:

$$x = aC_1^{-1}\int (\tau^2\wp \mp 1)(f \pm 2\tau\wp^2)^{1/2}(\tau f + 2\wp)^{-2}\,d\tau + C_3,$$
$$y = bC_1^7(f \pm 2\tau\wp^2)^{3/2}(\tau f + 2\wp)^{-2},$$

where $A = -\dfrac{1}{324}a^3b^{-1/3}\left(\dfrac{6b}{a}\right)^{2/3}$.

135. $y'''_{xxx} = Ay(y'_x)^{11}(y''_{xx})^{7/5}$.

Solution in the parametric form:

$$x = aC_1^{31}\int (f \pm 2\tau\wp^2)^{-3/2}(\tau f + 2\wp)^{13/6}\,d\tau + C_3,$$
$$y = bC_1^{27}(\tau^2\wp \mp 1)(f \pm 2\tau\wp^2)^{-1/2},$$

where $A = -20a^{10}b^{-12}\left(\dfrac{2a^2}{b}\right)^{2/5}$.

136. $y'''_{xxx} = Ay^5(y'_x)^3(y''_{xx})^{8/5}$.

Solution in the parametric form:

$$x = aC_1^{43}\int (\tau^2\wp \mp 1)^{-1/2}(f \pm 2\tau\wp^2)^{5/4}(\tau f + 2\wp)^{-4/3}\,d\tau + C_3,$$
$$y = bC_1^{16}(\tau f + 2\wp)^{-1/3},$$

where $A = 20a^2b^{-8}\left(\dfrac{2a^2}{b}\right)^{3/5}$.

137. $y'''_{xxx} = Ay^{-1/2}(y'_x)^{-4}(y''_{xx})^{4/5}$.

Solution in the parametric form:

$$x = aC_1^{47}\int (\tau^2\wp \mp 1)(f \pm 2\tau\wp^2)^{-3/2}(\tau f + 2\wp)^{4/3}\,d\tau + C_3,$$
$$y = bC_1^{54}(\tau^2\wp \mp 1)^2(f \pm 2\tau\wp^2)^{-1},$$

where $A = -320a^{-7}b^{11/2}\left(\dfrac{a^2}{4b}\right)^{4/5}$.

138. $y'''_{xxx} = Ay^{-5/2}(y''_{xx})^{11/5}$.

Solution in the parametric form:

$$x = aC_1^{-13} \int (f \pm 2\tau\wp^2)^{-3/2} (\tau f + 2\wp)^{1/3} \, d\tau + C_3,$$
$$y = bC_1^{14} (f \pm 2\tau\wp^2)^{-1} (\tau f + 2\wp)^{4/3},$$

where $A = \frac{5}{4} ab^{3/2} \left(\frac{a^2}{4b} \right)^{1/5}$.

► *In the solutions of equations 139–140, the following notation is used:*

$$U = \int \frac{\tau^{k-1} \, d\tau}{z(\tau)}, \qquad z = \begin{cases} \dfrac{1}{k+1} \tau^{k+1} + \dfrac{1}{k} \tau^k + C_2 & \text{if } k \neq 0, \ -1; \\ \tau + \ln|\tau| + C_2 & \text{if } k = 0; \\ \ln|\tau| - \dfrac{1}{\tau} + C_2 & \text{if } k = -1. \end{cases}$$

139. $y'''_{xxx} = Ay^\beta (y'_x)^{-1} (y''_{xx})^2, \qquad \beta \neq 0$.

Solution in the parametric form:

$$x = C_1 \int \tau^{\frac{1-\beta}{\beta}} \exp(-\tfrac{1}{2} U) \, d\tau + C_3, \qquad y = b\tau^{1/\beta},$$

where $k = 1/\beta, \ A = -2b^{-\beta}$.

140. $y'''_{xxx} = Ay^{-1} (y'_x)^\gamma y''_{xx}, \qquad \gamma \neq 1$.

Solution in the parametric form:

$$x = aC_1 \int \tau^{\frac{2-\gamma}{\gamma-1}} z^{-1} e^U \, d\tau + C_3, \qquad y = e^U,$$

where $k = \dfrac{2}{\gamma - 1}, \quad A = a^{\gamma-1} b^{1-\gamma}$.

► *In the solutions of equations 141–170, the following notation is used:*

$$R = \sqrt{\pm(4\tau^3 - 1)}, \qquad I_1 = 2\tau I \mp R, \qquad I_2 = \tau^{-1}(RI_1 - 1),$$
$$I_3 = 4\tau I_1^2 \mp I_2^2, \qquad I_4 = I_2 I_3 - 8I_1^2, \qquad I_5 = 2RI - \tau^2,$$

where $I = \int \tau R^{-1} \, d\tau + C_2$ *is the incomplete elliptic integral of the second kind in the form of Weierstrass.*

141. $y'''_{xxx} = A(y'_x)^{-7}$.

Solution in the parametric form:

$$x = aC_1^4 \int \tau^{-3/2} R^{-1} \, d\tau + C_3, \qquad y = bC_1^5 \tau^{-1} I_1,$$

where $A = \mp 3a^{-10} b^8$.

142. $y'''_{xxx} = Ay^{-4}y'_x(y''_{xx})^3.$

Solution in the parametric form:

$$x = aC_1^{-1}\int \tau^{-3/2}I_1^{-1/2}\,d\tau + C_3, \qquad y = bC_1^4\tau^{-1},$$

where $A = \pm 24a^4b.$

143. $y'''_{xxx} = Ayy'_x(y''_{xx})^{7/4}.$

Solution in the parametric form:

$$x = aC_1^{11}\int \tau^{5/2}I_1^{-1/2}R^{-1}\,d\tau + C_3, \qquad y = bC_1^6R,$$

where $A = \mp\frac{2}{3}b^{-2}\Big(\mp\frac{a^2}{3b}\Big)^{3/4}.$

144. $y'''_{xxx} = A(y'_x)^3(y''_{xx})^{5/4}.$

Solution in the parametric form:

$$x = aC_1^{13}\int \tau^{-2}R^{-3/2}\,d\tau + C_3, \qquad y = bC_1^{10}\tau^{-1}I_1,$$

where $A = -4a^2b^{-3}\Big(\mp\frac{a^2}{3b}\Big)^{1/4}.$

145. $y'''_{xxx} = Ay(y'_x)^{-7}.$

Solution in the parametric form:

$$x = aC_1^7\int I_1^{-3/2}R^{-1}\,d\tau + C_3, \qquad y = bC_1^{10}\tau I_1^{-1},$$

where $A = \mp 3a^{-10}b^7.$

146. $y'''_{xxx} = Ay^{-4}(y'_x)^3(y''_{xx})^3.$

Solution in the parametric form:

$$x = aC_1\int \tau^{-1/2}I_1^{-3/2}I_2R^{-1}\,d\tau + C_3, \qquad y = bC_1^6I_1^{-1},$$

where $A = \pm 24a^6b^{-1}.$

147. $y'''_{xxx} = Ay(y''_{xx})^{7/4}.$

Solution in the parametric form:

$$x = aC_1^7\int I_1^2R^{-1}\,d\tau + C_3, \qquad y = bC_1^2I_2,$$

where $A = -4a^{-1}b^{-1}\Big(\pm\frac{a^2}{6b}\Big)^{3/4}.$

148. $y'''_{xxx} = Ay^{-1/2}(y'_x)^3(y''_{xx})^{5/4}$.

Solution in the parametric form:

$$x = aC_1^{11}\int \tau I_1^{-3}I_2^{-1/2}R^{-1}\,d\tau + C_3, \qquad y = bC_1^{10}\tau^2 I_1^{-2},$$

where $A = \frac{1}{2}a^2b^{-5/2}\left(\pm\frac{a^2}{6b}\right)^{5/4}$.

149. $y'''_{xxx} = Ay^{-1/2}(y'_x)^{-5/3}(y''_{xx})^3$.

Solution in the parametric form:

$$x = aC_1^{-1}\int \tau I_1^{-1/2}I_2R^{-1}\,d\tau + C_3, \qquad y = bC_1^8 I_2^2,$$

where $A = \mp\frac{1}{27}a^2b^{-1/2}\left(\frac{12b}{a}\right)^{2/3}$.

150. $y'''_{xxx} = Ay^{-1/2}(y'_x)^{-4/3}(y''_{xx})^3$.

Solution in the parametric form:

$$x = aC_1\int I_1^{-5/2}I_2I_3R^{-1}\,d\tau + C_3, \qquad y = bC_1^{10}I_1^{-3}I_2^2,$$

where $A = \mp\frac{16}{27}a^2b^{-1/2}\left(\pm\frac{3b}{a}\right)^{1/3}$.

151. $y'''_{xxx} = Ay^{-1/2}(y'_x)^3(y''_{xx})^{8/7}$.

Solution in the parametric form:

$$x = aC_1^{37}\int I_1^{7/4}I_2^{-1/2}I_3R^{-1}\,d\tau + C_3, \qquad y = bC_1^{32}I_3^2,$$

where $A = \mp 7\cdot 2^{-10}a^2b^{-5/2}\left(\frac{a^2}{6b}\right)^{1/7}$.

152. $y'''_{xxx} = Ay(y''_{xx})^{13/7}$.

Solution in the parametric form:

$$x = aC_1^{13}\int I_1^{-5/2}R^{-1}\,d\tau + C_3, \qquad y = bC_1^5 I_1^{-3/2}I_2,$$

where $A = \frac{7}{2}a^{-1}b^{-1}\left(\frac{a^2}{6b}\right)^{6/7}$.

153. $y'''_{xxx} = Ay(y'_x)^{-13}$.

Solution in the parametric form:

$$x = aC_1^{13}\int I_1^{3/4}R^{-1}\,d\tau + C_3, \qquad y = bC_1^{16}I_3,$$

where $A = \pm 3\cdot 2^{25}a^{-16}b^{13}$.

154. $y'''_{xxx} = Ay^{-7}(y'_x)^3(y''_{xx})^3.$

Solution in the parametric form:

$$x = aC_1^{-1}\int I_1^{-1/2}I_2I_3^{-1/2}R^{-1}\,d\tau + C_3, \qquad y = bC_1^3I_1^{1/2},$$

where $A = \mp 12a^6b^2$.

155. $y'''_{xxx} = Ay^3(y'_x)^{-13}.$

Solution in the parametric form:

$$x = aC_1^{11}\int I_1^{3/4}I_3^{-3/2}R^{-1}\,d\tau + C_3, \qquad y = bC_1^{16}I_3^{-1},$$

where $A = \pm 3\cdot 2^{25}a^{-16}b^{11}$.

156. $y'''_{xxx} = Ay^{-7}(y'_x)^7(y''_{xx})^3.$

Solution in the parametric form:

$$x = aC_1\int I_1^{-1/2}I_3^{-3/2}I_4R^{-1}\,d\tau + C_3, \qquad y = bC_1^5I_1^{1/2}I_3^{-1},$$

where $A = \mp 192a^{10}b^{-2}$.

157. $y'''_{xxx} = Ay^{-7/6}(y''_{xx})^{2/3}.$

Solution in the parametric form:

$$x = aC_1^9\int I_1^{-5/2}I_3^5R^{-1}\,d\tau + C_3, \qquad y = bC_1^{10}I_1^{-3}I_3^6,$$

where $A = \pm 54a^{-3}b^{13/6}\left(\dfrac{2a^2}{9b}\right)^{2/3}$.

158. $y'''_{xxx} = Ay^{-1/2}(y'_x)^{-4/3}(y''_{xx})^{7/3}.$

Solution in the parametric form:

$$x = aC_1^3\int I_1^{-5/2}I_3^{-1}I_4R^{-1}\,d\tau + C_3, \qquad y = bC_1^2I_1^{-3}I_4^2,$$

where $A = 8b^{1/2}\left(\dfrac{2a}{3}\right)^{1/3}$.

159. $y'''_{xxx} = Ay^{-3/4}(y'_x)^3(y''_{xx})^{8/7}.$

Solution in the parametric form:

$$x = aC_1^{67}\int I_1^{7/4}I_3^{-5}I_4^{-1/2}R^{-1}\,d\tau + C_3, \qquad y = bC_1^{64}I_3^{-4},$$

where $A = \mp 7\cdot 2^{-13}a^2b^{-9/4}\left(\dfrac{a^2}{12b}\right)^{1/7}$.

160. $y'''_{xxx} = Ay(y'_x)^{-1/2}(y''_{xx})^{13/7}$.

Solution in the parametric form:

$$x = aC_1^{19} \int I_1^{-5/2} I_3^4 R^{-1}\, d\tau + C_3, \qquad y = bC_1^3 I_1^{-3/2} I_4,$$

where $A = \frac{7}{2} a^{-1} b^{-1} \left(\pm \frac{3b}{a}\right)^{1/2} \left(\frac{a^2}{12b}\right)^{6/7}$.

161. $y'''_{xxx} = A(y'_x)^2$.

Solution in the parametric form:

$$x = aC_1^{-1} \int R^{-1}\, d\tau + C_3, \qquad y = bC_1 \int \tau R^{-1}\, d\tau + C_2,$$

where $A = \pm 6a^{-1}b^{-1}$.

162. $y'''_{xxx} = Ay^{1/2} y'_x (y''_{xx})^3$.

Solution in the parametric form:

$$x = aC_1^7 \int \tau I^{-1/2}\, d\tau + C_3, \qquad y = bC_1^8 \tau^2,$$

where $A = \mp 24a^4 b^{-7/2}$.

163. $y'''_{xxx} = Ayy'_x (y''_{xx})^4$.

Solution in the parametric form:

$$x = aC_1^5 \int \tau^2 I^{-1/2} R^{-1}\, d\tau + C_3, \qquad y = bC_1^6 R,$$

where $A = -\frac{1}{162} a^6 b^{-5}$.

164. $y'''_{xxx} = A(y'_x)^3 (y''_{xx})^{-1}$.

Solution in the parametric form:

$$x = aC_1^{-1} \int \tau R^{-3/2}\, d\tau + C_3, \qquad y = bC_1^2 \int \tau R^{-1}\, d\tau + C_2,$$

where $A = 9a^{-2}b^{-1}$.

165. $y'''_{xxx} = Ay^{-7/2}(y'_x)^2$.

Solution in the parametric form:

$$x = aC_1^5 \int I^{-3/2} R^{-1}\, d\tau + C_3, \qquad y = bC_1^2 I^{-1},$$

where $A = \mp 6a^{-1}b^{5/2}$.

166. $y'''_{xxx} = Ay^{1/2}(y'_x)^{-6}(y''_{xx})^3.$

Solution in the parametric form:

$$x = aC_1^7 \int \tau I^{-3/2} I_5 R^{-1}\, d\tau + C_3, \qquad y = bC_1^6 \tau^2 I^{-1},$$

where $A = \mp 48a^{-3}b^{7/2}$.

167. $y'''_{xxx} = Ay(y'_x)^{-9/5}(y''_{xx})^4.$

Solution in the parametric form:

$$x = aC_1^{11} \int \tau^2 I^{-1/4} R^{-1}\, d\tau + C_3, \qquad y = bC_1^{16} I_5,$$

where $A = -\dfrac{2}{1125} a^4 b^{-3} \left(\dfrac{12b}{a}\right)^{4/5}.$

168. $y'''_{xxx} = Ay^{-7/5}(y'_x)^3(y''_{xx})^{-1}.$

Solution in the parametric form:

$$x = aC_1 \int \tau I^{3/2} I_5^{-1/2} R^{-1}\, d\tau + C_3, \qquad y = bC_1 I^{5/2},$$

where $A = \frac{36}{5} a^{-2} b^{2/5}$.

169. $y'''_{xxx} = Ay^{-1/3}(y''_{xx})^{9/7}.$

Solution in the parametric form:

$$x = aC_1 \int \tau^2 I^{-5/2} R^{-1}\, d\tau + C_3, \qquad y = bC_1^9 \tau^3 I^{-3/2},$$

where $A = \dfrac{7}{2} a^{-1} b^{1/3} \left(\dfrac{a^2}{18b}\right)^{2/7}.$

170. $y'''_{xxx} = Ay^{-1/2}(y'_x)^{1/3}(y''_{xx})^{12/7}.$

Solution in the parametric form:

$$x = aC_1^{23} \int \tau^{1/2} I^{7/4} I_5 R^{-1}\, d\tau + C_3, \qquad y = bC_1^{32} I_5^2,$$

where $A = -\dfrac{7}{4} a^{-1} b^{1/2} \left(\dfrac{a}{3b}\right)^{1/3} \left(\dfrac{a^2}{18b}\right)^{5/7}.$

► *In the solutions of equations 171–172, the following notation is used:*

$$f = \exp\left(\int \frac{d\tau}{\sqrt{z}}\right), \qquad z = \begin{cases} C_2 + \dfrac{1}{4}\tau^2 + \dfrac{2B}{k+1}\tau^{k+1} & \text{if } k \neq -1, \\ C_2 + \dfrac{1}{4}\tau^2 + 2B \ln|\tau| & \text{if } k = -1. \end{cases}$$

171. $y'''_{xxx} = Ay^{-\frac{\gamma+5}{4}}(y'_x)^\gamma$.

Solution in the parametric form:

$$x = aC_1^3 \int \tau^{-1/2} f^{3/4} z^{-1/2}\, d\tau + C_3, \qquad y = bC_1^4 f,$$

where $k = \dfrac{\gamma-1}{2}$, $A = \dfrac{1}{2} B a^{2(k-1)} b^{\frac{3}{2}(1-k)}$.

172. $y'''_{xxx} = Ay^{-\gamma-2}(y'_x)^\gamma (y''_{xx})^3$.

Solution in the parametric form:

$$x = a \int (\tau z^{-1/2} + 2)\, d\tau + C_3, \qquad y = C_1 \tau f^{1/2},$$

where $k = -\gamma - 2$, $A = -2^{3-k} a^{1-k} B$.

173. $y'''_{xxx} = Ay^{\frac{\gamma-1}{2}}(y'_x)^\gamma (y''_{xx})^{3/2}$.

Solution in the parametric form:

$$x = aC_1^3 \int \frac{\left(V - \sqrt{V^2+4}\right)}{(\tau U)^{3/4}\sqrt{V^2+4}}\, d\tau + C_3, \qquad y = bC_1^2 \tau^{1/2} U^{1/2},$$

where

$$U = \exp\left(\int \frac{V\, d\tau}{\tau\sqrt{V^2+4}}\right), \quad V = \begin{cases} \tau^{-1/2}\left(C_2 + \dfrac{B}{\gamma+1}\tau^{\frac{\gamma+1}{2}}\right) & \text{if } \gamma \neq -1, \\ \tau^{-1/2}\left(C_2 + \frac{1}{2}B \ln|\tau|\right) & \text{if } \gamma = -1, \end{cases}$$

$$A = 2^{3/2} a b^{-\frac{\gamma}{2}-1} B \left(-\frac{2a}{b}\right)^{\gamma-1}.$$

174. $y'''_{xxx} = Ay^\beta (y'_x)^\gamma (y''_{xx})^{\frac{\gamma+4\beta+5}{\gamma+2\beta+3}}$, $\quad\beta,\ \gamma \neq -1$.

Solution in the parametric form:

$$x = aC_1^{\gamma+\beta+2} \int \tau^{-3/2} U^{-\frac{\gamma+4\beta+5}{2(\gamma+1)}} z^{-1}\, d\tau + C_3, \qquad y = bC_1^{\gamma+1} U,$$

where $A = a^{\gamma-1} b^{-\beta-\gamma}\left(\dfrac{2a^2}{b}\right)^{\frac{2(\beta+1)}{\gamma+2\beta+3}} B$, $U = \exp\left(\displaystyle\int \frac{d\tau}{\tau z}\right)$, $z = z(\tau)$ is the solution of the algebraic equation

$$(z+k-1)(z+k)^{\frac{k}{1-k}} = \left(C_2 + \frac{2B}{\gamma+2\beta+3}\tau^{\frac{\gamma+1}{2}}\right)\tau^{\frac{1}{k-1}}, \qquad k = -\frac{2(\beta+1)}{\gamma+1}.$$

175. $y'''_{xxx} = Ay^{-1}(y'_x)^{-1}(y''_{xx})^\delta$, $\quad\delta \neq 1,\ 2$.

Solution in the parametric form:

$$x = aC_1^{\delta-3} \int \tau^{k-1} U^{\frac{2-k}{2k}} z^{-1}\, d\tau + C_3, \qquad y = bC_1^{2\delta-4} k(kz-\tau)^{-1} U^{1/k},$$

where $k = \dfrac{\delta-1}{\delta-2}$, $A = \dfrac{1-k}{2} a^{-4} b^3 \left(-\dfrac{2a^2}{b}\right)^\delta B$, $U = \exp\left(\displaystyle\int \frac{d\tau}{z}\right)$, $z = z(\tau)$ is the solution of the transcendental equation

$$\ln|kz - \tau| - \frac{\tau}{kz-\tau} = \frac{1}{k}\tau^k + C_2.$$

176. $y'''_{xxx} = Ay^{-1}(y'_x)^{-1}(y''_{xx})^2$.

Solution in the parametric form:

$$x = \pm C_1 \int e^\tau z^{-1/2} U^{-1/2}\, d\tau + C_3, \qquad y = \pm \tfrac{1}{2} e^\tau,$$

where $z = \mp A\tau + e^\tau + C_2$, $U = \exp\Bigl(\pm A \int \frac{d\tau}{z}\Bigr)$.

177. $y'''_{xxx} = Ay^{-1}(y'_x)^{-1} y''_{xx}$.

Solution in the parametric form:

$$x = C_1 \int e^{\tau/2} U\, d\tau + C_3, \qquad y = \pm C_1 z U,$$

where $z = \pm A\tau + e^\tau + C_2$, $U = \exp\Bigl(\mp \int \frac{d\tau}{z}\Bigr)$.

3.2.5. Some Transformations

Let us consider some transformations of the equation $y'''_{xxx} = Ax^\alpha y^\beta (y'_x)^\gamma (y''_{xx})^\delta$.

1. In the special case $\gamma = \delta = 0$, the transformation

$$x = \frac{1}{t}, \qquad y = \frac{w}{t^2}$$

reduces the equation

$$y'''_{xxx} = Ax^\alpha y^\beta$$

to an equation of the similar form

$$w'''_{ttt} = -At^{-\alpha-2\beta-4} w^\beta.$$

2. In the special case $\alpha = \delta = 0$, the transformation

$$x = -\int \frac{d\tau}{[z(\tau)]^{3/2}}, \qquad y = \frac{1}{z(\tau)}$$

reduces the equation

$$y'''_{xxx} = Ay^\beta (y'_x)^\gamma$$

to an equation of the similar form

$$z'''_{\tau\tau\tau} = Az^{-\frac{2\beta+\gamma+5}{2}} (w'_\tau)^\gamma.$$

3. In the special case $\beta = 0$, the substitution

$$u(x) = y'_x$$

reduces the equation

$$y'''_{xxx} = Ax^\alpha (y'_x)^\gamma (y''_{xx})^\delta$$

to the generalized Emden–Fowler equation

$$u''_{xx} = Ax^{\alpha}u^{\gamma}(u'_x)^{\delta}$$

(see Section 2.3 and Section 2.5).

4. In the special case $\alpha = 0$, the substitution

$$v(y) = (y'_x)^2$$

reduces the equation

$$y'''_{xxx} = Ay^{\beta}(y'_x)^{\gamma}(y''_{xx})^{\delta}$$

to the generalized Emden—Fowler equation

$$v''_{yy} = A \cdot 2^{1-\delta}y^{\beta}v^{\frac{\gamma-1}{2}}(v'_y)^{\delta}$$

(see Section 2.3 and Section 2.5).

3.3. Equations of the Form $y'''_{xxx} = f(y)g(y'_x)h(y''_{xx})$

3.3.1. Equations Containing Power Functions

1. $\boldsymbol{y'''_{xxx} = (ay^2 + by + c)^{-5/4}}$.

This is a special case of equation 3.5.2.17 with $f(w) = 1$.

2. $\boldsymbol{y'''_{xxx} = (Ay^n + By^m)y'_x}$.

This is a special case of equation 3.5.2.1 with $f(y) = Ay^n + By^m$.

3. $\boldsymbol{y'''_{xxx} = (Ay^n + By^m)\big[a(y'_x)^3 + by'_x\big]}$.

This is a special case of equation 3.5.2.4 with $f(y) = b(Ay^n + By^m)$, $g(y) = a(Ay^n + By^m)$.

4. $$\boldsymbol{y'''_{xxx} = y^{-2}\left[-\frac{(m+1)}{(m+3)^2}(y'_x)^3 + A(y'_x)^{2m+1}\right], \qquad m \neq -3,\ m \neq -1.}$$

The substitution $w(y) = (y'_x)^2$ leads to an equation of the form 2.4.2.4:

$$w''_{yy} = y^{-2}\left[-\frac{2(m+1)}{(m+3)^2}w + 2Aw^m\right].$$

5. $\boldsymbol{y'''_{xxx} = y^{-2}\left[\frac{15}{8}(y'_x)^3 + A(y'_x)^{-13}\right]}$.

The substitution $w(y) = (y'_x)^2$ leads to an equation of the form 2.4.2.35:

$$w''_{yy} = y^{-2}\left(\tfrac{15}{4}w + 2Aw^{-7}\right).$$

6. $y'''_{xxx} = y^{-2}\big[3(y'_x)^3 + A(y'_x)^{-7}\big]$.

The substitution $w(y) = (y'_x)^2$ leads to an equation of the form 2.4.2.31:

$$w''_{yy} = y^{-2}(6w + 2Aw^{-4}).$$

7. $y'''_{xxx} = y^{-2}\big[6(y'_x)^3 + A(y'_x)^{-4}\big]$.

The substitution $w(y) = (y'_x)^2$ leads to an equation of the form 2.4.2.64:

$$w''_{yy} = y^{-2}(12w + 2Aw^{-5/2}).$$

8. $y'''_{xxx} = y^{-2}\big[(y'_x)^3 + A(y'_x)^{-3}\big]$.

The substitution $w(y) = (y'_x)^2$ leads to an equation of the form 2.4.2.6:

$$w''_{yy} = y^{-2}(2w + 2Aw^{-2}).$$

9. $y'''_{xxx} = y^{-2}\big[-\tfrac{3}{32}(y'_x)^3 + A(y'_x)^{-7/3}\big]$.

The substitution $w(y) = (y'_x)^2$ leads to an equation of the form 2.4.2.26:

$$w''_{yy} = y^{-2}(-\tfrac{3}{16}w + 2Aw^{-5/3}).$$

10. $y'''_{xxx} = y^{-2}\big[-\tfrac{9}{200}(y'_x)^3 + A(y'_x)^{-7/3}\big]$.

The substitution $w(y) = (y'_x)^2$ leads to an equation of the form 2.4.2.10:

$$w''_{yy} = y^{-2}(-\tfrac{9}{100}w + 2Aw^{-5/3}).$$

11. $y'''_{xxx} = y^{-2}\big[\tfrac{3}{8}(y'_x)^3 + A(y'_x)^{-7/3}\big]$.

The substitution $w(y) = (y'_x)^2$ leads to an equation of the form 2.4.2.12:

$$w''_{yy} = y^{-2}(\tfrac{3}{4}w + 2Aw^{-5/3}).$$

12. $y'''_{xxx} = y^{-2}\big[\tfrac{63}{8}(y'_x)^3 + A(y'_x)^{-7/3}\big]$.

The substitution $w(y) = (y'_x)^2$ leads to an equation of the form 2.4.2.66:

$$w''_{yy} = y^{-2}(\tfrac{63}{4}w + 2Aw^{-5/3}).$$

13. $y'''_{xxx} = y^{-2}\big[-\tfrac{5}{72}(y'_x)^3 + A(y'_x)^{-9/5}\big]$.

The substitution $w(y) = (y'_x)^2$ leads to an equation of the form 2.4.2.29:

$$w''_{yy} = y^{-2}(-\tfrac{5}{36}w + 2Aw^{-7/5}).$$

14. $y'''_{xxx} = y^{-2}\big[-\tfrac{1}{9}(y'_x)^3 + A\big]$.

The substitution $w(y) = (y'_x)^2$ leads to an equation of the form 2.4.2.14:

$$w''_{yy} = y^{-2}(-\tfrac{2}{9}w + 2Aw^{-1/2}).$$

15. $y'''_{xxx} = y^{-2}\left[-\frac{2}{25}(y'_x)^3 + A\right].$

The substitution $w(y) = (y'_x)^2$ leads to an equation of the form 2.4.2.8:

$$w''_{yy} = y^{-2}(-\tfrac{4}{25}w + 2Aw^{-1/2}).$$

16. $y'''_{xxx} = y^{-2}\left[10(y'_x)^3 + A\right].$

The substitution $w(y) = (y'_x)^2$ leads to an equation of the form 2.4.2.33:

$$w''_{yy} = y^{-2}(20w + 2Aw^{-1/2}).$$

17. $y'''_{xxx} = y^{-2}\left[-\frac{6}{49}(y'_x)^3 + A(y'_x)^2\right].$

The substitution $w(y) = (y'_x)^2$ leads to an equation of the form 2.4.2.37:

$$w''_{yy} = y^{-2}(-\tfrac{12}{49}w + 2Aw^{1/2}).$$

18. $y'''_{xxx} = y^{-2}\left[A(y'_x)^5 - \frac{3}{25}(y'_x)^3\right].$

The substitution $w(y) = (y'_x)^2$ leads to an equation of the form 2.4.2.60:

$$w''_{yy} = y^{-2}(2Aw^2 - \tfrac{6}{25}w).$$

19. $y'''_{xxx} = y^{-2}\left[A(y'_x)^5 + \frac{3}{25}(y'_x)^3\right].$

The substitution $w(y) = (y'_x)^2$ leads to an equation of the form 2.4.2.62:

$$w''_{yy} = y^{-2}(2Aw^2 + \tfrac{6}{25}w).$$

20. $y'''_{xxx} = y^{-4/3}(Ay'_x + B).$

The substitution $w(y) = (y'_x)^2$ leads to an equation of the form 2.4.2.40:

$$w''_{yy} = y^{-4/3}(2A + 2Bw^{-1/2}).$$

21. $y'''_{xxx} = (Ay^4 + By^3)(y'_x)^{-13}.$

The substitution $w(y) = (y'_x)^2$ leads to an equation of the form 2.4.2.39:

$$w''_{yy} = (2Ay^4 + 2By^3)w^{-7}.$$

22. $y'''_{xxx} = (Ay^2 + B)(y'_x)^{-9}.$

The substitution $w(y) = (y'_x)^2$ leads to an equation of the form 2.4.2.16:

$$w''_{yy} = (2Ay^2 + 2B)w^{-5}.$$

23. $y'''_{xxx} = (Ay^{-1} + By^{-2})(y'_x)^{-3}.$

The substitution $w(y) = (y'_x)^2$ leads to an equation of the form 2.4.2.28:

$$w''_{yy} = (2Ay^{-1} + 2By^{-2})w^{-2}.$$

24. $y'''_{xxx} = (Ay^{-7/3} + By^{-10/3})(y'_x)^{-7/3}$.

The substitution $w(y) = (y'_x)^2$ leads to an equation of the form 2.4.2.48:

$$w''_{yy} = (2Ay^{-7/3} + 2By^{-10/3})w^{-5/3}.$$

25. $y'''_{xxx} = (Ay^{-4/3} + By^{-10/3})(y'_x)^{-7/3}$.

The substitution $w(y) = (y'_x)^2$ leads to an equation of the form 2.4.2.49:

$$w''_{yy} = (2Ay^{-4/3} + 2By^{-10/3})w^{-5/3}.$$

26. $y'''_{xxx} = (Ay^{-4/3} + By^{-7/3})(y'_x)^{-7/3}$.

The substitution $w(y) = (y'_x)^2$ leads to an equation of the form 2.4.2.24:

$$w''_{yy} = (2Ay^{-4/3} + 2By^{-7/3})w^{-5/3}.$$

27. $y'''_{xxx} = (Ay^{-2/3} + By^{-4/3})(y'_x)^{-7/3}$.

The substitution $w(y) = (y'_x)^2$ leads to an equation of the form 2.4.2.90:

$$w''_{yy} = (2Ay^{-2/3} + 2By^{-4/3})w^{-5/3}.$$

28. $y'''_{xxx} = (A + By^{-2/3})(y'_x)^{-7/3}$.

The substitution $w(y) = (y'_x)^2$ leads to an equation of the form 2.4.2.89:

$$w''_{yy} = (2A + 2By^{-2/3})w^{-5/3}.$$

29. $y'''_{xxx} = (Ay^2 + B)(y'_x)^{-7/3}$.

The substitution $w(y) = (y'_x)^2$ leads to an equation of the form 2.4.2.47:

$$w''_{yy} = (2Ay^2 + 2B)w^{-5/3}.$$

30. $y'''_{xxx} = (Ay^2 + By)(y'_x)^{-7/3}$.

The substitution $w(y) = (y'_x)^2$ leads to an equation of the form 2.4.2.46:

$$w''_{yy} = (2Ay^2 + 2By)w^{-5/3}.$$

31. $y'''_{xxx} = (Ay^n + By^k)y'_x(y''_{xx})^m$.

This is a special case of equation 3.5.4.15 with $f(y) = Ay^n + By^k$.

TABLE 3.2
Solvable equations of the form
$$y'''_{xxx} = Ae^y (y'_x)^\gamma (y''_{xx})^\delta$$

δ	γ	Equation
arbitrary ($\delta \neq 1$)	1	3.3.2.1
0	3	3.3.2.9
1	1	3.3.2.2
$\frac{3}{2}$	0	3.3.2.3
$\frac{3}{2}$	3	3.3.2.7
2	arbitrary ($\gamma \neq -1$)	3.3.2.11
2	-1	3.3.2.13

TABLE 3.3
Solvable equations of the form
$$y'''_{xxx} = Ay^\beta y'_x \exp[(y'_x)^2](y''_{xx})^\delta$$

δ	β	Equation
arbitrary ($\delta \neq 2$)	0	3.3.2.4
1	arbitrary ($\beta \neq -1$)	3.3.2.12
1	-1	3.3.2.14
$\frac{3}{2}$	$-\frac{1}{2}$	3.3.2.5
$\frac{3}{2}$	1	3.3.2.8
2	0	3.3.2.6
3	1	3.3.2.10

TABLE 3.4
Other solvable equations of the type considered

Form of equation	Equation
$y'''_{xxx} = Ae^y y'_x \exp[(y'_x)^2](y''_{xx})^\delta$	3.3.2.20
$y'''_{xxx} = A(y'_x)^\gamma \exp(y''_{xx}), \quad \gamma \neq -1$	3.3.2.15
$y'''_{xxx} = A(y'_x)^{-1} \exp(y''_{xx})$	3.3.2.16
$y'''_{xxx} = Ay^\beta y'_x \exp(y''_{xx}), \quad \beta \neq -1$	3.3.2.17
$y'''_{xxx} = Ay^{-1} y'_x \exp(y''_{xx})$	3.3.2.18
$y'''_{xxx} = Ae^y y'_x \exp[(y'_x)^2 + y''_{xx}]$	3.3.2.19

3.3.2. Equations Containing Exponential Functions

Tables 3.2–3.4 represent the equations whose solutions are given in this subsection.

▶ *In the solutions of equations 1–6, the following notation is used:*

$$E = \int \exp(\tau^2)\,d\tau + C_2, \qquad G = \int \tau^{-1}(1 \pm \tau)^k\,d\tau + C_2,$$
$$F = 2\tau E - \exp(\tau^2), \qquad H = \int \tau^{-1} \exp(\mp\tau)\,d\tau + C_2.$$

1. $y'''_{xxx} = Ae^y y'_x (y''_{xx})^\delta, \qquad \delta \neq 1.$

Solution in the parametric form:

$$x = aC_1 \int \tau^{-1} G^{-1/2}\,d\tau + C_3, \qquad y = \ln(bC_1^{2\delta-2}\tau),$$

where $k = \dfrac{1}{1-\delta}$, $A = \pm\dfrac{1}{2(1-\delta)} a^{-2} b^{-1} (2a^2)^\delta$.

2. $y'''_{xxx} = Ae^y y'_x y''_{xx}$.

Solution in the parametric form:

$$x = C_1 \int \tau^{-1} H^{-1/2}\, d\tau + C_3, \qquad y = \ln\Bigl(\mp\frac{\tau}{A}\Bigr).$$

3. $y'''_{xxx} = Ae^y (y''_{xx})^{3/2}$.

Solution in the parametric form:

$$x = C_1 \int E^{-1}\, d\tau + C_3, \qquad y = \tau^2 + \ln\bigl(\sqrt{2}\, A^{-1} E^{-1}\bigr).$$

4. $y'''_{xxx} = Ay'_x \exp\bigl[(y'_x)^2\bigr](y''_{xx})^\delta, \qquad \delta \neq 2.$

Solution in the parametric form:

$$x = aC_1 \int \tau^{-1}(1 \pm \tau)^{\frac{1}{\delta-2}} \bigl[\ln(bC_1^{\delta-2}\tau)\bigr]^{-1/2}\, d\tau + C_3, \qquad y = aC_1 G,$$

where $k = \dfrac{1}{\delta - 2}$, $A = \pm\dfrac{1}{2-\delta} a^{\delta-3} b^{-1}$.

5. $y'''_{xxx} = Ay^{-1/2} y'_x \exp\bigl[(y'_x)^2\bigr](y''_{xx})^{3/2}$.

Solution in the parametric form:

$$x = 4C_1 \int EF\bigl[\tau^2 + \ln(aE^{-1})\bigr]^{-1/2}\, d\tau + C_3, \quad y = C_1 F^2, \qquad A = -\sqrt{2}\, a^{-1}.$$

6. $y'''_{xxx} = Ay'_x \exp\bigl[(y'_x)^2\bigr](y''_{xx})^2$.

Solution in the parametric form:

$$x = C_1 \int \tau^{-1} \exp(\mp\tau)\Bigl[\ln\Bigl(\pm\frac{\tau}{A}\Bigr)\Bigr]^{-1/2}\, d\tau + C_3, \qquad y = C_1 H.$$

▶ *In the solutions of equations 7–8, the following notation is used:*

$$E = \sqrt{\tau(\tau+1)} - \ln\bigl[C_2\bigl(\sqrt{\tau} + \sqrt{\tau+1}\bigr)\bigr], \qquad R = \sqrt{\frac{\tau+1}{\tau}},$$

$$F = 1 - \sqrt{\frac{\tau+1}{\tau}}\ln\bigl[C_2\bigl(\sqrt{\tau} + \sqrt{\tau+1}\bigr)\bigr].$$

7. $y'''_{xxx} = Ae^y (y'_x)^3 (y''_{xx})^{3/2}$.

Solution in the parametric form:

$$x = aC_1 \int R^{-1} E^{-1} F^{-1/2}\, d\tau + C_3, \qquad y = -\ln(bC_1^{-3} E),$$

where $A = 2^{3/2} a^3 b$.

8. $y'''_{xxx} = Ayy'_x \exp\bigl[(y'_x)^2\bigr](y''_{xx})^{3/2}$.

Solution in the parametric form:

$$x = -\tfrac{1}{2} bC_1 \int \tau^{-2} R^{-1} E\bigl[\ln(aC_1^{-3/2} E^{-1})\bigr]^{-1/2}\, d\tau + C_3, \qquad y = bC_1 F,$$

where $A = -4a^{-1} b^{-3/2}$.

▶ *In the solutions of equations 9–10, the following notation is used:*

$$Z = \begin{cases} C_1 J_0(\tau) + C_2 Y_0(\tau) & \text{for the upper sign,} \\ C_1 I_0(\tau) + C_2 K_0(\tau) & \text{for the lower sign,} \end{cases}$$

where J_0 and Y_0 are Bessel functions, I_0 and K_0 are modified Bessel functions.

9. $y'''_{xxx} = Ae^y (y'_x)^3.$

Solution in the parametric form:

$$x = 2C_1 \int \tau^{-1} Z^{-1/2}\, d\tau + C_3, \qquad y = \ln(\mp\tfrac{1}{8}A^{-1}\tau^2).$$

10. $y'''_{xxx} = Ayy'_x \exp\big[(y'_x)^2\big](y''_{xx})^3.$

Solution in the parametric form:

$$x = C_1 \int Z'_\tau \left[\ln\left(\pm\frac{\tau^2}{A}\right)\right]^{-1/2} d\tau + C_3, \qquad y = C_1 Z.$$

11. $y'''_{xxx} = Ae^y (y'_x)^\gamma (y''_{xx})^2, \qquad \gamma \neq -1.$

Solution in the parametric form:

$$x = a \int \tau^{-1/2} f^{-1} \exp\left(\frac{U}{\gamma+1}\right) d\tau + C_3, \qquad y = U,$$

where $A = \frac{1}{2}a^{\gamma+1}k$, $U = \int \frac{d\tau}{f} + C_1$, $f = f(\tau)$ is the solution of the transcendental equation

$$\ln(\lambda f - \tau) - \frac{\tau}{\lambda f - \tau} = \frac{k}{\lambda}\tau^\lambda + C_2, \qquad \lambda = \frac{\gamma+1}{2}.$$

12. $y'''_{xxx} = Ay^\beta y'_x \exp\big[(y'_x)^2\big] y''_{xx}, \qquad \beta \neq -1.$

Solution in the parametric form:

$$x = a \int f^{-1}\left(f - \frac{\tau}{\beta+1}\right) U^{-1/2} \exp\left(-\frac{U}{\beta+1}\right) d\tau + C_3, \qquad y = a\tau \exp\left(-\frac{U}{\beta+1}\right),$$

where $A = a^{-\beta-1}k$, $U = \int \frac{d\tau}{f} + C_1$, $f = f(\tau)$ is the solution of the transcendental equation

$$\ln(\lambda f - \tau) - \frac{\tau}{\lambda f - \tau} = -\frac{k}{\lambda}\tau^\lambda + C_2, \qquad \lambda = \beta + 1.$$

13. $y'''_{xxx} = Ae^y (y'_x)^{-1} (y''_{xx})^2.$

Solution in the parametric form:

$$x = C_1 \int W^{-1/2}\, d\tau + C_3, \qquad y = \tau,$$

where $W = \exp\left(\int \frac{d\tau}{\tau - 2Ae^\tau + C_2}\right).$

14. $y'''_{xxx} = Ay^{-1}y'_x \exp\left[(y'_x)^2\right]y''_{xx}$.

Solution in the parametric form:

$$x = C_1 \int \tau^{-1/2}(\tau + Ae^\tau + C_2)^{-1}W\,d\tau + C_3, \qquad y = C_1 W,$$

where $W = \exp\left(\int \frac{d\tau}{\tau + Ae^\tau + C_2}\right)$.

▶ *In the solutions of equations 15–19, the following notation is used:*

$$V = C_1 - \frac{1}{\lambda}(m+1)(\tau+1)e^{-\tau}, \qquad M = C_1 - \frac{1}{\lambda}(\tau+2)e^{-\tau/2},$$

$$W = \begin{cases} C_2 - \int \ln\left(C_1 - \frac{\lambda}{n+1}\tau^{n+1}\right) d\tau & \text{if } n \neq -1, \\ C_2 - \int \ln(C_1 - \lambda \ln|\tau|)\,d\tau & \text{if } n = -1, \end{cases}$$

$$N = \ln M - \frac{1}{2\lambda}\int e^{-\tau/2}M^{-1}\,d\tau - \frac{1}{2} + C_2.$$

15. $y'''_{xxx} = A(y'_x)^\gamma \exp(y''_{xx})$, $\quad \gamma \neq -1$.

Solution in the parametric form:

$$x = \frac{1}{\lambda}\int e^{-\tau}V^{-\frac{2m+1}{2m+2}}\,d\tau + C_3, \quad y = \frac{2}{\lambda}\int e^{-\tau}V^{-\frac{m}{m+1}}\,d\tau + C_2,$$

where $m = \frac{1}{2}(\gamma - 1)$, $A = 2^{-\gamma}\lambda$.

16. $y'''_{xxx} = A(y'_x)^{-1}\exp(y''_{xx})$.

Solution in the parametric form:

$$x = \frac{1}{\lambda}\int \exp\left(-\tau + \frac{1}{2}V\right)d\tau + C_3, \qquad y = \frac{2}{\lambda}\int \exp(-\tau + V)\,d\tau + C_2,$$

where $m = 0$, $A = 2\lambda$.

17. $y'''_{xxx} = Ay^\beta y'_x \exp(y''_{xx})$, $\quad \beta \neq -1$.

Solution in the parametric form:

$$x = \int W^{-1/2}\,d\tau + C_3, \qquad y = 2\tau,$$

where $n = \beta$, $A = 2^{-\beta-1}\lambda$.

18. $y'''_{xxx} = Ay^{-1}y'_x \exp(y''_{xx})$.

Solution in the parametric form:

$$x = \int W^{-1/2}\,d\tau + C_3, \qquad y = 2\tau,$$

where $n = -1$, $A = \lambda$.

19. $y'''_{xxx} = Ae^y y'_x \exp\left[(y'_x)^2 + y''_{xx}\right]$.

Solution in the parametric form:

$$x = \frac{1}{2\lambda}\int e^{-\tau/2}M^{-1}N^{-1/2}\,d\tau + C_3, \qquad y = \frac{1}{2\lambda}\int e^{-\tau/2}M^{-1}\,d\tau + C_2,$$

where $A = \lambda$.

20. $y'''_{xxx} = Ae^y y'_x \exp[(y'_x)^2](y''_{xx})^\delta$.

Solution in the parametric form:

$$x = \int \tau^{-\delta}z^{-1}\left(\ln\frac{z}{A} - U - C_1\right)^{-1/2} d\tau + C_3, \qquad y = U,$$

where

$$U = \int \frac{d\tau}{z\tau^\delta}, \qquad z = \begin{cases} \dfrac{1}{2-\delta}\tau^{2-\delta} + \dfrac{1}{1-\delta}\tau^{1-\delta} + C_2 & \text{if } \delta \neq 2,\ \delta \neq 1, \\ \tau + \ln|\tau| + C_2 & \text{if } \delta = 1, \\ \ln|\tau| - \dfrac{1}{\tau} + C_2 & \text{if } \delta = 2. \end{cases}$$

3.3.3. Other Equations

1. $y'''_{xxx} = Ayy'_x\{\cosh[\omega(y'_x)^2]\}^{-2}y''_{xx}$.

The substitution $y'_x = \sqrt{u(y)}$ leads to a second order equation of the form 2.7.4.1: $u''_{yy} = Ay[\cosh(\omega u)]^{-2}u'_y$.

2. $y'''_{xxx} = Ayy'_x\{\sinh[\omega(y'_x)^2]\}^{-2}y''_{xx}$.

The substitution $y'_x = \sqrt{u(y)}$ leads to a second order equation of the form 2.7.4.2: $u''_{yy} = Ay[\sinh(\omega u)]^{-2}u'_y$.

3. $y'''_{xxx} = Ayy'_x\cosh[\omega(y'_x)^2]\,(y''_{xx})^{3/2}$.

The substitution $y'_x = \sqrt{u(y)}$ leads to a second order equation of the form 2.7.4.3: $u''_{yy} = \dfrac{A}{\sqrt{2}}y\cosh(\omega u)(u'_y)^{3/2}$.

4. $y'''_{xxx} = Ayy'_x\sinh[\omega(y'_x)^2]\,(y''_{xx})^{3/2}$.

The substitution $y'_x = \sqrt{u(y)}$ leads to a second order equation of the form 2.7.4.4: $u''_{yy} = \dfrac{A}{\sqrt{2}}y\sinh(\omega u)(u'_y)^{3/2}$.

5. $y'''_{xxx} = A\cosh(\omega y)\,(y'_x)^3(y''_{xx})^{3/2}$.

The substitution $y'_x = \sqrt{u(y)}$ leads to a second order equation of the form 2.7.4.5: $u''_{yy} = \dfrac{A}{\sqrt{2}}\cosh(\omega y)u(u'_y)^{3/2}$.

6. $y'''_{xxx} = A\sinh(\omega y)\,(y'_x)^3(y''_{xx})^{3/2}.$

The substitution $y'_x = \sqrt{u(y)}$ leads to a second order equation of the form 2.7.4.6: $u''_{yy} = \dfrac{A}{\sqrt{2}}\sinh(\omega y)u(u'_y)^{3/2}$.

7. $y'''_{xxx} = A[\cosh(\omega y)]^{-2}(y'_x)^3(y''_{xx})^2.$

The substitution $y'_x = \sqrt{u(y)}$ leads to a second order equation of the form 2.7.4.7: $u''_{yy} = \frac{1}{2}A[\cosh(\omega y)]^{-2}u(u'_y)^2$.

8. $y'''_{xxx} = A[\sinh(\omega y)]^{-2}(y'_x)^3(y''_{xx})^2.$

The substitution $y'_x = \sqrt{u(y)}$ leads to a second order equation of the form 2.7.4.8: $u''_{yy} = \frac{1}{2}A[\sinh(\omega y)]^{-2}u(u'_y)^2$.

9. $y'''_{xxx} = A\cosh^n(\omega y)\,y'_x(y''_{xx})^m.$

This is a special case of equation 3.5.4.15 with $f(y) = A\cosh^n(\omega y)$.

10. $y'''_{xxx} = A\sinh^n(\omega y)\,y'_x(y''_{xx})^m.$

This is a special case of equation 3.5.4.15 with $f(y) = A\sinh^n(\omega y)$.

11. $y'''_{xxx} = A\tanh^n(\omega y)\,y'_x(y''_{xx})^m.$

This is a special case of equation 3.5.4.15 with $f(y) = A\tanh^n(\omega y)$.

12. $y'''_{xxx} = A\coth^n(\omega y)\,y'_x(y''_{xx})^m.$

This is a special case of equation 3.5.4.15 with $f(y) = A\coth^n(\omega y)$.

13. $y'''_{xxx} = A\ln^n(\omega y)\,y'_x(y''_{xx})^m.$

This is a special case of equation 3.5.4.15 with $f(y) = A\ln^n(\omega y)$.

14. $y'''_{xxx} = Ayy'_x\{\cos[\omega(y'_x)^2]\}^{-2}y''_{xx}.$

The substitution $y'_x = \sqrt{u(y)}$ leads to a second order equation of the form 2.7.5.1: $u''_{yy} = Ay[\cos(\omega u)]^{-2}u'_y$.

15. $y'''_{xxx} = Ayy'_x\{\sin[\omega(y'_x)^2]\}^{-2}y''_{xx}.$

The substitution $y'_x = \sqrt{u(y)}$ leads to a second order equation of the form 2.7.5.2: $u''_{yy} = Ay[\sin(\omega u)]^{-2}u'_y$.

16. $y'''_{xxx} = A[\cos(\omega y)]^{-2}(y'_x)^3(y''_{xx})^2.$

The substitution $y'_x = \sqrt{u(y)}$ leads to a second order equation of the form 2.7.5.3: $u''_{yy} = \frac{1}{2}A[\cos(\omega y)]^{-2}u(u'_y)^2$.

17. $y'''_{xxx} = A[\sin(\omega y)]^{-2}(y'_x)^3(y''_{xx})^2.$

The substitution $y'_x = \sqrt{u(y)}$ leads to a second order equation of the form 2.7.5.4: $u''_{yy} = \frac{1}{2}A[\sin(\omega y)]^{-2}u(u'_y)^2.$

18. $y'''_{xxx} = Ayy'_x\cos[\omega(y'_x)^2]\,(y''_{xx})^{3/2}.$

The substitution $y'_x = \sqrt{u(y)}$ leads to a second order equation of the form 2.7.5.5: $u''_{yy} = \dfrac{A}{\sqrt{2}}y\cos(\omega u)(u'_y)^{3/2}.$

19. $y'''_{xxx} = Ayy'_x\sin[\omega(y'_x)^2]\,(y''_{xx})^{3/2}.$

The substitution $y'_x = \sqrt{u(y)}$ leads to a second order equation of the form 2.7.5.6: $u''_{yy} = \dfrac{A}{\sqrt{2}}y\sin(\omega u)(u'_y)^{3/2}.$

20. $y'''_{xxx} = A\cos(\omega y)\,(y'_x)^3(y''_{xx})^{3/2}.$

The substitution $y'_x = \sqrt{u(y)}$ leads to a second order equation of the form 2.7.5.7: $u''_{yy} = \dfrac{A}{\sqrt{2}}\cos(\omega y)u(u'_y)^{3/2}.$

21. $y'''_{xxx} = A\sin(\omega y)\,(y'_x)^3(y''_{xx})^{3/2}.$

The substitution $y'_x = \sqrt{u(y)}$ leads to a second order equation of the form 2.7.4.8: $u''_{yy} = \dfrac{A}{\sqrt{2}}\sin(\omega y)u(u'_y)^{3/2}.$

22. $y'''_{xxx} = A\cos^n(\omega y)\,y'_x(y''_{xx})^m.$

This is a special case of equation 3.5.4.15 with $f(y) = A\cos^n(\omega y)$.

23. $y'''_{xxx} = A\sin^n(\omega y)\,y'_x(y''_{xx})^m.$

This is a special case of equation 3.5.4.15 with $f(y) = A\sin^n(\omega y)$.

24. $y'''_{xxx} = A\tan^n(\omega y)\,y'_x(y''_{xx})^m.$

This is a special case of equation 3.5.4.15 with $f(y) = A\tan^n(\omega y)$.

25. $y'''_{xxx} = A\cot^n(\omega y)\,y'_x(y''_{xx})^m.$

This is a special case of equation 3.5.4.15 with $f(y) = A\cot^n(\omega y)$.

26. $y'''_{xxx} = A(\arcsin y)^n y'_x(y''_{xx})^m.$

This is a special case of equation 3.5.4.15 with $f(y) = A(\arcsin y)^n$.

27. $y'''_{xxx} = A(\arctan y)^n y'_x(y''_{xx})^m.$

This is a special case of equation 3.5.4.15 with $f(y) = A(\arctan y)^n$.

3.4. Some Nonlinear Equations with Arbitrary Parameters

3.4.1. Equations Containing Power Functions

1. $y'''_{xxx} = ay^{-5/2} + by^{-7/2}$.

Using the transformation given in 3.5.2.21, we reduce this equation to a nonhomogeneous constant-coefficient linear equation.

Solution in the parametric form ($b \neq 0$):

$$x = \int \frac{d\tau}{[\varphi(\tau)]^{3/2}} + C_3, \quad y = \frac{1}{\varphi(\tau)},$$

where

$$\varphi = -\frac{a}{b} + C_1 e^{-k\tau} + C_2 e^{k\tau/2} \sin\frac{k\tau\sqrt{3}}{2}, \qquad k = b^{1/3}.$$

2. $y'''_{xxx} = axy^{-5/2} + bx^3y^{-7/2}$.

The transformation $x = 1/t$, $y = w/t^2$ leads to an equation of the form 3.4.1.1: $w'''_{ttt} = -aw^{-5/2} - bw^{-7/2}$.

3. $y'''_{xxx} = (y + ax^2 + bx + c)^n$.

The substitution $z = y + ax^2 + bx + c$ leads to the equation $z'''_{xxx} = z^n$ whose solvable cases are outlined in Section 3.2.

4. $y'''_{xxx} = (ax + b)^n (cx + d)^{-n-2m-4} y^m$.

The transformation $\xi = \dfrac{ax+b}{cx+d}$, $w = \dfrac{y}{(cx+d)^2}$ leads to a simpler equation (see Section 3.2):

$$w'''_{\xi\xi\xi} = \Delta^{-3}\xi^n w^m, \qquad \text{where} \quad \Delta = ad - bc.$$

5. $y'''_{xxx} = x(ay^2 + bx^2y + cx^4)^{-5/4}$.

This is a special case of equation 3.5.2.18 with $f(\xi) \equiv 1$.

6. $y'''_{xxx} = bx^{2n-1}(x - a)^{-3}y^{-n}$.

This is a special case of equation 3.5.1.9 with $f(\xi) = b\xi^{-n}$.

7. $y'''_{xxx} = ax^{-n-2}y^n y'_x - ax^{-n-3}y^{n+1}$.

This is a special case of equation 3.5.2.2 with $f(\xi) = a\xi^n$.

8. $y'''_{xxx} = ax^{-2n-4}y^n y'_x - 2ax^{-2n-5}y^{n+1}$.

This is a special case of equation 3.5.2.3 with $f(\xi) = a\xi^n$.

9. $y'''_{xxx} = \lambda y^{-3}y'_x + ay^{-5/2} + by^{-7/2}$.

The transformation $x = \int [\varphi(\tau)]^{-3/2}\, d\tau$, $y = [\varphi(\tau)]^{-1}$ leads to a constant coefficient equation: $\varphi'''_{\tau\tau\tau} - \lambda\varphi'_\tau + b\varphi + a = 0$.

10. $y'''_{xxx} = -x^{-2}y'_x + x^{-3}y + ax^{1/2}y^{-5/2}.$

This is a special case of equation 3.5.2.20 with $f(\xi) = a$.

11. $y'''_{xxx} = -x^{-2}y'_x + x^{-3}y + ax^{-3/4}y^{-5/4}.$

This is a special case of equation 3.5.2.19 with $f(\xi) = a$.

12. $y'''_{xxx} = ay^n y'_x + by^m (y'_x)^3.$

This is a special case of equation 3.5.2.4 with $f(y) = ay^n$, $g(y) = by^m$.

13. $y'''_{xxx} = by^n (y'_x)^3 + a(y'_x)^{-5}.$

This is a special case of equation 3.5.2.9 with $f(y) = by^n$.

14. $y'''_{xxx} = (ax^2 + bx + c)^{-\frac{m+5}{4}} (y'_x)^m.$

This is a special case of equation 3.5.2.17 with $f(\xi) = \xi^m$.

15. $y'''_{xxx} = ax(xy'_x - y)^n.$

This is a special case of equation 3.5.2.13 with $f(\xi) = a\xi^n$.

16. $y'''_{xxx} = ax^{-n-2}(xy'_x - y)^n.$

This is a special case of equation 3.5.2.14 with $f(\xi) = a\xi^n$.

17. $y'''_{xxx} = ax^{-n-4}(xy'_x - 2y)^n.$

This is a special case of equation 3.5.2.16 with $f(\xi) = a\xi^n$.

18. $y'''_{xxx} = ax^{2n-7}y^{-n}(xy'_x - 2y)^3.$

This is a special case of equation 3.5.2.6 with $f(\xi) = a\xi^{-n}$.

19. $y'''_{xxx} = ax^{n-5}y^{-n}(xy'_x - y)^3.$

This is a special case of equation 3.5.2.5 with $f(\xi) = a\xi^{-n}$.

20. $y'''_{xxx} = ax^n y^m (xy'_x - 2y)^l.$

The transformation $t = x^{-1}$, $z = yx^{-2}$ leads to the equation

$$z'''_{ttt} = -a(-1)^l t^{-n-2m-l-4} z^m (z'_t)^l$$

which is discussed in Section 3.2.

21. $xy'''_{xxx} + 3y''_{xx} = ax^n y^n.$

The substitution $w(x) = xy$ leads to the equation $w'''_{xxx} = aw^n$ which is discussed in Section 3.2.

22. $xy'''_{xxx} = -\frac{3}{2}y''_{xx} + ax^{-n-2}y^{2n}(2xy'_x - y).$

This is a special case of equation 3.5.3.13 with $f(\xi) = a\xi^{2n}$.

23. $xy'''_{xxx} = -\frac{3}{2}y''_{xx} + ax^{-n-3}y^{2n}(2xy'_x - y)^3.$

This is a special case of equation 3.5.3.16 with $f(\xi) = a\xi^{2n}$.

24. $xy'''_{xxx} + y''_{xx} = ax^{-n-3}(xy'_x - y)^n.$

This is a special case of equation 3.5.3.15 with $f(\xi) = a\xi^n$.

25. $xy'''_{xxx} + (1-a)y''_{xx} = bx^{2a}(xy'_x - y)^n.$

This is a special case of equation 3.5.3.6 with $f(\xi) = b\xi^n$.

26. $x^2y'''_{xxx} + 6xy''_{xx} + 6y'_x = ax^{2n}y^n.$

The substitution $w(x) = x^2 y$ leads to the equation $w'''_{xxx} = aw^n$ which is discussed in Section 3.2.

27. $yy'''_{xxx} + \frac{1}{2}y'_x y''_{xx} = ax + b.$

The transformation $x = x(t)$, $y = (x'_t)^2$ leads to a nonhomogeneous constant-coefficient fourth-order linear equation of the form 4.1.2.2: $2x''''_{tttt} = ax + b$.

28. $yy'''_{xxx} + \frac{1}{2}y'_x y''_{xx} = Ax^{-5/3}.$

The transformation $x = x(t)$, $y = (x'_t)^2$ leads to the equation of the form 4.2.1.1: $2x''''_{tttt} = Ax^{-5/3}$.

29. $yy'''_{xxx} + \frac{1}{2}y'_x y''_{xx} = k\sqrt{y}\,y''_{xx} + my'_x + a\sqrt{y} + bx + c.$

The transformation $x = x(t)$, $y = (x'_t)^2$ leads to a constant coefficient linear equation of the fourth order: $2x''''_{tttt} = \pm 2kx'''_{ttt} + 2mx''_{tt} \pm ax'_t + bx + c$, where "+" corresponds to $x'_t > 0$, and "−" corresponds to $x'_t < 0$.

30. $yy'''_{xxx} + 3y'_x y''_{xx} + ax^n yy'_x = bx^m.$

This is a special case of equation 3.5.3.19 with $f(x) = ax^n$, $g(x) = bx^m$.

31. $yy'''_{xxx} + 3y'_x y''_{xx} + a\bigl[yy''_{xx} + (y'_x)^2\bigr] = bx^n.$

This is a special case of equation 3.5.3.20 with $f(x) = bx^n$.

32. $yy'''_{xxx} + (3y'_x + 2ay)y''_{xx} + 2a(y'_x)^2 + a^2yy'_x = bx^n.$

This is a special case of equation 3.5.3.33 with $f(x) = e^{ax}$, $g(x) = bx^n e^{ax}$.

33. $yy'''_{xxx} + (3y'_x + ax^n y)y''_{xx} + ax^n(y'_x)^2 = 0.$

This is a special case of equation 3.5.3.21 with $f(x) = ax^n$.

34. $(y+a)y'''_{xxx} + by'_x y''_{xx} + cy^n y'_x = 0.$

This is a special case of equation 3.5.3.25 with $f(y) = cy^n$.

35. $(y+ax+b)y'''_{xxx}+3(y'_x+a)y''_{xx}=cx^n.$

This is a special case of equation 3.5.3.27 with $f(x)=cx^n$.

36. $x(yy'''_{xxx}+3y'_xy''_{xx})+a\big[yy''_{xx}+(y'_x)^2\big]=bx^n.$

This is a special case of equation 3.5.3.28 with $f(x)=bx^n$.

37. $x^2yy'''_{xxx}+x(3xy'_x+2ay)y''_{xx}+2ax(y'_x)^2+a(a-1)yy'_x=bx^n.$

This is a special case of equation 3.5.3.33 with $f(x)=x^a$, $g(x)=bx^{n+a-2}$.

38. $y^2y'''_{xxx}-3yy'_xy''_{xx}+2(y'_x)^3=ax^ny^3.$

This is a special case of equation 3.5.3.29 with $f(x)=ax^n$.

39. $y^2y'''_{xxx}+3myy'_xy''_{xx}+m(m-1)(y'_x)^3=ax^ky^{2-m}.$

This is a special case of equation 3.5.3.30 with $f(x)=ax^k$, $n=m+1$.

40. $2y'_xy'''_{xxx}-(y''_{xx})^2=\lambda(y'_x)^2+ay^2+by+c.$

Differentiating with respect to x and dividing by y'_x, we arrive at a constant coefficient linear equation: $2y''''_{xxxx}=2\lambda y''_{xx}+2ay+b$.

41. $2y'_xy'''_{xxx}-3(y''_{xx})^2=ay^{4-2n}(y'_x)^n.$

This is a special case of equation 3.5.4.6 with $f(\xi)=a\xi^n$.

42. $2y'_xy'''_{xxx}-3(y''_{xx})^2=ax^{2n-8}y^{4-2n}(y'_x)^n.$

This is a special case of equation 3.5.4.8 with $f(\xi)=a\xi^n$.

43. $2y'_xy'''_{xxx}-3(y''_{xx})^2=ax^{n-4}y^{4-2n}(y'_x)^n.$

This is a special case of equation 3.5.4.7 with $f(\xi)=a\xi^n$.

44. $2y'_xy'''_{xxx}-3(y''_{xx})^2=ax^n(y'_x)^2+by^m(y'_x)^4.$

This is a special case of equation 3.5.4.3 with $f(x)=ax^n$, $g(y)=by^m$.

45. $2y'_xy'''_{xxx}-3(y''_{xx})^2=ay^n(y'_x)^4+bx^{-1}y^m(y'_x)^{7/2}.$

This is a special case of equation 3.5.4.5 with $f(y)=ay^n$, $g(y)=by^m$.

46. $2y'_xy'''_{xxx}-3(y''_{xx})^2=ax^n(y'_x)^2+bx^my^{-1}(y'_x)^{5/2}.$

This is a special case of equation 3.5.4.4 with $f(x)=ax^n$, $g(x)=bx^m$.

47. $2y'_xy'''_{xxx}-(y''_{xx})^2=\lambda x^n(y'_x)^2+ay^2+2by+c.$

This is a special case of equation 3.5.4.1 with $f(x)=-\lambda x^n$.

48. $xy'_x y'''_{xxx} - 3x(y''_{xx})^2 + 3y'_x y''_{xx} = axy^n (y'_x)^4 + by^m (y'_x)^5.$

This is a special case of equation 3.5.4.10 with $f(y) = ay^n$, $g(y) = by^m$.

49. $y'''_{xxx} = ax^{-2n-5}(xy'_x - y)^n (y''_{xx})^3.$

This is a special case of equation 3.5.4.13 with $f(\xi) = a\xi^n$.

50. $y'''_{xxx} = ax^{-4n-5}(xy'_x - y)^n (y''_{xx})^3.$

This is a special case of equation 3.5.4.12 with $f(\xi) = a\xi^n$.

51. $y'''_{xxx} = \left[ax^{-5} + bx^3(xy'_x - y)^n\right](y''_{xx})^3.$

This is a special case of equation 3.5.4.11 with $f(\xi) = b\xi^n$.

52. $y'''_{xxx} = \left[ax(y'_x)^n + by(y'_x)^m + c(y'_x)^k\right](y''_{xx})^3 + s(y'_x)^l (y''_{xx})^2.$

This is a special case of equation 3.5.4.14 with $f(\xi) = a\xi^n$, $g(\xi) = b\xi^m$, $h(\xi) = c\xi^k$, $\varphi(\xi) = s\xi^l$.

53. $xy'''_{xxx} + y''_{xx} = ax^n (xy'_x - y)^m (y''_{xx})^n.$

This is a special case of equation 3.5.5.5 with $f(\xi) = a\xi^m$, $g(\xi) = \xi^n$.

54. $y'''_{xxx} = Ax^n (y'_x)^m (xy'_x - y)^l (y''_{xx})^k.$

The Legendre transformation $x = w'_t$, $y = tw'_t - w$ leads to the equation $w'''_{ttt} = -At^m w^l (w'_t)^n (w''_{tt})^{3-k}$ which is discussed in Section 3.2.

55. $y'''_{xxx} = ax(xy'_x - y)^n (y''_{xx})^2 + bx(xy'_x - y)^m (y''_{xx})^k.$

This is a special case of equation 3.5.4.16 with $f(\xi) = a\xi^n$, $g(\xi) = b\xi^m$.

56. $yy'''_{xxx} = y'_x y''_{xx} + ay^{-n-m}(y'_x)^n (y''_{xx})^m \left[yy''_{xx} - (y'_x)^2\right].$

This is a special case of equation 3.5.5.7 with $f(\xi) = a\xi^n$, $g(\xi) = \xi^m$.

57. $(y'''_{xxx})^2 = a(x^2 y''_{xx} - 2xy'_x + 2y) + by''_{xx} + c.$

Differentiating with respect to x, we obtain

$$y'''_{xxx}(2y''''_{xxxx} - ax^2 - b) = 0$$

Equating the second factor to zero and integrating, we find the solution

$$y = \frac{ax^6}{720} + \frac{bx^4}{48} + C_3 x^3 + C_2 x^2 + C_1 x + C_0,$$

The integration constants C_i and parameters a, b, and c are related by

$$36C_3^2 = 2aC_0 + 2bC_2 + c.$$

This constraint is obtained by substituting the above solution into the original equation. In addition, to the first factor corresponds the solution $y = \widetilde{C}_2 x^2 + \widetilde{C}_1 x + \widetilde{C}_0$, where constants $\widetilde{C}_i$ are related by $2a\widetilde{C}_0 + 2b\widetilde{C}_2 + c = 0$.

3.4.2. Equations Containing Exponential Functions

1. $y'''_{xxx} = a(y + be^x + c)^n - be^x.$

The substitution $w = y + be^x + c$ leads to the equation $w'''_{xxx} = aw^n$ whose solvable cases are outlined in Section 3.2.

2. $y'''_{xxx} = ae^{\lambda y}y'_x + be^{\mu y}(y'_x)^3.$

This is a special case of equation 3.5.2.4 with $f(y) = ae^{\lambda y}$, $g(y) = be^{\mu y}$.

3. $y'''_{xxx} = ae^{\lambda y}y'_x + by^n(y'_x)^3.$

This is a special case of equation 3.5.2.4 with $f(y) = ae^{\lambda y}$, $g(y) = by^n$.

4. $y'''_{xxx} = ay^n y'_x + be^{\lambda y}(y'_x)^3.$

This is a special case of equation 3.5.2.4 with $f(y) = ay^n$, $g(y) = be^{\lambda y}$.

5. $y'''_{xxx} = be^{\lambda y}(y'_x)^3 + a(y'_x)^{-5}.$

This is a special case of equation 3.5.2.9 with $f(y) = be^{\lambda y}$.

6. $y'''_{xxx} = 2\lambda^2(y'_x)^3 + ae^{\lambda m y}(y'_x)^{m-5}.$

This is a special case of equation 3.5.2.23 with $f(\xi) = a\xi^{m-6}$.

7. $y'''_{xxx} = -3y''_{xx} + ae^{mx}y^m y'_x + ae^{mx}y^{m+1} + 2y.$

This is a special case of equation 3.5.3.3 with $f(\xi) = a\xi^m$.

8. $y'''_{xxx} + 3\lambda y'_x y''_{xx} + \lambda^2(y'_x)^3 = ae^{\beta x - \lambda y}.$

This is a special case of equation 3.5.3.2 with $f(x) = ae^{\beta x}$.

9. $y'''_{xxx} + 3\lambda y'_x y''_{xx} + \lambda^2(y'_x)^3 = ax^n e^{-\lambda y}.$

This is a special case of equation 3.5.3.2 with $f(x) = ax^n$.

10. $xy'''_{xxx} + (1 - ax)y''_{xx} = be^{2ax}(xy'_x - y)^n.$

This is a special case of equation 3.5.3.7 with $f(\xi) = b\xi^n$.

11. $yy'''_{xxx} + 3y'_x y''_{xx} = ae^{\lambda x}.$

Solution: $y^2 = C_2x^2 + C_1x + C_0 + 2a\lambda^{-3}e^{\lambda x}.$

12. $yy'''_{xxx} + 3y'_x y''_{xx} + ae^{\lambda x}yy'_x = be^{\mu x}.$

This is a special case of equation 3.5.3.19 with $f(x) = ae^{\lambda x}$, $g(x) = be^{\mu x}$.

13. $yy'''_{xxx} + 3y'_x y''_{xx} + ae^{\lambda x}yy'_x = bx^n.$

This is a special case of equation 3.5.3.19 with $f(x) = ae^{\lambda x}$, $g(x) = bx^n$.

14. $yy'''_{xxx} + 3y'_x y''_{xx} + ax^n yy'_x = be^{\lambda x}.$

This is a special case of equation 3.5.3.19 with $f(x) = ax^n$, $g(x) = be^{\lambda x}$.

15. $yy'''_{xxx} + 3y'_x y''_{xx} + a\big[yy''_{xx} + (y'_x)^2\big] = be^{\lambda x}.$

This is a special case of equation 3.5.3.20 with $f(x) = be^{\lambda x}$.

16. $yy'''_{xxx} + (3y'_x + 2ay)y''_{xx} + 2a(y'_x)^2 + a^2 yy'_x = be^{\lambda x}.$

This is a special case of equation 3.5.3.33 with $f(x) = e^{ax}$, $g(x) = be^{(\lambda+a)x}$.

17. $yy'''_{xxx} + (3y'_x + ae^{\lambda x}y)y''_{xx} + ae^{\lambda x}(y'_x)^2 = 0.$

This is a special case of equation 3.5.3.21 with $f(x) = ae^{\lambda x}$.

18. $(y + a)y'''_{xxx} + by'_x y''_{xx} + ce^{\lambda y}y'_x = 0.$

This is a special case of equation 3.5.3.25 with $f(y) = ce^{\lambda y}$.

19. $(y + ax + b)y'''_{xxx} + 3(y'_x + a)y''_{xx} = ke^{\lambda x}.$

Solution: $(y + ax + b)^2 = C_2 x^2 + C_1 x + C_0 + 2k\lambda^{-3}e^{\lambda x}$.

20. $y^2 y'''_{xxx} - 3yy'_x y''_{xx} + 2(y'_x)^3 = ae^{\lambda x}y^3.$

Solution: $\ln|y| = C_2 x^2 + C_1 x + C_0 + a\lambda^{-3}e^{\lambda x}$.

21. $y^2 y'''_{xxx} + 3myy'_x y''_{xx} + m(m-1)(y'_x)^3 = ae^{\lambda x}y^{2-m}.$

This is a special case of equation 3.5.3.30 with $f(x) = ae^{\lambda x}$, $n = m + 1$.

22. $x^2 yy'''_{xxx} + x(3xy'_x + 2ay)y''_{xx} + 2ax(y'_x)^2 + a(a-1)yy'_x = be^{\lambda x}.$

This is a special case of equation 3.5.3.33 with $f(x) = x^a$, $g(x) = bx^{a-2}e^{\lambda x}$.

23. $2y'_x y'''_{xxx} - (y''_{xx})^2 = ke^{\lambda x}(y'_x)^2 + ay^2 + 2by + c.$

This is a special case of equation 3.5.4.1 with $f(x) = -ke^{\lambda x}$.

24. $2y'_x y'''_{xxx} - 3(y''_{xx})^2 = ae^{\lambda x}(y'_x)^2 + be^{\mu y}(y'_x)^4.$

This is a special case of equation 3.5.4.3 with $f(x) = ae^{\lambda x}$, $g(y) = be^{\mu y}$.

25. $2y'_x y'''_{xxx} - 3(y''_{xx})^2 = ae^{\lambda x}(y'_x)^2 + by^m(y'_x)^4.$

This is a special case of equation 3.5.4.3 with $f(x) = ae^{\lambda x}$, $g(y) = by^m$.

26. $2y'_x y'''_{xxx} - 3(y''_{xx})^2 = ax^n(y'_x)^2 + be^{\mu y}(y'_x)^4.$

This is a special case of equation 3.5.4.3 with $f(x) = ax^n$, $g(y) = be^{\mu y}$.

27. $2y'_x y'''_{xxx} - 3(y''_{xx})^2 = ae^{\lambda x}(y'_x)^2 + be^{\mu x}y^{-1}(y'_x)^{5/2}.$

This is a special case of equation 3.5.4.4 with $f(x) = ae^{\lambda x}$, $g(x) = be^{\mu x}$.

28. $2y'_x y'''_{xxx} - 3(y''_{xx})^2 = ae^{\lambda x}(y'_x)^2 + bx^n y^{-1}(y'_x)^{5/2}$.

This is a special case of equation 3.5.4.4 with $f(x) = ae^{\lambda x}$, $g(x) = bx^n$.

29. $2y'_x y'''_{xxx} - 3(y''_{xx})^2 = ax^n(y'_x)^2 + be^{\lambda x}y^{-1}(y'_x)^{5/2}$.

This is a special case of equation 3.5.4.4 with $f(x) = ax^n$, $g(x) = be^{\lambda x}$.

30. $2y'_x y'''_{xxx} - 3(y''_{xx})^2 = ae^{\lambda y}(y'_x)^4 + bx^{-1}e^{\mu y}(y'_x)^{7/2}$.

This is a special case of equation 3.5.4.5 with $f(y) = ae^{\lambda y}$, $g(y) = be^{\mu y}$.

31. $2y'_x y'''_{xxx} - 3(y''_{xx})^2 = ae^{\lambda y}(y'_x)^4 + bx^{-1}y^n(y'_x)^{7/2}$.

This is a special case of equation 3.5.4.5 with $f(y) = ae^{\lambda y}$, $g(y) = by^n$.

32. $2y'_x y'''_{xxx} - 3(y''_{xx})^2 = ay^n(y'_x)^4 + bx^{-1}e^{\lambda y}(y'_x)^{7/2}$.

This is a special case of equation 3.5.4.5 with $f(y) = ay^n$, $g(y) = be^{\lambda y}$.

3.4.3. Equations Containing Hyperbolic Functions

1. $y'''_{xxx} = a\cosh^n(\lambda y)y'_x + b\cosh(\mu y)(y'_x)^3$.

This is a special case of equation 3.5.2.4 with $f(y) = a\cosh^n(\lambda y)$, $g(y) = b\cosh(\mu y)$.

2. $y'''_{xxx} = ay^n y'_x + b\cosh(\mu y)(y'_x)^3$.

This is a special case of equation 3.5.2.4 with $f(y) = ay^n$, $g(y) = b\cosh(\mu y)$.

3. $y'''_{xxx} = a\sinh^n(\lambda y)y'_x + b\sinh(\mu y)(y'_x)^3$.

This is a special case of equation 3.5.2.4 with $f(y) = a\sinh^n(\lambda y)$, $g(y) = b\sinh(\mu y)$.

4. $y'''_{xxx} = b\cosh(\lambda y)(y'_x)^3 + a(y'_x)^{-5}$.

This is a special case of equation 3.5.2.9 with $f(y) = b\cosh(\lambda y)$.

5. $y'''_{xxx} = \frac{1}{2}\lambda^2(y'_x)^3 + a(\cosh\lambda y)^{-m-3}(y'_x)^{2m+1}$.

This is a special case of equation 3.5.2.24 with $f(\xi) = a\xi^{2m}$.

6. $y'''_{xxx} = \frac{1}{2}\lambda^2(y'_x)^3 + a(\sinh\lambda y)^{-m-3}(y'_x)^{2m+1}$.

This is a special case of equation 3.5.2.25 with $f(\xi) = a\xi^{2m}$.

7. $yy'''_{xxx} + 3y'_x y''_{xx} = a\cosh(\lambda x)$.

Solution: $y^2 = C_2x^2 + C_1x + C_0 + 2a\lambda^{-3}\sinh(\lambda x)$.

8. $yy'''_{xxx} + 3y'_x y''_{xx} = a\sinh(\lambda x)$.

Solution: $y^2 = C_2x^2 + C_1x + C_0 + 2a\lambda^{-3}\cosh(\lambda x)$.

9. $yy'''_{xxx} + 3y'_x y''_{xx} = a\cosh^n(\lambda x)$.

This is a special case of equation 3.5.3.17 with $f(x) = a\cosh^n(\lambda x)$.

10. $yy'''_{xxx} + 3y'_x y''_{xx} = a\sinh^n(\lambda x)$.

This is a special case of equation 3.5.3.17 with $f(x) = a\sinh^n(\lambda x)$.

11. $yy'''_{xxx} + 3y'_x y''_{xx} = a\tanh^n(\lambda x)$.

This is a special case of equation 3.5.3.17 with $f(x) = a\tanh^n(\lambda x)$.

12. $yy'''_{xxx} + 3y'_x y''_{xx} + ax^n yy'_x = b\cosh^m(\lambda x)$.

This is a special case of equation 3.5.3.19 with $f(x) = ax^n$, $g(x) = b\cosh^m(\lambda x)$.

13. $yy'''_{xxx} + 3y'_x y''_{xx} + ax^n yy'_x = b\sinh^m(\lambda x)$.

This is a special case of equation 3.5.3.19 with $f(x) = ax^n$, $g(x) = b\sinh^m(\lambda x)$.

14. $yy'''_{xxx} + 3y'_x y''_{xx} + ax^n yy'_x = b\tanh^m(\lambda x)$.

This is a special case of equation 3.5.3.19 with $f(x) = ax^n$, $g(x) = b\tanh^m(\lambda x)$.

15. $yy'''_{xxx} + 3y'_x y''_{xx} + a\big[yy''_{xx} + (y'_x)^2\big] = b\cosh^n(\lambda x)$.

This is a special case of equation 3.5.3.20 with $f(x) = b\cosh^n(\lambda x)$.

16. $yy'''_{xxx} + 3y'_x y''_{xx} + a\big[yy''_{xx} + (y'_x)^2\big] = b\sinh^n(\lambda x)$.

This is a special case of equation 3.5.3.20 with $f(x) = b\sinh^n(\lambda x)$.

17. $yy'''_{xxx} + 3y'_x y''_{xx} + a\big[yy''_{xx} + (y'_x)^2\big] = b\tanh^n(\lambda x)$.

This is a special case of equation 3.5.3.20 with $f(x) = b\tanh^n(\lambda x)$.

18. $yy'''_{xxx} + (3y'_x + ay\cosh^n x)y''_{xx} + a\cosh^n x(y'_x)^2 = 0$.

This is a special case of equation 3.5.3.21 with $f(x) = a\cosh^n x$.

19. $yy'''_{xxx} + (3y'_x + ay\sinh^n x)y''_{xx} + a\sinh^n x(y'_x)^2 = 0$.

This is a special case of equation 3.5.3.21 with $f(x) = a\sinh^n x$.

20. $yy'''_{xxx} + (3y'_x + ay\tanh^n x)y''_{xx} + a\tanh^n x(y'_x)^2 = 0$.

This is a special case of equation 3.5.3.21 with $f(x) = a\tanh^n x$.

21. $(y+a)y'''_{xxx} + by'_x y''_{xx} + c\cosh^n(\lambda y)y'_x = 0$.

This is a special case of equation 3.5.3.25 with $f(y) = c\cosh^n(\lambda y)$.

22. $(y+a)y'''_{xxx} + by'_x y''_{xx} + c\sinh^n(\lambda y)y'_x = 0$.

This is a special case of equation 3.5.3.25 with $f(y) = c\sinh^n(\lambda y)$.

23. $(y+a)y'''_{xxx} + by'_x y''_{xx} + c\tanh^n(\lambda y)y'_x = 0.$

This is a special case of equation 3.5.3.25 with $f(y) = c\tanh^n(\lambda y)$.

24. $y^2 y'''_{xxx} + 3myy'_x y''_{xx} + m(m-1)(y'_x)^3 = a\cosh^k(\lambda x)y^{2-m}.$

This is a special case of equation 3.5.3.30 with $f(x) = a\cosh^k(\lambda x)$, $n = m+1$.

25. $y^2 y'''_{xxx} + 3myy'_x y''_{xx} + m(m-1)(y'_x)^3 = a\sinh^k(\lambda x)y^{2-m}.$

This is a special case of equation 3.5.3.30 with $f(x) = a\sinh^k(\lambda x)$, $n = m+1$.

26. $y^2 y'''_{xxx} + 3myy'_x y''_{xx} + m(m-1)(y'_x)^3 = a\tanh^k(\lambda x)y^{2-m}.$

This is a special case of equation 3.5.3.30 with $f(x) = a\tanh^k(\lambda x)$, $n = m+1$.

27. $2y'_x y'''_{xxx} - 3(y''_{xx})^2 = a\cosh^n(\lambda x)(y'_x)^2 + by^m(y'_x)^4.$

This is a special case of equation 3.5.4.3 with $f(x) = a\cosh^n(\lambda x)$, $g(y) = by^m$.

28. $2y'_x y'''_{xxx} - 3(y''_{xx})^2 = ax^n(y'_x)^2 + b\cosh^m(\lambda y)(y'_x)^4.$

This is a special case of equation 3.5.4.3 with $f(x) = ax^n$, $g(y) = b\cosh^m(\lambda y)$.

29. $2y'_x y'''_{xxx} - 3(y''_{xx})^2 = a\tanh^n(\lambda x)(y'_x)^2 + b\tanh^m(\mu y)(y'_x)^4.$

This is a special case of equation 3.5.4.3 with $f(x) = a\tanh^n(\lambda x)$, $g(y) = b\tanh^m(\mu y)$.

30. $2y'_x y'''_{xxx} - 3(y''_{xx})^2 = a\tanh^n(\lambda x)(y'_x)^2 + by^m(y'_x)^4.$

This is a special case of equation 3.5.4.3 with $f(x) = a\tanh^n(\lambda x)$, $g(y) = by^m$.

31. $2y'_x y'''_{xxx} - 3(y''_{xx})^2 = ax^n(y'_x)^2 + b\tanh^m(\lambda y)(y'_x)^4.$

This is a special case of equation 3.5.4.3 with $f(x) = ax^n$, $g(y) = b\tanh^m(\lambda y)$.

32. $2y'_x y'''_{xxx} - 3(y''_{xx})^2 = a\cosh^n(\lambda x)(y'_x)^2 + b\cosh^m(\mu x)y^{-1}(y'_x)^{5/2}.$

This is a special case of equation 3.5.4.4 with $f(x) = a\cosh^n(\lambda x)$, $g(x) = b\cosh^m(\mu x)$.

33. $2y'_x y'''_{xxx} - 3(y''_{xx})^2 = a\cosh^n(\lambda x)(y'_x)^2 + bx^m y^{-1}(y'_x)^{5/2}.$

This is a special case of equation 3.5.4.4 with $f(x) = a\cosh^n(\lambda x)$, $g(x) = bx^m$.

34. $2y'_x y'''_{xxx} - 3(y''_{xx})^2 = ax^n(y'_x)^2 + b\cosh^m(\lambda x)y^{-1}(y'_x)^{5/2}.$

This is a special case of equation 3.5.4.4 with $f(x) = ax^n$, $g(x) = b\cosh^m(\lambda x)$.

35. $2y'_x y'''_{xxx} - 3(y''_{xx})^2 = a\tanh^n(\lambda x)(y'_x)^2 + b\tanh^m(\mu x)y^{-1}(y'_x)^{5/2}.$

This is a special case of equation 3.5.4.4 with $f(x) = a\tanh^n(\lambda x)$, $g(x) = b\tanh^m(\mu x)$.

36. $2y'_x y'''_{xxx} - 3(y''_{xx})^2 = a\tanh^n(\lambda x)(y'_x)^2 + bx^m y^{-1}(y'_x)^{5/2}.$

This is a special case of equation 3.5.4.4 with $f(x) = a\tanh^n(\lambda x)$, $g(x) = bx^m$.

37. $2y'_x y'''_{xxx} - 3(y''_{xx})^2 = ax^n (y'_x)^2 + b\tanh^m(\lambda x) y^{-1} (y'_x)^{5/2}.$

This is a special case of equation 3.5.4.4 with $f(x) = ax^n$, $g(x) = b\tanh^m(\lambda x)$.

38. $2y'_x y'''_{xxx} - 3(y''_{xx})^2 = a\cosh^n(\lambda y)(y'_x)^4 + bx^{-1}\cosh^m(\mu y)(y'_x)^{7/2}.$

This is a special case of equation 3.5.4.5 with $f(y) = a\cosh^n(\lambda y)$, $g(y) = b\cosh^m(\mu y)$.

39. $2y'_x y'''_{xxx} - 3(y''_{xx})^2 = a\cosh^n(\lambda y)(y'_x)^4 + bx^{-1} y^m (y'_x)^{7/2}.$

This is a special case of equation 3.5.4.5 with $f(y) = a\cosh^n(\lambda y)$, $g(y) = by^m$.

40. $2y'_x y'''_{xxx} - 3(y''_{xx})^2 = ay^n (y'_x)^4 + bx^{-1}\cosh^m(\lambda y)(y'_x)^{7/2}.$

This is a special case of equation 3.5.4.5 with $f(y) = ay^n$, $g(y) = b\cosh^m(\lambda y)$.

41. $2y'_x y'''_{xxx} - 3(y''_{xx})^2 = a\tanh^n(\lambda y)(y'_x)^4 + bx^{-1}\tanh^m(\mu y)(y'_x)^{7/2}.$

This is a special case of equation 3.5.4.5 with $f(y) = a\tanh^n(\lambda y)$, $g(y) = b\tanh^m(\mu y)$.

42. $2y'_x y'''_{xxx} - 3(y''_{xx})^2 = a\tanh^n(\lambda y)(y'_x)^4 + bx^{-1} y^m (y'_x)^{7/2}.$

This is a special case of equation 3.5.4.5 with $f(y) = a\tanh^n(\lambda y)$, $g(y) = by^m$.

43. $2y'_x y'''_{xxx} - 3(y''_{xx})^2 = ay^n (y'_x)^4 + bx^{-1}\tanh^m(\lambda y)(y'_x)^{7/2}.$

This is a special case of equation 3.5.4.5 with $f(y) = ay^n$, $g(y) = b\tanh^m(\lambda y)$.

3.4.4. Equations Containing Logarithmic Functions

1. $y'''_{xxx} = a\ln^n(\lambda y) y'_x + b\ln^m(\mu y)(y'_x)^3.$

This is a special case of equation 3.5.2.4 with $f(y) = a\ln^n(\lambda y)$, $g(y) = b\ln^m(\mu y)$.

2. $y'''_{xxx} = a\ln^n(\lambda y) y'_x + by^m (y'_x)^3.$

This is a special case of equation 3.5.2.4 with $f(y) = a\ln^n(\lambda y)$, $g(y) = by^m$.

3. $y'''_{xxx} = ay^n y'_x + b\ln^m(\lambda y)(y'_x)^3.$

This is a special case of equation 3.5.2.4 with $f(y) = ay^n$, $g(y) = b\ln^m(\lambda y)$.

4. $y'''_{xxx} = ay^{-5/2}(2\ln y'_x - \ln y).$

This is a special case of equation 3.5.2.11 with $f(\xi) = 2a\ln\xi$.

5. $y'''_{xxx} = ay^{-5/4}(4\ln y'_x - \ln y).$

This is a special case of equation 3.5.2.12 with $f(\xi) = 4a\ln\xi$.

6. $x^3 y'''_{xxx} = a(\ln y - \ln x)(xy'_x - y).$

This is a special case of equation 3.5.2.2 with $f(\xi) = a\ln\xi$.

7. $x^5 y'''_{xxx} = a(\ln y - 2\ln x)(xy'_x - 2y)$.

This is a special case of equation 3.5.2.3 with $f(\xi) = a\ln\xi$.

8. $y'''_{xxx} = -3y''_{xx} + a(x + \ln y)^n (y'_x + y) + 2y$.

This is a special case of equation 3.5.3.3 with $f(\xi) = a\ln^n \xi$.

9. $xy'''_{xxx} = b(xy'_x - y + a\ln x)y''_{xx}$.

This is a special case of equation 3.5.3.8 with $f(\xi) = b\xi$.

10. $yy'''_{xxx} + 3y'_x y''_{xx} = a\ln^n(bx)$.

This is a special case of equation 3.5.3.17 with $f(x) = a\ln^n(bx)$.

11. $yy'''_{xxx} + 3y'_x y''_{xx} + ax^n yy'_x = b\ln^m(\lambda x)$.

This is a special case of equation 3.5.3.19 with $f(x) = ax^n$, $g(x) = b\ln^m(\lambda x)$.

12. $yy'''_{xxx} + 3y'_x y''_{xx} + a\big[yy''_{xx} + (y'_x)^2\big] = b\ln^n(\lambda x)$.

This is a special case of equation 3.5.3.20 with $f(x) = b\ln^n(\lambda x)$.

13. $(y + a)y'''_{xxx} + by'_x y''_{xx} + c\ln^n(\lambda y)y'_x = 0$.

This is a special case of equation 3.5.3.25 with $f(y) = c\ln^n(\lambda y)$.

14. $y^2 y'''_{xxx} + 3myy'_x y''_{xx} + m(m-1)(y'_x)^3 = a\ln^k(bx)y^{2-m}$.

This is a special case of equation 3.5.3.30 with $f(x) = a\ln^k(bx)$, $n = m + 1$.

15. $2y'_x y'''_{xxx} - 3(y''_{xx})^2 = ay^4(\ln y'_x - 2\ln y)$.

This is a special case of equation 3.5.4.6 with $f(\xi) = a\ln\xi$.

16. $2y'_x y'''_{xxx} - 3(y''_{xx})^2 = a\ln^n(\lambda x)(y'_x)^2 + b\ln^m(\mu y)(y'_x)^4$.

This is a special case of equation 3.5.4.3 with $f(x) = a\ln^n(\lambda x)$, $g(y) = b\ln^m(\mu y)$.

17. $2y'_x y'''_{xxx} - 3(y''_{xx})^2 = a\ln^n(\lambda x)(y'_x)^2 + by^m(y'_x)^4$.

This is a special case of equation 3.5.4.3 with $f(x) = a\ln^n(\lambda x)$, $g(y) = by^m$.

18. $2y'_x y'''_{xxx} - 3(y''_{xx})^2 = ax^n(y'_x)^2 + b\ln^m(\lambda y)(y'_x)^4$.

This is a special case of equation 3.5.4.3 with $f(x) = ax^n$, $g(y) = b\ln^m(\lambda y)$.

19. $2y'_x y'''_{xxx} - 3(y''_{xx})^2 = a\ln^n(\lambda x)(y'_x)^2 + b\ln^m(\mu x)y^{-1}(y'_x)^{5/2}$.

This is a special case of equation 3.5.4.4 with $f(x) = a\ln^n(\lambda x)$, $g(x) = b\ln^m(\mu x)$.

20. $2y'_x y'''_{xxx} - 3(y''_{xx})^2 = a\ln^n x\,(y'_x)^2 + bx^m y^{-1}(y'_x)^{5/2}$.

This is a special case of equation 3.5.4.4 with $f(x) = a\ln^n x$, $g(x) = bx^m$.

21. $2y'_x y'''_{xxx} - 3(y''_{xx})^2 = ax^n (y'_x)^2 + b\ln^m(\lambda x) y^{-1} (y'_x)^{5/2}.$

This is a special case of equation 3.5.4.4 with $f(x) = ax^n$, $g(x) = b\ln^m(\lambda x)$.

22. $2y'_x y'''_{xxx} - 3(y''_{xx})^2 = a\ln^n(\lambda y)(y'_x)^4 + bx^{-1}\ln^m(\mu y)(y'_x)^{7/2}.$

This is a special case of equation 3.5.4.5 with $f(y) = a\ln^n(\lambda y)$, $g(y) = b\ln^m(\mu y)$.

23. $2y'_x y'''_{xxx} - 3(y''_{xx})^2 = ay^n (y'_x)^4 + bx^{-1}\ln^m(\lambda y)(y'_x)^{7/2}.$

This is a special case of equation 3.5.4.5 with $f(y) = ay^n$, $g(y) = b\ln^m(\lambda y)$.

24. $2y'_x y'''_{xxx} - 3(y''_{xx})^2 = a\ln^n(\lambda y)(y'_x)^4 + bx^{-1}y^m (y'_x)^{7/2}.$

This is a special case of equation 3.5.4.5 with $f(y) = a\ln^n(\lambda y)$, $g(y) = by^m$.

3.4.5. Equations Containing Trigonometric Functions

1. $y'''_{xxx} = ay^n y'_x + b\cos(\lambda y)(y'_x)^3.$

This is a special case of equation 3.5.2.4 with $f(y) = ay^n$, $g(y) = b\cos(\lambda y)$.

2. $y'''_{xxx} = ay^n y'_x + b\sin(\lambda y)(y'_x)^3.$

This is a special case of equation 3.5.2.4 with $f(y) = ay^n$, $g(y) = b\sin(\lambda y)$.

3. $y'''_{xxx} = ay^n y'_x + b\tan(\lambda y)(y'_x)^3.$

This is a special case of equation 3.5.2.4 with $f(y) = ay^n$, $g(y) = b\tan(\lambda y)$.

4. $y'''_{xxx} = a\cos^n(\lambda y) y'_x + by^m (y'_x)^3.$

This is a special case of equation 3.5.2.4 with $f(y) = a\cos^n(\lambda y)$, $g(y) = by^m$.

5. $y'''_{xxx} = a\sin^n(\lambda y) y'_x + by^m (y'_x)^3.$

This is a special case of equation 3.5.2.4 with $f(y) = a\sin^n(\lambda y)$, $g(y) = by^m$.

6. $y'''_{xxx} = a\tan^n(\lambda y) y'_x + by^m (y'_x)^3.$

This is a special case of equation 3.5.2.4 with $f(y) = a\tan^n(\lambda y)$, $g(y) = by^m$.

7. $y'''_{xxx} = a\cos^n(\lambda y + \varepsilon) y'_x + b\cos(\mu y + \delta)(y'_x)^3.$

This is a special case of equation 3.5.2.4 with $f(y) = a\cos^n(\lambda y + \varepsilon)$, $g(y) = b\cos(\mu y + \delta)$.

8. $y'''_{xxx} = b\cos^n(\lambda y)(y'_x)^3 + a(y'_x)^{-5}.$

This is a special case of equation 3.5.2.9 with $f(y) = b\cos^n(\lambda y)$.

9. $y'''_{xxx} = b\tan^n(\lambda y)(y'_x)^3 + a(y'_x)^{-5}.$

This is a special case of equation 3.5.2.9 with $f(y) = b\tan^n(\lambda y)$.

10. $y'''_{xxx} = -\frac{1}{2}\lambda^2(y'_x)^3 + a(\cos\lambda y)^{-m-3}(y'_x)^{2m+1}.$

This is a special case of equation 3.5.2.28 with $f(\xi) = a\xi^{2m}$.

11. $y'''_{xxx} = -\frac{1}{2}\lambda^2(y'_x)^3 + a(\sin\lambda y)^{-m-3}(y'_x)^{2m+1}.$

This is a special case of equation 3.5.2.29 with $f(\xi) = a\xi^{2m}$.

12. $yy'''_{xxx} + 3y'_x y''_{xx} = a\cos(\lambda x).$

Solution: $y^2 = C_2 x^2 + C_1 x + C_0 - 2a\lambda^{-3}\sin(\lambda x)$.

13. $yy'''_{xxx} + 3y'_x y''_{xx} = a\sin^n(\lambda x).$

This is a special case of equation 3.5.3.17 with $f(x) = a\sin^n(\lambda x)$.

14. $yy'''_{xxx} + 3y'_x y''_{xx} = a\tan^n(\lambda x).$

This is a special case of equation 3.5.3.17 with $f(x) = a\tan^n(\lambda x)$.

15. $yy'''_{xxx} + 3y'_x y''_{xx} + a\sin(\lambda x)yy'_x = b\sin^n(\mu x).$

This is a special case of equation 3.5.3.19 with $f(x) = a\sin(\lambda x)$, $g(x) = b\sin^n(\mu x)$.

16. $yy'''_{xxx} + 3y'_x y''_{xx} + a\sin(\lambda x)yy'_x = bx^n.$

This is a special case of equation 3.5.3.19 with $f(x) = a\sin(\lambda x)$, $g(x) = bx^n$.

17. $yy'''_{xxx} + 3y'_x y''_{xx} + ax^n yy'_x = b\cos^m(\lambda x).$

This is a special case of equation 3.5.3.19 with $f(x) = ax^n$, $g(x) = b\cos^m(\lambda x)$.

18. $yy'''_{xxx} + 3y'_x y''_{xx} + ax^n yy'_x = b\sin^m(\lambda x).$

This is a special case of equation 3.5.3.19 with $f(x) = ax^n$, $g(x) = b\sin^m(\lambda x)$.

19. $yy'''_{xxx} + 3y'_x y''_{xx} + ax^n yy'_x = b\tan^m(\lambda x).$

This is a special case of equation 3.5.3.19 with $f(x) = ax^n$, $g(x) = b\tan^m(\lambda x)$.

20. $yy'''_{xxx} + 3y'_x y''_{xx} + a\big[yy''_{xx} + (y'_x)^2\big] = b\cos^n(\lambda x).$

This is a special case of equation 3.5.3.20 with $f(x) = b\cos^n(\lambda x)$.

21. $yy'''_{xxx} + 3y'_x y''_{xx} + a\big[yy''_{xx} + (y'_x)^2\big] = b\tan^n(\lambda x).$

This is a special case of equation 3.5.3.20 with $f(x) = b\tan^n(\lambda x)$.

22. $yy'''_{xxx} + (3y'_x + ay\sin^n x)y''_{xx} + a\sin^n x(y'_x)^2 = 0.$

This is a special case of equation 3.5.3.21 with $f(x) = a\sin^n x$.

23. $yy'''_{xxx} + (3y'_x + ay\tan^n x)y''_{xx} + a\tan^n x(y'_x)^2 = 0.$

This is a special case of equation 3.5.3.21 with $f(x) = a\tan^n x$.

24. $(y+a)y'''_{xxx} + by'_x y''_{xx} + c\cos^n(\lambda y)y'_x = 0.$

This is a special case of equation 3.5.3.25 with $f(y) = c\cos^n(\lambda y)$.

25. $(y+a)y'''_{xxx} + by'_x y''_{xx} + c\tan^n(\lambda y)y'_x = 0.$

This is a special case of equation 3.5.3.25 with $f(y) = c\tan^n(\lambda y)$.

25. $y^2 y'''_{xxx} + 3myy'_x y''_{xx} + m(m-1)(y'_x)^3 = a\cos^k(\lambda x)y^{2-m}.$

This is a special case of equation 3.5.3.30 with $f(x) = a\cos^k(\lambda x)$, $n = m+1$.

26. $y^2 y'''_{xxx} + 3myy'_x y''_{xx} + m(m-1)(y'_x)^3 = a\tan^k(\lambda x)y^{2-m}.$

This is a special case of equation 3.5.3.30 with $f(x) = a\tan^k(\lambda x)$, $n = m+1$.

27. $2y'_x y'''_{xxx} - 3(y''_{xx})^2 = a\cos(\lambda x+\varepsilon)(y'_x)^2 + b\cos(\mu y+\delta)(y'_x)^4.$

This is a special case of equation 3.5.4.3 with $f(x) = a\cos(\lambda x+\varepsilon)$, $g(y) = b\cos(\mu y+\delta)$.

28. $2y'_x y'''_{xxx} - 3(y''_{xx})^2 = a\cos(\lambda x)(y'_x)^2 + by^m(y'_x)^4.$

This is a special case of equation 3.5.4.3 with $f(x) = a\cos(\lambda x)$, $g(y) = by^m$.

29. $2y'_x y'''_{xxx} - 3(y''_{xx})^2 = ax^n(y'_x)^2 + b\cos(\lambda y)(y'_x)^4.$

This is a special case of equation 3.5.4.3 with $f(x) = ax^n$, $g(y) = b\cos(\lambda y)$.

30. $2y'_x y'''_{xxx} - 3(y''_{xx})^2 = a\tan^n(\lambda x)(y'_x)^2 + b\tan^m(\mu y)(y'_x)^4.$

This is a special case of equation 3.5.4.3 with $f(x) = a\tan^n(\lambda x)$, $g(y) = b\tan^m(\mu y)$.

31. $2y'_x y'''_{xxx} - 3(y''_{xx})^2 = a\tan^n(\lambda x)(y'_x)^2 + by^m(y'_x)^4.$

This is a special case of equation 3.5.4.3 with $f(x) = a\tan^n(\lambda x)$, $g(y) = by^m$.

32. $2y'_x y'''_{xxx} - 3(y''_{xx})^2 = ax^n(y'_x)^2 + b\tan^m(\lambda y)(y'_x)^4.$

This is a special case of equation 3.5.4.3 with $f(x) = ax^n$, $g(y) = b\tan^m(\lambda y)$.

33. $2y'_x y'''_{xxx} - 3(y''_{xx})^2 = a\sin^n(\lambda x)(y'_x)^2 + b\sin^m(\mu x)y^{-1}(y'_x)^{5/2}.$

This is a special case of equation 3.5.4.4 with $f(x) = a\sin^n(\lambda x)$, $g(x) = b\sin^m(\mu x)$.

34. $2y'_x y'''_{xxx} - 3(y''_{xx})^2 = a\cos^n(\lambda x)(y'_x)^2 + bx^m y^{-1}(y'_x)^{5/2}.$

This is a special case of equation 3.5.4.4 with $f(x) = a\cos^n(\lambda x)$, $g(x) = bx^m$.

35. $2y'_x y'''_{xxx} - 3(y''_{xx})^2 = ax^n(y'_x)^2 + b\cos^m(\lambda x)y^{-1}(y'_x)^{5/2}.$

This is a special case of equation 3.5.4.4 with $f(x) = ax^n$, $g(x) = b\cos^m(\lambda x)$.

36. $2y'_x y'''_{xxx} - 3(y''_{xx})^2 = a\tan^n(\lambda x)(y'_x)^2 + b\tan^m(\mu x)y^{-1}(y'_x)^{5/2}.$

This is a special case of equation 3.5.4.4 with $f(x) = a\tan^n(\lambda x)$, $g(x) = b\tan^m(\mu x)$.

37. $2y'_x y'''_{xxx} - 3(y''_{xx})^2 = a\tan^n(\lambda x)(y'_x)^2 + bx^m y^{-1}(y'_x)^{5/2}$.

This is a special case of equation 3.5.4.4 with $f(x) = a\tan^n(\lambda x)$, $g(x) = bx^m$.

38. $2y'_x y'''_{xxx} - 3(y''_{xx})^2 = ax^n(y'_x)^2 + b\tan^m(\lambda x)y^{-1}(y'_x)^{5/2}$.

This is a special case of equation 3.5.4.4 with $f(x) = ax^n$, $g(x) = b\tan^m(\lambda x)$.

39. $2y'_x y'''_{xxx} - 3(y''_{xx})^2 = a\cos^n(\lambda y)(y'_x)^4 + bx^{-1}\cos^m(\mu y)(y'_x)^{7/2}$.

This is a special case of equation 3.5.4.5 with $f(y) = a\cos^n(\lambda y)$, $g(y) = b\cos^m(\mu y)$.

40. $2y'_x y'''_{xxx} - 3(y''_{xx})^2 = a\cos^n(\lambda y)(y'_x)^4 + bx^{-1}y^m(y'_x)^{7/2}$.

This is a special case of equation 3.5.4.5 with $f(y) = a\cos^n(\lambda y)$, $g(y) = by^m$.

41. $2y'_x y'''_{xxx} - 3(y''_{xx})^2 = ay^n(y'_x)^4 + bx^{-1}\cos^m(\lambda y)(y'_x)^{7/2}$.

This is a special case of equation 3.5.4.5 with $f(y) = ay^n$, $g(y) = b\cos^m(\lambda y)$.

42. $2y'_x y'''_{xxx} - 3(y''_{xx})^2 = a\tan^n(\lambda y)(y'_x)^4 + bx^{-1}\tan^m(\mu y)(y'_x)^{7/2}$.

This is a special case of equation 3.5.4.5 with $f(y) = a\tan^n(\lambda y)$, $g(y) = b\tan^m(\mu y)$.

43. $2y'_x y'''_{xxx} - 3(y''_{xx})^2 = a\tan^n(\lambda y)(y'_x)^4 + bx^{-1}y^m(y'_x)^{7/2}$.

This is a special case of equation 3.5.4.5 with $f(y) = a\tan^n(\lambda y)$, $g(y) = by^m$.

44. $2y'_x y'''_{xxx} - 3(y''_{xx})^2 = ay^n(y'_x)^4 + bx^{-1}\tan^m(\lambda y)(y'_x)^{7/2}$.

This is a special case of equation 3.5.4.5 with $f(y) = ay^n$, $g(y) = b\tan^m(\lambda y)$.

3.5. Nonlinear Equations Containing Arbitrary Functions

3.5.1. Equations of the Form $F(x,y)y'''_{xxx} + G(x,y) = 0$

1. $y'''_{xxx} = f(y)$.

The substitution $w(y) = (y'_x)^2$ leads to a second order equation: $w''_{yy} = \pm 2f(y)w^{-1/2}$. In particular, with $f(y) = ay^n$ the obtained equation is the Emden—Fowler equation which is discussed in Section 2.3.

2. $y'''_{xxx} = f(x)y^{-1}$.

Having integrated the equation, we have

$$yy''_{xx} - \frac{1}{2}(y'_x)^2 = \int f(x)\,dx + C.$$

The substitution $y = w^2$ reduces the latter equation to the form

$$w''_{xx} = \frac{1}{2}\left[\int f(x)\,dx + C\right]w^{-3}.$$

3. $y'''_{xxx} = x^{-3} f(y)$.

The substitution $t = \ln|x|$ leads to an autonomous equation of the form 3.5.5.9: $y'''_{ttt} - 3y''_{tt} + 2y'_t = f(y)$.

4. $y'''_{xxx} = x^{-2} f(yx^{-1})$.

The transformation $t = \ln|x|$, $w = yx^{-1}$ leads to an autonomous equation of the form 3.5.5.9: $w'''_{ttt} - w'_t = f(w)$.

5. $y'''_{xxx} = x^{-4} f(yx^{-2})$.

The transformation $t = x^{-1}$, $w = yx^{-2}$ leads to an autonomous equation of the form 3.5.1.1: $w'''_{ttt} = -f(w)$.

6. $y'''_{xxx} = f(y + ax^3 + bx^2 + cx + k)$.

The substitution $w = y + ax^3 + bx^2 + cx + k$ leads to an autonomous equation of the form 3.5.1.1: $w'''_{xxx} = f(w) + 6a$.

7. $y'''_{xxx} = f(y + ae^{\lambda x}) - a\lambda^3 e^{\lambda x}$.

The substitution $w = y + ae^{\lambda x}$ leads to an autonomous equation of the form 3.5.1.1: $w'''_{xxx} = f(w)$.

8. $(y + ax^2 + bx + c)y'''_{xxx} = f(x)$.

The substitution $w = y + ax^2 + bx + c$ leads to an autonomous equation of the form 3.5.1.2: $ww'''_{xxx} = f(x)$.

9. $x(x-a)^3 y'''_{xxx} = f(yx^{-2}), \qquad a \neq 0$.

The transformation $\xi = \ln\left|\dfrac{x-a}{x}\right|$, $w = \dfrac{y}{x^2}$ leads to an autonomous equation of the form 3.5.5.9: $w'''_{\xi\xi\xi} - 3w''_{\xi\xi} + 2w'_\xi = a^{-3} f(w)$.

10. $(ax^2 + bx + c)^2 y'''_{xxx} = f\left(\dfrac{y}{ax^2 + bx + c}\right)$.

The transformation

$$\xi = \int \frac{dx}{ax^2 + bx + c}, \qquad w = \frac{y}{ax^2 + bx + c}$$

leads to an autonomous equation of the form 3.5.5.9: $w'''_{\xi\xi\xi} + (4ac - b^2)w'_\xi = f(w)$.

11. $(ay + be^x)y'''_{xxx} + be^x y = f(x)$.

Integrating, we obtain

$$(ay + be^x)y''_{xx} - \tfrac{1}{2}a(y'_x)^2 - be^x y'_x + be^x y = \int f(x)\,dx + C.$$

12. $y'''_{xxx} = F(x, y)$.

The transformation $x = \dfrac{1}{t}$, $y = \dfrac{w}{t^2}$ leads to an equation of the analogous form: $w'''_{ttt} = -t^{-4} F\left(\dfrac{1}{t}, \dfrac{w}{t^2}\right)$.

3.5.2. Equations of the Form $F(x, y, y'_x)y'''_{xxx} + G(x, y, y'_x) = 0$

1. $y'''_{xxx} = f(y)y'_x.$

Solution:

$$C_3 \pm x = \int \left[C_2 y + C_1 + 2\int F(y)\,dy\right]^{-1/2} dy, \qquad \text{where} \quad F(y) = \int f(y)\,dy.$$

2. $y'''_{xxx} = x^{-3} f\left(\dfrac{y}{x}\right)(xy'_x - y).$

The transformation $z = \dfrac{y}{x}$, $w = \left(y'_x - \dfrac{y}{x}\right)^2$ leads to a second order linear equation: $w''_{zz} = 2f(z) + 2$.

Integrating the latter twice, we obtain a homogeneous first-order equation:

$$y'_x = z \pm \left[z^2 + C_2 z + C_1 + 2\int_{z_0}^{z} (z - t) f(t)\,dt\right]^{1/2},$$

where $z = y/x$, z_0 is an arbitrary number.

3. $y'''_{xxx} = x^{-5} f\left(\dfrac{y}{x^2}\right)(xy'_x - 2y).$

The transformation $t = \dfrac{1}{x}$, $z = \dfrac{y}{x^2}$ leads to the equation $z'''_{ttt} = f(z)z'_t$. The substitution $w(z) = (z'_t)^2$ next yields a second order linear equation: $w''_{zz} = 2f(z)$, whose solution has the form

$$w = C_2 z + C_1 + 2\int_{z_0}^{z} (z - \xi) f(\xi)\,d\xi, \qquad z_0 \text{ is an arbitrary number.}$$

4. $y'''_{xxx} = f(y)y'_x + g(y)(y'_x)^3.$

The substitution $z(y) = (y'_x)^2$ leads to a second order linear equation: $z''_{yy} = 2g(y)z + 2f(y)$.

5. $y'''_{xxx} = x^{-5} f\left(\dfrac{y}{x}\right)(xy'_x - y)^3.$

The transformation $z = \dfrac{y}{x}$, $w = \left(y'_x - \dfrac{y}{x}\right)^2$ leads to a second order linear equation: $w''_{zz} = 2f(z)w + 2$.

6. $y'''_{xxx} = x^{-7} f\left(\dfrac{y}{x^2}\right)(xy'_x - 2y)^3.$

This is a special case of equation 3.5.2.8.

7. $y'''_{xxx} = x^{-3} f\left(\dfrac{y}{x}\right)(xy'_x - y) + x^{-5} g\left(\dfrac{y}{x}\right)(xy'_x - y)^3.$

This is a special case of equation 3.5.3.14 with $k = -1$.

The transformation $t = \ln x$, $z = y/x$ followed by the substitution $w(z) = (z'_t)^2$ yields a second order linear equation: $w''_{zz} = 2g(z)w + 2f(z) + 2$.

8. $y'''_{xxx} = x^{-5} f\Big(\dfrac{y}{x^2}\Big)(xy'_x - 2y) + x^{-7} g\Big(\dfrac{y}{x^2}\Big)(xy'_x - 2y)^3.$

The transformation $t = \dfrac{1}{x}$, $z = \dfrac{y}{x^2}$ followed by the substitution $w(z) = (z'_t)^2$ leads to a second order linear equation: $w''_{zz} = 2g(z)w + 2f(z)$.

9. $y'''_{xxx} = f(y)(y'_x)^3 + a(y'_x)^{-5}.$

The substitution $z(y) = (y'_x)^2$ leads to Yermakov's equation 2.9.1.12: $z''_{yy} = 2f(y)z + 2az^{-3}$.

10. $y'''_{xxx} = f(y'_x).$

Solution in the parametric form:

$$x = \int_{C_2}^{\tau} \frac{d\tau}{\varphi(\tau)}, \quad y = \int_{C_3}^{\tau} \frac{\tau\, d\tau}{\varphi(\tau)}, \qquad \text{where} \quad \varphi = \pm\Big[C_1 + 2\int f(\tau)\, d\tau\Big]^{1/2}.$$

11. $y'''_{xxx} = y^{-5/2} f\Big(\dfrac{y'_x}{\sqrt{y}}\Big).$

The substitution $w(y) = (y'_x)^2$ leads to an equation of the form 2.9.1.5:

$$w''_{yy} = y^{-3} F\Big(\frac{w}{y}\Big), \qquad \text{where} \quad F(\xi) = \pm 2\xi^{-1/2} f(\pm\xi^{1/2}).$$

12. $y'''_{xxx} = y^{-5/4} f\Big(\dfrac{y'_x}{y^{1/4}}\Big).$

The substitution $w(y) = (y'_x)^2$ leads to an equation of the form 2.9.1.6:

$$w''_{yy} = y^{-3/2} F\Big(\frac{w}{\sqrt{y}}\Big), \qquad \text{where} \quad F(\xi) = \pm\xi^{-1/2} f(\pm\xi^{1/2}).$$

13. $y'''_{xxx} = x f(xy'_x - y).$

The substitution $z = xy'_x - y$ leads to a second order equation of the form 2.9.2.4 with $n = 1$: $xz''_{xx} = z'_x + x^3 f(z)$.

14. $y'''_{xxx} = x^{-2} F\Big(y'_x - \dfrac{y}{x}\Big).$

The substitution $z = xy'_x - y$ leads to a second order equation: $xz''_{xx} = z'_x + F(z/x)$, which is the special case of the equation 2.9.4.12 with $n = -1$, $m = 1$, $k = -1$, $F(\xi) = \xi f(\xi)$.

15. $y'''_{xxx} = x^{-3} f(xy'_x - y).$

The transformation $t = \ln|x|$, $z = xy'_x - y$ leads to a second order autonomous equation: $z''_{tt} - 2z'_t = f(z)$, which is reduced, with the aid of the substitution $w(z) = \frac{1}{2}z'_t$, the Abel equation $ww'_z - w = \frac{1}{4}f(z)$ (for some functions f, the solutions of the latter equation are given in Subsection 1.3.1).

16. $y'''_{xxx} = x^{-4} f\Big(y'_x - 2\dfrac{y}{x}\Big).$

The transformation $t = \dfrac{1}{x}$, $z = \dfrac{y}{x^2}$ yields $z'''_{ttt} = -f(-z'_t)$. The substitution $w = -z'_t$ leads to a second order equation of the form 2.9.1.1: $w''_{tt} = f(w)$.

17. $y'''_{xxx} = (ay^2 + by + c)^{-5/4} f\bigg(\dfrac{y'_x}{(ay^2 + by + c)^{1/4}}\bigg).$

The substitution $w(y) = (y'_x)^2$ leads to an equation of the form 2.9.1.9:

$$w''_{yy} = w^{-3} F\bigg(\frac{w}{\sqrt{ay^2 + by + c}}\bigg), \qquad \text{where} \quad F(\xi) = \pm 2\xi^{5/2} f(\pm\xi^{1/2}).$$

18. $y'''_{xxx} = x(ay^2 + bx^2 y + cx^4)^{-5/4} f\bigg(\dfrac{xy'_x - 2y}{(ay^2 + bx^2 y + cx^4)^{1/4}}\bigg).$

The transformation $t = \dfrac{1}{x}$, $z = \dfrac{y}{x^2}$ leads to an equation of the form 3.5.2.17:

$$z'''_{ttt} = -(az^2 + bz + c)^{-5/4} f\bigg(\frac{-z'_t}{(az^2 + bz + c)^{1/4}}\bigg).$$

19. $y'''_{xxx} = -x^{-2} y'_x + x^{-3} y + x^{-3/4} y^{-5/4} f\bigg(\dfrac{xy'_x - y}{x^{3/4} y^{1/4}}\bigg).$

The transformation $t = \ln x$, $z = y/x$ followed by the substitution $w(z) = (z'_t)^2$ leads to a second order equation of the form 2.9.1.6:

$$w''_{zz} = z^{-3/2} F(wz^{-1/2}), \qquad \text{where} \quad F(\xi) = \pm 2\xi^{-1/2} f(\pm\sqrt{\xi}\,).$$

20. $y'''_{xxx} = -x^{-2} y'_x + x^{-3} y + x^{1/2} y^{-5/2} f\bigg(\dfrac{xy'_x - y}{\sqrt{xy}}\bigg).$

The transformation $t = \ln x$, $z = y/x$ followed by the substitution $w(z) = (z'_t)^2$ leads to a second order equation of the form 2.9.1.5:

$$w''_{zz} = z^{-3} F(w/z), \qquad \text{where} \quad F(\xi) = \pm 2\xi^{-1/2} f(\pm\sqrt{\xi}\,).$$

21. $y'''_{xxx} = f(y,\, y'_x).$

This is a special case of the equation 3.5.5.9. The transformation

$$x = \int [\varphi(\tau)]^{-3/2} d\tau, \quad y = [\varphi(\tau)]^{-1} \tag{1}$$

leads to an analogous equation with respect to function $\varphi = \varphi(\tau)$:

$$\varphi'''_{\tau\tau\tau} = -\varphi^{-5/2} f(\varphi^{-1},\, -\varphi^{-1/2}\varphi'_\tau).$$

Note two important cases of transforming equations of a special form:

$$y'''_{xxx} = f(y) \quad \xrightarrow{\text{transformation (1)}} \quad \varphi'''_{\tau\tau\tau} = -\varphi^{-5/2} f(\varphi^{-1}),$$

$$y'''_{xxx} = Ay^n \quad \xrightarrow{\text{transformation (1)}} \quad \varphi'''_{\tau\tau\tau} = -A\varphi^{-n-5/2}.$$

22. $y'''_{xxx} = \dfrac{1}{x^4} f\Big(\dfrac{y}{x^2},\ y'_x - 2\dfrac{y}{x}\Big).$

The transformation $t = \dfrac{1}{x}$, $z = \dfrac{y}{x^2}$ leads to an equation of the form 3.5.5.9: $z'''_{ttt} = -f(z,\ -z'_t)$, which admits, with the aid of the substitution $w(z) = (z'_t)^2$, lowering of its order: $w''_{zz} = \mp 2w^{-1/2} f(z,\ \mp w^{1/2})$.

23. $y'''_{xxx} = 2\lambda^2 (y'_x)^3 + e^{6\lambda y} f(e^{\lambda y} y'_x) y'_x.$

This is a special case of equation 3.5.2.32 with $\psi(y) = e^{-2\lambda y}$.

24. $y'''_{xxx} = \frac{1}{2}\lambda^2 (y'_x)^3 + (\cosh \lambda y)^{-3} f\Big(\dfrac{y'_x}{\sqrt{\cosh \lambda y}}\Big) y'_x.$

This is a special case of equation 3.5.2.32 with $\psi(y) = \cosh \lambda y$.

25. $y'''_{xxx} = \frac{1}{2}\lambda^2 (y'_x)^3 + (\sinh \lambda y)^{-3} f\Big(\dfrac{y'_x}{\sqrt{\sinh \lambda y}}\Big) y'_x.$

This is a special case of equation 3.5.2.32 with $\psi(y) = \sinh \lambda y$.

26. $y'''_{xxx} = (\sinh y)^{-2} (y'_x)^3 + (\tanh y)^3 f\big(y'_x \sqrt{\tanh y}\,\big) y'_x.$

This is a special case of equation 3.5.2.32 with $\psi(y) = \coth y$.

27. $y'''_{xxx} = -(\cosh y)^{-2} (y'_x)^3 + (\coth y)^3 f\big(y'_x \sqrt{\coth y}\,\big) y'_x.$

This is a special case of equation 3.5.2.32 with $\psi(y) = \tanh y$.

28. $y'''_{xxx} = -\frac{1}{2}\lambda^2 (y'_x)^3 + (\cos \lambda y)^{-3} f\Big(\dfrac{y'_x}{\sqrt{\cos \lambda y}}\Big) y'_x.$

This is a special case of equation 3.5.2.32 with $\psi(y) = \cos \lambda y$.

29. $y'''_{xxx} = -\frac{1}{2}\lambda^2 (y'_x)^3 + (\sin \lambda y)^{-3} f\Big(\dfrac{y'_x}{\sqrt{\sin \lambda y}}\Big) y'_x.$

This is a special case of equation 3.5.2.32 with $\psi(y) = \sin \lambda y$.

30. $y'''_{xxx} = (\sin y)^{-2} (y'_x)^3 + (\tan y)^3 f\big(y'_x \sqrt{\tan y}\,\big) y'_x.$

This is a special case of equation 3.5.2.32 with $\psi(y) = \cot y$.

31. $y'''_{xxx} = (\cos y)^{-2} (y'_x)^3 + (\cot y)^3 f\big(y'_x \sqrt{\cot y}\,\big) y'_x.$

This is a special case of equation 3.5.2.32 with $\psi(y) = \tan y$.

32. $y'''_{xxx} = \dfrac{1}{2} \dfrac{\psi''_{yy}}{\psi} (y'_x)^3 + \psi^{-3} f\Big(\dfrac{y'_x}{\sqrt{\psi}}\Big) y'_x, \qquad \psi = \psi(y).$

The substitution $z = (y'_x)^2$ leads to a second order equation of the form 2.9.1.14:

$$z''_{yy} = \frac{\psi''_{yy}}{\psi} z + 2\psi^{-3} f\Big(\pm\sqrt{\frac{z}{\psi}}\Big).$$

33. $[ay + f(x)]y'''_{xxx} + g(y)y'_x + f'''_{xxx}(x)y + h(x) = 0.$

The equation admits the first integral:

$$[ay + f(x)]y''_{xx} - \tfrac{1}{2}a(y'_x)^2 - f'_x(x)y'_x + f''_{xx}(x)y + \int g(y)\,dy + \int h(x)\,dx = C.$$

34. $y'''_{xxx} = F(x, y, y'_x).$

Let $F \neq \varphi(x)y'_x + \psi(x)y + \chi(x)$, i.e., the equation is a nonlinear one. Then, its order can be lowered by one if the right-hand side of the equation has the following form:

$$F(x,y,y'_x) = f^{-2}E\Big\{\Phi(u,w) + \int\big[2ff'''_{xxx}w + (f^2f''''_{xxxx} + 2ff'_xf'''_{xxx})u - (2f'''_{xxx}g + fg'''_{xxx})E^{-1} - (2ff'_xf'''_{xxx} + f^2f''''_{xxxx} - 2kff'''_{xxx})V\big]\,dx\Big\}, \qquad (1)$$

where

$$E = \exp\Big(-k\int f^{-1}\,dx\Big), \quad V = \int f^{-2}gE^{-1}\,dx,$$
$$u = f^{-1}E^{-1}y + V, \quad w = f^{-1}E^{-1}(fy'_x - f'_xy + g) - kV;$$

$\Phi = \Phi(u,w)$, $f = f(x)$, $g = g(x)$ are arbitrary functions, k is an arbitrary constant.

The integral in (1) may always be expressed in terms of E and V. The following cases are possible:

1) For $f'''_{xxx} \neq 0$,

$$F = f^{-2}E\Phi(u,\ w) + f^{-2}\big[2ff''_{xx} - (f'_x)^2\big]y'_x + f^{-3}\big[f^2f'''_{xxx} - 2ff'_xf''_{xx} + (f'_x)^3\big]y + f^{-3}\big\{\big[ff''_{xx} - (f'_x)^2\big]g + f(f'_xg'_x - fg''_{xx}) + k(fg'_x - f'_xg - kg)\big\} + k^3f^{-2}EV.$$

2) For $f = ax^2 + bx + c$, $a \neq 0$,

$$F = f^{-2}E\Phi(u,w) + f^{-3}\big[ff'_xg'_x - ff''_{xx}g - f^2g''_{xx} + k(fg'_x - f'_xg - kf)\big] + k(k^2 + \Delta)V,$$

where $\Delta = 4ac - b^2$.

3) For $f = ax + b$, $a \neq k$, $a \neq -\frac{1}{2}k$,

$$F = f^{-k/a-2}\Phi(u,w) - f^{-3}\big[f^2g''_{xx} - (a+k)fg'_x + k(a+k)g\big] - k(a^2 - k^2)f^{-k/a-2}U,$$

where

$$u = f^{k/a-1}y + U, \quad w = f^{k/a}y'_x - f^{k/a-1}g + (a-k)U, \quad U = \int f^{k/a-2}g\,dx.$$

4) For $f = kx + b$,

$$F = f^{-3}\big[\Phi(u,w) - f^2g''_{xx} + 2kfg'_x - 2k^2\big],$$

where

$$u = y + W, \qquad w = fy'_x + g, \qquad W = \int f^{-1}g\,dx.$$

5) For $f = -\frac{1}{2}kx + b$,

$$F = \Phi(u, w) + \frac{(kx-2b)^2 g''_{xx} + k(kx-2b)g'_x + 2k^2 g}{(kx-2b)^3} + 6kQ,$$

where

$$u = (kx-2b)^{-3}y - 2Q, \quad w = (kx-2b)^{-2}y'_x - 2(kx-2b)^{-3}g - 6kQ, \quad Q = \int \frac{g\,dx}{(kx-2b)^4}.$$

6) For $f = ax + b$, $g = c = \text{const}$, $k = 0$,

$$F = \frac{1}{(ax+b)^2}\Phi\Big(\frac{ay-c}{ax+b},\ y'_x\Big).$$

7) For $f = kx + b$, $g = 0$,

$$F = \frac{1}{(kx+b)^3}\Phi\big(y,\ (kx+b)y'_x\big).$$

In all these cases, the transformation

$$t = \int f^{-1}dx, \qquad u = f^{-1}E^{-1}y + V$$

leads to an autonomous equation:

$$u'''_{ttt} - 3ku''_{tt} + 3k^2u'_t - k^3u = \Phi(u,\ u'_t - ku),$$

which is reducible, with the aid of the substitution $z(u) = u'_t$, to a second order equation:

$$z^2z''_{uu} + z(z'_u)^2 - 3kzz'_u + 3k^2z - k^3u = \Phi(u,\ z - ku).$$

Remark. The original equation can be reduced, with the aid of point transformations, to an autonomous form *only* for function F of the form (1).

3.5.3. Equations of the Form $F(x,y,y'_x)y'''_{xxx} + G(x,y,y'_x)y''_{xx} + H(x,y,y'_x) = 0$

1. $\boldsymbol{y'''_{xxx} + ay''_{xx} + by'_x + cy = e^{\lambda x}f(ye^{-\lambda x})}$.

The substitution $w(x) = ye^{-\lambda x}$ leads to an autonomous equation of the form 3.5.5.9:

$$w'''_{xxx} + (3\lambda + a)w''_{xx} + (3\lambda^2 + 2a\lambda + b)w'_x + (\lambda^3 + a\lambda^2 + b\lambda + c)w = f(w).$$

2. $\boldsymbol{y'''_{xxx} + 3\lambda y'_x y''_{xx} + \lambda^2(y'_x)^3 = f(x)e^{-\lambda y}}$.

Solution:

$$e^{\lambda y} = C_2x^2 + C_1x + C_0 + \frac{\lambda}{2}\int_{x_0}^{x}(x-t)^2 f(t)\,dt,$$

where x_0 is an arbitrary number.

3. $y'''_{xxx} = -3y''_{xx} + 2y + f(e^x y)(y'_x + y).$

The transformation $z = e^x y$, $w = e^{2x}(y'_x + y)^2$ leads to a second order linear equation: $w''_{zz} = 2f(z) + 6$. Integrating the latter, we find the solution:

$$\int \frac{dz}{\sqrt{3z^2 + C_2 z + C_1 + 2\Phi(z)}} = \pm x + C_3,$$

where

$$z = e^x y, \quad \Phi(z) = \int \left[\int f(z)\,dz \right] dz.$$

4. $xy'''_{xxx} + 3y''_{xx} = f(xy).$

The substitution $w(x) = xy$ leads to an autonomous equation of the form 3.5.1.1: $w'''_{xxx} = f(w)$.

5. $xy'''_{xxx} = f(xy'_x - y)y''_{xx}.$

The substitution $z = xy'_x - y$ leads to a second order equation of the form 2.9.2.7: $xz''_{xx} = [f(z) + 1]z'_x$.

6. $xy'''_{xxx} + (1 - a)y''_{xx} = x^{2a} f(xy'_x - y).$

The substitution $z = xy'_x - y$ leads to a second order equation of the form 2.9.2.4: $xz''_{xx} = az'_x + x^{2a+1} f(z)$.

7. $xy'''_{xxx} + (1 - ax)y''_{xx} = e^{2ax} f(xy'_x - y).$

The substitution $z = xy'_x - y$ leads to a second order equation of the form 2.9.2.17: $z''_{xx} - az'_x = e^{2ax} f(z)$.

8. $xy'''_{xxx} = f(xy'_x - y + a\ln x)y''_{xx}.$

The substitution $z = xy'_x - y$ leads to a second order equation of the form 2.9.2.22: $xz''_{xx} = \left[f\left(\ln(x^a e^z)\right) + 1 \right] z'_x$.

9. $x^2 y'''_{xxx} + 6xy''_{xx} + 6y'_x = f(x^2 y).$

The substitution $w(x) = x^2 y$ leads to an autonomous equation of the form 3.5.1.1: $w'''_{xxx} = f(w)$.

10. $x^3 y'''_{xxx} + x^2 y''_{xx} = f(xy'_x - y).$

The transformation $t = \ln|x|$, $z = xy'_x - y$ leads to an autonomous equation of the form 2.9.6.2: $z''_{tt} - z'_t = f(z)$, which is reduced, with the aid of the substitution $w = z'_t$, to the Abel equation $ww'_z - w = f(w)$ (see Subsection 1.3.1).

11. $x^3 y'''_{xxx} + ax^2 y''_{xx} + bxy'_x = f(y).$

The substitution $t = \ln|x|$ leads to an autonomous equation of the form 3.5.5.9: $y'''_{ttt} + (a - 3)y''_{tt} + (b - a + 2)y'_t = f(y)$.

12. $x^3y'''_{xxx} + ax^2y''_{xx} + bxy'_x = f(x^m e^{\lambda y})$.

The transformation $t = \ln x$, $\lambda w = \lambda y + mt$ leads to an autonomous equation of the form 3.5.5.9:

$$w'''_{ttt} + (a-3)w''_{tt} + (b-a+2)w'_t = f(e^{\lambda w}) + \frac{m}{\lambda}(b-a+2).$$

13. $x^3y'''_{xxx} = -\frac{3}{2}x^2y''_{xx} + f\left(\frac{y}{\sqrt{x}}\right)(2xy'_x - y)$.

The transformation $t = \dfrac{y}{\sqrt{x}}$, $z = \dfrac{1}{x}\left(xy'_x - \tfrac{1}{2}y\right)^2$ leads to a second order linear equation $2z''_{tt} = 8f(t) + 1$ whose solution has the form

$$z = \tfrac{1}{4}t^2 + C_2t + C_1 + 4\int_{t_0}^{t}(t-\xi)f(\xi)\,d\xi, \qquad t_0 \text{ is an arbitrary number.}$$

Passing on to variables x, $t = yx^{-1/2}$, we obtain an equation with separation of variables.

14. $x^3y'''_{xxx} = -3(k+1)x^2y''_{xx} + k(k+1)(2k+1)y$
$\qquad + f(x^ky)(xy'_x + ky) + x^{2k}g(x^ky)(xy'_x + ky)^3$.

The transformation $t = \ln x$, $z = x^ky$ followed by the substitution $w(z) = (z'_t)^2$ yields a second order linear equation: $w''_{zz} = 2g(z)w + 2f(z) + 6k^2 + 6k + 2$.

15. $x^4y'''_{xxx} + x^3y''_{xx} = f\left(y'_x - \frac{y}{x}\right)$.

The substitution $w(x) = xy'_x - y$ leads to an equation of the form 2.9.1.5: $w''_{xx} = x^{-3}f(w/x)$.

16. $x^4y'''_{xxx} = -\frac{3}{2}x^3y''_{xx} + f\left(\frac{y}{\sqrt{x}}\right)(2xy'_x - y)^3$.

The transformation $t = \dfrac{y}{\sqrt{x}}$, $z = \dfrac{1}{x}(xy'_x - \tfrac{1}{2}y)^2$ leads to a second order linear equation: $2z''_{tt} = 16f(t)z + \frac{1}{2}$.

17. $yy'''_{xxx} + 3y'_xy''_{xx} = f(x)$.

Solution: $y^2 = C_2x^2 + C_1x + C_0 + \displaystyle\int_{x_0}^{x}(x-t)^2f(t)\,dt$, $\quad x_0$ is an arbitrary number.

18. $y(y'''_{xxx} + 3ay''_{xx} + 2a^2y'_x) = f(x)$.

Having integrated the equation, we obtain

$$2yy''_{xx} + 2ayy'_x - (y'_x)^2 = e^{-2ax}\left[2\int e^{2ax}f(x)\,dx + C\right].$$

19. $yy'''_{xxx} + 3y'_xy''_{xx} + f(x)yy'_x = g(x)$.

The substitution $w = yy'_x$ leads to a second order linear equation: $w''_{xx} + f(x)w = g(x)$.

20. $yy'''_{xxx} + 3y'_x y''_{xx} + a\big[yy''_{xx} + (y'_x)^2\big] = f(x).$

Solution:

$$y^2 = C_3 e^{-ax} + C_2 x + C_1 + 2\int_{x_0}^{x} (x-t)e^{-at}F(t)\,dt,$$

where $F(t) = \int e^{at} f(t)\,dt$, x_0 is an arbitrary number.

21. $yy'''_{xxx} + [3y'_x + f(x)y]y''_{xx} + f(x)(y'_x)^2 = 0.$

Solution:

$$y^2 = C_3 x + C_2 + C_1 \int_{x_0}^{x} (x-t)e^{-F(t)}\,dt, \qquad \text{where} \quad F(t) = \int f(t)\,dt.$$

22. $yy'''_{xxx} + (3y'_x + 2ay)y''_{xx} + 2a(y'_x)^2 + a^2 yy'_x = f(x).$

Integrating the equation twice, we obtain a first order equation with separation of variables:

$$e^{ax} yy'_x = C_2 x + C_1 + \int_{x_0}^{x} (x-t)e^{at} f(t)\,dt.$$

23. $yy'''_{xxx} + [3y'_x + f(x)y]y''_{xx} + f(x)(y'_x)^2 + g(x)yy'_x + h(x) = 0.$

The substitution $w = yy'_x$ leads to a second order linear equation: $w''_{xx} + f(x)w'_x + g(x)w + h(x) = 0$.

24. $(y+a)y'''_{xxx} + by'_x y''_{xx} = f(x).$

Having integrated the equation, we obtain

$$(y+a)y''_{xx} + \tfrac{1}{2}(b-1)(y'_x)^2 = \int f(x)\,dx + C.$$

With $b \neq -1$, the substitution $y = w^{\frac{2}{b+1}} - a$ leads to the equation

$$w''_{xx} = \frac{b+1}{2}\left[\int f(x)\,dx + C\right] w^{\frac{b-3}{b+1}}$$

(for $C = 0$ and $f(x) = \lambda x^n$, see Section 2.3).

25. $(y+a)y'''_{xxx} + by'_x y''_{xx} + f(y)y'_x = 0.$

Having integrated the equation, we obtain a second order equation:

$$(y+a)y''_{xx} + \tfrac{1}{2}(b-1)(y'_x)^2 + \int f(y)\,dy = C,$$

which is reduced, with the aid of the substitution $w(y) = (y'_x)^2$, to a first order linear equation:

$$(y+a)w'_y + (b-1)w + 2\int f(y)\,dy = 2C.$$

26. $(y+a)y'''_{xxx} + by'_x y''_{xx} + f(y)y'_x = g(x).$

Having integrated the equation, we obtain

$$(y+a)y''_{xx} + \tfrac{1}{2}(b-1)(y'_x)^2 + \int f(y)\,dy = \int g(x)\,dx + C.$$

27. $(y+ax+b)y'''_{xxx} + 3(y'_x + a)y''_{xx} = f(x).$

Solution:

$$(y+ax+b)^2 = C_2x^2 + C_1x + C_0 + \int_{x_0}^{x}(x-t)^2 f(t)\,dt,$$

where x_0 is an arbitrary number.

28. $x(yy'''_{xxx} + 3y'_x y''_{xx}) + a\big[yy''_{xx} + (y'_x)^2\big] = f(x).$

Solution:

$$y^2 = C_3x^{2-a} + C_2x + C_1 + 2\int_{x_0}^{x}(x-t)t^{-a}F(t)\,dt,$$

where $F(t) = \int t^{a-1}f(t)\,dt$, x_0 is an arbitrary number.

29. $y^2y'''_{xxx} - 3yy'_x y''_{xx} + 2(y'_x)^3 = f(x)y^3.$

Solution:

$$\ln|y| = C_2x^2 + C_1x + C_0 + \tfrac{1}{2}\int_{x_0}^{x}(x-t)^2 f(t)\,dt,$$

where x_0 is an arbitrary number.

30. $y^2y'''_{xxx} + 3(n-1)yy'_x y''_{xx} + (n-1)(n-2)(y'_x)^3 = f(x)y^{3-n}, \qquad n \neq 0.$

Solution:

$$y^n = C_2x^2 + C_1x + C_0 + \frac{n}{2}\int_{x_0}^{x}(x-t)^2 f(t)\,dt,$$

where x_0 is an arbitrary number.

31. $y^2y'''_{xxx} = -3y^2y''_{xx} + 2y^3 + y^2 f(e^x y)(y'_x + y) + g(e^x y)(y'_x + y)^3.$

The substitution $z(x) = e^x y$, followed by lowering of the equation order and the substitution $w(z) = (z'_x)^2$, leads to a second order linear equation: $w''_{zz} = 2z^{-2}g(z)w + 2f(z) + 6$.

32. $y\big(fy'''_{xxx} + \tfrac{3}{2}f'_x y''_{xx} + \tfrac{1}{2}f''_{xx}y'_x\big) = g(x), \qquad f = f(x).$

Having integrated the equation, we obtain

$$2fyy''_{xx} + f'_x yy'_x - f(y'_x)^2 = 2\int g(x)\,dx + C.$$

33. $fyy'''_{xxx} + (3fy'_x + 2f'_x y)y''_{xx} + 2f'_x(y'_x)^2 + f''_{xx}yy'_x = g, \qquad f = f(x),\ g = g(x).$

Integrating the equation twice, we obtain a first order linear equation with separation of variables:

$$f(x)yy'_x = C_2x + C_1 + \int_{x_0}^{x}(x-t)g(t)\,dt.$$

3.5.4. Equations of the Form $F(x, y, y'_x)y'''_{xxx} + \sum\limits_{\alpha} G_\alpha(x, y, y'_x)(y''_{xx})^\alpha = 0$

1. $2y'_x y'''_{xxx} - (y''_{xx})^2 + f(x)(y'_x)^2 = ay^2 + 2by + c.$

Differentiating both sides of the equation with respect to x and dividing by y'_x, we arrive at a fourth order linear equation: $y''''_{xxxx} + fy''_{xx} + \frac{1}{2}f'_x y'_x = ay + b$.

2. $2y'_x y'''_{xxx} - (y''_{xx})^2 - \lambda y'_x y''_{xx} + F(x)(y'_x)^2 = e^{\lambda x}(ay^2 + 2by + c).$

Multiplying both sides by $e^{-\lambda x}$, we arrive at an equation of the form 3.5.4.17 with $f(x) = e^{-\lambda x}$, $g(x) = e^{-\lambda x}F(x)$.

3. $2y'_x y'''_{xxx} - 3(y''_{xx})^2 = f(x)(y'_x)^2 + g(y)(y'_x)^4.$

Solution:

$$\int \frac{dy}{u^2(y)} = \int \frac{dx}{w^2(x)} + C,$$

where $u = u(y)$ and $w = w(x)$ are the general solutions of the second order linear equations

$$4u''_{yy} - g(y)u = 0 \quad \text{and} \quad 4w''_{xx} + f(x)w = 0.$$

4. $2y'_x y'''_{xxx} - 3(y''_{xx})^2 = f(x)(y'_x)^2 + g(x)y^{-1}(y'_x)^{5/2}.$

The substitution $w(x) = \dfrac{y}{\sqrt{y'_x}}$ leads to a nonhomogeneous second-order linear equation: $4w''_{xx} + f(x)w + g(x) = 0$.

5. $2y'_x y'''_{xxx} - 3(y''_{xx})^2 = f(y)(y'_x)^4 + x^{-1}g(y)(y'_x)^{7/2}.$

Taking y as the independent variable, we obtain an equation of the form 3.5.4.4 for $x = x(y)$:

$$2x'_y x'''_{yyy} - 3(x''_{yy})^2 = -f(y)(x'_y)^2 - g(y)x^{-1}(x'_y)^{5/2}.$$

6. $2y'_x y'''_{xxx} - 3(y''_{xx})^2 = y^4 f\left(\dfrac{y'_x}{y^2}\right).$

The substitution $w(x) = \dfrac{y}{\sqrt{y'_x}}$ leads to the second order autonomous equation of the form 2.9.1.1: $w''_{xx} = F(w)$, where $F(w) = -\frac{1}{4}w^5 f(w^{-2})$.

7. $2y'_x y'''_{xxx} - 3(y''_{xx})^2 = x^{-4}y^4 f\left(\dfrac{xy'_x}{y^2}\right).$

The substitution $w(x) = \dfrac{y}{\sqrt{y'_x}}$ leads to the second order equation of the form 2.9.1.6: $w''_{xx} = x^{-3/2}F(wx^{-1/2})$, where $F(\xi) = -\frac{1}{4}\xi^5 f(\xi^{-2})$.

8. $2y'_x y'''_{xxx} - 3(y''_{xx})^2 = x^{-8}y^4 f\left(\dfrac{x^2 y'_x}{y^2}\right).$

The substitution $w(x) = \dfrac{y}{\sqrt{y'_x}}$ leads to the second order equation of the form 2.9.1.5: $w''_{xx} = x^{-3}F(wx^{-1})$, where $F(\xi) = -\frac{1}{4}\xi^5 f(\xi^{-2})$.

9. $2xy'_x y'''_{xxx} - x(y''_{xx})^2 + ny'_x y''_{xx} + F(x)(y'_x)^2 = x^{1-n}(ay^2 + 2by + c).$

Multiplying both sides by x^{n-1}, we arrive at an equation of the form 3.5.4.17 with $f(x) = x^n$, $g(x) = x^{n-1}F(x)$.

10. $xy'_x y'''_{xxx} - 3x(y''_{xx})^2 + 3y'_x y''_{xx} = xf(y)(y'_x)^4 + g(y)(y'_x)^5.$

Taking y as the independent variable, we obtain an equation of the form 3.5.3.23 for $x = x(y)$:

$$xx'''_{yyy} + 3x'_y x''_{yy} = -g(y) - f(y)xx'_y.$$

11. $y'''_{xxx} = \left[x^3 f(xy'_x - y) + ax^{-5}\right](y''_{xx})^3.$

The Legendre transformation $x = w'_t$, $y = tw'_t - w$ leads to an equation of the form 3.5.2.9: $w'''_{ttt} = -f(w)(w'_t)^3 - a(w'_t)^{-5}$.

12. $y'''_{xxx} = x^{-5} f\left(\dfrac{xy'_x - y}{x^4}\right)(y''_{xx})^3.$

The Legendre transformation $x = w'_t$, $y = tw'_t - w$ leads to an equation of the form 3.5.2.12: $w'''_{ttt} = w^{-5/4}F(w'_t w^{-1/4})$, where $F(\xi) = -\xi^{-5}f(\xi^{-4})$.

13. $y'''_{xxx} = x^{-5} f\left(\dfrac{xy'_x - y}{x^2}\right)(y''_{xx})^3.$

The Legendre transformation $x = w'_t$, $y = tw'_t - w$ leads to an equation of the form 3.5.2.11: $w'''_{ttt} = w^{-5/2}F(w'_t w^{-1/2})$, where $F(\xi) = -\xi^{-5}f(\xi^{-2})$.

14. $y'''_{xxx} = [xf(y'_x) + yg(y'_x) + h(y'_x)](y''_{xx})^3 + \varphi(y'_x)(y''_{xx})^2.$

The Legendre transformation $x = w'_t$, $y = tw'_t - w$ leads to a linear equation:

$$w'''_{ttt} = -\varphi(t)w''_{tt} - [f(t) + tg(t)]w'_t + g(t)w - h(t).$$

15. $y'''_{xxx} = f(y)y'_x(y''_{xx})^m.$

Solution with $m \neq 1$:

$$C_3 \pm x = \int \left\{2\int \left[(1-m)F(y) + C_2\right]^{\frac{1}{1-m}} dy + C_1\right\}^{-\frac{1}{2}} dy, \quad F(y) = \int f(y)\,dy.$$

Solution with $m = 1$:

$$C_3 \pm x = \int \left\{C_2 \int e^{F(y)}\,dy + C_1\right\}^{-\frac{1}{2}} dy, \quad F(y) = \int f(y)\,dy.$$

16. $y'''_{xxx} = xf(xy'_x - y)(y''_{xx})^2 + xg(xy'_x - y)(y''_{xx})^k.$

The Legendre transformation $x = w'_t$, $y = tw'_t - w$ leads to the equation

$$w'''_{ttt} = -f(w)w'_t w''_{tt} - g(w)w'_t(w''_{tt})^{3-k}.$$

Lowering its order with the substitution $z(w) = w''_{tt}$ ($z'_w = w'''_{ttt}/w'_t$), we have the Bernoulli equation $z'_w = -f(w)z - g(w)z^{3-k}$.

17. $2fy'_x y'''_{xxx} - f(y''_{xx})^2 + f'_x y'_x y''_{xx} + g(y'_x)^2 = ay^2 + 2by + c, \quad f = f(x),\ g = g(x).$

Differentiating the both sides of the equation with respect to x and dividing by y'_x, we arrive at a fourth order linear equation:

$$fy''''_{xxxx} + \tfrac{3}{2}f'_x y'''_{xxx} + (g + \tfrac{1}{2}f''_{xx})y''_{xx} + \tfrac{1}{2}g'_x y'_x = ay + b.$$

3.5.5. Equations of the Form $F(x, y, y'_x, y''_{xx})y'''_{xxx} + G(x, y, y'_x, y''_{xx}) = 0$

1. $y'''_{xxx} = f(y''_{xx})$.

Solution in the parametric form:

$$x = \int_{C_1}^{\tau} \frac{d\tau_1}{f(\tau_1)}, \qquad y = \int_{C_2}^{\tau} \frac{d\tau_1}{f(\tau_1)} \int_{C_3}^{\tau_1} \frac{\tau_2\, d\tau_2}{f(\tau_2)}.$$

2. $y'''_{xxx} = f(y)y'_x g(y''_{xx})$.

Integrating the equation and substituting $w(y) = \frac{1}{2}(y'_x)^2$, we arrive at a first order equation:

$$\int \frac{d\xi}{g(\xi)} = \int f(y)\, dy + C, \qquad \text{where} \quad \xi = w'_y.$$

Solving this equation for w'_y, we obtain an equation with separation of variables.

3. $y'''_{xxx} = f(y)g(y'_x)h(y''_{xx})$.

The substitution $w(y) = \frac{1}{2}(y'_x)^2$ leads to a second order equation:

$$w''_{yy} = f(y)\varphi(w)h(w'_y), \qquad \text{where} \quad \varphi(w) = \pm\frac{g(\pm\sqrt{2w}\,)}{\sqrt{2w}},$$

whose solvable cases for some functions f, g, and h are outlined in Section 2.7.

4. $y'''_{xxx} = xf(xy'_x - y)g(y''_{xx})$.

The Legendre transformation $x = w'_t$, $y = tw'_t - w$, where $w = w(t)$, leads to an equation of the form 3.5.5.2: $w'''_{ttt} = -f(w)w'_t g\Big(\dfrac{1}{w''_{tt}}\Big)(w''_{tt})^3$.

5. $xy'''_{xxx} + y''_{xx} = f(xy'_x - y)g(xy''_{xx})$.

The substitution $w(x) = xy'_x - y$ leads to an equation of the form 2.9.4.2: $w''_{xx} = f(w)g(w'_x)$.

6. $y'''_{xxx} = f(x)g(x^2y''_{xx} - 2xy'_x + 2y)$.

The substitution $w(x) = x^2y''_{xx} - 2xy'_x + 2y$ leads to a first order equation with separation of variables: $w'_x = x^2 f(x)g(w)$.

7. $y'''_{xxx} = \dfrac{y'_x y''_{xx}}{y} + \left[y''_{xx} - \dfrac{(y'_x)^2}{y}\right] f\Big(\dfrac{y'_x}{y}\Big) g\Big(\dfrac{y''_{xx}}{y}\Big)$.

The transformation $t = \dfrac{y'_x}{y}$, $w = \dfrac{y''_{xx}}{y}$ leads to a first order equation with separation of variables: $w'_t = f(t)g(w)$.

8. $y'''_{xxx} = F(x, y'_x, y''_{xx})$.

The substitution $u(x) = y'_x$ leads to a second order equation: $u''_{xx} = F(x, u, u'_x)$.

9. $y'''_{xxx} = F(y, y'_x, y''_{xx})$.

Autonomous equation.

The substitution $w(y) = (y'_x)^2$ leads to a second order equation:

$$w''_{yy} = \pm\frac{2}{\sqrt{w}}F\big(y,\ \pm\sqrt{w},\ \tfrac{1}{2}w'_y\big).$$

10. $y'''_{xxx} = yF\left(\dfrac{y'_x}{y},\ \dfrac{y''_{xx}}{y}\right)$.

The transformation $t = \dfrac{y'_x}{y}$, $w = \dfrac{y''_{xx}}{y}$ leads to a first order equation: $(w - t^2)w'_t = -tw + F(t, w)$.

11. $y'''_{xxx} = F(x,\ xy'_x - y,\ xy''_{xx})$.

The substitution $z = xy'_x - y$ leads to a second order equation: $z''_{xx} = x^{-1}z'_x + xF(x, z, z'_x)$.

12. $y'''_{xxx} = x^{-k-3}F(x^k y,\ x^{k+1}y'_x,\ x^{k+2}y''_{xx})$.

Homogeneous equation in the extended sense.

The transformation $t = \ln x$, $z = x^k y$ leads to an autonomous equation:

$$z'''_{ttt} = 3(k+1)z''_{tt} - (3k^2 + 6k + 2)z'_t + k(k+1)(k+2)z \\ +F\big(z,\ z'_t - kz,\ z''_{tt} - (2k+1)z'_t + k(k+1)z\big).$$

The substitution $w(z) = (z'_t)^2$ leads to a second order equation:

$$w''_{zz} = \pm 3(k+1)w^{-1/2}w'_z - 6k^2 - 12k - 4 \pm 2k(k+1)(k+2)zw^{-1/2} \\ \pm 2w^{-1/2}F\big(z,\ \pm w^{1/2} - kz,\ \tfrac{1}{2}w'_z \mp (2k+1)w^{1/2} + k(k+1)z\big).$$

13. $y'''_{xxx} = yx^{-3}F\left(x^k y^m,\ \dfrac{xy'_x}{y},\ \dfrac{x^2 y''_{xx}}{y}\right)$.

Homogeneous equation in the extended sense.

The transformation $t = x^k y^m$, $z = \dfrac{xy'_x}{y}$ leads to a second order equation.

14. $y'''_{xxx} = yx^{-3}F\left(\dfrac{xy'_x}{y},\ \dfrac{x^2 y''_{xx}}{y}\right)$.

This is a special case of equation 3.5.5.13.

The transformation $z = \dfrac{xy'_x}{y}$, $w = \dfrac{x^2 y''_{xx}}{y}$ leads to a first order equation:

$$(w + z - z^2)w'_z = 2w - zw + F(z, w).$$

15. $y'''_{xxx} = F(x, y, y'_x, y''_{xx})$.

The Legendre transformation $x = w'_t$, $y = tw'_t - w$ leads to the equation

$$w'''_{ttt} = -F\Big(w'_t,\ tw'_t - w,\ t,\ \frac{1}{w''_{tt}}\Big)(w''_{tt})^3.$$

16. $y'''_{xxx} = e^{-\alpha x} F(e^{\alpha x} y,\ e^{\alpha x} y'_x,\ e^{\alpha x} y''_{xx})$.

The substitution $z = e^{\alpha x} y$ leads to an autonomous equation of the form 3.5.5.9:

$$z'''_{xxx} - 3\alpha z''_{xx} + 3\alpha^2 z'_x - \alpha^3 z = F(z,\ z'_x - \alpha z,\ z''_{xx} - 2\alpha z'_x + \alpha^2 z).$$

17. $y'''_{xxx} = yF\left(e^{\alpha x} y,\ \dfrac{y'_x}{y},\ \dfrac{y''_{xx}}{y}\right)$.

Exponential homogeneous equation.

The transformation $z = e^{\alpha x} y$, $w = \dfrac{y'_x}{y}$ leads to a second order equation:

$$z^2(w+\alpha)^2 w''_{zz} + z^2(w+\alpha)(w'_z)^2 + z(w+\alpha)(4w+\alpha)w'_z + w^3 = F\bigl(z,\ w,\ z(w+\alpha)w'_z + w^2\bigr).$$

18. $y'''_{xxx} = x^{-3} F(x^m e^y,\ xy'_x,\ x^2 y''_{xx})$.

Exponential homogeneous equation.

The transformation $z = x^m e^y$, $w = xy'_x + m$ leads to a second order equation:

$$z^2 w^2 w''_{zz} + z^2 w(w'_z)^2 + zw^2 w'_z - 3zww'_z + 2w - 2m = F(z,\ w - m,\ zww'_z - w + m).$$

Chapter 4
Fourth Order Differential Equations

4.1. Linear Equations

4.1.1. Preliminary Comments

1. A nonhomogeneous linear equation of the fourth order has the form

$$f_4 y''''_{xxxx} + f_3 y'''_{xxx} + f_2 y''_{xx} + f_1 y'_x + f_0 y = g(x), \qquad f_k = f_k(x). \tag{1}$$

Let $y_0 = y_0(x)$ be a nontrivial particular solution of the corresponding homogeneous equation (with $g \equiv 0$). Then, the substitution

$$y = y_0(x) \int z(x)\, dx \tag{2}$$

leads to a linear equation of the third order:

$$f_4 y_0 z''' + (4f_4 y_0' + f_3 y_0) z'' + (6f_4 y_0'' + 3f_3 y_0' + f_2 y_0) z' + (4f_4 y_0''' + 3f_3 y_0'' + 2f_2 y_0' + f_1 y_0) z = g, \tag{3}$$

where prime denotes differentiation with respect to x.

2. Let $y_1 = y_1(x)$ and $y_2 = y_2(x)$ be two nontrivial linearly-independent particular solutions of equation (1) with $g \equiv 0$. Then, the substitution

$$y = y_1 \int y_2 w\, dx - y_2 \int y_1 w\, dx \tag{4}$$

yields a second order linear equation:

$$f_4 \Delta_1 w'' + (3f_4 \Delta_2 + f_3 \Delta_1) w' + [f_4(3\Delta_3 + 2\varepsilon) + 2f_3 \Delta_2 + f_2 \Delta_1] w = g, \tag{5}$$

where

$$\Delta_1 = y_1' y_2 - y_1 y_2', \quad \Delta_2 = y_1'' y_2 - y_1 y_2'', \quad \Delta_3 = y_1''' y_2 - y_1 y_2''', \quad \varepsilon = y_1'' y_2' - y_1' y_2''.$$

4.1.2. Equations Containing Power Functions

1. $\quad y''''_{xxxx} + ay = 0.$

1°. Solution with $a = 0$:

$$y = C_1 + C_2 x + C_3 x^2 + C_4 x^3.$$

2°. Solution with $a = 4k^4 > 0$:

$$y = C_1 \cosh kx \cos kx + C_2 \cosh kx \sin kx + C_3 \sinh kx \cos kx + C_4 \sinh kx \sin kx.$$

3°. Solution with $a = -k^4 < 0$:

$$y = C_1 \cos kx + C_2 \sin kx + C_3 \cosh kx + C_4 \sinh kx.$$

2. $\boldsymbol{y''''_{xxxx} + \lambda y = ax^3 + bx^2 + cx + s, \qquad \lambda \neq 0.}$

Solution: $y = \dfrac{1}{\lambda}(ax^3 + bx^2 + cx + s) + w(x)$, where $w(x)$ is the general solution of the equation 4.1.2.1: $w''''_{xxxx} + \lambda w = 0$.

3. $\boldsymbol{y''''_{xxxx} = axy + b.}$

This is a special case of equation 5.1.2.4 with $n = 4$.

4. $\boldsymbol{y''''_{xxxx} = ax^m y.}$

For $m = -2, -4, -6, -8$, and -9, see equations 4.1.2.34, 4.1.2.42, 4.1.2.47, 4.1.2.48, and 4.1.2.53, respectively.

The transformation $x = t^{-1}$, $y = wt^{-3}$ leads to an equation of the similar form: $w''''_{tttt} = at^{-m-8}w$.

5. $\boldsymbol{y''''_{xxxx} + ay'_x + by = 0.}$

This is a special case of equation 4.1.2.24.

6. $\boldsymbol{y''''_{xxxx} + 2ay'_x - a^2x^2y = 0.}$

This is a special case of equation 4.1.2.13 with $n = 1$.

7. $\boldsymbol{y''''_{xxxx} + 4axy'_x + (2a - a^2x^4)y = 0.}$

This is a special case of equation 4.1.2.13 with $n = 2$.

8. $\boldsymbol{y''''_{xxxx} + ax(2b - 3a - a^2x^2)y'_x + b(2a - b + a^2x^2)y = 0.}$

The substitution $w = y''_{xx} - axy'_x + by$ leads to a second order linear equation of the form 2.1.2.28: $w''_{xx} + axw'_x + (2a - b + a^2x^2)w = 0$.

9. $\boldsymbol{y''''_{xxxx} + ax^m y'_x - 3ax^{m-1}y = 0.}$

Particular solution: $y_0 = x^3$.

The substitution $z = xy'_x - 3y$ leads to a third order equation of the form 3.1.2.7: $z'''_{xxx} + ax^m z = 0$.

10. $\boldsymbol{y''''_{xxxx} + ax^m y'_x + amx^{m-1}y = 0.}$

Integrating yields a third order equation: $y'''_{xxx} + ax^m y = C$.

11. $\boldsymbol{y''''_{xxxx} + ax^m y'_x + a(m+3)x^{m-1}y = 0.}$

The transformation $x = t^{-1}$, $y = wt^{-3}$ leads to an equation of the form 4.1.2.10: $w''''_{tttt} + bt^n w'_t + bnt^{n-1}w = 0$, where $b = -a$, $n = -m - 6$.

12. $y''''_{xxxx} + bx^m y'_x - a(a^3 + bx^m)y = 0.$

This is a special case of equation 4.1.5.1 with $f = bx^m$.

13. $y''''_{xxxx} + 2anx^{n-1}y'_x + a[n(n-1)x^{n-2} - ax^{2n}]y = 0.$

The substitution $w = y''_{xx} + ax^n y$ leads to a second order equation of the form 2.1.2.7: $w''_{xx} - ax^n w = 0$.

14. $y''''_{xxxx} + (ax^n + b^3)y'_x + abx^n y = 0.$

Particular solution: $y_0 = e^{-bx}$.

15. $y''''_{xxxx} + (ax^{n+1} + bx^n)y'_x - ax^n y = 0.$

Particular solution: $y_0 = ax + b$.

16. $y''''_{xxxx} + 2ay''_{xx} + a^2 y = 0.$

1°. Solution with $a = k^2 > 0$:

$$y = (C_1 + C_2 x)\cos(kx) + (C_3 + C_4 x)\sin(kx).$$

2°. Solution with $a = -k^2 < 0$:

$$y = (C_1 + C_2 x)\exp(kx) + (C_3 + C_4 x)\exp(-kx).$$

17. $y''''_{xxxx} + (a+b)y''_{xx} + aby = 0.$

The case of $a = b$ is given in 4.1.2.16. Let $a \neq b$.

1°. Solution with $a = \alpha^2 > 0,\ b = \beta^2 > 0$:

$$y = C_1\cos(\alpha x) + C_2\sin(\alpha x) + C_3\cos(\beta x) + C_4\sin(\beta x).$$

2°. Solution with $a = \alpha^2 > 0,\ b = -\beta^2 < 0$:

$$y = C_1\cos(\alpha x) + C_2\sin(\alpha x) + C_3\exp(\beta x) + C_4\exp(-\beta x).$$

3°. Solution with $a = -\alpha^2 < 0,\ b = \beta^2 > 0$:

$$y = C_1\exp(\alpha x) + C_2\exp(-\alpha x) + C_3\cos(\beta x) + C_4\sin(\beta x).$$

4°. Solution with $a = -\alpha^2 < 0,\ b = -\beta^2 < 0$:

$$y = C_1\exp(\alpha x) + C_2\exp(-\alpha x) + C_3\exp(\beta x) + C_4\exp(-\beta x).$$

18. $y''''_{xxxx} - 2a^2 y''_{xx} + a^4 y - \lambda(ax - b)(y''_{xx} - a^2 y) = 0.$

This equation is met with in the turbulence theory. Assuming

$$z(x) = y''_{xx} - a^2 y, \qquad (1)$$

yields a second order linear equation of the form 2.1.2.12:

$$z''_{xx} - a^2 z - \lambda(ax - b)z = 0. \qquad (2)$$

Given the boundary conditions

$$y(0) = y'_x(0) = 0, \quad y(1) = y'_x(1) = 0, \qquad (3)$$

we obtain

$$2ay = e^{ax} \int_0^x e^{-ax} z\,dx - e^{-ax} \int_0^x e^{ax} z\,dx$$

The latter is the solution of equation (1) that satisfies the first pair of the boundary conditions (3). In order to satisfy the second pair of the boundary conditions, the solution $z(x)$ of equation (2) must meet the requirements

$$\int_0^1 e^{-ax} z\,dx = \int_0^1 e^{ax} z\,dx = 0.$$

19. $y''''_{xxxx} + ax^n y''_{xx} + b(ax^n - b)y = 0.$

1°. Particular solutions with $b > 0$: $y_1 = \cos\bigl(x\sqrt{b}\bigr)$, $y_2 = \sin\bigl(x\sqrt{b}\bigr)$.

2°. Particular solutions with $b < 0$: $y_1 = \exp\bigl(-x\sqrt{-b}\bigr)$, $y_2 = \exp\bigl(x\sqrt{-b}\bigr)$.

The substitution $w = y''_{xx} + by$ leads to a second order linear equation: $w''_{xx} + (ax^n - b)w = 0$.

20. $y''''_{xxxx} + ax^{n+1} y''_{xx} - 4ax^n y'_x + 6ax^{n-1} y = 0.$

Particular solutions: $y_1 = x^2$, $y_2 = x^3$.

The substitution $w = x^2 y''_{xx} - 4xy'_x + 6y$ leads to a second order linear equation of the form 2.1.2.7: $w''_{xx} + ax^{n+1} w = 0$.

21. $y''''_{xxxx} + 10ax^n y''_{xx} + 10anx^{n-1} y'_x + [3an(n-1)x^{n-2} + 9a^2 x^{2n}]y = 0.$

This is a special case of equation 4.1.5.26 with $f = ax^n$.

22. $y''''_{xxxx} + (ax^n + b)y''_{xx} + abx^n y = 0.$

1°. Particular solutions with $b > 0$: $y_1 = \cos\bigl(x\sqrt{b}\bigr)$, $y_2 = \sin\bigl(x\sqrt{b}\bigr)$.

2°. Particular solutions with $b < 0$: $y_1 = \exp\bigl(-x\sqrt{-b}\bigr)$, $y_2 = \exp\bigl(x\sqrt{-b}\bigr)$.

The substitution $w = y''_{xx} + by$ leads to a second order linear equation of the form 2.1.2.7: $w''_{xx} + ax^n w = 0$.

23. $y''''_{xxxx} + ay''_{xx} + bx^n y'_x + bnx^{n-1} ny = sx^m.$

Integration yields a third order equation: $y'''_{xxx} + ay''_{xx} + bx^n y = s\int x^m\,dx + C$.

24. $y''''_{xxxx} + a_3 y'''_{xxx} + a_2 y''_{xx} + a_1 y'_x + a_0 y = 0.$

For $a_0 = 0$, the substitution $w(x) = y'_x$ leads to a third order equation. Let $a_0 \neq 0$ and $P(\lambda) = \lambda^4 + a_3\lambda^3 + a_2\lambda^2 + a_1\lambda + a_0$ be the characteristic polynomial.

1°. Let P be factorizable, so that

$$P(\lambda) = (\lambda - \lambda_1)(\lambda - \lambda_2)(\lambda - \lambda_3)(\lambda - \lambda_4),$$

where λ_1, λ_2, λ_3, and λ_4 are real numbers. The following cases are possible:

a) λ_i are all different, then

$$y = C_1 e^{\lambda_1 x} + C_2 e^{\lambda_2 x} + C_3 e^{\lambda_3 x} + C_4 e^{\lambda_4 x};$$

b) $\lambda_1 = \lambda_2$; λ_3 and λ_4 are different and not equal to λ_1, then

$$y = (C_1 + C_2 x)e^{\lambda_1 x} + C_3 e^{\lambda_3 x} + C_4 e^{\lambda_4 x};$$

c) $\lambda_1 = \lambda_2 = \lambda_3 \neq \lambda_4$, then

$$y = (C_1 + C_2 x + C_3 x^2)e^{\lambda_1 x} + C_4 e^{\lambda_4 x};$$

d) $\lambda_1 = \lambda_2 = \lambda_3 = \lambda_4$, then

$$y = (C_1 + C_2 x + C_3 x^2 + C_4 x^3)e^{\lambda_1 x}.$$

2°. Let

$$P(\lambda) = (\lambda - \lambda_1)(\lambda - \lambda_2)(\lambda^2 + 2b_1\lambda + b_0),$$

where λ_1 and λ_2 are real numbers, and $b_1^2 - b_0 < 0$. If

a) $\lambda_1 \neq \lambda_2$, then

$$y = C_1 e^{\lambda_1 x} + C_2 e^{\lambda_2 x} + e^{-b_1 x}[C_3 \cos(\mu x) + C_4 \sin(\mu x)], \qquad \mu = \sqrt{b_0 - b_1^2};$$

b) $\lambda_1 = \lambda_2$, then

$$y = (C_1 + C_2 x)e^{\lambda_1 x} + e^{-b_1 x}[C_3 \cos(\mu x) + C_4 \sin(\mu x)], \qquad \mu = \sqrt{b_0 - b_1^2}.$$

3°. Let us assume that

$$P(\lambda) = (\lambda^2 + 2b_1\lambda + b_0)(\lambda^2 + 2\beta_1\lambda + \beta_0),$$

where $b_1^2 - b_0 < 0$ and $\beta_1^2 - \beta_0 < 0$. If

a) $(b_1 - \beta_1)^2 + (b_0 - \beta_0)^2 \neq 0$, then

$$y = e^{-b_1 x}[C_1 \cos(\mu x) + C_2 \sin(\mu x)] + e^{-\beta_1 x}[C_3 \cos(\nu x) + C_4 \sin(\nu x)],$$

where $\mu = \sqrt{b_0 - b_1^2}$, $\nu = \sqrt{\beta_0 - \beta_1^2}$;

b) $b_1 = \beta_1$ and $b_0 = \beta_0$, then

$$y = e^{-b_1 x}[(C_1 + C_2 x)\cos(\mu x) + (C_3 + C_4 x)\sin(\mu x)], \qquad \mu = \sqrt{b_0 - b_1^2}.$$

25. $y''''_{xxxx} + 4axy'''_{xxx} + 6a^2x^2y''_{xx} + 4a^3x^3y'_x + a^4x^4y = 0.$

Solution:

$$y = \sum_{i=1}^{4} C_i \exp(\lambda_i x - \tfrac{1}{2}ax^2),$$

where λ_i are the roots of the biquadratic equation $\lambda^4 - 6a\lambda^2 + 3a^2 = 0$.

26. $y''''_{xxxx} + (ax+b)y'''_{xxx} + [b(a+c)x + c]y''_{xx} + b^2cxy'_x - b^2cy = 0.$

Particular solutions: $y_1 = x$, $y_2 = e^{-bx}$.

27. $y''''_{xxxx} = ax^ny'''_{xxx} + by'_x - abx^ny.$

Particular solutions: $y_k = \exp(\lambda_k x)$ $(k = 1,\ 2,\ 3)$, where λ_k are the roots of the cubic equation $\lambda^3 - b = 0$.

28. $y''''_{xxxx} + ax^{n+3}y'''_{xxx} - 3ax^{n+2}y''_{xx} + 6ax^{n+1}y'_x - 6ax^ny = 0.$

Particular solutions: $y_1 = x$, $y_2 = x^2$, $y_3 = x^3$.

The substitution $w = x^3y'''_{xxx} - 3x^2y''_{xx} + 6xy'_x - 6y$ leads to a first order linear equation: $w'_x + ax^{n+3}w = 0$.

29. $y''''_{xxxx} + ax^ny'''_{xxx} + bx^{m+1}y''_{xx} - 2bx^my'_x + 2bx^{m-1}y = 0.$

Particular solutions: $y_1 = x$, $y_2 = x^2$.

The substitution $w = x^2y''_{xx} - 2xy'_x + 2y$ leads to a second order linear equation: $xw''_{xx} + (ax^{n+1} - 2)w'_x + bx^{m+2}w = 0$.

30. $y''''_{xxxx} + ax^ny'''_{xxx} + bx^my''_{xx} + acx^ny'_x + c(bx^m - c)y = 0.$

1°. Particular solutions with $c > 0$: $y_1 = \cos(x\sqrt{c})$, $y_2 = \sin(x\sqrt{c})$.

2°. Particular solutions with $c < 0$: $y_1 = \exp(-x\sqrt{-c})$, $y_2 = \exp(x\sqrt{-c})$.

The substitution $w = y''_{xx} + cy$ leads to a second order linear equation: $w''_{xx} + ax^nw'_x + (bx^m - c)w = 0$.

31. $y''''_{xxxx} + ax^ny'''_{xxx} + (bx^m + c)y''_{xx} + acx^ny'_x + bcx^my = 0.$

1°. Particular solutions with $c > 0$: $y_1 = \cos(x\sqrt{c})$, $y_2 = \sin(x\sqrt{c})$.

2°. Particular solutions with $c < 0$: $y_1 = \exp(-x\sqrt{-c})$, $y_2 = \exp(x\sqrt{-c})$.

The substitution $w = y''_{xx} + cy$ leads to a second order linear equation: $w''_{xx} + ax^nw'_x + bx^mw = 0$.

32. $xy''''_{xxxx} + 4y'''_{xxx} + axy = 0.$

The substitution $w(x) = xy$ leads to a constant coefficient equation of the form 4.1.2.1: $w''''_{xxxx} + aw = 0$.

33. $xy''''_{xxxx} - 4my'''_{xxx} + axy = 0, \qquad m = 1,\ 2,\ 3,\ \ldots$

Solution:

$$y = x^{4m+3}\left(\frac{1}{x^3}\frac{d}{dx}\right)^m\left(\frac{w}{x^3}\right),$$

where $w = w(x)$ is the general solution of the constant coefficient equation 4.1.2.1 $w''''_{xxxx} + aw = 0$.

34. $x^2 y''''_{xxxx} = ay.$

This is a special case of equation 5.1.2.23 with $n = 2$.

35. $x^2 y''''_{xxxx} - 2(ax^2 + 6)y''_{xx} + a(ax^2 + 4)y = 0.$

Particular solutions: $y_1 = x^{-1/2} I_{1/2}(x\sqrt{a})$, $y_2 = x^{-1/2} K_{1/2}(x\sqrt{a})$, where $I_{1/2}$ and $K_{1/2}$ are modified Bessel functions.

36. $x^2 y''''_{xxxx} + 6xy'''_{xxx} + 6y''_{xx} - \lambda^2 y = 0.$

The equation of transverse vibrations of a pointed bar.

Solution:

$$y = \frac{1}{\sqrt{x}}\left[C_1 J_1\left(2\sqrt{\lambda x}\right) + C_2 Y_1\left(2\sqrt{\lambda x}\right) + C_3 I_1\left(2\sqrt{\lambda x}\right) + C_4 K_1\left(2\sqrt{\lambda x}\right)\right],$$

where J_1 and Y_1 are Bessel functions, I_1 and K_1 are modified Bessel functions.

37. $x^2 y''''_{xxxx} + 2(a + 2)xy'''_{xxx} + (a + 1)(a + 2)y''_{xx} - b^4 y = 0.$

Solution:

$$y = x^{-a/2}[C_1 J_a(\xi) + C_2 Y_a(\xi) + C_3 I_a(\xi) + C_4 K_a(\xi)],$$

where $\xi = 2b\sqrt{x}$, J_a and Y_a are Bessel functions, I_a and K_a are modified Bessel functions.

38. $x^2 y''''_{xxxx} + 8xy'''_{xxx} + 12y''_{xx} + ax^2 y = 0.$

The substitution $w(x) = x^2 y$ leads to a constant coefficient equation of the form 4.1.2.1: $w''''_{xxxx} + aw = 0$.

39. $x^2 y''''_{xxxx} + 8xy'''_{xxx} + 12y''_{xx} = ax^3 y + b.$

The substitution $w(x) = x^2 y$ leads to an equation of the form 4.1.2.3: $w''''_{xxxx} = axw + b$.

40. $x^2 y''''_{xxxx} + axy'''_{xxx} + (bx^{n+1} + c)y''_{xx} + (a - 4)bx^n y'_x + b(c - 2a + 6)x^{n-1} y = 0.$

The substitution $w(x) = x^2 y''_{xx} + (a - 4)xy'_x + (c - 2a + 6)y$ leads to a second order equation of the form 2.1.2.7: $w''_{xx} + bx^{n-1} w = 0$.

41. $x^3 y''''_{xxxx} + 2x^2 y'''_{xxx} - xy''_{xx} + y'_x - a^4 x^3 y = 0.$

Solution:

$$y = C_1 J_0(ax) + C_2 Y_0(ax) + C_3 I_0(ax) + C_4 K_0(ax),$$

where J_0 and Y_0 are Bessel functions, I_0 and K_0 are modified Bessel functions.

42. $x^4 y''''_{xxxx} = ay.$

Solution:

$$y = C_1 x^{k_1} + C_2 x^{k_2} + C_3 x^{k_3} + C_4 x^{k_4},$$

where $k_{1,2} = \frac{3}{2} \pm \sqrt{\frac{5}{4} + \sqrt{a+1}}$, $k_{3,4} = \frac{3}{2} \pm \sqrt{\frac{5}{4} - \sqrt{a+1}}$.

43. $x^4 y''''_{xxxx} + A_3 x^3 y'''_{xxx} + A_2 x^2 y''_{xx} + A_1 x y'_x + A_0 y = 0.$

The Euler equation.

The substitution $t = \ln|x|$ leads to a constant coefficient equation of the form 4.1.2.24:

$$y''''_{tttt} + (A_3 - 6)y'''_{ttt} + (11 - 3A_3 + A_2)y''_{tt} + (2A_3 - A_2 + A_1 - 6)y'_t + A_0 y = 0.$$

44. $x^4 y''''_{xxxx} - 2n(n+1)x^2 y''_{xx} + 4n(n+1)x y'_x + [ax^4 + n(n+1)(n+3)(n-2)]y = 0,$ where n is a positive integer.

Solution:

$$y = x^{-n} \sum_{\nu=1}^{4} C_\nu \exp(\lambda_\nu x) P_\nu(x), \qquad a \neq 0,$$

where λ_ν are four different roots of the equation $\lambda^4 + a = 0$, and P_ν is some definite polynomial of the degree $\leq 4n$. For $a = 0$, we have the Euler equation 4.1.2.43.

45. $x^4 y''''_{xxxx} + 2(2-n)x^3 y'''_{xxx} + (1-n)(2-n)x^2 y''_{xx} - a^4 x^{2n} y = 0.$

Solution:

$$y = \sqrt{x}\,[C_1 J_{1/n}(\xi) + C_2 Y_{1/n}(\xi) + C_3 I_{1/n}(\xi) + C_4 K_{1/n}(\xi)],$$

where $\xi = \dfrac{2a}{n} x^{n/2}$, J_ν and Y_ν are Bessel functions, I_ν and K_ν are modified Bessel functions.

46. $x^4 y''''_{xxxx} + 6x^3 y'''_{xxx} + [4x^4 + (7 - a^2 - b^2)x^2]y''_{xx}$
$\qquad + x(16x^2 + 1 - a^2 - b^2)y'_x + (8x^2 + a^2 b^2)y = 0.$

Solution with $ab \neq 0$:

$$y = C_1 J_\mu(x) J_\nu(x) + C_2 J_\mu(x) Y_\nu(x) + C_3 Y_\mu(x) J_\nu(x) + C_4 Y_\mu(x) Y_\nu(x),$$

where J_μ and Y_μ are Bessel functions, $2\mu = a + b$, $2\nu = a - b$.

47. $x^6 y''''_{xxxx} = ay.$

This is a special case of equation 5.1.2.24 with $n = 2$.

48. $x^8 y''''_{xxxx} = ay.$

The transformation $x = t^{-1}$, $y = wt^{-3}$ yields a constant coefficient equation: $w''''_{tttt} = aw$.

49. $x^8 y''''_{xxxx} + 4x^7 y'''_{xxx} = ay.$

The substitution $w(x) = xy$ leads to an equation of the form 4.1.2.48: $x^8 w''''_{xxxx} = aw$.

50. $(ax+b)^4 (cx+d)^4 y''''_{xxxx} = ky.$

The transformation

$$\xi = \ln \frac{ax+b}{cx+d}, \qquad w = \frac{y}{(cx+d)^3}$$

leads to a constant coefficient equation.

51. $(ax^2 + bx + c)^4 y''''_{xxxx} = ky.$

The transformation

$$\xi = \int \frac{dx}{ax^2 + bx + c}, \qquad w = \frac{y}{(ax^2 + bx + c)^{3/2}}$$

leads to a constant coefficient equation:

$$w''''_{\xi\xi\xi\xi} - \tfrac{5}{2} D w''_{\xi\xi} + \left(\tfrac{9}{16} D^2 - k\right) w = 0, \qquad \text{where} \quad D = b^2 - 4ac.$$

52. $(ax + b)^2 (cx + d)^6 y''''_{xxxx} = ky.$

The transformation

$$\xi = \frac{ax + b}{cx + d}, \qquad w = \frac{y}{(cx + d)^3}$$

leads to an equation of the form 4.1.2.34:

$$\xi^2 w''''_{\xi\xi\xi\xi} = k \Delta^{-4} w, \qquad \text{where} \quad \Delta = ad - bc.$$

53. $x^9 y''''_{xxxx} = ay + bx^4.$

The transformation $x = t^{-1}$, $y = wt^{-3}$ leads to an equation of the form 4.1.2.3: $w''''_{tttt} = atw + b$.

54. $(ax + b)^9 y''''_{xxxx} = (cx + d) y.$

The transformation

$$\xi = \frac{cx + d}{ax + b}, \qquad w = \frac{y}{(ax + b)^3}$$

leads to an equation of the form 4.1.2.3:

$$w''''_{\xi\xi\xi\xi} = \Delta^{-4} \xi w, \qquad \text{where} \quad \Delta = ad - bc.$$

4.1.3. Equations Containing Exponential, Hyperbolic, and Logarithmic Functions

1. $y''''_{xxxx} + a^3 y'_x + be^{ax}(a^2 - be^{ax}) y = 0.$

The substitution $w = y''_{xx} + ay'_x + be^{ax} y$ leads to a second order linear equation of the form 2.1.3.10: $w''_{xx} - aw'_x + (a^2 - be^{ax}) w = 0$.

2. $y''''_{xxxx} + ae^{\lambda x} y'_x - (abe^{\lambda x} + b^4) y = 0.$

Particular solution: $y_0 = e^{bx}$.

3. $y''''_{xxxx} + 2a\lambda e^{\lambda x} y'_x + a(\lambda^2 e^{\lambda x} - ae^{2\lambda x}) y = 0.$

The substitution $w = y''_{xx} + ae^{\lambda x} y$ leads to a second order linear equation of the form 2.1.3.1: $w''_{xx} - ae^{\lambda x} w = 0$.

4. $y''''_{xxxx} + (ae^{\lambda x} + b^3) y'_x + abe^{\lambda x} y = 0.$

Particular solution: $y_0 = e^{-bx}$.

5. $\boldsymbol{y''''_{xxxx} + (ax+b)e^{\lambda x}y'_x - ae^{\lambda x}y = 0.}$

Particular solution: $y_0 = ax + b$.

6. $\boldsymbol{y''''_{xxxx} + ae^{\lambda x}y''_{xx} - b(ae^{\lambda x} + b)y = 0.}$

1°. Particular solutions with $b > 0$: $y_1 = \exp(-x\sqrt{b})$, $y_2 = \exp(x\sqrt{b})$.

2°. Particular solutions with $b < 0$: $y_1 = \cos(x\sqrt{-b})$, $y_2 = \sin(x\sqrt{-b})$.

The substitution $w = y''_{xx} - by$ leads to a second order linear equation of the form 2.1.3.10: $w''_{xx} + (ae^{\lambda x} + b)w = 0$.

7. $\boldsymbol{y''''_{xxxx} + (a + be^{\lambda x})y''_{xx} + abe^{\lambda x}y = 0.}$

1°. Particular solutions with $a > 0$: $y_1 = \cos(x\sqrt{a})$, $y_2 = \sin(x\sqrt{a})$.

2°. Particular solutions with $a < 0$: $y_1 = \exp(-x\sqrt{-a})$, $y_2 = \exp(x\sqrt{-a})$.

The substitution $w = y''_{xx} + ay$ leads to a second order linear equation of the form 2.1.3.1: $w''_{xx} + be^{\lambda x}w = 0$.

8. $\boldsymbol{y''''_{xxxx} + 10ae^{\lambda x}y''_{xx} + 10a\lambda e^{\lambda x}y'_x + (3a\lambda^2 e^{\lambda x} + 9a^2 e^{2\lambda x})y = 0.}$

This is a special case of equation 4.1.5.26 with $f(x) = ae^{\lambda x}$.

9. $\boldsymbol{y''''_{xxxx} + ay'''_{xxx} + be^{\lambda x}y'_x + abe^{\lambda x}y = 0.}$

Particular solution: $y_0 = e^{-ax}$.

10. $\boldsymbol{y''''_{xxxx} = ae^{\lambda x}y'''_{xxx} + by'_x - abe^{\lambda x}y.}$

Particular solutions: $y_k = e^{\beta_k x}$ $(k = 1,\ 2,\ 3)$, where β_k are the roots of the cubic equation $\beta^3 - b = 0$.

11. $\boldsymbol{y''''_{xxxx} + ae^{\lambda x}y'''_{xxx} + be^{\mu x}y''_{xx} + ace^{\lambda x}y'_x + c(be^{\mu x} - c)y = 0.}$

1°. Particular solutions with $c > 0$: $y_1 = \cos(x\sqrt{c})$, $y_2 = \sin(x\sqrt{c})$.

2°. Particular solutions with $c < 0$: $y_1 = \exp(-x\sqrt{-c})$, $y_2 = \exp(x\sqrt{-c})$.

The substitution $w = y''_{xx} + cy$ leads to a second order linear equation:

$$w''_{xx} + ae^{\lambda x}w'_x + (be^{\mu x} - c)w = 0.$$

12. $\boldsymbol{y''''_{xxxx} + ae^{\lambda x}y'''_{xxx} + (be^{\mu x} + c)y''_{xx} + ace^{\lambda x}y'_x + bce^{\mu x}y = 0.}$

1°. Particular solutions with $c > 0$: $y_1 = \cos(x\sqrt{c})$, $y_2 = \sin(x\sqrt{c})$.

2°. Particular solutions with $c < 0$: $y_1 = \exp(-x\sqrt{-c})$, $y_2 = \exp(x\sqrt{-c})$.

The substitution $w = y''_{xx} + cy$ leads to a second order linear equation: $w''_{xx} + ae^{\lambda x}w'_x + be^{\mu x}w = 0$.

13. $\boldsymbol{y''''_{xxxx} + ax^3e^{\lambda x}y'''_{xxx} - 3ax^2e^{\lambda x}y''_{xx} + 6axe^{\lambda x}y'_x - 6ae^{\lambda x}y = 0.}$

Particular solutions: $y_1 = x$, $y_2 = x^2$, $y_3 = x^3$.

The substitution $w = x^3y'''_{xxx} - 3x^2y''_{xx} + 6xy'_x - 6y$ leads to a first order linear equation: $w'_x + ax^3e^{\lambda x}w = 0$.

14. $(ae^x + b)y''''_{xxxx} = ae^x y.$

Particular solution: $y_0 = ae^x + b.$

15. $(ax^m + be^x + c)y''''_{xxxx} = be^x y, \qquad m = 1,\ 2,\ 3.$

Particular solution: $y_0 = ax^m + be^x + c.$

16. $(ax^m e^x + b)y''''_{xxxx} = by, \qquad m = 0,\ 1,\ 2,\ 3.$

Particular solution: $y_0 = ax^m + be^{-x}.$

17. $y''''_{xxxx} + b\exp(\lambda x^n)\, y''_{xx} + a[b\exp(\lambda x^n) - a]y = 0.$

This is a special case of equation 4.1.5.5 with $f(x) = b\exp(\lambda x^n)$.

18. $y''''_{xxxx} + [a + b\exp(\lambda x^n)]y''_{xx} + ab\exp(\lambda x^n)\, y = 0.$

This is a special case of equation 4.1.5.6 with $f(x) = b\exp(\lambda x^n)$.

19. $y''''_{xxxx} + b\sinh^n(\lambda x)\, y''_{xx} + a[b\sinh^n(\lambda x) - a]y = 0.$

This is a special case of equation 4.1.5.5 with $f(x) = b\sinh^n(\lambda x)$.

20. $y''''_{xxxx} + [a + b\sinh^n(\lambda x)]y''_{xx} + ab\sinh^n(\lambda x)\, y = 0.$

This is a special case of equation 4.1.5.6 with $f(x) = b\sinh^n(\lambda x)$.

21. $y''''_{xxxx} + b\cosh^n(\lambda x)\, y''_{xx} + a[b\cosh^n(\lambda x) - a]y = 0.$

This is a special case of equation 4.1.5.5 with $f(x) = b\cosh^n(\lambda x)$.

22. $y''''_{xxxx} + [a + b\cosh^n(\lambda x)]y''_{xx} + ab\cosh^n(\lambda x)\, y = 0.$

This is a special case of equation 4.1.5.6 with $f(x) = b\cosh^n(\lambda x)$.

23. $x^2 y''''_{xxxx} + 2axy'_x - a[1 + ax^2\ln^2(bx)]y = 0.$

The substitution $w = y''_{xx} + a\ln(bx)\, y$ leads to a second order linear equation: $w''_{xx} - a\ln(bx)w = 0$.

24. $y''''_{xxxx} + a\ln^n(\lambda x)(x^3 y'''_{xxx} - 3x^2 y''_{xx} + 6xy'_x - 6y) = 0.$

This is a special case of equation 4.1.5.15 with $f(x) = a\ln^n(\lambda x)$.

4.1.4. Equation Containing Trigonometric Functions

1. $y''''_{xxxx} + 2ab\cos(bx)\, y'_x - a[b^2\sin(bx) + a\sin^2(bx)]y = 0.$

The substitution $w = y''_{xx} + a\sin(bx)\, y$ leads to a second order linear equation of the form 2.1.6.3: $w''_{xx} - a\sin(bx)w = 0$.

2. $y''''_{xxxx} + a\sin^n(\lambda x)\, y'_x + b[a\sin^n(\lambda x) - b^3]y = 0.$

Particular solution: $y_0 = e^{-bx}.$

3. $y''''_{xxxx} + [a\sin^n(\lambda x) + b^3]y'_x + ab\sin^n(\lambda x)\,y = 0.$

Particular solution: $y_0 = e^{-bx}$.

4. $y''''_{xxxx} + a\tan^n(\lambda x)\,y'_x + b[a\tan^n(\lambda x) - b^3]y = 0.$

Particular solution: $y_0 = e^{-bx}$.

5. $y''''_{xxxx} + [a\tan^n(\lambda x) + b^3]y'_x + ab\tan^n(\lambda x)\,y = 0.$

Particular solution: $y_0 = e^{-bx}$.

6. $y''''_{xxxx} + a\sin^n(\lambda x)\,y''_{xx} + b[a\sin^n(\lambda x) - b]y = 0.$

The substitution $w = y''_{xx} + by$ leads to a second order linear equation: $w''_{xx} + [a\sin^n(\lambda x) - b]w = 0$.

7. $y''''_{xxxx} + [a + b\sin^n(\lambda x)]y''_{xx} + ab\sin^n(\lambda x)\,y = 0.$

The substitution $w = y''_{xx} + ay$ leads to a second order linear equation: $w''_{xx} + b\sin^n(\lambda x)\,w = 0$.

8. $y''''_{xxxx} + b\tan^n(\lambda x)\,y''_{xx} + a[b\tan^n(\lambda x) - a]y = 0.$

This is a special case of equation 4.1.5.5 with $f(x) = b\tan^n(\lambda x)$.

9. $y''''_{xxxx} + [a + b\tan^n(\lambda x)]y''_{xx} + ab\tan^n(\lambda x)\,y = 0.$

This is a special case of equation 4.1.5.6 with $f(x) = b\tan^n(\lambda x)$.

10. $y''''_{xxxx} + a\sin^n(\lambda x)\,(x^3 y'''_{xxx} - 3x^2 y''_{xx} + 6xy'_x - 6y) = 0.$

This is a special case of equation 4.1.5.15 with $f(x) = a\sin^n(\lambda x)$.

11. $y''''_{xxxx} = a\sin^n(\lambda x)\,y'''_{xxx} + by'_x - ab\sin^n(\lambda x)\,y.$

Particular solutions: $y_k = e^{\beta_k x}$ $(k = 1,\ 2,\ 3)$, where β_k are the roots of the cubic equation $\beta^3 - b = 0$.

12. $y''''_{xxxx} = a\tan^n(\lambda x)\,y'''_{xxx} + by'_x - ab\tan^n(\lambda x)\,y.$

Particular solutions: $y_k = e^{\beta_k x}$ $(k = 1,\ 2,\ 3)$, where β_k are the roots of the cubic equation $\beta^3 - b = 0$.

13. $x^2 y''''_{xxxx} + a\sin^n(\lambda x)\,(x^2 y''_{xx} - 4xy'_x + 6y) = 0.$

The substitution $w = x^2 y''_{xx} - 4xy'_x + 6y$ leads to a second order linear equation: $w''_{xx} + a\sin^n(\lambda x)\,w = 0$.

14. $\sin^4 x\, y''''_{xxxx} + 2\sin^3 x \cos x\, y'''_{xxx} + \sin^2 x\,(\sin^2 x - 3)\, y''_{xx} + \sin x \cos x\,(2\sin^2 x + 3)\, y'_x + (a^4 \sin^4 x - 3)y = 0.$

The equation of a loaded rigid spherical shell.

If $a^4 = 1 - \lambda^2$ then the equation can be written as

$$\mathbf{LL}(y) - \lambda^2 y = 0, \qquad \text{where} \quad \mathbf{L} \equiv \frac{d^2}{dx^2} + \cot x \frac{d}{dx} - \cot^2 x.$$

This equation falls into two second order equations:

$$\mathbf{L}(y) + \lambda y = 0, \qquad \mathbf{L}(y) - \lambda y = 0,$$

which differ only in the sign of parameter λ. The transformation $\xi = \sin^2 x$, $w = y/\sin x$ reduces the latter equations to the hypergeometric equations 2.1.2.158:

$$\xi(\xi - 1)w''_{\xi\xi} + (\tfrac{5}{2}\xi - 2)w'_\xi + \tfrac{1}{4}(1 \mp \lambda)w = 0.$$

15. $(a\cos x + b)y''''_{xxxx} = a\cos x\, y.$

Particular solution: $y_0 = a\cos x + b$.

16. $(ax^m + b\cos x)y''''_{xxxx} = b\cos x\, y, \qquad m = 1,\ 2,\ 3.$

Particular solution: $y_0 = ax^m + b\cos x$.

17. $(a\sin x + b)y''''_{xxxx} = a\sin x\, y.$

Particular solution: $y_0 = a\sin x + b$.

18. $(ax^m + b\sin x)y''''_{xxxx} = b\sin x\, y, \qquad m = 1,\ 2,\ 3.$

Particular solution: $y_0 = ax^m + b\sin x$.

4.1.5. Equations Containing Arbitrary Functions

Notation: f, g, and h are arbitrary functions of x; a, b, and c are parameters.

1. $y''''_{xxxx} + fy'_x - a(f + a^3)y = 0.$

Particular solution: $y_0 = e^{ax}$.

2. $y''''_{xxxx} + (f + a^3)y'_x + afy = 0.$

Particular solution: $y_0 = e^{-ax}$.

3. $y''''_{xxxx} + xfy'_x - 3fy = 0.$

Particular solution: $y_0 = x^3$.

The substitution $z = xy'_x - 3y$ leads to a third order equation: $z'''_{xxx} + xfz = 0$.

4. $y''''_{xxxx} + (ax + b)fy'_x - afy = 0.$

Particular solution: $y_0 = ax + b$.

5. $y''''_{xxxx} + fy''_{xx} + a(f - a)y = 0$.

1°. Particular solutions with $a > 0$: $y_1 = \cos(x\sqrt{a})$, $y_2 = \sin(x\sqrt{a})$.

2°. Particular solutions with $a < 0$: $y_1 = \exp(-x\sqrt{-a})$, $y_2 = \exp(x\sqrt{-a})$.

The substitution $w = y''_{xx} + ay$ leads to a second order equation: $w''_{xx} + (f - a)w = 0$.

6. $y''''_{xxxx} + (f + a)y''_{xx} + afy = 0$.

1°. Particular solutions with $a > 0$: $y_1 = \cos(x\sqrt{a})$, $y_2 = \sin(x\sqrt{a})$.

2°. Particular solutions with $a < 0$: $y_1 = \exp(-x\sqrt{-a})$, $y_2 = \exp(x\sqrt{-a})$.

The substitution $w = y''_{xx} + ay$ leads to a second order equation: $w''_{xx} + fw = 0$.

7. $y''''_{xxxx} + f(x)(x^2 y''_{xx} - 4xy'_x + 6y) = 0$.

Particular solutions: $y_1 = x^2$, $y_2 = x^3$.

The substitution $w = x^2 y''_{xx} - 4xy'_x + 6y$ leads to a second order linear equation: $w''_{xx} + x^2 fw = 0$.

8. $y''''_{xxxx} + (ax^2 + bx + c)fy''_{xx} - 2afy = 0$.

Particular solution: $y_0 = ax^2 + bx + c$.

9. $y''''_{xxxx} + fy'''_{xxx} + xgy'_x - 2gy = 0$.

Particular solution: $y_0 = x^2$.

10. $y''''_{xxxx} + fy'''_{xxx} - 2a^2 y''_{xx} - a^2 fy'_x + a^4 y = 0$.

Particular solutions: $y_1 = e^{-ax}$, $y_2 = e^{ax}$.

11. $y''''_{xxxx} + fy'''_{xxx} + gy''_{xx} + afy'_x + a(g - a)y = 0$.

1°. Particular solutions with $a > 0$: $y_1 = \cos(x\sqrt{a})$, $y_2 = \sin(x\sqrt{a})$.

2°. Particular solutions with $a < 0$: $y_1 = \exp(-x\sqrt{-a})$, $y_2 = \exp(x\sqrt{-a})$.

The substitution $w = y''_{xx} + ay$ leads to a second order linear equation: $w''_{xx} + fw'_x + (g - a)w = 0$.

12. $y''''_{xxxx} + fy'''_{xxx} + (g + a)y''_{xx} + afy'_x + agy = 0$.

1°. Particular solutions with $a > 0$: $y_1 = \cos(x\sqrt{a})$, $y_2 = \sin(x\sqrt{a})$.

2°. Particular solutions with $a < 0$: $y_1 = \exp(-x\sqrt{-a})$, $y_2 = \exp(x\sqrt{-a})$.

The substitution $w = y''_{xx} + ay$ leads to a second order equation: $w''_{xx} + fw'_x + gw = 0$.

13. $y''''_{xxxx} + fy'''_{xxx} + gy''_{xx} + xhy'_x - hy = 0$.

Particular solution: $y_0 = x$.

14. $y''''_{xxxx} + f(x)y'''_{xxx} + g(x)(x^2 y''_{xx} - 2xy'_x + 2y) = 0$.

Particular solutions: $y_1 = x$, $y_2 = x^2$.

The substitution $z = x^2 y''_{xx} - 2xy'_x + 2y$ leads to a second order equation: $xz''_{xx} + (xf - 2)z'_x + x^3 gz = 0$.

15. $y''''_{xxxx} + f(x)(x^3 y'''_{xxx} - 3x^2 y''_{xx} + 6xy'_x - 6y) = 0.$

Particular solutions: $y_1 = x, \quad y_2 = x^2, \quad y_3 = x^3.$

The substitution $w = x^3 y'''_{xxx} - 3x^2 y''_{xx} + 6xy'_x - 6y$ leads to a first order linear equation: $w'_x + x^3 f w = 0$.

16. $y''''_{xxxx} = f y'''_{xxx} + a y'_x - a f y.$

Particular solutions: $y_k = e^{\lambda_k x}$ $(k = 1,\ 2,\ 3)$, where λ_k are the roots of the cubic equation $\lambda^3 - a = 0$.

17. $y''''_{xxxx} = (f - a) y'''_{xxx} + (af - b) y''_{xx} + (bf - c) y'_x + c f y.$

Particular solutions: $y_k = e^{\lambda_k x}$ $(k = 1,\ 2,\ 3)$, where λ_k are the roots of the cubic equation $\lambda^3 + a\lambda^2 + b\lambda + c = 0$.

18. $y''''_{xxxx} + (f + a) y'''_{xxx} + (af + g + axg) y''_{xx} + a^2 x g y'_x - a^2 g y = 0.$

Particular solutions: $y_1 = x, \quad y_2 = e^{-ax}.$

19. $y''''_{xxxx} + (f_3 + a) y'''_{xxx} + (f_2 + a f_3) y''_{xx} + (f_1 + a f_2) y'_x + a f_1 y = 0,$
where $f_k = f_k(x)$ $(k = 1,\ 2,\ 3)$.

Particular solution: $y_0 = e^{-ax}.$

20. $x y''''_{xxxx} + 4 y'''_{xxx} + a x y = f(x).$

The substitution $w(x) = xy$ leads to a nonhomogeneous constant-coefficient linear equation: $w''''_{xxxx} + aw = f(x)$.

21. $x y''''_{xxxx} + x f y'_x - [(x + 1) f + x + 4] y = 0.$

Particular solution: $y_0 = x e^x.$

22. $x^2 y''''_{xxxx} + a x y'''_{xxx} + (x^2 f + b) y''_{xx} + (a - 4) x f y'_x + (b - 2a + 6) f y = 0.$

The substitution $w = x^2 y''_{xx} + (a - 4) x y'_x + (b - 2a + 6) y$ leads to a second order equation: $w''_{xx} + f w = 0$.

23. $x^4 y''''_{xxxx} + a x^3 y'''_{xxx} + x f y'_x + (a - 3) f y = 0.$

Particular solution: $y_0 = x^{3-a}.$

24. $y''''_{xxxx} + f y'_x + f'_x y = g.$

Integrating yields $y'''_{xxx} + fy = \int g\,dx + C$.

25. $y''''_{xxxx} + 2 f'_x y'_x + (f''_{xx} - f^2) y = 0.$

The substitution $w = y''_{xx} + fy$ leads to a second order equation: $w''_{xx} - fw = 0$.

26. $y''''_{xxxx} + 10 f y''_{xx} + 10 f'_x y'_x + (3 f''_{xx} + 9 f^2) y = 0.$

Solution:

$$y = C_1 w_1^3 + C_2 w_1^2 w_2 + C_3 w_1 w_2^2 + C_4 w_2^3,$$

where w_1 and w_2 are nontrivial linearly-independent solutions of the second order equation $w''_{xx} + fw = 0$.

27. $y''''_{xxxx} + (f+g)y''_{xx} + 2f'_x y'_x + (f''_{xx} + fg)y = 0.$

The substitution $w = y''_{xx} + fy$ leads to a second order equation: $w''_{xx} + gw = 0$.

28. $y''''_{xxxx} + 6fy'''_{xxx} + (4f'_x + 11f^2 + 10g)y''_{xx} + (f''_{xx} + 7ff'_x + 6f^3 + 30fg + 10g'_x)y'_x$
$+ 3(2f'_x g + 5fg'_x + 6f^2 g + g''_{xx} + 3g^2)y = 0.$

Solution:

$$y = C_1 w_1^3 + C_2 w_1^2 w_2 + C_3 w_1 w_2^2 + C_4 w_2^3,$$

where w_1 and w_2 form a fundamental set of solutions of the second order equation $w''_{xx} + fw'_x + gw = 0$.

29. $(fy''_{xx})''_{xx} = 0.$

The equation of transverse vibrations of a bar.

Solution: $y = C_1 + C_2 x + \displaystyle\int_{x_0}^{x} \frac{x-t}{f(t)}(C_3 + C_4 t)\,dt.$

30. $y''''_{xxxx} + fy'_x + (f\tan x - 1)y = 0.$

Particular solution: $y_0 = \cos x$.

31. $y''''_{xxxx} + fy'_x - (1 + f\cot x)y = 0.$

Particular solution: $y_0 = \sin x$.

32. $y''''_{xxxx} = f(x)y.$

The transformation $x = t^{-1}$, $y = wt^{-3}$ leads to an equation of the similar form: $w''''_{tttt} = t^{-8} f(1/t)w$.

33. $y''''_{xxxx} = f\left(\dfrac{ax+b}{cx+d}\right)\dfrac{y}{(cx+d)^8}.$

The transformation $\xi = \dfrac{ax+b}{cx+d}$, $w = \dfrac{y}{(cx+d)^3}$ leads to a simpler equation:

$$w''''_{\xi\xi\xi\xi} = \Delta^{-4} f(\xi) w, \qquad \text{where} \quad \Delta = ad - bc.$$

34. $y''''_{xxxx} + f(x)y'_x + g(x)y + h(x) = 0.$

The transformation $x = t^{-1}, y = wt^{-3}$ leads to an equation of the similar form:

$$w''''_{tttt} - t^{-6} f\left(\frac{1}{t}\right) w'_t + \left[3t^{-7} f\left(\frac{1}{t}\right) + t^{-8} g\left(\frac{1}{t}\right)\right] w + t^{-5} h\left(\frac{1}{t}\right) = 0.$$

4.1.6. Asymptotic Solutions

This subsection presents asymptotic solutions, as $\varepsilon \to 0$ ($\varepsilon > 0$), of some fourth-order linear ordinary differential equations containing arbitrary functions (sufficiently smooth), with the independent variable being a real number.

1. Consider the equation

$$\varepsilon^4 y''''_{xxxx} - f(x)y = 0 \tag{1}$$

on a closed interval $a \le x \le b$. With the condition $f > 0$, the leading terms of the asymptotic expansions of the fundamental system of solutions, as $\varepsilon \to 0$, are given by the formulae

$$y_1 = [f(x)]^{-3/8} \exp\left\{-\frac{1}{\varepsilon}\int [f(x)]^{1/4}\,dx\right\}, \quad y_2 = [f(x)]^{-3/8} \exp\left\{\frac{1}{\varepsilon}\int [f(x)]^{1/4}\,dx\right\},$$

$$y_3 = [f(x)]^{-3/8} \cos\left\{\frac{1}{\varepsilon}\int [f(x)]^{1/4}\,dx\right\}, \quad y_4 = [f(x)]^{-3/8} \sin\left\{\frac{1}{\varepsilon}\int [f(x)]^{1/4}\,dx\right\}.$$

2. Now consider the "biquadratic" equation

$$\varepsilon^4 y''''_{xxxx} - 2\varepsilon^2 g(x) y''_{xx} - f(x)y = 0. \tag{2}$$

Introduce the notation

$$D(x) = [g(x)]^2 + f(x).$$

In the region where the conditions $f(x) \neq 0$ and $D(x) \neq 0$ are satisfied, the leading terms of the asymptotic expansions of the fundamental system of solutions of equation (2) are described by the formulae

$$y_k = [\lambda_k(x)]^{-1/2}[D(x)]^{-1/4} \exp\left\{\frac{1}{\varepsilon}\int \lambda_k(x)\,dx - \frac{1}{2}\int \frac{[\lambda_k(x)]'_x}{\sqrt{D(x)}}\,dx\right\}; \quad k = 1,\ 2,\ 3,\ 4.$$

where

$$\lambda_1(x) = \sqrt{g(x) + \sqrt{D(x)}}, \qquad \lambda_2(x) = -\sqrt{g(x) + \sqrt{D(x)}},$$

$$\lambda_3(x) = \sqrt{g(x) - \sqrt{D(x)}}, \qquad \lambda_4(x) = -\sqrt{g(x) - \sqrt{D(x)}}.$$

4.2. Nonlinear Equations

4.2.1. Equation Containing Power Functions

1. $\quad y''''_{xxxx} = Ay^{-5/3}$.

Multiply both sides of the equation by $y^{5/3}$ and differentiate the resulting expression with respect to x. We have

$$3yy^{(5)}_x + 5y'_x y''''_{xxxx} = 0.$$

Integrating the latter equation three times, we obtain a chain of equalities:

$$3yy''''_{xxxx} + 2y'_x y'''_{xxx} - (y''_{xx})^2 = 2C_2, \tag{1}$$

$$3yy'''_{xxx} - y'_x y''_{xx} = 2C_2 x + C_1, \tag{2}$$

$$3yy''_{xx} - 2(y'_x)^2 = C_2 x^2 + C_1 x + C_0, \tag{3}$$

where C_0, C_1, and C_2 are arbitrary constants. By eliminating the highest derivatives from (1)–(3) with the help of the original equation, we obtain a first order equation:

$$(2Py'_x - 3P'_x y)^2 = 9(C_1^2 - 4C_0C_2)y^2 - 2P^3 + 54APy^{4/3},$$

where $P = C_2x^2 + C_1x + C_0$. The substitution $y = (P/w)^{3/2}$ leads to an equation with separation of variables whereof integration finally yields

$$\int \left[9(C_1^2 - 4C_0C_2) + 54Aw - 2w^3\right]^{-1/2} \frac{dw}{w} \pm \int \frac{dx}{3P} = C_3.$$

2. $\boldsymbol{y''''_{xxxx} = Ay^m.}$

By integrating, we obtain ($m \neq -1$)

$$2y'_x y'''_{xxx} - \left(y''_{xx}\right)^2 = \frac{2A}{m+1} y^{m+1} + \frac{4}{3}C,$$

where C is an arbitrary constant. The substitution $w(y) = \left(y'_x\right)^{3/2}$ leads to a second order equation:

$$w''_{yy} = \left(\frac{3A}{2m+2} y^{m+1} + C\right) w^{-5/3}.$$

The value $C = 0$ corresponds to the Emden—Fowler equation whose integrable cases are specified in Section 2.3 for some values of m (to those cases correspond three-parameter families of particular solutions of the original equation).

3. $\boldsymbol{y''''_{xxxx} = Ax^{-3m-5}y^m.}$

The transformation $x = t^{-1}$, $y = t^{-3}w(t)$ leads to an equation of the form 4.2.1.2: $w''''_{xxxx} = Aw^m$.

4. $\boldsymbol{y''''_{xxxx} = Ax^{-\frac{3m+5}{2}} y^m.}$

This is a special case of equation 4.2.3.3 with $f(w) = Aw^m$.

5. $\boldsymbol{y''''_{xxxx} = (ay + bx^k)^m, \qquad k = 0,\ 1,\ 2,\ 3.}$

The substitution $aw = ay + bx^k$ leads to an equation of the form 4.2.1.2: $w''''_{xxxx} = a^m w^m$.

6. $\boldsymbol{x^{3m+1}(ax+b)^4 y''''_{xxxx} = cy^m.}$

This is a special case of equation 4.2.3.5 with $f(w) = cw^m$.

7. $\boldsymbol{y''''_{xxxx} = (ax^2 + bx + c)^{-\frac{3m+5}{2}} y^m.}$

This is a special case of equation 4.2.3.6 with $f(w) = w^m$.

8. $\boldsymbol{y''''_{xxxx} - \frac{5}{2}ay''_{xx} + \frac{9}{16}a^2 y = by^{-5/3}.}$

The transformation $\xi = e^{x\sqrt{a}}$, $w(\xi) = \xi^{3/2}y$ leads to an equation of the form 4.2.1.1: $w''''_{\xi\xi\xi\xi} = a^{-2}bw^{-5/3}$.

9. $xy''''_{xxxx} + 4y'''_{xxx} = Ax^{-5/3}y^{-5/3}.$

The substitution $w(x) = xy$ leads to an equation of the form 4.2.1.1: $w''''_{xxxx} = Aw^{-5/3}$.

10. $xy''''_{xxxx} + 2y'''_{xxx} = a(xy'_x - y)^m.$

The substitution $w(x) = xy'_x - y$ leads to a third order equation: $w'''_{xxx} = aw^m$ (Section 3.2 presents its solutions for $m = -\frac{7}{2}, -\frac{5}{2}, -2, -\frac{4}{3}, -\frac{7}{6}, -\frac{1}{2}$, 0, and 1).

11. $x^2y''''_{xxxx} + 8xy'''_{xxx} + 12y''_{xx} = ax^{-10/3}y^{-5/3}.$

The substitution $w(x) = x^2y$ leads to an equation of the form 4.2.1.1: $w''''_{xxxx} = aw^{-5/3}$.

12. $x^4y''''_{xxxx} + 6x^3y'''_{xxx} + 7x^2y''_{xx} + xy'_x = ay^{-5/3}.$

The substitution $t = \ln|x|$ leads to an equation of the form 4.2.1.1: $y''''_{xxxx} = ay^{-5/3}$.

13. $yy''''_{xxxx} = ay'_xy'''_{xxx}.$

Having integrated this equation, we obtain the third order equation $y'''_{xxx} = Cy^a$ whose solvable cases are specified in Section 3.2.

14. $yy''''_{xxxx} + 4y'_xy'''_{xxx} + 3(y''_{xx})^2 = ax^n.$

This is a special case of equation 4.2.3.22 with $f(x) = ax^n$.

15. $yy''''_{xxxx} + 4y'_xy'''_{xxx} + 3(y''_{xx})^2 = ay^{-10/3}.$

The substitution $w = y^2$ leads to an equation of the form 4.2.1.1: $w''''_{xxxx} = 2aw^{-5/3}$.

16. $yy''''_{xxxx} + \frac{3}{2}y'_xy'''_{xxx} + \frac{1}{2}(y''_{xx})^2 = (ax+b)y^{-1/2}.$

The transformation $x = x(t)$, $y = (x'_t)^2$ leads to a constant-coefficient fifth-order linear equation: $2x^{(5)}_t = ax + b$.

17. $y^3y''''_{xxxx} = 4y^2y'_xy'''_{xxx} + 3y^2(y''_{xx})^2 - 6(y'_x)^4.$

This is a special case of equation 4.2.3.27 with $f \equiv 0$.

Solution in the parametric form:

$$x = \pm\int\frac{dx}{\sqrt{2\xi^4 + C_2\xi + C_1}} + C_3, \qquad y = C_4\exp\left(\pm\int\frac{\xi\,d\xi}{\sqrt{2\xi^4 + C_2\xi + C_1}}\right).$$

18. $y''_{xx}y''''_{xxxx} = a(y'''_{xxx})^2.$

Solution:

$$y = \begin{cases} C_0 + C_1x + (C_2 + C_3x)^{\frac{3-2a}{1-a}} & \text{if } a \neq 1, \\ C_0 + C_1x + C_2\exp(C_3x) & \text{if } a = 1. \end{cases}$$

19. $y''_{xx}y''''_{xxxx} - \frac{1}{2}(y'''_{xxx})^2 = \alpha(xy'_x - y) + \beta y'_x + \gamma.$

Differentiating with respect to x yields

$$y''_{xx}\left(y_x^{(5)} - \alpha x - \beta\right) = 0.$$

By equating the expression in the parentheses to zero and integrating it, we find the solution:

$$y = \alpha\frac{x^6}{6!} + \beta\frac{x^5}{5!} + C_4x^4 + C_3x^3 + C_2x^2 + C_1x + C_0.$$

The constants C_k and parameters α, β, and γ are related by the constraint

$$48C_2C_4 - 18C_3^2 = -\alpha C_0 + \beta C_1 + \gamma$$

obtained by means of substituting the solution into the original equation. In addition, there exists the solution

$$y = \widetilde{C}_1x + \widetilde{C}_0, \qquad \text{where} \quad \alpha\widetilde{C}_0 - \beta\widetilde{C}_1 - \gamma = 0.$$

20. $y''''_{xxxx} = ay^k y'_x (y'''_{xxx})^s.$

This is a special case of equation 4.2.3.29 with $f(y) = Ay^k$, $g(w) = w^s$. For $k = -1$ and $s = 1$, see equation 4.2.1.13.

The first integral has the form:

$$\frac{1}{1-s}(y'''_{xxx})^{1-s} - \frac{A}{k+1}y^{k+1} = C \quad \text{if } k \neq -1,\ s \neq 1; \tag{1}$$

$$\ln y'''_{xxx} - \frac{A}{k+1}y^{k+1} = C \quad \text{if } k \neq -1,\ s = 1; \tag{2}$$

$$\frac{1}{1-s}(y'''_{xxx})^{1-s} - A\ln y = C \quad \text{if } k = -1,\ s = 1. \tag{3}$$

For $C = 0$, equality (1) is changing to the equation

$$y'''_{xxx} = \left[\frac{A(1-s)}{k+1}\right]^{\frac{1}{1-s}} y^{\frac{k+1}{1-s}}$$

which is discussed in Section 3.2 (the solutions given there generate 3-parametric families of particular solutions of the original equation for $k = (1-s)\beta - 1$, where $\beta = -\frac{7}{2}, -\frac{5}{2}, -2, -\frac{4}{3}, -\frac{7}{6}, -\frac{1}{2}$, 0, and 1).

4.2.2. Equations Containing Exponential, Hyperbolic, Logarithmic, and Trigonometric Functions

1. $y''''_{xxxx} = ae^{\lambda y}.$

This is a special case of equation 4.2.3.1 with $f(y) = ae^{\lambda y}$.

2. $y''''_{xxxx} = a(y + be^x)^{-5/3} - be^x.$

The substitution $w = y + be^x$ leads to an equation of the form 4.2.1.1: $w''''_{xxxx} = aw^{-5/3}$.

3. $y''''_{xxxx} = a(y + be^x)^m - be^x.$

The substitution $w = y + be^x$ leads to an equation of the form 4.2.1.2: $w''''_{xxxx} = aw^m$.

4. $y''''_{xxxx} - 4\lambda y'''_{xxx} + 6\lambda^2 y''_{xx} - 4\lambda^3 y'_x + \lambda^4 y = a\exp(\frac{8}{3}\lambda x)y^{-5/3}.$

The substitution $w(x) = ye^{-\lambda x}$ leads to an equation of the form 4.2.1.1: $w''''_{xxxx} = aw^{-5/3}$.

5. $y''''_{xxxx} - 4\lambda y'''_{xxx} + 6\lambda^2 y''_{xx} - 4\lambda^3 y'_x + \lambda^4 y = ae^{\lambda(1-m)x}y^m.$

The substitution $w(x) = ye^{-\lambda x}$ leads to an equation of the form 4.2.1.2: $w''''_{xxxx} = aw^m$.

6. $yy''''_{xxxx} + 4y'_x y'''_{xxx} + 3(y''_{xx})^2 = ae^{\lambda x}.$

Solution: $y^2 = C_3x^3 + C_2x^2 + C_1x + C_0 + 2a\lambda^{-4}e^{\lambda x}$.

7. $yy''''_{xxxx} + 4y'_x y'''_{xxx} + 3(y''_{xx})^2 = a\cosh(\lambda x).$

Solution: $y^2 = C_3x^3 + C_2x^2 + C_1x + C_0 + 2a\lambda^{-4}\cosh(\lambda x)$.

8. $yy''''_{xxxx} + 4y'_x y'''_{xxx} + 3(y''_{xx})^2 = a\tanh^m(\lambda x).$

This is a special case of equation 4.2.3.22 with $f(x) = a\tanh^m(\lambda x)$.

9. $y''''_{xxxx} = a\ln^m(by).$

This is a special case of equation 4.2.3.1 with $f(y) = a\ln^m(by)$.

10. $y''''_{xxxx} = ax^{-5}(\ln y - 3\ln x).$

This is a special case of equation 4.2.3.2 with $f(w) = a\ln w$.

11. $y''''_{xxxx} = ax^{-5/2}(2\ln y - 3\ln x).$

This is a special case of equation 4.2.3.3 with $f(w) = 2a\ln w$.

12. $yy''''_{xxxx} + 4y'_x y'''_{xxx} + 3(y''_{xx})^2 = a\ln^m(\lambda x).$

This is a special case of equation 4.2.3.22 with $f(x) = a\ln^m(\lambda x)$.

13. $y''''_{xxxx} = a\cos^m(\lambda y).$

This is a special case of equation 4.2.3.1 with $f(y) = a\cos^m(\lambda y)$.

14. $y''''_{xxxx} = a\tan^m(\lambda y).$

This is a special case of equation 4.2.3.1 with $f(y) = a\tan^m(\lambda y)$.

15. $yy''''_{xxxx} + 4y'_x y'''_{xxx} + 3(y''_{xx})^2 = a\cos(\lambda x).$

Solution: $y^2 = C_3x^3 + C_2x^2 + C_1x + C_0 + 2a\lambda^{-4}\cos(\lambda x)$.

16. $yy''''_{xxxx} + 4y'_x y'''_{xxx} + 3(y''_{xx})^2 = a\tan^m(\lambda x).$

This is a special case of equation 4.2.3.22 with $f(x) = a\tan^m(\lambda x)$.

4.2.3. Equations Containing Arbitrary Functions

1. $y''''_{xxxx} = f(y)$.

By integrating, we obtain

$$2y'_x y'''_{xxx} - (y''_{xx})^2 = 2\int f(y)\,dy + 2C.$$

The substitution $w(y) = |y'_x|^{3/2}$ leads to a second order equation:

$$w''_{yy} = \frac{3}{2}\left[\int f(y)\,dy + C\right] w^{-5/3}.$$

2. $y''''_{xxxx} = x^{-5} f(yx^{-3})$.

The transformation $x = t^{-1}$, $y = wt^{-3}$ leads to an equation of the form 4.2.3.1: $w''''_{tttt} = f(w)$.

3. $y''''_{xxxx} = x^{-5/2} f(yx^{-3/2})$.

The transformation $x = e^t$, $y = x^{3/2}w$ leads to an equation of the form 4.2.3.14: $w''''_{tttt} - \frac{5}{2}w''_{tt} = -\frac{9}{16}w + f(w)$.

4. $y''''_{xxxx} = f(y + \alpha x^3 + \beta x^2 + \gamma x + \delta)$.

The substitution $w = y + \alpha x^3 + \beta x^2 + \gamma x + \delta$ leads to an equation of the form 4.2.3.1: $w''''_{xxxx} = f(w)$.

5. $x(ax+b)^4 y''''_{xxxx} = f(yx^{-3})$.

The transformation $\xi = \ln\dfrac{ax+b}{x}$, $w = \dfrac{y}{x^3}$ leads to an autonomous equation of the form 4.2.3.34.

6. $y''''_{xxxx} = (ax^2+bx+c)^{-5/2} f\left(\dfrac{y}{(ax^2+bx+c)^{3/2}}\right)$.

1°. The transformation

$$\xi = \int \frac{dx}{ax^2+bx+c}, \qquad w = \frac{y}{(ax^2+bx+c)^{3/2}}$$

leads to an autonomous equation of the form 4.2.3.14 for $w = w(\xi)$:

$$w''''_{\xi\xi\xi\xi} - \tfrac{5}{2}\Delta w''_{\xi\xi} + \tfrac{9}{16}\Delta^2 w = f(w), \qquad \text{where} \quad \Delta = b^2 - 4ac.$$

Therefore, having integrated the latter equation, we obtain

$$w'_\xi w'''_{\xi\xi\xi} - \tfrac{1}{2}(w''_{\xi\xi})^2 - \tfrac{5}{4}\Delta(w'_\xi)^2 = -\tfrac{9}{32}\Delta^2 w^2 + \int f(w)\,dw + C.$$

The substitution $z(w) = |w'_\xi|^{3/2}$ leads to a second order equation:

$$z''_{ww} = \tfrac{15}{8}\Delta z^{-1/3} + \tfrac{3}{2}\left[-\tfrac{9}{32}\Delta^2 w^2 + \int f(w)\,dw + C\right] z^{-5/3}.$$

2°. The first integral of the original equation has the form

$$(Py'_x - \tfrac{3}{2}P'_x y)y'''_{xxx} - \tfrac{1}{2}P(y''_{xx})^2 + \tfrac{1}{2}P'_x y'_x y''_{xx} + 3ayy''_{xx} - 2a(y'_x)^2 = \int f(w)\,dw + C,$$

where $P = ax^2 + bx + c$, $w = yP^{-3/2}$.

7. $y''''_{xxxx} = f(y + ae^x) - ae^x.$

The substitution $w = y + ae^x$ leads to an autonomous equation of the form 4.2.3.1: $w''''_{xxxx} = f(w)$.

8. $y''''_{xxxx} = f(y + a\cosh x) - a\cosh x.$

The substitution $w = y + a\cosh x$ leads to an autonomous equation of the form 4.2.3.1: $w''''_{xxxx} = f(w)$.

9. $y''''_{xxxx} = f(y + a\sinh x) - a\sinh x.$

The substitution $w = y + a\sinh x$ leads to an autonomous equation of the form 4.2.3.1: $w''''_{xxxx} = f(w)$.

10. $y''''_{xxxx} = f(y + a\cos x) - a\cos x.$

The substitution $w = y + a\cos x$ leads to an autonomous equation of the form 4.2.3.1: $w''''_{xxxx} = f(w)$.

11. $y''''_{xxxx} = f(y + a\sin x) - a\sin x.$

The substitution $w = y + a\sin x$ leads to an autonomous equation of the form 4.2.3.1: $w''''_{xxxx} = f(w)$.

12. $y''''_{xxxx} = f(y)y'_x + g(x).$

By integrating, we find

$$y'''_{xxx} = \int f(y)\,dy + \int g(x)\,dx + C.$$

For $g(x) \equiv 0$, the order of this equation can lowered by one with the help of the substitution $w(y) = y'_x$.

13. $y''''_{xxxx} = x^{-4}f(xy'_x - y).$

The transformation $t = \ln|x|$, $w = xy'_x - y$ leads to a third order autonomous equation of the form 3.5.5.9: $w'''_{ttt} - 5w''_{tt} + 6w'_t = f(w)$.

14. $y''''_{xxxx} + ay''_{xx} = f(y).$

Having integrated this equation, we obtain

$$2y'_x y'''_{xxx} - (y''_{xx})^2 + a(y'_x)^2 = 2\int f(y)\,dy + 2C,$$

where C is an arbitrary constant. The substitution $w(y) = |y'_x|^{3/2}$ leads to a second order equation:

$$w''_{yy} = -\tfrac{3}{4}aw^{-1/3} + \tfrac{3}{2}\left[\int f(y)\,dy + C\right]w^{-5/3}.$$

15. $y''''_{xxxx} = x^{-2} f(xy'_x - y) y''_{xx}$.

The substitution $t = \ln|x|$, $w = xy'_x - y$ leads to a third order equation:

$$w'''_{ttt} - 5w''_{tt} + 6w'_t = f(w) w'_t.$$

Integrating it, we obtain a second order autonomous equation:

$$w''_{tt} - 5w'_t + 6w = \int f(w)\, dw + C.$$

The substitution $z(w) = \frac{1}{5} w'_t$ leads to the Abel equation of the second kind:

$$zz'_w - z = \tfrac{1}{25}\Big[-6w + \int f(w)\, dw + C\Big]$$

(see Section 1.3).

16. $y''''_{xxxx} = x^m f(x^2 y''_{xx} - 2xy'_x + 2y)$.

The substitution $w = x^2 y''_{xx} - 2xy'_x + 2y$ leads to a second order equation: $xw''_{xx} - 2w'_x = x^{m+3} f(w)$.

For $m = -4$, the substitution $z(w) = \frac{1}{3} xw'_x$ leads to the Abel equation of the second kind: $zz'_w - z = \frac{1}{9} f(w)$ (see Subsection 1.3.1).

17. $y''''_{xxxx} + ay'''_{xxx} + by''_{xx} + cy'_x = e^{\lambda x} f(ye^{-\lambda x})$.

The substitution $w(x) = ye^{-\lambda x}$ leads to an autonomous equation:

$$\begin{aligned} w''''_{xxxx} &+ (4\lambda + a) w'''_{xxx} + (6\lambda^2 + 3a\lambda + b) w''_{xx} \\ &+ (4\lambda^3 + 3a\lambda^2 + 2b\lambda + c) w'_x + (\lambda^4 + a\lambda^3 + b\lambda^2 + c\lambda) w = f(w), \end{aligned}$$

which can be reduced to a third order equation by means of the substitution $z(w) = w'_x$. For $a = -4\lambda$ and $c = 8\lambda^3 - 2b\lambda$, the above equation coincides, to a precision of the notation, with the equation 4.2.3.14 and can be reduced to a second order equation.

18. $xy''''_{xxxx} + 4y'''_{xxx} = f(xy)$.

The substitution $w(x) = xy$ leads to an equation of the form 4.2.3.1: $w''''_{xxxx} = f(w)$.

19. $x^2 y''''_{xxxx} + 8xy'''_{xxx} + 12y''_{xx} = f(x^2 y)$.

The substitution $w(x) = x^2 y$ leads to an autonomous equation of the form 4.2.3.1: $w''''_{xxxx} = f(w)$.

20. $x^4 y''''_{xxxx} + a_3 x^3 y'''_{xxx} + a_2 x^2 y''_{xx} + a_1 xy'_x = f(y)$.

The substitution $t = \ln|x|$ leads to an autonomous equation:

$$y''''_{tttt} + (a_3 - 6) y'''_{ttt} + (11 - 3a_3 + a_2) y''_{tt} + (2a_3 - a_2 + a_1 - 6) y'_t = f(y),$$

the order of which can lowered with the help of the substitution $w(y) = y'_t$. For $a_3 = 6$ and $a_1 = a_2 - 6$, the latter equation coincides, to a precision of the notation, with the equation 4.2.3.14 and can be reduced to a second order equation.

21. $x^4 y''''_{xxxx} + a x^3 y'''_{xxx} + b x^2 y''_{xx} + c x y'_x + s y = x^{-k} f(y x^k)$.

The transformation $t = \ln x$, $w = y x^k$ leads to an autonomous equation of the form 4.2.3.34.

22. $y y''''_{xxxx} + 4 y'_x y'''_{xxx} + 3 (y''_{xx})^2 = f(x)$.

Solution: $y^2 = C_3 x^3 + C_2 x^2 + C_1 x + C_0 + \frac{1}{3} \int_{x_0}^{x} (x-t)^3 f(t)\, dt$.

23. $y y''''_{xxxx} + a y'_x y'''_{xxx} + (a-1)(y''_{xx})^2 = f(x)$.

Having integrated this equation, we find

$$y y''_{xx} + \frac{a-2}{2} (y'_x)^2 = C_1 x + C_0 + \int_{x_0}^{x} (x-t) f(t)\, dt.$$

24. $y y''''_{xxxx} + (4 y'_x + f y) y'''_{xxx} + 3 (y''_{xx})^2 + 3 f y'_x y''_{xx} + g = 0, \qquad f = f(x),\ g = g(x)$.

The substitution $w = (y y'_x)''_{xx}$ leads to a first order linear equation: $w'_x + f w + g = 0$.
Solution:

$$y^2 = C_2 x^2 + C_1 x + C_0 + \int_{x_0}^{x} (x-t)^2 w(t)\, dt,$$

where $w(x) = e^{-F(x)} \bigl[C_3 - \int e^{F(x)} g(x)\, dx\bigr]$, $F(x) = \int f(x)\, dx$; x_0 is any number.

25. $y y''''_{xxxx} + (4 y'_x + f y) y'''_{xxx} + 3 (y''_{xx})^2 + (3 f y'_x + g y) y''_{xx} + g (y'_x)^2 + h y y'_x + s = 0$, where $f = f(x)$, $g = g(x)$, $h = h(x)$, $s = s(x)$.

The substitution $w = y y'_x$ leads to a nonhomogeneous third-order linear equation: $w'''_{xxx} + f w''_{xx} + g w'_x + h w + s = 0$.

26. $(y + a x + b) y''''_{xxxx} + 4 (y'_x + a) y'''_{xxx} + 3 (y''_{xx})^2 = f(x)$.

Solution: $(y + a x + b)^2 = C_3 x^3 + C_2 x^2 + C_1 x + C_0 + \frac{1}{3} \int_{x_0}^{x} (x-t)^3 f(t)\, dt$.

27. $y y''''_{xxxx} = 4 y'_x y'''_{xxx} + 3 (y''_{xx})^2 - 6 \dfrac{(y'_x)^4}{y^2} + \bigl[y y''_{xx} - (y'_x)^2\bigr] f\Bigl(\dfrac{y'_x}{y}\Bigr)$.

The transformation $\xi = \dfrac{y'_x}{y}$, $w = \dfrac{y''_{xx}}{y} - \Bigl(\dfrac{y'_x}{y}\Bigr)^2$ leads to a second order linear equation for w^2: $(w^2)''_{\xi\xi} = 24 \xi^2 + 2 f(\xi)$. Integrating yields

$$w^2 = C_2 \xi + C_1 + 2 \xi^4 + 2 \int_{\xi_0}^{\xi} (\xi - t) f(t)\, dt.$$

Taking into account that $\xi'_x = w$, $y'_x = \xi y$, $y'_\xi = \xi y / w$, we find the solution in the parametric form:

$$x = \int \frac{d\xi}{w} + C_3, \qquad y = C_4 \exp\Bigl(\int \frac{\xi\, d\xi}{w}\Bigr),$$

where

$$w = \pm \sqrt{C_2 \xi + C_1 + 2 \xi^4 + 2 \int_{\xi_0}^{\xi} (\xi - t) f(t)\, dt}.$$

28. $y''_{xx}y''''_{xxxx} - 3(y'''_{xxx})^2 = f(xy'_x - y)(y''_{xx})^5.$

The Legendre transformation $x = u'_t$, $y = tu'_t - u$ leads to an equation of the form 4.2.3.1: $u''''_{tttt} = -f(u)$.

29. $y''''_{xxxx} = f(y)y'_x g(y'''_{xxx}).$

By integrating, we obtain a third order autonomous equation:

$$\int \frac{dw}{g(w)} = \int f(y)\,dy + C, \qquad \text{where} \quad w = y'''_{xxx},$$

the order of which can be lowered by means of the substitution $z(y) = y'_x$.

30. $xy''''_{xxxx} + 2y'''_{xxx} = (xy''_{xx})^{-5} f\left(\dfrac{xy''_{xx}}{\sqrt{xy'_x - y}}\right).$

The substitution $w(x) = xy'_x - y$ leads to a third order equation of the form 3.5.2.11:

$$w'''_{xxx} = w^{-5/2} F\left(\frac{w'_x}{\sqrt{w}}\right), \qquad \text{where} \quad F(\xi) = \xi^{-5} f(\xi).$$

31. $x^2y''''_{xxxx} + 2xy'''_{xxx} = f(x^2y''_{xx} - 2xy'_x + 2y)g(x^2y'''_{xxx}).$

The substitution $w(x) = x^2y''_{xx} - 2xy'_x + 2y$ leads to a second order equation of the form 2.9.4.2: $w''_{xx} = f(w)g(w'_x)$.

32. $y''''_{xxxx} = f(x)g(x^3y'''_{xxx} - 3x^2y''_{xx} + 6xy'_x - 6y).$

The substitution $w(x) = x^3y'''_{xxx} - 3x^2y''_{xx} + 6xy'_x - 6y$ leads to a first order equation with separation of variables: $w'_x = x^3 f(x)g(w)$.

33. $y''''_{xxxx} = f(x,\ y'_x,\ y''_{xx},\ y'''_{xxx}).$

The substitution $w(x) = y'_x$ leads to a third order equation: $w'''_{xxx} = f(x,\ w,\ w'_x,\ w''_{xx})$.

34. $y''''_{xxxx} = f(y,\ y'_x,\ y''_{xx},\ y'''_{xxx}).$

Autonomous equation.

The substitution $w(y) = (y'_x)^2$ leads to a third order equation:

$$ww'''_{yyy} + \tfrac{1}{2}w'_y w''_{yy} = 2f(y,\ \pm\sqrt{w},\ \tfrac{1}{2}w'_y,\ \pm\tfrac{1}{2}\sqrt{w}\,w''_{yy}).$$

35. $y''''_{xxxx} = yf\left(\dfrac{y'_x}{y},\ \dfrac{y''_{xx}}{y},\ \dfrac{y'''_{xxx}}{y}\right).$

The transformation $\xi = \dfrac{y'_x}{y}$, $w = \dfrac{y''_{xx}}{y} - \left(\dfrac{y'_x}{y}\right)^2$ leads to a second order equation:

$$w^2w''_{\xi\xi} + w(w'_\xi)^2 + 4\xi ww'_\xi + 3w^2 + 6\xi^2 w + \xi^4 = f(\xi,\ w + \xi^2,\ ww'_\xi + 3\xi w + \xi^3).$$

36. $y''''_{xxxx} = yx^{-4} f\left(x^k y^m,\ \dfrac{xy'_x}{y},\ \dfrac{x^2y''_{xx}}{y},\ \dfrac{x^3y'''_{xxx}}{y}\right).$

The homogeneous equation in the extended sense.

The transformation $t = x^k y^m$, $z = \dfrac{xy'_x}{y}$ leads to a third order equation.

37. $$y''''_{xxxx} = yx^{-4} f\left(\frac{xy'_x}{y}, \frac{x^2 y''_{xx}}{y}, \frac{x^3 y'''_{xxx}}{y}\right).$$

The transformation $z = \dfrac{xy'_x}{y}$, $w = \dfrac{x^2 y''_{xx}}{y}$ leads to a second order equation.

38. $$y''''_{xxxx} = x^{-4} f(x^m e^{\alpha y}, xy'_x, x^2 y''_{xx}, x^3 y'''_{xxx}).$$

The transformation $z = x^m e^{\alpha y}$, $w = xy'_x$ leads to a third order equation.

39. $$y''''_{xxxx} = yf\left(e^{\alpha x} y^m, \frac{y'_x}{y}, \frac{y''_{xx}}{y}, \frac{y'''_{xxx}}{y}\right).$$

The transformation $z = e^{\alpha x} y^m$, $w = y'_x/y$ leads to a third order equation.

Chapter 5
Higher Order Differential Equations

5.1. Linear Equations

5.1.1. Preliminary Comments

In this Chapter, we denote higher derivatives by $y_x^{(n)}$ that stands for $\dfrac{d^n y}{dx^n}$.

1. The general solution of a homogeneous linear equation of the nth order

$$f_n(x)y_x^{(n)} + f_{n-1}y_x^{(n-1)} + \cdots + f_1(x)y'_x + f_0(x)y = 0 \tag{1}$$

has the form

$$y = C_1 y_1(x) + C_2 y_2(x) + \cdots + C_n y_n(x), \tag{2}$$

where $y_1(x)$, $y_2(x)$, ..., $y_n(x)$ make up a fundamental set of solutions (y_k are linearly-independent solutions; $y_k \not\equiv 0$); C_1, C_2, ..., C_n are arbitrary constants.

2. Let $y_0 = y_0(x)$ be a nontrivial particular solution of equation (1). Then, the substitution $y = y_0(x)\int z(x)\,dx$ leads to a linear equation of the $(n-1)$th order for function $z(x)$.

Given m linearly-independent solutions $y_1(x)$, $y_2(x)$, ..., $y_m(x)$ of equation (1), its order can be lowered down to $(n-m)$ by the following technique. The substitution $y = y_m(x)\int z(x)\,dx$ leads to an $(n-1)$th order equation for $z(x)$, with the following linearly-independent solutions known:

$$z_1 = \left(\frac{y_1}{y_m}\right)'_x, \quad z_2 = \left(\frac{y_2}{y_m}\right)'_x, \quad \ldots, \quad z_{m-1} = \left(\frac{y_{m-1}}{y_m}\right)'_x.$$

Furthermore, the substitution $z = z_{m-1}(x)\int w(x)\,dx$ yields an $(n-2)$th order equation, etc. Thus, the above procedure applied m times results in an $(n-m)$th order homogeneous linear equation.

3. A nonhomogeneous linear equation of the nth order has the form

$$f_n(x)y_x^{(n)} + f_{n-1}y_x^{(n-1)} + \cdots + f_1(x)y'_x + f_0(x)y = g(x). \tag{3}$$

The general solution of equation (3) is the sum of its particular solution and the general solution of the corresponding homogeneous equation (1).

Let $\{y_1(x), \ldots, y_n(x)\}$ be a fundamental set of solutions of the homogeneous differential equation (1), and $W(x)$ is the Wronskian:

$$W(y_1, y_2, \ldots, y_n) = \begin{vmatrix} y_1(x) & \cdots & y_n(x) \\ y_1'(x) & \cdots & y_n'(x) \\ \cdots & \cdots & \cdots \\ y_1^{(n-1)}(x) & \cdots & y_n^{(n-1)}(x) \end{vmatrix}, \tag{4}$$

where $y_k^{(m)}(x) = \dfrac{d^m y_k}{dx^m}$, $m = 1, 2, \ldots, n-1$; $k = 1, 2, \ldots, n$. Denote by $W_\nu(x)$ determinant (4) wherein the ν th column is replaced by the column $0, 0, \ldots, 0, g$ (from top to bottom). Then, the general solution of the nonhomogeneous linear equation (3) can be written as

$$y = \sum_{\nu=1}^{n} C_\nu y_\nu(x) + \sum_{\nu=1}^{n} y_\nu(x) \int \frac{W_\nu(x)\,dx}{f_n(x)W(x)}.$$

5.1.2. Equations Containing Power Functions

1. $\boldsymbol{y_x^{(6)} + ay = 0.}$

1°. For $a = 0$,

$$y = C_1 + C_2 x + C_3 x^2 + C_4 x^3 + C_5 x^4 + C_6 x^5.$$

2°. For $a = k^6 > 0$,

$$y = C_1 \cos kx + C_2 \sin kx + \cos\frac{kx}{2}(C_3 \cosh\xi + C_4 \sinh\xi) \\ + \sin\frac{kx}{2}(C_5 \cosh\xi + C_6 \sinh\xi), \qquad \xi = \frac{kx\sqrt{3}}{2}.$$

3°. For $a = -k^6 < 0$,

$$y = C_1 \cosh kx + C_2 \sinh kx + \cosh\frac{kx}{2}(C_3 \cos\xi + C_4 \sin\xi) \\ + \sinh\frac{kx}{2}(C_5 \cos\xi + C_6 \sin\xi), \qquad \xi = \frac{kx\sqrt{3}}{2}.$$

2. $\boldsymbol{y_x^{(2n)} = a^{2n} y.}$

Solution:

$$y = C_1 e^{ax} + C_2 e^{-ax} + \sum_{k=1}^{n-1} e^{\varphi_k}(A_k \cos\theta_k + B_k \sin\theta_k),$$

where $\varphi_k = ax\cos\dfrac{k\pi}{n}$, $\theta_k = ax\sin\dfrac{k\pi}{n}$; C_1, C_2, A_k, B_k $(k = 1, 2, \ldots, n-1)$ are arbitrary constants.

3. $\boldsymbol{y_x^{(n)} + a_{n-1} y_x^{(n-1)} + \dots + a_1 y'_x + a_0 y = 0.}$

The homogeneous constant-coefficient linear equation.

To solve this equation determine the n roots of the characteristic polynomial

$$P(\lambda) = \lambda^n + a_{n-1}\lambda^{n-1} + \dots + a_1\lambda + a_0.$$

The general solution is determined by these characteristic roots. Several cases are possible:

1. The roots are all real and different. Denote them by $\lambda_1, \lambda_2, \dots, \lambda_n$. Then, the general solution of the original equation is

$$y = C_1 \exp(\lambda_1 x) + C_2 \exp(\lambda_2 x) + \dots + C_n \exp(\lambda_n x).$$

2. There are $m \le n$ equal real roots, $\lambda_1 = \lambda_2 = \dots = \lambda_m$, while the other roots are real and different. In this case, the general solution is

$$y = \exp(\lambda_1 x)(C_1 + C_2 x + \dots + C_m x^{m-1}) \\ + C_{m+1} \exp(\lambda_{m+1} x) + C_{m+2} \exp(\lambda_{m+2} x) + \dots + C_n \exp(\lambda_n x).$$

3. There are m equal pairs ($2m \le n$) of complex conjugate roots, $\lambda = \alpha \pm i\beta$, while the other roots are real and different. Then, the general solution has the form

$$y = \exp(\alpha x)\cos(\beta x)(A_1 + A_2 x + \dots + A_m x^{m-1}) \\ + \exp(\alpha x)\sin(\beta x)(B_1 + B_2 x + \dots + B_m x^{m-1}) \\ + C_{2m+1} \exp(\lambda_{2m+1} x) + C_{2m+2} \exp(\lambda_{2m+2} x) + \dots + C_n \exp(\lambda_n x),$$

where $A_1, \dots, A_m, B_1, \dots, B_m$ are arbitrary constants.

4. In the general case, there are r different roots $\lambda_1, \lambda_2, \dots, \lambda_r$ of multiplicities $m_1, m_2, \dots, m_r$, respectively. Hence, the characteristic polynomial can be factorized:

$$P(\lambda) = (\lambda - \lambda_1)^{m_1} (\lambda - \lambda_2)^{m_2} \dots (\lambda - \lambda_r)^{m_r},$$

where $m_1 + m_2 + \dots + m_r = n$. Then, the general solution of the original equation is given by the formula

$$y = \sum_{k=1}^{r} \exp(\lambda_k x)(C_{k,0} + C_{k,1} x + \dots + C_{k,m_k-1} x^{m_k-1}),$$

where $C_{k,l}$ are arbitrary constants.

If $P(\lambda)$ has complex conjugate roots, in the above solution the real and imaginary parts should be taken, in view of the formula: $\exp(\alpha \pm i\beta) = e^{\alpha}(\cos\beta \pm i\sin\beta)$.

4. $\boldsymbol{y_x^{(n)} = axy + b, \qquad a > 0.}$

Solution:

$$y = \sum_{\nu=0}^{n} C_\nu \varepsilon_\nu \int_0^\infty \exp\left[\varepsilon_\nu x t - \frac{t^{n+1}}{a(n+1)}\right] dt,$$

where $\varepsilon_\nu = \exp\left(\dfrac{2\pi\nu i}{n+1}\right)$, $\displaystyle\sum_{\nu=0}^{n} C_\nu = \frac{b}{a}$, $i^2 = -1$.

5. $\boldsymbol{y_x^{(n)} + ax^\nu y_x' + a\nu x^{\nu-1} y = 0.}$

This equation can be reduced to an $(n-1)$th order equation: $y_x^{(n-1)} + ax^\nu y = C$, where C is an arbitrary constant.

6. $\boldsymbol{y_x^{(n)} + ax^{k+1} y_x' - a(n-1)x^k y = 0.}$

The substitution $z = xy_x' - (n-1)y$ leads to an $(n-1)$th order equation: $z_x^{(n-1)} + ax^{k+1}z = 0$.

7. $\boldsymbol{y_x^{(n)} + ax^{k+1} y_x' + a(k+n)x^k y = 0.}$

The transformation $x = t^{-1}$, $y = wt^{1-n}$ leads to an equation of the form 5.1.2.5: $w_t^{(n)} + bt^\nu w_t' + b\nu t^{\nu-1} w = 0$, where $b = a(-1)^{n+1}$, $\nu = 1 - k - 2n$.

8. $\boldsymbol{y_x^{(n)} + ax^k y_x^{(m)} - (ab^m x^k + b^n)y = 0.}$

Particular solution: $y_0 = e^{bx}$.

9. $\boldsymbol{y_x^{(n)} + (ax^k - b^{n-m})y_x^{(m)} - ab^m x^k y = 0.}$

Particular solution: $y_0 = e^{bx}$.

10. $\boldsymbol{y_x^{(n)} + (ax^{m+1} + bx^m)y_x' - ax^m y = 0.}$

Particular solution: $y_0 = ax + b$.

11. $\boldsymbol{y_x^{(n)} + ay_x^{(n-1)} + bx^m y_x' + abx^m y = 0.}$

Particular solution: $y_0 = e^{-ax}$.

12. $\boldsymbol{xy_x^{(n)} - nmy_x^{(n-1)} + axy = 0, \qquad n = 2,\ 3,\ 4,\ \dots, \qquad m = 1,\ 2,\ 3,\ \dots}$

Solution:

$$y = x^{(m+1)n-1}\left(x^{1-n}\frac{d}{dx}\right)^m (x^{1-n}w),$$

where w is the general solution of the constant coefficient equation $w_x^{(n)} + aw = 0$.

13. $\boldsymbol{xy_x^{(n)} + ny_x^{(n-1)} = axy + b.}$

The substitution $w = xy$ leads to a constant coefficient equation: $w_x^{(n)} = aw + b$.

14. $\boldsymbol{xy_x^{(n)} + ny_x^{(n-1)} = ax^2 y + b.}$

The substitution $w = xy$ leads to an equation of the form 5.1.2.4: $w_x^{(n)} = axw + b$.

15. $\boldsymbol{xy_x^{(n)} + (n - m - 1)y_x^{(n-1)} + ax^k y_x' - amx^{k-1} y = 0.}$

Particular solution: $y_0 = x^m$.

16. $\boldsymbol{xy_x^{(n)} + ax^k y_x^{(m)} - (ax^k + amx^{k-1} + x + n)y = 0.}$

Particular solution: $y_0 = xe^x$.

17. $xy_x^{(n)} = \sum_{\nu=0}^{n-1}[(aA_{\nu+1} - A_\nu)x + A_{\nu+1}]y_x^{(\nu)}$,

where $A_n = 1$, $A_0 = 0$; a and A_ν are arbitrary numbers ($\nu = 1, 2, \ldots, n-1$).

Denote $f(\lambda) = \sum_{\nu=0}^{n-1} A_{\nu+1}\lambda^\nu$. Let all the roots $\lambda_1, \lambda_2, \ldots, \lambda_{n-1}$ of the equation $f(\lambda) = 0$ be different, and $f(a) \neq 0$. Then, the solution is as follows:

$$y = C_1 e^{\lambda_1 x} + C_2 e^{\lambda_2 x} + \ldots + C_{n-1} e^{\lambda_{n-1} x} + C_n e^{ax}\left[x - \frac{f'_a(a)}{f(a)}\right].$$

18. $\sum_{\nu=0}^{n}(a_\nu x + b_\nu)y_x^{(\nu)} = 0.$

The Laplace equation.
Particular solutions:

$$y_k = \int_{\alpha_k}^{\beta_k} \frac{1}{P(t)} \exp\left[xt + \int \frac{Q(t)}{P(t)}\,dt\right] dt,$$

where $P(t) = \sum_{\nu=0}^{n} a_\nu t^\nu$, $Q(t) = \sum_{\nu=0}^{n} b_\nu t^\nu$; α_k and β_k are found from the condition

$$\exp\left(xt + \int \frac{Q(t)}{P(t)}\,dt\right)\bigg|_{\alpha_k}^{\beta_k} = 0.$$

In many cases, the path of integration should be chosen on the complex plane.

19. $x^2 y_x^{(n)} + 2nxy_x^{(n-1)} + n(n-1)y_x^{(n-2)} = ax^2 y + b.$

The substitution $w = x^2 y$ leads to a constant coefficient equation: $w_x^{(n)} = aw + b$.

20. $x^2 y_x^{(n)} + 2nxy_x^{(n-1)} + n(n-1)y_x^{(n-2)} = ax^3 y + b.$

The substitution $w = x^2 y$ leads to a equation of the form 5.1.2.4: $w_x^{(n)} = axw + b$.

21. $x(x+m)y_x^{(n)} + x(ax^k - x - n)y_x^{(m)} - a(x+m)x^k y = 0.$

Particular solution: $y_0 = xe^x$.

22. $x^{2n} y_x^{(n)} = ay.$

The transformation $x = t^{-1}$, $y = wt^{1-n}$ leads to a constant coefficient equation: $w_t^{(n)} = (-1)^n aw$.

23. $x^n y_x^{(2n)} = ay.$

Solution:

$$y = x^{n/2} \sum_{k=1}^{n}\left[C_{k1} I_n\left(2\beta_k\sqrt{x}\right) + C_{k2} K_n\left(2\beta_k\sqrt{x}\right)\right],$$

where I_n and K_n are modified Bessel functions; $\beta_1, \beta_2, \ldots, \beta_n$ are the roots of the equation $\beta^n = \sqrt{a}$.

24. $x^{3n} y_x^{(2n)} = ay.$

The transformation $x = t^{-1}$, $y = wt^{1-2n}$ leads to an equation of the form 5.1.2.23: $t^n w_t^{(2n)} = aw$.

25. $x^{n+1/2} y_x^{(2n+1)} = ay.$

Solution:

$$y = x^{(2n+1)/4} \sum_{k=0}^{2n} C_k \big[J_{-n-1/2}(2\beta_k \sqrt{x}\,) + i J_{n+1/2}(2\beta_k \sqrt{x}\,) \big],$$

where β_0, β_1, ..., β_{2n} are the roots of the equation $\beta^{2n+1} = -ai$; $i^2 = -1$.

26. $x^{3n+3/2} y_x^{(2n+1)} = ay.$

The transformation $x = t^{-1}$, $y = wt^{-2n}$ leads to an equation of the form 5.1.2.25: $t^{n+1/2} w_t^{(2n+1)} = -aw$.

27. $a_n x^n y_x^{(n)} + a_{n-1} x^{n-1} y_x^{(n-1)} + \cdots + a_1 x y'_x + a_0 y = 0.$

The Euler equation.

If all the roots λ_k ($k = 1, 2, \ldots, n$) of the algebraic equation

$$\sum_{\nu=1}^{n} a_\nu \lambda(\lambda-1)\ldots(\lambda-\nu+1) = -a_0$$

are different, the general solution of the original differential equation has the form

$$y = C_1|x|^{\lambda_1} + C_2|x|^{\lambda_2} + \cdots + C_n|x|^{\lambda_n}.$$

In the general case, the substitution $t = \ln|x|$ leads to a constant coefficient equation of the form 5.1.2.3:

$$\sum_{\nu=1}^{n} a_\nu D(D-1)\ldots(D-\nu+1)y = -a_0 y, \qquad \text{where} \quad D = \frac{d}{dx}.$$

28. $x^{2n+1} y_x^{(n)} = ay + bx^n.$

The transformation $x = t^{-1}$, $y = wt^{1-n}$ leads to an equation of the form 5.1.2.4: $w_t^{(n)} = (-1)^n (atw + b)$.

29. $x^{2n+1} y_x^{(n)} + n x^{2n} y_x^{(n-1)} = axy.$

The substitution $w = xy$ leads to an equation of the form 5.1.2.22: $x^{2n} w_x^{(n)} = aw$.

30. $x^{2n+1} y_x^{(n)} + n x^{2n} y_x^{(n-1)} = ay.$

The substitution $w = xy$ leads to an equation of the form 5.1.2.28: $x^{2n+1} w_x^{(n)} = aw$.

31. $x^n y_x^{(2n)} + 2n x^{n-1} y_x^{(2n-1)} = ay.$

The substitution $w = xy$ leads to an equation of the form 5.1.2.23: $x^n w_x^{(2n)} = aw$.

32. $x^{3n}y_x^{(2n)} + 2nx^{3n-1}y_x^{(2n-1)} = ay.$

The substitution $w = xy$ leads to an equation of the form 5.1.2.24: $x^{3n}w_x^{(2n)} = aw$.

33. $x^{n+1}y_x^{(2n+1)} + (2n+1)x^n y_x^{(2n)} = a\sqrt{x}y.$

The substitution $w = xy$ leads to an equation of the form 5.1.2.25: $x^{n+1/2}w_x^{(2n+1)} = aw$.

34. $x^{3n+3/2}y_x^{(2n+1)} + (2n+1)x^{3n+1/2}y_x^{(2n)} = ay.$

The substitution $w = xy$ leads to an equation of the form 5.1.2.26: $x^{3n+3/2}w_x^{(2n+1)} = aw$.

35. $(ax+b)^{2n+1}y_x^{(n)} = (cx+d)y.$

The transformation

$$\xi = \frac{cx+d}{ax+b}, \qquad w = \frac{y}{(ax+b)^{n-1}}$$

leads to an equation of the form 5.1.2.4: $w_\xi^{(n)} = \Delta^{-n}\xi w$, where $\Delta = bc - ad$.

36. $(ax+b)^n(cx+d)^n y_x^{(n)} = ky.$

1°. The transformation

$$\xi = \ln\frac{ax+b}{cx+d}, \qquad w = \frac{y}{(cx+d)^{n-1}}$$

leads to a constant coefficient equation.

2°. The transformation

$$\zeta = \frac{ax+b}{cx+d}, \qquad w = \frac{y}{(cx+d)^{n-1}}$$

leads to the Euler equation 5.1.2.27: $\zeta^n w_\zeta^{(n)} = k\Delta^{-n}w$, where $\Delta = ad - bc$.

37. $(ax^2+bx+c)^n y_x^{(n)} = ky.$

The transformation

$$\xi = \int\frac{dx}{ax^2+bx+c}, \qquad w = y(ax^2+bx+c)^{\frac{1-n}{2}}$$

leads to a constant coefficient equation.

38. $(ax+b)^n(cx+d)^{3n}y_x^{(2n)} = ky.$

The transformation

$$\xi = \frac{ax+b}{cx+d}, \qquad w = \frac{y}{(cx+d)^{2n-1}}$$

leads to an equation of the form 5.1.2.23: $\xi^n w_\xi^{(2n)} = k\Delta^{-2n}w$, where $\Delta = ad - bc$.

39. $(ax+b)^{n+1/2}(cx+d)^{3n+3/2}y_x^{(2n+1)} = ky.$

The transformation

$$\xi = \frac{ax+b}{cx+d}, \qquad w = \frac{y}{(cx+d)^{2n}}$$

leads to an equation of the form 5.1.2.25: $\xi^{n+1/2}w_\xi^{(2n+1)} = k\Delta^{-2n-1}w$, where $\Delta = ad - bc$.

40. $P_{n-1}(x)y_x^{(n)} + P_{n-2}(x)y_x^{(n-1)} + \cdots + P_1(x)y''_{xx} + (a_1x+b_1)y'_x - ma_1y = 0,$
where P_ν are polynomials of the degree $\le \nu$, m is a positive integer, $a_1 \ne 0$.

A particular solution of this equation is the polynomial of degree m which can be written as

$$y_0 = \sum_{k=0}^{m}\Bigl(-\frac{1}{a_1}\Bigr)^k [x^m I x^{-m-1}(P_{n-1}D^n + \cdots + P_1D^2 + b_1D)]^k x^m,$$

where $D = \dfrac{d}{dx}$, $Ix^\nu = \dfrac{x^{\nu+1}}{\nu+1}$ with $\nu \ne -1$.

41. $[a_nx^n + P_{n-1}(x)]y_x^{(n)} + \cdots + [a_1x + P_0(x)]y'_x + a_0y = 0,$
where P_ν are polynomials of the degree $\le \nu$.

Assume that for some integer $m \ge 0$,

$$\sum_{\nu=0}^{n} C_m^\nu \nu!\, a_\nu = 0, \qquad \text{where} \quad C_m^\nu = \frac{m!}{\nu!\,(m-\nu)!},$$

and m is the least of the numbers satisfying this condition. Then, there exists a solution in the form of a polynomial of degree m such that no polynomial of a smaller degree satisfies the original equation.

5.1.3. Equations Containing Exponential Functions

1. $y_x^{(n)} + (ax+b)e^{\lambda x}y'_x - ae^{\lambda x}y = 0.$

Particular solution: $y_0 = ax + b$.

2. $y_x^{(n)} + (ae^{\lambda x} - b^{n-m})y_x^{(m)} - ab^m e^{\lambda x}y = 0.$

Particular solution: $y_0 = e^{bx}$.

3. $y_x^{(n)} + ay_x^{(n-1)} + be^{\lambda x}y'_x + abe^{\lambda x}y = 0.$

Particular solution: $y_0 = e^{-ax}$.

4. $y_x^{(n)} + ae^{\lambda x}y_x^{(m)} - (ab^m e^{\lambda x} + b^n)y = 0.$

Particular solution: $y_0 = e^{bx}$.

5. $$y_x^{(n)} = \sum_{k=0}^{n-1} (A_{k+1}e^{\lambda x} + bA_{k+1} - A_k)y_x^{(k)},$$

where $A_n = 1$, $A_0 = 0$; b and A_k are arbitrary numbers ($k = 1, 2, \ldots, n-1$).

Particular solutions: $y_m = e^{\mu_m x}$, where μ_m are the roots of the polynomial equation $\sum_{k=0}^{n-1} A_{k+1}\mu^k = 0$.

6. $$xy_x^{(n)} + axe^{\lambda x}y_x^{(m)} - [a(x+m)e^{\lambda x} + x + n]y = 0.$$

Particular solution: $y_0 = xe^x$.

7. $$xy_x^{(n)} + (n - m - 1)y_x^{(n-1)} + axe^{\lambda x}y'_x - ame^{\lambda x}y = 0.$$

Particular solution: $y_0 = x^m$.

8. $$x(x+m)y_x^{(n)} + x(ae^{\lambda x} - x - n)y_x^{(m)} - a(x+m)e^{\lambda x}y = 0.$$

Particular solution: $y_0 = xe^x$.

9. $$(ax^m + be^x + c)y_x^{(n)} = be^x y, \qquad m = 0,\ 1,\ 2,\ \ldots,\ n-1.$$

Particular solution: $y_0 = ax^m + be^x + c$.

10. $$(ax^m e^x + b)y_x^{(n)} = (-1)^n by, \qquad m = 0,\ 1,\ 2,\ \ldots,\ n-1.$$

Particular solution: $y_0 = ax^m + be^{-x}$.

11. $$\left(ae^x + \sum_{k=0}^{n-1} b_k x^k\right)y_x^{(n)} = ae^x y.$$

Particular solution: $y_0 = ae^x + \sum_{k=0}^{n-1} b_k x^k$.

12. $$y_x^{(n)} + a\cosh^k x\, y_x^{(m)} - (ab^m \cosh^k x + b^n)y = 0.$$

Particular solution: $y_0 = e^{bx}$.

13. $$y_x^{(n)} + (a\cosh^k x - b^{n-m})y_x^{(m)} - ab^m \cosh^k x\, y = 0.$$

Particular solution: $y_0 = e^{bx}$.

14. $$y_x^{(n)} + (ax+b)\cosh^m(\lambda x)y'_x - a\cosh^m(\lambda x)y = 0.$$

Particular solution: $y_0 = ax + b$.

15. $$xy_x^{(n)} + ax\cosh^k x\, y_x^{(m)} - [a(x+m)\cosh^k x + x + n]y = 0.$$

Particular solution: $y_0 = xe^x$.

16. $$y_x^{(n)} + a\sinh^k x\, y_x^{(m)} - (ab^m \sinh^k x + b^n)y = 0.$$

Particular solution: $y_0 = e^{bx}$.

17. $y_x^{(n)} + (a \sinh^k x - b^{n-m}) y_x^{(m)} - ab^m \sinh^k x \, y = 0.$

Particular solution: $y_0 = e^{bx}$.

18. $y_x^{(n)} + (ax + b) \sinh^m(\lambda x) y'_x - a \sinh^m(\lambda x) y = 0.$

Particular solution: $y_0 = ax + b$.

19. $xy_x^{(n)} + ax \sinh^k x \, y_x^{(m)} - [a(x + m) \sinh^k x + x + n] y = 0.$

Particular solution: $y_0 = xe^x$.

20. $y_x^{(n)} + a \tanh^k x \, y_x^{(m)} - (ab^m \tanh^k x + b^n) y = 0.$

Particular solution: $y_0 = e^{bx}$.

21. $y_x^{(n)} + (a \tanh^k x - b^{n-m}) y_x^{(m)} - ab^m \tanh^k x \, y = 0.$

Particular solution: $y_0 = e^{bx}$.

22. $y_x^{(n)} + (ax + b) \tanh^m(\lambda x) y'_x - a \tanh^m(\lambda x) y = 0.$

Particular solution: $y_0 = ax + b$.

23. $xy_x^{(n)} + ax \tanh^k x \, y_x^{(m)} - [a(x + m) \tanh^k x + x + n] y = 0.$

Particular solution: $y_0 = xe^x$.

24. $y_x^{(n)} + a \coth^k x \, y_x^{(m)} - (ab^m \coth^k x + b^n) y = 0.$

Particular solution: $y_0 = e^{bx}$.

25. $y_x^{(n)} + (a \coth^k x - b^{n-m}) y_x^{(m)} - ab^m \coth^k x \, y = 0.$

Particular solution: $y_0 = e^{bx}$.

26. $y_x^{(n)} + (ax + b) \coth^m(\lambda x) \, y'_x - a \coth^m(\lambda x) \, y = 0.$

Particular solution: $y_0 = ax + b$.

27. $xy_x^{(n)} + ax \coth^k x \, y_x^{(m)} - [a(x + m) \coth^k x + x + n] y = 0.$

Particular solution: $y_0 = xe^x$.

5.1.4. Equations Containing Trigonometric Functions

1. $y_x^{(n)} + a \sin^k x \, y_x^{(m)} - (ab^m \sin^k x + b^n) y = 0.$

Particular solution: $y_0 = e^{bx}$.

2. $y_x^{(n)} + (a \sin^k x - b^{n-m}) y_x^{(m)} - ab^m \sin^k x \, y = 0.$

Particular solution: $y_0 = e^{bx}$.

3. $y_x^{(n)} + ay_x^{(n-1)} + b\sin^m(\lambda x)y'_x + ab\sin^m(\lambda x)y = 0.$

Particular solution: $y_0 = e^{-ax}$.

4. $y_x^{(n)} + (ax+b)\sin^m(\lambda x)y'_x - a\sin^m(\lambda x)y = 0.$

Particular solution: $y_0 = ax + b$.

5. $y_x^{(n)} + a\cos^k x\, y_x^{(m)} - (ab^m\cos^k x + b^n)y = 0.$

Particular solution: $y_0 = e^{bx}$.

6. $y_x^{(n)} + (a\cos^k x - b^{n-m})y_x^{(m)} - ab^m\cos^k x\, y = 0.$

Particular solution: $y_0 = e^{bx}$.

7. $y_x^{(n)} + ay_x^{(n-1)} + b\cos^m(\lambda x)y'_x + ab\cos^m(\lambda x)y = 0.$

Particular solution: $y_0 = e^{-ax}$.

8. $y_x^{(n)} + (ax+b)\cos^m(\lambda x)y'_x - a\cos^m(\lambda x)y = 0.$

Particular solution: $y_0 = ax + b$.

9. $y_x^{(n)} + a\tan^k x\, y_x^{(m)} - (ab^m\tan^k x + b^n)y = 0.$

Particular solution: $y_0 = e^{bx}$.

10. $y_x^{(n)} + (a\tan^k x - b^{n-m})y_x^{(m)} - ab^m\tan^k x\, y = 0.$

Particular solution: $y_0 = e^{bx}$.

11. $y_x^{(n)} + ay_x^{(n-1)} + b\tan^m(\lambda x)y'_x + ab\tan^m(\lambda x)y = 0.$

Particular solution: $y_0 = e^{-ax}$.

12. $y_x^{(n)} + (ax+b)\tan^m(\lambda x)y'_x - a\tan^m(\lambda x)y = 0.$

Particular solution: $y_0 = ax + b$.

13. $xy_x^{(n)} + ax\sin^k(\lambda x)y_x^{(m)} - [a(x+m)\sin^k(\lambda x) + x + n]y = 0.$

Particular solution: $y_0 = xe^x$.

14. $xy_x^{(n)} + ax\cos^k(\lambda x)y_x^{(m)} - [a(x+m)\cos^k(\lambda x) + x + n]y = 0.$

Particular solution: $y_0 = xe^x$.

15. $xy_x^{(n)} + ax\tan^k(\lambda x)y_x^{(m)} - [a(x+m)\tan^k(\lambda x) + x + n]y = 0.$

Particular solution: $y_0 = xe^x$.

16. $(ax^m + b\sin x)y_x^{(n)} = b\sin\left(x + \dfrac{\pi n}{2}\right)y, \qquad m = 0,\ 1,\ 2,\ \ldots,\ n-1.$

Particular solution: $y_0 = ax^m + b\sin x$.

17. $$\left(a\sin x + \sum_{k=0}^{n-1} b_k x^k\right) y_x^{(n)} = a\sin\left(x + \frac{\pi n}{2}\right) y.$$

Particular solution: $y_0 = a\sin x + \sum_{k=0}^{n-1} b_k x^k$.

18. $$(ax^m + b\cos x) y_x^{(n)} = b\cos\left(x + \frac{\pi n}{2}\right) y, \qquad m = 0,\ 1,\ 2,\ \ldots,\ n-1.$$

Particular solution: $y_0 = ax^m + b\cos x$.

19. $$\left(a\cos x + \sum_{k=0}^{n-1} b_k x^k\right) y_x^{(n)} = a\cos\left(x + \frac{\pi n}{2}\right) y.$$

Particular solution: $y_0 = a\sin x + \sum_{k=0}^{n-1} b_k x^k$.

5.1.5. Equations Containing Arbitrary Functions

1. $y_x^{(n)} = f(x)$.

Solution:

$$y = \sum_{\nu=0}^{n-1} C_\nu x^\nu + \int_{x_0}^{x} \frac{(x-t)^{n-1}}{(n-1)!} f(t)\,dt,$$

where x_0 may be chosen arbitrarily.

2. $y_x^{(n)} + xf(x)y'_x - mf(x)y = 0$.

If $m = 0,\ 1,\ 2,\ \ldots$, or $n-1$, the equation has a particular solution $y_0 = x^m$, and the substitution $z = xy'_x - my$ leads to an $(n-1)$ th order equation:

$$D^{n-m-1}\left(\frac{z_x^{(m)}}{x}\right) + f(x)z = 0, \qquad \text{where} \quad D = \frac{d}{dx}.$$

In particular, for $m = n-1$, we have $z_x^{(n-1)} + xf(x)z = 0$.

3. $y_x^{(n)} + (ax+b)f(x)y'_x - af(x)y = 0$.

Particular solution: $y_0 = ax + b$.

4. $y_x^{(n)} + f(x)(x^2 y''_{xx} - 2xy'_x + 2y) = 0$.

Particular solutions: $y_1 = x$, $y_2 = x^2$.

The substitution $z = x^2 y''_{xx} - 2xy'_x + 2y$ leads to a linear equation of the $(n-2)$ th order.

5. $y_x^{(n)} + f(x)y_x^{(m)} - [a^n + a^m f(x)]y = 0$.

Particular solution: $y_0 = e^{ax}$.

6. $y_x^{(n)} + (f - a^{n-m})y_x^{(m)} - a^m f y = 0, \qquad f = f(x).$

Particular solution: $y_0 = e^{ax}$.

7. $y_x^{(n)} + a y_x^{(n-1)} + f y_x' + a f y = 0, \qquad f = f(x).$

Particular solution: $y_0 = e^{-ax}$.

8. $y_x^{(n)} + f(x) y_x^{(n-1)} + g(x) y_x^{(n-2)} + h(x) = 0.$

The substitution $w(x) = y_x^{(n-2)}$ leads to a linear equation of the second order: $w''_{xx} + f(x)w'_x + g(x)w + h(x) = 0$.

9. $y_x^{(n)} + a_{n-1} y_x^{(n-1)} + \cdots + a_1 y_x' + a_0 y = f(x).$

The nonhomogeneous constant-coefficient linear equation.

The general solution of this equation is the sum of the general solution of the corresponding homogeneous equation (see 5.1.2.3) and any particular solution of the nonhomogeneous equation.

If all the roots of the polynomial

$$P(\lambda) = \lambda^n + a_{n-1}\lambda^{n-1} + \cdots + a_1\lambda + a_0$$

are different, the original equation has the general solution

$$y = \sum_{\nu=1}^{n} C_\nu e^{\lambda_\nu x} + \sum_{\nu=1}^{n} \frac{e^{\lambda_\nu x}}{P'_\lambda(\lambda_\nu)} \int f(x) e^{-\lambda_\nu x}\, dx$$

(with complex roots, the real part should be taken).

In Table 5.1 are listed the forms of particular solutions corresponding to some special forms of functions on the right-hand side of the linear nonhomogeneous equation.

10. $x^n y_x^{(n)} + b_{n-1} x^{n-1} y_x^{(n-1)} + \cdots + b_1 x y_x' + b_0 y = f(x).$

The substitution $x = ae^t$ $(a \neq 0)$ leads to an equation of the form 5.1.5.9.

11. $y_x^{(n)} + f(x) \sum_{k=0}^{n-1} (-1)^k k!\, C_{n-1}^k x^{n-k-1} y_x^{(n-k-1)} = 0,$

where $C_m^k = \dfrac{m!}{k!\,(m-k)!}$ are binomial coefficients.

Particular solutions: $y_m = x^m$, where $m = 1,\ 2,\ \ldots,\ n-1$.

The substitution $z = \sum_{k=0}^{n-1} (-1)^k k!\, C_{n-1}^k x^{n-k-1} y_x^{(n-k-1)}$ leads to a first order linear equation: $z'_x + x^{n-1} f(x) z = 0$. Having solved this equation, we obtain an $(n-1)$th order equation of the form 5.1.5.10 for function $y(x)$.

12. $y_x^{(n)} = \sum_{k=0}^{n-1} (a_{k+1} f - a_k) y_x^{(k)},$

where $f = f(x)$; $a_n = 1$, $a_0 = 0$; a_k are arbitrary numbers $(k = 1,\ 2,\ \ldots,\ n-1)$.

Particular solutions: $y_k = e^{\lambda_k x}$ $(k = 1,\ 2,\ \ldots,\ n-1)$, where λ_k are the roots of the polynomial equation $\sum_{k=0}^{n-1} a_{k+1}\lambda^k = 0$.

TABLE 5.1
The forms of particular solutions of the nonhomogeneous constant-coefficient linear equation $y_x^{(n)} + a_{n-1}y_x^{(n-1)} + \cdots + a_1 y_x' + a_0 y = f(x)$ which correspond to some special forms of function $f(x)$.

The form of function $f(x)$	Roots of the characteristic equation $\lambda^n + a_{n-1}\lambda^{n-1} + \cdots + a_1\lambda + a_0 = 0$	The form of a particular solution $y = \widetilde{y}(x)$
$P_m(x)$	Zero is not a root of the characteristic equation (i.e., $a_0 \neq 0$)	$\widetilde{P}_m(x)$
	Zero is a root of the characteristic equation (multiplicity r)	$x^r \widetilde{P}_m(x)$
$P_m(x)e^{\alpha x}$ (α is a real number)	α is not a root of the characteristic equation	$\widetilde{P}_m(x)e^{\alpha x}$
	α is a root of the characteristic equation (multiplicity r)	$x^r \widetilde{P}_m(x)e^{\alpha x}$
$P_m(x)\cos\beta x + Q_n(x)\sin\beta x$	$i\beta$ is not a root of the characteristic equation	$\widetilde{P}_\nu(x)\cos\beta x + \widetilde{Q}_\nu(x)\sin\beta x$
	$i\beta$ is a root of the characteristic equation (multiplicity r)	$x^r[\widetilde{P}_\nu(x)\cos\beta x + \widetilde{Q}_\nu(x)\sin\beta x]$
$[P_m(x)\cos\beta x + Q_n(x)\sin\beta x]e^{\alpha x}$	$\alpha + i\beta$ is not a root of the characteristic equation	$[\widetilde{P}_\nu(x)\cos\beta x + \widetilde{Q}_\nu(x)\sin\beta x]e^{\alpha x}$
	$\alpha + i\beta$ is a root of the characteristic equation (multiplicity r)	$x^r[\widetilde{P}_\nu(x)\cos\beta x + \widetilde{Q}_\nu(x)\sin\beta x]e^{\alpha x}$

Notation: P_m and Q_n are polynomials of the degrees m and n with given coefficients; $\widetilde{P}_m$, $\widetilde{P}_\nu$, and $\widetilde{Q}_\nu$ are polynomials of the degrees m and ν whose coefficients are determined as a result of substituting the particular solution into the basic equation; $\nu = \max(m, n)$; $i^2 = -1$.

13. $xy_x^{(n)} + xfy_x^{(m)} - [(x+m)f + x + n]y = 0, \qquad f = f(x).$

Particular solution: $y_0 = xe^x$.

14. $x(x+m)y_x^{(n)} + x(f - x - n)y_x^{(m)} - (x+m)fy = 0, \qquad f = f(x).$

Particular solution: $y_0 = xe^x$.

15. $x^k y_x^{(n)} + xf(x)y_x' - mf(x)y = 0, \qquad m = 0,\ 1,\ 2,\ \ldots,\ n-1.$

Particular solution: $y_0 = x^m$.

16. $x^n y_x^{(n)} + (n - m - 1)x^{n-1}y_x^{(n-1)} + xfy_x' - mfy = 0, \qquad f = f(x).$

Particular solution: $y_0 = x^m$.

17. $x^n y_x^{(n)} + x^m f y_x^{(m)} - (n!C_a^n + m!C_a^m f)y = 0,$

where $f = f(x)$, $C_a^n = \dfrac{\Gamma(a+1)}{n!\,\Gamma(a-n+1)}$ are binomial coefficients, $\Gamma(a)$ is the gamma-function.

Particular solution: $y_0 = x^a$.

18. $x^m y_x^{(n)} = \sum\limits_{k=0}^{n-1} [x^m(a_{k+1}f - a_k) + a_{k+1}]y_x^{(k)},$

where $f = f(x)$; $a_n = 1$, $a_0 = 0$; m and a_k are arbitrary numbers ($k = 1, 2, \dots, n-1$).

Particular solutions: $y_k = e^{\lambda_k x}$ ($k = 1, 2, \dots, n-1$), where λ_k are the roots of the polynomial equation $\sum_{k=0}^{n-1} a_{k+1}\lambda^k = 0$.

19. $\sum\limits_{k=2}^{n} f_k(x)y_x^{(k)} = g(x)(xy_x' - y).$

Particular solution: $y_0 = x$.

The substitution $w(x) = xy_x' - y$ leads to an $(n-1)$ th order equation.

20. $\sum\limits_{k=3}^{n} f_k(x)y_x^{(k)} = g(x)(x^2 y_{xx}'' - 2xy_x' + 2y).$

Particular solutions: $y_1 = x$, $y_2 = x^2$.

The substitution $w(x) = x^2 y_{xx}'' - 2xy_x' + 2y$ leads to an $(n-2)$ th order equation.

21. $\sum\limits_{k=4}^{n} f_k(x)y_x^{(k)} = g(x)(x^3 y_{xxx}''' - 3x^2 y_{xx}'' + 6xy_x' - 6y).$

Particular solutions: $y_1 = x$, $y_2 = x^2$, $y_3 = x^3$.

The substitution $w(x) = x^3 y_{xxx}''' - 3x^2 y_{xx}'' + 6xy_x' - 6y$ leads to an $(n-3)$ th order equation.

22. $\sum\limits_{k=m+1}^{n} f_k(x)y_x^{(k)} + g(x)\sum\limits_{k=0}^{m} (-1)^k k!\, C_m^k x^{m-k} y_x^{(m-k)} = 0,$

where $C_m^k = \dfrac{m!}{k!\,(m-k)!}$ are binomial coefficients.

Particular solutions: $y_s = x^s$, where $s = 1, 2, \dots, m$.

The substitution $z = \sum_{k=0}^{m} (-1)^k k!\, C_m^k x^{m-k} y_x^{(m-k)}$ leads to an $(n-m)$ th order equation:

$$\sum_{k=m+1}^{n} f_k(x) D^{(k-m-1)}\left(\frac{z_x'}{x^m}\right) + g(x)z = 0, \qquad \text{where} \quad D = \frac{d}{dx}.$$

23. $\sum\limits_{k=0}^{n} (f_k - af_{k+1})y_x^{(k)} = 0,$

where $f_k = f_k(x)$ ($k = 1, 2, \dots, n$), $f_{n+1} \equiv f_0 \equiv 0$.

Particular solution: $y_0 = e^{ax}$.

24. $\displaystyle\sum_{k=0}^{n} x^k[f_k + (k-m)f_{k+1}]y_x^{(k)} = 0,$
where $f_k = f_k(x)$ $(k = 1,\ 2,\ \ldots,\ n)$, $f_{n+1} \equiv f_0 \equiv 0$.

Particular solution: $y_0 = x^m$.

25. $f y_x^{(n)} - f_x^{(n)} y = 0, \qquad f = f(x).$

Particular solution: $y_0 = f(x)$.

26. $f y_x^{(2n+1)} + f_x^{(2n+1)} y = g(x), \qquad f = f(x).$

The first integral has the form

$$\sum_{k=0}^{2n} (-1)^k f_x^{(2n-k)} y_x^{(k)} = \int g(x)\,dx + C.$$

27. $\sin x\, y_x^{(n)} + \sin x\, f(x) y_x^{(m)} - \Big[\sin\Big(x + \dfrac{\pi n}{2}\Big) + f(x)\sin\Big(x + \dfrac{\pi m}{2}\Big)\Big] y = 0.$

Particular solution: $y_0 = \sin x$.

28. $\cos x\, y_x^{(n)} + \cos x\, f(x) y_x^{(m)} - \Big[\cos\Big(x + \dfrac{\pi n}{2}\Big) + f(x)\cos\Big(x + \dfrac{\pi m}{2}\Big)\Big] y = 0.$

Particular solution: $y_0 = \cos x$.

29. $y_x^{(n)} = f(x) y.$

The transformation $x = t^{-1}$, $y = wt^{1-n}$ leads to an equation of the similar form: $w_t^{(n)} = (-1)^n t^{-2n} f\Big(\dfrac{1}{t}\Big) w$.

30. $y_x^{(n)} = (cx+d)^{-2n} f\Big(\dfrac{ax+b}{cx+d}\Big) y.$

The transformation $\xi = \dfrac{ax+b}{cx+d}$, $w = \dfrac{y}{(cx+d)^{n-1}}$ leads to the equation $w_\xi^{(n)} = \Delta^{-n} f(\xi) w$, where $\Delta = ad - bc$.

31. $y_x^{(n)} + f(x) y_x' + g(x) y + h(x) = 0.$

The transformation $x = t^{-1}$, $y = wt^{1-n}$ leads to an equation of the similar form:

$$w_t^{(n)} + (-1)^n t^{-2n} \Big\{ -t^2 f\Big(\frac{1}{t}\Big) w_t' + \Big[(n-1) t f\Big(\frac{1}{t}\Big) + g\Big(\frac{1}{t}\Big)\Big] w + t^{n-1} h\Big(\frac{1}{t}\Big) \Big\} = 0.$$

32. $y_x^{(n+2)} + f(x)[x^2 y_{xx}'' - 2nx y_x' + n(n+1) y] = 0.$

The substitution $w(x) = x^2 y_{xx}'' - 2nx y_x' + n(n+1) y$ leads to an nth order equation: $w_x^{(n)} + x^2 f(x) w = 0$.

33. $xy_x^{(n)} + ny_x^{(n-1)} = x^{1-2n} f\Big(\dfrac{1}{x}\Big) y + x^{-n-1} g\Big(\dfrac{1}{x}\Big).$

The transformation $t = x^{-1}$, $w = yx^{2-n}$ leads to an nth order equation: $w_t^{(n)} = (-1)^n[f(t)w + g(t)]$.

34. $x^2 y_x^{(n+2)} + \alpha x y_x^{(n+1)} + \beta y_x^{(n)}$

$$+ f(x)[x^2 y''_{xx} + (\alpha - 2n)xy'_x + (\beta - \alpha n + n^2 + n)y] = 0.$$

The substitution

$$w(x) = x^2 y''_{xx} + (\alpha - 2n)xy'_x + (\beta - \alpha n + n^2 + n)y$$

leads to an nth order equation: $w_x^{(n)} + f(x)w = 0$.

5.1.6. Asymptotic Solutions

This subsection presents asymptotic solutions, as $\varepsilon \to 0$ ($\varepsilon > 0$), of some fifth-order linear ordinary differential equations containing arbitrary functions (sufficiently smooth), with the independent variable being a real number.

1. Consider an equation of the form

$$\varepsilon^n y_x^{(n)} - f(x)y = 0. \tag{1}$$

on a closed interval $a \le x \le b$. Assume that $f \ne 0$. Then, the leading terms of the asymptotic expansions of the fundamental system of solutions, as $\varepsilon \to +0$, has the form

$$y_m = \big[f(x)\big]^{-\frac{1}{2} + \frac{1}{2n}} \exp\bigg\{\frac{\omega_m}{\varepsilon} \int \big[f(x)\big]^{\frac{1}{n}}\,dx\bigg\}\big[1 + O(\varepsilon)\big],$$

where $\omega_1, \omega_2, \ldots, \omega_n$ are the roots of the equation $\omega^n = 1$:

$$\omega_m = \cos\Big(\frac{2\pi m}{n}\Big) + i\sin\Big(\frac{2\pi m}{n}\Big), \qquad m = 1,\, 2,\, \ldots,\, n.$$

2. Consider an equation of the form

$$\varepsilon^n y_x^{(n)} + \varepsilon^{n-1} f_{n-1}(x) y_x^{(n-1)} + \cdots + \varepsilon f_1(x) y'_x + f_0(x) y = 0. \tag{2}$$

on a closed interval $a \le x \le b$. Let $\lambda_m = \lambda_m(x)$ $(m = 1, 2, \ldots, n)$ be the roots of the characteristic equation

$$P(x, \lambda) \equiv \lambda^n + f_{n-1}(x)\lambda^{n-1} + \cdots + f_1(x)\lambda + f_0(x) = 0.$$

Let all the roots of the characteristic equation be different on the interval $a \le x \le b$, i.e., the conditions $\lambda_m(x) \ne \lambda_k(x)$ if $m \ne k$ are satisfied, which is equivalent to the fulfillment of the conditions $P_\lambda(x, \lambda_m) \ne 0$. Then, the leading terms of the asymptotic expansions of the fundamental system of solutions, as $\varepsilon \to +0$, are given by the formulae

$$y_m = \exp\bigg\{\frac{1}{\varepsilon}\int \lambda_m(x)\,dx - \frac{1}{2}\int [\lambda_m(x)]'_x \frac{P_{\lambda\lambda}(x, \lambda_m(x))}{P_\lambda(x, \lambda_m(x))}\,dx\bigg\},$$

where

$$P_\lambda(x, \lambda) \equiv \frac{\partial P}{\partial \lambda} = n\lambda^{n-1} + (n-1)f_{n-1}(x)\lambda^{n-2} + \ldots + 2\lambda f_2(x) + f_1(x),$$

$$P_{\lambda\lambda}(x, \lambda) \equiv \frac{\partial^2 P}{\partial \lambda^2} = n(n-1)\lambda^{n-2} + (n-1)(n-2)f_{n-1}(x)\lambda^{n-3} + \ldots + 6\lambda f_3(x) + 2f_2(x).$$

5.2. Nonlinear Equations

5.2.1. Equations Containing Power Functions

1. $yy_x^{(5)} + 5y'_x y''''_{xxxx} + 10y''_{xx} y'''_{xxx} = ax^n.$

This is a special case of equation 5.2.6.1 with $f(x) = ax^n$.

2. $y_x^{(6)} = Ay^{-7/5}.$

Multiplying both sides by $y^{7/5}$ and differentiating with respect to x, we obtain the equation $5yy_x^{(7)} + 7y'_x y_x^{(6)} = 0$. Having integrated the latter three times, we arrive at a chain of equalities:

$$5yy_x^{(6)} + 2y'_x y_x^{(5)} - 2y''_{xx} y''''_{xxxx} + (y'''_{xxx})^2 = 2C_2, \qquad (1)$$

$$5yy_x^{(5)} - 3y'_x y''''_{xxxx} + y''_{xx} y'''_{xxx} = 2C_2 x + C_1, \qquad (2)$$

$$5yy''''_{xxxx} - 8y'_x y'''_{xxx} + \tfrac{9}{2}(y''_{xx})^2 = C_2 x^2 + C_1 x + C_0, \qquad (3)$$

where C_0, C_1, and C_2 are arbitrary constants. Eliminating the highest derivatives from (1)–(3), with the aid of the original equation, we can obtain a third order equation which can be reduced to a second order equation (see equation 5.2.1.4 with $n = 3$).

3. $yy_x^{(6)} + 6y'_x y_x^{(5)} + 15y''_{xx} y''''_{xxxx} + 10(y'''_{xxx})^2 = ax^n.$

This is a special case of equation 5.2.6.4 with $f(x) = ax^n$.

4. $y_x^{(2n)} = Ay^{\frac{1+2n}{1-2n}}.$

Multiply both sides by $y^{\frac{2n+1}{2n-1}}$ and differentiate with respect to x. As a result we obtain

$$(2n-1)yy_x^{(2n+1)} + (2n+1)y'_x y_x^{(2n)} = 0.$$

Three integrals containing arbitrary constants C_0, C_1, and C_2 are presented in 5.2.6.22 wherein we should let $f \equiv 0$. Eliminating the highest derivatives from those integrals and the original equation, we may always obtain a $(2n-3)$ th order equation. With the aid of the transformation

$$t = \int \frac{dx}{P}, \quad w = yP^{\frac{1-2n}{2}}, \qquad \text{where} \quad P = C_2 x^2 + C_1 x + C_0,$$

the latter equation can be reduced to the autonomous form 5.2.6.40. Therefore, the substitution $z(w) = w'_t$ finally leads to a $(2n-4)$ th order equation with respect to $z = z(w)$.

5. $y_x^{(2n)} = Ay^k, \qquad k \neq -1.$

Having integrated, we arrive at

$$\sum_{m=1}^{n-1} (-1)^m y_x^{(m)} y_x^{(2n-m)} + \frac{1}{2}(-1)^n \big[y_x^{(n)}\big]^2 = -\frac{A}{k+1} y^{k+1} + C,$$

where C is an arbitrary number. Further, the order of the obtained autonomous equation can be lowered by the substitution $w(y) = y'_x$.

6. $y_x^{(n)} = ax^{-n}y^m.$

This is a special case of equation 5.2.6.10 with $f(y) = ay^m$.

7. $y_x^{(n)} = ax^k y^m.$

1°. The transformation $x = t^{-1}$, $y = t^{1-n}w(t)$ leads to an equation of the similar form: $w_t^{(n)} = (-1)^n A t^{-k-(n-1)m-n-1} w^m$.

2°. The transformation $\xi = x^{n+k}y^{m-1}$, $z = xy'_x/y$ leads to an $(n-1)$ th order equation.

8. $yy_x^{(2n+1)} = ax^n + b.$

This is a special case of equation 5.2.6.16 with $f(x) = ax^n + b$.

9. $y_x^{(n)} = x^{m-nm-n-1}(ay + bx^{n-1})^m.$

This is a special case of equation 5.2.6.11 with $f(w) = (aw + b)^m$.

10. $y_x^{(2n)} = x^{\frac{m-2nm-2n-1}{2}}\left(ay + bx^{\frac{2n-1}{2}}\right)^m.$

This is a special case of equation 5.2.6.12 with $f(w) = (aw + b)^m$.

11. $y_x^{(n)} = (ay + bx^k)^m; \qquad k = 1,\ 2,\ \ldots,\ n-1.$

The substitution $aw = ay + bx^k$ leads to an autonomous equation: $w_x^{(n)} = a^m w^m$ (see 5.2.1.4, 5.2.1.5, and 5.2.6.40).

12. $y_x^{(n)} = (ax^2 + bx + c)^{\frac{m-nm-n-1}{2}} y^m.$

This is a special case of equation 5.2.6.21 with $f(w) = w^m$.

13. $y_x^{(n)} = (ax + b)^{-n}(cx + d)^{m-nm-1}y^m.$

This is a special case of equation 5.2.6.20 with $f(w) = w^m$.

14. $yy_x^{(2n+1)} = ay'_x y_x^{(2n)}.$

The equation admits two different (with $a \neq -1$) first integrals:

$$y_x^{(2n)} = \widetilde{C}_1 y^a,$$

$$yy_x^{(2n)} + (a+1)\sum_{m=1}^{n-1}(-1)^m y_x^{(m)} y_x^{(2n-m)} + \tfrac{1}{2}(-1)^n(a+1)\big[y_x^{(n)}\big]^2 = \widetilde{C}_2,$$

where $\widetilde{C}_1$ and $\widetilde{C}_2$ are arbitrary constants. Eliminating the highest derivative from the integrals, we arrive at a $(2n-1)$ th order autonomous equation:

$$\sum_{m=1}^{n-1}(-1)^m y_x^{(m)} y_x^{(2n-m)} + \tfrac{1}{2}(-1)^n\big[y_x^{(n)}\big]^2 = C_1 y^{a+1} + C_2,$$

where $C_1 = -\dfrac{\widetilde{C}_1}{a+1}$, $C_2 = \dfrac{\widetilde{C}_2}{a+1}$. The order of the obtained equation next can be lowered by the standard substitution $w(y) = y'_x$.

15. $y_x^{(n-2)} y_x^{(n)} = a\big(y_x^{(n-1)}\big)^2.$

Solution:

$$y = \begin{cases} C_0 + C_1 x + \cdots + C_{n-3} x^{n-3} + (C_{n-2} + C_{n-1} x)^{n-2+\frac{1}{1-a}} & \text{if } a \neq 1, \\ C_0 + C_1 x + \cdots + C_{n-3} x^{n-3} + C_{n-2} \exp(C_{n-1} x) & \text{if } a = 1. \end{cases}$$

16. $y_x^{(n)} = a x^{m-n} y^{1-m} (y'_x)^m.$

This is a special case of equation 5.2.6.15 with $f(w) = aw^m$.

17. $y_x^{(n+1)} = a y^k y'_x \big(y_x^{(n)}\big)^m.$

This is a special case of equation 5.2.6.17 with $f(y) = y^{-k}$, $g(w) = aw^m$.

18. $y_x^{(n)} = a x^m (x y'_x - y)^k (y''_{xx})^l.$

The substitution $w(x) = x y'_x - y$ leads to an $(n-1)$th order equation:

$$\frac{d^{n-2}}{dx^{n-2}}\left(\frac{w'_x}{x}\right) = a x^{m-l} w^k (w'_x)^l.$$

19. $y_x^{(n)} = a x^{m_1} y^{m_2} (y'_x)^{m_3} \cdots \big(y_x^{(n-1)}\big)^{m_{n+1}}.$

Homogeneous equation in the extended sense.

The transformation $\xi = x^\lambda y^\mu$, $w = \dfrac{x y'_x}{y}$, where

$$\lambda = n + m_1 - m_3 - 2m_4 - \cdots - (n-1) m_{n+1}, \quad \mu = m_2 + m_3 + \cdots + m_{n+1} - 1,$$

leads to an $(n-1)$th order equation.

20. $x y_x^{(n)} + n y_x^{(n-1)} = a x^m y^m.$

This is a special case of equation 5.2.6.23 with $f(w) = aw^m$.

21. $x^2 y_x^{(n)} + 2n x y_x^{(n-1)} + n(n-1) y_x^{(n-2)} = a x^{2m} y^m.$

This is a special case of equation 5.2.6.24 with $f(w) = aw^m$.

22. $(2n-1) y y_x^{(2n+1)} + (2n+1) y'_x y_x^{(2n)} = a x^m.$

This is a special case of equation 5.2.6.22 with $f(x) = ax^m$.

23. $\left(\sqrt{y}\,\dfrac{d}{dx}\right)^{n-1} (y'_x) = ax + b.$

The transformation $x = x(t)$, $y = (x'_t)^2$ leads to a constant coefficient linear equation: $2x_t^{(n+1)} = ax + b$.

24. $2\displaystyle\sum_{m=1}^{n-1} (-1)^m y_x^{(m)} y_x^{(2n-m)} + (-1)^n \big[y_x^{(n)}\big]^2 + \lambda (y'_x)^2 = ay^2 + by + c.$

Differentiating both sides with respect to x and dividing by y'_x, we arrive at a constant coefficient equation: $2y_x^{(2n)} - 2\lambda y''_{xx} + 2ay + b = 0$.

25. $\displaystyle 2\sum_{m=1}^{n-1}(-1)^m y_x^{(m)} y_x^{(2n-m)} + (-1)^n\big[y_x^{(n)}\big]^2 = \alpha(xy'_x - y) + \beta y'_x + \gamma.$

Differentiating both sides of the equation with respect to x, we have

$$y''_{xx}\big[2y_x^{(2n-1)} - \alpha x - \beta\big] = 0. \tag{1}$$

Equating the second factor to zero, we find

$$y = \frac{1}{2}\frac{\alpha x^{2n}}{(2n)!} + \frac{1}{2}\frac{\beta x^{2n-1}}{(2n-1)!} + \sum_{k=0}^{2n-2} C_k x^k.$$

Integration constants C_k and parameters α, β, and γ are related by the equality

$$2\sum_{m=2}^{n-1}(-1)^m m!\,(2n-m)!\,C_m C_{2n-m} + (-1)^n (n!)^2 C_n^2 = \beta C_1 - \alpha C_0 + \gamma,$$

which is obtained as a result of substituting the above solution into the original equation.

In addition, there is the solution corresponding to equating the first factor in (1) to zero:

$$y = \widetilde{C}_1 x + \widetilde{C}_0, \qquad \text{where} \quad \beta\widetilde{C}_1 - \alpha\widetilde{C}_0 + \gamma = 0.$$

26. $\displaystyle 2\sum_{m=1}^{n-1}(-1)^m y_x^{(m)} y_x^{(2n-m)} + (-1)^n\big[y_x^{(n)}\big]^2 + s(y''_{xx})^2 = \alpha(xy'_x - y) + \beta y'_x + \gamma,$

where n is an integer greater than or equal to 3.

With $s = 0$ see 5.2.1.25. Let $s \neq 0$. Differentiating the equation with respect to x, we have

$$y''_{xx}\big[2y_x^{(2n-1)} + 2sy'''_{xxx} - \alpha x - \beta\big] = 0.$$

Equating the second factor to zero and integrating, we obtain

$$y = \frac{\alpha x^4}{48s} + \frac{\beta x^3}{12s} + C_2 x^2 + C_1 x + C_0 + \iiint w\,dx\,dx\,dx,$$

where $w = w(x)$ is the general solution of a constant coefficient equation of the form 5.1.2.2: $w_x^{(2n-4)} + sw = 0$. The constants of integration are related by an equality which is found as a result of substituting the obtained solution into the original equation.

In addition, there is the solution $y = \widetilde{C}_1 x + \widetilde{C}_0$, where the constants of integration are related by $\beta\widetilde{C}_1 - \alpha\widetilde{C}_0 + \gamma = 0$.

27. $\displaystyle \sum_{m=1}^{n} a_m\left\{2\sum_{\nu=1}^{m-1}(-1)^\nu y_x^{(\nu)} y_x^{(2m-\nu)} + (-1)^m\big[y_x^{(m)}\big]^2\right\} = \alpha y^2 + 2\beta y + \gamma.$

Differentiating with respect to x, we arrive at a constant coefficient linear equation: $\sum_{m=1}^{n} a_m y_x^{(2m)} + \alpha y + \beta = 0$.

5.2.2. Equations Containing Exponential Functions

1. $yy_x^{(5)} + 5y'_x y''''_{xxxx} + 10y''_{xx} y'''_{xxx} = ae^{\lambda x}$.

This is a special case of equation 5.2.6.1 with $f(x) = ae^{\lambda x}$.

2. $yy_x^{(6)} + 6y'_x y_x^{(5)} + 15y''_{xx} y''''_{xxxx} + 10(y'''_{xxx})^2 = ae^{\lambda x}$.

This is a special case of equation 5.2.6.4 with $f(x) = ae^{\lambda x}$.

3. $y_x^{(2n)} = ae^{\lambda y}$.

This is a special case of equation 5.2.6.6 with $f(y) = ae^{\lambda y}$.

4. $y_x^{(n)} = ax^{-n} e^{\lambda y}$.

This is a special case of equation 5.2.6.10 with $f(y) = ae^{\lambda y}$.

5. $y_x^{(n)} = ax^k e^{\alpha y}$.

This is a special case of equation 5.2.6.31 with $f(w) = aw$, $m = k + n$.

6. $y_x^{(n)} = Ae^{\alpha x} y^m$.

This is a special case of equation 5.2.2.11 with $m = m_1$ and $m_2 = m_3 = \dots = m_n = 0$.

7. $yy_x^{(2n+1)} = ae^{\lambda x} + b$.

This is a special case of equation 5.2.6.16 with $f(x) = ae^{\lambda x} + b$.

8. $y_x^{(n)} = ae^{by} e^{cx^m}, \qquad m = 1,\ 2,\ \dots,\ n-1.$

The substitution $bw = by + cx^m$ leads to an autonomous equation: $w_x^{(n)} = ae^{bw}$, which, for even n, admits lowering of its order by two (see 5.2.2.3).

9. $(2n-1)yy_x^{(2n+1)} + (2n+1)y'_x y_x^{(2n)} = ae^{\lambda x}$.

This is a special case of equation 5.2.6.22 with $f(x) = ae^{\lambda x}$.

10. $y_x^{(n+1)} = ae^{\lambda y} y'_x \big(y_x^{(n)}\big)^m$.

This is a special case of equation 5.2.6.17 with $f(y) = e^{-\lambda y}$, $g(w) = aw^m$.

11. $y_x^{(n)} = Ae^{\alpha x} y^{m_1} (y'_x)^{m_2} \dots \big(y_x^{(n-1)}\big)^{m_n}$.

The substitution $w(x) = ye^{\beta x}$, where $\beta = \dfrac{\alpha}{m_1 + m_2 + \dots + m_n - 1}$, leads to an autonomous equation of the form 5.2.6.40.

12. $y_x^{(n)} = Ae^{\alpha y} x^{m_1} (y'_x)^{m_2} (y''_{xx})^{m_3} \dots \big(y_x^{(n-1)}\big)^{m_n}$.

The transformation $z = x^\sigma e^{\alpha y}$, $w = xy'_x$, where $\sigma = n + m_1 - m_2 - 2m_3 - 3m_4 - \dots - (n-1)m_n$, leads to an $(n-1)$ th order equation.

5.2.3. Equations Containing Hyperbolic Functions

1. $yy_x^{(5)} + 5y'_x y''''_{xxxx} + 10y''_{xx} y'''_{xxx} = a\cosh^m(\lambda x).$

This is a special case of equation 5.2.6.1 with $f(x) = a\cosh^m(\lambda x)$.

2. $yy_x^{(5)} + 5y'_x y''''_{xxxx} + 10y''_{xx} y'''_{xxx} = a\sinh^m(\lambda x).$

This is a special case of equation 5.2.6.1 with $f(x) = a\sinh^m(\lambda x)$.

3. $yy_x^{(5)} + 5y'_x y''''_{xxxx} + 10y''_{xx} y'''_{xxx} = a\tanh^m(\lambda x).$

This is a special case of equation 5.2.6.1 with $f(x) = a\tanh^m(\lambda x)$.

4. $yy_x^{(5)} + 5y'_x y''''_{xxxx} + 10y''_{xx} y'''_{xxx} = a\coth^m(\lambda x).$

This is a special case of equation 5.2.6.1 with $f(x) = a\coth^m(\lambda x)$.

5. $yy_x^{(6)} + 6y'_x y_x^{(5)} + 15y''_{xx} y''''_{xxxx} + 10(y'''_{xxx})^2 = a\cosh^m(\lambda x).$

This is a special case of equation 5.2.6.4 with $f(x) = a\cosh^m(\lambda x)$.

6. $yy_x^{(6)} + 6y'_x y_x^{(5)} + 15y''_{xx} y''''_{xxxx} + 10(y'''_{xxx})^2 = a\sinh^m(\lambda x).$

This is a special case of equation 5.2.6.4 with $f(x) = a\sinh^m(\lambda x)$.

7. $yy_x^{(6)} + 6y'_x y_x^{(5)} + 15y''_{xx} y''''_{xxxx} + 10(y'''_{xxx})^2 = a\tanh^m(\lambda x).$

This is a special case of equation 5.2.6.4 with $f(x) = a\tanh^m(\lambda x)$.

8. $yy_x^{(6)} + 6y'_x y_x^{(5)} + 15y''_{xx} y''''_{xxxx} + 10(y'''_{xxx})^2 = a\coth^m(\lambda x).$

This is a special case of equation 5.2.6.4 with $f(x) = a\coth^m(\lambda x)$.

9. $y_x^{(2n)} = a\cosh^m(\lambda y).$

This is a special case of equation 5.2.6.6 with $f(y) = a\cosh^m(\lambda y)$.

10. $y_x^{(2n)} = a\sinh^m(\lambda y).$

This is a special case of equation 5.2.6.6 with $f(y) = a\sinh^m(\lambda y)$.

11. $y_x^{(2n)} = a\tanh^m(\lambda y).$

This is a special case of equation 5.2.6.6 with $f(y) = a\tanh^m(\lambda y)$.

12. $y_x^{(2n)} = a\coth^m(\lambda y).$

This is a special case of equation 5.2.6.6 with $f(y) = a\coth^m(\lambda y)$.

13. $y_x^{(n)} = ax^{-n}\cosh^m(\lambda y).$

This is a special case of equation 5.2.6.10 with $f(y) = a\cosh^m(\lambda y)$.

14. $y_x^{(n)} = ax^{-n}\sinh^m(\lambda y).$

This is a special case of equation 5.2.6.10 with $f(y) = a\sinh^m(\lambda y)$.

15. $y_x^{(n)} = ax^{-n}\tanh^m(\lambda y).$

This is a special case of equation 5.2.6.10 with $f(y) = a\tanh^m(\lambda y)$.

16. $y_x^{(n)} = ax^{-n}\coth^m(\lambda y).$

This is a special case of equation 5.2.6.10 with $f(y) = a\coth^m(\lambda y)$.

17. $yy_x^{(2n+1)} = a\cosh^m(\lambda x).$

This is a special case of equation 5.2.6.16 with $f(x) = a\cosh^m(\lambda x)$.

18. $yy_x^{(2n+1)} = a\sinh^m(\lambda x).$

This is a special case of equation 5.2.6.16 with $f(x) = a\sinh^m(\lambda x)$.

19. $yy_x^{(2n+1)} = a\tanh^m(\lambda x).$

This is a special case of equation 5.2.6.16 with $f(x) = a\tanh^m(\lambda x)$.

20. $yy_x^{(2n+1)} = a\coth^m(\lambda x).$

This is a special case of equation 5.2.6.16 with $f(x) = a\coth^m(\lambda x)$.

21. $(2n-1)yy_x^{(2n+1)} + (2n+1)y'_x y_x^{(2n)} = a\cosh^m(\lambda x).$

This is a special case of equation 5.2.6.22 with $f(x) = a\cosh^m(\lambda x)$.

22. $(2n-1)yy_x^{(2n+1)} + (2n+1)y'_x y_x^{(2n)} = a\sinh^m(\lambda x).$

This is a special case of equation 5.2.6.22 with $f(x) = a\sinh^m(\lambda x)$.

23. $(2n-1)yy_x^{(2n+1)} + (2n+1)y'_x y_x^{(2n)} = a\tanh^m(\lambda x).$

This is a special case of equation 5.2.6.22 with $f(x) = a\tanh^m(\lambda x)$.

24. $(2n-1)yy_x^{(2n+1)} + (2n+1)y'_x y_x^{(2n)} = a\coth^m(\lambda x).$

This is a special case of equation 5.2.6.22 with $f(x) = a\coth^m(\lambda x)$.

25. $y_x^{(n+1)} = a\cosh^k(\lambda y)\, y'_x \big(y_x^{(n)}\big)^m.$

This is a special case of equation 5.2.6.17 with $f(y) = \cosh^{-k}(\lambda y)$, $g(w) = aw^m$.

26. $y_x^{(n+1)} = a\sinh^k(\lambda y)\, y'_x \big(y_x^{(n)}\big)^m.$

This is a special case of equation 5.2.6.17 with $f(y) = \sinh^{-k}(\lambda y)$, $g(w) = aw^m$.

27. $y_x^{(n+1)} = a\tanh^k(\lambda y)\, y'_x \big(y_x^{(n)}\big)^m.$

This is a special case of equation 5.2.6.17 with $f(y) = \tanh^{-k}(\lambda y)$, $g(w) = aw^m$.

28. $y_x^{(n+1)} = a\coth^k(\lambda y)\, y'_x \big(y_x^{(n)}\big)^m.$

This is a special case of equation 5.2.6.17 with $f(y) = \coth^{-k}(\lambda y)$, $g(w) = aw^m$.

5.2.4. Equations Containing Logarithmic Functions

1. $yy_x^{(5)} + 5y'_x y''''_{xxxx} + 10y''_{xx} y'''_{xxx} = a\ln^m(bx).$

This is a special case of equation 5.2.6.1 with $f(x) = a\ln^m(bx)$.

2. $yy_x^{(6)} + 6y'_x y_x^{(5)} + 15y''_{xx} y''''_{xxxx} + 10(y'''_{xxx})^2 = a\ln^m(bx).$

This is a special case of equation 5.2.6.4 with $f(x) = a\ln^m(bx)$.

3. $y_x^{(2n)} = a\ln^m(by).$

This is a special case of equation 5.2.6.6 with $f(y) = a\ln^m(by)$.

4. $yy_x^{(2n+1)} = a\ln^m(bx).$

This is a special case of equation 5.2.6.16 with $f(x) = a\ln^m(bx)$.

5. $y_x^{(n)} = y(\alpha x + m\ln y + b).$

This is a special case of equation 5.2.6.30 with $f(w) = \ln w + b$.

6. $y_x^{(n)} = x^{-n}(\alpha y + m\ln x + b).$

This is a special case of equation 5.2.6.31 with $f(w) = \ln w + b$.

7. $y_x^{(n)} = ax^{-n}\ln^m(by).$

This is a special case of equation 5.2.6.10 with $f(y) = a\ln^m(by)$.

8. $y_x^{(n)} = ax^{-n-1}[\ln y + (1-n)\ln x].$

This is a special case of equation 5.2.6.11 with $f(w) = a\ln w$.

9. $y_x^{(n)} = ax^{-n-k}(\ln y + k\ln x).$

This is a special case of equation 5.2.6.13 with $f(w) = a\ln w$.

10. $y_x^{(n)} = ayx^{-n}(m\ln y + k\ln x).$

This is a special case of equation 5.2.6.14 with $f(w) = a\ln w$.

11. $y_x^{(2n)} = ax^{-\frac{2n+1}{2}}[2\ln y + (1-2n)\ln x].$

This is a special case of equation 5.2.6.12 with $f(w) = 2a\ln w$.

12. $y_x^{(n)} = (ax^2+c)^{-\frac{n+1}{2}}[2\ln y + (1-n)\ln(ax^2+c)].$

This is a special case of equation 5.2.6.21 with $b = 0$, $f(w) = 2\ln w$.

13. $y_x^{(n)} = be^{\alpha x}(\ln y - \alpha x).$

This is a special case of equation 5.2.6.29 with $f(w) = b\ln w$.

14. $(2n-1)yy_x^{(2n+1)}+(2n+1)y'_x y_x^{(2n)}=a\ln^m(bx).$

This is a special case of equation 5.2.6.22 with $f(x)=a\ln^m(bx)$.

15. $y_x^{(n+1)}=a\ln^k(by)\,y'_x\bigl(y_x^{(n)}\bigr)^m.$

This is a special case of equation 5.2.6.17 with $f(y)=\ln^{-k}(by)$, $g(w)=aw^m$.

16. $y_x^{(n+1)}=ay^m y'_x\ln y_x^{(n)}.$

This is a special case of equation 5.2.6.17 with $f(y)=y^{-m}$, $g(w)=a\ln w$.

5.2.5. Equations Containing Trigonometric Functions

1. $yy_x^{(5)}+5y'_x y''''_{xxxx}+10y''_{xx}y'''_{xxx}=a\cos^m(\lambda x).$

This is a special case of equation 5.2.6.1 with $f(x)=a\cos^m(\lambda x)$.

2. $yy_x^{(5)}+5y'_x y''''_{xxxx}+10y''_{xx}y'''_{xxx}=a\sin^m(\lambda x).$

This is a special case of equation 5.2.6.1 with $f(x)=a\sin^m(\lambda x)$.

3. $yy_x^{(5)}+5y'_x y''''_{xxxx}+10y''_{xx}y'''_{xxx}=a\tan^m(\lambda x).$

This is a special case of equation 5.2.6.1 with $f(x)=a\tan^m(\lambda x)$.

4. $yy_x^{(5)}+5y'_x y''''_{xxxx}+10y''_{xx}y'''_{xxx}=a\cot^m(\lambda x).$

This is a special case of equation 5.2.6.1 with $f(x)=a\cot^m(\lambda x)$.

5. $yy_x^{(6)}+6y'_x y_x^{(5)}+15y''_{xx}y''''_{xxxx}+10(y'''_{xxx})^2=a\cos^m(\lambda x).$

This is a special case of equation 5.2.6.4 with $f(x)=a\cos^m(\lambda x)$.

6. $yy_x^{(6)}+6y'_x y_x^{(5)}+15y''_{xx}y''''_{xxxx}+10(y'''_{xxx})^2=a\sin^m(\lambda x).$

This is a special case of equation 5.2.6.4 with $f(x)=a\sin^m(\lambda x)$.

7. $yy_x^{(6)}+6y'_x y_x^{(5)}+15y''_{xx}y''''_{xxxx}+10(y'''_{xxx})^2=a\tan^m(\lambda x).$

This is a special case of equation 5.2.6.4 with $f(x)=a\tan^m(\lambda x)$.

8. $yy_x^{(6)}+6y'_x y_x^{(5)}+15y''_{xx}y''''_{xxxx}+10(y'''_{xxx})^2=a\cot^m(\lambda x).$

This is a special case of equation 5.2.6.4 with $f(x)=a\cot^m(\lambda x)$.

9. $y_x^{(2n)}=a\cos^m(\lambda y).$

This is a special case of equation 5.2.6.6 with $f(y)=a\cos^m(\lambda y)$.

10. $y_x^{(2n)}=a\sin^m(\lambda y).$

This is a special case of equation 5.2.6.6 with $f(y)=a\sin^m(\lambda y)$.

11. $y_x^{(2n)} = a \tan^m(\lambda y).$

This is a special case of equation 5.2.6.6 with $f(y) = a \tan^m(\lambda y)$.

12. $y_x^{(2n)} = a \cot^m(\lambda y).$

This is a special case of equation 5.2.6.6 with $f(y) = a \cot^m(\lambda y)$.

13. $y_x^{(n)} = a x^{-n} \cos^m(\lambda y).$

This is a special case of equation 5.2.6.10 with $f(y) = a \cos^m(\lambda y)$.

14. $y_x^{(n)} = a x^{-n} \sin^m(\lambda y).$

This is a special case of equation 5.2.6.10 with $f(y) = a \sin^m(\lambda y)$.

15. $y_x^{(n)} = a x^{-n} \tan^m(\lambda y).$

This is a special case of equation 5.2.6.10 with $f(y) = a \tan^m(\lambda y)$.

16. $y_x^{(n)} = a x^{-n} \cot^m(\lambda y).$

This is a special case of equation 5.2.6.10 with $f(y) = a \cot^m(\lambda y)$.

17. $y y_x^{(2n+1)} = a \cos^m(\lambda x).$

This is a special case of equation 5.2.6.16 with $f(x) = a \cos^m(\lambda x)$.

18. $y y_x^{(2n+1)} = a \sin^m(\lambda x).$

This is a special case of equation 5.2.6.16 with $f(x) = a \sin^m(\lambda x)$.

19. $y y_x^{(2n+1)} = a \tan^m(\lambda x).$

This is a special case of equation 5.2.6.16 with $f(x) = a \tan^m(\lambda x)$.

20. $y y_x^{(2n+1)} = a \cot^m(\lambda x).$

This is a special case of equation 5.2.6.16 with $f(x) = a \cot^m(\lambda x)$.

21. $(2n-1) y y_x^{(2n+1)} + (2n+1) y'_x y_x^{(2n)} = a \cos^m(\lambda x).$

This is a special case of equation 5.2.6.22 with $f(x) = a \cos^m(\lambda x)$.

22. $(2n-1) y y_x^{(2n+1)} + (2n+1) y'_x y_x^{(2n)} = a \sin^m(\lambda x).$

This is a special case of equation 5.2.6.22 with $f(x) = a \sin^m(\lambda x)$.

23. $(2n-1) y y_x^{(2n+1)} + (2n+1) y'_x y_x^{(2n)} = a \tan^m(\lambda x).$

This is a special case of equation 5.2.6.22 with $f(x) = a \tan^m(\lambda x)$.

24. $(2n-1) y y_x^{(2n+1)} + (2n+1) y'_x y_x^{(2n)} = a \cot^m(\lambda x).$

This is a special case of equation 5.2.6.22 with $f(x) = a \cot^m(\lambda x)$.

25. $y_x^{(n+1)} = a\cos^k(\lambda y)\, y'_x \left(y_x^{(n)}\right)^m$.

This is a special case of equation 5.2.6.17 with $f(y) = \cos^{-k}(\lambda y)$, $g(w) = aw^m$.

26. $y_x^{(n+1)} = a\sin^k(\lambda y)\, y'_x \left(y_x^{(n)}\right)^m$.

This is a special case of equation 5.2.6.17 with $f(y) = \sin^{-k}(\lambda y)$, $g(w) = aw^m$.

27. $y_x^{(n+1)} = a\tan^k(\lambda y)\, y'_x \left(y_x^{(n)}\right)^m$.

This is a special case of equation 5.2.6.17 with $f(y) = \tan^{-k}(\lambda y)$, $g(w) = aw^m$.

28. $y_x^{(n+1)} = a\cot^k(\lambda y)\, y'_x \left(y_x^{(n)}\right)^m$.

This is a special case of equation 5.2.6.17 with $f(y) = \cot^{-k}(\lambda y)$, $g(w) = aw^m$.

5.2.6. Equations Containing Arbitrary Functions

1. $yy_x^{(5)} + 5y'_x y''''_{xxxx} + 10y''_{xx} y'''_{xxx} = f(x)$.

Solution:

$$y^2 = C_4x^4 + C_3x^3 + C_2x^2 + C_1x + C_0 + \frac{1}{12}\int_{x_0}^{x} (x-t)^4 f(t)\,dt,$$

where x_0 is an arbitrary number.

2. $yy_x^{(5)} + ay'_x y''''_{xxxx} + (3a-5)y''_{xx} y'''_{xxx} = f(x)$.

Integrating the equation three times, we obtain

$$yy''_{xx} + \frac{a-3}{2}(y'_x)^2 = C_2x^2 + C_1x + C_0 + \frac{1}{2}\int_{x_0}^{x} (x-t)^2 f(t)\,dt,$$

where x_0 is an arbitrary number.

3. $(a+y)y_x^{(5)} + by'_x y''''_{xxxx} + cy''_{xx} y'''_{xxx} = f(x)$.

Integrating, we obtain

$$(a+y)y''''_{xxxx} + (b-1)y'_x y'''_{xxx} + \frac{1}{2}(1-b+c)(y''_{xx})^2 = \int f(x)\,dx + C.$$

4. $yy_x^{(6)} + 6y'_x y_x^{(5)} + 15y''_{xx} y''''_{xxxx} + 10(y'''_{xxx})^2 = f(x)$.

Solution: $y^2 = C_5x^5 + C_4x^4 + C_3x^3 + C_2x^2 + C_1x + C_0 + \dfrac{1}{60}\displaystyle\int_{x_0}^{x} (x-t)^5 f(t)\,dt.$

5. $y_x^{(6)} = (ax^2+bx+c)^{-7/2} f\left((ax^2+bx+c)^{-5/2}\right)$.

This is a special case of equation 5.2.6.21 with $n=6$.

6. $y_x^{(2n)} = f(y)$.

The first integral of the equation is

$$\sum_{m=1}^{n-1}(-1)^m y_x^{(m)} y_x^{(2n-m)} + \tfrac{1}{2}(-1)^n\bigl[y_x^{(n)}\bigr]^2 + \int f(y)\,dy = C.$$

Next, the order of the obtained equation can be lowered by the substitution $w(y) = y'_x$.

7. $y_x^{(n)} = f\bigl(y_x^{(n-1)}\bigr)$.

Having set $u(x) = y_x^{(n-1)}$, we obtain $u'_x = f(u)$. Further, find u from the relation $x = \int \frac{du}{f(u)} + C_1$. Then, the $(n-1)$-fold integration yields y.

The solution in the parametric form is written as

$$x = \int_{C_1}^{u} \frac{du}{f(u)}, \quad y = \int_{C_2}^{u} \frac{du_1}{f(u_1)} \int_{C_3}^{u_1} \frac{du_2}{f(u_2)} \cdots \int_{C_{n-1}}^{u_{n-3}} \frac{du_{n-2}}{f(u_{n-2})} \int_{C_n}^{u_{n-2}} \frac{u_{n-1}\,du_{n-1}}{f(u_{n-1})}.$$

8. $y_x^{(n)} = f\bigl(y_x^{(n-2)}\bigr)$.

Setting $u(x) = y_x^{(n-2)}$, we obtain the equation $u''_{xx} = f(u)$ whose solution has the form

$$x = \int \frac{du}{\varphi(u)} + C_2, \qquad \text{where} \quad \varphi(u) = \pm\Bigl[C_1 + 2\int f(u)\,du\Bigr]^{1/2}.$$

Expressing u in terms of x and integrating the resulting relation $(n-2)$ times, we find y.

The solution in the parametric form is written as

$$x = \int_{C_2}^{u} \frac{du}{\varphi(u)}, \quad y = \int_{C_3}^{u} \frac{du_1}{\varphi(u_1)} \int_{C_4}^{u_1} \frac{du_2}{\varphi(u_2)} \cdots \int_{C_{n-1}}^{u_{n-4}} \frac{du_{n-3}}{\varphi(u_{n-3})} \int_{C_n}^{u_{n-3}} \frac{u_{n-2}\,du_{n-2}}{\varphi(u_{n-2})}.$$

9. $y_x^{(n)} = f(y + ax^m), \qquad m = 0,\ 1,\ 2,\ \ldots,\ n-1.$

The substitution $w = y + ax^m$ lead to an autonomous equation: $w_x^{(n)} = f(w)$, which, for even n, admits lowering of its order by two (see 5.2.6.6).

10. $y_x^{(n)} = x^{-n} f(y)$.

The substitution $t = \ln|x|$ leads to an autonomous equation of the form 5.2.6.40.

11. $y_x^{(n)} = x^{-n-1} f(x^{1-n} y)$.

The transformation $x = t^{-1}$, $y = t^{1-n} w$ leads to an autonomous equation: $w_t^{(n)} = (-1)^n f(w)$, whose order, for even n, can be lowered by two (see 5.2.6.6).

12. $y_x^{(2n)} = x^{-\frac{2n+1}{2}} f\Bigl(x^{\frac{1-2n}{2}} y\Bigr)$.

The transformation $x = e^t$, $y = x^{\frac{2n-1}{2}} w(t)$ leads to an autonomous equation of the form 5.2.6.25, whose order can be lowered by two.

13. $y_x^{(n)} = x^{-n-k} f(yx^k)$.

This is a special case of equation 5.2.6.44.

The transformation $t = \ln x$, $w = x^k y$ leads to an autonomous equation of the form 5.2.6.40.

14. $y_x^{(n)} = yx^{-n} f(x^k y^m)$.

The transformation $t = x^k y^m$, $w = \dfrac{xy'_x}{y}$ leads to an $(n-1)$th order equation.

15. $y_x^{(n)} = yx^{-n} f\left(\dfrac{xy'_x}{y}\right)$.

The transformation $z = \dfrac{xy'_x}{y}$, $w = \dfrac{x^2 y''_{xx}}{y}$ leads to an $(n-2)$th order equation.

16. $yy_x^{(2n+1)} = f(x)$.

Having integrated the equation, we obtain

$$2\sum_{m=0}^{n-1}(-1)^m y_x^{(m)} y_x^{(2n-m)} + (-1)^n \left[y_x^{(n)}\right]^2 = 2\int f(x)\,dx + C,$$

where the notation $y_x^{(0)} \equiv y$ is used.

17. $f(y) y_x^{(n+1)} = y'_x g\left(y_x^{(n)}\right)$.

Having integrated the equation, we obtain

$$\int \frac{dw}{g(w)} = \int \frac{dy}{f(y)} + C, \qquad \text{where} \quad w = y_x^{(n)}.$$

Next, the order of this equation can be lowered by the substitution $z(y) = y'_x$.

18. $y_x^{(n)} = f(x, y)$.

The transformation $x = t^{-1}$, $y = t^{1-n} w(t)$ leads to an equation of the similar form: $w_t^{(n)} = (-1)^n t^{-n-1} f(t^{-1}, t^{1-n} w)$.

19. $y_x^{(n)} = f\left(x, y_x^{(n-2)}, y_x^{(n-1)}\right)$.

The substitution $w(x) = y_x^{(n-2)}$ leads to a second order equation: $w''_{xx} = f(x, w, w'_x)$.

20. $(ax+b)^n (cx+d) y_x^{(n)} = f\left(\dfrac{y}{(cx+d)^{n-1}}\right)$.

The transformation $\xi = \ln \dfrac{ax+b}{cx+d}$, $w = \dfrac{y}{(cx+d)^{n-1}}$ leads to an autonomous equation of the form 5.2.6.40.

21. $y_x^{(n)} = (ax^2+bx+c)^{-\frac{1+n}{2}} f\Bigl(y(ax^2+bx+c)^{\frac{1-n}{2}}\Bigr).$

1°. The transformation

$$t = \int \frac{dx}{ax^2+bx+c}, \qquad w = y(ax^2+bx+c)^{\frac{1-n}{2}} \tag{1}$$

leads to an autonomous equation with respect to $w = w(t)$, which admits lowering of its order by the substitution $z(w) = w'_t$.

2°. Let $n = 2m$ be an even integer ($m = 1, 2, 3, \ldots$). In this case, transformation (1) yields an equation of the form 5.2.6.25, whose order can be lowered by two.

Setting $P = ax^2+bx+c$, $y = wP^{\frac{2m-1}{2}}$ and multiplying both sides of the original equation by $w'_x = P^{-\frac{1+2m}{2}}\Bigl(Py'_x + \dfrac{1-2m}{2}P'_x y\Bigr)$, we obtain

$$\Bigl(Py'_x + \frac{1-2m}{2}P'_x y\Bigr) y_x^{(2m)} = f(w)w'_x.$$

Integrating both sides of this equality with respect to x (the left-hand side is integrated by parts), we have

$$\sum_{k=0}^{m-2}(-1)^k \psi_x^{(k)} y_x^{(2m-1-k)} + (-1)^{m-1}\int \psi_x^{(m-1)} y_x^{(m+1)}\,dx = \int f(w)\,dw + C, \tag{2}$$

where

$$\psi_x^{(k)} = \frac{d^k}{dx^k}\Bigl(Py'_x + \frac{1-2m}{2}P'_x y\Bigr) = Py_x^{(k+1)} + \Bigl(k-m+\frac{1}{2}\Bigr)P'_x y_x^{(k)} + ak(k-2m)y_x^{(k-1)}.$$

(remind that $n = 2m$). It can be shown that the integrand on the left-hand side of (2) is the total differential. Finally, we arrive at the first integral

$$\begin{aligned}
&\sum_{k=0}^{m-2}(-1)^k\Bigl[Py_x^{(k+1)} + \Bigl(k-m+\frac{1}{2}\Bigr)P'_x y_x^{(k)} + ak(k-2m)y_x^{(k-1)}\Bigr] y_x^{(2m-1-k)}\\
&\quad + (-1)^{m-1}\Bigl\{\frac{1}{2}P\bigl[y_x^{(m)}\bigr]^2 - \frac{1}{2}P'_x y_x^{(m-1)} y_x^{(m)}\\
&\quad + a(1-m^2)y_x^{(m-2)} y_x^{(m)} + \frac{am^2}{2}\bigl[y_x^{(m-1)}\bigr]^2\Bigr\}\\
&\quad = \int f(w)\,dw + C.
\end{aligned}$$

22. $(2n-1)yy_x^{(2n+1)} + (2n+1)y'_x y_x^{(2n)} = f(x).$

Having integrated the equation, we have

$$(2n-1)yy_x^{(2n)} + 2\sum_{i=1}^{n-1}(-1)^{i+1}y_x^{(i)}y_x^{(2n-i)} + (-1)^{n+1}\bigl[y_x^{(n)}\bigr]^2 = \int f(x)\,dx + 2C_2.$$

The second integration yields

$$\sum_{i=0}^{n-1}(2n-1-2i)(-1)^i y_x^{(i)} y_x^{(2n-1-i)} = 2C_2x + C_1 + \int_{x_0}^{x}(x-t)f(t)\,dt.$$

The third integration leads to a $(2n-2)$ th order equation:

$$\begin{aligned}
&\sum_{i=0}^{n-2}(i+1)(2n-i-1)(-1)^i y_x^{(i)} y_x^{(2n-2-i)} + \frac{1}{2}(-1)^{n-1}n^2\bigl[y_x^{(n-1)}\bigr]^2\\
&\qquad = C_2x^2 + C_1x + C_0 + \frac{1}{2}\int_{x_0}^{x}(x-t)^2 f(t)\,dt.
\end{aligned}$$

23. $xy_x^{(n)} + ny_x^{(n-1)} = f(xy)$.

The substitution $w(x) = xy$ leads to the autonomous equation $w_x^{(n)} = f(w)$ (see 5.2.6.6 and 5.2.6.40).

24. $x^2y_x^{(n)} + 2nxy_x^{(n-1)} + n(n-1)y_x^{(n-2)} = f(x^2y)$.

The substitution $w(x) = x^2y$ leads to the autonomous equation $w_x^{(n)} = f(w)$ (see 5.2.6.6 and 5.2.6.40).

25. $\sum\limits_{m=1}^{n} a_m y_x^{(2m)} = f(y)$.

The first integral has the form

$$\sum_{m=1}^{n} a_m \left\{ \sum_{\nu=1}^{m-1} (-1)^\nu y_x^{(\nu)} y_x^{(2m-\nu)} + \frac{1}{2}(-1)^m \left[y_x^{(m)}\right]^2 \right\} + \int f(y)\,dy = C,$$

where C is an arbitrary constant. Further, the order of the obtained equation next be lowered by the substitution $w(y) = y'_x$.

26. $\sum\limits_{m=1}^{n} a_m x^m y_x^{(m)} = f(y)$.

The substitution $t = \ln|x|$ leads to an autonomous equation of the form 5.2.6.40.

27. $y \sum\limits_{m=0}^{n} a_m y_x^{(2m+1)} = f(x)$.

Having integrated the equation, we obtain

$$\sum_{m=0}^{n} a_m \left\{ 2\sum_{\nu=0}^{m-1} (-1)^\nu y_x^{(\nu)} y_x^{(2m-\nu)} + (-1)^m \left[y_x^{(m)}\right]^2 \right\} = 2\int f(x)\,dx + C,$$

where $y_x^{(0)}$ stands for y.

28. $\sum\limits_{m=0}^{n} a_m y_x^{(m)} y_x^{(2n+1-m)} = f(x)$.

The first integral has the form

$$2\sum_{m=0}^{n-1} A_m y_x^{(m)} y_x^{(2n-m)} + A_n \left[y_x^{(n)}\right]^2 = 2\int f(x)\,dx + C,$$

where

$$A_m = \sum_{k=0}^{m} (-1)^{m+k} a_k = a_m - a_{m-1} + a_{m-2} - a_{m-3} + \cdots.$$

If the condition $A_n = 2\sum_{m=0}^{n-1}(-1)^{n-1+m} A_m$ is satisfied, the obtained equation can be integrated two times more (see, in particular, equation 5.2.6.22).

29. $y_x^{(n)} = e^{\alpha x} f(ye^{-\alpha x})$.

The substitution $w(x) = ye^{-\alpha x}$ leads to an autonomous equation of the form 5.2.6.40.

30. $y_x^{(n)} = yf(e^{\alpha x} y^m)$.

The transformation $z = e^{\alpha x} y^m$, $w(z) = y'_x/y$ leads to an $(n-1)$ th order equation.

31. $y_x^{(n)} = x^{-n} f(x^m e^{\alpha y})$.

The transformation $z = x^m e^{\alpha y}$, $w(z) = xy'_x$ leads to an $(n-1)$ th order equation.

32. $y_x^{(n)} = f(y + ae^{\lambda x}) - a\lambda^n e^{\lambda x}$.

The substitution $w(x) = y + ae^{\lambda x}$ leads to an autonomous equation: $w_x^{(n)} = f(w)$ (see 5.2.6.6 and 5.2.6.40).

33. $y_x^{(2n)} = f(y + a\cosh x) - a\cosh x$.

The substitution $w(x) = y + a\cosh x$ leads to an autonomous equation: $w_x^{(2n)} = f(w)$ (see 5.2.6.6).

34. $y_x^{(2n)} = f(y + a\sinh x) - a\sinh x$.

The substitution $w(x) = y + a\sinh x$ leads to an autonomous equation of the form 5.2.6.6: $w_x^{(2n)} = f(w)$.

35. $y_x^{(2n+1)} = f(y + a\cosh x) - a\sinh x$.

The substitution $w(x) = y + a\cosh x$ leads to an autonomous equation of the form 5.2.6.40: $w_x^{(2n+1)} = f(w)$.

36. $y_x^{(2n+1)} = f(y + a\sinh x) - a\cosh x$.

The substitution $w(x) = y + a\sinh x$ leads to an autonomous equation of the form 5.2.6.40: $w_x^{(2n+1)} = f(w)$.

37. $y_x^{(n)} = f(y + a\cos x) - a\cos\Bigl(x + \dfrac{\pi n}{2}\Bigr)$.

The substitution $w(x) = y + a\cos x$ leads to an autonomous equation: $w_x^{(n)} = f(w)$ (see 5.2.6.6 and 5.2.6.40).

38. $y_x^{(n)} = f(y + a\sin x) - a\sin\Bigl(x + \dfrac{\pi n}{2}\Bigr)$.

The substitution $w(x) = y + a\sin x$ leads to an autonomous equation: $w_x^{(n)} = f(w)$ (see 5.2.6.6 and 5.2.6.40).

39. $F(x,\ y'_x,\ y''_{xx},\ \dots,\ y_x^{(n)}) = 0$.

The substitution $w(x) = y'_x$ leads to an $(n-1)$ th order equation:

$$F(x,\ w,\ w'_x,\ \dots,\ w_x^{(n-1)}) = 0.$$

40. $F(y,\ y'_x,\ y''_{xx},\ \dots,\ y_x^{(n)}) = 0.$

Autonomous equation.

The substitution $w(y) = y'_x$ leads to an $(n-1)$th order equation. The derivatives of the original equation and the transformed one are related by the formulae

$$y''_{xx} = ww'_y, \quad y'''_{xxx} = w^2 w''_{yy} + w(w'_y)^2, \quad \dots, \quad y_x^{(n)} = w\bigl(y_x^{(n-1)}\bigr)'_y.$$

41. $F(x,\ xy'_x - y,\ y''_{xx},\ y'''_{xxx},\ \dots,\ y_x^{(n)}) = 0.$

The substitution $w(x) = xy'_x - y$ leads to an $(n-1)$th order equation:

$$F\left(x,\ w,\ \frac{w'_x}{x},\ \frac{d}{dx}\left(\frac{w'_x}{x}\right), \dots,\ \frac{d^{n-2}}{dx^{n-2}}\left(\frac{w'_x}{x}\right)\right) = 0.$$

42. $F(x,\ x^2y''_{xx} - 2xy'_x + 2y,\ y'''_{xxx},\ \dots,\ y_x^{(n)}) = 0.$

The substitution $w(x) = x^2y''_{xx} - 2xy'_x + 2y$ leads to an $(n-2)$th order equation:

$$F\left(x,\ w,\ \frac{w'_x}{x^2},\ \dots,\ \frac{d^{n-3}}{dx^{n-3}}\left(\frac{w'_x}{x^2}\right)\right) = 0.$$

43. $\displaystyle\sum_{k=0}^{m}(-1)^k k!\, C_m^k x^{m-k} y_x^{(m-k)} = F(x,\ y_x^{(m+1)},\ \dots,\ y_x^{(n)}),$

where $C_m^k = \dfrac{m!}{k!\,(m-k)!}$ are binomial coefficients.

The substitution $w(x) = \sum_{k=0}^{m}(-1)^k k!\, C_m^k x^{m-k} y_x^{(m-k)}$ leads to an $(n-m)$th order equation; the derivatives on the right-hand side are calculated in consecutive manner using the formula $y_x^{(m+1)} = x^{-m} w'_x$.

44. $F(x^k y,\ x^{k+1}y'_x,\ \dots,\ x^{k+n}y_x^{(n)}) = 0.$

Homogeneous equation in the extended sense.

The transformation $t = \ln x$, $w = x^k y$ leads to an autonomous equation of the form 5.2.6.40.

45. $F\left(\dfrac{xy'_x}{y},\ \dfrac{x^2y''_{xx}}{y},\ \dots,\ \dfrac{x^n y_x^{(n)}}{y}\right) = 0.$

Homogeneous equation in the extended sense.

The transformation $z = \dfrac{xy'_x}{y}$, $w = \dfrac{x^2y''_{xx}}{y}$ leads to an $(n-2)$th order equation.

46. $F\left(x^k y^m,\ \dfrac{xy'_x}{y},\ \dfrac{x^2y''_{xx}}{y},\ \dots,\ \dfrac{x^n y_x^{(n)}}{y}\right) = 0.$

Homogeneous equation in the extended sense.

The transformation $t = x^k y^m$, $z = \dfrac{xy'_x}{y}$ leads to an $(n-1)$th order equation.

47. $F\left(e^{\alpha x}y,\ e^{\alpha x}y'_x,\ e^{\alpha x}y''_{xx},\ \ldots,\ e^{\alpha x}y_x^{(n)}\right)=0.$

Exponential homogeneous equation.

The substitution $w(x)=e^{\alpha x}y$ leads to an autonomous equation of the form 5.2.6.40.

48. $F\left(e^{\alpha x}y^m,\ \dfrac{y'_x}{y},\ \dfrac{y''_{xx}}{y},\ \ldots,\ \dfrac{y_x^{(n)}}{y}\right)=0.$

Exponential homogeneous equation.

The transformation $z=e^{\alpha x}y^m$, $w=\dfrac{y'_x}{y}$ leads to an $(n-1)$th order equation.

49. $F\left(x^m e^{\alpha y},\ xy'_x,\ x^2y''_{xx},\ \ldots,\ x^n y_x^{(n)}\right)=0.$

Exponential homogeneous equation.

The transformation $z=x^m e^{\alpha y}$, $w=xy'_x$ leads to an $(n-1)$th order equation.

Supplement 1 Some Elementary Functions and Their Properties

In Supplement 1, n is a positive integer, unless otherwise specified.

1.1. Trigonometric Functions

1.1.1. Simplest Relations

$$\sin^2 x + \cos^2 x = 1, \qquad \tan x \cot x = 1,$$
$$\sin(-x) = -\sin x, \qquad \cos(-x) = \cos x,$$
$$\tan x = \frac{\sin x}{\cos x}, \qquad \cot x = \frac{\cos x}{\sin x},$$
$$\tan(-x) = -\tan x, \qquad \cot(-x) = -\cot x.$$

1.1.2. Relations Between Trigonometric Functions of Identical Argument

$$\sin x = \pm\sqrt{1-\cos^2 x} = \pm\frac{\tan x}{\sqrt{1+\tan^2 x}} = \pm\frac{1}{\sqrt{1+\cot^2 x}},$$
$$\cos x = \pm\sqrt{1-\sin^2 x} = \pm\frac{1}{\sqrt{1+\tan^2 x}} = \pm\frac{\cot x}{\sqrt{1+\cot^2 x}},$$
$$\tan x = \pm\frac{\sin x}{\sqrt{1-\sin^2 x}} = \pm\frac{\sqrt{1-\cos^2 x}}{\cos x} = \frac{1}{\cot x},$$
$$\cot x = \pm\frac{\sqrt{1-\sin^2 x}}{\sin x} = \pm\frac{\cos x}{\sqrt{1-\cos^x}} = \frac{1}{\tan x}.$$

1.1.3. Reduction Formulae

$$\sin(x \pm n\pi) = (-1)^n \sin x, \qquad \cos(x \pm n\pi) = (-1)^n \cos x,$$
$$\sin\Bigl(x \pm \frac{2n+1}{2}\pi\Bigr) = \pm(-1)^n \cos x, \qquad \cos\Bigl(x \pm \frac{2n+1}{2}\pi\Bigr) = \mp(-1)^n \sin x,$$
$$\tan(x \pm n\pi) = \tan x, \qquad \cot(x \pm n\pi) = \cot x,$$
$$\tan\Bigl(x \pm \frac{2n+1}{2}\pi\Bigr) = -\cot x, \qquad \cot\Bigl(x \pm \frac{2n+1}{2}\pi\Bigr) = -\tan x,$$

$$\sin\left(x \pm \frac{\pi}{4}\right) = \frac{\sqrt{2}}{2}(\sin x \pm \cos x), \qquad \cos\left(x \pm \frac{\pi}{4}\right) = \frac{\sqrt{2}}{2}(\cos x \mp \sin x),$$
$$\tan\left(x \pm \frac{\pi}{4}\right) = \frac{\tan x \pm 1}{1 \mp \tan x}, \qquad \cot\left(x \pm \frac{\pi}{4}\right) = \frac{\cot x \mp 1}{1 \pm \cot x}.$$

1.1.4. Addition Formulae

$$\sin(x \pm y) = \sin x \cos y \pm \sin y \cos x, \qquad \cos(x \pm y) = \cos x \cos y \mp \sin x \sin y,$$
$$\tan(x \pm y) = \frac{\tan x \pm \tan y}{1 \mp \tan x \tan y}, \qquad \cot(x \pm y) = \frac{1 \mp \tan x \tan y}{\tan x \pm \tan y}.$$

1.1.5. Addition and Subtraction of Trigonometric Functions

$$\begin{aligned}
\sin x \pm \sin y &= 2\sin\left(\frac{x \pm y}{2}\right)\cos\left(\frac{x \mp y}{2}\right),\\
\cos x + \cos y &= 2\cos\left(\frac{x + y}{2}\right)\cos\left(\frac{x - y}{2}\right),\\
\cos x - \cos y &= -2\sin\left(\frac{x + y}{2}\right)\sin\left(\frac{x - y}{2}\right),
\end{aligned}$$

$$a\cos x + b\sin x = r\sin(x + \varphi) = r\cos(x - \psi),$$

where $r = \sqrt{a^2 + b^2}$, $\sin\varphi = a/r$, $\cos\varphi = b/r$, $\sin\psi = b/r$, $\cos\psi = a/r$,

$$\sin^2 x - \sin^2 y = \cos^2 y - \cos^2 x = \sin(x + y)\sin(x - y),$$

$$\sin^2 x - \cos^2 y = -\cos(x + y)\cos(x - y),$$

$$\tan x \pm \tan y = \frac{\sin(x \pm y)}{\cos x \cos y}, \qquad \cot x \pm \cot y = -\frac{\sin(x \pm y)}{\sin x \sin y},$$

1.1.6. Product of Trigonometric Functions

$$\begin{aligned}
\sin x \sin y &= \tfrac{1}{2}[\cos(x - y) - \cos(x + y)],\\
\cos x \cos y &= \tfrac{1}{2}[\cos(x - y) + \cos(x + y)],\\
\sin x \cos y &= \tfrac{1}{2}[\sin(x - y) + \sin(x + y)].
\end{aligned}$$

1.1.7. Powers of Trigonometric Functions

$$\begin{aligned}
\cos^2 x &= \tfrac{1}{2}\cos 2x + \tfrac{1}{2}, & \sin^2 x &= -\tfrac{1}{2}\cos 2x + \tfrac{1}{2},\\
\cos^3 x &= \tfrac{1}{4}\cos 3x + \tfrac{3}{4}\cos x, & \sin^3 x &= -\tfrac{1}{4}\sin 3x + \tfrac{3}{4}\sin x,\\
\cos^4 x &= \tfrac{1}{8}\cos 4x + \tfrac{1}{2}\cos 2x + \tfrac{3}{8}, & \sin^4 x &= \tfrac{1}{8}\cos 4x - \tfrac{1}{2}\cos 2x + \tfrac{3}{8},\\
\cos^5 x &= \tfrac{1}{16}\cos 5x + \tfrac{5}{16}\cos 3x + \tfrac{5}{8}\cos x, & \sin^5 x &= \tfrac{1}{16}\sin 5x - \tfrac{5}{16}\sin 3x + \tfrac{5}{8}\sin x,
\end{aligned}$$

$$\cos^{2n} x = \frac{1}{2^{2n-1}} \sum_{k=0}^{n-1} C_{2n}^k \cos[2(n-k)x] + \frac{1}{2^{2n}} C_{2n}^n,$$

$$\cos^{2n+1} x = \frac{1}{2^{2n}} \sum_{k=0}^{n} C_{2n+1}^k \cos[(2n-2k+1)x],$$

$$\sin^{2n} x = \frac{1}{2^{2n-1}} \sum_{k=0}^{n-1} (-1)^{n-k} C_{2n}^k \cos[2(n-k)x] + \frac{1}{2^{2n}} C_{2n}^n,$$

$$\sin^{2n+1} x = \frac{1}{2^{2n}} \sum_{k=0}^{n} (-1)^{n-k} C_{2n+1}^k \sin[(2n-2k+1)x],$$

where $C_m^k = \dfrac{m!}{k!(m-k)!}$ are binomial coefficients.

1.1.8. Trigonometric Functions of Multiple Arguments

$$\cos 2x = 2\cos^2 x - 1, \qquad \sin 2x = 2\sin x \cos x,$$

$$\cos 3x = -3\cos x + 4\cos^3 x, \qquad \sin 3x = 3\sin x - 4\sin^3 x,$$

$$\cos 4x = 1 - 8\cos^2 x + 8\cos^4 x, \qquad \sin 4x = 4\cos x(\sin x - 2\sin^3 x),$$

$$\cos 5x = 5\cos x - 20\cos^3 x + 16\cos^5 x, \qquad \sin 5x = 5\sin x - 20\sin^3 x + 16\sin^5 x,$$

$$\cos(2nx) = 1 + \sum_{k=1}^{n} (-1)^k \frac{n^2(n^2-1)\dots[n^2-(k-1)^2]}{(2k)!} 4^k \sin^{2k} x,$$

$$\cos[2(n+1)x] = \cos x \biggl\{ 1 + \sum_{k=1}^{n} (-1)^k \frac{[(2n+1)^2-1][(2n+1)^2-3^2]\dots[(2n+1)^2-(2k-1)^2]}{(2k)!} \sin^{2k} x \biggr\},$$

$$\sin(2nx) = 2n\cos x \biggl[\sin x + \sum_{k=1}^{n} (-1)^k \frac{(n^2-1)(n^2-2^2)\dots(n^2-k^2)}{(2k-1)!} 4^k \sin^{2k-1} x \biggr],$$

$$\sin[2(n+1)x] = (2n+1) \biggl\{ \sin x + \sum_{k=1}^{n} (-1)^k \frac{[(2n+1)^2-1][(2n+1)^2-3^2]\dots[(2n+1)^2-(2k-1)^2]}{(2k+1)!} \sin^{2k+1} x \biggr\},$$

$$\tan 2x = \frac{2\tan x}{1-\tan^2 x}, \qquad \tan 3x = \frac{3\tan x - \tan^3 x}{1 - 3\tan^2 x}, \qquad \tan 4x = \frac{4\tan x - 4\tan^3 x}{1 - 6\tan^2 x + \tan^4 x}.$$

1.1.9. Euler and de Moivre Formulae, Relation to Hyperbolic Functions

$$e^{z+ix} = e^z(\cos x + i\sin x), \qquad (\cos x + i\sin x)^n = \cos(nx) + i\sin(nx), \qquad i^2 = -1,$$

$$\sin(ix) = i\sinh x, \qquad \cos(ix) = \cosh x, \qquad \tan(ix) = i\tanh x, \qquad \cot(ix) = -i\coth x.$$

1.1.10. Differentiation and Integration Formulae

$$\frac{d\sin x}{dx} = \cos x, \qquad \frac{d\cos x}{dx} = -\sin x, \qquad \frac{d\tan x}{dx} = \frac{1}{\cos^2 x}, \qquad \frac{d\cot x}{dx} = -\frac{1}{\sin^2 x},$$

$$\int \sin x\,dx = -\cos x + C,$$

$$\int \cos x\,dx = \sin x + C,$$

$$\int \tan x\,dx = -\ln|\cos x| + C,$$

$$\int \cot x\,dx = \ln|\sin x| + C,$$

$$\int \sin^{2n} x\,dx = \frac{1}{2^{2n}} C_{2n}^{n} x + \frac{(-1)^n}{2^{2n-1}} \sum_{k=0}^{n-1} (-1)^k C_{2n}^{k} \frac{\sin[(2n-2k)x]}{2n-2k} + C,$$

$$\int \sin^{2n+1} x\,dx = \sum_{k=0}^{n} \frac{(-1)^{k+1}}{2k+1} C_n^k \cos^{2k+1} x + C,$$

$$\int \cos^{2n} x\,dx = \frac{1}{2^{2n}} C_{2n}^{n} x + \frac{1}{2^{2n-1}} \sum_{k=0}^{n-1} C_{2n}^{k} \frac{\sin[(2n-2k)x]}{2n-2k} + C,$$

$$\int \cos^{2n+1} x\,dx = \sum_{k=0}^{n} \frac{(-1)^{k}}{2k+1} C_n^k \sin^{2k+1} x + C,$$

$$\int \tan^{2n} x\,dx = (-1)^n x - \sum_{k=1}^{n} \frac{(-1)^k (\tan x)^{2n-2k+1}}{2n-2k+1} + C,$$

$$\int \tan^{2n+1} x\,dx = (-1)^{n+1} \ln|\cos x| - \sum_{k=1}^{n} \frac{(-1)^k (\tan x)^{2n-2k+2}}{2n-2k+2} + C,$$

$$\int \cot^{2n} x\,dx = (-1)^n x + \sum_{k=1}^{n} \frac{(-1)^k (\cot x)^{2n-2k+1}}{2n-2k+1} + C,$$

$$\int \cot^{2n+1} x\,dx = (-1)^n \ln|\sin x| + \sum_{k=1}^{n} \frac{(-1)^k (\cot x)^{2n-2k+2}}{2n-2k+2} + C,$$

where $C_m^k = \dfrac{m!}{k!\,(m-k)!}$ are binomial coefficients.

1.2. Hyperbolic Functions

1.2.1. Simplest Relations

$$\sinh x = \frac{e^x - e^{-x}}{2}, \qquad \cosh x = \frac{e^x + e^{-x}}{2},$$

$$\tanh x = \frac{e^x - e^{-x}}{e^x + e^{-x}}, \qquad \coth x = \frac{e^x + e^{-x}}{e^x - e^{-x}},$$

$$
\begin{aligned}
&\cosh^2 x - \sinh^2 x = 1, && \tanh x \cdot \coth x = 1,\\
&\sinh(-x) = -\sinh x, && \cosh(-x) = \cosh x,\\
&\tanh x = \frac{\sinh x}{\cosh x}, && \coth x = \frac{\cosh x}{\sinh x},\\
&\tanh(-x) = -\tanh x, && \coth(-x) = -\coth x.
\end{aligned}
$$

1.2.2. Relations Between Hyperbolic Functions of Identical Argument

$$
\begin{aligned}
\sinh x &= \pm\sqrt{\cosh^2 x - 1} = \pm\frac{\tanh x}{\sqrt{1-\tanh^2 x}} = \pm\frac{1}{\sqrt{\coth^2 x - 1}},\\
\cosh x &= \sqrt{\sinh^2 x + 1} = \pm\frac{1}{\sqrt{1-\tanh^2 x}} = \pm\frac{\coth x}{\sqrt{\coth^2 x - 1}},\\
\tanh x &= \frac{\sinh x}{\sqrt{\sinh^2 x + 1}} = \pm\frac{\sqrt{\cosh^2 x - 1}}{\cosh x} = \frac{1}{\coth x},\\
\coth x &= \frac{\sqrt{\sinh^2 x + 1}}{\sinh x} = \pm\frac{\cosh x}{\sqrt{\cosh^2 x - 1}} = \frac{1}{\tanh x}.
\end{aligned}
$$

1.2.3. Addition Formulae

$$
\begin{aligned}
\sinh(x \pm y) &= \sinh x \cosh y \pm \sinh y \cosh x,\\
\cosh(x \pm y) &= \cosh x \cosh y \pm \sinh x \sinh y,
\end{aligned}
$$

$$
\tanh(x \pm y) = \frac{\tanh x \pm \tanh y}{1 \pm \tanh x \tanh y}, \qquad \coth(x \pm y) = \frac{\coth x \coth y \pm 1}{\coth y \pm \coth x}.
$$

1.2.4. Addition and Subtraction of Hyperbolic Functions

$$
\begin{aligned}
\sinh x \pm \sinh y &= 2\sinh\left(\frac{x \pm y}{2}\right)\cosh\left(\frac{x \mp y}{2}\right),\\
\cosh x + \cosh y &= 2\cosh\left(\frac{x+y}{2}\right)\cosh\left(\frac{x-y}{2}\right),\\
\cosh x - \cosh y &= 2\sinh\left(\frac{x+y}{2}\right)\sinh\left(\frac{x-y}{2}\right),
\end{aligned}
$$

$$
\sinh^2 x - \sinh^2 y = \cosh^2 x - \cosh^2 y = \sinh(x+y)\sinh(x-y),
$$

$$
\sinh^2 x + \cosh^2 y = \cosh(x+y)\cosh(x-y),
$$

$$
\tanh x \pm \tanh y = \frac{\sinh(x \pm y)}{\cosh x \cosh y}, \qquad \coth x \pm \coth y = \pm\frac{\sinh(x \pm y)}{\sinh x \sinh y},
$$

1.2.5. Product of Hyperbolic Functions

$$
\begin{aligned}
\sinh x \sinh y &= \tfrac{1}{2}[\cosh(x+y) - \cosh(x-y)],\\
\cosh x \cosh y &= \tfrac{1}{2}[\cosh(x+y) + \cosh(x-y)],\\
\sinh x \cosh y &= \tfrac{1}{2}[\sinh(x+y) + \sinh(x-y)].
\end{aligned}
$$

1.2.6. Powers of Hyperbolic Functions

$$\cosh^2 x = \tfrac{1}{2}\cosh 2x + \tfrac{1}{2}, \qquad \sinh^2 x = \tfrac{1}{2}\cosh 2x - \tfrac{1}{2},$$
$$\cosh^3 x = \tfrac{1}{4}\sinh 3x + \tfrac{3}{4}\sinh x, \qquad \sinh^3 x = \tfrac{1}{4}\sinh 3x - \tfrac{3}{4}\sinh x,$$
$$\cosh^4 x = \tfrac{1}{8}\cosh 4x + \tfrac{1}{2}\cosh 2x + \tfrac{3}{8}, \qquad \sinh^4 x = \tfrac{1}{8}\cosh 4x - \tfrac{1}{2}\cosh 2x + \tfrac{3}{8},$$

$$\cosh^5 x = \tfrac{1}{16}\sinh 5x + \tfrac{5}{16}\sinh 3x + \tfrac{5}{8}\sinh x,$$
$$\sinh^5 x = \tfrac{1}{16}\sinh 5x - \tfrac{5}{16}\sinh 3x + \tfrac{5}{8}\sinh x,$$

$$\cosh^{2n} x = \frac{1}{2^{2n-1}}\sum_{k=0}^{n-1} C_{2n}^k \cosh[2(n-k)x] + \frac{1}{2^{2n}}C_{2n}^n,$$

$$\cosh^{2n+1} x = \frac{1}{2^{2n}}\sum_{k=0}^{n} C_{2n+1}^k \cosh[(2n-2k+1)x],$$

$$\sinh^{2n} x = \frac{1}{2^{2n-1}}\sum_{k=0}^{n-1} (-1)^k C_{2n}^k \cosh[2(n-k)x] + \frac{(-1)^n}{2^{2n}}C_{2n}^n,$$

$$\sinh^{2n+1} x = \frac{1}{2^{2n}}\sum_{k=0}^{n} (-1)^k C_{2n+1}^k \sinh[(2n-2k+1)x],$$

where $C_m^k = \dfrac{m!}{k!\,(m-k)!}$ are binomial coefficients.

1.2.7. Hyperbolic Functions of Multiple Argument

$$\cosh 2x = 2\cosh^2 x - 1, \qquad \sinh 2x = 2\sinh x\cosh x,$$
$$\cosh 3x = -3\cosh x + 4\cosh^3 x, \qquad \sinh 3x = 3\sinh x + 4\sinh^3 x,$$
$$\cosh 4x = 1 - 8\cosh^2 x + 8\cosh^4 x, \qquad \sinh 4x = 4\cosh x(\sinh x + 2\sinh^3 x),$$

$$\cosh 5x = 5\cosh x - 20\cosh^3 x + 16\cosh^5 x,$$
$$\sinh 5x = 5\sinh x + 20\sinh^3 x + 16\sinh^5 x.$$

$$\cosh(nx) = 2^{n-1}\cosh^n x + \frac{n}{2}\sum_{k=0}^{[n/2]} \frac{(-1)^{k+1}}{k+1} C_{n-k-2}^{k-2} 2^{n-2k-2}(\cosh x)^{n-2k-2},$$

$$\sinh(nx) = \sinh x \sum_{k=0}^{[(n-1)/2]} 2^{n-k-1} C_{n-k-1}^k (\cosh x)^{n-2k-1},$$

where $C_m^k = \dfrac{k!}{m!\,(k-m)!}$ are binomial coefficients, $[A]$ stands for the integer part of number A.

1.2.8. Relation to Trigonometric Functions

$$\sinh(ix) = i\sin x, \quad \cosh(ix) = \cos x, \quad \tanh(ix) = i\tan x, \quad \coth(ix) = -i\cot x,$$

where $i^2 = -1$.

1.2.9. Differentiation and Integration Formulae

$$\frac{d\sinh x}{dx} = \cosh x, \quad \frac{d\cosh x}{dx} = \sinh x, \quad \frac{d\tanh x}{dx} = \frac{1}{\cosh^2 x}, \quad \frac{d\coth x}{dx} = -\frac{1}{\sinh^2 x},$$

$$\int \sinh x\,dx = \cosh x + C,$$

$$\int \cosh x\,dx = \sinh x + C,$$

$$\int \tanh x\,dx = \ln\cosh x + C,$$

$$\int \coth x\,dx = \ln|\sinh x| + C,$$

$$\int \sinh^{2n} x\,dx = \frac{(-1)^n}{2^{2n}} C_{2n}^n x + \frac{1}{2^{2n-1}} \sum_{k=0}^{n-1} (-1)^k C_{2n}^k \frac{\sinh[(2n-2k)x]}{2n-2k} + C,$$

$$\int \sinh^{2n+1} x\,dx = \sum_{k=0}^{n} \frac{(-1)^{n+k}}{2k+1} C_n^k \cosh^{2k+1} x + C,$$

$$\int \cosh^{2n} x\,dx = \frac{1}{2^{2n}} C_{2n}^n x + \frac{1}{2^{2n-1}} \sum_{k=0}^{n-1} C_{2n}^k \frac{\cosh[(2n-2k)x]}{2n-2k} + C,$$

$$\int \cosh^{2n+1} x\,dx = \sum_{k=0}^{n} \frac{1}{2k+1} C_n^k \sinh^{2k+1} x + C,$$

$$\int \tanh^{2n} x\,dx = x - \sum_{k=1}^{n} \frac{(\tanh x)^{2n-2k+1}}{2n-2k+1} + C,$$

$$\int \tanh^{2n+1} x\,dx = \ln\cosh x - \sum_{k=1}^{n} \frac{(\tanh x)^{2n-2k+2}}{2n-2k+2} + C,$$

$$\int \coth^{2n} x\,dx = x - \sum_{k=1}^{n} \frac{(\coth x)^{2n-2k+1}}{2n-2k+1} + C,$$

$$\int \coth^{2n+1} x\,dx = \ln|\sinh x| - \sum_{k=1}^{n} \frac{(\coth x)^{2n-2k+2}}{2n-2k+2} + C,$$

where $C_m^k = \dfrac{m!}{k!\,(m-k)!}$ are binomial coefficients.

1.3. Inverse Trigonometric Functions

1.3.1. Simplest Relations

Principal values of inverse trigonometric functions are defined by the inequalities

$$-\frac{\pi}{2} \le \arcsin x \le \frac{\pi}{2}, \qquad 0 \le \arccos x \le \pi, \qquad \text{where} \quad -1 \le x \le 1,$$

$$-\frac{\pi}{2} < \arctan x < \frac{\pi}{2}, \qquad 0 < \operatorname{arccot} x < \pi, \qquad \text{where} \quad -\infty < x < +\infty.$$

$$\arcsin(-x) = -\arcsin x, \qquad \arccos(-x) = \pi - \arccos x,$$
$$\arctan(-x) = -\arctan x, \qquad \operatorname{arccot}(-x) = \pi - \operatorname{arccot} x,$$

$$\sin(\arcsin x) = x, \qquad \cos(\arccos x) = x,$$
$$\tan(\arctan x) = x, \qquad \cot(\operatorname{arccot} x) = x,$$

$$\arcsin(\sin x) = \begin{cases} x - 2n\pi & \text{if } 2n\pi - \frac{1}{2}\pi \le x \le 2n\pi + \frac{1}{2}\pi, \\ -x + 2(n+1)\pi & \text{if } (2n+1)\pi - \frac{1}{2}\pi \le x \le 2(n+1)\pi + \frac{1}{2}\pi, \end{cases}$$

$$\arccos(\cos x) = \begin{cases} x - 2n\pi & \text{if } 2n\pi \le x \le (2n+1)\pi, \\ -x + 2(n+1)\pi & \text{if } (2n+1)\pi \le x \le 2(n+1)\pi, \end{cases}$$

$$\arctan(\tan x) = x - n\pi \quad \text{if } n\pi - \tfrac{1}{2}\pi < x < n\pi + \tfrac{1}{2}\pi,$$
$$\operatorname{arccot}(\cot x) = x - n\pi \quad \text{if } n\pi < x < (n+1)\pi.$$

1.3.2. Relation Between Inverse Trigonometric Functions

$$\arcsin x + \arccos x = \frac{\pi}{2}, \qquad \arctan x + \operatorname{arccot} x = \frac{\pi}{2}.$$

$$\arcsin x = \begin{cases} \arccos\sqrt{1-x^2} & \text{if } 0 \le x \le 1, \\ -\arccos\sqrt{1-x^2} & \text{if } -1 \le x \le 0, \\ \arctan\dfrac{x}{\sqrt{1-x^2}} & \text{if } -1 < x < 1, \\ \operatorname{arccot}\dfrac{\sqrt{1-x^2}}{x} - \pi & \text{if } -1 \le x < 0, \end{cases}$$

$$\arccos x = \begin{cases} \arcsin\sqrt{1-x^2} & \text{if } 0 \le x \le 1, \\ \pi - \arcsin\sqrt{1-x^2} & \text{if } -1 \le x \le 0, \\ \arctan\dfrac{\sqrt{1-x^2}}{x} & \text{if } 0 < x \le 1, \\ \operatorname{arccot}\dfrac{x}{\sqrt{1-x^2}} & \text{if } -1 \le x \le 1. \end{cases}$$

$$\arctan x = \begin{cases} \arcsin\dfrac{x}{\sqrt{1+x^2}} & \text{if } x \text{ – any}, \\ \arccos\dfrac{1}{\sqrt{1+x^2}} & \text{if } x \ge 0, \\ -\arccos\dfrac{1}{\sqrt{1+x^2}} & \text{if } x \le 0, \\ \operatorname{arccot}\dfrac{1}{x} & \text{if } x > 0. \end{cases}$$

$$\operatorname{arccot} x = \begin{cases} \arcsin\dfrac{1}{\sqrt{1+x^2}} & \text{if } x > 0, \\ \pi - \arcsin\dfrac{1}{\sqrt{1+x^2}} & \text{if } x < 0, \\ \arctan\dfrac{1}{x} & \text{if } x > 0, \\ \pi + \arctan\dfrac{1}{x} & \text{if } x < 0. \end{cases}$$

1.3.3. Addition and Subtraction of Inverse Trigonometric Functions

$$\arcsin x + \arcsin y = \arcsin\left(x\sqrt{1-y^2} + y\sqrt{1-x^2}\,\right) \qquad \text{if} \quad x^2+y^2 \le 1,$$
$$\arccos x + \arccos y = \arccos\left[xy - \sqrt{(1-x^2)(1-y^2)}\,\right] \qquad \text{if} \quad x+y \ge 0,$$
$$\arccos x - \arccos y = -\arccos\left[xy + \sqrt{(1-x^2)(1-y^2)}\,\right] \qquad \text{if} \quad x-y \ge 0,$$
$$\arctan x + \arctan y = \arctan\frac{x+y}{1-xy} \qquad \text{if} \quad xy < 1,$$
$$\arctan x - \arctan y = \arctan\frac{x-y}{1+xy} \qquad \text{if} \quad xy > -1.$$

1.3.4. Differentiation and Integration Formulae

$$\frac{d}{dx}\arcsin x = \frac{1}{\sqrt{1-x^2}}, \qquad \frac{d}{dx}\arccos x = -\frac{1}{\sqrt{1-x^2}},$$
$$\frac{d}{dx}\arctan x = \frac{1}{1+x^2}, \qquad \frac{d}{dx}\operatorname{arccot} x = -\frac{1}{1+x^2},$$

$$\int \arcsin x\,dx = x\arcsin x + \sqrt{1-x^2} + C,$$
$$\int \arccos x\,dx = x\arccos x - \sqrt{1-x^2} + C,$$
$$\int \arctan x\,dx = x\arctan x - \frac{1}{2}\ln(x^2+1) + C,$$
$$\int \operatorname{arccot} x\,dx = x\operatorname{arccot} x + \frac{1}{2}\ln(x^2+1) + C.$$

1.4. Inverse Hyperbolic Functions

1.4.1. Relation to Logarithmic Functions and Simplest Relations

$$\operatorname{Arsinh} x = \ln\left(x + \sqrt{x^2+1}\,\right), \qquad \operatorname{Arcosh} x = \pm\ln\left(x + \sqrt{x^2-1}\,\right),$$
$$\operatorname{Artanh} x = \frac{1}{2}\ln\frac{1+x}{1-x}, \qquad \operatorname{Arcoth} x = \frac{1}{2}\ln\frac{1+x}{x-1},$$

$$\operatorname{Arsinh}(-x) = -\operatorname{Arsinh} x, \qquad \operatorname{Arcosh}(-x) = \operatorname{Arcosh} x,$$
$$\operatorname{Artanh}(-x) = -\operatorname{Artanh} x, \qquad \operatorname{Arcoth}(-x) = -\operatorname{Arcoth} x,$$

1.4.2. Relations Between Inverse Hyperbolic Functions

$$\operatorname{Arsinh} x = \operatorname{Arcosh}\sqrt{x^2+1} = \operatorname{Artanh}\frac{x}{\sqrt{x^2+1}},$$
$$\operatorname{Arcosh} x = \operatorname{Arsinh}\sqrt{x^2-1} = \operatorname{Artanh}\frac{\sqrt{x^2-1}}{x},$$
$$\operatorname{Artanh} x = \operatorname{Arsinh}\frac{x}{\sqrt{1-x^2}} = \operatorname{Arcosh}\frac{1}{\sqrt{1-x^2}}x = \operatorname{Arcoth}\frac{1}{x}.$$

1.4.3. Addition and Subtraction of Inverse Hyperbolic Functions

$$
\begin{aligned}
\operatorname{Arsinh} x \pm \operatorname{Arsinh} y &= \operatorname{Arsinh}\bigl(x\sqrt{1+y^2} \pm y\sqrt{1+x^2}\,\bigr),\\
\operatorname{Arcosh} x \pm \operatorname{Arcosh} y &= \operatorname{Arcosh}\bigl[xy \pm \sqrt{(x^2-1)(y^2-1)}\,\bigr],\\
\operatorname{Arsinh} x \pm \operatorname{Arcosh} y &= \operatorname{Arsinh}\bigl[xy \pm \sqrt{(x^2+1)(y^2-1)}\,\bigr],\\
\operatorname{Artanh} x \pm \operatorname{Artanh} y &= \operatorname{Artanh}\frac{x \pm y}{1 \pm xy},\\
\operatorname{Artanh} x \pm \operatorname{Arcoth} y &= \operatorname{Artanh}\frac{xy \pm 1}{y \pm x}.
\end{aligned}
$$

1.4.4. Differentiation and Integration Formulae

$$
\frac{d}{dx}\operatorname{Arsinh} x = \frac{1}{\sqrt{x^2+1}}, \qquad \frac{d}{dx}\operatorname{Arcosh} x = \frac{\pm 1}{\sqrt{x^2-1}},
$$

$$
\frac{d}{dx}\operatorname{Artanh} x = \frac{1}{1-x^2} \quad (x^2<1), \qquad \frac{d}{dx}\operatorname{Arcoth} x = \frac{1}{1-x^2} \quad (x^2>1),
$$

$$
\begin{aligned}
\int \operatorname{Arsinh} x\,dx &= x\operatorname{Arsinh} x - \sqrt{x^2+1} + C,\\
\int \operatorname{Arcosh} x\,dx &= x\operatorname{Arcosh} x \mp \sqrt{x^2-1} + C,\\
\int \operatorname{Artanh} x\,dx &= x\operatorname{Artanh} x + \frac{1}{2}\ln(1-x^2) + C,\\
\int \operatorname{Arcoth} x\,dx &= x\operatorname{Arcoth} x + \frac{1}{2}\ln(x^2-1) + C.
\end{aligned}
$$

1.5. Some Conventional Symbols

1.5.1. Factorial

$$
\begin{aligned}
&0! = 1! = 1, \quad n! = 1\cdot 2\cdot 3\ldots(n-1)n, \qquad n = 2,\ 3,\ 4,\ \ldots,\\
&(2n)!! = 2\cdot 4\cdot 6\ldots(2n-2)(2n) = 2^n n!,\\
&(2n+1)!! = 1\cdot 3\cdot 5\ldots(2n-1)(2n+1) = \frac{2^{n+1}}{\sqrt{\pi}}\Gamma\Bigl(n+\frac{3}{2}\Bigr),\\
&n!! = \begin{cases} (2k)!! & \text{if } n = 2k,\\ (2k+1)!! & \text{if } n = 2k+1, \end{cases} \qquad 0!! = 1.
\end{aligned}
$$

1.5.2. Binomial Coefficients

$$
C_a^b = \frac{\Gamma(a+1)}{\Gamma(b+1)\Gamma(a-b+1)},
$$

where $\Gamma(x)$ is the gamma-function.

$$
C_a^k = (-1)^k\frac{(-a)_k}{k!} = \frac{a(a-1)\ldots(a-k+1)}{k!} \qquad k = 1,\ 2,\ 3,\ldots
$$

$$C_n^k = \frac{n!}{k!(n-k)!}, \qquad k = 1,\ 2,\ 3,\ \dots,\ n,$$

$$C_a^0 = 1, \quad C_n^k = 0 \quad \text{for } k = -1,\ -2,\ -3,\ \dots \quad \text{or} \quad k > n,$$

$$C_a^{b+1} = \frac{a}{b+1} C_{a-1}^b = \frac{a-b}{b+1} C_a^b, \quad C_a^b + C_a^{b+1} = C_{a+1}^{b+1},$$

$$C_{-1/2}^n = \frac{(-1)^n}{2^{2n}} C_{2n}^n = (-1)^n \frac{(2n-1)!!}{(2n)!!},$$

$$C_{1/2}^n = \frac{(-1)^{n-1}}{n 2^{2n-1}} C_{2n-2}^{n-1} = \frac{(-1)^{n-1}}{n} \frac{(2n-3)!!}{(2n-2)!!},$$

$$C_{n+1/2}^{2n+1} = (-1)^n 2^{-4n-1} C_{2n}^n, \quad C_{2n+1/2}^n = 2^{-2n} C_{4n+1}^{2n},$$

$$C_n^{1/2} = \frac{2^{2n+1}}{\pi C_{2n}^n}, \quad C_n^{n/2} = \frac{2^{2n}}{\pi} C_n^{(n-1)/2}.$$

1.5.3. Pochhammer Symbol

$$(a)_n = a(a+1)\dots(a+n-1) = \frac{\Gamma(a+n)}{\Gamma(a)} = (-1)^n \frac{\Gamma(1-a)}{\Gamma(1-a-n)},$$

$$(a)_0 = 1, \quad (a)_{n+k} = (a)_n (a+n)_k, \quad (n)_k = \frac{(n+k-1)!}{(n-1)!},$$

$$(a)_{-n} = \frac{\Gamma(a-n)}{\Gamma(a)} = \frac{(-1)^n}{(1-a)_n}, \quad \text{where } a \neq 1,\ 2,\ \dots,\ n; \ k = 1,\ 2,\ 3,\ \dots,$$

$$(1)_n = n!, \quad (1/2)_n = 2^{-2n} \frac{(2n)!}{n!}, \quad (3/2)_n = 2^{-2n} \frac{(2n+1)!}{n!},$$

$$(a+mk)_{nk} = \frac{(a)_{mk+nk}}{(a)_{mk}}, \quad (a+n)_n = \frac{(a)_{2n}}{(a)_n}, \quad (a+n)_k = \frac{(a)_k (a+k)_n}{(a)_n}.$$

Supplement 2
Some Special Functions

In Supplement 2, n is a positive integer, unless otherwise specified.

2.1. Gamma-function

The gamma-function $\Gamma(z)$ is an analytic function of z everywhere, except for the points $z = 0,\ -1,\ -2,\ \dots$

For $\operatorname{Re} z > 0$,

$$\Gamma(z) = \int_0^\infty t^{z-1} e^{-t}\, dt.$$

For $-(n+1) < \operatorname{Re} z < -n$, where n is a nonnegative integer,

$$\Gamma(z) = \int_0^\infty \left[e^{-t} - \sum_{m=0}^{n} \frac{(-1)^m}{m!} \right] t^{z-1}\, dt.$$

The gamma-function possesses the following properties:

$$\Gamma(z+1) = z\Gamma(z), \qquad \Gamma(n+1) = n!, \qquad \Gamma(1) = \Gamma(2) = 1,$$

$$\Gamma(z)\Gamma(-z) = -\frac{\pi}{z\sin(\pi z)}, \quad \Gamma(z)\Gamma(1-z) = \frac{\pi}{\sin(\pi z)}, \quad \Gamma\Bigl(\frac{1}{2}+z\Bigr)\Gamma\Bigl(\frac{1}{2}-z\Bigr) = \frac{\pi}{\cos(\pi z)},$$

$$\Gamma(2z) = \frac{2^{2z-1}}{\sqrt{\pi}}\Gamma(z)\Gamma\Bigl(z+\frac{1}{2}\Bigr), \qquad \Gamma(3z) = \frac{3^{3z-1/2}}{2\pi}\Gamma(z)\Gamma\Bigl(z+\frac{1}{3}\Bigr)\Gamma\Bigl(z+\frac{2}{3}\Bigr),$$

$$\Gamma(nz) = (2\pi)^{(1-n)/2} n^{nz-1/2} \prod_{k=0}^{n-1} \Gamma\Bigl(z+\frac{k}{n}\Bigr),$$

$$\Gamma\Bigl(\frac{1}{2}\Bigr) = \sqrt{\pi}, \qquad \Gamma\Bigl(n+\frac{1}{2}\Bigr) = \frac{\sqrt{\pi}}{2^n}(2n-1)!!,$$

$$\Gamma\Bigl(-\frac{1}{2}\Bigr) = -2\sqrt{\pi}, \quad \Gamma\Bigl(\frac{1}{2}-n\Bigr) = (-1)^n \frac{2^n\sqrt{\pi}}{(2n-1)!!},$$

$$\frac{\Gamma(z+n)}{\Gamma(z)} = (z)_n, \quad \frac{\Gamma(z+1)}{\Gamma(w+1)\Gamma(z-w+1)} = C_z^w.$$

2.2. Bessel functions J_ν and Y_ν

2.2.1. Basic Formulae

The Bessel functions of the first and second kind J_ν and Y_ν (function Y_ν is also called the Neumann function) are solutions of the Bessel equation 2.1.2.121 and are defined by the formulae

$$J_\nu(x) = \sum_{k=0}^{\infty} \frac{(-1)^k (x/2)^{\nu+2k}}{k!\,\Gamma(\nu+k+1)}, \quad Y_\nu(x) = \frac{J_\nu(x)\cos\pi\nu - J_{-\nu}(x)}{\sin\pi\nu}.$$

The formulae for $Y_\nu(x)$ is valid for $\nu \neq 0, \pm1, \pm2, \ldots$; (see below for the integral representation of $Y_\nu(x)$, as well as equation 2.1.2.121 for $\nu = 0, \pm1, \pm2, \ldots$).

Function $Z_\nu(x) = C_1 J_\nu(x) + C_2 Y_\nu(x)$ is referred to as the cylindrical function.

The Bessel functions possess the following properties:

$$2\nu Z_\nu(x) = x[Z_{\nu-1}(x) + Z_{\nu+1}(x)],$$

$$\frac{d}{dx} Z_\nu = \frac{1}{2}[Z_{\nu-1}(x) - Z_{\nu+1}(x) = \pm\frac{1}{x}[\nu Z_\nu(x) - Z_{\nu\pm1}],$$

$$\frac{d}{dx}[x^\nu Z_\nu(x)] = x^\nu Z_{\nu-1}(x), \qquad \frac{d}{dx}[x^{-\nu} Z_\nu(x)] = -x^{-\nu} Z_{\nu+1}(x),$$

$$\left(\frac{d}{x\,dx}\right)^n [x^\nu J_\nu(x)] = x^{\nu-n} J_{\nu-n}(x), \qquad \left(\frac{d}{x\,dx}\right)^n [x^{-\nu} J_\nu(x)] = (-1)^n x^{-\nu-n} J_{\nu+n}(x),$$

$$J_{-n}(x) = (-1)^n J_n(x), \quad Y_{-n}(x) = (-1)^n Y_n(x) \qquad n = 0,\ 1,\ 2,\ \ldots$$

2.2.2. Bessel functions for $\nu = \pm n \pm \frac{1}{2}$; $n = 0, 1, 2, \ldots$

$$J_{1/2}(x) = \sqrt{\frac{2}{\pi x}}\sin x, \qquad J_{-1/2}(x) = \sqrt{\frac{2}{\pi x}}\cos x,$$

$$J_{3/2}(x) = \sqrt{\frac{2}{\pi x}}\left(\frac{1}{x}\sin x - \cos x\right), \qquad J_{-3/2}(x) = \sqrt{\frac{2}{\pi x}}\left(-\frac{1}{x}\cos x - \sin x\right),$$

$$J_{n+1/2}(x) = \sqrt{\frac{2}{\pi x}}\left[\sin\left(x - \frac{n\pi}{2}\right)\sum_{k=0}^{[n/2]} \frac{(-1)^k (n+2k)!}{(2k)!\,(n-2k)!\,(2x)^{2k}}\right.$$
$$\left. + \cos\left(x - \frac{n\pi}{2}\right)\sum_{k=0}^{[(n-1)/2]} \frac{(-1)^k (n+2k+1)!}{(2k+1)!\,(n-2k-1)!\,(2x)^{2k+1}}\right],$$

$$J_{-n-1/2}(x) = \sqrt{\frac{2}{\pi x}}\left[\cos\left(x + \frac{n\pi}{2}\right)\sum_{k=0}^{[n/2]} \frac{(-1)^k (n+2k)!}{(2k)!\,(n-2k)!\,(2x)^{2k}}\right.$$
$$\left. - \sin\left(x + \frac{n\pi}{2}\right)\sum_{k=0}^{[(n-1)/2]} \frac{(-1)^k (n+2k+1)!}{(2k+1)!\,(n-2k-1)!\,(2x)^{2k+1}}\right],$$

$$Y_{1/2}(x) = -\sqrt{\frac{2}{\pi x}}\cos x, \qquad Y_{-1/2}(x) = \sqrt{\frac{2}{\pi x}}\sin x,$$

$$Y_{n+1/2}(x) = (-1)^{n+1} J_{-n-1/2}(x), \qquad Y_{-n-1/2}(x) = (-1)^n J_{n+1/2}(x),$$

2.2.3. Wronskians an Similar Formulae

Notation: $W(f,g) = fg'_x - f'_x g$.

$$W(J_\nu, J_{-\nu}) = -\frac{2}{\pi x}\sin(\pi\nu), \qquad W(J_\nu, Y_\nu) = \frac{2}{\pi x},$$

$$J_\nu(x)J_{-\nu+1}(x) + J_{-\nu}(x)J_{\nu-1}(x) = \frac{2\sin(\pi\nu)}{\pi x}, \quad J_\nu(x)Y_{\nu+1}(x) - J_{\nu+1}(x)Y_\nu(x) = -\frac{2}{\pi x}.$$

2.2.4. Integral Representation

Functions J_ν and Y_ν may be expressed in terms of definite integrals (with $x > 0$):

$$\pi J_\nu(x) = \int_0^\pi \cos(x\sin\theta - \nu\theta)\,d\theta - \sin\pi\nu \int_0^\infty \exp(-x\sinh t - \nu t)\,dt,$$

$$\pi Y_\nu(x) = \int_0^\pi \sin(x\sin\theta - \nu\theta)\,d\theta - \int_0^\infty (e^{\nu t} + e^{-\nu t}\cos\pi\nu)e^{-x\sinh t}\,dt.$$

For $|\nu| < \frac{1}{2}$, $x > 0$,

$$J_\nu(x) = \frac{2^{1+\nu}x^{-\nu}}{\pi^{1/2}\Gamma(\frac{1}{2}-\nu)}\int_1^\infty \frac{\sin(xt)\,dt}{(t^2-1)^{\nu+1/2}}, \quad Y_\nu(x) = -\frac{2^{1+\nu}x^{-\nu}}{\pi^{1/2}\Gamma(\frac{1}{2}-\nu)}\int_1^\infty \frac{\cos(xt)\,dt}{(t^2-1)^{\nu+1/2}}.$$

For $\nu = 0$, $x > 0$,

$$J_0(x) = \frac{2}{\pi}\int_0^\infty \sin(x\cosh t)\,dt, \qquad Y_0(x) = -\frac{2}{\pi}\int_0^\infty \cos(x\cosh t)\,dt.$$

2.2.5. Integrals with Bessel Functions on Closed Intervals

$$\int_0^x x^\lambda J_\nu(x)\,dx = \frac{x^{\lambda+\nu+1}}{2^\nu(\lambda+\nu+1)\,\Gamma(\nu+1)}F\Big(\frac{\lambda+\nu+1}{2},\ \frac{\lambda+\nu+3}{2},\ \nu+1;\ -\frac{x^2}{4}\Big),$$

where $\operatorname{Re}(\lambda+\nu) > -1$, $F(a,b,c;x)$ is the hypergeometric series (see equation 2.1.2.158),

$$\int_0^x x^\lambda Y_\nu(x)\,dx = -\frac{\cos(\nu\pi)\,\Gamma(-\nu)}{2^\nu\pi(\lambda+\nu+1)}x^{\lambda+\nu+1}F\Big(\frac{\lambda+\nu+1}{2},\ \nu+1,\ \frac{\lambda+\nu+3}{2},\ -\frac{x^2}{4}\Big)$$
$$- \frac{2^\nu\Gamma(\nu)}{\lambda-\nu+1}x^{\lambda-\nu+1}F\Big(\frac{\lambda-\nu+1}{2},\ 1-\nu,\ \frac{\lambda-\nu+3}{2},\ -\frac{x^2}{4}\Big),$$

where $\operatorname{Re}\lambda > |\operatorname{Re}\nu| - 1$.

2.2.6. Asymptotic Expansion, as $|x| \to \infty$

$$J_\nu(x) = \sqrt{\frac{2}{\pi x}}\bigg\{\cos\Big(\frac{4x-2\nu\pi-\pi}{4}\Big)\bigg[\sum_{m=0}^{M-1}(-1)^m(\nu,2m)(2x)^{-2m} + O(|x|^{-2M})\bigg]$$
$$- \sin\Big(\frac{4x-2\nu\pi-\pi}{4}\Big)\bigg[\sum_{m=0}^{M-1}(-1)^m(\nu,2m+1)(2x)^{-2m-1} + O(|x|^{-2M-1})\bigg]\bigg\},$$

$$Y_\nu(x) = \sqrt{\frac{2}{\pi x}} \left\{ \sin\left(\frac{4x - 2\nu\pi - \pi}{4}\right) \left[\sum_{m=0}^{M-1} (-1)^m (\nu, 2m)(2x)^{-2m} + O(|x|^{-2M}) \right] \right.$$
$$\left. + \cos\left(\frac{4x - 2\nu\pi - \pi}{4}\right) \left[\sum_{m=0}^{M-1} (-1)^m (\nu, 2m+1)(2x)^{-2m-1} + O(|x|^{-2M-1}) \right] \right\},$$

where $(\nu, m) = \dfrac{1}{2^{2m} m!}(4\nu^2 - 1)(4\nu^2 - 3^2)\dots[4\nu^2 - (2m-1)^2] = \dfrac{\Gamma(\frac{1}{2} + \nu + m)}{m!\,\Gamma(\frac{1}{2} + \nu - m)}$.

2.3. Modified Bessel Functions I_ν and K_ν

2.3.1. Basic Formulae

The modified Bessel functions of the first and second kind I_ν and K_ν (function K_ν is also called the Basset function) are solutions the modified Bessel equation 2.1.2.122 and are defined by the formulae

$$I_\nu(x) = \sum_{k=0}^{\infty} \frac{(x/2)^{2k+\nu}}{k!\,\Gamma(\nu + k + 1)}, \qquad K_\nu(x) = \frac{\pi}{2} \frac{I_{-\nu} - I_\nu}{\sin \pi\nu},$$

(see equation 2.1.2.122 with $\nu = 0,\ 1,\ 2,\ \dots$ for the representation of $K_\nu(x)$).

The modified Bessel functions possess the following properties:

$$K_{-\nu}(x) = K_\nu(x); \qquad I_{-n}(x) = (-1)^n I_n(x), \quad n = 0,\ 1,\ 2,\ \dots$$

$$2\nu I_\nu(x) = x[I_{\nu-1}(x) - I_{\nu+1}(x)], \qquad 2\nu K_\nu(x) = -x[K_{\nu-1}(x) - K_{\nu+1}(x)],$$

$$\frac{d}{dx} I_\nu(x) = \frac{1}{2}[I_{\nu-1}(x) + I_{\nu+1}(x)], \qquad \frac{d}{dx} K_\nu(x) = -\frac{1}{2}[K_{\nu-1}(x) + K_{\nu+1}(x)],$$

2.3.2. Modified Bessel Functions for $\nu = \pm n \pm \frac{1}{2}$; $n = 0,\ 1,\ 2,\ \dots$

$$I_{1/2}(x) = \sqrt{\frac{2}{\pi x}} \sinh x, \qquad I_{-1/2}(x) = \sqrt{\frac{2}{\pi x}} \cosh x,$$

$$I_{3/2}(x) = \sqrt{\frac{2}{\pi x}} \left(-\frac{1}{x} \sinh x + \cosh x \right), \quad I_{-3/2}(x) = \sqrt{\frac{2}{\pi x}} \left(\frac{1}{x} \cosh x + \sinh x \right),$$

$$I_{n+1/2}(x) = \frac{1}{\sqrt{2\pi x}} \left[e^x \sum_{k=0}^{n} \frac{(-1)^k (n+k)!}{k!\,(n-k)!\,(2x)^k} - (-1)^n e^{-x} \sum_{k=0}^{n} \frac{(n+k)!}{k!\,(n-k)!\,(2x)^k} \right],$$

$$I_{-n-1/2}(x) = \frac{1}{\sqrt{2\pi x}} \left[e^x \sum_{k=0}^{n} \frac{(-1)^k (n+k)!}{k!\,(n-k)!\,(2x)^k} + (-1)^n e^{-x} \sum_{k=0}^{n} \frac{(n+k)!}{k!\,(n-k)!\,(2x)^k} \right],$$

$$K_{\pm 1/2}(x) = \sqrt{\frac{\pi}{2x}} e^{-x}, \qquad K_{\pm 3/2}(x) = \frac{\pi}{2x} \left(1 + \frac{1}{x} \right) e^{-x},$$

$$K_{n+1/2}(x) = K_{-n-1/2}(x) = \sqrt{\frac{\pi}{2x}} e^{-x} \sum_{k=0}^{n} \frac{(n+k)!}{k!\,(n-k)!\,(2x)^k}.$$

2.3.3. Wronskians and Similar Formulae

Notation: $W(f,g) = fg'_x - f'_x g$.

$$W(I_\nu, I_{-\nu}) = -\frac{2}{\pi x}\sin(\pi\nu), \qquad W(I_\nu, K_\nu) = -\frac{1}{x},$$

$$I_\nu(x)I_{-\nu+1}(x) + I_{-\nu}(x)I_{\nu-1}(x) = -\frac{2\sin(\pi\nu)}{\pi x}, \qquad I_\nu(x)K_{\nu+1}(x) + I_{\nu+1}(x)K_\nu(x) = \frac{1}{x}.$$

2.3.4. Integral Representation

Functions I_ν and K_ν may be expressed in terms of definite integrals (with $x>0$, $\nu>-\frac{1}{2}$):

$$I_\nu(x) = \frac{x^\nu}{\pi^{1/2}2^\nu\Gamma(\nu+\frac{1}{2})}\int_{-1}^{1}\exp(-xt)(1-t^2)^{\nu-1/2}\,dt,$$

$$K_\nu(x) = \int_0^\infty \exp(-x\cosh t)\cosh(\nu t)\,dt.$$

2.3.5. Integrals with Modified Bessel Functions on Closed Intervals

$$\int_0^x x^\lambda I_\nu(x)\,dx = \frac{x^{\lambda+\nu+1}}{2^\nu(\lambda+\nu+1)\,\Gamma(\nu+1)}F\Big(\frac{\lambda+\nu+1}{2},\ \frac{\lambda+\nu+3}{2},\ \nu+1;\ \frac{x^2}{4}\Big),$$

where $\operatorname{Re}(\lambda+\nu) > -1$, $F(a,b,c;x)$ is the hypergeometric series (see equation 2.1.2.158),

$$\int_0^x x^\lambda K_\nu(x)\,dx = \frac{2^{\nu-1}\Gamma(\nu)}{\lambda-\nu+1}x^{\lambda-\nu+1}F\Big(\frac{\lambda-\nu+1}{2},\ 1-\nu,\frac{\lambda-\nu+3}{2},\ \frac{x^2}{4}\Big) + \frac{2^{-\nu-1}\Gamma(-\nu)}{\lambda+\nu+1}x^{\lambda+\nu+1}F\Big(\frac{\lambda+\nu+1}{2},\ 1+\nu,\frac{\lambda+\nu+3}{2},\ \frac{x^2}{4}\Big),$$

where $\operatorname{Re}\lambda > |\operatorname{Re}\nu| - 1$.

2.3.6. Asymptotic Expansion, as $x \to +\infty$

$$I_\nu(x) = \frac{e^x}{\sqrt{2\pi x}}\left[1+\sum_{m=1}^{M}(-1)^m\frac{(4\nu^2-1)(4\nu^2-3^2)\dots[4\nu^2-(2m-1)^2]}{m!\,(8x)^m} + O(x^{-M-1})\right],$$

$$K_\nu(x) = \sqrt{\frac{\pi}{2x}}\,e^{-x}\left[1+\sum_{m=1}^{M}\frac{(4\nu^2-1)(4\nu^2-3^2)\dots[4\nu^2-(2m-1)^2]}{m!\,(8x)^m} + O(x^{-M-1})\right].$$

2.4. Degenerate Hypergeometric Functions

2.4.1. Definitions

The degenerate hypergeometric functions $\Phi(a,b;x)$ and $\Psi(a,b;x)$ are solutions of the degenerate hypergeometric equation 2.1.2.65.

If $b \neq 0, -1, -2, -3, \ldots$, function $\Phi(a, b; x)$ is expressed in terms of Kummer's series:

$$\Phi(a, b; x) = 1 + \sum_{k=1}^{\infty} \frac{(a)_k}{(b)_k} \frac{x^k}{k!},$$

where $(a)_k = a(a+1)\ldots(a+k-1)$, $(a)_0 = 1$. Function $\Psi(a, b; x)$ is defined as follows:

$$\Psi(a, b; x) = \frac{\Gamma(1-b)}{\Gamma(a-b+1)} \Phi(a, b; x) + \frac{\Gamma(b-1)}{\Gamma(a)} x^{1-b} \Phi(a-b+1,\ 2-b;\ x).$$

Table 2.1 represents some special cases where Φ is expressed in terms of simpler functions.

2.4.2. Basic Properties

Kummer's transformation:

$$\Phi(a, b; x) = e^x \Phi(b-a, b; -x), \qquad \Psi(a, b; x) = x^{1-b} \Psi(1+a-b, 2-b; x).$$

Linear relations for function Φ:

$$\begin{aligned}
&(b-a)\Phi(a-1, b; x) + (2a-b+x)\Phi(a, b; x) - a\Phi(a+1, b; x) = 0,\\
&b(b-1)\Phi(a, b-1; x) - b(b-1+x)\Phi(a, b; x) + (b-a)x\Phi(a, b+1; x) = 0,\\
&(a-b+1)\Phi(a, b; x) - a\Phi(a+1, b; x) + (b-1)\Phi(a, b-1; x) = 0,\\
&b\Phi(a, b; x) - b\Phi(a-1, b; x) - x\Phi(a, b+1; x) = 0,\\
&b(a+x)\Phi(a, b; x) - (b-a)x\Phi(a, b+1; x) - ab\Phi(a+1, b; x) = 0,\\
&(a-1+x)\Phi(a, b; x) + (b-a)\Phi(a-1, b; x) - (b-1)\Phi(a, b-1; x) = 0,
\end{aligned}$$

Linear relations for function Ψ:

$$\begin{aligned}
&\Psi(a-1, b; x) - (2a-b+x)\Psi(a, b; x) + a(a-b+1)\Psi(a+1, b; x) = 0,\\
&(b-a-1)\Psi(a, b-1; x) - (b-1+x)\Psi(a, b; x) + x\Psi(a, b+1; x) = 0,\\
&\Psi(a, b; x) - a\Psi(a+1, b; x) - \Psi(a, b-1; x) = 0,\\
&(b-a)\Psi(a, b; x) - x\Psi(a, b+1; x) + \Psi(a-1, b; x) = 0,\\
&(a+x)\Psi(a, b; x) + a(b-a-1)\Psi(a+1, b; x) - x\Psi(a, b+1; x) = 0,\\
&(a-1+x)\Psi(a, b; x) - \Psi(a-1, b; x) + (a-c+1)\Psi(a, b-1; x) = 0.
\end{aligned}$$

Differentiation formulae:

$$\frac{d}{dx}\Phi(a, b; x) = \frac{a}{b}\Phi(a+1, b+1; x), \qquad \frac{d}{dx}\Psi(a, b; x) = -a\Psi(a+1, b+1; x),$$

$$\frac{d^n}{dx^n}\Phi(a, b; x) = \frac{(a)_n}{(b)_n}\Phi(a+n, b+n; x), \qquad \frac{d^n}{dx^n}\Psi(a, b; x) = (-1)^n (a)_n \Psi(a+n, b+n; x),$$

Wronskian:

$$W(\Phi, \Psi) = \Phi\Psi'_x - \Phi'_x\Psi = -\frac{\Gamma(b)}{\Gamma(a)} x^{-b} e^x.$$

2.4.3. Integral Representation

$$\Phi(a,b;x) = \frac{\Gamma(b)}{\Gamma(a)\,\Gamma(b-a)} \int_0^1 e^{xt} t^{a-1}(1-t)^{b-a-1}\,dt, \qquad (\text{if } b > a > 0)$$

$$\Psi(a,b;x) = \frac{1}{\Gamma(a)} \int_0^\infty e^{-xt} t^{a-1}(1+t)^{b-a-1}\,dt, \qquad (\text{if } a > 0, x > 0)$$

where $\Gamma(a)$ is the gamma-function.

2.4.4. Integrals with Degenerate Hypergeometric Functions

$$\int \Phi(a,b;x)\,dx = \frac{b-1}{a-1}\Psi(a-1,b-1;x)+C, \quad \int \Phi(a,b;x)\,dx = \frac{1}{1-a}\Psi(a-1,b-1;x)+C,$$

$$\int x^n \Phi(a,b;x)\,dx = n! \sum_{k=1}^{n+1} \frac{(-1)^{k+1}(1-b)_k x^{n-k+1}}{(1-a)_k (n-k+1)!} \Phi(a-k, b-k; x) + C,$$

$$\int x^n \Psi(a,b;x)\,dx = n! \sum_{k=1}^{n+1} \frac{(-1)^{k+1} x^{n-k+1}}{(1-a)_k (n-k+1)!} \Phi(a-k, b-k; x) + C.$$

2.4.5. Asymptotic Expansion, as $|x| \to \infty$

$$\Phi(a,b;x) = \frac{\Gamma(b)}{\Gamma(a)} e^x x^{a-b} \left[\sum_{n=0}^{N} \frac{(c-a)_n (1-a)_n}{n!} x^{-n} + O(x^{-N-1})\right], \qquad x > 0,$$

$$\Phi(a,b;x) = \frac{\Gamma(b)}{\Gamma(b-a)} (-x)^{-a} \left[\sum_{n=0}^{N} \frac{(a)_n (a-b+1)_n}{n!} (-x)^{-n} + O(|x|^{-N-1})\right], \qquad x < 0,$$

$$\Psi(a,b;x) = x^{-a} \left[\sum_{n=0}^{N} (-1)^n \frac{(a)_n (a-b+1)_n}{n!} x^{-n} + O(|x|^{-N-1})\right], \qquad -\infty < x < +\infty.$$

2.5. Legendre Functions

2.5.1. Definitions

The associated Legendre functions $P_\nu^\mu(z)$ and $Q_\nu^\mu(z)$ of the first and second kind are linearly-independent solutions the Legendre equation (see 2.1.2.212):

$$(1-z^2)y''_{zz} - 2zy'_z + [\nu(\nu+1) - \mu^2(1-z^2)^{-1}]y = 0,$$

where parameters ν, μ and variable z may be arbitrary real or complex numbers.

With $|1-z| < 2$, the following formulae may be used:

$$P_\nu^\mu(z) = \frac{1}{\Gamma(1-\mu)} \left(\frac{z+1}{z-1}\right)^{\mu/2} F\left(-\nu,\ 1+\nu,\ 1-\mu,\ \frac{1-z}{2}\right),$$

$$Q_\nu^\mu(z) = A\left(\frac{z-1}{z+1}\right)^{\frac{\mu}{2}} F\left(-\nu,\ 1+\nu,\ 1+\mu,\ \frac{1-z}{2}\right) + B\left(\frac{z+1}{z-1}\right)^{\frac{\mu}{2}} F\left(-\nu,\ 1+\nu,\ 1-\mu,\ \frac{1-z}{2}\right),$$

$$A = e^{i\mu\pi} \frac{\Gamma(-\mu)\,\Gamma(1+\nu+\mu)}{2\,\Gamma(1+\nu-\mu)}, \quad B = e^{i\mu\pi} \frac{\Gamma(\mu)}{2}, \quad i^2 = -1,$$

where $F(a,b,c;z)$ is the hypergeometric series (see equation 2.1.2.158).

With $|z|>1$, the following formulae may be used:

$$P_\nu^\mu(z) = \frac{2^{-\nu-1}\Gamma(-\frac{1}{2}-\nu)}{\sqrt{\pi}\,\Gamma(-\nu-\mu)} z^{-\nu+\mu-1}(z^2-1)^{-\mu/2} F\Big(\frac{1+\nu-\mu}{2}, \frac{2+\nu-\mu}{2}, \frac{2\nu+3}{2}, \frac{1}{z^2}\Big)$$
$$+ \frac{2^\nu\Gamma(\frac{1}{2}+\nu)}{\sqrt{\pi}\,\Gamma(1+\nu-\mu)} z^{\nu+\mu}(z^2-1)^{-\mu/2} F\Big(-\frac{\nu+\mu}{2}, \frac{1-\nu-\mu}{2}, \frac{1-2\nu}{2}, \frac{1}{z^2}\Big),$$

$$Q_\nu^\mu(z) = e^{i\pi\mu}\frac{\sqrt{\pi}\,\Gamma(\nu+\mu+1)}{2^{\nu+1}\Gamma(\nu+\frac{3}{2})} z^{-\nu-\mu-1}(z^2-1)^{\mu/2} F\Big(\frac{2+\nu+\mu}{2}, \frac{1+\nu+\mu}{2}, \frac{2\nu+3}{2}, \frac{1}{z^2}\Big).$$

The functions

$$P_\nu(z) \equiv P_\nu^0(z), \qquad Q_\nu(z) \equiv Q_\nu^0(z)$$

are called the Legendre functions and are solutions of the Legendre equation 2.1.2.148.

The modified associated Legendre functions on the cut $z=x$, $-1<x<1$, are defined by the formulae

$$P_\nu^\mu(x) = \frac{1}{2}\Big[e^{\frac{1}{2}i\mu\pi}P_\nu^\mu(x+i0) + e^{-\frac{1}{2}i\mu\pi}P_\nu^\mu(x-i0)\Big],$$

$$Q_\nu^\mu(x) = \frac{1}{2}e^{-i\mu\pi}\Big[e^{-\frac{1}{2}i\mu\pi}Q_\nu^\mu(x+i0) + e^{\frac{1}{2}i\mu\pi}Q_\nu^\mu(x-i0)\Big].$$

2.5.2. Trigonometric Expansions

With $-1<x<1$, the modified associated Legendre functions can be expressed in terms of trigonometric series:

$$P_\nu^\mu(\cos\theta) = \frac{2^{\mu+1}}{\sqrt{\pi}}\frac{\Gamma(\nu+\mu+1)}{\Gamma(\nu+\frac{3}{2})}(\sin\theta)^\mu \sum_{k=0}^\infty \frac{(\frac{1}{2}+\mu)_k(1+\nu+\mu)_k}{k!\,(\nu+\frac{3}{2})_k}\sin[(2k+\nu+\mu+1)\theta],$$

$$Q_\nu^\mu(\cos\theta) = \sqrt{\pi}\,2^\mu\frac{\Gamma(\nu+\mu+1)}{\Gamma(\nu+\frac{3}{2})}(\sin\theta)^\mu \sum_{k=0}^\infty \frac{(\frac{1}{2}+\mu)_k(1+\nu+\mu)_k}{k!\,(\nu+\frac{3}{2})_k}\cos[(2k+\nu+\mu+1)\theta],$$

where $0<\theta<\pi$.

2.5.3. Some Relations

$$P_\nu^\mu(z) = P_{-\nu-1}^\mu(z), \quad P_\nu^n(z) = \frac{\Gamma(\nu+n+1)}{\Gamma(\nu-n+1)}P_\nu^{-n}(z) \qquad n=0,\ 1,\ 2,\ \ldots$$

$$Q_\nu^\mu(z) = \frac{\pi}{2\sin(\mu\pi)}e^{i\pi\mu}\Big[P_\nu^\mu(z) - \frac{\Gamma(1+\nu+\mu)}{\Gamma(1+\nu-\mu)}P_\nu^{-\mu}(z)\Big].$$

For $-1<x<1$,

$$P_{\nu+1}^\mu(x) = \frac{2\nu+1}{\nu-\mu+1}xP_\nu^\mu(x) - \frac{\nu+\mu}{\nu-\mu+1}P_{\nu-1}^\mu(x),$$
$$P_{\nu+1}^\mu(x) = P_{\nu-1}^\mu(x) - (2\nu+1)(1-x^2)^{1/2}P_\nu^{\mu-1}(x),$$
$$\frac{d}{dx}P_\nu^\mu(x) = \frac{\nu x}{x^2-1}P_\nu^\mu(x) - \frac{\nu+\mu}{x^2-1}P_{\nu-1}^\mu(x).$$

For $0 < x < 1$,

$$P_\nu^\mu(-x) = P_\nu^\mu(x)\cos[\pi(\nu+\mu)] - 2\pi^{-1}Q_\nu^\mu(x)\sin[\pi(\nu+\mu)],$$
$$Q_\nu^\mu(-x) = -Q_\nu^\mu(x)\cos[\pi(\nu+\mu)] - \tfrac{1}{2}\pi P_\nu^\mu(x)\sin[\pi(\nu+\mu)].$$

Wronskians:

$$W(P_\nu, Q_\nu) = \frac{1}{1-x^2}, \qquad W(P_\nu^\mu, Q_\nu^\mu) = \frac{k}{1-x^2},$$

where $k = 2^{2\mu}\dfrac{\Gamma\left(\frac{\nu+\mu+1}{2}\right)\Gamma\left(\frac{\nu+\mu+2}{2}\right)}{\Gamma\left(\frac{\nu-\mu+1}{2}\right)\Gamma\left(\frac{\nu-\mu+2}{2}\right)}$.

For $n = 0,\ 1,\ 2,\ \ldots,$

$$P_\nu^n(x) = (-1)^n(1-x^2)^{n/2}\frac{d^n}{dx^n}P_\nu(x), \qquad Q_\nu^n(x) = (-1)^n(1-x^2)^{n/2}\frac{d^n}{dx^n}Q_\nu(x).$$

The Legendre polynomials $P_n(x)$ and the Legendre function $Q_n(x)$ are given by the formulae

$$P_n(x) = \frac{1}{n!\,2^n}\frac{d^n}{dx^n}(x^2-1)^n,$$
$$Q_n(x) = \frac{1}{2}P_n(x)\ln\frac{1+x}{1-x} - \sum_{m=1}^{n}\frac{1}{m}P_{m-1}(x)P_{n-m}(x).$$

Functions $P_n = P_n(x)$ can be conveniently calculated by the recurrence relations

$$P_0 = 1, \quad P_1 = x, \quad P_2 = \frac{1}{2}(3x^2-1), \quad \ldots, \quad P_{n+1} = \frac{2n+1}{n+1}xP_n - \frac{n}{n+1}P_{n-1}.$$

Three leading functions $Q_n = Q_n(x)$ are

$$Q_0 = \frac{1}{2}\ln\frac{1+x}{1-x}, \quad Q_1 = \frac{x}{2}\ln\frac{1+x}{1-x} - 1, \quad Q_2 = \frac{3x^2-1}{4}\ln\frac{1+x}{1-x} - \frac{3}{2}x.$$

2.5.4. Integral Representation

For $n = 0,\ 1,\ 2,\ \ldots$

$$P_\nu^n(z) = \frac{\Gamma(\nu+n+1)}{\pi\Gamma(\nu+1)}\int_0^\pi \left(z + \cos t\sqrt{z^2-1}\right)^\nu \cos(nt)\,dt, \quad \operatorname{Re} z > 0,$$
$$Q_\nu^n(z) = (-1)^n\frac{\Gamma(\nu+n+1)}{2^{\nu+1}\Gamma(\nu+1)}(z^2-1)^{-n/2}\int_0^\pi (z+\cos t)^{n-\nu-1}(\sin t)^{2\nu+1}\,dt, \quad \operatorname{Re}\nu > -1,$$

Note that in the latter formulae $z \neq x$, $-1 < x < 1$.

2.6. The Weierstrass function ℘

2.6.1. Definitions

The Weierstrass function $\wp = \wp(z, g_2, g_3)$ is defined implicitly by the elliptic integral

$$z = \int_\infty^\wp \frac{dt}{\sqrt{4t^3 - g_2t - g_3}}$$

and satisfies the first order differential equation

$$(\wp_z')^2 = 4\wp^3 - g_2\wp - g_3.$$

2.6.2. Some Properties

Below $\wp(z)$ stands for $\wp(z, g_2, g_3)$.

$$\wp(z) = \wp(-z), \qquad \wp(z_1 + z_2) = -\wp(z_1) - \wp(z_2) + \frac{1}{4}\left[\frac{\wp'(z_1) - \wp'(z_2)}{\wp(z_1) - \wp(z_2)}\right]^2.$$

In the vicinity of the point $z = 0$, the Weierstrass function can be expanded into the series

$$\wp(z) = \frac{1}{z^2} + \frac{g_2}{20}z^2 + \frac{g_3}{28}z^4 + \frac{g_2^2}{1200}z^6 + \frac{3g_2g_3}{6160}z^8 + \cdots.$$

References

Abramowitz, M., and Stegun, I.A., *Handbook of Mathematical Functions with Formulas, Graphs and Mathematical Tables*, Washington, National Bureau of Standards Applied Mathematics, 1964.

Bateman, H., and Erdélyi, A., *Higher Transcendental Functions, Vol. 1*, New York, McGraw-Hill Book Comp., 1953.

Bateman, H., and Erdélyi, A., *Higher Transcendental Functions, Vol. 2*, New York, McGraw-Hill Book Comp., 1953.

Bateman, H., and Erdélyi, A., *Higher Transcendental Functions, Vol. 3*, New York, McGraw-Hill Book Comp., 1955.

Bercovich, L.M., *Factorization and Transformations of Ordinary Differential Equations*, Saratov, Saratov University Publ., 1989.

Bluman, G.W., and Cole, J.D. *Similarity Methods for Differential Equations,* New York, Springer—Verlag, 1974.

Cole, J.D., *Perturbation Methods in Applied Mathematics*, Waltham, Blaisdell Publ. Comp., 1968.

Fedoryuk, M.V., *Asymptotic Methods for Linear Ordinary Differential Equations*, Moscow, Nauka, 1983 (in Russian).

Gradshteyn, I.S., and Ryzhik, I.M. *Tables of Integrals, Series, and Products*, New York, Academic Press, 1980.

Gromak, V.I., and Lukashevich, N.A. *Analytical Properties of Solutions of Painlevé Equations*, Minsk, Universitetskoye, 1990 (in Russian).

Hartman, P., *Ordinary Differential Equations*, New York, Jonh Wiley & Sons, 1964.

Hill, J.M., *Solution of differential equations by means of one-parameter groups*, Mass.: Pitman, Marshfield, 1982.

Ibragimov, N.H., *CRC Handbook of Lie Group to Differential Equations,* Vol. 1, Boca Raton, CRC Press, 1993.

Ince, E.L., *Ordinary Differential Equations*, New York, Dover Publications, 1964.

Kamke, E., *Handbook on Ordinary Differential Equations*, Moscow, Nauka, 1976 (in Russian).

Korn, G.A., and Korn, T.M., *Mathematical Handbook for Scientists and Engineers*, New York, McGraw-Hill Book Comp., 1961.

Matveev, N.M., *Methods of Integration of Ordinary Differential Equations*, Moscow, Vysshaya Shkola, 1967 (in Russian).

Murphy, G.M., *Ordinary Differential Equations and Their Solutions*, New York, D. Van Nostrand, 1960.

Nayfeh, A.H., *Perturbation Methods*, New York, John Wiley & Sons, 1973.

Nayfeh, A.H., *Introduction to Perturbation Techniques*, New York, John Wiley & Sons, 1981.

Olver, P.J., *Application of Lie Group to Differential Equations,* New York, Springer—Verlag, 1986.

Ovsiannikov, L.V., *Group Analysis of Differential Equations,* New York, Academic Press, 1982.

Prudnikov, A.P., Brychkov, Yu.A., and Marichev, O.I., *Integrals and Series, Vol. 1, Elementary Functions*, New York, Gordon & Breach Sci. Publ., 1986.

Prudnikov, A.P., Brychkov, Yu.A., and Marichev, O.I., *Integrals and Series, Vol. 2, Special Functions*, New York, Gordon & Breach Sci. Publ., 1986.

Prudnikov, A.P., Brychkov, Yu.A., and Marichev, O.I., *Integrals and Series, Vol. 3, More Special Functions*, New York, Gordon & Breach Sci. Publ., 1988.

Zaitsev, V.F. and Polyanin, A.D., *Discrete-Group Methods for Integrating Equations of Nonlinear Mechanics*, Boca Raton, CRC Press—Begel House, 1994.

Zaitsev, V.F. and Polyanin, A.D., *Handbook on Nonlinear Differential Equations*, Moscow, Nauka, 1993 (in Russian).

Zwillinger, D., *Handbook of Differential Equations*, San Diego, Academic Press, 1989.

Index

A

B

C

D

E

F

G

H

I

J

K

L

M

N

P

Q

R

S

T

V

W

Y